ALLYL COMPOUNDS AND THEIR POLYMERS

(Including Polyolefins)

HIGH POLYMERS

A SERIES OF MONOGRAPHS ON THE CHEMISTRY, PHYSICS, AND

TECHNOLOGY OF HIGH POLYMERIC SUBSTANCES

Editorial Board

VOLUME XXVIII

ALLYL COMPOUNDS
AND THEIR POLYMERS
(Including Polyolefins)

CALVIN E. SCHILDKNECHT

Consultant, and Professor, Gettysburg College

WILLEY-INTERSCIENCE

A Division of John Wiley & Sons, Inc.

New York • London • Sydney • Toronto

Library of Congress Cataloging in Publication Data

Schildknecht, Calvin Everett, 1910–
Allyl compounds and their polymers (including polyolefins)

(High polymers, v. 28)
1. Allyl compounds. 2. Polymers and polymerization.
I. Title. II. Series.

TP248.A56S35   547'.01   72–1363
ISBN 0–471–39380–0

Printed in the United States of America

10 9 8 7 6 5 4 3 2 1

To the late Mabel (Duthey) Reiner

and

Althea (Schneider) Schildknecht

who made this book possible

PREFACE

This book surveys for the first time the preparations and properties of allyl compounds and their polymers together with their applications in plastics, fibers, synthetic rubbers, and adhesives. Numerous monomeric allyl and related compounds useful in flavors, perfumes, pharmaceuticals, and biocides are included. In general the allyl compounds are less reactive in polymerizations than the better-known vinyl compounds, but are more reactive in other ways.

The book supplements my book of 1952, Vinyl and Related Polymers, which surveyed principally the polymers from ethylene and substituted ethylenic compounds where a group activating polymerization is attached directly at the double bond. Only 30 allyl compounds were mentioned in the Vinyl book. An interesting challenge was offered by the great number of allyl compounds where a third carbon atom of a group having low electron-displacing activity is attached to the ethylenic group. The relatively low reactivity in polymerization and other difficulties presented by allyl monomers have now been overcome in compounds such as the 1-alkenes, diallyl carbonates, and diallyl phthalates permitting production of heat-resistant, quality synthetic polymers of high utility in electronic, space, medical, optical, and other applications. New interest in the sluggish polymerizations of allyl compounds comes from the outstanding research and development of stereoregulated polymerizations of allylic 1-olefins beginning in 1954; from the use of diallyl and triallyl compounds in crosslinking and in graft copolymerizations; as well as from the discovery of cyclopolymerization of polyfunctional allyl compounds.

In contrast to the vinyl monomers, numerous allyl compounds occur in nature and many are of interest in reactions and applications other than polymerization. Studies of their reactivity in additions, substitutions, and rearrangements have been important in the history of theoretical organic chemistry. The allyl compounds exhibit an amazing range of pungent odors, of toxicity, and biological activity which are challenging for correlation with functional groups and structure. They include the allyl compounds of onions and mustard which, although toxic, have been eaten with pleasure in low concentrations by humans since prehistoric times.

Literature studies for this book were begun in 1965 by the late Mabel D. Reiner and myself. A collection of allyl compounds was made and some gaps and uncertainties in the literature were checked by laboratory tests with the assistance of college students. In order to provide a book of practical size it became necessary to include little theory and only the most significant references. Patent references have been chosen for their examples; legal questions of patent claims and priority are beyond the scope of the book. A graph giving publication dates of patents and a list of complete company names is given in the appendix.

The introductory chapter defines allyl compounds and discusses their reactivities in comparison to typical vinyl compounds such as styrenes, acrylic monomers, vinyl halides, and vinyl esters. The second chapter on polymerizations of 1-alkenes (alpha-olefins other than ethylene) was most challenging because of the tremendous recent volume of literature on the epoch-making development of stereoregulated polymerizations, frequently with conflicting claims or omissions relating to priority. I have made ionic stereoregulation a special field for study, first with vinyl ether polymers in 1947. I have made every effort to present an impartial account of the art for synthesis of stereoregular polyolefins and their useful compositions based upon records available, from examination of samples, and from discussions here and in Europe. Some later chapters were difficult because so little research has been reported on polymerizations. Many of these allyl compounds may be expected to yield high polymers by the use of new catalysts such as modified Ziegler-Natta and Grignard systems.

I continue to believe, as in my preface of Volume X of the High Polymer Series, that the struggles, successes, and failures of technology eventually contribute to theory. Advances in theory based upon the work of more than a century with allyl compounds, and hopefully with new experiments giving greater attention to monomer purity and to kinetics, should be the subject of a future book.

I wish to acknowledge encouragement in polymer research and writing from Herman Mark and C. S. (Speed) Marvel since the 1940s. I have received help with this book from my teacher, E. Emmet Reid of Baltimore, who first pointed out to me in 1932 some peculiarities of allyl compounds and industrial polymers. His 100th birthday on June 27, 1972, was a special day for organic chemistry. My manuscript had the advantage of criticism and assistance from a number of pioneers in research of addition polymerizations and their application to manufacture of useful products. These include Franklin Strain, William Simpson, Arthur D. F. Toy, Edwin J. Vandenberg, William R. Dial, Edmund F. Jordan, Jr., George B. Butler, William E. Hanford, Robert N. Haward, Paul Lagally, Walter M. Thomas, the late Leo A. Wall, and the late Edward L. Kropa.

Assistance is also acknowledged to the following: David Aelony, Massimo Baer, Paul D. Bartlett, H. Richard Basso, Walter Bauer, Harry H. Beacham, Doris Bernstein, Katherine A. Birula, Graham J. Blake, Albert Bloom, John C. Boccalini, Marshall H. Cohen, Harold W. Coles, John A. Cornell, James W. Crossin, Donna E. Cullison, Helen H. Darrah, Aldo DeBenedictis, Charles H. Fisher, Rudolph F. Fischer, Werner Freudenberg, Junji Furukawa, Kiyoshi Fujii, Carlin F. Gibbs, C. E. Gleim, Robert H. Goldsmith, Frank P. Greenspan, Howard L. Gerhart, David W. Hartman, Toshinobu Higashimura, Mervin F. Hoover, Dirk T. A.

Huibers, Lester M. Jampolsky, Michael Kareha, Wolf Karo, John L. Kinnin, Edward C. Leonard, Jr., Jules Lindau, James S. Long, Robert L. Markus, Robert N. Meals, Willem J. Mijs, Arnold E. Molzon, Herman D. Noether, Alex G. Oblad, L. Raymond Olson, Janet Ommert, Emery Parker, Solomon H. Pinner, Peter H. Plesch, Al C. Poshkus, Malcolm M. Renfrew, Gerhard Rehse, George Roedel, David E. Schildknecht, August Schwalbach, Joel A. Seckar, James W. Shenck, Thomas S. Shoupe, Ozias F. Solomon, Irving Skeist, Curt W. Smith, Howard W. Starkweather, Jr., James Starner, John K. Stille, Eugene A. Talley, James L. Thomas, Ivor H. Updegraff, Emil A. Vitales, Melvin J. Wagner, John E. Watson, Glenn S. Weiland, Robert F. Williams, Ruth A. Willis, Albert J. Winters, Arthur N. Wrigley, Bunichiro Yamada, Ronald H. Yocum, Morris Zief, American Cyanamid Company, AMP, Inc., Armstrong Cork Company, Borden Chemical Company, Dow Chemical Company, FMC Corporation, Fritzsche Brothers, The B. F. Goodrich Company, Goodyear Tire & Rubber Company, Hoffmann-LaRoche, Morton Chemical Co., Pennwalt Corporation, Pfaltz & Bauer, Inc., Pierce Chemical Co., PPG Industries, Sartomer Resins, Shell Development Company, Stauffer Chemical Company, Libraries of The Chemists' Club and U.S. Patent Office. Some contributions of unpublished work are acknowledged with the references. For the final form of the book I am indebted to my wife, Althea J. Schildknecht.

CALVIN E. SCHILDKNECHT

RD3, Gettysburg, Pennsylvania 17325
March 1, 1973

CONTENTS

ALLYL COMPOUNDS AND THEIR POLYMERS
(Including Polyolefins)

1. INTRODUCTION TO POLYMERIZATIONS OF
VINYL AND ALLYL COMPOUNDS

This chapter outlines the scope and reactivity of allyl compounds in relation to other ethylenic compounds, especially the better-known vinyl compounds, including styrenes, acrylic monomers, vinyl halides, and vinyl esters. The numerous allyl compounds have been relatively unfamiliar to scientists and reviews of polymerizability have given allyl compounds little attention (1). In general, these compounds are more difficult to polymerize; they are highly reactive in ways other than polymerization, and their nomenclature frequently has been misleading. Some statements made in this chapter are documented more completely in later chapters.

The odors and biological properties of allyl and other ethylenic compounds, like their polymerizations and other addition reactions, seem to be related to inductive effects and resonance, as well as to highly reactive allylic atoms or groups (alpha hydrogen, chlorine, etc.). This is a promising area for further research and generalizations. Many allyl compounds occur in plants, and some may affect the health and behavior of animals even in low concentrations. Onions provide an old example; the sexy synthetic allyl esters in some modern perfumes are new examples.

The polymerization reactions of different allyl monomers offer interesting comparisons both of reactivities and of useful synthetic products, such as heat-resistant plastics (polypropylenes, diallyl phthalate polymers, and diallyl carbonate polymers), as well as fibers, rubbers, and adhesives that have made possible advanced techniques in medical, optical, electronic, aircraft, space, and computer technology. The monoallyl compounds do not homopolymerize readily by conventional free radical polymerizations. However, the monomers bearing two or more allyl groups often yield formable, soluble prepolymers by controlled polymerizations with peroxide or other free radical initiators. Such prepolymer plastics, suitable for molding with catalysts to form thermoset or

1

cured polymer shapes, are a feature of the allyl polymers. In contrast, most polyfunctional vinyl, acrylic, and styrene derivatives do not readily form fusible polymers. A number of the diallyl monomers form polymers with unique cyclic segments.

The study of the literature in writing this book led to a classification of ethylenic compounds in which the 1-olefins are shown to belong among the allyl compounds with low reactivity of the double bonds in polymerization and high reactivity of allylic or alpha hydrogen atoms. Closely related to these features is the wide utility of allyl compounds as intermediates for organic synthesis, often involving addition and substitution reactions. The allyl compounds emphasized here are those compounds containing the unsubstituted terminal $CH_2=$ or $CF_2=$ group [as in $CH_2=CHCH_2Y$ and $CH_2=C(CH_3)CH_2Y$], where the ethylenic group has little activation from resonance or polarity from attached electron-releasing or electron-repelling groups.

Vinyl-type monomers bearing the $CH_2=C\langle$ or $CF_2=C\langle$ groups may be divided into four groups:

1. Vinyl monomers including vinyl esters, acrylic monomers, and styrenes in which electronic displacing and/or resonating groups facilitate polymerization of both the monofunctional and polyfunctional compounds to form polymers and copolymers of high molecular weight, many of which have proved useful.

2. Monomers of lower polymerization activity lacking strong electron-displacing groups directly attached to the ethylenic nucleus and having no reactive allylic hydrogen or halogen on the third carbon. These compounds, such as $CH_2=CHC(CH_3)_3$ and $CH_2=CHC(CH_3)_2OH$, have not received much attention in polymerizations until recently. That they do not homopolymerize readily shows that degradative chain transfer at allylic hydrogen atoms is not solely responsible for the inability of allyl compounds to form homopolymers by conventional free radical and ionic initiations.

3. Typical allyl monomers that polymerize with difficulty because of lack of inductive and/or resonating substituents attached directly to the ethylenic nucleus and because of competing degradative reactions of H or X attached to the third carbon. In general, only the polyfunctional monomers give polymers of high molecular weight by radical initiation. However, Ziegler-type catalysts seem unique in forming high polymers from a number of monoallyl compounds, such as 1-alkenes $CH_2=CHR$.

4. Nonexistent compounds such as the vinyl and iso-

propenyl alcohols and amines, which would have strong
electron-repelling OH and NH_2 groups substituted into
ethylene. The enol forms such as vinyl alcohol CH_2=CHOH
lack resonance stabilization, in contrast to highly
reactive but existing phenol and aniline.

   In the familiar styrene, acrylic, and conjugated diene
monomers, a group is attached to the ethylenic nucleus
which attracts electron density and may provide resonance.
This withdrawal and delocalization of the π-bond electrons
of the double bond activates these monomers for polymerization by free radical methods (e.g., heating with
peroxide or azo catalysts), which are most often employed
in commercial polymerizations.

CH_2=CH⟨◯⟩ CH_2=CHCN CH_2=CHCCl=CH_2
     styrene   acrylonitrile   chloroprene

   For example, polymerization of many liquid vinyl monomers may be carried out with 1% added benzoyl peroxide or
azo initiator by heating at 50 to 70°C in sealed tubes,
preferably with precautions to exclude air. Liquid
styrene slowly becomes viscous from dissolved polymer;
finally it is transformed into a solid, clear polymer
mass. Acrylonitrile heated with peroxide under nitrogen
with exclusion of air becomes turbid from insoluble
polymer and later may polymerize violently from local
overheating in the monomer-polymer slurry.

$$CH_2=CHCN \xrightarrow[\text{catalyst}]{\text{radical}} --CH_2\underset{\underset{CN}{|}}{CH}-CH_2\underset{\underset{CN}{|}}{CH}-- + 17 \text{ kcal/mole}$$

In contrast, most monoallyl monomers do not yield solid
polymers when they are heated with free radical catalysts.
The less efficient allyl polymerizations, however, may
yield solid polymers when polyfunctional monomers such as
diallyl phthalates (bearing electron-withdrawing groups
on the third carbon) are heated at higher temperatures and
with larger concentrations of peroxide. Free radical
polymerizations may also be promoted by heating and by
ultraviolet light or other higher energy radiation. Many
of the vinyl monomers also respond to polymerization by
Lewis-basic or acidic catalysts at moderate or low
temperatures (ionic polymerizations).
   There seems to be an optimum electron withdrawal for
radical homopolymerization of vinyl-type compounds,
since monomers such as CH_2=C(CN)$_2$, CH_2=C(CN)COOCH$_3$, and
CH_2=CHNO$_2$ bearing very strongly electron-attracting

groups (meta-directing in benzene) do not polymerize
readily with peroxide catalyst or in UV light. This may
result from repulsive forces for the following reasons:
these monomers copolymerize readily with monomers bearing
electron-donating groups, and they homopolymerize readily
in presence of basic catalysts which by complexing may
reduce their polarity. Resonance-activated, relatively
nonpolar monomers such as styrene may be polymerized by
many methods of polymerization: free radical, cationic
(Lewis-acid catalysts), anionic (basic catalysts), or by
Ziegler-Natta catalysts.

Commercial radical polymerizations are most often
carried out by heating with small amounts of peroxide,
persulfate, or azo catalysts in the following types of
systems: (a) bulk or by casting (without solvent),
(b) in solution, (c) in aqueous suspension, or (d)
aqueous emulsion (2).

Vinyl monomers bearing electron-donating groups such as
the following do not homopolymerize to polymers of high
molecular weight by radical methods; instead, they
respond to Lewis-acidic catalysts (usually at low temperatures in order to control the reactions):

$$CH_2=CHOR \qquad CH_2=C{\overset{\displaystyle CH_3}{\underset{\displaystyle OR}{\big<}}} \qquad CH_2=C(CH_3)_2$$

   vinyl alkyl ethers   isopropenyl ethers   isobutylene

It is most interesting that a number of monomers that
would bear very strong electron-donating groups are unstable and do not exist. These include the vinyl and
isopropenyl alcohols, mercaptans, and amines. When
attempts are made to synthesize these monomers, the
aldehyde, ketone, thioketone, and imine isomers generally
result. This behavior is consistent with the greater
stability of enols bearing electron-attracting groups
(e.g., in acetoacetic and malonic ester). The vinylidene
glycols and diamines also are nonexistent, in contrast to
vinylidene ethers $CH_2=C(OR)_2$. Allyl and methallyl
alcohols, mercaptans, and amines exist, but they are
highly reactive. The instability of vinyl alcohol is
represented:

     nonexistent $CH_2=CH \longrightarrow CH_3\underset{O}{CH}$
     vinyl alcohol HO $\overset{\parallel}{O}$ acetaldehyde

The driving forces in this reaction are the tendency for
the proton to become more positive by moving to carbon
and the tendency for oxygen to become more negative by

forming a double bond (3). The positive charge on the
alpha carbon atom attracts a π-electron pair from oxygen
and the negative charge of the beta carbon atom attracts
the proton of the hydroxyl group. Both in monomers and
in polymers there is growing evidence that greater
stability is imparted by dispersal of electron density
from unsaturated centers by electron-attracting groups,
as well as by resonance (4).

Allyl compounds are ethylenic compounds having lower
reactivity in polymerization than typical vinyl compounds
and generally higher reactivity in other reactions (5),
which often compete with polymerization reactions. The
word allyl is derived from allyl sulfide compounds found
in plants of the Allium or onion family; the name has
been most frequently used for the radical $CH_2=CH\overset{.}{C}H_2$. That
monoallyl compounds have not yielded homopolymers of high
molecular weight by conventional methods of free radical
and ionic polymerization has been attributed principally
to the high reactivity of allylic hydrogen or halogen
atoms on the third carbon in terminating growth of chain
molecules (degradative chain transfer), as taught by
Bartlett and co-workers (6). Thus the monomer itself acts
as an inhibitor of the formation of long-chain polymer,
and only liquid low polymer results. The small increase
in viscosity of such polymerizing systems makes the reactions appear slower than they actually are.

This book emphasizes the importance of the second factor
in the reluctant polymerization of allyl compounds; namely,
the lack of activating electron-displacing and/or
resonance-promoting groups attached to the ethylene
nucleus. Such groups not only activate ethylenic double
bonds of vinyl monomers toward addition polymerization,
but they also may prevent activity of hydrogen on a third
carbon in degradative chain transfer (7). Thus methyl
methacrylate $CH_2=\underset{\underset{COOCH_3}{|}}{C}CH_3$ does not behave as an allyl com-
pound, although formally it appears to possess allylic
hydrogen atoms. Resonance and electron attraction by the
$COOCH_3$ group prevents methyl methacrylate from reacting
as an allyl compound.

When there is no reactive hydrogen or chlorine on the
third carbon for degradative chain transfer, as in
$CH_2=CHC(CH_3)_3$, radical polymerizations occur somewhat
more readily; reactivity, however, is still relatively
low. For example, $CH_2=CH-C(CH_3)_2NCO$ copolymerized with
ethylene more readily than did allyl isocyanate, and
homopolymers were obtained by heating with benzoyl peroxide (8). Compounds of the type of $CH_2=CHC(CH_3)_2OCH_2NR_2$
copolymerized with methyl acrylate on heating at 65°C

with azobisisobutyronitrile (azobis) (9). The copolymer
contained 79% acrylate ester units. Note that ethylene,
lacking both activating groups and allylic hydrogen atoms,
requires forcing conditions of high pressure or transition metal catalysts of the Ziegler type for practical
homopolymerizations.

We classify 1-olefins $CH_2=CHCH_2R$, allylidene $CH_2=CH\overset{.}{C}H$,
and also $CH_2=C\overset{.}{C}H$ compounds as allyl compounds. In these
compounds the ethylenic nucleus lacks strong activating
groups, and allylic hydrogen atoms are present on the
third carbon for chain transfer (10). Allylic hydrogen
atoms are underlined in the following examples:

$CH_2=CHC\underline{H}_3$ $CH_2=CHC\underline{H}_2Cl$ $CH_2=CHC\underline{H}(OOCCH_3)_2$

propylene   allyl chloride   allylidene diacetate

Removal of allyl hydrogen or chlorine atoms can terminate
a growing chain radical, and the resonance-stabilized
radicals formed have low efficiency in propagating a new
polymer chain: $R + CH_2=CHCH_3 \longrightarrow RH + CH_2\text{---}CH\text{---}CH_2$.

Study of the literature as represented by this book
supports such a classification of allyl compounds by
showing that such monofunctional monomers generally do
not form homopolymers of high molecular weight by addition
polymerizations using conventional catalyst systems. However, special complex catalyst systems, such as Ziegler-
Natta transition metal catalysts, can yield homopolymers
of controlled steric structure, especially from 1-alkenes
and monomers of lowest polarity. Many polyfunctional
allyl compounds, however, can form high-molecular-weight
homopolymers by suitable radical initiation, in spite of
macroradical wastage by degradative chain transfer and by
cyclization. This, for example, is true of polyfunctional
allyl esters and other polyfunctional monomers bearing
electron-attracting groups on the third carbon atom.
These compounds respond to convenient bulk polymerization
with peroxide or azo initiators for making useful cast
forms and molding plastics.

Convenient but misleading names such as vinyl cyclohexane, vinyl dioxolane, and methylene cyclohexane should
not conceal the fact that such compounds are actually
allylic in structure and behavior. Nature does not
recognize chemical classifications, and it is not surprising that borderline cases exist. Among the compounds
that behave somewhat like allylic compounds but are not
discussed in detail here are α-methylstyrene (11), isopropenyl phenyl ketone (12), and isobutylene (13). A
hydrogen atom apparently can be removed from these
monomers rather easily by a growing macroradical to form

a resonance-stabilized radical of low reactivity, which does not readily continue free radical chain growth. However, these compounds form homopolymers by conventional ionic systems, in contrast to typical allyl compounds.

Compounds bearing CX_3 (trihalomethyl groups) seem to represent a special case. For example, $CH_2=CHCF_3$ polymerizes at high pressures with lower ratios of transfer to propagation than are found in the case of propylene (14). The CF_3 is a relatively stable electron-attracting group. However, high-melting polymers reported from irradiation of $CH_2=CHCCl_3$ were attributed to isomerization polymerization giving a 1,3-chain structure (15). Compounds having the structure $CH_2=CHCF_2CF_2R$ also seem to give radical polymerizations more readily than typical monoallyl monomers (16). Fluoroalkenes are discussed in Chapter 4.

Beta-substituted ethylenic compounds such as trans-β-methylstyrene and cis-2-butene have lower reactivities in polymerizations than vinyl and allyl compounds. Such ethylenic compounds lacking $CH_2=$, $CF_2=$, and $CHF=$ groups might be called vinylic and allylic, respectively. Note that some beta-substituted polar monomers such as maleic anhydride, free from bulky substituents, readily copolymerize with monomers of opposite polarity.

The polymerization characteristics of different groups of vinyl-type compounds are now outlined as a background for the polymerizations of allyl compounds. Other reactions besides polymerizations must be considered also in preparing, purifying, handling, and polymerizing the monomers to obtain products of reproducible structure and properties for commercial use.

FREE RADICAL-CHAIN POLYMERIZATIONS

Characteristics of initiation and inhibition by small amounts of added agents and the kinetics of these polymerizations on heating and in light suggested application of the theory of chain reactions involving reactive odd electron or radical species (17). Since the 1930s this theory has given good service in interpretation of vinyl polymerizations other than those initiated by added Lewis acids and Lewis bases. Free-radical-producing initiators and light can promote polymerizations with high yields. Small amounts of molecular oxygen, arylamines, phenols, and allylic compounds may inhibit or retard these polymerizations. Typical kinetics and mechanisms have been interpreted in terms of basic reactions of initiation, propagation, chain transfer, and termination (18). A propagation step in the formation of chain molecules of linear vinyl chloride polymer initiated by a peroxide or azo catalyst may be represented

$$R(CH_2\overset{\bullet}{C}H)_n CH_2\overset{\bullet}{C}H + CH_2=CHCl \longrightarrow R(CH_2\overset{\bullet}{C}H)CH_2\overset{\bullet}{C}H + 17 \text{ kcal/mole}$$
$$\quad\quad Cl \quad\quad Cl \quad (\text{gas or liquid}) \quad\quad Cl_{n+1} \; Cl$$

   In typical free radical polymerizations, all the
hundreds to thousands of individual propagation steps for
building up a macromolecule of 10^5 to 10^7 mol. wt. may be
completed within a second or even less. The active
radical species have short life except at high conversion
to polymer, when trapped radicals may terminate very
slowly because of high local viscosity. At low conversion
there is already some completed polymer of high molecular
weight dissolved or precipitated in the unreacted monomer.
This contrasts with condensation polymerizations forming
polyesters and polyamides, where the whole mass passes
through mixtures of dimer, trimer, and so on. In the
actual polymerizations of vinyl chloride in bulk, the
polymers formed are somewhat branched and they precipitate
as formed.
   Acrylic, styrene, and vinyl monomers are familiar because most of them polymerize easily by free radical chain
reactions with heating or irradiation, and many of their
polymers have achieved commercial importance. Their
behavior can be correlated in terms of three major
qualities of the groups substituted into ethylene:
(a) inductive effects or polarity (electron-attracting
versus electron-repelling groups), (b) resonance or conjugation, and (c) steric effects. Electron-attracting
and conjugated groups normally promote polymerization by
radical initiation as well as by anionic methods (Lewis-
basic or electron-donating catalysts). Electron-donating
groups promote polymerization by cationic methods
(Lewis-acid or electrophilic catalysts). It is interesting
that most volatile acrylic and styrene monomers having
pronounced conjugation and unsubstituted alpha-hydrogen
atoms have strong odors, whereas vinyl esters, vinyl
halides, and methacrylic monomers have relatively mild
odors. Vinyl and acrylic compounds rarely occur naturally,
but many allyl compounds are found in nature.
   Quantitative values for polarity and for reactivity
(attributed largely to resonance) can be expressed,
respectively, as the e and Q values of Alfrey and Price
for the different monomers (19). These are related to the
monomer reactivity ratio obtained from radical copolymerizations in bulk or solution (usually with peroxide
catalyst, in the range of 50 to 80°C) to low conversion,
followed by analysis of the resulting copolymers. In
general, the less reactive monomers give the most reactive

radicals. Thus styrene radicals and acrylonitrile monomer are relatively reactive. In radical copolymerization, the reactivity ratio (r_1 = 0.52) represents the relative rate of acrylonitrile monomer to that of styrene monomer in adding to an acrylonitrile radical end of a growing chain. The reactivity ratio of styrene (r_2 = 0.03) represents the relative rate of styrene monomer to that of acrylonitrile monomer in adding to a growing chain bearing a styrene radical. Reactivity ratios predict relative rates and compositions of copolymers in radical systems but not actual laboratory rates of reaction.

Very bulky, nonconjugated alpha substituents in ethylenic compounds generally prevent formation by radical initiation of homopolymers of high molecular weight, and copolymerizations may also occur reluctantly; for example, α-isopropyl acrylates and α-cyclohexyl acrylates (20). Steric effects also prevent most 2-substituted ethylenic compounds, such as 1,2-diphenyl ethylene compounds, from readily forming polymers of high molecular weight. Recently cyclobutene and related cyclic 1,2-disubstituted ethylenes have given low yields of saturated or unsaturated polymers depending upon the transition metal catalyst used (21).

The composition and structure of polymers and copolymers prepared by radical-chain reactions are relatively little affected by specific peroxide or azo catalysts or by types of radiation employed. This is in contrast to ionic polymerizations, where in some cases reactivity ratios and polymer structure, including stereoisomerism, can be varied by choice of catalyst-solvent system. Incompletely soluble and heterogeneous ionic catalysts may control each monomer addition step rather than merely functioning as an initiator.

Examples of the relatively high e and Q values of acrylic monomers resulting from electron-attracting and conjugated substituents follow:

		e	Q
Acrylic acid	$CH_2{=}CHCOOH$	0.77	1.15
Ethyl acrylate	$CH_2{=}CHCOOC_2H_5$	0.22	0.52
Methacrylic acid	$CH_2{=}\overset{\overset{\displaystyle CH_3}{\vert}}{C}COOH$	0.65	2.34
Methyl methacrylate	$CH_2{=}\overset{\overset{\displaystyle CH_3}{\vert}}{C}COOCH_3$	0.40	0.74
Acrylamide	$CH_2{=}CHCONH_2$	1.30	1.18
Acrylonitrile	$CH_2{=}CHCN$	1.20	0.60
Methacrylonitrile	$CH_2{=}\overset{\overset{\displaystyle CH_3}{\vert}}{C}CN$	0.81	1.12

Esterification of the free acids reduces both e and Q values. Association of the free acids and amides by hydrogen bonding, especially in polar solvents, is a complicating factor. In general, acrylic monomers tend to polymerize rapidly. The acrylic monomers (but not the methacrylic) tend to give branched polymers unless regulators such as mercaptans are added. Lewis acids tend to retard or inhibit radical polymerization of acrylic and methacrylic ester monomers, as well as vinyl halides and isopropenyl acetylene (22). This might be called ionic inhibition. It is suggested that the group of Lewis-acidic monomers including acrylics, methylene malonics, nitroalkenes, and itaconic monomers may be called acethenes.

   Styrene is given arbitrary values of e = -0.80 and Q = 1.0 as a reference monomer. Under similar conditions of radical initiation, styrene polymerizes more slowly than methyl methacrylate and ethyl acrylate. The reactivities of styrene derivatives are greatly affected by the substituents as shown in the following:

	e	Q
o-Methylstyrene	-0.78	0.90
m-Methylstyrene	-0.72	0.91
p-Methylstyrene	-0.98	1.27
α-Methylstyrene	-1.27	0.98
m-Divinyl benzene	-1.77	3.35
p-Cyanostyrene	-0.21	1.86
p-Chlorostyrene	-0.33	1.28
p-Nitrostyrene	+0.39	1.63

   The reactivity of different styrene derivatives follows approximately the orientation effects of substituents in benzene substitution and sigma values in the Hammett equation (23). Strongly meta-directing and electron-attracting groups (as in the nitrostyrenes) give high e values. The other styrene monomers here tend to co-polymerize readily with acrylic monomers such as acrylo-nitrile, generally giving alternating monomer units in the macromolecule chains. Such alternating copolymerizations between reactive monomers having opposite inductive effects often occur even without added free radical initiator (cross-initiation). The freedom from catalyst residues can have practical advantages. These reactive monomer combinations can be regarded as self-initiating. The tendency to alternation increases as the product of the reactive ratios $r_1 \times r_2$ approaches zero. When random addition of comonomer units occurs readily, $r_1 \times r_2$ approaches 1.

Vinyl halides and vinyl esters are characterized by low
Q values (little conjugation or resonance). Fluorine
raises e values remarkably:

		e	Q
Vinyl chloride	$CH_2=CHCl$	0.20	0.04
Vinylidene chloride	$CH_2=CCl_2$	0.36	0.22
Vinyl fluoride	$CH_2=CHF$	1.28	0.01
Ethylene	$CH_2=CH_2$	-0.20	0.02
Vinyl acetate	$CH_2=CHOCOCH_3$	-0.22	0.03
Vinyl n-butyrate	$CH_2=CHOCOC_4H_9$	-0.26	0.04
Vinyl trifluoroacetate	$CH_2=CHOCOCF_3$	1.06	0.03

These monomers homopolymerize and copolymerize with each
other by free radical methods but do not give high polymers readily by conventional methods of cationic or
anionic polymerization. With monomers of high e and Q
values they often resist copolymerization even with high
temperatures. For example, mixtures of vinyl acetate and
styrene are self-inhibited under ordinary conditions. A
number of such monomers undergo chain transfer readily
with monomer and with polymer and tend to give branched
polymers. In low reactivity, these monomers approach in
some systems the behavior of vinyl ethers and monoallyl
monomers.

A group of N-vinyl monomers has still lower e values,
but some homopolymerize in selected free radical systems
because of higher Q values:

	e	Q
2-Vinyl pyridine	-0.50	1.30
4-Vinyl pyridine	-0.20	0.82
N-Vinyl pyrrolidone	-1.14	0.14
N-Vinyl carbazole	-1.40	0.41
N-Vinyl phthalimide	-1.52	0.36

These monomers may respond also to cationic polymerization,
and added bases act as inhibitors against polymerization.
Quaternary salts of the vinyl pyridines polymerize much
more readily than the free amines.

The conjugated dienes are divinyl and diisopropenyl
compounds showing high Q values and low e values. They
polymerize most readily in heterogeneous systems, such as
by emulsion free radical polymerization, and rather
unpredictably in proliferous or popcorn polymerization
upon active surfaces. They often show only slow radical
polymerization in homogeneous solutions or in bulk.

	e	Q
1,3-Butadiene or divinyl	-1.05	2.39
Isoprene or vinyl isopropenyl	-1.22	3.33
Chloroprene or 2-chloro-1,3-butadiene	-0.02	7.26
Hexafluoro-1,3-butadiene	0.47	0.93

Some monomers bearing strongly electron-attractive groups do not homopolymerize readily by free radical initiation. However, the first two compounds listed below homopolymerize even with such relatively weak Lewis-basic initiators as water, and they polymerize very readily with added basic comonomers such as vinyl ethers and N-vinyl compounds.

		e	Q
Methyl α-cyanoacrylate	$\overset{CN}{CH_2=\overset{|}{C}COOCH_3}$	2.10	High
Vinylidene cyanide	$CH_2=C(CN)_2$	2.58	20.13
Maleic anhydride	$\underset{O}{\overset{CH=CH}{O=CC=O}}$	2.25	0.23
Maleonitrile	$\overset{CH=CH}{CNCN}$	2.32	0.42
Fumaronitrile	$\overset{CN}{\underset{CN}{CH=CH}}$	1.96	0.80

In the cases of the cyanoacrylates and vinylidene cyanide, strong Lewis acids are useful inhibitors against polymerization. The 1,2-substituted maleic and fumaric monomers do not easily give homopolymers of high molecular weight, but they copolymerize very readily with basic monomers.

Figure 1.1 shows that e values are approximately proportional to the Hammett sigma values for monomer substituents. Vinylidene cyanide lies outside the range of e values +1.6 to -1.0 and sigma +0.7 to -0.1, where radical homopolymerization occurs to products of high molecular weight. Isobutylene and other monomers with low e values also do not homopolymerize readily by radical methods.

The high reactivities of acrylic, styrene, and con-jugated diene monomers in free-radical-initiated poly-

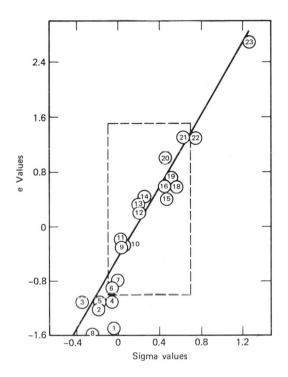

Fig. 1.1. Approximate parallel between e values and Hammett sigma
values in comparing polarity and polymerizability of monomers.
Radical homopolymerizations are possible with monomers within the
broken line: (1) methyl vinyl sulfide; (2) p-dimethylaminostyrene;
(3) isobutylene; (4) p-methoxystyrene; (5) α-methylstyrene;
(6) p-methylstyrene; (7) styrene; (8) ethyl vinyl ether; (9) p-
chlorostyrene; (10) p-iodostyrene; (11) p-bromostyrene; (12) vinyl
chloride; (13) p-cyanostyrene; (14) p-nitrostyrene; (15) methyl
methacrylate; (16) vinylidene chloride; (18) methyl acrylate;
(19) methyl vinyl ketone; (20) methacrylonitrile; (21) acrylonitrile; (22) methyl vinyl sulfone; (23) vinylidene cyanide. Sigma
values of disubstituted compounds were taken as the algebraic sum
of the sigma values of the two substituents. From J. Furukawa and
T. Tsuruta, J. Polym. Sci. _36_, 275 (1959)

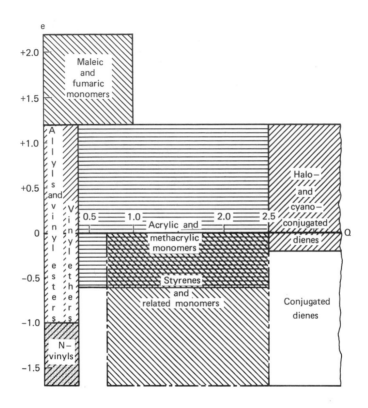

Fig. 1.2. Comparative reactivities of different families of
monomers are shown by e values (polarity), plotted vertically, and
Q values (reactivity from resonance), plotted horizontally: Lewis-
acid monomers at the top; Lewis-basic monomers at the bottom;
monomers with resonance (conjugation) to the right. Allyl monomers,
like alkyl vinyl ethers and most vinyl esters, have low resonance
and e values between -1.7 and +1.2. The ranges of the reactive
acrylic monomers overlap those of common styrene derivatives.
Vinylidene cyanide, cyanoacrylates, and α-cyanostyrene would be off
the diagram to the upper right. Conjugated dienes have very high
reactivity in polymerization compared with nonconjugated dienes such
as diallyl. C. E. and D. E. Schildknecht.

14

merizations are presented in Fig. 1.2. With a few exceptions, the values of e (polarity or inductive effects)
and Q (reactivity from resonance) reported for the
numerous monomers lie within the zones indicated. The
e-Q area of the acrylics and methacrylics overlaps that
of styrene and its derivatives. The allyl monomers,
vinyl halides, vinyl esters, vinyl ethers, and a number
of N-vinyl compounds overlap each other in a band of low
resonance from e values of about -1.7 to +1.2. The maleic
and fumaric monomers have higher e values and considerably
more resonance. Vinylidene cyanide, cyanoacrylates, and
α-cyanostyrene do not respond to radical homopolymerization; their e and Q values place them off the graph, above
and to the right. Nitro groups can introduce into such
monomers as nitroethylene Lewis acidity and resonance
comparable to that found in cyanoacrylics. Since there
are few data on nitro monomers, they are not included in
the plot. Also omitted are aminostyrenes where Lewis
basicity causes inhibition in free radical systems.

   Note that a very wide range of monomers covering most
of the chart respond to anionic polymerization by Lewis-
basic catalysts such as alkali metals and metal alkyls.
However, the allyl monomers, vinyl halides, vinyl esters,
and vinyl ethers do not homopolymerize by basic catalysts.
These monomers generally have mild odors--in contrast to
the strong odors of acrylic, styrene, conjugated dienes,
and other high-resonance monomers. The relatively mild
odor of methacrylates $CH_2=C(CH_3)COOR$ might be attributed
to the electron-releasing methyl group opposing the
resonating, electron-attracting COOR group or to the
steric effect. Note that the functional groups rather
than the overall molecular size or shape determine the
powerful odor of acrylate esters. There is no great
difference in stench from methyl acrylate to butyl
acrylates. A few allyl sulfur compounds have comparable
high olefactory potency. Some recent theories of odor,
in publications of J. E. Amoore and R. W. Moncrief, have
neglected effects of functional groups.

   Besides monomer structure and catalyst type, several
other factors may influence addition polymerizations,
especially those by free radical-chain reactions. One
is acceleration by reduced chain termination resulting
from high viscosity or immobilization of radicals in a
solid phase (trapped radical or gel effect). Another is
chemical participation of solvent or other nonmonomeric
compounds in polymerization (telomerization).

   The concept of acceleration of free radical polymerization and increase in polymer molecular weight resulting
from high viscosity preventing termination of macro-

radicals [Trommsdorff effect (24)] has proved to be very
useful. The termination rate constant may be inversely
proportional to solvent viscosity (25), but for stiff
chains the effect may be less. We suggest that the
accelerated polymerization of monomers such as methyl
methacrylate in poor solvents (e.g., lower alcohols) may
involve inaccessibility of macroradical ends within
clumped chain molecules. When a monomer is a poor solvent
for its polymer, as in the case of acrylonitrile, bulk
polymerizations become very rapid.

TELOMERIZATIONS

   Knowledge of vinyl-type polymerizations has been
advanced through studies of chemical participation of
solvents and other added saturated compounds in free
radical polymerizations. In reactions with inhibitors
such as aromatic amines or nitro compounds, phenols, and
mercaptans, a growing macroradical is believed to abstract
hydrogen, thus terminating the chain and giving a
resonance-stabilized inhibitor radical incapable of
propagating a new macromolecule. Reactive solvents and
other telogens may give terminated polymers (telomers),
plus new radicals that may be capable of propagating new
polymer chains. Staudinger and Schwalbach observed that
low-molecular-weight polymers of vinyl acetate prepared
in chloroform solution contained chemically bound
chlorine (26). Hanford and Joyce suggested the names
telogen and telomer in systems giving products of low
molecular weight; the use of these terms has gradually
been extended, however, to terminated polymers of higher
molecular weights. In the high-pressure aqueous polymerization of ethylene in the presence of halogen compounds,
Hanford and Joyce (27) found that carbon-halogen bonds
were attacked, giving liquid to waxy halogen-terminated
telomers. Ethylene and CCl_4 gave mixtures of tetra-
chloroalkanes $Cl_3(CH_2CH_2)_nCl$. Chloroform was unusual
among halogen compounds in that hydrogen was attacked and
the telomers contained terminal CCl_3 groups which, on
hydrolysis, gave carboxyl groups. Ethers and related
compounds were particularly reactive in telomerization
with ethylene, forming waxlike materials of 20 or more
monomer units (28). Telomerizations have been used in
preparation of commercial low-molecular-weight fluoropolymers but have been rather slow in reaching industrial
importance. Recently telomerization of ethylene with
trimethyl borate, followed by methanolysis, has been
investigated for the synthesis of long-chain alcohols (29).
Some of the more interesting possibilities remaining for
exploration include the syntheses of medium and high-

molecular-weight polymers terminated by stable polar groups imparting surface activity, antistatic, and stabilizing properties. The concept of telomers merges with that of chain transfer regulators of the mercaptan type which have long been used in commercial butadiene and acrylic polymerizations.

Relative rates of formation of low-molecular-weight telomers between $CH_2=CHCH_2Y$ compounds and CCl_4 under gamma radiation were studied in Russia (30). Electron-withdrawing Y groups caused faster reaction. At lower temperatures the ratio approached 1:1. Telomers of allyl alcohol with CCl_4 lost HCl on distillation.

It is not surprising that polar and resonance factors are important in chain transfer, as shown by values of chain transfer constants C defined by the formula $1/DP = C/M + 1/DP_0$ (where M is monomer concentration and DP is degree of polymerization or average number of monomer units per polymer molecule) (31). With carbon tetrabromide at 60°C, chain transfer constants were reported as follows (32): vinyl acetate, 39; styrene, 2.2; methyl acrylate, 0.41; methyl methacrylate, 0.27. Thus relatively basic monomers including allyl compounds undergo chain transfer most readily, whereas conjugated and acidic monomers, which give less reactive radicals, do not abstract halogen or hydrogen readily. Tertiary hydrogen atoms, such as in isopropyl compounds and acrylate ester polymers, are more reactive in chain transfer than primary and secondary hydrogen atoms. Jordan and co-workers made studies of chain transfer constants for styrene, methyl methacrylate, vinyl acetate, and diethyl maleate polymerized with N-allyl stearamide at 90°C (33). An empirical relation was found between the DP of the polymer and the mole fraction of allylic monomer entering into the copolymer. Ethylene with peroxide catalysts is about equally reactive with $CHCl_3$ and with CCl_4 as telogens; vinyl acetate at 60°C is about 60 times as reactive with CCl_4 as with $CHCl_3$ (34).

Most of the work on telomerization of vinyl-type monomers used relatively large proportions of CCl_4 or other telogens, and the halogen was believed to become attached only to the ends of telomer chain molecules. The possibility of telomers acting as telogens to give halogen atoms in addition to those at chain ends seems to have been overlooked. Apparently this occurs when alkyl vinyl ethers "copolymerize" or telocopolymerize with CCl_4 or $CHCl_3$ at 50°C with peroxide initiation or in UV light (35). Styrene telocopolymerized more readily with minor proportions of CCl_4, $CHCl_3$, or C_2Cl_6 with photochemical initiation than by peroxide initiation with heating.

Degradative chain transfer, which opposes monoallyl compounds forming high polymers in radical systems, may be regarded as self-telomerization. Telomers from the reaction of allyl compounds with halogen compounds and other common telogens have been little studied.

IONIC POLYMERIZATIONS

Early in this century, ionic polymerizations such as sodium-catalyzed polymerization of conjugated dienes were studied before initiation by peroxides became widely known. Later special attention was given to ionic poly-merization of those ethylenic compounds which did not respond to free radical polymerization to give products of high molecular weight. Polymerizations of Lewis-basic or electron-rich ethylenic monomers such as isobutylene, alkyl vinyl ethers, and α-methylstyrene initiated by strong Lewis-acids such as BF_3 and $AlCl_3$ are characterized by violence when initiated from ordinary temperatures. The development of controlled conditions at lower tempera-tures by Michael Otto and others was necessary for preparation of high polymers free of side products. It was suggested in the 1930s by Chalmers (36), Whitmore (37) and others that these were chain-type reactions leading to chain polymers by way of growing ionic species.

Cationic and anionic polymerizations include a wider range of behavior and conditions than the free radical reactions. The inhibitors of radical polymerizations have little effect. Among substances that retard or pre-vent ionic polymerizations are those which destroy acids or basic catalysts (e.g., more than traces of water in cationic polymerizations). Low energies of activation, low heats of polymerization, speed of reaction, and other characteristics of ionic polymerizations have been dis-cussed (37); but recently more diverse types have been developed, especially complex heterogeneous catalyst systems such as Ziegler-Natta polymerizations for syn-thesis of stereoregular polymers. Some of these are now considered to be slow step-wise ionic polymerizations.

The behavior of ethylenic monomers in ionic polymeri-zation follows the e and Q values (Fig. 1.2). Monomers of low e value and low Q value respond best to cationic polymerization. Monomers of high e and high Q values polymerize by basic catalysts. Ionic polymerizations generally give greater difficulties in controlling the synthesis of reproducible polymers of high molecular weight than do free radical polymerizations. When they can be controlled, rates of cationic polymerization may increase linearly with monomer and catalyst concentra-tions (Figs. 1.3 and 1.4).

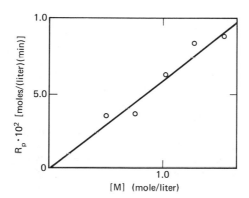

Fig. 1.3. Rates of polymerization of vinyl isobutyl ether in
ethylene dichloride increased approximately linearly with monomer
concentration. Initial catalyst concentration of iodine was 0.175
mole/liter and the temperature was 30°C. S. Okamura, N. Kanoh,
and T. Higashimura, Makromol. Chem. <u>47</u>, 37 (1961).

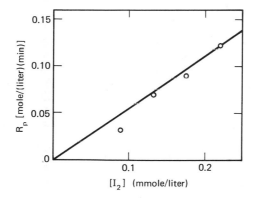

Fig. 1.4. As in Fig. 1.3, the increase in rate is nearly linear
with initial catalyst concentration. S. Okamura, N. Kanoh, and
T. Higashimura, Makromol. Chem. <u>47</u>, 37 (1961).

Behavior in copolymerization is one of the best criteria
of ionic mechanisms. By free radical initiation, mixtures
of two different monomers normally form copolymeric products containing units from both monomers; often this is
not true with ionic systems. Different monomers may
require initiators of different Lewis acidities or
basicities, so that one monomer may homopolymerize without
participation of the second monomer. Thus one alkyl vinyl
ether such as isopropyl vinyl ether can homopolymerize in
the presence of another less reactive vinyl ether. However, ionic copolymerizations are indeed possible under
favorable conditions, as in the preparation of butyl
rubber from isobutylene and isoprene. Monomer copolymerization reactivity ratios are not generally useful in
different types of ionic systems. The composition of
copolymers obtained from ionic systems depends more on
specific catalysts, as well as solvents, which influence
polarity and acidity. In general, there is less tendency
for alternation of units in the copolymer molecules and
more tendency to form blocks of successive identical units.

TABLE 1.1

Growing Macromolecules of Long Life in
Proliferous Cationic Polymerization of
Vinyl n-Butyl Ether at $-78°C$[a]

Reaction time, min	$\eta sp/C$	Conditions
2	2.4	Cold $BF_3 \cdot (C_2H_5)_2O$ added to
5	4.5	10 g monomer, 40 g liquid propane, and 40 g solid CO_2.
10	6.7	Viscosities from solutions of 0.20 g polymer per
15	7.3	100 ml benzene at 25°C

[a]C. E. Schildknecht, A. O. Zoss, and F. Grosser,
Ind. Eng. Chem. 41, 2893 (1949).

Growing macroions can have long life (minutes or longer),
as in the slow proliferous cationic polymerization of
vinyl ethers (Table 1.1) or in the polymerization of
styrene by organoalkali metal initiators. In typical
ionic polymerizations, both initiation and propagation
may be regulated by added catalysts. This was demonstrated by the discovery of stereoregulation in ionic

polymerizations of monomers of the type $CH_2=CHY$ (e.g., vinyl ethers and propylene). Many of these ionic polymerizations involve growth of polymer masses directly from heterogeneous catalyst systems. By contrast, in radical polymerizations direct growth of polymer masses is seldom encountered in the synthesis of linear polymers.

Ionic polymerizations not only make possible greater variation in monomer reactivity than in radical copolymerizations, but certain electrophilic catalyst systems (in which orbitals of metalloids or transition metal atoms apparently coordinate with monomers as ligands) permit steric control in polymerizations. This area has been called coordination polymerization, stereospecific polymerization, or stereopolymerization; often it is called Ziegler-Natta polymerization because of the commercial success of such catalyst systems in polymerization of l-alkenes.

The first stereopolymerizations forming either stereoregular, crystallizable polymers or random, amorphous polymers by choice of ionic catalysts and other conditions were observed with alkyl vinyl ethers (38), propylene (39), and styrene (39). The more stereoregular DDDD-polymers result from slow growth or proliferous polymerization upon relatively weak Lewis-acid complexes, such as boron fluoride etherates in the case of vinyl ethers, and upon reaction products of alkyl aluminum compounds and titanium halides (Ziegler-Natta catalysts) in the case of propylene and styrene. Later other organometallic catalysts were found to yield stereoregular styrene, acrylic, and methacrylic polymers as well as cis-1,4-diene polymers. Generally more rapid and more homogeneous ionic polymerizations tend to give more amorphous polymers. Of the activated transition metal compound catalysts, titanium favors stereoregular l-olefin polymers, whereas vanadium favors amorphous rubberlike polymers and copolymers. Both in the alkyl vinyl ethers (40) and the alkyl acrylates (41), stereoregular crystallizable polymers form more readily from isobutyl monomers than from ethyl and isopropyl derivatives. X-ray data are the best criteria of crystallinity and stereoregularity in polymers that can crystallize. Figure 1.5 shows x-ray patterns of amorphous and crystalline stereoregular polymers. However, some stereoregular polymers do not crystallize since, as in the case of polyisobutene, the steric factors prevent orderly packing in a crystal lattice. For this reason crystallinity cannot always be used as a criterion of regularity of structure. Few allyl polymers other than polyolefins have been sufficiently regular for crystallization. Polypropylene plastics are the outstanding isotactic stereoregular allyl polymers of industrial importance (see

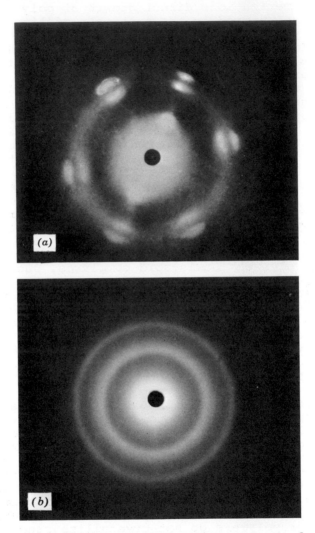

Fig. 1.5. Characteristic x-ray diffraction patterns of stereoregular polymers: (a) spot pattern of drawn (fibered) polymer of
vinyl n-butyl ether; S. T. Gross, C. E. Schildknecht, and A. O. Zoss
(1948) unpublished; (b) halo-type pattern from quenched isotactic
propylene film obtained by rapid cooling or quenching from melt
(smectic form).

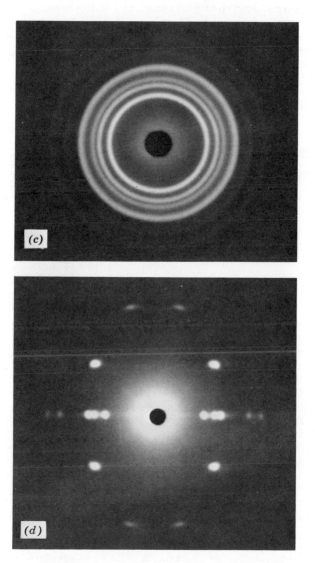

Fig. 1.5. (continued). (c) sharp line type diagram obtained by
annealing the film in (b) by heating for several hours below the
melting point; (d) spot pattern of isotactic polypropylene oriented
by drawing or fibering. X-ray patterns (b), (c), and (d) by
Marjorie D. Chris, Hercules, Inc. (1972) unpublished.

23

Chapter 2). In radical polymerizations, polymer structure cannot be controlled specifically by selection of catalysts, solvents and heterogeneity. However, in some cases lower temperatures of radical polymerization promote more DLDL or syndiotactic structure. Stereoregulated polymerizations of vinyl isobutyl ether either to give normally crystalline isotactic polymer or to form amorphous, atactic polymer were described in 1948 by Schildknecht, Gross, Zoss, Davidson and Lambert. The reactions may be represented as follows:

CH_2=CHOR $\xrightarrow[\text{slow}]{BF_3 \cdot Et_2 O}$

Stereoregular normally crystalline polyvinyl ether. (Crystalline part actually in helical conformation.)

CH_2=CHOR $\xrightarrow[\text{rapid}]{BF_3}$

Normally amorphous polyvinyl ether. (Not forming crystals with helical conformation.)

Most polymerizations initiated by high-energy radiations at room temperature and above are of the free radical type. However, in recent years some radiation-promoted polymerizations at low temperatures and in particular solvents have been discovered to have characteristics of cationic polymerization with inhibition by water and amines. Positive ions formed seem to have sufficiently long duration for cationic polymerizations, especially in isobutylene, cyclopentadiene, α-methyl styrene, styrene, β-pinene, and isobutyl vinyl ether (42). Isobutylene has been polymerized ionically by use of UV light, and use of an electric field is said to eliminate need for an added gegenion or heterogeneous surfaces (43). The advantages of keeping the ions apart suggests a Trommsdorff-like effect in ionic polymerization.

Irradiation of maleic anhydride in the presence of dioxane, aldehydes, or ketones initiates radical-cationic homopolymerization of isobutyl vinyl ether (44). In this and most examples of charge-transfer polymerizations, strongly colored ion-radicals or charge-transfer complexes participate, and polymer molecular weights tend to be rather low. Of particular interest are polymerizations of such basic monomers as N-vinyl carbazole by electron

acceptors often accompanied by the formation of deeply
colored complexes. Thus N-vinyl carbazole treated with
p-chloroanil, acetonitrile, or tetracyanoethylene, or
allowed to stand in carbon tetrachloride solution, slowly
polymerizes from room temperatures (45). The infrared
spectrum of the polymer isolated from the colored solutions is similar to that of polymer prepared using boron
fluoride-etherate as catalyst. Vinyl carbazole also was
polymerized by metal ion oxidizing agents possibly
initiated as follows (46):

$$Fe^{3+} \;+\; \overset{\cdot\cdot}{\underset{\underset{HC=CH_2}{|}}{N}} \;\longrightarrow\; Fe^{2+} \;+\; \overset{\cdot\,+}{\underset{\underset{HC=CH_2}{|}}{N}} \;\longrightarrow\; \overset{+}{\underset{\underset{HC-\overset{\cdot}{C}H_2}{|}}{N}}$$

Dilute aqueous solutions of perchloric acid initiated
exothermic polymerization of N-vinyl carbazole in orange-
colored solutions (47).
     The relative rates of polymerization of geometrical
isomeric monomers are of interest. Cis-stilbene was
reported to polymerize to oligomers faster than trans-
stilbene at -78°C with titanium tetrachloride-trichloro-
acetic acid (48). Cis-propenyl n-butyl ether was
reported more reactive than the trans isomer in polymerization at -78°C with boron fluoride-etherate catalyst
(49). However, the trans-alkenyl vinyl ethers more
readily stereopolymerized cationically to crystallizable
polymers (50).
     Cationic telomerization of butadiene with acetic acid
was described (51). Styrene and acetic acid in the
presence of boron fluoride-acetic acid complex for 20 hr
at 25°C gave 90% of telomers of the type $H(CH_2CHR)_nOAC$,
along with 10% of styrene dimer (52). The liquid telomers
consisted of n = 1, 10%; n = 2, 38%; n = 3, 20%, and
n > 3, 31%. The same workers telomerized styrene
cationically with acetonitrile and with formaldehyde in
acetic acid. Ethylene has been telomerized with aromatic
hydrocarbons in the presence of modified Ziegler-type
catalysts and halocycloalkanes (53).
     By cationic systems copolymerization of some vinyl-type
monomers has been observed with aldehydes and with cyclic
compounds. For example, styrene and 1,3-dioxolane in
toluene were reacted with boron trifluoride-etherate
catalyst to form block copolymers having rather long
sequences of each monomer unit (54).
     By certain cationic or anionic systems, it has been
possible to prepare homopolymers of moderately high
molecular weight from certain β- or 2-substituted ethylenic compounds. By acidic catalysts, homopolymers have

been prepared from indene, cumarone, acenaphthylene (55),
and 2-methoxystyrene (56). Anionic conditions using
alkali metal organic catalysts have given polymers from
ethyl (56) and from t-butyl crotonate (57). The latter
produced stereoregular polymers when phenyl magnesium
bromide acted for 2 hr at 27°C. Tributyl phosphine at
-15°C promoted vinyl-type polymerization of crotonaldehyde along with some cyclization (58). Products of
number average molecular weight above 3000 were obtained.

OTHER REACTIONS OF VINYL MONOMERS

Electron-donating alkyl and alkoxy groups substituted
into ethylene promote rapid addition of bromine to double
bonds. Electron-attracting and conjugated groups as in
acrylic, methacrylic, maleic, and fumaric compounds very
greatly retard addition of bromine. Propylene adds
hydrogen very much faster than ethylene under similar
conditions over contact catalysts. Styrene and vinyl
ethers are readily hydrogenated. Hydrogen chloride
generally adds only slowly to ethylenic compounds. Hydrogen bromide adds more rapidly and in the Markovnikov (59)
way to many unsymmetrical olefinic hydrocarbons, as well
as to chloro- and bromosubstituted olefins. Bauer
observed reverse (or Posner) addition of hydrogen halides
to vinyl halides in presence of oxidizing agents (60):

$$CH_2=CHCl + HCl \longrightarrow CH_2ClCH_2Cl$$

When the direction of addition of hydrogen halides to
unsymmetrical alkenes and haloalkenes can be reversed by
addition of peroxides or oxygen, the mechanism is believed
to be free radical (61). Reversed addition of hydrogen
bromide to allyl bromide and to diallyl in the absence of
oxygen was favored by photochemical reaction (62).
Active hydrogen compounds generally add slowly and in
the reverse way to ethylenic compounds bearing strong
electron-attracting groups, as in the case of acrylic and
methacrylic monomers. The alkali salts may be somewhat
more reactive. Substitution of halogen at or next to the
double bond of an olefinic hydrocarbon retards addition
reactions; as a rule, however, addition still occurs in
the Markovnikov way. Nevertheless, small proportions of
n-propyl iodide were obtained from propylene and hydrogen
iodide (63). The addition of mercuric acetate in
methanol solution to ethylene derivatives such as styrene,
divinyl benzene, alkyl vinyl ethers, N-vinyl carbazole,
allyl alcohol, allyl ethers, and allyl esters occurs
readily to give compounds of the type $\underset{\underset{CH_3O}{|}\;\underset{HgAc}{|}}{CH_2CH-R}$ (64). This

reaction occurs too slowly with acrylic, methacrylic, maleate, and itaconate esters for analytical purposes.

Posner, using thiophenol and benzyl mercaptan, discovered that mercaptans have a prevailing tendency to add in the reverse way to ethylenic double bonds (65). Note that peroxide-catalyzed free radical reactions are generally inhibited by high concentrations of mercaptans and that such additions seem to be ionic in character. Mercaptans add only very slowly to acidic monomers such as maleic and fumaric acids (66). Bisulfite salts generally add to olefins in the Posner way, but the products are often mixtures. Vinyl-type monomers bearing strong electron-attracting groups, such as acrylonitrile, readily undergo nucleophilic reverse addition with alkaline catalysts (67); for example

$$CH_2=CHCN + ROH \xrightarrow{\text{KOH}} ROCH_2CH_2CN$$

Epoxidations of olefinic hydrocarbons or of oleic acid with perbenzoic acid or peracetic acid are complete in a few hours, but crotonic acid and acrylic compounds react slowly. Olefinic monomers often hydrolyze or saponify more rapidly than their saturated analogs. Lower alkyl vinyl ethers hydrolyze to alcohol and acetaldehyde slowly in water and much faster in dilute mineral acid. Vinyl acetate hydrolyzes in dilute mineral acid or mercuric solutions and more rapidly in alkaline solutions. Acrylic and methacrylic ester monomers are saponified only by strong aqueous alkali. Styrenes, vinyl ethers, acrylic, and methacrylic monomers may react with oxygen in light with the formation of hydroperoxides and other oxidation products. However, some monomers bearing hydroxyl, mercaptans, or amine groups are self-stabilized against such autoxidation. Small concentrations of added phenols, arylamines, or sulfur compounds can of course retard oxidation as well as polymerization. Additions of soluble Lewis acids to methyl methacrylate and to some other acidic monomers retard both oxidation and polymerization (68).

NONVINYL POLYMERIZATIONS OF VINYL COMPOUNDS

Explorations of unusual reaction conditions have revealed several mechanisms besides simple addition by which macromolecules may form from ethylenic compounds; especially notable are reactions involving migration of active hydrogen, isomerization polymerizations, with cyclization, and by attack of multiple bonds other than vinyl.

Hydrogen-transfer or H-polymerizations may be regarded as related to chain transfer, where an active hydrogen atom is displaced at each polymerization step. A single

ethylenic monomer can supply both the active hydrogen and
the double bond, as in the H-polymerization of acrylamide
to form a polyamide discovered by Breslow and co-workers
(69):

$$CH_2=CHCONH_2 \xrightarrow{\text{NaOR}} --CH_2CH_2CONH--$$

Sodium t-butylate catalyst was added to a solution of
acrylamide in dry pyridine containing an arylamine as
inhibitor of vinyl polymerization. Polymers formed as a
separate phase.

Much research remains in designing anionic conditions
to obtain by H-polymerization products of really high
molecular weight. Copolymers of mixed vinyl-type and
H-type structure were indicated by polymerization of
acrylamide in presence of peroxide and base (70). In
Japan (71), the H-polymerization of acrylic and other
monomers has received attention. Methacrylamide and
crotonamides respond more sluggishly than acrylamide.
Substituents in acrylamide derivatives decreased both the
rate and the molecular weight of the polymer. A substituent on the nitrogen showed the greatest effect, on the
beta carbon next, and on the alpha carbon least. These
reactions are anionic with nucleophilic catalysts, but
free-radical-initiated examples are also known, at least
with allyl compounds (see Chapter 3). The addition of
these active hydrogen compounds seems to occur predominantly in the Posner manner.

Yokota and co-workers studied monomers containing amide
hydrogens of different acidities (72). Using sodium t-
butoxide at 105°C, N-cyclohexylacrylamide only gave a
vinyl-type polymer, whereas H-polymer structures predominated from N-benzyl, N-phenylethyl, N-p-anisyl,
N-p-tolyl, N-phenyl, and N-m-chlorophenyl acrylamides.
Polymers of mixed vinyl and H-polymer structure confirmed
by infrared were obtained by lithium butyl catalysis from
p-vinyl benzamide (73). Addition of lithium chloride
suppressed the H-polymerization, perhaps by reducing the
basicity of the amide groups by complex formation.

H-type homopolymerizations have also been reported with
vinyl sulfonamide (74) and methyl vinyl ketone (75).
Hydrogen-transfer polymerizations may occur when divinyl,
diacrylic, or diallyl compounds react with diamines,
diols, dithiols, or other bifunctional compounds bearing
reactive hydrogens:

$$CH_2=CH-A-CH=CH_2 + HBH \longrightarrow --CH_2CH_2-A-CH_2CH_2B--$$

Examples include base-catalyzed reactions of divinyl
sulfone with polyfunctional hydrogen donors, such as

malonate esters, glycols, and thioglycols (76); divinyl sulfone plus urea (77); diacrylic monomers plus diamines (78); diacrylates plus diisopropylamine plus hydrogen sulfide (79). Copolymerizations of allylic dienes with dimercaptans are discussed in Chapter 3.

When the compounds of functionality greater than two are used, nucleophilic additions of active hydrogen compounds can give crosslinked products. Thus cellulose fibers can be crosslinked by divinyl sulfone for shrink-proofing (80). A triacrylyl triazine derivative reacted with H_2S in alkaline solution to give crosslinked polymers (81).

The participation of cyclization in polymerization of ethylenic monomers was discovered with diallyl o-phthalate by Simpson (Chapter 11), and the phenomenon has been encountered with many diallyl compounds. Earlier Rothrock, Strain, and co-workers had prepared soluble prepolymers from diallyl esters which may have contained cyclic structures (Chapters 10 and 27). Radical polymerization in solution of some diacrylic monomers, such as acrylic anhydride, can give soluble polymers containing rings (82). These may open upon hydrolyzing to form acrylic acid polymers. Jones proposed the name cyclopolymerization. He also observed cyclopolymerization of alloocimene by boron fluoride-etherate. Both double bonds were consumed. Cyclizations of preformed chains may also occur, such as in rapid polymerizations of acrylonitrile to colored polymers. Opening of some $C \equiv N$ groups in acrylonitrile polymerizations and, under special conditions, participation of carbonyl double bonds in vinyl-type addition polymerizations has been suspected. Vinyl-type monomers may even participate occasionally in condensation polymerizations such as the reactions of p-xylylene dichloride with styrene or methyl methacrylate, giving heat-resistant polymers believed to contain 1,4-ring units in the main chain (83).

RADICAL POLYMERIZATIONS OF ALLYL AND RELATED COMPOUNDS

The monoallyl and monoallylidene compounds do not form homopolymers of high molecular weight on heating with radical catalysts; moreover, they retard radical polymerizations of most vinyl, styrene, and acrylic polymerizations On heating with peroxide catalysts, typical monoallyl compounds such as allyl acetate form oils or viscous liquid low polymers of DP up to about 20. The explanation of Bartlett and co-workers has been generally accepted that reaction of radicals with allylic hydrogen atoms causes self-termination (allylic degradative transfer or autoinhibition) (84).

$$--CH_2\overset{\bullet}{C}H \;+\; CH_2=CH \longrightarrow --CH_2CH_2 \;+\; CH_2=CH$$
$$\quad\;\; CH_2Y \qquad\quad CH_2Y \qquad\qquad\;\; CH_2Y \qquad\;\; \cdot CHY$$

growing macroradical	allyl monomer	terminated polymer	resonance-stabilized radical

For each polymerization experiment, there was a linear relation between peroxide consumed and allyl acetate low polymer formed. At high initial catalyst concentrations, however, this ratio fell off. Degradative transfer was confirmed by faster rates and higher polymer molecular weights from deuterated allyl acetate. Ultrahigh pressures reduce degradative chain transfer and increase polymer molecular weights somewhat, as noted for propylene and fluoroalkenes in Chapters 2 and 4.

The ability of allyl compounds to terminate chain reactions can be used to control or retard radical-type polymerizations, oxidations, and degradations. Allyl acetate was suggested as a stabilizer of diallyl phthalate prepolymer (85). Allyl t-butylphenyl ether was added as a temporary inhibitor of unsaturated polyester-monomer mixtures (86).

Degradative chain transfer has been somewhat over-emphasized with regard to allyl compounds in polymerization. Low reactivity of the allyl group from absence of strong polarity and resonance merits more consideration. Values of e and Q from radical copolymerization data show the similarity of monoallyl compounds to alkyl vinyl ethers (87).

	e	Q
$CH_2=CH_2$	-0.20	0.02
$CH_2=CHCH_3$	-0.78	0.002
$CH_2=CH(CH_2)_2CH_3$	-0.63	0.07
$CH_2=C(CH_3)_2$	-0.96	0.03
$CH_2=CHOC_2H_5$	-1.17	0.03
$CH_2=CHO(CH_2)_3CH_3$	-1.20	0.09
$CH_2=CHOCH_2CH(CH_3)_2$	-1.77	0.02
$CH_2=CHOC(CH_3)_3$	-1.58	0.15
$CH_2=CHCH_2Cl$	+0.11	0.06
$CH_2=CHCH_2OH$	+0.36	0.05
$CH_2=CHCH_2OCOCH_3$	-1.13	0.03

	e	Q
$CH_2=\overset{CH_3}{\overset{|}{C}}CH_2Cl$	-0.91	0.12
$CH_2=\overset{CH_3}{\overset{|}{C}}CH_2OCOCH_3$	-1.33	0.04
$CH_2=\overset{Cl}{\overset{|}{C}}CH_2OCOCH_3$	-1.12	0.53
$CH_2=\overset{Cl}{\overset{|}{C}}CH_2Cl$	+0.44	0.27
$CH_2=\overset{Cl}{\overset{|}{C}}CH_2OH$	+0.56	0.24

The allyl and methallyl chlorides and acetates resemble
vinyl alkyl ethers in having low Q values and relatively
low e values. When heated with peroxide catalysts, vinyl
alkyl ethers like the monoallyl compounds give only
viscous liquid polymers of low DP. It seems probable
that degradative chain transfer by attack of hydrogen
atoms alpha to the ether oxygen atoms limits the DP of
polymers from the vinyl ethers, as in the case of the
allyl compounds. The monoallyl compounds and vinyl alkyl
ethers actually polymerize faster than one expects because
the low DP of the products imparts little change in
viscosity. However, when the monoallyl compounds are
distilled at atmospheric pressure, polymerization often
causes wide boiling ranges and large undistillable liquid
residues.

In contrast to the vinyl and acrylic monomers, relatively
few of the monoallyl compounds have been studied very
thoroughly in polymerizations. Uniform polymers of high
molecular weight have been considered unattainable. However, the recent success of Ziegler-type catalysts with
1-alkenes should encourage further research.

There have been few studies of relative rates of polymerization of different monoallyl compounds. However,
compounds of the type $CH_2=CHCH_2Y$ gave the following
relative polymer yields by irradiating with the same dose
of gamma rays from ^{60}Co where Y was Cl = 1.77, $OOCCH_3$ = 44,
OC_2H_5 = 26.6, OH = 25.5, CN = 24.7, COOH = 22.7,
NH_2 = 10.6 (88).

Under free radical conditions, most monoallyl compounds,
including propylene and higher olefins, copolymerize
readily with SO_2 to give solid products of relatively high
molecular weight. Sulfur dioxide behaves as an acidic
monomer where an unshared electron pair of S apparently
acts like a π-electron pair of a monomer double bond.
These polysulfones have found little use because of
chemical instability to heat and to aqueous alkalis. The
sulfone polymer reaction was discovered by Solonina, who
made SO_2 copolymers from allyl alcohol and from allyl
phenyl ether (89). Unsaturated compounds bearing electron-
attracting groups such as COOH, COOR, and halogen did not
copolymerize with SO_2. Monoallyl compounds of negative e
values may copolymerize readily with acidic monomers of
high e value, such as maleic anhydride.

Allyl compounds having two or more allyl groups often
can be polymerized under free radical conditions to form
solid polymers even with the waste of growing macro-
radicals by chain transfer. The latter and cyclization
promote formation of soluble prepolymer syrups in the
early stages of free radical polymerization of polyfunctional allyl monomers. That radicals may be lost to

chain growth through cyclization was first observed with
diallyl phthalates by Simpson in England and was later
studied extensively by Butler and his students at the
University of Florida. The polyfunctional vinyl and
acrylic monomers give crosslinked polymers quickly; they
do not lend themselves as well as the polyfunctional
monomers to preparation of soluble and formable prepolymers, nor have they shown as many examples of cyclopolymerization. Occurrence of cyclic chain segments from
polyfunctional allyl compounds is discussed in Chapters 11,
25, and 26. Control of rates of polymerization of diallyl-
o-phthalate by peroxide initiator and by inhibitors is
shown in Figs. 1.6 and 1.7. In practice, the concentrations of peroxides (e.g., 4%) necessary for complete
polymerization are higher than those useful for polymerization of common vinyl monomers. Homogeneous polymerizations of ethylenic monomers to form soluble polymers may
be followed by the increase in refractive index resulting
from change from double bonds to single bonds:

$$CH_2{=}CHCH_2Y \longrightarrow {-}{-}CH_2\underset{\underset{\displaystyle CH_2Y}{|}}{CH}{-}{-}$$

There is a linear relation between the percentage of
polymer and the increase in n_D when complex side reactions
are not occurring. Figure 1.8 illustrates change in
refractive index in the bulk polymerization of diallyl-
o-phthalate.
   On warming with peroxide catalysts, divinyl ether
polymerizes to brittle crosslinked polymers, whereas
diallyl ether with benzoyl peroxide under ultraviolet
gives soluble, soft solid polymers (90). The allyl
monomers that have been used most successfully in commercial polymers are the polyfunctional ones which can be
formed in partially polymerized states, followed by
completion of polymerization or curing to thermoset
structures by heating in the presence of peroxide initiators. Thus diallyl diglycol carbonates are used for cast
optical plastics and the diallyl phthalate prepolymers are
used in fiber- and filler-reinforced thermoset articles.
The relatively low reactivity of the allyl groups permits
formation of fusible prepolymers. Another advantage of
low reactivity is control of copolymerizations when small
proportions of polyfunctional allyl monomers are added as
curing or crosslinking agents with other monomer or
polymer mixtures. Typical allyl monomers show little
tendency for thermal polymerization without added initiators. If initiators are absent, some of them can be heated
to surprisingly high temperatures with little reaction.

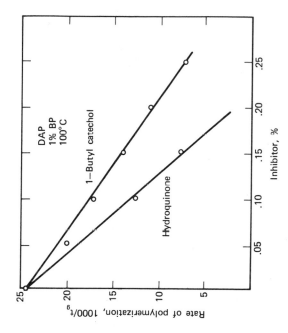

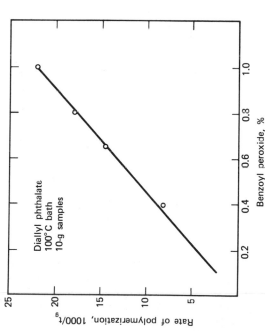

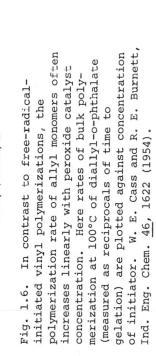

Fig. 1.7. Plot of the retardation of
the polymerization rate by addition of
phenolic inhibitors. W. E. Cass and
R. E. Burnett, Ind. Eng. Chem. 46,
1622 (1954).

Fig. 1.6. In contrast to free-radical-
initiated vinyl polymerizations, the
polymerization rate of allyl monomers often
increases linearly with peroxide catalyst
concentration. Here rates of bulk polymerization at 100°C of diallyl-o-phthalate
(measured as reciprocals of time to
gelation) are plotted against concentration
of initiator. W. E. Cass and R. E. Burnett,
Ind. Eng. Chem. 46, 1622 (1954).

33

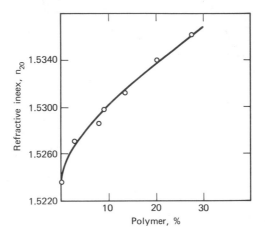

Fig. 1.8. Increase in refractive index in the early syrup phase of
polymerization of diallyl isophthalate, achieved by heating with
peroxide catalyst. Alex Schulthess, Promotionsarbeit, Zurich ETH
(1960).

The following values of e and Q show the low reactivity
of some diallyl monomers, nonconjugated dienes, and divinyl ether monomers:

	e	Q
Diallyl-o-phthalate	+0.36	0.04
Diallyl melamine	-0.95	0.02
Diallyl phenyl phosphonate	-0.07	0.05
Diallyl n-butyl phosphonate	-0.07	0.05
Divinyl ether	-1.28	0.04
Vinyl cyclohexene	-1.64	0.06

A typical styrene or methacrylate ester monomer requires
20 ppm of phenolic-type inhibitor in order to prevent
polymerization on storage for a year at room temperature.
Diallyl esters without added inhibitor often do not polymerize appreciably on storage for a year under air.
Casting sheets from methyl methacrylate using 0.1% diacetyl peroxide at 50 to 60°C compares in rate with

o-diallyl phthalate cast at 100 to 110°C using 4% t-butyl
perbenzoate catalyst. Occasionally the relatively slow
rates of polymerization of allyl monomers present technical
advantages. Rates may be accelerated somewhat by adding
maleic anhydride, acrylic or methacrylic comonomers to
polyfunctional allyl monomers in cast polymerizations.

Methallyl compounds $CH_2{=}CCH_2Y$ have been less thoroughly
investigated than allyl compounds. One reason is the
ready isomerization of methallyl alcohol to isobutyraldehyde in the presence of mineral acids. As would be expected from the electron-donating qualities of the methyl
group, methallyl compounds seem to polymerize less readily
than corresponding allyl compounds in free radical systems.
The lower e values of methallyl compounds compared with
corresponding allyl compounds should make them respond
better in cationic polymerization. Exploratory work by
the author gives some evidence of this. Compounds of the
type $YCH{=}CHCH_3$ do not polymerize readily by conventional
methods. Crotyl and cinnamyl esters are less reactive in
polymerization than allyl esters (91).

The so-called air-drying polymerization of certain
unsaturated hydrocarbons, long-chain esters, and ethers
containing peroxidizable allylic hydrogen atoms seems to
involve radical processes quite different from homogeneous conventional radical-initiated polymerizations.
The thickening and solidification of linseed oil, one of
the very oldest polymerization processes used by man,
remains incompletely understood. Oxidation at allylic
hydrogen atoms leads to hydroperoxide groups, which
initiate formation of prepolymer, followed by crosslinking. Such unsaturated liquids that form skins of
polymer in contact with air, especially in the presence
of traces of transition metal compounds (driers such as
cobalt naphthenate), generally do not respond well to
heating with added radical initiators out of contact with
air. For air-oxidative polymerization to occur readily,
a molecule seems to require at least two nonconjugated
double bonds associated with hydrocarbon or ether groups
as in drying oils, sorbitol tetraallyl ether, polyallyl
ethers of carbohydrates, and certain diallylidene cyclic
acetal esters. Most polyfunctional synthetic allyl
monomers investigated have proved too slow in rates of
air-drying polymerization for large-scale use in coatings
(see Chapters 14 and 15). However, research continues.

Molecular oxygen retards polymerization of allyl esters,
diallyl glycol carbonates, diallyl phthalates, diallyl
adipate, and triallyl cyanurate, but the effect seems less
than in the case of typical acrylic and methacrylic com-

pounds. Dissolved high polymers, as in syrups, reduce the
inhibiting effects of air; however, surfaces unprotected
from oxygen may remain soft and tacky. Certain polyfunctional allyl ethers with air-drying properties have been
added to vinyl-type monomer systems for the purpose of
overcoming inhibition by air, especially in coatings (92).
In a number of cast polymers prepared by the author from
different polyfunctional allyl compounds by photopolymerization with benzoyl peroxide in beakers under air, the
thickness of the top soft layers was less than encountered
with methyl methacrylate.

In radical copolymerizations with vinyl halides,
styrenes, and acrylic monomers in the range of 40 to 80°C
the allyl compounds often show low reactivity, and some
are added as inhibitors or regulators. In attempted copolymerizations with allyl or diallyl compounds, products
close to homopolymer composition may result in some cases.
Special procedures may be desirable in order to prepare
chemically homogeneous copolymers (interpolymers) (93).

Additions of very small amounts of bifunctional allyl
monomers in suspension polymerizations of vinyl chloride
were reported to give modified polymers of higher
molecular weight (94). A large addition of diallyl ester
was found necessary for modifying styrene in copolymerization at 60°C with azobis initiator (95). The chain
transfer constants of the diallyl esters, however, were
considerably lower than those of diallyl acetals.

Comparatively little has been reported about the copolymerizations of methallyl compounds with vinyl monomers.
Recently, using azobis as initiator, exploratory copolymerizations with minor proportions of different monomethallyl compounds were carried out at 60°C in toluene to
low conversions (96). With methyl methacrylate, greatest
retardation was observed from methallyl acetate, methallyl
phenyl ether, and methallyl benzene. Methallyl alcohol,
methallyl cyanide, and methallyl acetone produced less
retardation and did not reduce polymer viscosity excessively. With styrene under similar conditions, additions
of methallyl alcohol and methallyl phenyl ether caused
greatest reduction in rates, but copolymer viscosities
were not as low as expected.

Certain compounds bearing hydroperoxidizable allylic
hydrogen atoms can graft copolymerize readily with styrenes
and acrylic monomers. An example seems to be the graft
copolymerizations of shellac and its derivatives dispersed
in aqueous emulsion (97). The shellac ammonium salt may
function both as promoter and dispersing agent to give
latices of extremely small particle size with styrene or
acrylate esters. Such products are used in household
polishes.

A number of monoallyl compounds, as well as isobutylene, have been found to copolymerize more readily with acrylic monomers in the presence of zinc chloride or other Lewis acids, which seem to act as complexing agents (98). The increased reactivity of the allyl comonomers may be attributed to decreased degradative chain transfer (99). Complexing of allyl compounds with electrophilic $ZnCl_2$ may remove electron density from the double bond, making the allylic hydrogen atoms less reactive (more like the hydrogen atoms in CH_3 of methyl methacrylate). That the added Lewis acid complexes with the acrylic monomer and modifies its reactivity seems more probable than that it promotes formation of a reactive complex between both monomers. Boron-containing Lewis acids which were strong enough to form a complex with acrylonitrile formed 1:1 molar copolymers with 1-hexene (100). When carried out in methylene chloride or in benzene there was no evidence of a charge-transfer intermediate.

Attempts have been made to prepare polymers by poly-recombination from allyl aromatic compounds encouraged by the stability of allylic radicals:

$$C_6H_5CH_2CH=CH_2 \xrightarrow[\substack{\text{peroxide} \\ \text{conc.}}]{\text{high}} \quad \begin{array}{cc} CH_2=CH & CH=CH_2 \\ | & | \\ --C & C-- \\ | & | \\ C_6H_5 & C_6H_5 \end{array}$$

For example, Sorin treated p-allyl anisole at 200°C under nitrogen by adding t-butyl peroxide portionwise (101). Crosslinked and oligomer fractions of uncertain structure were obtained.

IONIC POLYMERIZATIONS OF ALLYL COMPOUNDS

The 1-alkenes and other monoallyl compounds do not polymerize readily with alkali metal and other conventional Lewis-base catalysts. Usual Lewis-acid catalysts, such as boron trifluoride or its etherate, generally give liquid low polymers only. On heating, ionic catalysts may isomerize or decompose the monomers. Allyl compounds may terminate chain growth in ionic polymerization of other compounds. Chain transfer may form resonating unreactive allyl ions (102).

$$-------C+ \ + \ C=C-CH \longrightarrow -------CH \ + \ \overset{\delta+}{C}\cdots C\cdots\overset{\delta+}{C}$$

Special heterogeneous catalysts containing activated transition metal compounds make possible the synthesis of solid high polymers from 1-olefins and some other allyl compounds. These Ziegler-Natta or coordination catalysts permit polymerization of monoallyl compounds that are not possible by conventional ionic and radical systems as discussed in Chapter 2.

After the feasibility of stereoregulation in ionic polymerization had been demonstrated with isobutyl vinyl ether and with methyl vinyl ether, solid crystalline 1-olefin polymers were prepared in 1954 in laboratories in Italy and the United States. All decisions on priority have not yet been made by patent offices. Natta, Pino, Mazzanti, and co-workers have published most disclosures on the heterogeneous catalysts from polymeric $TiCl_3$ and $Al(C_2H_5)_2Cl$ or AlR_3 which can form predominantly stereoregular polymers from such 1-alkenes as propylene (103). Although Natta has maintained that his catalyst systems are of the anionic type, we suggest that they must be more cationic in nature, since the catalysts are electrophilic. Furthermore, the methyl or other alkyl group of the monomer is electron donating, making the 1-olefins Lewis basic and only responsive to electrophilic ionic catalysts (104). Only catalysts relatively insoluble in the diluent-monomer solution have formed normally-crystalline 1-olefin polymers. Stereopolymerizations of 1-olefins using activated transition metal catalysts are discussed in Chapter 2.

Allyl compounds bearing polar groups (e.g., allylamines, alcohols, and esters) seem to react or complex too strongly with most coordination or Ziegler-Natta catalysts for practical formation of high polymers. Except for a few systems of industrial interest, relatively little research has been reported on ionic copolymerizations. One of the latter is the terpolymerization of ethylene, propylene, and nonconjugated dienes such as diallyl to amorphous elastomers. These terpolymer rubbers (EPDM) are are discussed in the next chapter.

Recently allyl halides have been reported to copolymerize with 1-olefins using BF_3 or $AlCl_3$ in alkyl chloride solvent (105). Allyl chloride in alkyl chloride could first be reacted with Lewis-acid catalyst to give a deeply colored solution. This was added to a solution of alkene in ethyl chloride at low temperatures to form copolymers of high molecular weight. A number of allyl compounds noted in this book give viscous liquid polymers under the action of Lewis acids such as BF_3 or $AlCl_3$. In most cases conditions have not been developed for obtaining good yields of solid high polymers. Allyl compounds seldom if ever respond to basic catalysts (anionic polymerization).

POLYMERIZATION OF ALLYL COMPOUNDS
BY NONVINYL MECHANISMS

Allyl compounds have provided examples of hydrogen-transfer or H-polymerization to products of low or intermediate molecular weight. The ready addition of mercaptans to olefins was demonstrated by Posner and Nicolet (106). It was found difficult to suppress the Posner type of addition in reaction of allyl lauryl sulfide with lauryl mercaptan, but additions of sulfur favored Markovnikov addition (107). In early research, Hofmann had observed gradually rising boiling temperatures in the distillation of allyl mercaptan. Polymerization of allyl mercaptan and related ethylenic compounds on heating was attributed by von Braun and Murjahn (108) to the reaction of SH groups with the double bond:

$$CH_2=CHCH_2SH \longrightarrow --CH_2CH_2CH_2S--$$

Treatment of allyl mercaptan with sodium ethoxide followed by carbon tetrachloride gave brown-colored polymers (109). Such polymerizations of allyl mercaptan seem to occur by free radical mechanisms, but the unique odor of the monomer has discouraged research. Homopolymers of high molecular weight have not been reported.

Marvel and his students succeeded in making high polymers by free radical emulsion polymerization of diallyl with dimercaptans, and they reported interesting characteristics of the reactions (110).

$$CH_2=CHCH_2CH_2CH=CH_2 + HS(CH_2)_6SH \longrightarrow --(CH_2)_6S(CH_2)_6S--$$

These polymerizations are discussed in Chapter 3. Both diallyl and diallyl ether have been heated or irradiated with phenyl phosphine to give linear polyphosphines (111). Diallyl ether and phenyl phosphine heated for 90 hr at 70°C with benzoyl peroxide formed a tough elastomer.

Another example of isomerization-polymerization is the intramolecular chloride shift polymerization of the allyl compound 3-chloro-3-methyl-1-butene $H_2C=CHCClCH_3$ (112).
$$CH_3$$
Aluminum chloride was used as catalyst in liquid ethyl chloride at -130 to -30°C, giving soft amorphous polymers having about equal proportions of 1,3 and 1,2 structures:

$$--CH_2CH-\underset{\underset{CH_3}{|}}{\overset{\overset{CH_3}{|}}{C}}-- \qquad\qquad --CH_2CH--$$
$$\underset{Cl \ \ CH_3}{} \qquad\qquad\qquad CH_3\underset{\underset{Cl}{|}}{C}CH_3$$

Polymerization of allyl benzene using BF_3 or $AlCl_3$ gave polymers believed to have benzene rings in the main chain; this judgment was based on infrared, nuclear magnetic resonance (NMR), and oxidation studies (113). In CS_2 at -30°C with $AlBr_3$ catalyst, allylbenzene gave polymer of structure

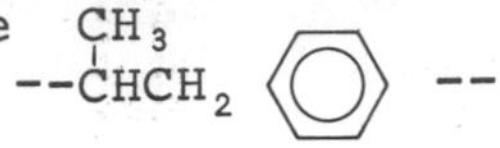

$$--\overset{\displaystyle CH_3}{\underset{|}{C}}HCH_2 \bigcirc --$$

having DP of about 32 (114).

Cyclization in free radical polymerizations of diallyl compounds to soluble polymers has been noted. Cyclization has been found to contribute to numerous ionic polymerizations of diallyl compounds. Among these compounds are nonconjugated diolefinic hydrocarbons (115) and 4-vinyl cyclohexene by Ziegler catalysts (116), 2-allyl-1-methylene cyclohexane by BF_3 in methylene chloride at -70°C (117), and diallyl ammonium halide (see Chapter 22). Polymerizations of diallyl compounds to form soluble polymers may consist of two types of steps: simple vinyl-type addition of monomer, and addition with cyclization using up two double bonds (Fig. 1.9). Reviews of cyclopolymerization of nonconjugated diethylenic monomers have been published by Butler and by Schulz (118).

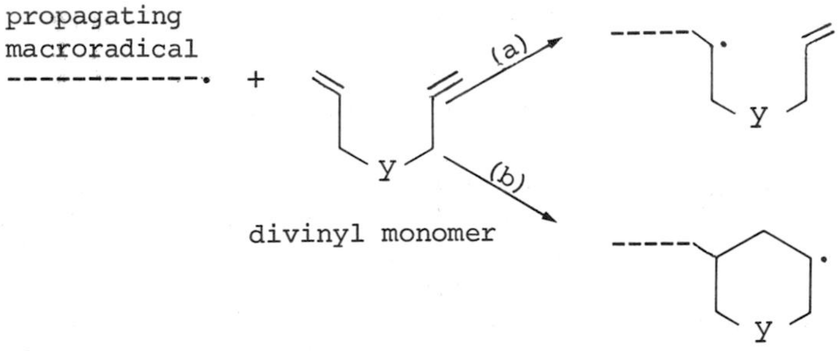

Fig. 1.9. Polymerization steps of a generalized diallyl compound with omission of C and H symbols. In (a) simple addition occurs at one double bond; in (b) addition occurs with cyclization. Partial cyclization can occur in ionic as well as in free radical polymerizations. Cf. G. B. Butler and S. Kimura, J. Macromol. Sci.-Chem. A5, 181 (1971).

OTHER REACTIONS OF ALLYL COMPOUNDS

The diversified and complex role of the allyl compounds in chemical research and production results in large part from the great reactivity of these compounds in ways other than polymerization. In this challenging area much remains to be discovered and confirmed, not only with regard to synthetic organic reactions but also in biological chemistry, e.g., odor, toxicity, and plant and animal behavior (perfumes). Older literature on reactions of many allyl compounds has been reviewed (119). Among the most important reactions of allyl compounds are rearrangements, additions to the double bond, substitutions, and oxidations. Although most allyl compounds exhibit biological activity and toxicity, few have proved to be dangerous. Man has been exposed for a long time to the toxic allyl sulfur compounds occurring naturally in onions and mustard.

Many allyl compounds, like benzyl compounds, are highly reactive because the allyl cation and radical are stabilized by resonance:

$$CH_2=CHCH_2{}^+ \longleftrightarrow \overset{+}{C}H_2-CH=CH_2 \quad or \quad \overset{1/2+}{CH_2} \, \text{---} \, C \, \text{---} \, \overset{1/2+}{CH_2}$$

resonating allyl cation

$$CH_2=CH\overset{\bullet}{C}H_2 \longleftrightarrow \overset{\bullet}{C}H_2-CH=CH_2 \quad or \quad CH_2 \, \text{---} \, CH \, \text{---} \, CH_2$$

resonating allyl radical

Vinyl chloride $CH_2=CHCl$ represents the opposite condition, in which the molecule is stable except for polymerization because it resonates as a whole. Also promoting reactivity, oxygen and sulfur in allyl compounds have unshared electrons which can promote ionic reactions (e.g., by attracting protons).

Academic research showing the high reactivity of allyl compounds in rearrangements and in substitution reactions was reviewed by DeWolfe and Young (120).

Allyl and methallyl alcohol (in contrast to nonexistent vinyl and isopropenyl alcohol) can be purified and stored. However, they can isomerize to aldehydes under the influence of acid catalysts. As would be expected from the presence of the electron-donating methyl group, methallyl alcohol isomerizes more easily than allyl alcohol:

$$CH_2=\underset{\underset{CH_3}{|}}{C}-CH_2OH \xrightarrow[H_2SO_4]{H^+} (CH_3)_2CHCHO$$

This rearrangement to isobutyraldehyde can prevent the preparation in good yield of methallyl esters by conven-

tional esterification with mineral acid catalysts, as
well as attempts to polymerize methallyl compounds with
acid catalysts. Preparations of allyl esters from allyl
alcohol and organic acids using acid catalysts are more
practical than esterifications with methallyl alcohol.

Allylic rearrangements may occur readily where a substituent moves from one end to the other of the resonating
allyl group. For example, either one of the following
pure bromides (obtained from butadiene plus hydrogen
bromide) isomerizes on standing to the equilibrium mixture indicated (121):

$$CH_2=CHCHBrCH_3 \qquad BrCH_2CH=CHCH_3$$
15% allyl compound   85% propenyl compound

The corresponding chlorine compounds apparently isomerize
less readily in this way. The peroxide ascaridole was
observed to catalyze this type of isomerization. The
isomerization of $CH_2CH=CHCH_2Cl$ and $CH_2=CHCH_2CH_2Cl$ to an
equilibrium mixture is catalyzed by ferric chloride,
indicating an ionic reaction. Changes in double-bond
positions may be called allylic shifts regardless of
whether true allyl compounds are involved. Solvolysis of
allyl compounds may lead to partial rearrangement,
especially in an SN_1 mechanism that occurs by an ionic
intermediate; SN_2 reactions may give less rearrangement.

Allyl ethers isomerize slowly to propenyl ethers at
150 to 175°C in the presence of Lewis-basic catalysts
(122). Thus with potassium t-butoxide as catalyst, 35%
isomerization of triallyl pentaerythritol occurred in an
hour. On heating with concentrated alkali, allylphenols
can isomerize slowly to propenyl phenols, but this reaction does not occur normally in the rearrangement of
allyl aryl ethers.

In the Claisen rearrangement, an allyl group moves from
an ether linkage to become attached largely at the
adjacent ortho position on an aromatic ring (if unsubstituted) (123). Thus allyl phenyl ether, on heating
without catalyst or with an arylamine, gives o-allyl
phenol (see Chapter 16). Allyl ethers of enols such as
o-allyl acetoacetate esters also can rearrange to C-allyl
compounds. Some allyl thioaryl ethers can give Claisen-
type rearrangements (124), but amine catalysts generally
are necessary. Allyl vinyl ether and allyl isopropenyl
ether can rearrange to allyl acetaldehyde and allyl
acetone, respectively (125).

Most allyl compounds, including 1-alkenes, add bromine
or chlorine rapidly at room temperature. Acrylic monomers,
maleate and fumarate esters, and other olefinic compounds

bearing strong electron-attracting groups react very
slowly with halogen under usual conditions.

Additions of halogens to allyl compounds were faster
in water or acetic acid than in carbon tetrachloride or
alkanes (126). Great differences in rates occur, depending on the halogenating agent, the rates increasing in
the series $I_2 <$ IBr $<$ ICl $<$ Br$_2 <$ BrCl. In rates of addition
in acetic acid at 25°C, CRR'=CH$_2$ reacted faster than
RCH=CH$_2$ (127). Normal butylethylene was slightly more
reactive than t-butylethylene; neopentylethylene was much
less reactive. The order of increasing reactivity of
allyl halides was ABr $<$ ACl $<$ AF. The following rates for
halogen addition were observed:

$CH_2=CH_2$: 1 $CH_2=CHCH_2Cl$: 2 $(ClCH_2)_2C=CH_2$: 0.00024

Presence of lithium chloride accelerated additions of
halogen to t-butylethylene.

Bromine in acetic acid reacted rapidly with allyl
chloride and allyl acetate but only slowly with maleic
and fumaric acid (128). Lithium bromide accelerated
rates of addition, as did hydrogen bromide. Adding water
impaired the action of HBr perhaps because water is a
Lewis base. Oxygen and light had little effect. In the
presence of bromide ion, second-order kinetics prevailed.
Relative rates of addition of bromine to allyl compounds
in methanol solution in the presence of 0.2 formal sodium
bromide correlated well with Hammett sigma values of
substituents attached to the third carbon (129). Values
of log K (rate constants) ranged from 3.3 for sigma = 0
to -0.82 for sigma = 1.74. Bromine has been added by the
Wijs reagent or by use of 0.5N Br$_2$ in CHCl$_3$ at 4°C for
10 min (130). The following allyl compounds were estimated: A acetate, A phthalate, A sucrose, A starch, and
triallyl and diallyl glycerol ethers. Siggia found that
Br$_2$ adds much faster to allyl formate, acetate, and
propionate in water or methanol than in acetic acid or
CCl$_4$ (131). However, oleates added Br$_2$ fast in acetic
acid or CCl$_4$. Sodium salts of maleic and fumaric acids
were somewhat more reactive to Br$_2$ than the free acids.

While BrCl reacts well enough with some unsaturated
acids, addition can be prevented by strong electron-
attracting groups such as COOR, CN, and F, conjugated
with double bonds (132). Allyl and fumarate double bonds
in unsaturated polyesters and their copolymers may be
estimated separately. After saponification with alkali
and neutralization, bromination without catalyst or by
the Wijs method gave the content of allylic double bonds.
Bromination in the presence of 0.2N HgSO$_4$ and 6N H$_2$SO$_4$

determined the total double-bond analysis. Mercuric
acetate reacted with propenyl compounds more readily than
with allyl compounds. Estimation of allyl compounds by
other reactions has been discussed (133).

In Markovnikov (134) or normal addition of active
hydrogen compounds to unsymmetrical olefins, the electron-
donating alkyl group promotes reaction and determines the
direction of addition:

$$R \rightarrow \overset{\delta-}{CH}=\overset{\delta+}{CH_2} + HX \longrightarrow RCHXCH_3$$

Such addition is now believed to occur by way of initial
formation of a π complex between a proton and the double
bond. The rather slow rates of addition increase with
acidity from HF to HI and may be accelerated by addition
of Lewis acids such as H_2SO_4, BF_3, $SnCl_4$, or I_2.

Posner studied the addition of thiophenol and benzyl
mercaptan to many alkenes and found the direction of
addition to be the reverse of that found by Markovnikov
for hydrogen halides (135).

$$RCH=CH_2 + HSR \longrightarrow RCH_2CH_2SR$$

Although Posner addition is entirely normal for mercaptans
and hydrogen sulfide, it is referred to hereafter as
reverse addition. Mercaptans are also very slow in addition to maleic acid, fumaric acid, and unsaturated compounds bearing strong electron-attracting groups (136).
Additions of H_2S to diallyl ether in UV light (137),
polymerizations of allyl mercaptan (138), and additions
of dimercaptans to double bonds occur in the reverse way.
Additions of iron, cobalt, or nickel have been observed
to promote reverse addition; light may also accelerate
addition of the reverse type. Allyl chloride with HBr
also has shown the peroxide effect, but the reversal of
addition in operating by a peroxide-catalyzed free radical
mechanism is not a general one. The 1-olefins rapidly
react by ionic mechanism with HX in the normal way, whereas with mercaptans, reverse addition occurs with acid
catalysts regardless of added peroxides.

Bauer observed that addition of HBr or HCl to CH≡CH
gave 1,2-dihaloethanes by reverse addition when oxygen or
oxidizing compounds were present (139). Kharasch and Mayo
studied the addition of hydrogen bromide to allyl bromide
(140). The addition of HCl was too slow for convenient
study and HI gave unstable products. Normal addition
occurred in the dark during several days if oxygen was
excluded. In the presence of air or with 1% added peroxide catalyst, however, reverse addition occurred in a
few hours.

Although less effective than when attached directly to the ethylene nucleus, different groups attached to the third carbon atom can influence rates of addition to allyl compounds (144). Thus under similar conditions the following in carbon tetrachloride solution decolorized dilute bromine solution almost instantly at room temperature in diffused light: allyl alcohol, diallyl amine, cyclohexene, 2-methyl-1-pentene, allyl ethylene glycol, allylurea, and safrole. The following required from 0.5 to 10 min for decolorization: allyl acetate, allyl propionate, allyl caproate, diallyl adipate, allyl bromide, allyl cyanide, and diallyl maleate. The following required more than 10 min: diallyl fumarate, diallyl-o-phthalate, and diallyl-m-phthalate. Allyl compounds do not react readily by nucleophilic addition in alkaline systems as do monomers bearing strong electron-attracting groups, e.g., acrylonitrile and other acrylic monomers.

Allyl monomers differ greatly in rates of photochemical oxidation under air. Allyl alcohols, mercaptans, and amines are self-stabilized against free radical oxidation. Small additions of numerous allyl compounds have been found to inhibit or retard radical oxidations and polymerizations (142), but the relations of these activities to self-oxidation and resonance stabilization of radical species remain to be completely clarified. The slightly delayed, very severe sting in the nose resulting from cautiously smelling vapors of allyl alcohol seems to result from oxidation to acrolein in the living tissues:

$$2CH_2=CHCH_2OH + O_2 \longrightarrow 2CH_2=CHCHO + 2H_2O$$

Exposure of nasal tissues of a dead cow to allyl alcohol resulted in positive tests for acrolein (143). Enzymatic oxidation may be involved. Propenyl compounds $CH_3CH=CHY$ are more stable toward oxidation, apparently.

Allylic substitution, which normally occurs along with some addition to the double bond, is discussed with the synthesis of allyl chloride from propylene and chlorine in Chapter 4. The well-known reagent N-bromosuccinimide and its use in allylic brominations originates in work of Wohl and Ziegler and has been reviewed by Djerassi (144). Reactions of N-bromosuccinimide with allylbenzene were accelerated by benzoyl peroxide addition or by UV light (145).

$$\langle\bigcirc\rangle CH_2CH=CH_2 + (CH_2CO)_2NBr \longrightarrow \langle\bigcirc\rangle CH_2-CH=CHBr, \text{ etc.}$$

Some addition occurred, as well as allylic rearrangement to cinnamyl halide. More of the latter occurred with the bromo compounds than with the chloroallyl derivatives. Reactions were carried out in carbon tetrachloride by refluxing for 2 hr.

A wide variety of allyl compounds have been epoxidized, at first using perbenzoic acid and later peracetic acid (146). Reactions under mild conditions in organic solvents give epoxy compounds (oxiranes) containing the glycidyl group -CH$_2$-CH-CH$_2$, and some of these have found
                        $\diagdown$O$\diagup$
applications in polymer products.

Allyl halides, allyl alcohol, and other allyl compounds are often chosen in syntheses because of their much greater reactivity than methyl, ethyl or propyl compounds. Menschutkin of St. Petersburg discovered at the beginning of this century that allyl bromide reacts much faster than methyl bromide with aromatic amines (147). At the other extreme, vinyl chloride is a very slow, impractical reagent for introducing vinyl groups.

The relations of the great reactivity of allyl compounds to their frequent pungent odors, to toxicity, and to other biological properties is a challenging field for research. There have been few directly comparative tests of toxicity. Allyl aliphatic compounds (alcohol and esters) were more toxic than n-propyl aliphatic compounds (liver lesions in rats (148). Allyl aromatic compounds were more toxic than propyl aromatics, but differences were less marked than in aliphatics. Unfortunately, all allyl compounds are sometimes falsely regarded as equally toxic. Latest information on toxicities should be obtained from suppliers and from Chemical Abstracts.

Physical evidence of allyl radicals (149) and of allylic carbonium ions (150) has been presented. A great number of unsaturated carbon compounds contain reactive allylic groups but are not allyl compounds containing =CH$_2$, which are discussed in this book. Among the latter are industrially important natural products such as shellac, linseed oil, and cardol (derived from cashew nut oil).

POLYFUNCTIONAL ALLYL COMPOUNDS

Divinyl benzene isomers, among the first readily available crosslinking monomers, are unstable in storage; they generally give brittle copolymers, and they do not provide double bonds of different reactivity. Monomers such as dimethacrylates of tri- or tetraethylene glycol and diallyl monomers show better stability as monomers and in their copolymers. Often they also provide more flexible crosslinks and less brittle copolymer products. In general, polyfunctional allyl and allyl-vinyl monomers are more useful because of better latitude in controlling

crosslinking and tighter final crosslinking. Such
monomers as allyl methacrylate, diallyl maleate, diallyl
fumarate, diallyl phthalates, triallyl cyanurate, and
triallyl isocyanurate in 1 to 5% concentration can be
employed to crosslink not only comonomers and partial
polymers but also substantially saturated polymers. Procedures involving suitable heating with peroxides or other
radical-forming compounds and/or radiation with radiation
sensitizers are adapted to the particular crosslinking
monomers. Among the fields of application are dental
plastics, finishes, synthetic rubbers and their latices,
electrical insulations, fiber-reinforced plastics, adhesives, flocculating agents, semipermeable membranes,
optical plastics, and crosslinked fibers.
    Polyfunctional allyl monomers such as the three isomeric
diallyl phthalates, diallyl diglycol carbonate, triallyl
isocyanurate, and diallyl adipate, can be brought to
formable, partial polymer stages before curing in useful
shapes. In general the polyfunctional styrene, acrylic,
and vinyl ester monomers polymerize too rapidly to arrest
the reactions at such partially polymerized forms suitable
for storage or shipment. The prepolymerizable polyfunctional allyl esters have made possible new techniques and
superior end-properties, especially in thermosetting
moldings and fiber-reinforced thermoset cast forms.
    It is interesting that one polyfunctional allyl monomer,
allyl cinnamate, had already been polymerized thermally
and photochemically in 1912 (151). Soluble prepolymer was
isolated and found to react only slowly with bromine. The
finished crosslinked polymers were brittle but were considered as amber substitutes. Allyl cinnamalacetate (152)
and allyl sorbate are early examples of trifunctional
monomers.
    Diallyl and triallyl ester monomers attracted attention
in experiments on high-energy radiation crosslinking of
polymers in England by Pinner and others (153). These
monomers have advantages of storage stability without
inhibitors, short propagation chain length, and high chain
transfer efficiency. By adding 25 parts or more of diallyl sebacate or triallyl cyanurate per 100 parts of
polyvinyl chloride, "attenuated network" structures were
obtained. Not all of the polyvinyl chloride is crosslinked directly but is penetrated by the network of
polyallyl ester. Even some polymers such as methyl methacrylate polymer, which normally degrade on irradiation,
may be converted to insoluble networks by polyfunctional
allyl monomers. Under certain conditions polyallyl
monomers can provide controllable high concentrations
of free radicals at sufficiently high temperatures that
saturated polymer chains can be attacked efficiently for

formation of crosslinks. At the same time, some of the
unreacted nonvolatile allyl monomers can act as stabilizers against oxidative degradation and discoloration.
Radiation curing of films by x-rays, high-energy electrons,
gamma radiation, and UV radiation account for much of the
recent interest in polyfunctional allyl compounds. However, high-temperature radical initiators are being
developed for accomplishing similar curing without
radiation.

Other recent developments in polyfunctional ethylenic
monomers include water-dispersible polyfunctional monomers
such as allyl vinyl sulfonate (154), which is of interest
for crosslinking semipermeable hydrophilic membranes,
stereoregular polymers from allyl acrylate (155) and from
methacrylate (156).

Related to polyfunctional monomers are polymers containing attached free ethylenic groups which can be
reactive in crosslinking or for synthesis of graft copolymers. Thickened drying oils, soluble diene copolymer
rubbers, diallyl phthalate prepolymers, and unsaturated
polyester resins are examples of such commercial polymers.
Patent literature shows numerous examples of partially
unsaturated prepolymers that show reactivity in crosslinking. For example, vinyl allyl ether can be cationically polymerized through the vinyl group (157). The
soluble prepolymers crosslink slowly with catalysts in
air. Polyvinyl acetate has been heated with allyl bromide
and sodium hydroxide in ketone solution in order to
prepare a polyvinyl allyl ether (158). Polymerization of
divinyl ethylene oxide through the epoxy groups gave
polymers with allylic structure (159). Allyl-containing
epoxy resins have been prepared (160).

A number of condensation polymers bearing free allyl
groups have been evaluated. Allyl melamine resins cure
only slowly in films (161). Allyl-terminated polyurethane
rubbers have been prepared (162). Allyl phenols have been
reacted with formaldehyde to give allyl-substituted
phenolic resins (163); phenol-formaldehyde resins have
been allylated, but these have had limited success (164).
Allyl phenols have been used for preparation of allyl-
substituted phenylene oxide polymers (165). Allyl-
containing aryl polycarbonates have been studied (166).
Sorbitol triallyl ether was reacted with 2,4- and 2,6-
toluene diisocyanate to form allyl-substituted polyurethanes (167).

Unsaturated epoxy compounds have been polymerized to
give unsaturated ether polymers. Thus allyl glycidyl
ether with alkyl aluminum catalyst in ether gave predominantly polymerization through the epoxy group, forming
amorphous elastomers (168). Homopolymers of 1,4-pentadiene

monoepoxide gave ether polymers bearing side allyl groups (169). Allylation of shellac by allyl alcohol has been described (170).

The synthetic polybutadienes containing 1,2-linkages have an allyl structure --CH$_2$CH-- . The graft copoly-
$$\overset{|}{C}H=CH_2$$
merization of butadiene-styrene rubber in solution with styrene provided the first practical high-impact styrene polymer molding plastics (171). Polybutadienes may be crosslinked by heating with peroxides (172). Some polyethylenes also contain small amounts of allyl groups. Certain polyethylenes containing unsaturation have been reported to increase in molecular weight on heating under vacuum with potassium hydroxide at 200 to 300°C (so-called macropolymerization) (173). Terminal vinyl or allyl groups in linear polyethylene made by the Phillips process can be reacted with dibutyl maleate in the presence of 0.2% t-butyl hydroperoxide in a mixer at 316°C (174).

References

1. C. E. Schildknecht, Vinyl and Related Polymers, Wiley, 1952; Ind. Eng. Chem. 50, 110 (1958); H. Luessi, Chimia 20, 379 (1966).
2. C. E. Schildknecht, Ed., Polymer Processes (High Polymers, Vol. X), Interscience, 1956.
3. T. M. Dyott unpublished; cf. J. A. Pople and M. Gordon, J. Am. Chem. Soc. 89, 4253 (1967).
4. C. E. Schildknecht and K. B. Williams, unpublished.
5. R. H. DeWolfe and W. G. Young, chapter in Chemistry of Alkenes, Vol. 1, S. Patai, Ed., Interscience, 1964.
6. P. D. Bartlett and R. Altschul, J. Am. Chem. Soc. 67, 816 (1945); P. D. Bartlett and F. A. Tate, ibid. 75, 91 (1953).
7. C. E. Schildknecht, Polym. Eng. Sci. 6, 240 (July 1966); Chimia 22, 261 (1968).
8. F. W. Hoover, U. S. 3,242,140 (Du Pont).
9. H. Naarmann and R. Fischer, Ger. 1,248,942 (Badische).
10. cf. C. E. Schildknecht, Polym. Eng. Sci. 6, 240 (1966); Polymer Preprints 13, #1, 253 (1972).
11. G. Smets and L. deHaes, Bull. Soc. Chim. Belg. 59, 13 (1950).
12. J. E. Mulvaney et al, J. Polym. Sci. Al 6, 1841 (1968).
13. C. E. Schildknecht, Vinyl and Related Polymers, Wiley, 1952.
14. L. A. Wall, private communication.
15. Brit. 888,730 (Dow-Corning).
16. Y. K. Kim, U. S. 3,503,945 (Dow)
17. W. Nernst, Z. Elektrochem. 24, 335 (1919); CA 14, 38 (1920); cf. M. Bodenstein and H. S. Taylor, Z. Angew. Chem. 28, III, 621 (1915); CA 10, 2077 (1916); Haber and Willstaetter, Ber. 64B, 2844 (1931); CA 26, 3774 (1932); cf. H. Staudinger,

Die Hochmolekularen Organischen Verbindungen, Springer, 1932.
18. F. W. Billmeyer, Textbook of Polymer Science, Interscience-
     Wiley, 1971.
19. T. Alfrey et al, Copolymerization, Interscience, 1952;
     L. J. Young, J. Polym. Sci. 54, 411 (1961); T. Alfrey and
     L. J. Young, in Copolymerization (High Polymers, Vol. XVIII),
     G. Ham, Ed., Interscience, 1964; cf. H. Luessi, Makromol. Chem.
     103, 47 (1967).
20. K. Chikanishi and T. Tsuruta, Makromol. Chem. 81, 198 (1965).
21. V. A. Kormer et al, J. Polym. Sci. A1, 10, 251 (1972).
22. R. J. Burch and C. E. Schildknecht, U. S. 3,117,167 (Airco).
23. C. Walling et al, J. Am. Chem. Soc. 70, 1537 (1948); M. Charton,
     J. Org. Chem. 30, 557 (1965).
24. E. Trommsdorff, H. Köhle and P. Lagally, Makromol. Chem. 1,
     176 (1948).
25. A. E. Nicholson and R. G. W. Norrish, Discussions Faraday Soc.
     22, 104 (1956); R. D. Burkhart, J. Polym. Sci. A3, 883 (1965).
26. H. Staudinger and A. Schwalbach, Ann. 480, 8 (1931);
     cf. F. R. Mayo, J. Am. Chem. Soc. 65, 2324 (1943).
27. R. M. Joyce, W. E. Hanford et al, J. Am. Chem. Soc. 70, 2529
     (1948); 72, 2213 (1950); cf. U. S. 2,440,800 (Du Pont),
     appl. 1942.
28. W. E. Hanford and J. R. Roland, U. S. 2,402,137 (Du Pont);
     M. D. Peterson and A. G. Weber, U. S. 2,395,292 (Du Pont);
     C. S. Marvel et al, J. Am. Chem. Soc. 69, 52 (1947).
29. W. T. House et al, Ind. Eng. Chem. Prod. Res. Develop. 5,
     331 (1966).
30. M. Okubo, CA 66, 80749 (1967).
31. F. R. Mayo et al, J. Am. Chem. Soc. 65, 2324 (1943); 70, 2373,
     3689, 3740 (1948).
32. N. Fuhrman and R. B. Mesrobian, J. Am. Chem. Soc. 76, 3281 (1954).
33. E. F. Jordan, B. Artymyshyn and A. N. Wrigley, J. Polym. Sci.
     A1, 6, 575 (1968).
34. F. M. Lewis and F. R. Mayo, J. Am. Chem. Soc. 76, 457 (1954).
35. C. E. Schildknecht, Chimia 22, 261 (1968); C. E. Schildknecht,
     K. B. Williams, and W. D. Kent, Polymer Preprints 12, 117
     (Sept. 1971).
36. W. Chalmers, Can. J. Res. 7, 113 and 472 (1932); J. Am. Chem.
     Soc. 56, 912 (1934); F. C. Whitmore, Ind. Eng. Chem. 26, 94 (1934).
37. C. E. Schildknecht, Vinyl and Related Polymers, Wiley, 1952;
     C. E. Schildknecht, Polymer Processes, Interscience, 1956;
     P. H. Plesch, The Chemistry of Cationic Polymerization,
     Macmillan, New York, 1963.
38. C. E. Schildknecht, A. O. Zoss, S. T. Gross, H. R. Davidson, and
     J. M. Lambert, Ind. Eng. Chem. 40, 2108 (1948); 41, 2981 (1949);
     cf. E. J. Vandenberg, J. Polym. Sci. C, 207 (1963); C. E.
     Schildknecht, C. H. Lee, and W. E. Maust, in Macromolecular
     Syntheses, Vol. II, J. R. Elliot, Ed., Wiley, 1966.
39. G. Natta et al, J. Am. Chem. Soc. 77, 1708 (1955); G. Natta,
     P. Pino and G. Mazzanti, U. S. 3,112,300 (Monte).

40. C. E. Schildknecht, Ind. Eng. Chem. 50, 107 (1958); Polymer
    Preprints 13, #1, 253 (1972); 13, #2, 1071 (1972).
41. M. L. Miller et al, J. Am. Chem. Soc. 80, 4115 (1958);
    J. Polym. Sci. 38, 63 (1959); Polymer 4, 75 (1963).
42. S. H. Pinner et al, Chem. and Ind. (London) 1274 (1957);
    M. Magat, Makromol. Chem. 35, 159 (1960); F. Williams et al,
    Polymer Preprints 7, #2, 479 (Sept. 1966).
43. E. W. Schlag and J. J. Sparapany, J. Am. Chem. Soc. 86, 1875
    (1964).
44. H. Yamaoka and K. Takakura, J. Polym. Sci. B 4, 509 (1966).
45. H. Scott et al, Tetrahedron Lett. 14, 1073 (1963); J. Polym.
    Sci. B 2, 689 (1964); J. W. Breitenbach and O. F. Olaj, J.
    Polym. Sci. B 2, 685 (1964); H. Nomori, J. Polym. Sci. B 4,
    261 (1966).
46. C-H. Wang, Chem. and Ind. (London) 751 (1964).
47. O. F. Solomon et al, J. Polym. Sci. B 2, 311 (1964).
48. P. H. Plesch and D. S. Brockman, J. Chem. Soc. 3563 (1958).
49. T. Higashimura et al, Kobunshi Kagaku 18, 561 (1961);
    Polymer Preprints 7, 409 (Sept. 1966).
50. G. Natta et al, Rend. Accad. Naz. Lincei (8) 28, 442 (1960).
51. E. L. Jenner and R. S. Schreiber, J. Am. Chem. Soc. 73, 4348
    (1951).
52. D. D. Coffman and E. L. Jenner, J. Am. Chem. Soc. 76, 2685 (1954).
53. D. V. Favis, U. S. 3,097,246 (Esso).
54. A. W. Campbell et al, J. Am. Chem. Soc. 58, 1051 (1936).
55. C. E. Schildknecht, Vinyl and Related Polymers, Wiley, 1952.
56. O. C. Bockman and C. Schuerch, Polym. Lett. 1, 145 (1963);
    cf. S. I. Miller, J. Am. Chem. Soc. 78, 6091 (1956).
57. G. Natta et al, U. S. 3,259,612 (Monte); M. L. Miller and J.
    Skogman, J. Polym. Sci. A 2, 4551 (1964); U. S. 3,274,168
    (Am. Cyan.)
58. J. N. Koral, J. Polym. Sci. 61, 537 (1962); U. S. 3,163,622
    (Am. Cyan.); Makromol. Chem. 62, 148 (1963).
59. W. Markovnikov, Compt. rend. 81, 670 (1875).
60. W. Bauer, Ger. 394,194 and U. S. 1,540,748 (Röhm & Haas);
    cf. Ger. 368,467; U. S. 1,414,852 (Röhm & Haas).
61. F. R. Mayo and C. Walling, Chem. Rev. 27, 351 (1940).
62. W. E. Vaughn et al, J. Org. Chem. 7, 477 (1942).
63. A. Michael and V. Leighton, J. prakt. Chem. [2] 60, 445 (1899).
64. R. W. Martin, Anal. Chem. 21, 922 (1949).
65. T. Posner, Ber. 38, 646 (1905).
66. E. J. Morgan and E. Friedman, Biochem. J. 32, 738 (1938).
67. H. A. Bruson, Cyanoethylation in Organic Reactions, Vol. 5,
    R. Adams, Ed., Wiley, 1949.
68. C. E. Schildknecht, unpublished.
69. D. S. Breslow, G. E. Hulse, and A. S. Matlack, J. Am. Chem.
    Soc. 79, 3760 (1957).
70. C. E. Schildknecht, H. Knutson, S. S. Stivala, 126th ACS meeting,
    New York, September 1954; Ind. Eng. Chem. 50, 107 (1958).

71. N. Ogata, J. Polym. Sci. 46, 271 (1960); Makromol. Chem. 40, 55 (1960; S. Okamura et al, Chem. High Polym.,Japan 19, 323 (1962); 20, 364 (1963); H. Tani et al, Makromol. Chem. 76, 82 (1964).

72. K. Yokota, M. Shimizu, Y. Yamashita, and Y. Ishii, Makromol. Chem. 77, 1 (1964).

73. T. Asahaera and N. Yoda, Polym. Lett. 4, 921 (1966).

74. D. S. Breslow, G. E. Hulse, and A. E. Matlack, J. Am. Chem. Soc. 79, 3760 (1957).

75. S. Iwatsuki, Y. Yamashita, and Y. Ishii, Polym. Lett. 1, 545 (1963).

76. D. L. Schoene, U. S. 2,493,364 and 2,505,366 (U. S. Rubber).

77. J. W. Schappel, U. S. 2,623,035 (Am. Viscose).

78. G. E. Hulse, U. S. 2,759,913 (Hercules).

79. J. G. Erickson, J. Polym. Sci. A 1, 519 (1966).

80. D. L. Schoene and V. S. Chambers, U. S. 2,524,399 (1950).

81. R. Wegler and A. Ballauf, Chem. Ber. 81, 530 (1948).

82. J. F. Jones (Goodrich), J. Polym. Sci. 33, 15 and 513 (1958).

83. P. J. Canterino and J. E. Cook, U. S. 3,193,538 (Phillips).

84. P. D. Bartlett and R. Altschul, J. Am. Chem. Soc. 67, 812 (1945); P. D. Bartlett and F. O. Tate, 75, 91 (1953).

85. G. F. D'Alelio, U. S. 2,339,058 (G. E.).

86. J. M. Howard, U. S. 3,371,129 (Am. Cyan.).

87. L. J. Young, in Polymer Handbook, J. Brandrup and E. H. Immergut, Ed., Interscience, 1966; cf. J. Polym. Sci. 54, 444 (1961).

88. E. A. Dolmatov and L. S. Polak, CA 64, 3691 (1966).

89. W. Solonina, J. Russ. Phys. Chem. Soc. 30, 826 (1896); CZ I, 249 (1899).

90. C. E. Schildknecht, unpublished.

91. F. Strain, U. S. 2,397,631 (PPG).

92. W. J. Maker, U. S. 2,852,487 (Glidden); W. Gumlich, P. Kränzlein, and G. Böhm, Ger. 1,019,421 (Hüls), CA 54, 8110 (1960); K. Raichle and W. Biedermann, Ger. 1,024,654 and Brit. 810,222 (Bayer); CA 53, 12704 (1959).

93. cf. C. E. Schildknecht, Chem. Eng. News 29, 1390 (1951); M. W. Perrin et al, Brit. 497,643 (ICI).

94. R. H. Martin, U. S. 2,996,484; U. S. 3,012,010-2; U. S. 3,043,814; U. S. 3,047,549 (Monsanto).

95. A. Matsumoto and M. Ouiva, J. Polym. Sci. A1, 10, 103 (1972).

96. R. L. Harville and S. F. Reid, J. Polym. Sci. A1, 8, 2535 (1970).

97. R. J. Frey and M. H. Roth, U. S. 2,961,420 (Monsanto); H. Schmalz and E. H. Hoffman, Ger. 1,151,381 (Resart); cf. U. S. 3,061,564 (Röhm und Haas)

98. G. E. Sernuik and R. M. Thomas, U. S. 3,183,217 and 3,278,503 (Esso); M. Imoto et al, Makromol. Chem. 82, 277 (1965); N. G. Gaylord et al, Polym. Lett. 6, 743 (1968); Macro-molecules 2, 442 (1969).

99. V. P. Zubov et al, J. Polym. Sci. C 23, 147 (1967).

100. R. E. Uschold, Macromolecules 4, 552 (1971).

101. S. L. Sosin et al, J. Polym. Sci. USSR 6, 1352 (1965);
      cf. A. Klages, Ber. 32, 1437 (1899).
102. J. P. Kennedy, J. Macromol. Sci. Chem. A1, 805 (1967).
103. G. Natta, P. Pino, and G. Mazzanti, Ital. 535,712 (June 1954);
      G. Natta et al, J. Am. Chem. Soc. 77, 1708 (1955) and 84,
      1488 (1962); J. Polym. Sci. 51, 387 (1961).
104. C. E. Schildknecht, Polym. Eng. Sci. 6, 240 (1966).
105. E. A. Hunter and C. L. Aldridge, U. S. 3,299,020 (Esso).
106. T. Posner, Ber. 38, 646 (1905).
107. S. O. Jones and E. E. Reid, J. Am. Chem. Soc. 60, 2452 (1938).
      (Emmet Reid at age 99, in 1971, still maintained that mercap-
      tans smell good.)
108. J. von Braun and R. Murjahn, Ber. 59B, 1207 (1926).
109. H. J. Backer and P. L. Stedehouder, Rec. Trav. Chim. Pays-Bas
      52, 453 (1933).
110. C. S. Marvel et al, J. Am. Chem. Soc. 70, 993 (1948) to
      75, 6318 (1953).
111. A. Y. Garner, U. S. 3,010,946 (Monsanto); cf. U. S. 2,671,080.
112. J. P. Kennedy et al, Makromol. Chem. 93, 191 (1966).
113. S. Murahashi et al, Chem. High Polymers (Tokyo) 23, 253, 354
      (1966).
114. E. B. Davidson, J. Polym. Sci. B 4, 175 (1966).
115. C. S. Marvel and J. K. Stille, J. Am. Chem. Soc. 80, 1740
      (1958); H. S. Makowski et al, J. Polym. Sci. A 2, 1549 (1964).
116. G. B. Butler and M. L. Miles, J. Polym. Sci. A 3, 1609 (1965).
117. G. B. Butler and M. L. Miles, Polym. Eng. Sci. 6, 71
      (Jan. 1966).
118. R. C. Schulz, Kolloid Z. 216-7, 309 (1967); G. B. Butler and
      S. Kimura, J. Macromol. Sci.-Chem. A 5, 181 (1971).
119. Allyl Chloride and Other Allyl Halides, Shell Chemical Co.,
      New York and San Francisco, 1949.
120. R. H. DeWolfe and W. G. Young, Chem. Rev. 56, 753 (1956);
      R. H. DeWolfe, chapter in Chemistry of Alkenes, Vol. 1,
      S. Patai, Ed., Interscience, 1964.
121. M. S. Kharasch et al, J. Org. Chem. 1, 393 (1936); S. Winstein
      and W. G. Young, J. Am. Chem. Soc. 58, 104 (1936).
122. T. J. Prosser, U. S. 3,168,575 (Hercules).
123. L. Claisen, Ber. 45B, 3157 (1912); D. S. Tarbell, Chem.
      Rev. 27, 497 (1940).
124. C. D. Hurd and Greengard, J. Am. Chem. Soc. 52, 3356 (1930).
125. C. D. Hurd and L. R. Pollack, J. Am. Chem. Soc. 60, 1905 (1938).
126. P. W. Robertson et al, J. Chem. Soc. 335 (1937) and 1509 (1939).
127. P. W. Robertson et al, J. Chem. Soc. 1014 (1952).
128. K. Nozaki and R. A. Ogg, J. Am. Chem. Soc. 64, 697 (1942).
129. J. E. Dubois and E. Goetz, Tetrahedron Lett. (5), 303 (1965).
130. H. M. Boyd and G. Roach, Anal. Chem. 19, 158 (1947).
131. S. Siggia et al, Anal. Chem. 35, 362 (1963).
132. R. Belcher and B. Fleet, Talanta 12, 677 (1965); CA 63, 2391
      (1965); A. I. Tarasov et al, Z. Anal. Chim. 21, 360 (1966);
      CA 65, 2366 (1966).

133. W. Karo in Encycl. Ind. Chem. Analysis, Vol. 5, F. D. Snell and C. L. Hilton, Ed., Interscience, 1967; A. Polgar and J. L. Jungnickel in Organic Analysis, Vol. 3, J. Mitchell et al, Ed., Interscience, 1956.

134. W. Markovnikov, Ann. $\underline{153}$, 256 (1870); Compt. Rend. $\underline{81}$, 670 (1875).

135. T. Posner, Ber. $\underline{38}$, 646 (1905); Review by K. Griesbaum, Angew. Chem. Int. Ed. $\underline{9}$, 273 (1970).

136. E. J. Morgan and E. Friedman, Biochem. J. $\underline{32}$, 733 (1938).

137. W. E. Vaughn and F. F. Rust, U. S. 2,522,589 (Shell); J. Org. Chem. $\underline{7}$, 472 (1942); cf. E. E. Reid and S. Jones, J. Am. Chem. Soc. $\underline{60}$, 2452 (1938).

138. J. Braun and T. Plate, Ber. $\underline{67}$, 281 (1934).

139. W. Bauer, U. S. 1,540,748 and Ger. 394,194 (Röhm & Haas), April 1922.

140. M. S. Kharasch and F. R. Mayo, J. Am. Chem. Soc. $\underline{55}$, 2468 (1933); $\underline{60}$, 3097 (1938); F. R. Mayo and C. Walling, Chem. Rev. $\underline{27}$, 351 (1940).

141. C. E. Schildknecht and D. J. Walborn, unpublished.

142. G. F. D'Alelio, U. S. 2,339,058 (G. E.).

143. C. E. Schildknecht and G. E. Carvell, unpublished.

144. A. Wohl, Ber. $\underline{52}$, 51 (1919); C. Djerassi, Chem. Rev. $\underline{43}$, 271 (1948).

145. E. A. Braude and E. S. Waight, J. Chem. Soc. 1116 (1952).

146. D. Swern, Chem. Rev. $\underline{45}$, 1 (1949); Org. Reactions $\underline{7}$, 378 (1953).

147. N. A. Menschutkin, CZ II 415 (1900).

148. J. M. Taylor et al, Toxicol. Appl. Pharmacol. $\underline{6}$ (4) 378 (1964).

149. C. L. Currie and D. A. Ramsay, J. Chem. Phys. $\underline{45}$/2, 488 (1966).

150. N. C. Deno et al, J. Am. Chem. Soc. $\underline{85}$, 2998 (1963).

151. C. Liebermann and M. Kardos, Ber. $\underline{46}$, 1065 (1912); cf. A. Kronstein, Ber. $\underline{46}$, 1812 (1913); P. Herold, U. S. 2,290,164 (Standard Catalytic).

152. F. F. Blicke, J. Am. Chem. Soc. $\underline{17}$, 1562 (1923).

153. S. H. Pinner et al, Plastics (London) $\underline{25}$, 35 (1960); Nature $\underline{182}$, 1108 and $\underline{184}$, 1303 (1959); J. Appl. Polym. Sci. $\underline{3}$, 338 (1960).

154. E. J. Goethals, J. Polym. Sci. B $\underline{4}$, 691 (1966).

155. M. Donati and M. Farino, Makromol. Chem. $\underline{60}$, 233 (1963).

156. D. M. Wiles and S. Brownstein, J. Polym. Sci. B $\underline{3}$, 951 (1965).

157. R. Paul et al, Bull. Soc. Chim. 1-2, 121 (1950); Brit. 659,288 (Rhône-Poulenc).

158. D. K. Alpern and W. Kimel, Paint, Varnish, & Plastics Div., ACS meeting, April 1947.

159. E. L. Stogryn and A. J. Passannante, U. S. 3,261,819 and 3,261,848 (Esso).

160. R. M. Christenson and W. C. Bean, U. S. 2,910,455 (PPG).

161. P. Zuppinger and G. Widmer, U. S. 2,885,382 (Ciba).

162. G. X. R. Boussu et al, U. S. 3,219,633 (Michelin).

163. H. Yoshioka and M. Kawano, Japanese Patent 3743 (1958); CA $\underline{53}$, 5704 (1959).

164.   N. Gaylord, U. S. 3,291,770 (Interchemical).
165.   C. J. Kurian and C. C. Price, J. Polym. Sci. 49, 267 (1961);
       A. S. Hay, French 1,322,152 (G.E.); CA 60, 685 (1964).
166.   R. Butterworth and J. A. Parker, U. S. 3,164,564 (Armstrong
       Cork.
167.   R. W. Hall, Brit. 1,034,816 (Distillers).
168.   E. J. Vandenberg, U. S. 3,065,213 (Hercules).
169.   E. L. Stogryn and A. J. Passannante, U. S. 3,261,874 (Esso).
170.   S. V. Puntambekar and T. K. Venkatachalam, Indian J. Technol.
       1, (6) 231-3 (1963).
171.   J. L. Amos, J. L. McCurdy, and O. R. McIntire,
       U. S. 2,694,692 (Dow), appl. 1954.
172.   S. P. Boutsicaris and R. A. Hayes, U. S. 3,595,851 (Firestone).
173.   J. M. Paushkin and G. P. Losev, J. Polym. Sci. no. 22, 501
       (1968).
174.   R. J. Zeitlin, J. Polym. Sci. C # 24, 269 (1968).

2.   POLYMERIZATIONS OF 1-ALKENES

The allylic 1-alkenes or α-olefin family, of which propylene is the first and most important member, are exceeded only by ethylene and styrene in volume of derived hydrocarbon products. In the chemical literature, the 1-alkenes have seldom been called allylic compounds, although leading organic chemists agree with the author that their chemical properties place them with the allyl compounds. The lower 1-alkenes are unique monomers, low in toxicity, low in price, and available in abundance from cracking of petroleum. Beside difficulties in preparing uniform linear polymers of high molecular weight, the flammability of the monomers and polymers has been a limitation to development of useful industrial materials.
The lack of activating substituents in ethylene and its low boiling point prevented industrial synthesis of polyethylene until the high-pressure process of Imperial Chemical Industries (ICI) in 1933, nearly a century after high polymers had been prepared from styrene and suggested as synthetic materials. Propylene as an allyl compound presented even more difficulty. That tough, heat-resistant polymers could be made from the lower 1-alkenes was unknown and unexpected before 1954. The discovery of new stereoregulated ionic polymerization processes, together with keen international competition in research and commercial development, led quickly to large-scale manufacture of propylene polymers for new plastics, fibers, and synthetic rubbers. From small production in 1957, the production in the United States alone exceeded 1.2 billion lb in 1969. Heat resistance substantially better than that of polyethylenes, along with strength and solvent resistance, has promoted wide use of normally crystalline propylene polymers in moldings, in films for packaging, and in fibers. The injection moldings and extrusions have replaced metals and inferior plastics in automotive, household, and medical components. Commercial uses of polymers from 1-butene, isobutylethylene, and other 1-alkenes is much smaller, but research and development continue. Although smaller in production than the high-propylene plastics and fibers, another major development

56

is synthetic rubber based upon ethylene-propylene-diene copolymers (EPDM) which have outstanding ozone resistance and long life. Different types containing 30% or more propylene units are being "tailored" to obtain properties required in specific applications such as automotive, household appliances, and building components.

Since the development of polyamide and polyester fibers by Carothers, Whinfield, and their co-workers in the 1930s and 1940s, there has been no explosion of laboratory activity and publications in organic chemistry on an international scale approaching that which followed disclosures of polymerizations of alkenes using transition metal-organometallic catalysts with control of polymer stereoisomerism. The discovery of normally-crystalline high polymers from 1-alkenes did not stand alone but was related to the discoveries of isomeric polyethylenes and isomeric vinyl ether polymers, which cannot be discussed in detail here (1). Karl Ziegler and Giulio Natta, who had become interested in this area already in 1951, led enthusiastic research teams in Germany and Italy. Their publications (2) stimulated world-wide activity, for which Ziegler and Natta received Nobel Prizes in 1963.

Many chemists and engineers in North America and Europe contributed to developing practical industrial processes for polymerizing and forming 1-olefin polymers. It is not surprising that controversies continue regarding priority of inventions and infringement of patents in an area so successful commercially and so basic scientifically as stereoregulation. Nomenclature is not yet uniformly accepted, and alternate names are employed: coordination polymerization, insertion polymerization, transition metal catalysis, Ziegler-Natta polymerization, and Ziegler polymerization. Unfortunately some publications, including reviews, have been incomplete or influenced by patent positions. This chapter cannot give all references, but those considered most significant.

The 1-alkenes as allyl compounds have been among the most difficult of monomers for commercial polymerization. Note that a number of industrial processes for polymerization of ethylene (e.g., the well-known ICI, Du Pont, Union Carbide, BASF, Phillips, and Standard Oil of Indiana processes) have not proved commercially feasible for the synthesis of propylene high polymers.

REACTIVITY OF 1-ALKENES

The 1-olefins considered here are those bearing allylic hydrogen atoms on the third carbon; they include propylene, 1-butene, 1-pentene, isopropylethylene, and isobutylethylene, but not isobutene or styrene. The most important

are gases, but beginning with 1-pentene they are normally
liquids having mildly pungent and slightly unpleasant
odors. The 1-alkenes boil slightly lower than the
related alkanes and also lower than isomers with more
central double bonds. Melting points are lower than for
the alkanes, and for straight-chain compounds the double
bonds increase the density about 0.015 g/cc over those of
alkanes. The terminal double bond increases refractive
index (n_D^{20}) about 0.011 (less than for more central double
bonds). The olefins burn with luminous flames. Slow
oxidation on storage in light may give peroxides, carbonyl
compounds, and viscous liquid polymeric residues. Cyclohexene with four allylic hydrogen atoms forms oxidation
products and viscous liquid polymers on standing in air
faster than most 1-alkenes (3). Oxidation of 1-hexene in
air can form $CH_2=CHCHCH_2CH_2CH_3$. Branching in alkenes can
$\qquad\qquad\qquad\qquad\quad$ ÓOH
promote oxidation to hydroperoxides. Oxidation of propylene to acrylic acid is in production in Japan and is
expected to become important in the U. S. and Europe. A
two-step process may be employed via acrolein (BP Chemicals
and Japan Catalytic) or direct reaction with oxygen over
complex catalysts, for example, Co, Fe, Mo, and Te
silicates (4).

The 1-olefins are prepared by cracking and isomerization
of natural gas and petroleum fractions, processes that are
beyond the scope of this book. Recently oxidative methods
of dehydrogenation of alkanes have been investigated. For
example, gaseous alkanes containing oxygen may be passed
over iron oxide catalysts at 600 to 800°F (5).

The electron-donating alkyl groups attached to the
ethylenic group of 1-olefins make the 1-alkenes more reactive than ethylene (e. g., in ionic addition reactions
with Lewis-acid or electrophilic catalysts). Propylene,
for example, adds chlorine and bromine rapidly, iodine
chloride adds more slowly, and iodine very slowly and
incompletely. Substitution may also occur, especially at
higher temperatures and in light. Hydrogen iodide adds
more rapidly than hydrogen chloride and hydrogen bromide.
Addition of active hydrogen compounds generally follows
Markovnikov's rule with some exceptions in the presence
of peroxides or light or in additions of mercaptans. In
the presence of UV light and sensitizers $NaHSO_3$ added to
1-dodecene in the reverse way giving largely sodium
1-dodecane sulfonate (6).

Propylene is hydrogenated over activated nickel catalysts
much faster than ethylene. However, steric effects
apparently cause decreasing rates of hydrogenation in the
order mono-, di-, and trialkyl-substituted ethylenes (7).

Thus rates of hydrogenation decreased in the order isopropylethylene, 2-pentene, and trimethylethylene (8).
Unfortunately, few studies have been made of comparative
rates of hydrogenation of different pure vinyl and allyl
compounds.

Alkenes may be characterized by solubility in sulfuric
acid solutions, which can involve physical solubility,
addition to double bonds, and polymerization. Propylene
is absorbed by 85% sulfuric acid, whereas concentrated
acid is required for ethylene. In concentrated sulfuric
acid, 1-butene dissolved faster than propylene (9). In
85% sulfuric acid, 2-butenes dissolved faster than
1-butene, whereas isobutylene dissolved 10 times as fast
as 2-butenes. Trimethylethylene was absorbed three to
four times as fast as isobutylene.

Early work showed that dissolved or liquefied propylene
(10) and 1-butene were polymerized to viscous liquid low
polymers by anhydrous acidic catalysts more readily than
was ethylene. Such polymerizations with anhydrous sulfuric acid, aluminum chloride, or acidic silicates
occurred much more readily with isobutylene and with
amylenes. The literature on polymerization of 1-olefins
to liquid polymers (e.g., up to hexamers) for use as
fuels and lubricants is not reviewed here. Catalysts
used recently include π-allyl complexes containing transition metals (11).

High polymers from the 1-alkenes have been prepared only
by special ionic polymerization systems, notably by coordination or Ziegler-Natta catalyst systems, which
surprisingly have often been regarded as anionic. After
the success of the free radical polymerization of ethylene
at high pressures in the 1930s, many attempts were made to
prepare propylene homopolymers of high molecular weight
under free radical conditions. There has been no success
in this, some statements in patents notwithstanding, and
these failures have fully confirmed the profound effect
of the electron-donating methyl group upon reactivity of
the ethylenic group. Moreover, no practical homopolymerizations of propylene or other 1-alkenes by electron-
donating catalysts such as alkali metals or alkali alkyls
(true anionic polymerization) have been developed.

Soft viscous polypropylenes of molecular weight as high
as 10,000 were made from propylene in methyl chloride
near -78°C using a dark red liquid complex of Al(OEt)$_3$
and AlCl$_3$ (12). These polymers, suggested as insulating
oils and tackifiers for rubber, offered no promise as
plastics. Soft solid polypropylenes, but only of average
molecular weight of 1500 and below, were obtained by
adding AlCl$_3$ in ethyl chloride to propylene in methyl-

cyclohexane near -78°C (13). Evidence was found for rearrangements in 1-butene and 2-butenes during polymerizations with gaseous BF_3 at 0 to 60°C (14).

With many of the Lewis-acid catalysts which polymerize isobutylene rapidly at low temperatures, propylene ordinarily reacts sluggishly and gives only viscous liquid low polymers. Fontana and Kidder of Socony laboratories explored conditions for preparing soft, solid higher polymers from propylene (15). Additions of HBr in ethyl chloride to 6 to 10 mole % propylene and 0.20 mole % $AlBr_3$ in liquid butane accelerated the rates of polymerization. Reaction times of 20 to 48 min near -78°C produced "plastic semisolid" polypropylenes. Rates were proportional to the promoter concentration and to that of the 1 : 1 molar complex [e.g., $H^+(\bar{Al}Br)_4$] and relatively insensitive to monomer concentration. The heat of polymerization was estimated as 16.5 kcal/mole. Fontana and Kidder favored a mechanism of stepwise synthesis instead of a typical chain reaction. Macromolecular propagations were suggested, with close association of the two ions:

$$CH_2{=}CHCH_3 + RCH_2\underset{\displaystyle CH_3}{CH}CH_2\overset{+}{\underset{\displaystyle CH_3}{CH}}\quad \overset{-}{Al}Br_4 \longrightarrow R(CH_2\underset{\displaystyle CH_3}{CH})_2CH_2\overset{+}{\underset{\displaystyle CH_3}{CH}}\quad \overset{-}{Al}Br_4$$

A gradual increase in polymer specific viscosity was observed during the early stages of polymerization, showing macroions of relatively long life, which were observed also in slow proliferous cationic polymerization of alkyl vinyl ethers (page 20).

Fontana and co-workers observed differences in the properties of their polypropylenes and attributed these to branching and molecular weight rather than to stereoisomerism (16). The "normal" polypropylenes of high viscosity index showed second-order transitions near -18°C; the apparently more branched polypropylenes of high viscosity index showed glass transition temperatures of about -30°C. The latter were prepared using slower propylene feed rates (17). Polypropylenes made by a slow, semicontinuous polymerization (up to 4 hr) with low ratio of propylene to HBr promoter had average molecular weights as high as 144,000. The polypropylene fractions of high molecular weight were said to be superficially similar to high-molecular-weight polyisobutylenes. When dissolved at higher concentrations in hydrocarbon oils, however, they tended to show gelation instead of stringiness and tack. The hydrogen bromide-promoted aluminum bromide catalyst remaining in the soft polypropylenes could be quenched and dissolved away by adding an excess of isopropanol (18).

The polymerization methods of Fontana gave polymers from 1-butene of higher molecular weights than from propylene. The additions of HBr to AlBr$_3$ catalyst suitable for 1-butene polymerization were considerably greater than the cocatalyst concentrations generally favoring isobutene polymerization to high polymers (19). Both systems gave only oily polymers from 2-butenes.

Some of the 1-butene polymers prepared in Magnolia Oil Company laboratories using HBr-promoted AlBr$_3$ were solids insoluble in solvents at ordinary temperatures (20). Good yields of high polymers, however, were not achieved; steric regularity and crystallinity were not observed. Propylene and 1-butene polymerizations using boron fluoride as catalyst gave only low-molecular-weight liquid products. Surprisingly, when petroleum refinery hydrocarbon streams of predominantly C$_4$ hydrocarbons were treated with chromium oxide-silica-alumina catalysts, they formed deposits of partially crystalline 1-butene polymers (21). Phillips Petroleum Company undertook the study of this type of polymerization system.

Zletz and co-workers of Standard Oil of Indiana used partially reduced molybdenum trioxide on alumina as catalyst at 130°C or above to prepare from propylene greaselike to normally solid polymers (22). In one example, 1 hr at 400°C and 150 psi with hydrogen was employed in the catalyst reduction step. Another patent disclosed normally solid resinous polymers (23). Rates of propylene polymerization were lower than those of ethylene over catalysts from activated metal oxides. No conclusive evidence of polymer structure from x-ray and infrared data was presented. Peters and Evering disclosed Group V A or VI A metal oxides along with aluminum hydride or aluminum alkyls as catalysts for synthesis of crystallizable polyethylene and polypropylene, for example, at 25 to 104°C and 100 to 300 psi (24). Commercially useful polypropylenes have not been available from these processes, and little has been revealed about the properties and structures of the products.

In 1953 Hogan and Banks applied for patents on polymerization of 1-olefins using heat-activated chromium oxide on silica-alumina, which developed into the industrially important Phillips process for high-density polyethylenes and ethylene-1-alkene copolymers (25). In one example, an activated catalyst of 3% chromic oxide on silica-alumina at 88°C was brought into contact with a solution of 20 mole % of 1-olefin and 80 mole % isobutane for 4 hr. Propylene gave about 90% conversion and 1-butene gave 75% conversion to "resinous" or semisolid products. Although 1-pentene and 1-hexene also were said to form resinous polymers, 1-octene gave only liquid dimers and trimers.

A sample of resinous polypropylene melting near 149°C
could be drawn out to fibers. Polypropylenes of up to
50,000 molecular weight could be obtained by hydrocarbon
extraction of low-molecular-weight fractions. The high
polymers were suggested for fabrication of rubberlike and
plastic articles. Solid polymer formed upon the solid
catalyst in fixed or moving beds or in suspension. The
polymers were dissolved off the catalyst by hot pentane
or isooctane. The solid polypropylene fractions of high
molecular weight were not soluble in methyl isobutyl
ketone at 93°C. The favorable higher softening temperatures of propylene high polymers compared with polyethylenes were disclosed (26). Preparation of the
catalysts was discussed (27). These catalyst types of
Standard Oil of Indiana and Phillips have not proved
practical for manufacture of normally-crystalline polypropylenes suitable for plastics.

  Ketley and Harvey later made propylene and 1-butene
polymers--apparently similar to those of Fontana--near
-78°C with $AlBr_3$ promoted by HBr (28). Few details of the
polymerization reactions were given. Infrared spectra of
the propylene polymers showed side ethyl groups (769 cm^{-1})
and propyl groups (739 cm^{-1}) attributed to inter- and
intramolecular hydride transfer during polymerization.
The growing carbonium ion might remove hydrogen from its
own chain or from another polymer chain. The authors did
not disclose the range of catalysts and polymerization
conditions investigated; they stated that the polymer
structure was independent of the nature of the catalyst
or cocatalyst used. Homopolymers were formed both in
n-butane and in CH_3Cl solvent, using promoted $AlBr_3$
catalyst. Only traces of polymer were obtained when
isobutane was the solvent, apparently because of rapid
chain transfer with the t-hydrogen atoms. Tetrachlorides
of titanium and tin, with water as cocatalyst, gave no
polypropylene; boron fluoride gave only oily, nonviscous
low polymers.

FREE RADICAL POLYMERIZATION OF PROPYLENE

  Solid polymers of high molecular weight have not been
prepared from propylene by conventional free radical
systems (29). However, polymerization of propylene by
radical initiation is accelerated by very high pressures.
Brown and Wall used pressures up to 16,000 atm to obtain
improved rates of homopolymerization (30). Gamma radiation was used for free radical initiation at 21, 48, and
83°C. Rates could be varied by the intensity of radiation
from ^{60}Co. The molecular weights of the propylene homopolymers formed were relatively low (DP = 50-100) and were

not simply related to the rates of polymerization. Inherent viscosities of solutions were 0.11 and lower. At
comparatively low radiation doses, the rate of polymerization was proportional to the square root of the intensity of radiation. The rate of polymerization increased
by a factor of 100 at room temperature by increasing
pressure from 5000 to 16,000 atm.

Wall (31) suggested that pressures of the order of 10^3
atm and higher may favor polymerization in the following
ways: (a) acceleration of the bimolecular propagation or
growth step (closer approach of reactants), (b) retardation
of termination and chain transfer as a result of increased
viscosity, (c) elevation of the ceiling or depropagation
temperature of the polymer, (d) elimination of inhibition
by impurities.

The amorphous propylene homopolymers prepared at high
pressures by Brown and Wall ranged from faintly colored,
viscous liquids similar in flow properties to polyisobutenes of low molecular weight to opalescent gums of
somewhat higher DP. The latter were so viscous that some
samples retained impressions for several days. All the
polymers were soluble in benzene at room temperature, but
some of the solutions showed mysterious opalescence.
There was no evidence of crystallinity or isotactic
structure. NMR spectra suggested predominantly alternating or syndiotactic structure in these free-radical-
initiated polypropylenes (32). In Monsanto laboratories
polypropylenes of number-average molecular weight 1000
were prepared at 27,000 atm and 130°C using t-butyl
peroxide as initiator (33).

PROPYLENE HIGH POLYMERS BY ZIEGLER-TYPE CATALYSTS

Ethylene polymerization by organometallic catalysts had
been reported in the 1940s. Hanford, Roland, and Young
obtained hard, waxy polymer melting at 119 to 121°C after
17 hr at 20 to 26°C and 900 atm by use of n-butyl lithium
in benzene (34). Roedel prepared orientable polyethylene
of high tensile strength by use of ethyl magnesium
bromide Grignard catalyst at 150°C and 900 atm (35).
Beginning about 1950, Ziegler at Mülheim-Ruhr, with
support of the German government, worked on metallorganic
compounds as catalysts for synthesis from ethylene of
liquid higher olefins; but occasionally, surprising solid
high polymers formed even at atmospheric pressure.
Ziegler and co-workers observed solid polyethylenes formed
slowly in experiments with Grignard catalysts at 60°C or
higher under moderate pressures (36). A suspension of
dibutyl magnesium in benzene also could be used (37).
Propylene formed largely liquid dimer at 150°C in the

presence of magnesium dialkyls. Small amounts of nickel
hydrogenation catalyst contaminating an autoclave reactor
were found to accelerate reaction rates, leading to the
discovery of transition metal-organometallic combinations
as practical catalysts for 1-olefin polymerization at
atmospheric pressures. Ziegler, Martin, and co-workers
studied additions of heavy metal salts systematically and
disclosed the remarkable catalytic activity of reaction
products of aluminum alkyls with transition metal halides
such as $TiCl_4$, VCl_4, and $ZrCl_4$ in the synthesis of substantially linear polyethylenes of high molecular
weight (38).

Patent applications from 1954 disclosed that solid
polypropylenes showing evidence of crystallinity had been
prepared in Du Pont laboratories by the use of activated
transition metal catalysts somewhat resembling those disclosed by Ziegler. In order for polymerization to occur,
it was believed necessary that the titanium from $TiCl_4$ be
reduced to bivalent titanium by the organometallic
reducing agent, such as lithium aluminum alkyl or phenyl
magnesium bromide. The name coordination polymerization
was applied because the monomer was believed to coordinate
with the complex catalysts. For example, Anderson,
Merckling, and co-workers prepared a catalyst from mixing
0.01 mole of $LiAl(n-hexyl)_4$ and 0.01 mole of $TiCl_4$ in 50
ml of cyclohexane in a shaking reactor (39). Propylene
at initial pressure of 300 psi was applied for 2 hr at
75°C. Exothermic reaction gave 45 g of homopolymer which
was washed with methanol, methanol-HCl, and acetone.
After drying, "x-ray determination showed the polymer to
be 15-25% crystalline." One polypropylene had an inherent
viscosity of 3.5 in decalin at 130°C. The polymers could
be molded at 150°C to flexible films and could be drawn
to fibers.

The Du Pont group did not disclose stereoregularity or
stereoisomerism to explain the different proportions of
normally-crystalline polypropylene fractions, as did
Schildknecht, Natta, and co-workers for isomeric polyvinyl ethers and polyolefins. Little was published about
the American work on normally-crystalline polypropylenes
in contrast to the extensive literature describing the
researches carried out in Europe.

The early Du Pont work was extended by Schreyer to
preparation of rubbery solid propylene polymers, using
catalyst prepared by reaction of $LiAlBu_4$ with VCl_4 (40).
The coordination catalysts prepared from $TiCl_4$ and $LiAlR_4$
or AlR_3 gave heat-resistant polymers from norbornene

 (41). The polymer, which was named poly-3-cyclo-

pentanylvinylene, had a crystal melting point 180 to
205°C. With titanium or other transition metal compounds,
the useful reducing agents were stated to include metal
alkyls, metal hydrides, alkali or alkaline earth metals,
and Grignard compounds. Copolymers of norbornene and
ethylene prepared at 1 atm and room temperature gave
clear, tough fibers and films (42). Ethylene was the
faster reacting monomer. Copolymers of propylene with
approximately 10% norbornene units were observed to be
35% crystalline by x-ray tests and could be molded at
200°C (43). It is interesting that norbornene slowly
becomes viscous by polymerization on storage in contact
with air as does cyclohexene (44).

Belgian patents, which are published without contest,
give a record of the research led by Ziegler at Mülheim
and by Natta at Milan. In contrast, some of the U. S.
patent applications of 1954-1955 remained in litigation
in 1972. In one example of Ziegler-activated transition
metal catalysis, 0.2 g of titanium tetrachloride was
added to 20 cc of aluminum triisopropyl, giving a black
dispersion (45). Ethylene was introduced under pressure
at 100°C. After 15 hr at 20 atm, there was formed a
paste of solid polyethylene in a solution of low-molecular-
weight polymers. After treatment with HCl in methanol,
and drying, the polymer could be molded at 150°C to form
films. In Ziegler polymerizations, reactants must be
substantially free of water, oxygen, and acetylene, and
reactions ordinarily are carried out in closed reactors
or autoclaves. Red-brown dispersions made with lower
ratios of aluminum alkyl to $TiCl_4$ gave polymers of lower
molecular weight (46). In another example, a solution of
$TiCl_4$ in hexane was added to $Al(C_2H_5)_2Cl$ and ethylene was
admitted (47). Active polymerization caused a spontaneous
rise from room temperature to 70°C.

On June 8 and July 27, 1954, Montecatini applied in
Italy for patents on polymerization of propylene and other
1-olefins to form crystalline polymers with catalysts
prepared by reacting compounds of metals of groups IV B
to VI B with organometallic compounds of groups I to III
(48). In one example, 4.75 g of $TiCl_4$ was added to a
solution of 5.8 g of $Al(C_2H_5)_3$ in purified diesel oil.
After an hour, 600 g of propylene was introduced into the
brown-black dispersion. Reaction at 70°C and 21 atm for
72 hr, followed by treatment with acetone and then by
HCl-methanol, gave 338 g of polypropylene that could be
molded at 140°C. Polypropylene films were less brittle
than those from highly crystalline polyethylene. Higher
temperatures of polymerization with a given complex
catalyst gave polymer of greater solubility in n-heptane,

e.g., lower stereoregularity (49). Preformed titanium
trichloride powder with aluminum alkyls gave slower
polymerization of propylene but greater polymer stereoregularity (50). Larger alkyl groups attached to
aluminum produced lower stereoregularity.

The discovery that 1-alkenes or α-olefins can yield
normally-crystalline, stereoregular high polymers under
the action of Ziegler-Natta catalysts was published in
America by Natta, Pino, Mazzanti, and co-workers on
March 20, 1955 (51). Earlier work on stereoregulated
cationic polymerizations of alkyl vinyl ethers was
recognized. Solid homopolymers of propylene, 1-butene,
and styrene made "by heterogeneous, solid catalysts" were
found to be normally crystalline with identity periods by
x-ray diffraction of 6.5 to 6.7 Å and crystal melting
points about 160, 127, and 230°C, respectively. Amorphous
fractions could be removed from the initial polymers by
extraction with hydrocarbon or ether solvents. It seemed
probable that "at least for long portions of the principal
chain, all the asymmetric carbon atoms have the same configuration", e.g., DDDD or isotactic (52). The identity
periods, like that reported earlier for crystalline vinyl
isobutyl ether polymer, were too short for a planar,
zig-zag conformation and were attributed to helices of
three monomer units in repetition. Each methyl group was
displaced 120° from its nearest methyl group. Infrared
spectra characteristics of the crystalline and amorphous
polymers were given. On melting the polymers, most of
the differences in IR spectra disappeared, although they
reappeared on cooling. Even in light of the researches
on polyvinyl ethers and the predictions of this writer
(53), it was indeed surprising that polymers of such high
crystal melting points could be prepared from propylene
and from styrene.

The principal U. S. patents of Natta, Pino, and
Mazzanti claim as a new substance high-molecular-weight
polypropylene consisting essentially of isotactic macromolecules insoluble in boiling n-heptane and in boiling
ether, and having substantially the type of stereoregular
structure illustrated in Fig. 2.1 (54). These informative
patents showed the remarkable contrast between the properties of the predominantly stereoregular polymers from
lower 1-alkenes and those of polyethylenes and polyisobutenes. A normally-crystalline polypropylene had density
of 0.92 g/ml and was insoluble in acetone, ethyl acetate,
and diethyl ether but soluble in hot toluene. An amorphous polypropylene fraction had density near 0.85 g/ml
and was soluble in all these solvents.

In example 1 of the patent, 1.8 g TiCl$_4$ in 50 ml of
anhydrous gasoline was added dropwise at 5 to 10°C to a

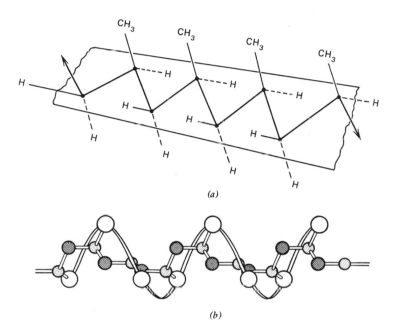

(a)

(b)

Fig. 2.1. (a) Representation of four monomer units of an isotactic polypropylene chain molecule in zig-zag conformation; G. Natta et al., U. S. 3,112,300, 1954. (b) Alternative helical conformation in crystalline polypropylene, proposed by Natta and Corradini, consisting of three monomer units per helix turn. The nine monomer units in (b) are represented using smallest circles (black) for CH, medium circles (gray) for CH_2, and large circles (white) for CH_3. In the helix each monomer unit and each CH_3 is rotated 120° with regard to its neighbor.

solution of 11.4 g of $AlEt_3$ in 150 ml of gasoline. The catalyst dispersion was then diluted with 50 ml of gasoline in a 2-liter steel autoclave. Into this was pumped 190 g of dry propylene, with agitating, at 55 to 60°C. After the pressure had fallen from 10 to 2 atm, 160 g of additional propylene was added. After 20 hr the residual gas was vented, and 95 g of methanol was added to decompose residual catalyst. The recovered solid polypropylene

was treated with dry HCl in diisopropyl ether for removal
of catalyst residues. There was obtained 180 g of dried
polymer having a wide range of molecular weights and containing both normally-crystalline and amorphous fractions.
Successive polypropylene fractions were removed by extraction with a series of solvents: acetone, ether, and
n-heptane (55). The fraction insoluble in boiling
n-heptane lost its crystallinity only above 150°C.

   Natta and co-workers also reported low melting crystalline homopolymer fractions from 1-pentene and 1-hexene.
Neither this patent nor variations of Ziegler-Natta
processes during the subsequent 15 years obtained wholly
stereospecific polymerization without forming some amorphous polymer fractions. Vanadium tetrachloride with
aluminum triethyl in n-heptane at 80°C gave a similar
polypropylene from which 28% of highly crystalline polymer
insoluble in n-heptane was recovered. The latter had an
intrinsic viscosity of 1.88, corresponding to a molecular
weight of 85,000. Viscosities were determined in hot
tetrahydronaphthalene.

   These polymerizations of 1-olefins and styrene were
believed by the Milan group to be examples of anionic
polymerization. The titanium halide was regarded generally as a cocatalyst or activator, whereas later the
aluminum organic compounds were regarded as activators of
the transition metal compounds. The percentage of insoluble fraction in boiling specific solvent continues to
be the best practical measure of the percentage of isotactic polymer as long as no crosslinking is present. In
the fractionation of crude polypropylenes by successively
more active solvents, it was generally supposed that the
most soluble fractions were atactic in structure; intermediate fractions were believed to be block copolymers of
atactic and isotactic sequences, and the undissolved
residues were regarded as substantially stereoregular (56).
Catalysts prepared by Natta and co-workers from reaction
of preformed titanium trichloride with organoaluminum
compounds also gave mixtures of stereoregular and amorphous polymers which could be separated by solvent
extraction (57). In one example, 1.2 g of finely divided
solid $TiCl_3$ (violet polymeric form) was present as a
slurry in a solution of 2.3 g of $Al(C_2H_5)_2Cl$ in 300 ml
of n-heptane.

LATER ZIEGLER-NATTA POLYMERIZATION SYSTEMS

   Progress toward commercial processes for manufacture of
1-olefin high polymers in Germany and in the United States
was promoted by joint efforts of Farbwerke Hoechst and
Hercules, who early licensed patents for Ziegler-Natta

catalysts. Vandenburg developed a two-component catalyst process (58). Titanium tetrachloride was reacted with AlR_3 or AlR_2Cl and the precipitate was separated and washed with hydrocarbon. After it had been aged or given a heat treatment, this very active catalyst precipitate was combined with additional AlR_3 or AlR_2Cl as a promoter, to give high yields of stereoregular polypropylene along with some amorphous, soluble polymer fractions. Rust and co-workers used a heat treatment of up to 3 hr at 40 to 150°C for the purified catalyst precipitate (59). As a second component $Al(C_2H_5)_2Cl$ or a sesquichloride was added for polymerization of 1-alkenes at -2 to +20°C. Yields as high as 93% of normally-crystalline polypropylene could be obtained by further addition of sodium chloride and polymerizing at 50°C for 5 hr at 15 psi gauge pressure (60). Later a catalyst precipitate of very small particle size (1-3 microns) was developed which formed polypropylene particles of about 10 times the catalyst diameter (61).

Regulators must be added to control polypropylene molecular weights to practical ranges for molding, calendering, and spinning to fibers. Additions of zinc diethyl were suggested by Natta, but hydrogen is employed generally in practice (62). Vandenberg found that reduction in molecular weight occurs approximately proportional to the pressure of hydrogen gas added and without adverse side effects (63). This principle is one of the most important in the commercial production of polyolefins.

The treatment of polyolefin slurries with alcohol and dilute acid was found to be unsatisfactory for removal of aluminum- and titanium-containing catalyst residues. The first polypropylene plastic granules attacked metal molds rather severely. Jacobi and Wolf bubbled ethylene oxide through the suspension of polyethylene obtained by Ziegler catalyst until the red-brown color was bleached (64). Precautions were taken to avoid excessive contact with oxygen and water. The polymer granules then were separated and washed by acetone or alcohol followed by water. For the removal of titanium residues, it was critical to add not more than one mole of water per mole of titanium halide residue. Such polymerizations were carried out with 3.2 parts of reaction product of titanium tetrachloride and ethyl aluminum sesquichloride (Al/Ti ratio = 0.5) together with 3.4 parts of $(C_2H_5)_2AlCl$ in petroleum hydrocarbon solvent (bp, 170-200°C) (65). The propylene and hydrogen reacted for 5 hr at 50°C. The polymer slurry in about 4 parts of liquid hydrocarbon was quenched by n-butanol and then stirred with addition of propylene oxide and the critical amount of water before filtering.

In another process of Mahlman and Spurlin, the polypropylene granules were slurried with n-butanol together with gluconic or tartaric acid and a little alkali in order to form soluble complexes from the aluminum and titanium residues (66).

Pioneer work on the synthesis of block copolymers by periodic change of feed, e.g., 5 min ethylene, 5 min propylene, was carried out by Bier and co-workers of Hoechst (67). Lives of active growing chains were indicated to be as long as 15 hr under some conditions. For copolymerization of ethylene with propylene by means of improved Ziegler-Natta catalysts, reactivity ratios were observed to be $r_1 = 3.0 \pm 5$ and $r_2 = 0.25 \pm 0.05$. The greater reactivity of ethylene is less pronounced under normal operating conditions because of the greater solubility of propylene in hydrocarbon diluents. Bier reported that 30% propylene in the feed gas gave about 10% propylene units in the copolymer product, which had stiffness similar to high-pressure branched polyethylenes. With 4% propylene in the copolymer, a modified polyethylene plastic was obtained comparable to an ethylene-low-1-butene plastic from the Phillips process. Experiments of Hoechst employed continuous flow of fresh gas mixtures. Figures 2.2 and 2.3 compare IR spectra of block copolymers of ethylene and propylene with those of propylene homopolymers. Alternate feeding of ethylene and propylene gave block copolymers of lower density than mixtures of the homopolymers (Fig. 2.4).

Hoechst and Hercules chemists showed that properties of the block copolymers of ethylene and propylene could be superior to those of blends of the homopolymers (especially higher ultimate extensibility and better toughness at low temperatures). More active catalysts, higher rates of polymerization, and longer periods of chain growth were possible. In polymerizations of propylene at 30, 50, and 70°C using $TiCl_3/AlR_2Cl$, initiation on the surface of violet polymeric $TiCl_3$ also involved the organoaluminum activator (68). Chain transfer and disproportionation were essentially absent. Practically no metal-free macromolecules were formed. Acceleration of rates at higher temperatures could involve faster diffusion of monomer to active catalyst sites.

Bier and co-workers obtained increased rates in propylene stereopolymerization by using $(C_2H_5)_2AlCl$ and alkyl aluminum fluoride along with titanium tetrachloride (69). There was less sacrifice of specificity than when AlR_3 activators were added to increase rates. Good yields were obtained in 3 hr at 50°C. Vanadium tetrachloride was reduced more rapidly by organoaluminum compounds than was titanium tetrachloride; vanadium was easily reduced

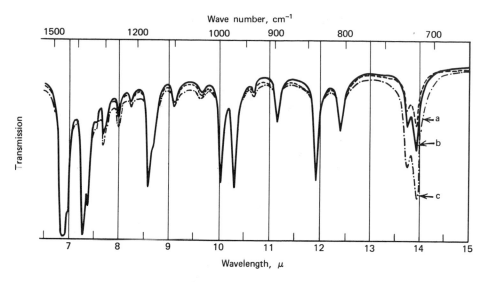

Fig. 2.2. Infrared spectra of three different block copolymers of ethylene and propylene (curves a, b, and c). Content of ethylene units and crystalline CH_2 sequences affect the sharp bands at 720 and 730 cm^{-1}. Absorptions at 995, 972, and 840 cm^{-1} are crystallization sensitive in polypropylenes according to G. Schnell, Ber. Bunsengesellschaft 70, 308 (1966).

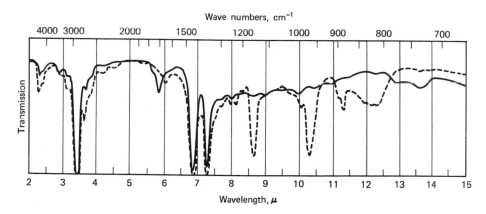

Fig. 2.3. Infrared spectra of largely amorphous propylene homo-polymers showing very little absorption at 720 and 730 cm^{-1}. The solid curve is that of a polypropylene made by conventional cationic polymerization. The broken curve is the spectrum of the soluble fraction of polypropylene obtained by Ziegler-Natta catalysis. G. Schnell, Ber. Bunsengesellschaft 70, 308 (1966).

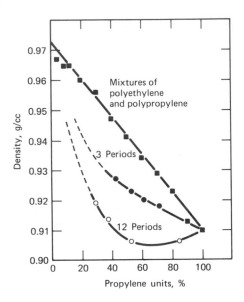

Fig. 2.4. Comparison of densities of blends of polyethylene and polypropylene with densities of block copolymers made with alternating periods of charging ethylene versus propylene to the reactor. Ziegler-type catalysts were used. Research from Hoechst laboratories reported by G. Schnell, Ber. Bunsengesellschaft 70, 308 (1966).

below a valence of three (70). AlR$_2$Cl compounds show advantages over AlR$_3$ in that titanium and vanadium compounds are not thereby reduced below the trivalent state favoring olefin polymerization. Although VCl$_4$ + Al(C$_2$H$_5$)$_2$Cl gave normally amorphous polypropylenes, TiCl$_3$ + Al(C$_2$H$_5$)$_2$Cl tended to give polypropylenes of greater stereoregularity and wider molecular weight range.

For successful polypropylene plastics, it was necessary to improve impact resistance at low temperatures. Molded bottles from early homopolymers could fracture when dropped at below 0°C. Superior blends of isotactic polypropylene with copolymers of propylene and ethylene (e.g., 93 : 7 molar ratio) could be made directly by change of monomer feeds) (71). In one experiment, ethylene, propylene, and hydrogen were reacted 45 min at 50°C using a catalyst made by injecting a suspension of 100 mmole TiCl$_3$ (in n-heptane) into a solution of 200 mmole of (C$_2$H$_5$)$_2$AlCl in 10 liters

of alkane diluent (bp 200-230°C). After preparation of
the ethylene-propylene copolymer, the temperature was
raised to 60°C and propylene containing 0.9 mole % hydrogen was fed for the second stage of polymerization. After
a granular product containing 87% propylene units was
obtained, n-butanol, ammonia, and gluconic acid were added
and the slurry was stirred for an hour under nitrogen.
Molding plastics containing crystalline block copolymers
of ethylene and propylene have been called allomers or
polyallomers by Eastman (72).

Natta and co-workers did research with catalysts based
on preformed solid titanium trichloride in different
polymeric forms (73). The violet modification was more
reactive than a brown form in making polypropylenes of
80 to 90% isotactic structure. Rates were proportional
to the propylene pressure and the amount of violet $TiCl_3$.
At 25°C in n-heptane, copolymerization reactivity ratios
were as follows using the same aluminum alkyl:

	r_1 Ethylene	r_2 Propylene	$r_1 \times r_2$
$TiCl_4$	37.4	0.3	1.1
$VOCl_3$	18.0	0.7	1.2
VCl_4	7.1	.09	0.6

The average life of the growing chains in these systems
ranged from 4 to 7 min.

The weaker or milder catalysts were found to be more
specific in preparing stereoregular polypropylenes, without excessive amorphous fractions. Along with titanium
trichloride, a better activator than $AlEt_3$ is $Al(C_2H_5)_2Cl$
(74). This activator also gives less reduction in rate
with time than triethyl aluminum. Temperatures up to 90°C
in inert liquid diluent were suitable, and no titanium was
reduced below the trivalent state. Mixtures of $Al(C_2H_5)_3$
and $Al(C_2H_5)_2Cl$ gave favorable results with $TiCl_3$ (75).
Catalysts of improved stereospecificity at 70°C were
obtained by adding to the transition metal halide-
aluminum organic catalysts small amounts of tertiary
amines or quaternary nitrogen compounds such as tetraethyl
ammonium iodide (76).

The precise mechanism by which the purple crystalline
modification of $TiCl_3$ in combination with trialkyl aluminum or dialkyl aluminum chloride can form isotactic polypropylene remains to be elucidated. However, a number of
characteristics of the systems most commonly employed
have been well established. Heterogeneity of the catalyst
system is essential for the stereopolymerization of 1-
olefins. The three-dimensional polymeric lattice of

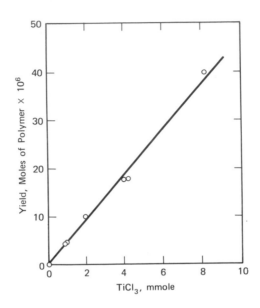

Fig. 2.5. The rate of polymerization of propylene is proportional to the concentration of active titanium trichloride (e.g., the number of catalyst sites, when activated by diethyl aluminum chloride). A. K. Ingberman, I. J. Levine, and R. J. Turbett, J. Polym. Sci. A 1 **4**, 2795 (1966).

titanium trichloride is important since the brown beta modification, which is a linear inorganic polymer, gives much lower rates and lower degrees of stereoregulation. The dialkyl aluminum halides seem to be only chemisorbed, or at least to react less with the titanium trichloride surface than do aluminum trialkyls (77). The active sites seem to be transition metal atoms at points of the titanium trichloride surface at which titanium atoms have been alkylated by the aluminum compound.

When titanium trichloride is used in heptane at 40°C with Et_2AlCl there may be formed less than 2% amorphous polypropylene fraction (77). In general, when triethyl aluminum is used under similar conditions rates are faster, stereoregulation is less efficient, and about 25% amorphous fraction may be formed. The fractions of amorphous polyolefin produced often have been reported too low in the literature because of inefficient recovery

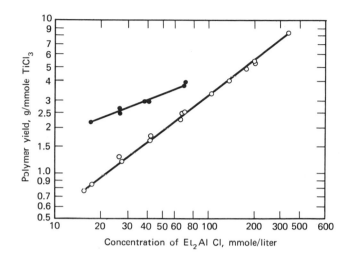

Fig. 2.6. Yields of polypropylene were proportional to the concentration of R_2AlCl used with 10 mmole of $TiCl_3$ in n-heptane at
40°C during 2.5 hr. The yields were higher with $TiCl_3$ prepared
from reaction of $TiCl_4$ with Bu_3Al (solid dots) than that from
reaction of $TiCl_4$ with Et_2AlCl. G. Schnell, Ber. Bunsengesellschaft, 70, 308 (1966).

methods. Adding a mixture of methanol-isopropanol 1:1
equal to the volume of polymerization solvent is effective in precipitating the amorphous fraction.

The polymerization rate may be proportional to the concentration of active sites, the concentration of propylene,
and the concentration of organoaluminum activator (Figs.
2.5 and 2.6). Finely divided titanium trichloride may be
added as such or prepared from titanium tetrachloride, for
example, by the equation $TiCl_4 + \frac{1}{3}R_3Al \longrightarrow \frac{1}{3}AlCl_3 +
TiCl_3 + R\cdot$. The fate of the radicals is uncertain, but
they are not initiators of polymerization. Little termination occurs at 40°C, and the polymer molecular weight
continues to increase with time unless terminating agents
are added. Because of the long life of the growing
chains, block copolymers are formed readily.

Increasing rates of polymerization are observed with
increasing concentrations of Et_2AlCl. The polymer yield
per unit of $TiCl_3$ may be fairly constant; however, it
tends to fall off at high $TiCl_3$ concentrations where

termination becomes more important and polymer molecular weights become lower. Similar plots were obtained for yields of polymer from toluene, chlorobenzene, and di-chlorobenzene.

Studies of polypropylene morphology showed that the physical form of the polymer particles closely resembles that of the $TiCl_3$ particles except for larger size (78). Each polymer flake may consist of thousands of roundish flakelets about 0.5 μ in diameter. The structure of the flakes was not altered by the treatments to remove catalyst residues. Some catalysts form flakes with open porous texture, others give more dense flakes. Such observations are consistent with growth of polymer chains on active sites of the $TiCl_3$ polymer surfaces. Fibrils of 200 to 1000 A° in width are often characteristic of the morphology of polyolefins prepared by Ziegler-Natta catalysts (79).

Of the many other variations from the early Ziegler and Natta polymerization techniques explored, a number became important in the manufacture of polypropylene plastics, fibers, and propylene copolymer synthetic rubbers. Early catalysts disclosed in patents gave poor yields of stereoregular olefin polymers of suitable molecular weight for plastics and fibers.

Laboratory small-scale polymerizations can be carried out in beverage bottles shaken or tumbled in constant temperature baths (e.g., at 50°C). Perforated metal bottle caps with self-sealing preextracted nitrile rubber liners permit injection of catalyst dispersions and removal of liquid samples by hypodermic-type syringes. More sophisticated apparatus and methods for polymeriza-tion and handling dangerously flammable metal alkyls have been described (80).

Tornquist and co-workers of Esso increased catalyst efficiency in propylene polymerization by addition of $AlCl_3$ (81). Titanium chloride and $AlCl_3$ were ball milled dry under nitrogen and then blended with $AlEt_3$ in inert xylene diluent. The improvement was believed to result not from increased surface but from changes in crystal structure. For example, 0.4 g of cocrystallized $TiCl_3 \cdot AlCl_3$ (molar ratio 1 : 0.2) was dry milled with chromium steel balls under nitrogen for 4 days at room temperature (Fig. 2.7). The product was suspended in xylene and activated with 0.5 g $AlEt_3$. More xylene was added to give 0.95 g of catalyst per liter, and propylene was bubbled through for 1 hr at 75°C. The polymerization was terminated by adding a mixture of isopropanol and acetylacetone, which dissolved out the catalyst residues. Tornquist reacted aluminum with $TiCl_4$ at 200°C for several hours to obtain $TiCl_{2.5}$ to $TiCl_{2.8}$ (82). This

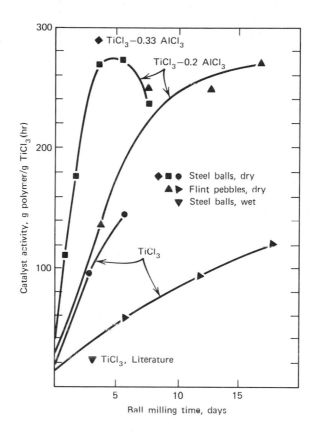

Fig. 2.7. Effects of ball milling and addition of $AlCl_3$ in promoting activity of catalysts based on $TiCl_3$ in polymerization of propylene. E. G. M. Tornquist, Ann. N.Y. Acad. Sci. <u>155</u>, 447-467 (1969).

product in paraffin oil was reacted with excess AlR_3 over that needed to reduce the $AlCl_3$ present to R_2AlCl at room temperature. In polymerizations, yields as high as 2000 to 6000 g of polypropylene per gram of titanium chloride were obtained at 80°C. Later in Europe aluminum alkyls were first treated with BF_3 before mixing with $VOCl_3$ or $TiCl_4$ to form catalysts of high activity (83).

Langer and Tornquist described "monomer-staged" polymerizations of propylene for obtaining higher polymer

crystallinity and catalyst efficiency (84). In one
example, four reactors were used in series with stepwise
increase in monomer pressure. Residence time was 1 hr in
each pressure reactor. The catalyst comprised
$2AlEt_3/TiCl_3 \cdot 1/3AlCl_3$ slurried in xylene at a concentration of 0.8 g/liter and the propylene concentration was
0.8% by weight. From the first reactor a polymer slurry
of 11.3% was pumped along with fresh xylene into the
second reactor, in which a propylene pressure of 10 psi
was maintained. Monomer pressures of 25 and 40 psi were
used in the third and fourth reactors, and more xylene
was added so that the final slurry had a concentration of
11.4%. The average catalyst efficiency was 125 g of
polymer/(hr)(g) compared with 72 g/(hr)(g) without monomer
staging.

Hercules and others developed cocrystallized catalysts
of $TiCl_3$ or VCl_3 along with $AlCl_3$ which may be given
milling, grinding, and heat treatments (85). For example,
0.05 to 0.9 mole of $AlCl_3$ and 0.15 to 0.5 mole of $AlRCl_2$
may be present per mole of $TiCl_3$ for stereopolymerization
of 1-olefins. BASF disclosed catalysts from alkali metal
hydrides, aluminum chloride, and titanium halide (86).
Other inorganic halides added include boron halides (87),
cuprous chloride (88), and cadmium chloride (89).

Besides aging of catalysts, various heat treatments
(e.g., 100-180°C) have been patented, particularly when
mixed catalysts of violet $TiCl_3$ containing $AlCl_3$ are
prepared from brown $TiCl_3$ (90). In one example a mixed
crystal precipitate containing $TiCl_3$, $AlCl_3$, and $EtAlCl_2$
prepared at -20°C was then heated 2 hr at 90°C (91). To
this mixture dispersed in n-heptane was added more $EtAlCl_2$
and a little $AlEt_3$; polymerization during 5 hr at 50°C
yielded polypropylene of reduced viscosity above 12.
Aging for 28 days at room temperature could replace the
heat treatment. Complex catalysts from reaction of $TiCl_4$
with aluminum alkyls or alkali hydrides have been heated
as high as 200 to 350°C for activation (92).

As in stereopolymerization of alkyl vinyl ethers, rapid
polymerization of 1-alkenes with more soluble, electrophilic catalysts tend to produce more amorphous polymers.
More weakly electrophilic complexes formed by adding
small amounts of ethers, amines, or other Lewis-basic
electron donors to Ziegler-Natta catalysts promote stereoregularity. Thus alkyl aluminum dichloride catalysts or
their mixtures with $AlCl_3$ alone are not the best stereoregulating catalysts for polymerization of 1-olefins. A
great number of patents have been granted on minor
additions of relatively weak Lewis bases having unshared
electron pairs. Adding too much of these makes the complex basic and completely destroys polymerization activity.

Thus the addition of 0.5 to 0.8 mole of hexamethyl phosphoramide to 1.0 mole of $C_2H_5AlCl_2$ in the presence of titanium trichloride gave polypropylene about 95% insoluble in methyl isobutyl ketone (93). A 1:1 molar ratio gave no polymerization. It is quite likely that in some early syntheses of stereoregular 1-olefin polymers, small amounts of water or other impurities were actually improving stereoregulation. Contact with a little oxygen reduced the activity of Ziegler-Natta catalysts and was shown to increase the tacticity of the polypropylene formed (94).

Instead of AlR_3 or AlR_2Cl, other activators along with transition metal halides have been proposed such as aluminum alkoxy alkyls (95). Grignard reagents (96), $(CH_2AlR)_n$ (97), and lithium aryl compounds (98). Du Pont obtained patent claims on 1-olefin polymerization, using as reducing activators zinc or any metal above zinc in the electromotive series (99).

Titanium trichloride-activated 1:1 mixtures of $Al(C_2H_5)_2Cl$ and $Al(C_2H_5)_2F$ gave polypropylene of high stereoregularity (100). Catalyst for 1-olefin polymerization was prepared by treating a titanium halide with high-energy radiation sufficient to generate ion pairs (101). Lithium hydride with diethyl aluminum chloride in hydrocarbon diluent at 60°C was then added to give polypropylene of about 70% crystallinity. (Although such statements have been usual, it would be more accurate to say 70% crystallizable or 70% normally crystalline.) Quenched polypropylene moldings reach maximum crystallinity only slowly.

A great many other disclosures of electron-donating (nucleophilic) third components for improving stereoregulation of transition halide-organometal catalysts have been suggested; these include quaternary nitrogen compounds, amides, imides, nitrogen heterocyclic compounds, ethers, esters, sulfur compounds, phosphites, and many others (102). Addition of allyl ethyl ether promoted highly crystalline polyolefin products (103). Boor studied the ability of 50 donors to destroy the most active, less stereoregulating catalyst sites, thereby increasing the yield of isotactic structure (104).

A number of heterogeneous catalysts have been devised without use of metal alkyl compounds, but rates are slow and the polypropylenes formed have not shown very high crystallinity. Youngman and Boor used titanium trichloride activated by a t-amine (105). Pino and Mazzanti made a catalyst from bis(cyclopentadienyl) titanium diphenyl and $TiCl_4$ in benzene which gave 50% crystalline polypropylene at 60°C (106). A complex of titanium borohydride with hexamethyl phosphoramide was effective (107).

Titanium trichloride with a reaction product of $AlCl_3$ and alkali compounds containing nitrogen, oxygen and sulfur gave polypropylene having stereoregularity (108). Recently MgR_2, Grignard reagents, and other magnesium compounds have been explored in catalysts for polymerization of 1-alkenes (109). Detailed mechanisms of stereoregulated polymerizations of 1-alkene remain to be established as shown by recent reviews (110).

The Sun Oil Company developed supported Ziegler catalysts instead of using slurries of insoluble catalyst. Titanium trichloride-based catalyst could be positioned in a tubular metal support (111). Solid, preformed $TiCl_3$ particles could be embedded in a solid carrier such as $NaAlX_4$ or metal alloy (112) or upon alumina (113). Propylene could be passed continuously over a fixed bed of catalyst (114). Zeolites or TiO_2 also were suggested as carriers of catalyst formed by reaction of $TiCl_4$ with AlR_3 (115). Eastman investigated addition of alkali metal fluorides and carbonates to Ziegler-Natta catalysts (116).

Processes for stereopolymerization of propylene in bulk without diluents have been developed (117). In a process disclosed by Phillips, an $AlEt_2Cl-TiCl_3$ complex catalyst, in the presence of hydrogen and liquid propylene, gave a product containing about 40% solids of which 3.7% was soluble amorphous polymer. The working up of the viscous mixture included the addition of acetone and propylene oxide. After polymerization, the temperature and pressure may be lowered at first only enough to precipitate the more crystallizable polypropylene fraction (118). At least one manufacturer uses bulk polymerization of propylene. With $Al(C_2H_5)_2Cl-3TiCl_3 \cdot AlCl_3$ as catalyst, the ratio of vinyl to vinylidene double bonds in polypropylene formed was found to be relatively high, 2.0 (119). Isotactic polypropylene has been made by $MoCl_5-Al_2F_6$ catalyst activated by heating above 120°C (120). Propylene was polymerized to mixtures of amorphous and crystalline polymers in low conversion by titanium metal and ethyl bromide (121). A catalyst from ball milling titanium powder with $AlBr_3$ gave from an ethylene-propylene mixture (16 hr at 90°C) a readily soluble copolymer and a residue of crystalline polyethylene (122). Partial replacement of alkyl groups of AlR_3 by chlorine in complex catalysts increased the polymerization rates of ethylene and propylene and also increased electrical conductivity of the catalyst dispersions (123).

Stereoregular homopolymers of propylene continue to be used in plastics, but new copolymers and copolymer blends are superior in low-temperature impact, clarity, and other properties. Films of unmodified isotactic polypropylene cooled slowly from the amorphous, melted state often are brittle and show beautiful radiating spherulites of crystals under the polarizing microscope (124). Finely dispersed "inert" solids and soaplike nucleating agents, such as long-chain organic salt, may be added to give better clarity by increasing the number and reducing the size of crystallites. Possibly the organic ends of the molecules of nucleating agents provide dispersion while the polar insoluble ends of the molecules furnish local insoluble nuclei upon which polymer crystals can grow. The high-propylene copolymers (discussed later), which have favorable crystallization and impact properties, continue to be called polypropylenes by plastics tech-nologists.

PROPYLENE POLYMER PLASTICS AND FIBERS

Stereoregular polypropylenes immediately gave promise of plastics and fibers more heat resistant than poly-ethylenes, but a number of obstacles presented themselves. Pure 1-alkenes and Ziegler-Natta catalysts form polyole-fins of very wide molecular-weight distribution and of average molecular weight too high for most practical molding of plastics and extrusion of fibers. Fortuitous impurities could not be relied upon for control of molecular weight. The unmodified propylene homopolymer moldings were brittle at winter temperatures. Moreover variations in the size and distribution of crystal aggre-gates (spherulites) caused uncontrolled differences in transparency and strength properties.

At first impact strength at low temperatures was improved by blending with 5 to 20% of such elastomers as polyisobutenes, butyl rubber, or EPDM rubber. However, the most satisfactory method for imparting flexibility at low temperatures is copolymerization, especially block copolymerization with small amounts of other olefins. Patents disclosed 5 to 15 volume % of ethylene copoly-merized with propylene to produce plastics of good impact strength at low temperatures (125). Another method involves copolymerization of propylene with minor propor-tions of a nonconjugated diolefin to give plastics of brittle points below -15°C (126). Ethylene-propylene molding copolymers with resistance to stress cracking were made by two polymerization steps (127).

Numerous suggestions were made of regulating agents for addition to transition metal catalyst systems in order to obtain uniform polyolefins of moderate molecular weight and favorable flow properties required for practical molding. Without a regulator, polypropylenes of reduced viscosity as high as 24.4 are obtained (128). The addition of hydrogen as a regulator has been widely adopted. For example, into a stainless steel reactor containing argon was introduced 10 g of $3TiCl_3 \cdot AlCl_3$ and 20 ml of a solution of 25% Et_2AlCl in cyclohexane (129). A mixture of 5 psi hydrogen and 135 psi propylene (550 g) was introduced, and polymerization proceeded for 30 min at 50 to 60°C. The reactor then was vented, the temperature reduced to 40°C, and ethylene introduced during 30 min. Repetition of these procedures followed by quenching by K_2CO_3-alcohol gave block copolymers which then were partially degraded near 350°C, in order to further improve low-temperature impact. In another process, 43 ppm of hydrogen based on propylene was preferred, along with xylene as solvent (130).

Note that soluble catalysts and homogeneous systems have not proved successful for preparing high-softening polypropylenes of predominantly stereoregular structure. Plastic polypropylenes have in all cases required hetero-geneous catalysts upon which solid propylene polymers swollen by monomer and diluent apparently grow as a solid phase. Soluble catalysts, however, have produced largely amorphous rubbery propylene polymers.

Important to the commercial success of polypropylene plastics has been the control of crystallinity and spherulite morphology. The addition of suitable nucleating agents to keep the spherulites small and more numerous generally improves clarity and other physical properties. Dispersed soaplike or other insoluble materials often act as nuclei for crystallization. Amounts up to 0.5% of fine alkali, alkaline earth, aluminum or titanium benzoate, or adipate salts (and also fluorides) have been effective (131). With a dispersed nucleating agent such as basic aluminum dibenzoate, crystallization temperatures were about 10°C higher than when fewer large spherulites of crystals formed (132). About 0.5% piperidinium t-butyl benzoate has been incorporated to improve clarity (133). For obtaining very high degrees of crystallinity (e.g., 90%), stereoregular polypropylenes may be annealed at 165°C for 3 hr in inert atmosphere, followed by slow cooling to room temperature (134). The ratio of IR absorption at 10.00 to 10.27 microns may be used to estimate the fraction of crystal-lized isotactic polymer. Estimates of the crystallinity of polypropylenes in the common plastics range from 50 to 70%. Adding 2 to 20% polyethylene or EP copolymer of low molecular weight, as well as orientation of films by drawing, can improve clarity (135). Specific nucleation has been obtained by oriented fibrous fillers and condi-tions of molding (136). Approximate properties of molded homopolymer and copolymer plastics are shown in Table 2.1.

TABLE 2.1

Properties of Stereoregular 1-Olefin Polymers

Properties[a]	Propylene homopolymer	Propylene copolymer plastic (high impact)	1-Butene polymer form I	1-Butene polymer form II	Isobutyl ethylene polymer, TPX
Density, g/cc	0.885-0.900	0.900-0.91	0.912	0.881	0.83
Shore durometer D hardness	72	70	65	39	74
Yield strength, psi	5000	4000	2100	640	--
Tensile strength, psi	5000	4000	4370	4275	4000
Elongation at break	>200	>400	325	350	15
Notched impact, ft-lb/in.	0.4	1.5-3.0	--	--	--
Crystal melting point, °C	174	--	124-135	--	240
Coefficient of expansion, in./(in.)(°C)	about 0.00006	about 0.00008	0.00015	--	0.00012
Specific heat, cal/(g)(°C)	0.45	0.45	--	--	0.52
Brittle point, °C	0-15	<-15	-25	--	--
Dielectric constant, 10^3 and 10^6 cp	2.1	2.2	2.25	--	2.12
Dissipation factor, 10^3 and 10^6 cp	0.0003	--	0.005	--	--
Water absorption, %	<0.01	<0.01	--	--	0.01

[a]A number of these properties vary with tacticity.

Optically clear, thin films of propylene polymers have
been achieved, and by 1972 thicker propylene polymer
moldings such as laboratory beakers and hospital ware
with improved clarity were being developed. Some of the
improvements by copolymerization and blending were
attained with sacrifice in heat resistance. In one case,
improved transparency was obtained by copolymerization
with 11% 1-butene and by quenching (137). Moldings from
present compositions may shrink 1 to 2% linearly during
10 days after cooling; the first 90% of the shrinkage
takes place during the first 6 hours as crystallization
proceeds. Examples of propylene copolymers giving good
clarity in films are Novolen KR and Novolen V of BASF.
From 10 to 30% of short fibers of glass, asbestos, graphite, or Wollastonite (calcium silicate) may improve
impact strength of propylene polymer molding plastics.
Polypropylenes modified by maleic anhydride are suitable
for metal coating (138); they also bond well to glass
fibers to give superior impact strength in moldings (139).
Maleated polypropylenes were neutralized with bases in
order to improve adhesion to fibers and metals (140).
The development of superior propylene copolymer plastics
and fibers has required intensive research into stabilization against thermal, photochemical, and oxidative
degradation (e.g., by incorporation of antioxidants, UV
absorbers, and other agents--note Figs. 2.8 and 2.9).
Carbonyl groups formed in light act as photosensitizers,
but salicylate esters and nickel chelates are effective
stabilizers (141). Initial autoxidation of a commercial
polypropylene in chlorobenzene solution at 110°C gave a
hydroperoxide of lower molecular weight (12,500 by
Mechrolab osmometer)(142). Infrared data suggested that
about 90% of the OOH groups are intramolecularly hydrogen
bonded (3378 $cm^{=1}$). The free hydroxyl groups absorb at
3553 cm^{-1}. Syndiotactic and isotactic segments of polypropylenes were believed to have different autoxidation
characteristics (143). Incorporation of 3% or more of a
good carbon black is the best stabilization against
degradation in light. Along with phenolic or phenol
sulfide antioxidant stabilizers, copper deactivators such
as oxamide derivatives are added for copper wire insulation (144).
A recent development is improvement of the strength
properties of polypropylene plastics by cold working
(e.g., hydrostatic extrusion near room temperature) (145).
About 50% of the polypropylene produced is used for
injection molding, especially for automotive parts.
Electrostatic spraying of "fluidized" fine powders of
propylene polymers can deposit coatings of .002 to .050
in thickness.

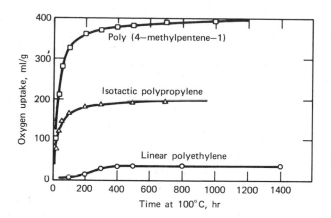

Fig. 2.8. Without antioxidants the thermal oxidations of propylene and isobutylethylene polymers are much more rapid than that of linear polyethylene. When selected antioxidant stabilizers are added, however, the branched polymers perform satisfactorily.
F. H. Winslow and W. Matreyek, Polym. Preprints 3, 232 (March 1960).

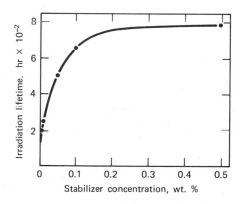

Fig. 2.9. An example of the improvement of polypropylenes by addition of stabilizers is the endurance to light obtained by incorporation of nickel dimethyl glyoxime. D J. Carlson and D. M. Wiles, IUPAC Macromol. Preprint I, 61 (July 1971); cf. M. L. Solder, U. S. 3,282,887 (Hercules).

About 30% of United States production of polypropylene is used as fibers. Fibers and monofilaments of stereo-regular propylene polymers such as Herculon I and II have found wide use in carpeting, ropes, and upholstery. Monofilaments may be made by melt spinning with rapid quenching in an inert liquid less than half an inch from the die (146). Draw ratios of 3:1 to 9:1 may be employed, followed by heat setting at 130 to 155°C. After elongations up to 75%, the drawn polypropylene monofilaments should have recovery of at least 85%. Composite or conjugate filaments may contain several propylene polymers of different molecular weights and with different modifiers of crystallinity (147). Thus "helical" crimp can be made to improve bulk and resilience. Polypropylene fibers useful in carpeting are made by fibrillation of oriented flattened tubes or sheets, e.g., through the splitting action of rubber belts (148) or rotating toothed rolls (149). Chevron Chemical Company makes polypropylene fibrillated yarns for synthetic grass. Nonwoven fabrics can be made from combinations with cellulosic fibers (150) or with polypropylene fibers alone (151). The latter may be partially fused in the process but maintain their identity after the webs of two fibers are compressed at 325 to 400°F. Formation of filaments by means of jet streams of gas has been most successful with molten glass, polyesters, and polyolefins. Mixtures of polypropylene (PP) and polyethylene terephthalate were used in a patent example (152). The Du Pont and Scott Paper companies are developing nonwoven textile products from so-called "spun-bonded" fibers such as Reemay polyester and Typar polypropylene. Synthetic "papers" based upon PP fibers are being developed by Mitsubishi Rayon, Hercules, Bakelite Xylonite, and others. For some of these promising products both the words textile and paper have misleading connotations. Among new composites, two sheets of polypropylene film have been laminated with a core of PP fibers (>75 crystalline) impregnated with amorphous PP (80% noncrystalline) (153).

Early tests of polypropylene fibers in clothing were not very promising. However, improved copolymer fibers are being developed which provide better dye-receptivity, better strength, better textures, and low static.

PROPERTIES AND STRUCTURES OF ISOMERIC POLYPROPYLENES

As in the case of branched polyethylenes prepared at high pressures by Fawcett and coinventors, the propylene polymers obtained by Ziegler-Natta or coordination type catalysts have complex structures imparting properties more useful than would the simpler structures at first

attributed to them. During the first decade after discovery there was great emphasis upon highly isotactic and
highly crystalline polypropylene structures (154). Bovey
and coworkers interpreted NMR spectra of isotactic propylenes (155). About 2% of random structure (racemic
dyads) was believed to occur at junctures of isotactic
blocks of opposite configuration. However, proton NMR
studies of proportions of isotactic, syndiotactic, and
atactic structures have not been regarded as conclusive
by many investigators because of overlapping of bands and
other uncertainties. Methods of synthesis and fractionation of the PP samples examined by physical chemists
often have not been indicated. References have been
given to published estimates of variation from perfect
stereoregularity of less than 2% (Natta) to 5 to 10%
(Flory) in "isotactic polypropylene" (156). This author
believes that both percentages may be too low for the
propylene polymers currently useful in plastics and
fibers. In research for further improvement in properties
more attention should be given to the less crystalline
domains in which the predominantly isotactic, crystallized
chain segments are dispersed.
When molten, largely isotactic polypropylenes or their
solutions are cooled slowly, spherulites comprising
radiating crystals form, but even single crystals obtained
from dilute solutions are believed to be only 75-95%
crystalline (157). One view is that the amorphous phase
comprises loops at the end of the chain folds comprising
crystals. New techniques have been developed to isolate
fractions of specially high crystallinity, such as
quenching partially crystalline polymer followed by
solvent extraction (158). On the other hand, patents
have described sterically rearranging or "randomizing"
isotactic polypropylenes by heating with peroxides and
certain plasticizers (159). It is difficult to prove
such steric changes because of simultaneous degradation
reactions. Reduction of PP crystallinity by high energy
radiation has been reported (160).
Besides the normal alpha crystalline form of isotactic
polypropylenes, Turner-Jones found that a second gamma
form is obtained under certain conditions (161). Single
crystals of high polymer formed from solution in paraffin
wax cooled to 100°C; also polymer of low molecular weight
gave gamma crystals (162). Propylene copolymers with
small amounts of ethylene crystallized in the gamma form
apparently because of interruptions in the isotactic
sequences (163). Thus fractions obtained from commercial
propylene block copolymer plastics at 35 to 70°C from
solutions in petroleum ether or xylene crystallized

entirely in the gamma form.   Curves for estimating per
cent of gamma form from x-ray data were given.

   Films of isotactic polypropylene cold draw and become
oriented on stretching, in contrast to soft, rubbery
polypropylenes.   Although samples of the latter are
usually called atactic they normally contain crystallizable fractions.   Densities of the crystalline and
amorphous phases were taken to be 0.93 g/cm^3 and 0.858 g/
cm^3 in estimating that four commercial polypropylenes had
66 to 69% normal crystallinity (164).   Commercial thin
films of stereoregular propylene polymers can be prepared
by rapid biaxial stretching of quenched extruded sheet
(165).   Only the monoclinic form was observed in dual
crystal orientation.

   Danusso and Gianotti showed that crystal melting points
of stereoregular polymers rise with temperatures used for
crystallization (166).   By repeated recrystallization at
rising temperatures, crystal melting points as high as
180 to 200°C were observed for isotactic polypropylene.

   Stereoregular polyolefins of high tacticity are only
soluble in solvents on heating, e.g., above 100°C, but
once dissolved their behavior in good solvents where
chains are extended seems not very different from that of
atactic polymers.   The following approximate constants in
the equation $\eta i = KM^a$ have been reported (167):

	K	a
Atactic polypropylene in decalin at 135°C	15.8×10^{-5}	0.77
Isotactic polypropylene in decalin at 135°C	11.0×10^{-5}	0.80
Isotactic poly-1-butene in decalin at 115°C	9.5×10^{-5}	0.73
Atactic polystyrene in toluene at 30°C	11.0×10^{-5}	0.72
Isotactic polystyrene in toluene at 30°C	11.0×10^{-5}	0.72

It is expected that in poor solvents such as hydrocarbon-
alcohol where there is more polymer-polymer interaction,
viscosity-molecular weight characteristics will be
different.   Slightly higher theta temperatures were
reported for atactic polypropylene in diphenyl ether than
for isotactic polymer (168).   However, isotactic 1-butene
polymers had higher theta temperatures and lower values
of second virial coefficients in solution (169).   Solution
studies were interpreted to indicate the presence of some
heterotactic links in alleged highly isotactic polyolefins
(170).   For a given intrinsic viscosity of polyolefins in

solution in good solvents, molecular weights increase with
short chain branching. Thus Coover and co-workers (171)
give the following values of molecular weight for solutions of $\eta i = 1.0$ in hydrocarbon solvent:

                High density polyethylene = 45,000
                Isotactic polypropylene   = 89,000
                Isotactic poly-1-butene   = 320,000

Atactic, substantially amorphous polypropylenes have
received much less attention than the normally crystallizable polymers and the amorphous EP terpolymer rubbers.
Completely atactic uniform amorphous polypropylenes of
high molecular weight and room-temperature solubility are
not easy to prepare. They are tacky, clear, and rubberlike. Crude atactic polypropylene fractions recovered
from solution in the hydrocarbon solvents used for
isotactic polypropylene synthesis generally have low
molecular weight and may contain combined oxygen or other
impurities (172). They also may contain branched chains
(173). Highly atactic polypropylene has been obtained by
a catalyst prepared from aluminum triisobutyl and titanium
dichloride dipropionate (174). Amorphous polymers of
intrinsic viscosity 1.5 to 2 were obtained by control of
the molar ratios of titanium to aluminum in the catalyst
(175).

A range of low-viscosity and low-molecular-weight polypropylenes (LVP) of partially isotactic structure have
become useful as modifiers of petroleum waxes; they serve
to raise blocking temperatures and to improve scuff and
abrasion resistance. Such polypropylenes may be prepared
by heating isotactic PP high polymers above 300°C (176).
The polymers supplied by Eastman are fairly hard, brittle
pellets of straw color, having viscosities at 190°C
ranging from 10 to 300 poises and number average molecular
weights 7000 to 9000 (177). When allowed to cool slowly
from the melt, the density was 0.9150 g/cc, whereas the
quenched LVP with minimum crystallization was 0.8866. In
tests with 2, 5, and 10% LVP in paraffin wax, cloud points
indicated lower compatibility than obtained with a low-
molecular-weight polyethylene in paraffin wax. The LVP
gives surprisingly glossy coatings on paper, apparently
because the crystals formed after quenching are very small.

Natta and co-workers reported the preparation of syndiotactic polypropylene with predominant alternating DLDL
structure, giving characteristic x-ray and IR data (178).
However, only very small laboratory samples have been
available for testing and their properties have not been
clearly defined. Like atactic PP, the syndiotactic PP was
relatively soft and elastic. In one example, PP believed
to show syndiotactic structure was prepared at -78°C in

toluene, using vanadium triacetyl acetonate and
$Al_2(C_2H_5)_3Cl_3$ for 15 hr (179). The low yield of PP had
x-ray crystal spacings different from those of isotactic
PP. In another example, a soluble catalyst from VCl_4 and
$AlEt_2Cl$ in n-heptane contacted propylene at -78°C for
18 hr (180). The active catalyst was believed to be
$C_2H_5VCl_2$. The infrared bands of the resulting polymer
were interpreted to indicate that the product was
"enriched of syndiotactic polymer." Polypropylene
samples believed to have syndiotactic structure from
infrared spectra had crystal melting points ranging from
65 to 131°C and densities 0.859 to 0.885 g/cc; however,
there was not enough material to permit evaluation (181).
Coordination catalysts modified by triaryl or trialkyl
phosphines were said to give minor proportions of syndiotactic together with isotactic polypropylene structures
(182).

Amorphous rubberlike polypropylenes have been used in
compositions for caulking (183), in adhesives (184), as
polymeric plasticizers for high-impact styrene copolymers
(185), and for improving the viscosity index and the
shear stability of lubricating oils (186). Copolymers
with ethylene or other comonomers generally have better
compatibility for useful blends than have amorphous homopolymers of propylene.

Reaction products of polypropylenes with maleic anhydride have been developed by Hercules. For example, a
polypropylene in inert organic solvent solution may be
heated with maleic anhydride in presence of a peroxide
catalyst (187). Copolymers containing 0.5 to 4.0% combined maleic anhydride units dissolve in xylene, from
which priming coats may be applied to aluminum surfaces
for promoting adhesion to polypropylene foam layers.

There have been several claims that propylene with
specific catalysts can undergo ionic isomerization polymerizations, giving different structures from $(CH_2CH)_n$.

$$CH_3$$

For example, 2.0 g of $Pd(CN)_2$ in 60 ml of dry benzene was
pressurized with 50 psi propylene at 60°C for 114 hr
(188). After the catalyst had been filtered off, an
excess of methanol was added to precipitate the polymer
from the filtrate. The x-ray pattern and the IR spectrum
of the polymer were said to be identical with a known
copolymer of 93% ethylene and 7% propylene units.

ETHYLENE-PROPYLENE COPOLYMER RUBBERS

Whereas heterogeneous transition metal catalysts such
as titanium halide-AlR_3 tend to form relatively crystalline block copolymers from ethylene-propylene mixtures,

soluble catalysts based on vanadium compounds such as
$VOCl_3$ or VCl_4 with aluminum organo-compounds can promote
formation of normally-amorphous, homogeneous copolymers
or interpolymers. The titanium-halide-insoluble catalysts
are used for manufacture by proliferous or growth polymerization of propylene homopolymer and copolymer plastics
and fibers; the more soluble vanadium-containing catalysts
are employed for synthetic elastomers such as ethylene-
propylene terpolymer rubbers (EPDM) which generally form
as viscous solutions. However, some suspension-type
polymerization systems have been developed recently for
preparation of EPDM rubber. Note that homogeneous cationic
polymerizations of alkyl vinyl ethers using soluble catalysts had been shown earlier to favor formation of
amorphous atactic vinyl ether polymers, in contrast to
heterogeneous growth polymerizations.

Ethylene and propylene were copolymerized in some early
work without full recognition of possibilities for synthetic rubber. Example 4 of a patent of Anderson,
Merckling, and co-workers describes the preparation of an
EP copolymer of density 0.91 g/ml which was tougher and
more transparent than polyethylene (189). The catalyst
believed to contain divalent titanium was obtained from
$TiCl_4$ and phenylmagnesium bromide; the polymerization was
carried out in cyclohexane during 2 hr at 30°C.

Gresham and Hunt prepared sulfur-vulcanizable EP copolymers by means of titanium catalysts (190). A mixture
of equal volumes of ethylene and propylene was bubbled
slowly into liquid 1,5-hexadiene; then the resulting mixed
gases, in molar ratio 4:4:1, were passed into an agitated
catalyst solution. The latter had been made by heating
0.38 g $LiAlH_4$, 6.5 g 1-decene, and 4 g xylene at 140°C,
followed by cooling and adding 1.3 g $TiCl_4$. Copolymerization caused the temperature to rise to 48°C. After
quenching with methanol and HCl, 38 g of dried copolymer
was recovered, showing an intrinsic viscosity of 1.0 in
benzene solution at 25°C. When mixed with sulfur, carbon
black, and accelerators, the rubbery copolymer could be
cured by 60 min at 153°C. Other dienes could be employed,
such as 1,4-hexadiene. More than 73% of ethylene units
in EP copolymers gave harder, nonrubbery products. Under
these conditions, with strong agitation and with xylene
as diluent, apparently the titanium-base catalyst did not
give proliferous polymerization to normally-crystalline
copolymers, but solution copolymerization to amorphous
terpolymer.

Adamek, Dudley, and Woodhams of Dunlop Laboratories in
Canada prepared a catalyst by adding 20 ml aluminum of
triisobutyl to 500 ml of anhydrous petroleum ether,
followed by a solution of 4 ml of titanium tetrachloride

dissolved in 75 ml of petroleum ether (191). With agitation, a mixture of ethylene and propylene gases was
introduced continuously while a solution was added of 5 g
dicyclopentadiene and 1 ml aluminum triisobutyl dissolved
in 500 ml of petroleum ether. Copolymerization occurred
in solution, giving increasing viscosity during 1 hr. By
final addition of methanol, a white, elastic terpolymer
was precipitated from the viscous solution. The patent
claims included as third monomers endocyclic diolefinic
hydrocarbons containing 7 to 10 carbon atoms in which the
bridge consists of one or two CH_2 groups. The first
provisional specification of this British patent of Dunlop presented examples of terpolymer rubbers by use both
of catalysts based on $TiCl_4$ and VCl_4. It is significant
that the modified patent specification gave only examples
using vanadium-containing catalysts. The rubber was best
cured by adding both sulfur and organic peroxide. For
example, 100 p of EP dicyclopentadiene terpolymer
(20:100:1 by weight) was mixed with 50 p of carbon black,
5 p of zinc oxide, 2 p of stearic acid, 1 p of dicumyl
peroxide, 2 p of sulfur, and 1.5 p accelerators, and
heated at 320°F for 30 min (192).

Research at Hercules laboratories at Wilmington showed
that uniform amorphous EP copolymers of narrow molecular
weight could be obtained by solution polymerization with
soluble catalysts based on vanadium compounds (193).
The soluble catalyst could be formed in situ during
reaction, and the copolymer remained largely in solution
as it was formed. Into an autoclave containing 500 ml of
chlorobenzene under nitrogen was passed separately and
continuously streams of (a) ethylene + propylene containing 68 mole % propylene, (b) a solution of 3.6 mmole/
liter of aluminum triisobutyl in heptane, and (c) 0.64
mmole/liter of liquid $VOCl_3$ in heptane. Copolymerization
began immediately and continued until a highly viscous
solution was obtained. Within 33 min, a homogeneous
copolymer or interpolymer of reduced viscosity 7.2 and
containing 24 mole % (or 33 wt %) propylene units had
formed at the rate of 30 g/(liter)(hr). The reaction was
quenched by adding 10 ml n-butanol with 200 ml heptane.
In other examples, the catalyst was prepared from
$Al(C_2H_5)_2Cl$ and $VO(OC_2H_5)_3$ just before use. Ratios of
Al/V were about 5 for catalysts based on VCl_4 or $VOCl_3$.
Uniform copolymers of reduced viscosities 3 to 5 were
made.

Natta, Mazzanti and Valvassori obtained amorphous EP
copolymers both by use of titanium and vanadium halides
(194). For example, the mixed monomer gases were passed
into an agitated suspension of $TiCl_3$ in a solution of

trihexyl aluminum in n-heptane at 75°C. A copolymer
containing 41% ethylene units showed an amorphous x-ray
pattern. In another patent to Montecatini, hydrocarbon-
soluble catalysts from VCl_4 or $VOCl_3$ and organometallic
reducing agents were employed to prepare amorphous linear
copolymers of ethylene and propylene (195). These elastomers showed characteristic IR absorptions at 13.65, 13.85,
10.67, and 10.3 microns. Molecular weights above 20,000
were determined using solution viscosities in hot tetralin
and the equation: intrinsic viscosity = $1.18 \times 10^{-3} \times M^{0.65}$.
Natta and co-workers made vulcanizable terpolymers from
ethylene, propylene, and vinyl or allyl bicycloheptenes at
about -20°C using catalysts from vanadium acetylacetonate
and $(C_2H_5)_2AlCl$ (196). Valvassori and Sartori used 1,7-
octadiene or methylene cyclooctenes as termonomers with
catalyst from $VOCl_3$ and $(C_2H_5)_2AlCl$, also at low temperatures (197). They considered the copolymerization
reactions to be coordinated anionic in type. A process
for bulk copolymerization of ethylene and propylene without diluent was patented, but control of the reaction
even at -50°C was difficult (198). Mixtures of ethylene
and propylene (5 to 95 molar) gave elastomers of about
equimolar content using catalyst from aluminum isobutyl
and VCl_4. Effects of some variables in the synthesis of
amorphous copolymers from ethylene and propylene using
$VOCl_3$ and $(C_2H_5)_2AlCl$ are shown in Fig. 2.10.
   Numerous other examples of syntheses of EP copolymer
elastomers using vanadium catalysts have been published.
EPDM rubbers containing about 40% propylene units were
prepared with AlR_3, an organic vanadium compound, and a
halogen-containing activator (199). For example, ethylene,
propylene, and dicyclopentadiene were copolymerized in
toluene solution by addition of aluminum isobutyl, tributyl vanadate, and benzyl chloride. A four-component
catalyst mixture was used in benzene solution at 23 to
38°C: $VOCl_3$, alkyl aluminum sesquichloride ($Et_3Al_2Cl_3$),
nitropropane, and a Lewis base such as acetone (200). A
continuous copolymerization process invented by Henderson
and Carnell used soluble vanadium compounds and organoaluminum compounds to produce solutions of EP copolymers
(201). Temperatures as low as -20 to +30°C were employed
for copolymerization of ethylene, propylene, and polyenes
(202).
   EP terpolymer rubbers were prepared using 23 different
nonconjugated dienes (203). These copolymers were
obtained by employing soluble catalysts from reaction of
$VOCl_3$, VCl_4, or $VO(OR)_3$ with R_3Al, R_2AlCl, or $R_3Al_2Cl_3$.
Allyl derivatives of 2-norbornene were among the dienes
evaluated. Soluble catalysts from $VOCl_3$ plus aluminum
alkyl sesquichloride in the presence of a trialkyl amine

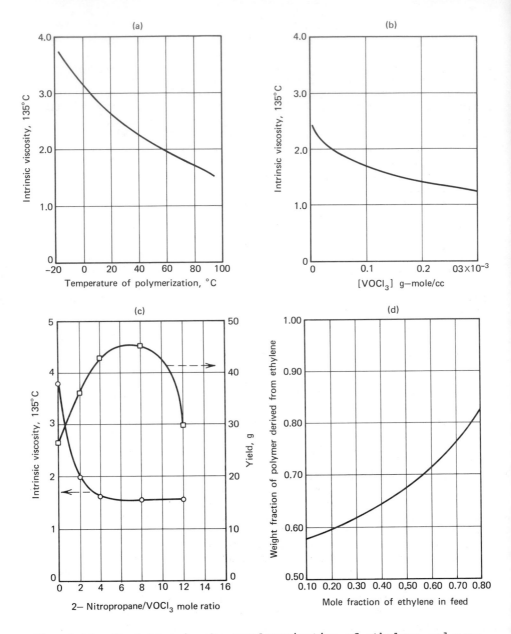

Fig. 2.10. Variables in the copolymerization of ethylene and propylene with vanadium-containing soluble catalyst: (a) intrinsic
viscosity as a function of polymerization temperature; (b) intrinsic
viscosity as a function of catalyst concentration; (c) intrinsic
viscosity and yield as a function of 2-nitropropane/$VOCl_3$ mole ratio;
(d) polymer composition as a function of monomer composition. E. K.
Esterbrook, T. J. Brett, Jr., F. C. Loveless, and D. N. Matthews,
IUPAC Meeting, Boston, Macromol. Preprints 712 (1971).

oxide were patented (204). Sulfur-curing EP terpolymers were made containing 2.8 to 5.0% dicyclopentadiene units (205). The catalyst was based on $VOCl_3$ + Et_2AlCl. In order to increase the proportion of dicyclopentadiene combining, the monolefins may be added continuously using VCl_4-$Al_2Et_3Cl_3$ as catalyst in heptane (206). Difficulties included yellow discoloration and shortness of catalyst life. Vinyl cyclohexene has lower reactivity than diallyl or dicyclopentadiene as a termonomer (207).

Ethylene-propylene terpolymer rubbers were prepared using low polymers of butadiene (e.g., mol. wt. 3800) as polyfunctional comonomers (208). A solution of ethyl aluminum sesquichloride and $VOCl_3$ in hydrocarbon was added as a catalyst. The rubbers produced were gel-free, readily soluble in hydrocarbons, and vulcanizable by conventional sulfur recipes without addition of peroxides. Ethyl aluminum chloride and $VOCl_3$, with PCl_3 as activator, were patented (209). Phillips disclosed a slurry-type copolymerization with $RAlCl_2$, VCl_4, and AlR_3 (210). Huels used a suspension process based on $VOCl_3$ and ethyl aluminum sesquichloride in CH_2Cl_2 as diluent (211). Their EP terpolymers with methyl butenyl norbornene contained 40 to 45% propylene units. Goodrich reported lower costs by employing a suspension process instead of the older solution methods. Dart Industries made EP terpolymers using bridged ring diene hydrocarbons with vanadium organophosphate catalysts (212).

The Polymer Corporation patented a continuous process for synthesis of ethylene-propylene-dicyclopentadiene terpolymer rubber (213). Two liquid streams were added continuously with rapid agitation to the reactor: (1) monomers with $VOCl_3$ in hexane and (2) $Al_2(C_2H_5)_3Cl_3$ (aluminum sesquichloride) in hexane. Using ethylene to propylene monomer ratios near unity, copolymers containing 76% ethylene units and about 1% unsaturation were formed. These showed no normal-crystallinity. Such conditions avoided 10 or more successive ethylene groups which were said to produce detectable crystallinity. In order to avoid handling of viscous solutions, Schaum and Hoerlein of Hoechst developed a suspension copolymerization process using methylene chloride and $VOCl_3$ (214).

Reactivity ratios for ethylene and propylene vary widely depending on the type of transition metal catalyst; for example, r_E = 5.9 to 20.3 and r_P = 0.01 to 0.03 (215). The nature of active catalyst sites and the mechanisms are incompletely understood (216). Still more complex catalysts are being studied, notably ones containing magnesium organic compounds which might be called Ziegler-Natta-Grignard systems. For example, Minchak of Goodrich used a reducible compound of titanium, an alkyl aluminum halide, and a diaryl magnesium compound (217).

Ethylene-propylene terpolymer rubbers with such third comonomers as 1,4-hexadiene, cyclooctadiene, methylene norbornene, or ethylidene norbornene are becoming important commercially. However, future growth depends on increasing rates of vulcanization and improving the tack and compatibility of the uncured elastomers. Resistance of the cured rubber to ozone, weather and chemicals is good. Properties of the new rubbers have been reviewed (218). Among the trade names of EPDM rubbers are Nordel (Du Pont), Vistalon (Esso), and Epcar (Goodrich). They are only partially compatible with butadiene-styrene copolymers, butadiene-acrylonitrile copolymers, and chloroprene polymer synthetic rubbers. EPDM or EPR rubber can be blended into isotactic polypropylene to improve spherulite morphology and low-temperature flexibility (219). As in the case of isotactic polypropylene, antioxidant stabilizers are essential for EPDM. Aldimines and ketimines stabilize against yellowing and degradation by autoxidation without interfering with cure by peroxides (220). Cure of EPDM by sulfur recipes requires up to 30 min at 150°C. Injection molding of Nordel EPDM rubber with curing at 290 to 350°F and 10,000 psi gives impact resistant automotive parts including bumpers and other safety devices (221). A range of EPDM elastomers of different molecular weight ranges and 30 to 85% propylene units have been supplied. For example, Goodrich has offered nine types of Epcar copolymers. An infrared band at 1378 cm^{-1} may be used to estimate the ratio of E/P units. Vulcanizable terpolymers containing less than 20% propylene units have been suggested for somewhat elastic fibers and films (222) but are not yet important commercially.

For ready copolymerization and relatively fast curing, EPD copolymer rubbers containing about 5% units from 5-ethylidene norbornene (ENB) have been favored. ENB is said to be outstanding because of the great difference in reactivity of its two double bonds. A Du Pont terpolymer rubber contained 53% ethylene units, 43% propylene units, and 3.2% 1,4-hexadiene units (223). Choice of polyfunctional termonomer is hoped to overcome limitations of slow cure, poor compression set, limited compatibility, and high gel content. Unsaturation in EP terpolymer rubbers may be estimated by pyridinium bromide or iodine chloride in a solvent such as CCl_4-CH_3OH (9 to 1) (224). However, dicyclopentadiene copolymers were sluggish in reaction. Patents disclosing other termonomers and other variables are listed here.

Up to 15% fulvene units: Kahle, U.S. 3,313,786 (Phillips)

Decatrienes, Schneider, U. S. 3,392,208-9 (Goodrich)
Cycloheptatrienes, Rust, German appl. (Hoechst);
  CA 75, 6668 (1971)
Alkenyl acetylenes, Valvassori, U. S. 3,313,787 (Monte)
Cyclobutene, alkenyl cyclohexene, Natta, U. S. 3,505,301-2
  (Monte)'
Diels-Alder adducts of triene + cyclopentadiene,
  Benedikter, U. S. 3,418,299 (Huels)
Diolefin + azo compound, Schleimer, U. S. 3,424,734 (Huels)
5-Propenyl-2-norbornene, Makowski, U. S. 3,427,360 (Esso)
Alkylidene or alkenyl norbornene + 1,4-hexadiene, Emde et
  al., U. S. 3,554,988 (Huels)
4-Ethylidene-1,6-heptadiene, Park, U. S. 3,480,599
  (Goodrich)
Alkenyl cyclopentenes, Van de Castle, U. S. 3,438,951
  (Esso)
Norbornene with chlorinated activator, Gumboldt,
  U. S. 3,531,447 (Hoechst)
Halohydrocarbon solvent, Sommer and Brandrup,
  U. S. 3,531,446 (Hoechst)
Addition of $SnCl_4$ or BF_3 etherate, Duck and Grieve,
  U. S. 3,536,678 (Int. Syn. Rubber)

A Du Pont patent disclosed methods of producing finely
divided crumbs of EP-nonconjugated diene rubber from
aqueous dispersions (225). For faster curing, EPDM rubber
was blended with copolymers of ethylene and 1,4-hexadiene
having 3 to 10 g-mole unsaturation per kilogram (226), or
natural or cis-1,4-polybutadiene rubber (223).
  The rate of production of ethylene-propylene diene
synthetic rubber (EPDM) in the United States in 1971 was
near 10^5 long tons per year. Production is also underway
in Canada, Europe, and Japan. The outstanding properties
of EPDM rubbers include better resistance to heat, ozone,
and weathering when compared with natural rubber and
butadiene-styrene rubbers. One specific use has been
integral insulation jackets for electrical wiring, replacing separate insulation and protective layers (227).
In order to upgrade their dielectric properties, from 20
to 30% of low-density polyethylene may be incorporated.
For electrical insulations, addition of 2.8% dicumyl
peroxide with 1% triallyl cyanurate produced crosslinking
fast enough for conventional continuous vulcanization.
However, faster cure remains an objective of research
(228). Along with tackifiers such as terpene polymers,
oils, or amorphous low polymers, the EPDM rubber blends
have promise as adhesives and sealants (229). Largest
uses of EPDM rubbers have been in automotive parts, tire
sidewalls, hose, wire covering, and coated fabrics. Graft
copolymers of styrene with EPDM rubber are being developed

as high-impact plastics (230). As much as 15% EPDM may
be used in suspension polymerizations with styrene to
yield impact-molding plastics with stability in sunlight
superior to that of plastics based on polybutadiene-
styrene graft copolymerizations.

1-BUTENE POLYMERS

Natta's group prepared partially crystalline, ether-
insoluble polybutenes of density d_{25}^{25} = 0.91 g/cc, crystal
melting point 126°C (231). Soluble amorphous fractions
of density 0.87 g/cc also were present. Poly-1-butenes
received relatively little attention until later, when
investigators began to appreciate their unique properties
such as flexibility with outstanding resistance to cold
flow. Blends of 1-butene isotactic polymers with polypropylene gave plastics with lower brittle temperatures
and higher impact resistance (232). Montecatini was
granted a U.S. patent with a composition of matter claim
for substantially isotactic 1-butene polymers (233).
Isotactic polymers of 1-butene can be prepared at 75 to
80°C at 800 psi using catalyst from the reaction of $TiCl_4$
with $AlEt_2Cl$ or with mixtures of $AlEt_2Cl$ and $AlEt_3$ (234).
Stereoregular 1-butene polymers were made also using
catalyst from mixing 0.9 g of powdered $TiCl_3$ (containing
some free titanium) dispersed in 245 g of n-heptane, 10 ml
of 0.87 molar $AlEt_3$ in n-heptane, and 5 ml of 0.87 molar
$AlEt_2Cl$ (235). After 250 g of cold liquid 1-butene and
nitrogen had been added to give 400 psi, the bomb reactor
was shaken 43 hr at 80°C. The solid violet-colored
polymer was washed 3 times with isopropanol and 3 times
with methyl ethyl ketone. A yield of 216 g of dried
polymer of melting point 106 to 110°C and molecular weight
about 200,000 was recovered. Montecatini laboratories
used violet preformed $TiCl_3$ solid or liquid $TiCl_4$ with
$AlEt_3$ or $AlEt_2Cl$ to obtain isotactic polymer of softening
point 110°C, together with amorphous soluble fractions
(236). Mixtures of isotactic and amorphous polymers gave
elastomeric films that could be sterilized at 100°C (237).
In the polymerizations with Ziegler-type catalysts,
addition of fine solids such as inorganic chlorides or
sulfates, diatomaceous earth or quartz was reported to
favor formation of normally amorphous polybutenes (238).
Poly-1-butene also was prepared using as catalyst a mixture of 5% metallic lithium on sand with $TiCl_3$ in naphtha
at 150°C for 6 hr (239). Polymerization from vapor phase
upon Ziegler-Natta catalyst gave stereoregular 1-butene
polymers (240). Cationic polymerizations of 1-butene were
studied in Canada (241), and variations in polymerizations
with Ziegler-type catalysts were tested (242).

Nagel and Gablicks of Petro-Tex have developed improved
systems for faster polymerization of 1-butene to stereopolymers of improved physical properties (243). In one
example, $AlEt_3$ in heptane solution was added in portions
to a mixture of $TiCl_4$ and Cu_2Cl_2 in n-heptane. The
catalyst was aged 20 min at 90°C before 1-butene was
introduced. Other added agents that were suggested included $SnCl_2$ and Hg_2Cl_2 (244). A later process used alkyl
aluminum chlorides plus $TiCl_3$-$AlCl_3$ and a little potassium
iodide (245). Hydrogen was found to accelerate polymerization of 1-butene in presence of $3TiCl_3 \cdot AlCl_3$ and various
cocatalysts (246).

On cooling from the melt, stereoregular 1-butene polymers show interesting examples of crystal polymorphism
(247). Under ordinary conditions of cooling, elastomeric
polymer having 4/1 helical chain structure of fourfold
symmetry and identity period 6.8 Å is formed first (as in
the stable crystal structure of 1-olefin polymers having
side groups larger than ethyl). However, this structure
slowly is transformed into a more stable, harder modification 1 characterized by 3/1 helical chains with identity
period of 6.5 Å similar to that occurring in normal isotactic polypropylenes. The transformation which may
require several days at room temperature may be accelerated
by high pressures, by catalyst residues, and by contacting
solvents or stearic acid. Poly-1-butene of about 60%
normal-crystallinity (x-ray) was found to require 150 hr
at 25°C after molding in order for the density to reach a
value of 0.91 g/ml (248). The change of density is about
4% in going from the unstable modification 2 (apparent t_m
of 120-126°C) to the stable modification 1 (t_m = 135-142°C).
Higher melting temperatures are shown by poly-1-butenes of
very high stereoregularity. In the range of 0 to 60°C,
the half-life of the transition from form 2 to 1 is of the
order of 10^4 min or less (249). A third form may deposit
from solutions. It is relatively stable at room temperature but changes to form 2 on heating at 90 to 100°C.
Form 1 may be identified by IR absorptions at 10.81, 11.78,
and 12.25μ and form 2 shows a band at 11.03μ (250).

Studies of crystallization of 1-butene copolymers with
minor proportions of higher alkenes are among those
bringing about new concepts of polymer crystallinity. It
had been assumed generally that the most stereoregular
homopolymers would crystallize most rapidly and completely.
Actually cocrystallinity often can occur where minor
amounts of comonomer units are accommodated in the crystal
lattice. Moreover, suitable minor comonomer units, and
apparently also small amounts of atactic connecting segments, may actually accelerate crystallization in some
cases. Turner-Jones found that small amounts of ethylene,

propylene, or 1-pentene copolymerized with 1-butene
accelerated crystallization to the stable form 1 (3/1
helix) from form 2 (4/1 helix) or directly from the melt
(251). She discovered that copolymers of 1-butene with
linear 1-alkenes of more than 5 carbon atoms in suitable
amount can remain stable in the crystal form 1. Units
from 1-hexene and 1-octene can enter the poly-1-butene 1
lattice but not units of higher 1-alkenes. Note also
that a number of stereoregular homopolymers such as polyisobutylene, isopropyl, n-butyl, and sec-butyl vinyl
ether polymers do not crystallize readily (252).

Copolymers of 1-butene with up to 8 mole % of C_{10} to C_{18}
1-olefin or with 5 to 9 mole % of propylene had desirable
low brittle temperatures (253). In one example, 5% 1-
dodecene copolymerized with 1-butene at 150°F during 1 hr.
The catalyst was obtained by reaction of $3TiCl_3 \cdot AlCl_3$,
$(C_2H_5)_2AlCl$, and $C_2H_5AlI_2$. The copolymer brittle point
was -4°C. Copolymers of 1-butene with 1 to 7% 1-hexene
made by organoaluminum-titanium chloride catalyst were 70%
insoluble in n-hexane (254). They were tough and tear
resistant, and were suggested for heavy-duty agricultural
bags.

Copolymers of 1-butene with 10 to 25% propylene, showing
tensile strengths of 3000 psi or more and elongations of
at least 500%, were made by catalyst from $(C_2H_5)_2AlCl$ and
the reaction product of $TiCl_4$ with aluminum (approximately
Ti_3AlCl_{12}) (255). Copolymers of promising resistance to
mechanical shock and good clarity were obtained from
1-butene and a minor proportion of 4,4-dimethyl-1-pentene
(256). The catalyst from $(C_2H_5)_2AlCl$ and $TiCl_3$ was prepared by reduction of $TiCl_4$ by $(C_2H_5)_3Al_2Cl_3$. Reaction
at 40°C for 2.25 hr gave copolymer containing 3.3% of
units from dimethyl pentene. A molding of the copolymer
retained form 1 structure even after 12 months storage
and vigorous beating with a hammer. Turner-Jones found
that mixtures of 1-butene and 3-methyl-1-butene tend to
form block copolymers in presence of $TiCl_3-(C_2H_5)_2AlCl$
catalyst (257). The 1-butene polymerized much faster and
random copolymers could not be obtained. Approximately
equimolar copolymers showed 50% crystallinity.

Depending on heat treatment, poly-1-butene films may
exhibit complex spherulite growths, greatly affecting
transparency, which are influenced by nucleating agents.
Lamellar single crystals can be formed from dilute solutions (258). Surprisingly, clear films become white
opaque on cold drawing. They become clearer again on
further elongation or upon relaxation (259). The
phenomenon may result from reversible opening of voids
between lamellae. Well-defined necking zones do not form
on drawing 1-butene polymers.

Stereoregular polymers from 1-butene, called Bu-Tuf, have been supplied by Petro-Tex Chemical Corporation of Houston, Texas, for use in piping, tubing, industrial packaging, and defense applications (260). The outstanding property compared with other plastics is high resistance to cold flow over the range of -10 to 190°F. Of course dimensional stability presupposes sufficient time (e.g., 4 to 10 days) for complete transformation into stable form 1. Bu-Tuf may be extruded, molded, calendered, and applied as coatings. The polymers are soluble in hydrocarbons or chlorinated hydrocarbon solvents above 140°F. Petro-Tex has offered several grades of polybutene pellets (e.g., densities 0.912 and 0.915 with brittle temperatures -25 and -20°C, respectively). Polymers of moderate tacticity have been offered for use as dimensionally stable thermoplastics and films of high tear strength. The 1-butene polymers supplied by Huels and others have less tendency to stress crack and less tendency to creep than polyethylenes and polypropylenes of similar flexibility. These polymers have found application especially in high-pressure piping and in storage tanks for corrosive liquids. They can be extruded at 130 to 180°C with sulfur and organic peroxide, followed by crosslinking at 180 to 240°C (261).

Crystalline block copolymers of 1-butene and propylene have favorable compatibility for improving the strength of hydrocarbon wax films (262). Copolymers of ethylene with 0.5 to 5% 1-butene have outstanding resistance to environmental stress cracking (263).

POLYMERIZATION OF BRANCHED 1-ALKENES

Early work showed that viscous liquid low polymers could be obtained from branched liquid alkenes by action of anhydrous acidic catalysts. Thomas and Reynolds polymerized isopropylethylene or 3-methyl-1-butene in ethyl chloride at -78°C using anhydrous aluminum chloride to give soft polymers of molecular weight about 6500 (264). The monomer also polymerized at about -60°C in the presence of boron fluoride to form polymers reported to resemble polyisobutylenes (265). Compounds of the type $CH_2 = CRR'$ (other than isobutene) do not give high polymers readily.

The polymers prepared at low temperatures from isopropylethylene or isobutylethylene using strong Lewis acids were discovered by Edwards and Chamberlain to have unexpected structures resulting from hydride shift or isomerization polymerization (266):

$$CH_2=CHCH(CH_3)_2 \quad \xrightarrow[\text{$-130°C$}]{\text{AlCl}_3} \quad --CH_2CH_2\underset{CH_3}{\overset{CH_3}{C}}--$$

$$\xrightarrow[\text{catalyst}]{\text{Ziegler}} \quad --CH_2\underset{CH(CH_3)_2}{CH}--$$

In contrast, heterogeneous polymerization using Ziegler-type catalysts can form stereoregular isopropylethylene polymers by vinyl-type polymerization. Differences in structure were proved by NMR spectra (Fig. 2.11).

Polymerizations at -130°C in ethyl chloride with $AlCl_3$, $AlBr_3$ or $TiCl_4$ catalyst gave almost completely 1,3-polymerization by isomerization to the more stable tertiary carbonium ion (following ideas of Whitmore). Infrared studies also revealed that the polymers from the low-temperature conventional cationic polymerization have structures different from those obtained by Ziegler-type conditions (267). These polymerizations have been further studied by Kennedy and co-workers (268). At higher temperatures with conventional Lewis acid catalysts, amorphous isobutylethylene polymers of mixed structure resulted. In these polymerizations the activation energy of isomerization apparently is smaller than that of 1,2-polymerization. The ratio of the proportion of 1,3 to 1,2 chain units was a linear function of the reciprocal temperature in the range +10 to -100°C (269). The activation energy changed from -2.5 to -0.35 kcal/mole in the range -80 to -130°C.

Homopolymerizations of isopropylethylene, using typical Ziegler-Natta catalysts, occur sluggishly. By adding about 5% of 1-butene along with 95% isopropylethylene, rates become faster and crystallizable block copolymers may be obtained. Activating additions of ethylene or propylene also may be employed (270). Addition of $AlCl_3$ to Ziegler catalyst improved polymerization rates. A decreasing order of reactivity of 1-olefins in the presence of a typical Ziegler catalyst was given as ethylene, propylene, 1-butene, 3-methyl butene and 3,3-dimethyl butene (271). The chromium oxide catalysts of Phillips do not polymerize isopropylethylene as they do isobutylethylene.

In benzene solution, 3-methyl-1-butene was copolymerized with smaller amounts of 4-methyl-1-pentene using aluminum isobutyl-vanadium trichloride catalyst (272). At 50°C reaction required 42 hr. Molded sheets of the copolymers were clear and tough. Isotactic homopolymers of 3-methyl-

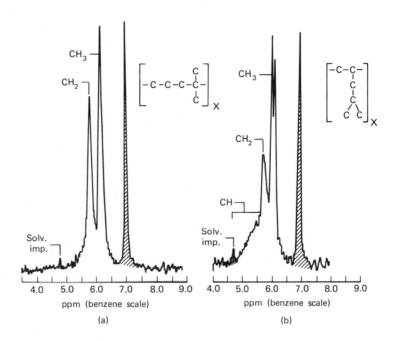

Fig. 2.11. Proton NMR spectra of two isomeric polymers obtained from isobutylethylene by using aluminum chloride catalyst (a) and Ziegler-type catalyst (b). The normal vinyl polymer structure gives the doublet methyl resonance and the broad resonance characteristic of the CH group in (b). The conditions used were as follows: solvent, 2-chlorothiophene; estimated concentration, <5%; internal reference, hexamethyldisiloxane; sample temperature, 120°C; frequency, 60 MHz. W. R. Edwards and N. F. Chamberlain, Polym. Preprints 3, 114 (September 1962).

1-butene have not been commercially attractive, in part because of polymer brittleness and high melting points. However, up to 5% has been blended with polyethylene terephthalate in heat-resistant molding plastics (273). Isopropylethylene polymer fibers were made experimentally by Celanese (274).

Isobutylethylene or 4-methyl-1-pentene is a low boiling, liquid, reactive monomer from which stereoregular polymers of attractive heat resistance can be prepared. The monomer can be obtained by dimerization of propylene over alkali or transition metal catalysts (275). Isobutylethylene polymerizes with catalysts from $TiCl_3$ and $AlEt_3$

even more readily than does propylene. Solid polymers
were prepared using chromium oxide on silica-alumina (276)
about the same time that Ziegler catalysts were developed.
   Polymers with high crystal melting points were prepared
from branched 1-olefins by Natta, Pino, and Mazzanti (277).
In the first example of the Belgian patent, 36 g of $TiCl_3$
was added to a solution of 5.8 g of $Al(C_2H_5)_3$ in 200 ml of
n-heptane. To this catalyst dispersion was added 92 g of
isopropylethylene in an autoclave. After 15 hr at 70°C
there was recovered 43 g of polymer. Extraction of the
crude polymer by solvents left a yield of 2.2% of partially crystalline polymer. The crystalline polymer
fractions from such branched monomers were found to have
relatively high melting points by x-ray studies.
   Watt described synthesis of isobutylethylene isotactic
polymers and gave earlier references (278). In n-heptane
at 40°C rates and stereoregulation were greatly affected
by the ratio of $AlEt_3$ to $TiCl_4$. Best yields of polymer
were realized by aluminum-to-titanium ratios of 1.2:2.0;
but, judging from insolubility in hot heptane, greatest
tacticity resulted from Al/Ti=1. Campbell made the
stereoregular polymers using a catalyst from lithium
aluminum tetradecyl in cyclohexane at room temperature
for 24 hr (279). Polymer of intrinsic viscosity 2.3
could be heated at 275°C for 6 hr under vacuum to reduce
the viscosity to 1.35. After extraction with petroleum
ether, this polyisobutylethylene could be melt spun to
fibers which showed sticking temperatures of 215°C. As
stabilizer, 0.3% N,N-p-phenylene bisphenyl amine was
added. Crystallizable polymers from isobutylethylene and
other branched 1-alkenes have been studied by other
laboratories (280).
   Polymerizations of isobutylethylene by $AlCl_3$ and other
conventional Lewis-acid catalysts at -78°C and below give
complex polymer structures by different proportions of
1,2 or vinyl, together with 1,3 and 1,4 polymerizations
(281). Polymers made at 20 to 60°C in heptane with $AlCl_3$,
$TiCl_3$ and mixed anhydrous acid catalysts gave 46 to 72%
of 1,4 structure by hydride-shift polymerization, along
with some vinyl-type segments (282).
   Stereoregular isobutylethylene polymers can be molded at
250°C and above. Although spherulites have been observed,
the crystalline polymers normally are quite clear. The
differences in density between the crystalline and amorphous polymers are small (283). One polyisobutylethylene
was said to be 78% crystalline, but most of the polymers
available for evaluation have been lower in crystalliza-
bility and tacticity. The birefringence of the polymers
has a minimum around 40°C, where the densities of the
amorphous and crystalline phases cross over with changing

temperature (284). Highly oriented fibers were made by
melt spinning at 150°C and then drawing elevenfold (285).
The polymer crystallizes in the tetragonal system with
cell dimensions a = b = 18.66 Å and C (fiber axis) = 13.8 Å.
Figure 2.12 shows the x-ray fiber pattern.

The polymers are characterized by comparatively high
tensile strength, stiffness, and surface hardness, but
their Achilles' heel is low impact at and below room
temperature. Although crystal melting temperatures are
high, heat-distortion temperatures as low as 58°C have
been observed for isobutylethylene polymers. Isopropylethylene polymers in contrast have given heat-distortion
temperatures near 155°C. The isobutylethylene polymers
normally are fairly transparent due to limited spherulite
growth. Surprisingly, however, spherulite crystal growth
could be promoted by copolymerization with a little propylene (sequential monomer addition). Pyrolysis of
isotactic isobutylethylene polymers at 300°C produced
isobutene and propane along with 2% monomer (286). Clear,
transparent molded polymers were obtained in Japan by
copolymerization with 0.45% of 1-hexene (287).

Isobutylethylene polymers were available from I.C.I.
and California Chemical Company beginning in 1964. They
can be injection molded at 260 to 320°C with a mold at
60°C to facilitate crystallization. Extrusion blowing
can be carried out at 275 to 290°C; ordinary extrusion
takes 260 to 275°C. The dielectric constant is about 2.2
at 25 and is nearly independent of frequency. The high
temperatures required for forming and limited impact
strength in the cold are limitations of the plastics.
The polymers absorb less than 0.1% water at 20°C. They
may show stress cracking in contact with detergents and
are swollen or crazed by some hydrocarbons and chlorinated hydrocarbons. The optical clarity and low density
of the molded TPX resins of I.C.I. give advantages in
laboratory, hospital, and electronic ware, withstanding
repeated sterilizations at 140°C (288). Scratches and
crazing of molded graduated cylinders impaired clarity
rather quickly in the author's laboratory.

Surprisingly, crystal melting points of stereoregular
polymers of branched 1-olefins (Table 2.2) can be raised
in some cases by sequential or block copolymerization
with small proportions of certain second olefins (289).
Terpolymers were prepared by sequential stereoregular
copolymerizations of branched 1-olefins with linear
1-olefins of 12 to 16 carbon atoms (290).

Amorphous high polymers were prepared from 4-methyl-1-
pentene using vanadium-containing catalysts (291). For
example, 16.4 g of monomer was treated with 0.073 g of
$VOCl_3 - (C_2H_5)_2AlCl$ in methyl cyclohexane, the molar ratio

TABLE 2.2
Properties of Stereoregular Polymers
from Branched Monomers

1-Alkene monomer	Melting point of crystalline phase, °C	Chain conformation	References, etc.
C C=CCC	310, high cryst.	4/1-helix	a,b; Hard, capable of high orientation
C C=CCCC	250, high cryst.	7/2-helix	a, b, c, d
C C=CCCCC	188 or 160	7/2-helix	a, b; Hard
C C=CCCCC	130 or 110, med. cryst.	3/1-helix	a, b; Somewhat rubbery
C C=CCCCCC	noncryst.	--	a
C C=CCCC C	>350, high cryst.	--	a, b, d; Hard
C C=CCCCC C	>350, med. cryst.	--	a, b; Hard
C C=CCCCCC	52, low cryst.	3/1-helix	b; Soft

[a]T. W. Campbell and A. C. Haven, J. Appl. Polym. Sci. 1, 75 (1959); cf. K. R. Dunham et al, J. Polym. Sci. A1, 751 (1963).

[b]R. L. Miller and L. E. Nielsen, J. Polym. Sci. 55, 644 (1961).

[c]B. G. Ranby et al, J. Polym. Sci. 58, 545 (1962).

[d]F. P. Reding, J. Polym. Sci. 21, 547 (1956).

of vanadium to aluminum being 2:1. After 30 min at 20°C,
the resulting polymer was precipitated and the catalyst
was quenched by adding isopropanol followed by dilute
hydrochloric acid. Higher ratios of 1,3 to 1,4 addition
were obtained by use of V to Al ratio of 2:1 than with
the 1:1 molar ratio. The 1:1 catalyst $VOCl_3-(C_2H_5)_2AlCl$
also was an active catalyst for isobutene polymerization.
The polymers from 4-methyl-1-pentene and from 1-butene
have been discussed in more detail (292). The olefin
monomers with branching on the third carbon atom, such as
isopropylethylene, sec-butylethylene, and t-butylethylene,
show relatively low reactivity with transition metal
halide-organoaluminum catalysts.

Copolymers of 4-methyl-1-hexene with minor proportions
of 1-olefins of 4 to 8 carbon atoms remained clear, and
they could be injection molded to form useful articles
(293). Polymers of 4-methyl-1-hexene and of 5-methyl-1-
hexene were blended with minor proportions of 1-olefin
homopolymers, all prepared using Ziegler-Natta type
catalysts (294).

Evidence of methyl migration or methide-shift polymerization of t-butylethylene at -130°C in the presence of
$AlCl_3$ was observed by Kennedy and co-workers (295). The
product was a white amorphous powder melting at about
60°C. Higher temperatures gave only trimers. Polymerizations of 4-methyl-1-hexene at low temperatures in ethyl
chloride initiated by $AlCl_3$ produced soluble, rubberlike
polymers (296). These had brittle points about -23°C.
Ultraviolet and NMR spectra showed the occurrence of
1,4-polymerization by hydride shift, but less than under
similar polymerization conditions with 4-methyl-1-pentene.
Isotactic polymers of 4,4-dimethyl-1-pentene melt only
above 400°C; since they are insoluble in solvents, it is
not feasible to make films and fibers by conventional
methods. Hydride-shift polymerization was believed to
occur in polymerizations of 3-phenyl-1-butene at -78°C
with aluminum chloride catalyst (297).

Monomers of the type $CH_2=CHCR_3$ lack allylic hydrogen
atoms; in free radical copolymerizations, therefore, they
might be expected to give products of higher molecular
weight. However, none of the branched 1-olefins discussed
in this chapter has given homopolymers of high molecular
weight by free radical polymerizations. A number of
purified 1-alkenes for research are supplied by the
Phillips Petroleum and Aldrich companies.

OPTICAL ACTIVITY AND STEREOISOMERISM IN POLYOLEFINS

Stereopolymerizations of branched alkenes have interest
for investigation of optical activity in relation to

structure of monomers and high polymers. In organic
molecules optical activity from stereoisomerism never
results from optically inactive materials except through
biosynthetic processes of living organisms. However,
inactive racemates and diastereoismers containing asymmetric carbon atoms of opposing D and L configurations
show striking differences in melting points, solubilities,
and other properties. Likewise, largely isotactic, syndiotactic or atatic polyolefins made from optically inactive
monomers do not show optical activity but show great
differences in many physical properties. A perfect,
hypothetical isotactic macromolecule would have all D or
all L configurations with one helix direction determined
by a possible different end group. The real isotactic
polyolefins prepared to date are believed to be block
copolymers with $(D)_n$ and $(L)_n$ segments including occa-
sional random short irregularities. It has been suggested
that in the polymer crystals, right helix segments may
prefer to be surrounded by left helices. Besides the
concept that isotactic polyolefins are diastereoisomers
(with compensating D and L structures), other reasons for
lack of optical activity include the presence of a plane
of symmetry passing through the stretched chain (if end
groups are ignored) and that every other C atom in the
main chain may be only "formally" asymmetric (since for
neighboring monomer units the two directions along the
chain are equivalent). Further elucidation of the
structure of different stereoisomeric polyolefins remains
a most intriguing challenge to thought and experiment;
this is also true of stereoisomeric polymers synthesized
from optically active monomers in Ziegler-Natta systems
which will now be discussed briefly.

Pino and co-workers prepared stereoregular polyolefins,
having asymmetric centers in the side chains, from optically active monomers (298). The observed optical rota-
tions of the polymers having asymmetric centers located
at α- or β- carbon atoms with respect to the main chain
were greater than those shown by model compounds of low
molecular weight. This enhancement of optical rotation
in the polymers decreased rapidly as the center of
asymmetry was located further from the main chain. The
temperature coefficient of change of molar rotation of
monomer units with change of temperature was negative and
was larger than expected from analogous small molecules.

Abe suggested that the enhancement arises from conforma-
tional rigidity of branched asymmetric centers close to
the backbone and that the contribution to optical activity
from the isotactic helical structure may not be important
(299). Theoretical calculations for some stereoregular
polyolefins such as poly-(S)-3-methyl-1-pentene agreed

with the order of magnitude of experiment, but not for
polymers of higher monomers such as (S)-5-methyl-1-heptene.
The optical activity of copolymers of (R)- and (S)-4-
methyl-1-hexene and from (R)- and (S)-5-methyl-1-heptene
was consistent with expectations from a crowded, partially
blocked condition along the main polymer chain.

    Bailey and Yates attributed the optical enhancement
effect in polyolefins to the existence of one sense of
helix rotation in preference to the other (300). They
polymerized 3-methyl-1-pentene of $[\alpha]_D^{25} = 33.49°$ and the
racemic monomer by Ziegler catalyst obtaining polymer
fractions which were only partially characterized. Birsh-
stein and Luisi predicted that in optically active stereoregular polymers one helical screw sense should predominate
(301). Pino and Luisi believed that the macromolecules of
isotactic polymer from (S)-4-methyl-1-hexene consist of
relatively long left-handed sequences alternating with
short segments having right-handed helical structure (302).
The optical activities of olefin copolymers gave linear
relations with optical activity of the monomers in some
cases but not in others (303). Pino and co-workers, using
racemic and optically active branched alkenes, found
Ziegler-Natta catalysts to be somewhat stereoselective
only when the center of asymmetry of the monomer was alpha
to the double bond (304). Natta and co-workers tried
optically active catalysts (305). Thermodynamics and
statistical theories have had limited success in this area
as yet compared to experimental investigations. Only a
beginning has been made in utilizing optical activity to
elucidate polymerization mechanisms and polymer structures.

OTHER ALLYLIC 1-ALKENES

    Whereas isobutylene does not behave as a typical meth-
allyl compound and it polymerizes readily by conventional
cationic polymerization at low temperatures, the related
methallyl compound $CH_2=C-CH_2CH_3$ has been little studied
                          |
                         CH_3
in polymerizations. Waterman obtained oily polymers using
$AlCl_3$ at about -80°C (306). Steric hindrance apparently
prevents formation of high polymers from this olefin by
conventional low-temperature cationic methods. With
sulfuric acid the monomer polymerizes faster than does
isopropylethylene, but high molecular weights are not
attained. Boron fluoride etherate gave only orange sticky
liquid polymers at room temperature. With $2AlEt_2Cl\cdot TiCl_3$
as catalyst at about 50°C, 1-pentene polymerized (307) but
2-methyl-1-butene did not. Compounds of the type $CH_2=CRR'$
and $CH_2=CHCRR'R''$ are sufficiently Lewis-basic from
electron-releasing alkyl groups to destroy the electro-

TABLE 2.3

Some Liquid Allylic 1-Alkenes

	BP, °C	mp, °C	n_D^{20}	References, etc.
$\underset{\displaystyle CH_2=\overset{\textstyle CH_3}{\underset{\textstyle }{C}}-CH_2CH_3}{}$	31	-138	1.3778	Sherril, JACS 58, 742 (1936)
$CH_2=\overset{CH_3}{C}-CH_2CH_2CH_3$	61	-135	1.3921	Schmitt, JACS 54, 754 (1932)
$CH_2=\overset{CH_3}{C}-CH_2CH_2CH_2CH_3$	91	--	1.4040	Wibaut, Rec. Trav. Chim. 72, 1037 (1953)
$CH_2=CHCH(CH_3)_2$	20	-168	1.3638	Norris, JACS 49, 2629 (1927); Waterman, Rec. Trav. Chim. 52, 517 (1933)
$CH_2=CHCH_2CH(CH_3)_2$	54	--	1.3825	Schmitt, JACS 54, 754 (1932)
$CH_2=CHC(CH_3)_3$	41	--	1.3763	Wibaut, Rec. Trav. Chim. 60, 241 (1941)
$CH_2=CHCH_2CH_2CH_3$	30	-165	1.3711	Adams, JACS 40, 1918 (1950)
$CH_2=CHCH_2C(CH_3)_3$	72	-138	1.3909	Robertson, J. Chem. Soc. 1014 (1952)
$CH_2=CHCH\overset{CH_3}{\underset{\textstyle }{-}}CH\overset{CH_3}{\underset{\textstyle }{-}}CH_2CH_3$	--	--	1.4103	Aldrich supplier
$CH_2=CH(CH_2)_3CH_3$	64	-139	1.3837	Wilkinson, J. Chem. Soc. 134, 3057 (1931)
$CH_2=CH(CH_2)_4CH_3$	94	--	1.3996	Wilkinson, J. Chem. Soc. 134, 3057 (1931)
$CH_2=CH(CH_2)_5CH_3$	122	-102	1.4087	Pomevantz, J. Res. Natl. Bur. Std. (U.S.) 52, 59 (1954)
$CH_2=CH(CH_2)_{15}CH_3$	145(8)[a]	+18	1.4448	Aubrey, J. Polym. Sci. A1, 1709 (1966); polymer of $\eta = 0.9$

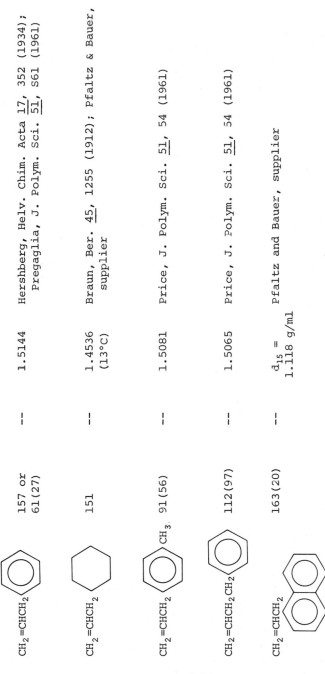

CH$_2$=CHCH$_2$ (phenyl)	157 or 61(27)	--	1.5144	Hershberg, Helv. Chim. Acta 17, 352 (1934); Pregaglia, J. Polym. Sci. 51, S61 (1961)
CH$_2$=CHCH$_2$ (cyclohexyl)	151	--	1.4536 (13°C)	Braun, Ber. 45, 1255 (1912); Pfaltz & Bauer, supplier
CH$_2$=CHCH$_2$—(aryl)CH$_3$	91(56)	--	1.5081	Price, J. Polym. Sci. 51, 54 (1961)
CH$_2$=CHCH$_2$CH$_2$ (phenyl)	112(97)	--	1.5065	Price, J. Polym. Sci. 51, 54 (1961)
CH$_2$=CHCH$_2$ (naphthyl)	163(20)	--	d$_{15}$ = 1.118 g/ml	Pfaltz and Bauer, supplier

aBoiling point expressed as 145(8) indicates: boils at 145°C at 8 mm.

111

philic Ziegler-Natta catalysts.   The first three olefins
in Table 2.3 are examples.
   Serniuk and Thomas were able to free-radical copolymerize
2-methyl-1-pentene with acrylonitrile by complexing with
zinc chloride (308).   In one example, 2 moles of acrylonitrile, 2 moles of the alkene containing hydroperoxides,
and 2 moles of $ZnCl_2$ were warmed to 63°C and stirred 3 hr.
The mixture was then repeatedly extracted with water to
remove $ZnCl_2$.   The copolymer recovered was soluble in
acetone and contained 13.39% nitrogen.   Without $ZnCl_2$ no
copolymer formed.   Tough films and fibers of light color
could be made.   Methyl acrylate also was copolymerized
with 2-methyl-1-pentene. See page 37 for further dis-
cussion of such methods for synthesis of alternating 1:1
molar interpolymers.
   Price prepared high-melting polymers from methyl-
substituted heptenes (309).   For example, isotactic
$$\begin{array}{c}\qquad\quad CH_3\quad\ CH_3 \\ \text{polymers from } CH_2=CHCHCH_2CCH_2CH_3 \text{ had crystal melting points} \\ \qquad\qquad\qquad\qquad\quad CH_3 \end{array}$$
about 422°C and were insoluble in hydrocarbons below 135°C.
Heat-resistant polymers were also prepared from 4-phenyl-
1-pentene (310).
   Stereoregular 1-pentene polymers of crystal mp about
75°C can be prepared by $TiCl_3$-$AlEt_2Cl$ catalysts under
conditions similar to those used for poly-1-butenes (311).
The polymers undergo complex changes in microcrystal-
linity (312).   Rubbery, soluble polymers were synthesized
from 1-hexene with a catalyst from $TiCl_3$ and $Al(C_2H_5)_2Cl$
(313).   The polymers, although believed to be isotactic,
show only slight crystallinity when oriented by stretching.
The polymers of linear 1-alkenes made by $TiCl_4$-$LiAlR_4$
catalyst gave a minimum melting point with the 1-heptene
polymer at 17°C (314).   Side chain crystallinity was shown
by polymers of 1-decene and 1-octadecene.   A heptane-
insoluble fraction was prepared from 1-octene which had a
crystal melting point of 115°C (315).   Polymers made from
C_6 to C_{12} 1-olefins in presence of aluminum alkyl chlorides
and $TiCl_4$ were suggested as synthetic lubricants (316).
   Campbell and Havens prepared polymers melting above 200°C
by Ziegler-type polymerization of allylcyclopentane and
allylcyclohexane (see Table 2.4) (317).   Polymerization of
allylbenzene using BF_3 at low temperatures seemed to give
some benzene rings in the main chains (318).   The monomer
did not polymerize readily with $AlBr_3$ or BF_3 etherate,
but it polymerized readily by using $TiCl_3$-$Al(C_2H_5)_3$ at
room temperature, forming high polymers of normal 1,2
structure (319).   Working with aluminum isobutyl-titanium-
trichloride catalyst, Price and co-workers prepared
crystalline high-melting polymers from a number of isomeric

TABLE 2.4

Polymers from Some Allylic Cyclic Hydrocarbons

Monomer, A is $CH_2=CHCH_2-$	mp, °C, (T_m) polymer crystalline phase	Polymer softening temperature, °C	Type of polymer, reference, etc.
A (cyclopentane ring)	225, med. cryst.	220	Hard polymer[a]
A (cyclohexane ring)	230, med. cryst.	180	Hard[a] (with re-arrangement)[c]
A (benzene ring)	230, med. cryst.	276	Hard, orientable[a]
A (benzene ring)CH_3		--	Polymers insoluble[b]
A (benzene ring with CH_3 and H_3C)	338	--	Polymers insoluble[b]
ACH_2 (benzene ring)	160, low cryst.	168	Slightly rubbery[a]
ACH_2CH_2 (benzene ring)	Noncryst.	Ca. 110	Rubbery[a]
$CH_2=CHCH$ (benzene ring) CH_3	>360, med. cryst.	338	Hard, untractable[a]

[a]Campbell and Havens, J. Appl. Polymer Sci. 1, 75 (1959).
[b]Price, J. Polym. Sci. 51, 541 (1961).
[c]Kleiner et al, Eur. Polym. J. 7, 1677 (1971).

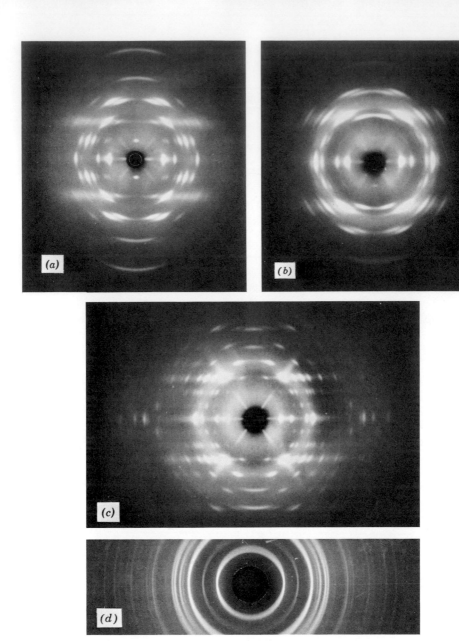

Fig. 2.12. X-Ray diffraction patterns showing complex order obtained in drawn (fibered) stereoregular polymers: (a) poly(4-methyl-1-pentene); (b) poly(vinylcyclohexane); (c) poly(vinylcyclopentane). In (d) a typical sharp ring pattern was obtained from isotactic 4,4-dimethyl-1- pentene polymer powder. H. D. Noether, Celanese Laboratories, Summit, N. J.; J. Polym. Sci. C 16, 725 (1967).

allyltoluenes and allylxylenes (320).   In general, methyl
groups substituted on the aromatic rings raised crystal
melting points.   Some of the polymers were insoluble in
decalin even at 135°C.   A soluble polymer fraction from
m-allyltoluene had $t_m = 290°C$ and showed η_{sp} in decalin at
135°C of 1.37.   When 1-allyl-1-cyclohexene was polymerized
in heptane by addition of 2:1 molar isobutyl aluminum-
titanium tetrachloride catalyst, it yielded a polymer
soluble in organic solvents and said to be 60% crystalline
by x-ray pattern (321).   Treatment of 1-allylindane with
$TiCl_4$ and $AlBr_3$ gave fluorescent polymers of low molecular
weight (322).
     Vinylcyclohexane and related monomers have been prepared
by pyrolysis of esters in presence of acetic or boric
acids (323).   Unsubstituted vinylcycloalkanes seem to
behave as allylic olefins.   Vinylcyclohexane gave stereoregular polymers by $TiCl_3$-AlR_3 catalyst at 50°C (324).
The polymers showed low crystallinity at first but became
highly crystalline after annealing at 150°C for several
hours.   The t_g was about 133°C, compared with 150°C for
hydrogenated isotactic polystyrene.   Highly isotactic
vinylcyclohexane polymer showed a crystal melting point
of 370°C.   Kennedy and co-workers polymerized vinylcyclohexane with $AlCl_3$ in the temperature range of -144
to +7°C (325).   NMR and IR data suggested 1,3-polymeriza-
tion by hydride shift.   By addition of aluminum bromide in
ethyl chloride at -78°C, vinylcyclohexane gave 36% conversion to polymer in 3 hr (326).   The product seemed to be a
structural copolymer resulting principally from 1,3-
addition.   One polymer melting about 93°C had a molecular
weight of 10,500 and showed an IR band at 745 cm^{-1},
characteristic of --CH_2CH_2--.   In Fig. 2.12 are shown
x-ray fiber patterns of isotactic polymers from vinylcyclohexane and from vinylcyclopentane.
     A sample of vinylcyclohexane monomer was observed to
become highly viscous after storage five years.   Amorphous
vinylcyclohexane polymers prepared by hydrogenation of
polystyrene have heat distortion temperatures up to 60°C
higher than those of atactic polystyrene (327).   Polymerization of allylbenzene by BF_3 or $AlCl_3$ gave only liquid
low polymers containing aromatic rings in the main chain
(328).
     Cationic polymerization of vinylcyclopropane occurred
predominantly by vinyl-type or 1,2-polymerization (329).
However, with radical initiation 1,5-type polymerization
occurred with ring opening.   Heptane-insoluble fractions
of vinylcyclopropane polymer prepared by Ziegler catalyst
showed crystallinity in drawn fibers (330).   Polymerizations of vinyl cycloalkanes were reviewed by Daly (331).

Methylene cycloalkanes are olefins that do not polymerize very readily to high polymers of uniform structure. Pinazzi, Rossi and co-workers studied methylenecyclobutane (332). In preliminary studies, cyclobutane groups were largely retained in the polymers when conventional Lewis-acid catalysts were used. Ziegler-Natta systems gave ring opening and also polymers of complex cyclic structure. Catalysts containing vanadium, chromium halides, and rhodium have given rather poor yields of isomeric polymers. Methylenecyclobutane copolymerized with acrylonitrile, methyl methacrylate, and styrene with radical initiation (333). The r_1 values were 0.65 to 1.0; r_2 was near zero. Methylenecyclopropane (334) and methylene cyclopentane (335) have been prepared.

Beta-pinene is a methylene bicycloalkane which, like isobutene and the olefins just mentioned, has two electron-donating substituents in ethylene. It forms polymers of moderate to high molecular weight by cationic polymerization. It was suggested that isomerization polymerization might occur, forming 1,4-linked cyclohexene rings in the main chain (336). Polymers of beta-pinene made by triisobutyl aluminum-titanium tetrachloride catalyst were reported to have identical structure to those prepared using ordinary Friedel-Crafts catalysts (337). Monomethylene- and monovinylcycloalkanes do not homopolymerize by free radical initiation. Treatment of 2-methylene-norbornene with $AlC_2H_5Cl_2$ in C_2H_5Cl at -30 to 100°C gave only viscous oils; norbornene yielded solid polymers (338). Similar catalyst with 2-vinylnorbornane for 3 days at -78°C gave a polymer of about 1440 mol. wt. that melted near 160°C. Infrared data indicated that 1,2- or vinyl-type polymerization had predominated. Soluble, solid polymers of similar molecular weight were prepared from 2-vinyl-5-norbornene. The same catalyst at -100°C polymerized 2-isopropenyl-5-norbornene to crosslinked products. Allyl-adamantanes were prepared and suggested for polymerizations (339).

Cyclic methallyl derivatives have been little studied. On warming with a little aluminum chloride etherate or sulfuric acid, methallylbenzene gave soft solid to viscous liquid low polymers (340). No polymerization was initiated by sodium.

OTHER COPOLYMERIZATIONS OF 1-ALKENES

In addition to copolymerizations forming propylene-ethylene copolymer plastics and ethylene-propylene terpolymer synthetic rubbers, many other copolymerizations of 1-alkenes by transition metal complex catalysts have been studied. When polar groups of comonomers are sufficiently distant from ethylenic double bonds, copolymerization may occur with 1-alkenes under the influence of Ziegler-Natta type catalysts. Thus propylene copolymerized with small amounts of p-naphthyl undecenyl thioether in the presence of a red-colored catalyst obtained by mixing diethyl aluminum chloride with titanium trichloride (341). The copolymers showed crystallinity. Alkyl vinyl thioethers and allyl thioethers were said to inhibit 1-olefin polymerization with Ziegler catalysts. Omega-haloalkenes copolymerized with 4-methyl-1-pentene in the presence of a Ziegler catalyst (342). The further removed the halogen atom from the double bond and the larger the halogen atom, the more readily did copolymerizations take place. Propylene was copolymerized with small amounts of 7-chloro-1-octene in heptane at 50°C using a catalyst from $TiCl_3$, $AlC_2H_5Cl_2$, and a small addition of amine (343). Up to 10 mole % of styrene or allylbenzene may be copolymerized with 1-alkenes, using compounds of vanadium and aluminum as catalysts (344). Copolymers of promising processing properties and heat resistance were obtained. Crystalline copolymers of propylene with about 7% styrene units were made using modified Ziegler catalyst at 50°C (345). These were suggested for dyeable fibers. Minor proportions of 1-olefins have been randomly copolymerized with formaldehyde in ionic systems (346). The thermal stability of the copolymers may be improved by capping with organic acids, vinyl ethers, epoxides, or isocyanates.

Hogan and co-workers of Phillips Petroleum Company have developed copolymers of ethylene with minor proportions of 1-butene or 1-hexene, particularly suited for extruded pipe and for molded bottles (347). The comonomer units supply random short branches that reduce crystallinity and density, and they impart greater elongation, impact strength, and resistance to environmental stress cracking compared with linear ethylene homopolymers. Small amounts of 1-butene units (e.g., 1 mole %) improve long-term load-bearing ability in pipes, and similar proportions of 1-hexene units give outstanding resistance to stress cracking.

In the Phillips "particle form process", the catalyst may be 2% CrO_3 on 98% silica (activated 5 hr at 1400°F). Water and oxygen must be excluded. In a laboratory polymerization in a stainless steel autoclave, 0.05 g of catalyst dispersed in liquid butane gave 250 g of copolymer after 60 min at 220°F. The higher 1-olefin may be added initially and the ethylene added gradually to keep the dissolved monomer concentration near 6% in the liquid diluent. The copolymer particles obtained were treated with antioxidant, dried, homogenized on a two-roll mill, and chopped.

It is surprising that 1 mole % of higher 1-olefins
accelerates the polymerization rate by the Phillips
chromium oxide catalyst as much as 40% over that of the
homopolymerization of ethylene. This seems to result
from generation of a more active catalyst complex, since
delayed addition of higher olefin did not cause similar
acceleration. Propylene copolymerizes with ethylene somewhat faster than do 1-butene and 1-hexene, but it does not
give as good copolymer properties. Additions of stronger
Lewis-basic monomers such as isobutene reacted strongly
with the acidic catalyst and stopped polymerization completely. The relations between copolymer density and
mole % of different 1-olefins appear in Fig. 2.13.   The
propylene copolymers showed high draw ratios with good
resistance to stress cracking. Linear ethylene homopolymers break without drawing down uniformly in this
test at 90°C. Copolymers of density about 0.94 g/cc are
especially suitable for piping. Copolymers with 4-methyl-
1-pentene or isobutylethylene also have good resistance to
stress cracking. In this work copolymers of similar melt
viscosity and molecular weight were compared. Copolymers
of 1-alkenes with minor proportions of conjugated diolefins
such as butadiene also can be made using chromium oxide
catalysts.(348). Furukawa and co-workers are developing
elastomeric alternating copolymers of 1-alkene and butadiene using modified Ziegler-Natta catalysts (349).

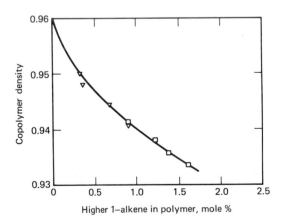

Fig. 2.13. Copolymers of ethylene with minor proportions of 1-hexene
(squares) or 4-methyl-1-pentene (triangles) show good resistance to
stress cracking in pipes; the comonomers reduce the density and normal-
crystallinity of the polymers. J. P. Hogan, B. E. Nasser, and R. T.
Werkman, IUPAC Meeting, Boston, Macromol. Reprints 710 (1971).

Copolymers of ethylene with small proportions of 1-dodecene have been studied by Sumitomo Chemical Company, and copolymers with 3 to 7% 1-octadecene were made by the Tokyo Kasei Company (350). Branching in these copolymers was estimated by IR absorption at 7.25μ. Long branches may favorably reduce viscosities of melts and solutions, perhaps by decreasing chain entanglement.

Polar allyl compounds such as allyl chloride and allyl amine prevent propylene polymerization by ordinary Ziegler-Natta catalysts. When the polar group is further removed from the double bond, there seems to be greater possibility for copolymerization. Thus 8-bromo-1-octene was copolymerized with propylene at 50 psi and 70°C for 4 hr (351). The copolymer, which was 66% insoluble in heptane, could be crosslinked by reaction with a poly-amine. With the purpose of obtaining copolymers chemically stable to heat and ultraviolet light, propylene was co-polymerzied using Ziegler-type catalysts with small amounts of t-N-alkenyl amines of the type of $RArN(CH_2)_nCH=CH_2$ (352). For example, N-methyl-N-undecenyl aniline in hydrocarbon diluent with $TiCl_3$ and Et_3Al was pressured with propylene at 50°C. After 3 hr n-butanol was added. The copolymer (mp 162°C) was 75% crystalline and had reduced viscosity 5.08 at 135°C in decohydronaphthalene. Long-chain 1-alkenes were copolymerized with linoleic acid with $AlCl_3$ catalyst in ethyl chloride solvent at 0 to 50°C (353). Copolymerization of 1-olefins with carbon disul-fide has been investigated using activated transition metal catalysts (354).

It has been proposed that polar monomers be pretreated with organoaluminum compounds in order to stabilize them for copolymerization in activated ionic systems (355). Treatment of an acrylate ester complex of diethyl aluminum chloride with $TiCl_3$ and propylene at 60°C gave a modified propylene polymer (356). Propylene and complexed methyl vinyl ketone gave a 1:1 molar copolymer (357). Some patent examples of 1-alkene copolymerization with polar monomers by Ziegler-type catalysts do not give uniform products of high molecular weight.

Contrary to some patents, aliphatic 1-alkenes do not copolymerize very readily with most other monomers by free radical methods. Even small additions of 1-alkenes can retard polymerization of vinyl, styrene, and acrylic monomers except for copolymerization with certain Lewis-acidic monomers such as maleic anhydride. Thus small proportions of 1-alkenes may be added as regulators to reduce the molecular weight of vinyl, styrene, or acrylic polymers. In this respect 1-alkenes generally act as typical allyl compounds, and only small proportions of the 1-alkene units generally enter into a copolymer structure.

From very active free radical systems, however, significant amounts of alkene units can be incorporated into
copolymers with certain vinyl monomers.

Ethylene mixtures with small proportions of propylene
and other 1-alkenes were polymerized at high pressure by
ICI and Du Pont in the 1930s. Recently Union Carbide
patented free-radical-initiated high-pressure copolymerization of ethylene with 3 to 40 mole % of propylene at
70,000 to 125,000 psi at 70°C to 250°C (358). Products
containing at least 1 mole % propylene units were claimed.

Vinyl chloride-propylene copolymers containing 2 to 6%
combined propylene units were developed by Heiberger and
co-workers by suspension polymerization using t-butyl
peroxypivalate as catalyst (359). The copolymers showed
favorable stability and molding properties, apparently
because there was less chain branching than in common
vinyl chloride polymers. The effect of propylene concentration in the monomer feed on propylene content and the
intrinsic viscosity of the copolymers is shown in Fig.
2.14. Copolymers of vinyl chloride containing 2.7% propylene units showed better flow and blow molding properties
than vinyl chloride homopolymers of similar values of
η_{sp}/C about 0.74 (360). The propylene units seem to im-

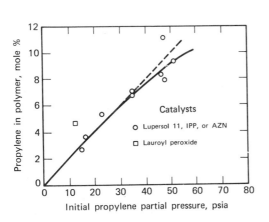 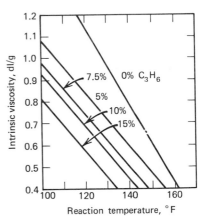

Fig. 2.14. Copolymerizations of vinyl chloride with propylene in
free-radical-initiated aqueous suspension. Lupersol 11 is t-butyl
peroxypivalate, IPP is isopropyl percarbonate, and AZN is azobis-
butyronitrile. M. J. R. Cantow, C. W. Cline, C. A. Heiberger, D. T.
A. Huibers, and R. Phillips, Mod. Plastics 46, 126 (June 1969).

prove stability to heat by blocking dehydrochlorination
(361). Blown filaments of vinyl chloride-ethylene-
propylene have been suggested for packing and insulation
(362). Copolymers of vinyl chloride with minor proportions of propylene have been available commercially (363).
Reactivity ratios for vinyl chloride radical copolymerization with various 1-alkenes were determined (364). Most
reactive was 1-pentene, while branched alkenes were less
reactive. Of the latter the methallyl monomers were most
reactive. Reding and co-workers of Union Carbide promoted
the copolymerization of vinyl chloride with branched
allylic olefins by high pressures, e.g., 15,000 psi at
60°C (365).

The butenes copolymerize by free radical mechanism less
readily with maleic anhydride than does isobutylene (366).
At 60°C in benzene with azobis initiator, the decreasing
rate of reaction and lowering DP of copolymer were reported: isobutylene > 1-butene > cis-2-butene > trans-2-
butene. The products were believed to have alternate 1:1
molar composition. A 1:1 molar copolymer of 1-hexene with
maleic anhydride was half-esterified by methanol and used
in aqueous dispersion for floor polishes (367).

Mixtures of fine 1-olefin-maleic anhydride copolymers
with metal oxides such as CaO were reacted during molding
with heat to densely crosslinked heat-resistant moldings
(368). Equimolar copolymers of 1-allyl naphthalene and
maleic anhydride were formed by heating at 160°C in
presence of 2% di-t-butyl peroxide (369). When excess
maleic anhydride was present, Diels-Alder addition to the
naphthalene ring occurred.

It has long been known that 1-olefins and many other
allyl compounds can form copolymers with sulfur dioxide
when irradiated by UV light or in presence of relatively
high concentrations of peroxides or azo initiators. Thus
propylene and 1-butene with SO_2 give viscous liquid to
glassy colorless solid polymers (370). Excess liquid SO_2
may serve as solvent. Generally thermal initiation under
pressure is not effective. Phillips developed emulsion
polymerization systems near 25°C for preparing 1-butene-
sulfone copolymers using $LiNO_3$, NH_4NO_3, or $AgNO_3$ as
catalysts (371). Copolymers of 1-butene and SO_2 made
with t-butyl hydroperoxide or UV light were soluble in
acetone but precipitated by methanol (372). They were
decomposed by gamma radiation even at 0°C. Among the
limitations that have discouraged industrial interest in
olefin-sulfone copolymers have been variable solubility
and crosslinking with chemical instability on heating.
So-called ceiling temperatures at which depolymerizations
become active have been reported: propylene-SO_2, 87°C;
1-butene-SO_2, 63°C; 2-butene-SO_2, 45°C; isobutylene-SO_2,

4°C.   Ethylene is said to react slower with SO_2 than butenes and propylene.

When preformed polymers of 1-alkenes such as polypropylene have been partially oxidized to form hydroperoxide groups (especially at reactive tertiary hydrogen atoms), they can be graft copolymerized with styrene or acrylic monomers (375).   This is one of the methods evaluated for making polyolefin fibers receptive to dyestuffs.   Grafting a hydroperoxide of polypropylene with diethaminoethyl methacrylate had little effect upon the normal crystallinity of the base polymer (374).   Grafting of polypropylene with styrene and maleic anhydride in a heated mixer has been disclosed (375).

References

1. H. F. Mark, Anal. Chem. 20, 104 (1948); Chimie et Physique des Molecules Geantes, Liège, 1950; H. F. Mark and N. G. Gaylord, Linear and Stereoregular Addition Polymers, Interscience, 1959; C. E. Schildknecht, Vinyl and Related Polymers, Wiley, 1952; Polym. Eng. Sci. 6, 240 (July 1966).
2. G. Natta, Science 147, 261 (January 15, 1965) (107 references).
3. H. Hock and A. Neuwirth, Ber. 72, 1562 (1939), confirmed by C. E. Schildknecht.
4. Chemical Week 41 (February 9, 1972); L. Jakubowicz et al, CA 76, 46668 (1972); cf. J. H. Murib, CA 78, 30459 (1973).
5. M. Z. Woskow, U.S. 3,476,824 (Petro-Tex); cf. P. Boutry et al, U.S. 3,577,477 (Inst. Francais du Pétrole); J. E. Connor et al, U.S. 3,586,733 (Atlantic Richfield).
6. C. L. Furrow and C. E. Stoops, Ind. Eng. Chem. Res. Develop. 7, 26 (1968).
7. S. V. Lebedev and M. Platonov, J. Chem. Soc. 127, 417 (1925).
8. H. S. Davis et al, J. Am. Chem. Soc. 54, 2340 (1932).
9. H. S. Davis and R. Schuler, J. Am. Chem. Soc. 52, 721 (1930).
10. F. Hofmann and W. Stegemann, Ger. 524,891 (I.G.); CA 25, 4282 (1931); cf. G. A. Mortimer and L. C. Arnold, J. Polym. Sci. A 2, 4247 (1964).
11. B. A. Englin et al (Moscow), CA 74, 126151 and 142414 (1971); S. G. Abasova et al (Moscow), CA 74, 142621 (1971).
12. C. A. Kraus and J. D. Calfee, U.S. 2,440,750 (Esso); cf. D. W. Young, U.S. 2,542,610 (Esso).
13. A. S. Hersberger and R. G. Heiligmann, U.S. 2,474,670 (Atlantic Refining).
14. R. L. Meier, J. Chem. Soc. 3656 (1950).
15. C. M. Fontana and G. A. Kidder, J. Am. Chem. Soc. 70, 3745 (1948).
16. C. M. Fontana et al, Ind. Eng. Chem. 44, 1695 and 2955 (1952); J. Phys. Chem. 63, 1167 (1959).
17. C. M. Fontana, U.S. 2,571,354; cf. U.S. 2,678,957 (Socony).

18. C. M. Fontana, in Cationic Polymerization, P. H. Plesch, Ed., Heffer, 1953.
19. D. J. Oriolo, U.S. 2,521,940 (Socony).
20. E. R. Boldecker and A. G. Oblad, Magnolia Oil Co. unpublished work.
21. M. P. Matuszak, U.S. 2,826,620 (Phillips).
22. A. Zletz et al, U.S. 2,692,257-9; cf. U.S. 2,478,006; Brit. 734,501 (Std. Oil Ind.) appl. 1951. Broad claims on polypropylenes to Zletz were announced by the Canadian Patent Office in 1969 but the patent has not issued.
23. Brit. 748,583 (Std. Oil Ind.) appl. 1952.
24. E. F. Peters and B. L. Evering, U.S. 2,824,089 and Brit. 786,014; cf. U.S. 2,898,326; U.S. 2,912,419; U.S. 2,936,291 (Std. Oil Ind.).
25. J. P. Hogan and R. L. Banks, U.S. 2,825,721; Brit. 790,195; Belg. 530,617, appl. 1953; cf. U.S. 2,826,620 (Phillips).
26. Belg. 535,082 (Phillips).
27. J. P. Hogan et al, Ind. Eng. Chem. 48, 1152 (1956).
28. A. D. Ketley and M. C. Harvey, J. Org. Chem. 26, 4649 (1961).
29. Viscous liquid and low-melting polymers: M. W. Perrin et al, Brit. 497,643 (ICI); also U.S. 2,396,677 (Du Pont) and U.S. 2,478,066 (Shell).
30. D. W. Brown and L. A. Wall, J. Phys. Chem. 67, 1016 (1963); cf. G. A. Mortimer and L. C. Arnold, J. Polym. Sci. A 2, 4247 (1964).
31. L. A. Wall, Report on High-Pressure Polymerization, National Bureau of Standards, Washington, D. C., July 1970.
32. P. Ehrlich and L. A. Wall, unpublished work.
33. G. A. Mortimer and L. C. Arnold, J. Polym. Sci. A 2, 4247 (1964).
34. W. E. Hanford, U.S. 2,377,779 (Du Pont) appl. Feb. 18, 1942.
35. M. J. Roedel, U.S. 2,475,520 (Du Pont) appl. May 22, 1944.
36. K. Ziegler and H.-G. Gellert, Ger. 883,067 appl. May 8, 1952 and 878,560 appl. May 29, 1952 (to Ziegler); CA 52, 12457; cf. Ziegler, U.S. 2,699,457.
37. K. Ziegler, H.-G. Gellert, and Helga Kuehlhorn, Ger. 889,229 (Ziegler), appl. Aug. 28, 1952.
38. K. Ziegler, E. Holzkamp, H. Breil, and H. Martin, Angew. Chem. 67, 541 (1955); cf. Ziegler, U.S. 3,579,493 (Ziegler); M. Fischer, Ger. 874,215 (Badische).
39. A. W. Anderson, N. G. Merckling, W. N. Baxter, I. M. Robinson, G. S. Stamatoff, W. L. Truett, and D. H. Payne, Brit. 777,538 and French 1,135,808 (Du Pont), appl. August and November 1954; A. W. Anderson and D. H. Payne, U.S. 3,254,139 (Du Pont), appl. November 1954; cf. D. H. Payne, Ger. 1,420,641 (Du Pont).
40. R. C. Schreyer, U.S. 2,962,451 (Du Pont) appl. September 1954.
41. A. W. Anderson and N. G. Merckling, U.S. 2,721,189 (Du Pont) appl. August 1954; I. M. Robinson, L. H. Rombach and W. L. Truett, U.S. 2,932,630 (Du Pont).
42. A. W. Anderson, N. G. Merckling, and P. H. Settlage, U.S. 2,799,668 (Du Pont).

43. G. R. McKay and P. H. Settlage, U.S. 2,934,527 (Du Pont); cf.
     Eleuterio, U.S. 3,074,918 (Du Pont).

44. C. E. Schildknecht, unpublished.

45. Belg. 533,362 (Ziegler) appl. Nov. 16, 1954; cf. Ziegler, Breil,
     Holzkamp, and Martin, Ger. 973,626, appl. Nov. 17, 1953.

46. Belg. 540,459 (Ziegler) appl. Aug. 16, 1954; cf. K. Ziegler and
     H. Martin, Ger. 1,039,055 (Ziegler).

47. Belg. 534,792 (Ziegler) appl. Jan. 11, 1955; cf. Belg. 534,888
     (Ziegler).

48. Belg. 538,782 (Monte and Ziegler); cf. Ital. 535,712 and 537,425;
     Belg. 543,259; 544,576; and 558,563 (Monte and Ziegler).

49. G. Natta, P. Pino and G. Mazzanti, Belg. 549,638 (Monte-Ziegler),
     appl. July 19, 1955.

50. Belg. 543,259 (Monte and Ziegler), appl. Dec. 3 and 16, 1954.

51. G. Natta, P. Pino, P. Corradini, F. Danusso, E. Mantica, G.
     Mazzanti, and G. Moraglio, J. Am. Chem. Soc. 77, 1708 (1955)
     (received for publication Dec. 10, 1954).

52. The terms isotactic, atactic and syndiotactic (DLDL, etc.) were
     suggested by Rosita Beati Natta.

53. C. E. Schildknecht, Vinyl and Related Polymers, Wiley, 1952.

54. G. Natta, P. Pino, and G. Mazzanti, U.S. 3,112,300-1 (Monte)
     appl. in Italy, June 8, 1954; cf. Chim. Ind. (Milan) 37, 888
     (1955); 38, 124 (1956); J. Am. Chem. Soc. 84, 1488 (1962).

55. G. Natta, P. Pino and G. Mazzanti, U.S. 3,438,956 (Monte).

56. G. Natta and G. Crespi, Belg. 550,093 (Monte and Ziegler)
     appl. Aug. 6, 1955; U.S. 3,175,999.

57. G. Natta et al, Belg. 562,918 (Monte and Ziegler) appl. Dec. 4,
     1956; cf. Belg. 585,563; 572,726; and 563,350.

58. E. J. Vandenberg, U.S. 3,058,963 (Hercules) appl. April 7, 1955;
     cf. U.S. 2,954,367; 3,058,973; 3,261,821 (Hercules).

59. K. Rust et al, Ger. 1,109,894 (Hoechst) appl. April 27, 1957;
     cf. U.S. 3,058,970 (Hercules).

60. W. W. Thomas, U.S. 2,909,510 (Hercules) appl. Nov. 20, 1958.

61. H. W. Blunt, U.S. 3,575,948 (Hercules).

62. G. Natta et al, U.S. 3,245,973; cf. U.S. 3,252,958 and
     3,227,700 (all to Montecatini).

63. E. J. Vandenberg, U.S. 3,051,690 (Hercules); Hercules Chemist
     No. 46, 7-11 (February 1963); cf. U.S. 3,146,223 (Shell);
     cf. U.S. 3,129,209 (Sun Oil).

64. B. Jacobi and O. Wolff, U.S. 2,974,132 (Hoechst); cf.
     French 1,314,673 (Hoechst).

65. E. A. Harris, U.S. 3,436,386 (Hercules).

66. B. H. Mahlman and H. M. Spurlin, U.S. 3,009,907 (Hercules).

67. G. Bier, G. Lehmann, and H. J. Leugering, Makromol. Chem. 44-46,
     347 (1961); 58, 1 (1962); G. Bier, Angew. Chem. 73, 186 (1961);
     G. Bier, A. Gumboldt, and G. Lehmann, Trans. J. Plastics Inst.
     (London) 28, 98-110 (1960); CA 55, 2171 (1961).

68. G. Bier, Makromol. Chem. 70, 44 (1964).

69. G. Bier et al, Ger. 1,109,895 (Hoechst) appl. Oct. 19, 1957.

70. G. Bier, A. Gumboldt, and G. Schleitzer, Makromol. Chem. 58, 43 (1962).

71. W. M. Schilling, U.S. 3,200,173 (Hercules) appl. Feb. 8, 1960; cf. Brit. 940,123 and U.S. 3,592,880 (Hoechst; M. Asada et al, Japan 71-32, 414 (Sumitomo), CA 76, 46698 (1972).

72. H. J. Hagemeyer et al, U.S. 3,600,463 (Eastman); cf. so-called stereosymmetric PP of Brit. Amended 921,635 (Eastman); CA 76, 34738 (1972); cf. U.S. 3,639,515 (Eastman).

73. G. Natta, J. Polym. Sci. 34, 21 (1959); cf. Makromol. Chem. 76, 54 (1964).

74. G. Natta et al, U.S. 3,141,872 (Monte); J. Polym. Sci. 51, 387 (1961).

75. Belg. 573,034 (Hoechst).

76. G. Natta et al, Belg. 570,168 (Monte and Ziegler) appl. Aug. 8, 1957.

77. A. K. Ingberman, I. J. Levine, and R. J. Turbett, J. Polym. Sci. A 1 4, 2781 (1966).

78. C. W. Hock, J. Polym. Sci. A 1 4, 3055 (1966).

79. P. Blais and R. S. Manley, J. Polym. Sci. A 1 6, 291 (1968).

80. D. B. Ludlum, A. W. Anderson, and C. E. Ashby, J. Am. Chem. Soc. 80, 1380 (1958); J. Boor, J. Polym. Sci. C 1, 237 (1963); A. K. Ingberman et al, J. Polym. Sci. A 1 4, 2781 (1966).

81. E. Tornquist and A. W. Langer, U. S. 3,032,510; E. Tornquist and C. W. Seelbach, U.S. 3,034,992 (Esso); cf. E. J. Vandenberg, U.S. 3,108,973 (Hercules).

82. E. Tornquist, U.S. 3,121,063; cf. U.S. 3,128,252 and 3,032,513 (Esso).

83. Fr. 1,561,994 (Scholven Chemie); CA 71, 102,428 (1969).

84. A. W. Langer and E. Tornquist, U.S. 3,047,558 (Esso); cf. L. D. Etherington, U.S. 3,155,640 (Esso); R. H. Schwaar and E. G. Foster, U.S. 3,502,633 (Shell).

85. E. J. Vandenberg, U.S. 3,108,973 and 3,261,821 (Hercules).

86. E. G. Kastning and K. Wisseroth, U.S. 3,075,958 (Badische).

87. R. G. Hay, U.S. 2,922,782 (Goodrich-Gulf).

88. P. A. Argabright and E. A. Schmall, U.S. 3,161,604 (Esso).

89. R. D. Lundberg and F. E. Bailey, U.S. 3,035,036 (Carbide).

90. A. W. Langer et al, U.S. 3,032,511; U.S. 3,063,798 (Esso). J. L. Jezl, U.S. 3,057,838; W. E. Thompson et al, U.S. 3,107,236 (Sun Oil).

91. E. J. Vandenberg, U.S. 3,108,973 (Hercules).

92. H. deVries et al, U.S. 3,068,216 (Stamicarbon).

93. H. W. Coover and F. B. Joyner, J. Polym. Sci. A 3, 2407 (1965); cf. U.S. 2,956,991 and twenty later U.S. patents to Eastman.

94. Y. Doi et al, CA 72, 122009 (1970).

95. J. W. McFarland, U.S. 3,012,996 (Du Pont).

96. D. Kaufman and B. H. McMullen, U.S. 3,031,440 (Natl. Lead); G. Bo and E. Fichet, Fr. 1,129,678 (Rhône-Poulenc) appl. August 1955.

97. J. Chatt and P. A. Small, U.S. 3,109,838 (ICI).

98. I. Kunz, U.S. 3,155,641 (Esso).

99.   A. W. Anderson, U.S. 3,050,471 (Du Pont).

100. K. Rust et al, U.S. 3,047,557 (Hoechst).

101. H. G. Schutze et al, U.S. 3,105,024 (Esso).

102. U.S. Patents as follows:
      2,912,424 (Eastman)   3,129,209 (Sun Oil)   3,153,641 (Monte)
      3,055,878 (Sun Oil)   3,135,724 (FMC)   3,216,987 (Avisum)
      3,099,647 (Sun Oil)   3,139,418 (Monte)   3,502,634 (Phillips)
      3,116,274 (Badische)   3,147,238 (Shell)   3,510,465 (Sumitomo)
      3,121,064 (Hoechst)   3,149,098 (Avisun)   3,669,948 (Chisso)

103. J. Sasaki et al, U.S. 3,536,686 (Mitsui).

104. J. Boor, Jr., J. Polym. Sci. B 3, 7 (1965) and 9, 617 (1971).

105. E. A. Youngman and J. Boor, J. Polym. Sci. B 4, 913 (1966);
      U.S. 3,400,112 (Shell).

106. P. Pino and G. Mazzanti, U.S. 3,000,870 (Monte).

107. E. B. Mirviss et al, U.S. 3,310,547 (Esso).

108. A. W. Langer and J. W. Harding, U.S. 3,418,304 (Esso).

109. M. deVries, Ger. appl. (Stamicarbon); CA 74, 64593 (1971);
      B. Diedrich and R. Klein, Ger. appl. (Hoechst); CA 75, 6670
      (1971); J. Stevens and G. Michel, Ger. Offen. 2,109,273,
      CA 76, 86420 (1972).

110. J. Boor et al, Preprint, Org. Coatings and Plastics Chem. 30,
      158 (May 1970); G. Allegra, Makromol. Chem. 145, 235 (1971);
      A. Gumboldt, Fortschr. Chem. Forsch. 16, 299 (1971).

111. R. M. Kennedy, U.S. 2,948,711 (Sun Oil).

112. J. L. Jezl et al, U.S. 2,980,662 (Sun Oil).

113. C. L. Thomas, U.S. 3,153,634 (Sun Oil); J. P. Herrmans et al,
      Ger. Offen. 2,059,822 (Solvay), CA 76, 15221 (1972).

114. C. L. Thomas, U.S. 3,047,551 (Sun Oil); cf. U.S. 3,652,527.

115. W. R. F. Guyer, U.S. 3,008,943 (Esso); cf. U.S. 3,639,378.

116. H. W. Coover, U.S. 3,038,892; H. J. Hagemeyer and M. B. Edwards,
      U.S. 3,143,537; Brit. 943,206 (Eastman).

117. H. G. Kirschner et al, U.S. 3,002,961 and Belg. 551,997
      (Hoechst); J. J. Moon, U.S. 3,257,372 (Phillips).

118. J. E. Cottle, U.S. 3,294,772 (Phillips).

119. G. Mazzanti et al, Makromol. Chem. 61, 63 (1963).

120. D. F. Hoeg, U.S. 3,138,578 (Grace).

121. A. S. Matlack and D. S. Breslow, J. Polym. Sci. A 3, 2853 (1965).

122. Brit. 839,695 (ICI); cf. Brit. 872,970 and 878,756 (Grace).

123. R. D. Bushick and R. S. Stearns, J. Polym. Sci. A 1 4, 215
      (1966).

124. F. J. Padden and H. D. Keith, J. Appl. Phys. 37, 4013 (1966).

125. For example, D. F. Hoeg, U.S. 3,374,213 (Rexall); E. Renaudo,
      U.S. 3,347,955 (Phillips); J. L. Jezl, U.S. 3,442,978 (Avisun).

126. R. Bacskai, U.S. 3,351,621 (Chevron); cf. M. R. Cines, U.S.
      3,527,579 (Phillips).

127. For example, B. Diedrich and K. D. Kreil, U.S. 3,592,880
      (Hoechst).

128. E. J. Vandenberg, U.S. 2,954,367; Brit. 795,182 (Hercules).

129. D. E. Hostetter, U.S. 3,435,095 (Rexall); cf. CA 70, 7854 (1969).

130. A. M. Jones et al, U.S. 3,414,637 (Esso).

131. C. J. Kuhre et al, SPE J. 20, 1113 (October 1964); H. V. Wood, U.S. 3,499,884 (Phillips); H. Naarmann, U.S. 3,513,146 (Badische).

132. H. N. Beck, J. Polym. Sci. A 2 4, 631 (1966); Belg. 622,919 and 633,715-6 (Shell).

133. F. B. Joyner and G. O. Cash, U.S. 3,408,341 (Eastman).

134. R. H. Hughes, J. Appl. Polym. Sci. 13, 417 (1969).

135. W. G. Frizelle et al, U.S. 3,536,644 (Natl. Distillers); R. L. Combs et al, U.S. 3,515,775 (Eastman); cf. U.S. 3,517,086 (Sumitomo).

136. S. Y. Hobbs, CA 76, 72920 (1972); M. R. Kantz et al, J. Appl. Polym. Sci. 16, 1249 (1972).

137. Brit. appl. ICI; CA 66, 66029 (1967); cf. K. Tsuboshima, U.S. 3,510,549 (Kohjin); K. Ohno, CA 78, 30781 (1973).

138. B. H. Mahlman, U.S. 3,483,276 (Hercules).

139. L. C. Cessna et al, SPE J. 25, 10, 35 (1969).

140. R. L. McConnell and F. B. Joyner, U.S. 3,579,486 (Eastman).

141. J. E. Guillet and F. J. Golemba, SPE J. 26, 88 (1970).

142. J. C. W. Chien, E. J. Vandenberg, and H. Jabloner, J. Polym. Sci. A 1 6, 381 (1968).

143. L. Dulog, E. Radlman, and W. Kern, Makromol. Chem. 80, 67 (1964).

144. R. H. Hansen et al, J. Polym. Sci. A 2, 587 (1964).

145. J. A. Kies, U.S. 3,507,941 (USA).

146. G. C. Oppenlander, U.S. 3,485,906 (Hercules).

147. G. C. Oppenlander, U.S. 3,509,013 (Hercules); G. Jurkiewitsch, U.S. 3,533,904 (Hercules).

148. F. Kalwaites and W. Sibbach, U.S. 3,501,565 (J & J); cf. F. W. Johnson and J. Leach, U.S. 3,544,404 (Burlington Industries); D. F. Stewart et al, U.S. 3,579,618 (Phillips).

149. L. M. Guenther, U.S. 3,496,259-60 (Chevron); cf. U.S. 3,576,931 (Celanese); U.S. 3,624,194 (Bemberg).

150. A. H. Drelich and M. R. Fechillas, U.S. 3,501,369 (J & J); cf. U.S. 3,630,816 (Chevron).

151. R. H. Saunders, U.S. 3,516,899 (Hercules); A. J. Herrman, U.S. 3,546,062 (Du Pont).

152. D. C. Prevorsek et al, U.S. 3,506,535 (Allied).

153. H. S. Barbehenn and R. F. Williams, U.S. 3,607,616 (Eastman).

154. G. Natta, P. Corradini et al, Makromol. Chem. 39, 238 (1960) and many other publications.

155. F. A. Bovey et al, Macromolecules 2, 619 (1969).

156. P. J. Flory and Y. Fugiwara, Macromolecules 2, 327 (1969); cf. A. Abe, Polymer J. 1, 232 (1970).

157. A. Peterlin, Polymer Preprints 11, #2, 1269 (1970).

158. K. Maeda and H. Kanetsuna, Japan 71-25,856; CA 76, 4366 (1972).

159. G. J. Listner and A. J. Sampson, U.S. 3,547,870 (J & J).

160. V. S. Tikhomiro, CA 76, 34695 (1972); cf. CA 78, 30859 (1973).

161. A. Turner-Jones, Makromol. Chem. 75, 134 (1964); J. L. Kardos et al, J. Polym. Sci. A 2, 4, 777 (1966).

162. M. Kojima, J. Polym. Sci. A 2 6, 1255 (1968).

163. A. Turner-Jones, Polymer 12, 487 (1971).

128   1-ALKENE POLYMERIZATIONS

164. C. R. Desper, Polymer Preprints 13, #1, 310 (1972).
165. G. C. Adams, Org. Coatings and Plastics Chem. Preprints 32, #1, 52 (1972).
166. F. Danusso and G. Gianotti, Makromol. Chem. 80, 1-12 (1964); CA 62, 7880 (1965).
167. J. Brandrup and E. H. Immergut, Polymer Handbook, Interscience 1965.
168. J. B. Kinsinger et al, J. Am. Chem. Soc. 81, 2908 (1959); J. Phys. Chem. 63, 2002 (1959).
169. W. R. Krigbaum et al, J. Phys. Chem. 65, 1984 (1961).
170. A. Abe, Polymer J. 1, 232 (1970); cf. G. Natta and P. Corradini, Makromol. Chem. 110, 291 (1967).
171. H. W. Coover et al, J. Appl. Polym. Sci. 13, 519 (1969).
172. G. Natta, P. Pino, and G. Mazzanti, U.S. 3,261,820 (Monte).
173. E. H. Immergut et al, J. Polym. Sci. 51, S57 (1961).
174. J. Kumamoto, J. Polym. Sci. A 3, 3355 (1965).
175. S. B. Lippincott et al, U.S. 2,976,271; cf. U.S. 2,996,492 (Esso).
176. J. E. Guillet, U.S. 2,835,959 (Eastman).
177. Technical Data Sheet, Low-Viscosity Polypropylene, Eastman Chemical Products; J. E. Guillet and H. W. Coover, SPE J. 16, #3, 311 (1960).
178. G. Natta, P. Corradini, and M. Peraldo, U.S. 3,258,455; cf. U.S. 3,257,370 (Monte); J. Polym. Sci. C 1, 411 (1964); J. Am. Chem. Soc. 84, 1488 (1962).
179. G. Natta et al, U.S. 3,305,538 (Monte); cf. J. Am. Chem. Soc. 84, 1488 (1962).
180. A. Zambelli et al, J. Polym. Sci. C 16, 2485 (1967); Makromol. Chem. 112, 160 (1968); 115, 73 (1968).
181. J. Boor and E. A. Youngman, J. Polym. Sci. A 1, 3, 577 (1965); A 1 4, 1861 (1966), and private communication.
182. D. D. Emrick, U.S. 3,268,627 and 3,364,190 (Std. Oil Ohio).
183. G. B. Sterling, U.S. 3,139,412 (Dow).
184. A. J. Reinert, U.S. 3,306,817 (Phillips) (bird repellents).
185. L. T. Dempsey and L. R. Babcock, U.S. 3,506,740 (Shell).
186. W. M. Sweeney et al, U.S. 3,522,180 (Texaco).
187. J. R. Lewis, U.S. 3,499,819 (Hercules).
188. A. D. Ketley, U.S. 3,535,302 (Grace).
189. A. W. Anderson, J. M. Bruce, N. G. Merckling and W. L. Truett, U.S. 3,541,074 (Du Pont) appl. August 1954.
190. W. F. Gresham and M. Hunt, U.S. 2,933,480 (Du Pont) appl. 1956.
191. S. Adamek, E. A. Dudley and R. T. Woodhams, Brit. 880,904 (Dunlop) appl. July 17 and Nov. 16, 1957; CA 58, 2557 (1963).
192. S. Adamek and A. D. Dingle, U.S. 3,033,835 (Dunlop) appl. April 1, 1959; cf. Brit. 923,411-2; CA 59, 4152 (1963).
193. Brit. 857,183 and 857,939 (Hercules), appl. March 2, 1959.
194. G. Natta, G. Mazzanti, and A. Valvassori, Belg. 569,816 (Monte), appl. July 29, 1957.
195. G. Natta, G. Mazzanti, and G. Boschi, U.S. 3,300,459 (Monte).
196. G. Natta et al, U.S. 3,489,733 (Monte) appl. Aug. 9, 1962;

     cf. U.S. 3,453,250; Italian 566,913 and 638,953 (Monte).

197. A. Valvassori and G. Sartori, U.S. 3,444,146-7 (Monte) appl.
     Dec. 13, 1963.

198. G. DiDrusco et al, U.S. 3,506,634; cf. U.S. 3,562,227 (Monte).

199. G. R. Kahle and O. G. Buck, U.S. 3,489,729 and 3,481,911
     (Phillips).

200. D. N. Matthews et al, U.S. 3,562,228 (Uniroyal).

201. J. F. Henderson and D. Carnell, U.S. 3,294,766 (Polymer Corp.).

202. A. Schrage and J. E. Schoenberg, U.S. 3,577,393 (Dart).

203. D. L. Christman and G. I. Keim, Macromolecules $\underline{1}$, 358 (1968).

204. D. N. Matthews and R. J. Kelly, U.S. 3,405,107; cf. U.S.
     3,437,645 (Uniroyal).

205. N. E. Tunnicliffe et al, Europ. Polym. J. $\underline{1}$, 259 (1965).

206. R. E. Cunningham, J. Polym. Sci. A 1, $\underline{5}$, 243 (1967).

207. S. W. Caywood, Rubber Chem. Technol. $\underline{44}$, #3, 653 (1971).

208. H. E. Diem. U.S. 3,494,983 (Goodrich-Gulf).

209. E. C. Loveless, U.S. 3,507,843 (Uniroyal).

210. R. J. Sonnenfeld, U.S. 3,539,541 (Phillips).

211. H. Schaum et al, U.S. 3,551,395 and 3,600,368 (Huels); cf.
     U.S. 3,622,548 (Huels).

212. J. R. Huerta et al, U.S. 3,595,842-4; 3,595,890 (Dart).

213. H. B. Mirza and S. Zaim, U.S. 3,600,364 (Polymer Corp.).

214. H. Schaum and G. Hoerlein, U.S. 3,551,395 (Hoechst); cf. E.
     Leblon et al, Ger. Offen. 2,109,924 (Solvay), CA $\underline{76}$, 60289 (1972).

215. C. Cozewith and G. VerStrate, Macromolecules $\underline{4}$, 482 (1971).

216. cf. G. Henrici-Olivé and S. Olivé, Adv. Polym. Sci. $\underline{6}$, 421 (1969).

217. R. J. Minchak, U.S. 3,624,056 (Goodrich).

218. R. L. Morgan, Rubber World 158, 61 and 65 (1968).

219. M. Kryszewski et al, J. Appl. Polym. Sci. $\underline{15}$, 1139 (1971).

220. M. Wismer and K. F. Schimmel, U.S. 3,582,520 (PPG).

221. R. E. Knox, Modern Plastics $\underline{49}$, 56 (February 1972); Rubber
     World $\underline{165}$, 31 (February 1972).

222. P. Longi et al, U.S. 3,658,770 and 3,660,364 (Monte).

223. M. A. Schoenbeck, U.S. 3,359,221 (Du Pont).

224. M. E. Tunnicliffe, J. Appl. Polym. Sci. $\underline{14}$, 827 (1970).

225. W. J. Keller and A. L. Moore, U.S. 3,350,370 (Du Pont).

226. J. L. Nyce and A. L. Shloss, U.S. 3,365,418 (Du Pont).

227. J. J. Murphy, SPE J. $\underline{25}$, 45 (1969).

228. D. G. McRitchie. U.S. 3,312,757 (Raybestos); C. A. Peri,
     U.S. 3,522,225 (Monte).

229. T. P. Flanagan, U.S. 3,492,372 (Natl. Starch); F. W. Bickel et
     al, U.S. 3,536,652-3 (McDonnell-Douglas); V. V. Raimondi,
     U.S. 3,539,525 (Std. Oil Ind.); P. Hamed, U.S. 3,354,107
     (Goodrich).

230. W. A. Bishop, U.S. 3,538,192 (Copolymer Rubber); also Mitsui,
     Toatsu, and other companies.

231. G. Natta et al, J. Am. Chem. Soc. $\underline{77}$, 1708 (1955).

232. G. Guzzetta and E. Ercoli, Belg. 560,523 (Monte and Ziegler).

233. G. Natta, P. Pino, and G. Mazzanti, U.S. 3,435,017 (Monte)
     appl. July 27, 1954.

234. C. W. Seelbach and L. T. Eby, U.S. 2,964,510; R. F. Leary and
     L. W. Bowman, U.S. 2,983,720 (Esso); K. O. Jung and H. Schneck,
     Makromol. Chem. 154, 227 (1972).
235. C. W. Seelbach and W. J. G. McCulloch, U.S. 2,893,984; cf.
     U.S. 2,925,392 (Esso).
236. G. Natta et al, U.S. 3,197,452; cf. U.S. 3,435,017 (Monte).
237. Belg. 561,458 (Monte and Ziegler).
238. C. Longiave and R. Castelli, U.S. 3,061,600 (Monte).
239. H. J. Hagemeyer and M.B.Edwards, Brit. 932,247 (Eastman).
240. M. K. Rosen and C. D. Mason, U.S. 3,580,898 (Allied).
241. M. H. Jones and M. P. Thorne, Can. J. Chem. 40, 1510 (1962);
     H. R. Allcock and A. M. Eastham, Can. J. Chem. 41, 932 (1963).
242. A. I. Medalia et al, J. Polym. Sci. 41, 241 (1959).
243. R. M. Nagel and M. Gabliks, U.S. 3,219,645 (Petro-Tex).
244. R. M. Nagel and M. Gabliks, U.S. 3,190,866 and 3,244,685
     (Petro-Tex).
245. R. M. Nagel, U.S. 3,314,930 (Petro-Tex).
246. C. D. Mason and R. J. Schaffhauser, J. Polym. Sci. B 9, 661 (1971)
247. G. Natta, P. Corradini, and I. W. Bassi, Makromol. Chem. 21, 240
     (1956); 35, 94 (1960); F. Danusso and G. Gianotti, Makromol.
     Chem. 80, 1-12 (1964); cf. CA 78, 4687 (1973).
248. J. Boor et al, J. Polym. Sci. A 1, 59 (1963); B 2, 903 (1964).
249. F. Danusso and G. Gianotti, J. Polym. Sci. B 3, 537 (1965).
250. T. Asada and S. Onogi, Preprints, Org. Coatings & Plastics Chem.
     32, #1, 68 (1972).
251. A. Turner-Jones, Polymer 7, 23 (1966); J. Polym. Sci. C #16,
     393 (1967).
252. C. E. Schildknecht, Polymer Preprints 13, #1, 253 (1972).
253. R. Eichenbaum and C. Geacintov, U.S. 3,489,732 and 3,562,357
     (Mobil).
254. J. W. Cleary, U.S. 3,296,232 (Phillips).
255. J. W. Cleary, U.S. 3,332,921 (Phillips).
256. Netherlands application of ICI; CA 66, 11305 (1967).
257. A. Turner-Jones, Polymer Letters 3, 591 (1965).
258. V. F. Holland and R. L. Miller, J. Appl. Phys. 35, 3241 (1964).
259. T. H. Shepherd, U.S. 3,574,044 (Princeton Chem. Res.).
260. T. Reed, Modern Plastics, October 1965 and February 1966; cf.
     A. J. Foglia, J. Appl. Polym. Sci. 11, 1 (1969).
261. F. Seifert et al, U.S. 3,546,326 (Huels).
262. J. E. Guillett and R. L. Combs, U.S. 3,519,586 (Eastman).
263. B. Diedrich et al, U.S. 3,513,143 (Hoechst).
264. R. M. Thomas and H. C. Reynolds, U.S. 2,387,784; Brit. 598,323
     (Esso).
265. R. F. Ruthruff, ACS meeting, Boston, 1939.
266. W. R. Edwards and N. F. Chamberlain, ACS Div. Org. Coatings and
     Plastics 22, 105 (September 1962); J. Polym. Sci. A 1, 2299 (1963)
267. J. P. Kennedy, R. M. Thomas et al, Makromol. Chem. 58, 28 (1962);
     J. Polym. Sci. C 4, 289 (1963); cf. W. V. Bush, U.S. 3,349,148
     (Shell).
268. J. P. Kennedy et al, J. Polym. Sci. A 2, 367, 1441, 2093 (1964);

W. R. Edwards, U.S. 3,317,501 (Esso); cf. U.S. 3,484,423 (Esso).

269. J. P. Kennedy and R. G. Squires, J. Polym. Sci. C #16, 1541 (1967).

270. A. D. Ketley, U.S. 3,251,819 (Grace); Polym. Lett. 1, 121 (1963).

271. G. Natta et al, Makromol. Chem. 30, 246 (1959).

272. F. P. Reding and C. W. McGary, U.S. 3,091,601 (Carbide); cf. Reding and McGary, Brit. 895,330 (Carbide); CA 57, 10050 (1962); J. DiPietro, U.S. 3,481,909 (Celanese); R. M. Schramm, U.S. 2,986,588 (Cal. Res.).

273. W. Herwig et al, U.S. 3,578,730 (Hoechst).

274. W. J. Polestak and K. G. Adams, U.S. 3,631,160 (Celanese).

275. A. A. Yeo et al, U.S. 3,084,206 and 3,182,096 (Brit. Pet.); J. B. Wilkes, U.S. 3,174,020-1 (Cal. Res.); A. W. Shaw et al, J. Org. Chem. 30, 3296 (1965).

276. J. P. Hogan and R. L. Banks, U.S. 2,825,721 and 2,606,940; Belg. 530,617 and 551,826 (all to Phillips).

277. G. Natta, P. Pino and G. Mazzanti, Belg. 549,891 (Monte and Ziegler), appl. July 1956; cf. Belg. 545,952.

278. W. R. Watt, J. Polym. Sci. 45, 509 (1960).

279. T. W. Campbell, U.S. 2,842,532 (Du Pont); J. Appl. Polym. Sci. 1, 78 (1959); 5, 184 (1960); cf. R. Bacskai et al, J. Polym. Sci. A 1, 10, 1529 (1972).

280. A. C. Havens, Belg. 548,964 (Du Pont); Belg. 557,115 (Carbide); T. Komaisha and M. Iwamoto, CA 74, 142631 (1971).

281. G. G. Wanless and J. P. Kennedy, Polymer 6, 111 (1965).

282. J. F. Goodrich and R. S. Porter, J. Polym. Sci. D 2, 353 (1964).

283. J. H. Griffith and B. G. Ranby, J. Polym. Sci. 44, 369 (1960); 58, 545 (1962); K. S. Chan et al, J. Polym. Sci. 61, S 29 (1962).

284. I. Kirshenbaum et al, Polym. Lett. 2, 897 (1964).

285. H. D. Noether, J. Polym. Sci. C #16, 737 (1967).

286. L. Reginato, Makromol. Chem. 132, 113 (1970).

287. T. Kamaishi and M. Iwamoto, Japan 71-20, 818 (Toray).

288. Modern Plastics, 104 (May 1967).

289. R. B. Isaacson et al, J. Appl. Polym. Sci. 8, 2789 (1964).

290. L. Bohn et al, U.S. 3,472,917 (Badische).

291. G. A. Pope, Brit. 1,105,324 (Dunlop); CA 68, 78735 (1968); L. Bohn et al, U.S. 3,472,917 (Hercules); K. J. Clark and M. E. B. Jones, U.S. 3,489,735 (ICI).

292. A. D. Caunt and J. B. Rose, in Supplementary Vol. Encycl. Chem. Tech., Kirk & Othmer, 2nd Ed., Wiley, New York, 1971.

293. K. J. Clark, U.S. 3,405,108 (ICI).

294. K. J. Clark, U.S. 3,505,430 (ICI).

295. J. P. Kennedy, J. J. Elliott, and B. E. Hudson, Makromol. Chem. 79, 109 (1964).

296. J. P. Kennedy, W. Naegele, and J. J. Elliott, J. Polym. Sci. B 3, 729 (1965); cf. A. D. Ketley, Polym. Lett. 2, 827 (1964).

297. J. P. Kennedy, C. A. Cohen, and W. Naegele, J. Polym. Sci. B 2, 1159 (1964).

298. P. Pino and G. P. Lorenzi, J. Am. Chem. Soc. 82, 4745 (1960); Makromol. Chem. 61, 207 (1963); Pino, Fortschr. Hochpolymer-

Forsch. $\underline{4}$, 393 (1965).

299. A. Abe, J. Am. Chem. Soc. $\underline{90}$, 2205 (1968); $\underline{92}$, 1136 (1970); cf. Abe and M. Goodman, J. Polym. Sci. A $\underline{1}$, 2193 (1963).

300. W. J. Bailey and E. T. Yates, J. Org. Chem. $\underline{25}$, 1800 (1960).

301. T. M. Birshtein and P. L. Luisi, Vysokomol. Soedin $\underline{6}$, 1238 (1964); CA $\underline{61}$, 10793 (1964); cf. G. Allegra et al, Makromol. Chem. $\underline{90}$, 60 (1966).

302. P. Pino, J. Phys. Chem. $\underline{72}$, 2405 (1968); P. L. Luisi and P. Pino, J. Phys. Chem. $\underline{72}$, 2400 (1968).

303. P. Pino et al, J. Polym. Sci. B $\underline{5}$, 307 (1967).

304. P. Pino et al, Makromol. Chem. $\underline{147}$, 53 (1971); cf. J. Polym. Sci. A 2, $\underline{9}$, 193 (1971); F. Ciardelli et al, Makromol. Chem. $\underline{147}$, 53 (1971).

305. G. Natta, Chim. Ind. (Milan) $\underline{43}$, 529 (1960).

306. H. I. Waterman et al, Rec. Trav. Chim. Pays-Bas, $\underline{52}$, 515 (1933); $\underline{53}$, 699 (1934).

307. E. F. Lutz and G. M. Bailey, J. Polym. Sci. A 1, $\underline{4}$, 1885 (1966).

308. G. E. Serniuk and R. M. Thomas, U.S. 3,183,217 (Esso), appl. 1961; cf. M. Hirooka, Polym. Lett. $\underline{10}$, 171 (1972).

309. J. A. Price, U.S. 3,071,568; 3,010,951 (Am. Viscose).

310. J. A. Price, U.S. 3,042,664 (Am. Viscose).

311. G. Natta, P. Pino and G. Mazzanti, U.S. 3,435,018 (Monte).

312. F. Danusso and G. Gianotti, Makromol. Chem. $\underline{61}$, 164 (1963); A. T.-Jones et al, J. Polym. Sci. B $\underline{1}$, 471 (1963).

313. C. F. Tu et al, Macromolecules $\underline{3}$, 206 (1970).

314. K. J. Clark et al, Chem. and Ind. 2010 (1962); CA $\underline{63}$, 1483 (1965).

315. Brit. 807,204 (Hercules), example 26; cf. D. Antonsen et al, Ind. Eng. Chem. $\underline{2}$, 224 (1963); Belg. 602,680 (Hoechst).

316. D. H. Antonsen and R. H. Johnson, U.S. 3,253,052; 3,259,667-8 (Sun Oil).

317. A. C. Havens, U.S. 3,257,367 (Du Pont); J. Appl. Polym. Sci. $\underline{1}$, 75 (1959); cf. A. D. Ketley and R. J. Ehrig, U.S. 3,294,771 (Grace); J. Polym. Sci. A 2, 4461 (1964).

318. S. Murahashi et al, Bull. Chem. Soc. Japan $\underline{37}$, 706 (1964); cf. J. P. Kennedy, J. Polym. Sci. A 2, 5171 (1964).

319. A. Shimizu et al, Bull. Chem. Soc. Japan $\underline{41}$, 953 (1968).

320. J. A. Price, M. R. Lytton, and B. G. Ranby, J. Polym. Sci. $\underline{51}$, 541 (1961).

321. S. I. Sadykhzade et al, CA $\underline{69}$, 87529 (1968).

322. J. Quere and E. Marechal, Bull. Soc. Chim. France 4087 (1969); CA $\underline{72}$, 44218 (1970).

323. V. S. Markevich et al, CA $\underline{76}$, 25759 (1972).

324. A. Abe and T. Hama, Polym. Lett. $\underline{7}$, 427 (1969); cf. S. T. Barsamyan et al, CA $\underline{68}$, 3287 (1968).

325. J. P. Kennedy et al, J. Polym. Sci. A 2, 5029 (1964); cf. V. I. Kleiner, CA $\underline{76}$, 100145 (1972).

326. A. D. Ketley and R. J. Ehrig, U.S. 3,297,672 (Grace).

327. J. F. Pendleton et al, Polymer Preprints $\underline{13}$, #1, 427 (1972); cf. A. Abe et al, J. Polym. Sci. B $\underline{7}$, 427 (1969).

328. S. Murahashi et al, CA 66, 65879 (1967).

329. T. Takahashi, J. Polym. Sci. A 1, 6, 403 and 3327 (1968).

330. H. D. Noether et al, J. Polym. Sci. A 1 7, 201 (1969).

331. W. H. Daly in Vol. 14, Encycl. Polym. Sci. Tech., N. M. Bikales, ed., Wiley, 1971.

332. C. P. Pinazzi and J. Brosse, Makromol. Chem. 122, 105 (1969); Polymer Preprints 13, #1, 447 (1972); R. Rossi et al, Macromolecules 5, 247 (1972).

333. K. Takomoto and M. Izubayashi, Makromol. Chem. 109, 81 (1967).

334. A. T. Blomquist et al, J. Am. Chem. Soc. 81, 2012 (1959).

335. K. Ziegler, Angew. Chem. 68, 729 (1956).

336. N. J. Roberts and A. R. Day, J. Am. Chem. Soc. 72, 1226 (1950); J. P. Kennedy and H. S. Makowski, J. Polym. Sci. C #22, 253 (1968).

337. C. S. Marvel, J. Hanley, and D. T. Longone, J. Polym. Sci. 40, 551 (1959).

338. J. P. Kennedy and H. S. Makowski, J. Polym. Sci. C #22, 247 (1968).

339. E. C. Capaldi and A. E. Borchert, U.S. 3,457,318 (Atlantic-Richfield).

340. C. E. Schildknecht, unpublished.

341. M. Feldhoff et al, U.S. 3,308,104 (Hercules).

342. K. J. Clark and T. Powell, Polymer 6, 531 (1965).

343. H. V. Holler and E. A. Youngman, U.S. 3,481,914 (Shell).

344. S. Zaim, U.S. 3,506,627 (Polymer Corp.).

345. H. Chabert et al, U.S. 3,529,038 (Rhône-Poulenc).

346. C. N. Wolt, U.S. 3,506,616 (Ethyl).

347. J. P. Hogan and R. L. Banks, U.S. 2,825,721; Belg. 530,617 (January 1955) (Phillips); J. P. Hogan et al, IUPAC (Boston) Macromol. Preprints 703 (1971); cf. J. E. Pritchard et al, Modern Plastics (October 1959).

348. M. R. Cines, U.S. 3,527,579 (Phillips).

349. J. Furukawa et al, J. Polym. Sci. A 1, 10, 681 (1972).

350. K. Shirayama et al, Makromol. Chem. 151, 97 (1972).

351. R. Bacskai, U.S. 3,311,600 (Chevron); cf. P. P. Spiegelman, U.S. 3,523,107 (Du Pont).

352. M. Feldhoff et al, U.S. 3,308,108 (Hercules).

353. T. J. Clough, U.S. 3,539,603 (Atlantic-Richfield).

354. H. M. Pitt and F. M. Pallos, U.S. 3,390,140 (Stauffer).

355. K. J. Clark, U.S. 3,492,277 (ICI).

356. K. Matsumura and O. Fukumoto, J. Polym. Sci. A 1 9, 471 (1971).

357. T. Diem et al, Makromol. Chem. 137, 61 (1970).

358. F. P. Reding and E. W. Wise, U.S. 3,197,449 (Carbide).

359. C. A. Heiberger (Airco) et al, Modern Plastics 46, 126 (June 1969); cf. A. R. Cain (Firestone), Polymer Preprints 11, 312 (February 1970); P. W. Ager and L. A. Graham, U.S. 3,607,986 (FMC); cf. R. Buening et al, CA 78, 30255 (1973).

360. W. Taylor and L. F. King, Polym. Eng. Sci. 10, 204 (1970).

361. R. V. Albarino et al, J. Polym. Sci. A 1 9, 1517 (1971).

362. H. Bartl et al, U.S. 3,597,370 (Bayer).

363. C. A. Heiberger et al, Polym. Eng. Sci. 9, 445 (1969); M. J.
      Cantow et al, Modern Plastics 46, 126 (1969); cf. U.S. 3,530,104;
      Ger. Offen. 2,006,775, CA 76, 25777 (1972).
364. A. R. Cain, CA 76, 4212 (1972); cf. A. Akimoto and T. Yoshida,
      J. Polym. Sci. 10, 993 (1972).
365. F. P. Reding et al, U.S. 3,256,256 (Carbide).
366. T. Otsu et al, J. Polym. Sci. B 2, 973 (1964).
367. D. L. Burdick et al, U.S. 3,532,656 and 3,488,311 (Gulf).
368. R. G. Hay and W. J. Heilman, U.S. 3,586,659 (Gulf).
369. R. G. Ismailov et al, CA 75, 6424 (1971).
370. F. E. Mathews and H. M. Elder, Brit. 11,635 (1914); cf. E.
      Sauter, Z. Krist. 83, 340 (1932); CA 27, 2362 (1933); cf.
      G. Vanhaeren and G. Butler, Polymer Preprints 6, 709 (September
      1965); R. Bacskai, ibid. 6, 687 (September 1965); M. A. Naylor
      and A. W. Anderson, J. Am. Chem. Soc. 76, 3962 (1954).
371. W. Crouch and J. E. Wicklatz, Ind. Eng. Chem. 47, 160 (1955);
      W. Crouch, U.S. 2,593,414, 2,602,787, and 2,686,171 (Phillips).
372. J. R. Brown and J. H. O'Donnell, Macromolecules 3, 266 (1970)
      and 5, 109 (1972); cf. M. A. Jobard, J. Polym. Sci. 29, 275
      (1958); cf. J. R. Brown et al, CA 78, 30358 (1973).
373. E. J. Vandenberg, U.S. 2,837,496 (Hercules); G. Natta et al,
      J. Polym. Sci. 34, 685 (1959); M. Baer, J. Appl. Polym. Sci. 16,
      1125 (1972); T. O'Neill, J. Polym. Sci. A 1, 10, 569 (1972).
374. H. Jabloner and R. H. Mumma, J. Polym. Sci. A 1, 10, 763 (1972).
375. N. G. Gaylord, Polym. Lett. 10, 95 (1972).

3. ALLYLIC DIOLEFINS

In contrast to conjugated dienes, the nonconjugated allylic diolefins have received little attention until recently in polymerization. Allylic polyfunctional unsaturated hydrocarbons are considered here as those bearing a terminal double bond and one or more double bonds not conjugated with the terminal allyl double bond. Although information is far from complete, the odors, rearrangements, polymerization behavior, and other addition reactions of many of these compounds indicate that they should be regarded as allyl compounds.

In general, radical-initiated homopolymerizations and copolymerizations do not occur readily, but polymerizations occur under the influence of electrophylic catalysts (e.g., cationic and especially Ziegler-Natta or coordination systems). Numerous examples of cyclopolymerizations have been reported, but few reactions have been studied closely to find the precise mechanisms involved.

The area of greatest industrial interest and widest research has been the copolymerization of ethylene, propylene, and minor proportions of polyfunctional allylic alkenes using complex vanadium-containing catalysts for the synthesis of EPDM vulcanizable elastomers, as discussed in the preceding chapter. An area of special scientific interest is the nonvinyl addition polymerization of dimercaptans with nonconjugated dienes, investigated by Marvel and his students. These reactions in free radical systems, giving polymers of low to moderate molecular weights, have been called hydrogen-transfer or H-polymerizations. They are examples of slow stepwise polymerization with radical initiation.

DIALLYL

Diallyl, or 1,5-hexadiene $(CH_2=CHCH_2)_2$, has been prepared in the laboratory from allyl bromide or iodide and sodium (1). It can be made conveniently from the action of magnesium metal on allyl chloride in ether (2). It was obtained from passing allyl chloride vapor over hot copper

135

(3) or zinc (4). Diallyl was also made by pyrolysis of
the diacetate of 1,6-hexanediol (5), from allyl acetate
and $Ni(CO)_5$ (6), and, in low yields, directly from allyl
halides (7). Diallyl has a fairly sharp allylic odor
(radishlike) in contrast to the more pleasant terpene or
camphorlike odor of dimethallyl. Diallyl is a water-
insoluble liquid that is miscible with hydrocarbons,
ethers, and many organic solvents. Like many allyl
compounds it isomerizes readily to propenyl products,
e.g., at 60°C over heavy metal catalysts on alumina (8).
    Diallyl has chemical properties contrasting sharply
with those of conjugated dienes. It does not homopolymerize readily in conventional radical-emulsion systems
nor by proliferous or popcorn polymerization. Diallyl
adds halogens or active hydrogen compounds readily.
With dry hydrogen bromide, a mixture of bromine compounds
was obtained by normal and reverse addition (9). Hydrogen bromide in acetic acid at room temperature gave
2,5-dibromohexane and some 5-bromo-1-hexene (10) in
contrast to reverse addition in UV light (11). Chloroform added to diallyl in presence of a peroxide to give
$H_2C=CHCH_2CH_2CH_2CH_2CCl_3$; carbon tetrachloride gave
$(Cl_3CCH_2CHClCH_2)_2$ (12).
    With ozone, diallyl formed an explosive, syrupy di-
ozonide (13). Mercury-sensitized irradiation of diallyl
gave allylcyclopropane and other products (14). Irradiation in air formed hydroperoxides (15). Diepoxides have
been prepared and polymerized (16). Diallyl reacts with
N-bromosuccinimide to give bromo-1,4-hexadienes (17). On
heating with catalysts, 1,5-hexadiene and derivatives may
rearrange to conjugated 2,4-hexadienes (18). Only one
double bond of diallyl reacted with maleic anhydride in
benzene at 160°C, instead of normal Diels-Alder addition
(19).

DIALLYL HYDROCARBON POLYMERIZATIONS

    Diallyl and related nonconjugated dienes generally homo-
polymerize only very slowly by radical initiation and by
heating. However, on heating with strong Lewis-acids such
as concentrated sulfuric acid, violent reactions with
complex products can occur. Heating alone above 200°C
gives slow polymerization. Apparently high polymers of
established structure have not been prepared using conventional catalyst systems. Small additions of certain
diallylic hydrocarbons do not excessively retard vinyl
polymerizations; indeed, they may be beneficial. With
little reduction of initial rates of polymerization,
addition of 0.05% 1,4-cyclohexadiene "damped" the Trommsdorff or gel effect in methyl methacrylate radical
polymerization (20).

In Du Pont laboratories in 1954 diallyl was polymerized by coordination catalysts to form colorless, elastomeric high polymers (21). A catalyst effective in one hour at room temperature in cyclohexane was prepared from lithium aluminum tetrabutyl and titanium tetrachloride. The same patent described polymerization of 4-vinylcyclohexene with similar catalyst at 128-140°C during 3 hr giving a solid, predominantly 1,4-polymer. It was molded at 200°C forming tough, stiff films. Such polymerizations, including those by Ziegler-Natta catalysts, may involve formation of donor-acceptor complexes between the monomer and an electrophilic catalyst component. Cationic polymerizations of such allyl monomers to products of high molecular weight cannot be accomplished by use of ordinary Lewis-acid catalysts.

Marvel and Vest were able to prepare soluble polymers containing cyclic recurring units from some 2,6-disubstituted-1,6-heptadienes by free radical initiation (22). Marvel and Stille prepared partially soluble polymers believed to contain carbocyclic rings from diallyl and other nonconjugated diolefins by the use of Ziegler-Natta transition metal catalysts (23). Polymers of highest DP (η_{sp} 0.4 in benzene) were obtained by high aluminum isobutyl to titanium chloride molar ratios (e.g., 3:1). A 1:1 molar ratio produced no polymerization. The polymer viscosities were not related to polymerization temperatures (30-60°C) nor to the proportion of heptane solvent. Infrared evidence indicated that from 4 to 10% of the monomer units in these polymers retained double bonds, along with cyclic units free of double bonds. The tough, white polymers had capillary melting points above 200°C. Transparent films could be deposited from benzene solutions from soluble portions of the diallyl polymers. A polymer from 1,6-heptadiene was dehydrogenated to form

segments of the type . Marvel and Stille, using a similar catalyst from aluminum triisobutyl and titanium trichloride, could only prepare polymers of very low molecular weight from dimethallyl.

Valvasori and co-workers prepared ether-soluble homopolymers from diallyl using $Al(C_2H_5)_2Cl$ with vanadium acetylacetonate as catalyst (24). Crystalline high polymers from diallyl having inherent viscosities up to 2 were prepared at 60°C using transition metal catalysts (25). The molded polymers were high melting and strong, yet flexible. The major chain segments were believed to contain cyclic units. Shrinkage during polymerization was unusually great (e.g., 38%). At 400°C the polymers volatilized less rapidly than linear polyethylenes.

Densities above 1.0 g/cc and the surprising flexibility of moldings confirmed a high proportion of 5-numbered rings along with some remaining allyl groups. Copolymers of ethylene with 1,5-hexadiene were crystalline and largely insoluble in boiling n-heptane (26). These block copolymers melted above 117°C.

Crystalline polymers were prepared from 1,5-hexadiene and from 1,6-heptadiene by Olson (27). The soluble polymers formed in heptane at 30°C during 18 hr with a catalyst from aluminum triisobutyl and titanium tetra-chloride were believed to have cyclic structures:

$$--CH_2\!-\!\bigpentagon\!-- \qquad\qquad --CH_2\!-\!\bighexagon\!--$$

In another example, the catalyst was aluminum isobutyl-vanadium trichloride. This also gave a crystalline polymer, but IR data showed 2.7% of vinyl groups corresponding to 7.9% 1,2-addition. Toluene-insoluble fractions were highly crystalline and substantially free of methyl and vinyl groups. Higher nonconjugated dienes such as $A(CH_2)_nA$, where n=4 to 12, formed soluble polymers apparently through reaction of only one double bond (28).

COPOLYMERIZATIONS OF DIALLYLIC HYDROCARBONS

Of greatest interest have been copolymerizations with larger proportions of ethylene and/or propylene, using transition metal halide catalysts of the Ziegler type to give curable plastics and elastomers. The ethylene-propylene-diallyl terpolymers prepared, for example, using vanadium halide-aluminum alkyl catalysts and cured with sulfur were among the most promising early ethylene-propylene terpolymer rubbers. Addition of small amounts of dimethacrylates along with high-temperature peroxides can accelerate cure and give less discoloration than sulfur curing (29). Curable polyethylenes have been developed by Phillips by copolymerization with small proportions of diallyl. Curable, heat-resistant EP ter-polymer rubbers using minor proportions of nonconjugated dienes have been developed by Du Pont, Dunlop, Esso, Goodrich, and many others (see Chapter 2).

Sulfur-vulcanizable elastomers were prepared by Gresham and Hunt from equal volumes of ethylene and propylene together with diallyl, using $LiAlH_4$-$TiCl_4$ catalyst (30). The temperature rose spontaneously during copolymerization from 35 to 48°C. Natta, Mazzanti, and co-workers made copolymers from ethylene, propylene, and diallyl in molar ratio 1:4:0.168 in n-heptane diluent and in the presence of a catalyst prepared from aluminum diethyl chloride and

vanadium triacetylacetonate (31). Delayed addition of
diallyl along with arylamine antioxidant was described in
one example. One of the amorphous terpolymer elastomers
had an intrinsic viscosity of 3.0 (tetralin at 30°C) and
prominent infrared bands at 6.08, 10, and 11 μ. Another
patent described a catalyst from aluminum diethyl chloride
and vanadium trichloride-trimethyl amine complex prepared
at -10 to -50°C. The ethylene-propylene mixture was added
continuously to diallyl and catalyst to form terpolymers
containing up to 5% of diallyl units (32).

Although the early terpolymer rubbers were based on
ethylene-propylene-diallyl, the less reactive 1,4-hexa-
diene, cyclopentadiene, and other nonconjugated dienes
were later found to be more desirable (33). Such EPDM
abrasion-resistant rubbers may be blended with natural or
butadiene-styrene rubber before curing with sulfur formulations (34). Copolymers of 4-methyl-1-pentene with 0.02
to 0.08 mole % alpha-omega diene made with Lewis-acid
catalysts at -35 to -70°C have been suggested as lubricating oil additives (35).

Ethylene-diallyl copolymers were prepared in Holland
using modifications of Ziegler processes (36). A terpolymer of 55% ethylene, 30% propylene, and 15% diallyl
units prepared by Huels could be vulcanized by sulfur or
peroxide formulations (37). Shell copolymerized diallyl
with a series of different olefins and nonconjugated
diolefins, using aluminum alkyl-titanium tetrachloride
catalysts (38). Oil-soluble copolymers of diallyl were
prepared by special $AlEt_3$-$TiCl_4$ catalyst further activated
by $AlEt_2Cl$ (39). These noncrystalline polymers were
believed to contain cyclic structures. Ethylene-diallyl
copolymers were prepared having block copolymer structures,
but the blocks probably were not perfectly homopolymeric
(40).

Copolymers of high diallyl to ethylene were prepared in
heptane with catalyst from $VOCl_3$ and CH_3AlBr_2 at 5°C (41).
They were suggested for packaging films and molded
plastics. Tacky vulcanizable rubbers were made at 5 to
50°C from mixtures of diallyl with butadiene or isoprene
by use of catalysts from reaction of aluminum isobutyl
with titanium tetrachloride (42).

Reactions of diallyl with maleic anhydride at high
temperatures were studied by Alder, but copolymers were
not characterized (43). Terpolymers of ethylene, maleic
anhydride, and different proportions of diallyl were
prepared by benzoyl peroxide catalysis at 70°C under
pressures of 20 atm (44).

Diallyl additions in minor proportions to free radical
polymerizations have interest for controlling branching,
crosslinking, and grafting. The monomer has been used in

synthesis of styrene graft copolymers for use in high-impact polyblend plastics. In butadiene emulsion systems, diallyl retards polymerization but is not a strong chain-transfer agent (45). In copolymerizations in bulk with vinyl acetate at 70°C in the presence of BP, diallyl produced less retardation of rate than did conjugated dienes, 1-hexene, or 2-hexene (46). Other radical co-polymerizations of diallyl have been reported with N-vinyl pyrrolidone (47) and with acrylonitrile (48).

Diallyl has been copolymerized with sulfur dioxide using peroxide and azo initiators at room temperature for 24 hours (49). The products contained two SO_2 units per diolefin unit. Fractions of about 25% or less retained solubility in ketone solvents or in sulfuric acid. Soluble fractions had molecular weights as high as 125,000 and inherent viscosities up to 1.5. The suggested structure for the soluble fractions was

OTHER ALLYLIC POLYENES

Dimethallyl or 2,5-dimethyl-1,5-hexadiene is a liquid with mild, camphor-like odor. It is insoluble in water and soluble in hydrocarbons. It shows strong IR bands at 6.9, 7.2, and 11.2 μ. This and other dienes can be identified by exhaustive oxidation by permanganate or dichromate giving organic acids, ketones, and carbon dioxide. On heating with bronze powder it isomerizes to the conjugated diene 2,5-dimethyl-2,4-hexadiene. Both diallyl and dimethallyl have been obtained in low yields from mixtures of propylene or isobutylene with vaporized hydrogen peroxide heated above 400°C (50). Less hazardous methods of preparation have been discussed (51). Dimeth-allyl has been prepared from methallyl chloride and magnesium (52) by dehydration of diols (53), also by reaction of methallyl chloride with $Ni(CO)_4$ (54). Di-methallyl can be made by procedures similar to those disclosed for preparing diallyl. It has been used for synthesis of pharmaceuticals.

Dimethallyl was polymerized with BF_3 and $TiCl_4$ as catalysts at -78 to +20°C (55). Reactions without solvent and with heptane or dichloroethane as solvents produced soluble polymers of low double-bond content.

Dimethallyl has been reported to give polymers of low DP and low melting point by use of Lewis-acid or Ziegler-type catalyst (56). Dimethallyl gave no polymer with sodium but polymerized near -78°C with BF_3 to give solid, non-adhesive polymers (57). With 5% benzoyl peroxide, diallyl showed little polymerization after long irradiation with UV light; long heating with 5% t-butyl perbenzoate at 80°C gave low yields of soluble, film-forming polymer (57).

Small proportions of dimethallyl have been copolymerized with isobutylene using $AlCl_3$ at low temperatures (58). Dimethallyl was copolymerized with larger proportions of acrylate esters by heating at reflux in benzene and other solvents containing t-BHP (59). The soluble copolymers obtained could be crosslinked by heating at 250°C. Co-polymerization occurs with maleic anhydride in dioxane under high-energy radiation (60). The crosslinked products were suggested as ion-exchange resins.

Crystallizable stereopolymers were reported from 1,4-pentadiene by using $Ti(OR)_4$ with AlR_3 as catalyst (61). Allyl cyclopentadiene (boiling point 26°C at 13 mm) was observed to polymerize on storage (62). The monomer could be reacted with maleic anhydride to form an allyl tetra-hydrophthalic anhydride. Allyl, methallyl, and allylmethyl derivatives of cyclopentadiene were polymerized using BF_3 (63). The three polymers were soluble in benzene, and NMR studies of their structures indicated greater proportions of 1,4-structure with increasing steric hindrance in the monomers.

Allylisopropenyl or 2-methyl-1,4-pentadiene has been prepared by dehydration of dimethyl allyl carbinol with sulfuric acid (64). On oxidation by permanganate it gave a mixture of formic, acetic, and oxalic acids. Maleic anhydride and 1,4-pentadiene formed copolymers of 2:1 molar composition by radical polymerization in benzene or xylene (65). The reactions were fairly slow and the inherent viscosity of the products was below 0.30 in DMF. Cyclopolymerization was suggested.

1,4-Hexadiene forms in good yield from reaction of ethylene and butadiene in the presence of organorhodium catalysts (66). On heating, the 1,4-compound isomerizes to 2,4-hexadiene. By passing 1,4-hexadiene over catalysts at 200°C, the proportions of cis to trans isomers can be controlled (67). The trans isomer (which is less avail-able) is preferred for preparing ethylene-propylene terpolymer elastomers. The formation of trans-1,4-hexadiene from ethylene and butadiene in the presence of $PdCl_2$ + R_2AlCl may be explained by formation of a syn-π allyl complex with butadiene by the bulky palladium atom (68). Similarly high trans-to-cis ratios were obtained using nickel compounds with R_2AlCl and a t-phosphine (69). A different catalyst, rhenium acetylacetonate with AlR_3, gives stereoselective synthesis of cis-1,4-hexadiene (70).

Other allylic dienes which have been much evaluated as termonomers for synthesis of EPDM elastomers have been derivatives of norbornenes. These are bicyclic compounds, as in the following formulas:

$=CH_2$ / CH₂   =CH$_2$ 5-methylene-2-norbornene (MNB)

CH₂   CH=CH$_2$ 5-vinyl-2-norbornene (VNB)

CH₂   =CHCH$_3$ 5-ethylidene-2-norbornene (ENB)

In the presence of coordination or Ziegler-Natta catalysts predominantly, one allylic double bond is reactive while the second double bond of the ring is effective in forming crosslinks during vulcanization of the ethylene-propylene-diene-terpolymer. Analysis of the unsaturation of the unvulcanized terpolymers may be based on the C=C stretching vibration near 6μ or upon the =C- out of plane deformation between 11 and 12μ in the IR spectra (71).

Tetraallyl methane has been prepared by reacting pentaerythrityl bromide with vinyl magnesium bromide in tetrahydrofuran, and the monomer has been copolymerized with acrylic acid (72). Tetraallylmethane was also made by reaction of triallylmethyl magnesium chloride with allyl bromide for 5 hr at room temperature (73).

Homopolymers and copolymers have been prepared from 5,7-dimethyl-1,6-octadiene using aluminum alkyl-vanadium oxychloride catalyst (74). Ziegler-Natta catalysis was used to prepare soluble polymers of intrinsic viscosity 0.46 from 3-vinyl-1,5-hexadiene (75). These were believed to contain 2,6-linked bicycloheptyl groups. Butler polymerized trans-1,3,8-nonatriene by means of Ziegler-Natta catalyst to obtain partially soluble polymers (76). The cis isomer gave only a low yield of crosslinked polymer. Low yields of partially soluble polymer were obtained from 1,3,7-octatriene. Cyclic structures were believed to contribute to solubility; however, all the soluble polymers had low molecular weights (e.g., η_{inh} 0.15 and lower).

Marvel and Gall provided additional evidence that 6-membered rings are present in partially cyclic polymers derived from diallyl compounds (77). Purified 2,5-diphenyl-1,5-hexadiene (mp 51°C) was polymerized with a

variety of catalysts, giving soluble polymers of low
molecular weight. The monomer in methylene chloride was
treated with BF_3 near -78°C and allowed to stand for an
hour. After quenching the residual catalyst by methanol
and drying, a polymer melting above 140°C was obtained.
Titanium tetrachloride catalyst formed polymer of lower
viscosity and melting point. Lithium phenyl as catalyst
gave a polymer that melted above 275°C and showed 5 to
10% residual unsaturation. Heating 2,5-diphenyl-1,5-
hexadiene for 5 days at 100°C formed a polymer melting in
the 240-270°C range. Ziegler-Natta catalyst resulted in
polymers of low inherent viscosity (0.07).

Persulfate-initiated polymerization of 1,2-dimethylene
cyclopentane at 50°C for 24 hr gave rubberlike soluble
polymers, largely by 1,4-addition according to infrared
spectra (78). Polymerization of cis-1,3-divinylcyclo-
hexane using Ziegler-Natta catalyst gave soluble polymers
having low unsaturation and believed to contain cyclic
structure (79). In polymerization behavior, 3-methylene-
cyclobutene seems to behave more as an allyl compound
than as a conjugated diene (80).

Polymerization of 1-methylene-4-vinylcyclohexane using
BF_3 in CH_2Cl_2 led to a solid, soluble polymer of softening
temperature about 150°C (81). NMR spectra showed one
residual double bond to two monomer units. The same
monomer by use of Ziegler-Natta catalyst in n-heptane gave
solid polymer which was only about 10% soluble in solvents.
Intrinsic polymer viscosities were only 0.11 and lower.
Heating the monomer two weeks at 68°C with BP gave no
increase of viscosity by free radical polymerization.
Polymerization largely to soluble polymers occurred with
2-allyl-1-methylenecyclohexane and BF_3 catalyst (82). By
Ziegler-Natta catalyst a low yield of soluble polymer was
obtained from 1,3,9,11-dodecatetraene (83).

Dimerization of butadiene at 140°C in the presence of
polymerization inhibitors gives 4-vinylcyclohexene (84).
Polymerizations of 4-vinylcyclohexene using BF_3, BF_3
etherate, as well as Ziegler catalyst gave low conversions
to partially crosslinked and partially soluble polymers of
low molecular weight (85). The polymers were considered
to have predominantly cyclic units and to be essentially
saturated. Polymers of high molecular weight were prepared at 50°C using a catalyst from $SbCl_5$ and $AlEt_2Cl$ in
n-hexane solution (86). See page 137 for polymerization
of 4-vinylcyclohexene using coordination catalysts. The
monomer was oxidized to 4-vinyl epoxycyclohexane which was
copolymerized by heating with maleic anhydride in benzene
using azobis as initiator (87). Copolymers of only low
molecular weight were obtained from 4-vinylcyclohexene
and maleic anhydride by heating in dioxane or acetone with

peroxide or azo catalyst (4 hr at 70°C) (88). Recently
3-vinylcyclopentene and 3-vinylcyclohexene have been
isomerized to ethylidene compounds by heating with sodium
in hexamethyl phosphoramide (89). Isomerization must
have occurred in the observed polymerization of 3-vinyl-
cyclopentene with sodium naphthalene as catalyst. Diallyl
benzenes have been polymerized by transition metal catalysts giving films melting above 230°C (90). Some additional references to diallylic hydrocarbons are given in
Table 3.1.

Note that 1,2-polybutadienes have allylic structure
--CH$_2$CH-- , and give some reactions of polyfunctional
      |
     CH=CH$_2$
allyl compounds. Viscous liquid butadiene polymers of low
molecular weight and about 85% vinyl unsaturation along
with curing agents are used for coatings, adhesives,
printed circuits, and in other electronic applications.
With specific terminating groups "telechelic" polybutadienes provide a variety of more complex polymers bearing
allyl groups. NMR spectra of 1,2-polybutadienes of
different tacticities are shown in Fig. 3.1. The much
simpler NMR spectrum of 1,4-dimethylene cyclohexane
appears in Fig. 3.2.

Polymerizations of allene or propyne CH$_2$=C=CH$_2$ are
interesting in comparison to those of allyl and conjugated
diolefins. Pyrolysis of isobutene, butene, or 1-butanol
may give allene (bp, -27°C; mp, -105°C) along with methylacetylene. However, allene is best made by reacting
1,1-dibromocyclopropane with sodium or magnesium. With
acid catalysts, allene ordinarily gives only liquid dimers
and other low polymers. Uncrosslinked allene polymers of
high molecular weight were prepared by Baker using TiCl$_4$
and VOCl$_3$ with AlR$_3$ and especially pure monomer (91). A
variety of solvents served for the polymerization. Polymers of intrinsic viscosity 1 to 5.5 (in bromobenzene at
110°C) showed crystal melting points near 122°C. Films
were crystalline, clear, and colorless, but brittle; they
were crosslinked by heating. Nickel-containing catalysts
gave crystalline polymers from allene having side methylene groups (92). Allene polymerized in methanol with
rhodium trichloride catalyst (70°C, 10 hr) gave a low
yield of polymer melting at 120°C (93). The η_{sp}/C was
1.0 using 0.5% solution in tetrachloroethylene at 65°C.
Polyallenes employed as crosslinking agents were made in
acetic acid solution using a catalyst containing a Pd
salt (94). These polymerizations were said to occur in
a head-to-head manner, giving pendant methylene groups.
Polyallene reacted with maleic anhydride in Diels-Alder
manner, and the products with ammonia or amines gave
water-soluble polymers (95). The isomerization of allyl-

TABLE 3.1

Properties of Some Nonconjugated Diene Hydrocarbons

	bp, °C	Refractive indices	References, etc.
Vinyl allyl or 1,4-pentadiene	26	$n_D^{20} = 1.3883$	Benson, Org. Syn. 38, 78 (1958); Pfaltz & Bauer (supplier)
Diallyl	60	$n_D^{20} = 1.4044$	Marvel, JACS 80, 1742 (1958); cf. Streiff et al, J. Res. Nat. Bur. Std. 45, 173 (1950).
Allyl propenyl	6?	--	Griner, Ann. Chim. [6] 26, 332 (1892)
Allyl isopropenyl	59	$n_D^{17} = 1.4081$	Saizew, Ann. 185, 156 (1877); Bacon, J.Chem.Soc. 1072 (1937)
Allyl methallyl	92	--	Perkins, J. Chem. Soc. 87, 658 (1905)
Dimethallyl	115	$n_D^{25} = 1.4310$; $n_D^{21} = 1.4399$	O. Schales, Ber. 70, 119 (1937); Mueller, Chem. Ber. 90, 543 (1957)
1,6-Heptadiene	90(75)	$n_D^{26} = 1.4142$	Henne and Greenlee, JACS 65, 2020 (1943); Marvel and Stille described purification
2,4-Dimethyl-pentadiene	92	$n_D^{12} = 1.4468$	Fellenberg, Ber. 37, 3579 (1904)
p-Diallyl benzene	94(12)	$n_D^{20} = 1.5250$	Quelet, Bull. Soc. Chim. [4] 45, 266 (1929): aniselike odor, copolymers; U.S. 2,687,383 (Koppers)
9,10-Diallyl anthracene	--	--	Kivac, CA 74, 76773 (1971); fluorescent in PMMA
5,5'-Diallyl diphenyl derivatives	--	--	Runeberg, Acta Chem. Scand. 12, 188 (1958); magnolia bark
Biphenylallyl $(CH_2=CCH_2)_2$	--	--	Kolobielski, JACS 79, 5820 (1957)
cis-1,3-Divinyl cyclohexane	164	$n_D^{20} = 1.4615$	Corfield, J. Macromol. Sci.-Chem. A5, 21 (1971)

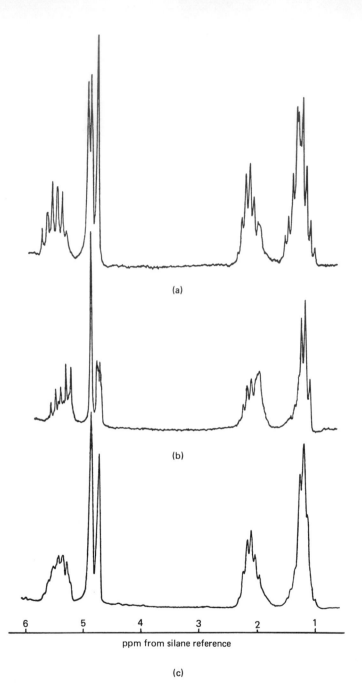

6   5   4   3   2   1

ppm from silane reference

(c)

Fig. 3.1.   Complex 100-MHz NMR spectra of (a) isotactic 1,2-poly-
butadiene; (b) syndiotactic 1,2-polybutadiene; (c) atactic 1,2-
polybutadiene (in o-dichlorobenzene at 120°C).   J. Zymonas and
H. J. Harwood, Polym. Preprints 12, 335 (September 1971).

146

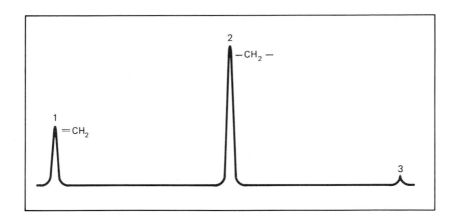

Fig. 3.2. The simple NMR spectrum of 1,4-dimethylenecyclohexane (10% in CCl_4) showing only two types of protons in $=CH_2$ and in $-CH_2-$ groups, together with the small peak at 3 of the reference tetramethylsilane. The frequency was 60 MHz. L. E. Ball and H. J. Harwood, Polym. Preprints 1, 65 (March 1961).

acetylene to vinylallene was studied in France (96). Allylacetylene (bp 42°C; $n_D^{19} = 1.4152$) and diallylacetylene have been little studied and no polymers of industrial interest have been reported (97).

POLYMERIZATION OF ALLYLIC DIENES WITH DITHIOLS

Von Braun and Murjahn had shown that allyl mercaptan, as well as cinnamyl mercaptan, slowly polymerize to low-molecular-weight polymers during distillation (98). Low polymers were prepared by Vaughn and Rust from diallyl and hydrogen sulfide at 0°C in light (99). Marvel and co-workers of the University of Illinois succeeded in preparing relatively high polymers by hydrogen-migration copolymerization of diallyl and other nonconjugated dienes with dimercaptans. They reported characteristics of these interesting polymerizations:

$$CH_2=CHCH_2CH_2CH=CH_2 + HS(CH_2)_nSH \longrightarrow --CH_2CH_2CH_2CH_2CH_2CH_2S(CH_2)_nS--$$

Free radical conditions catalyzed these addition polymerizations in the Posner or reverse manner, which seem to occur by a slow stepwise or "step growth" mechanism. Conjugated dienes also may give such polymerizations with dimercaptans if precautions are taken to prevent competing

vinyl-type addition polymerizations. Apparently anionic
polymerization conditions also can be used. There is
some evidence that, in contrast to typical vinyl-addition
polymerization, the separation of solid polymer from these
slow, stepwise polymerizations stops the growth of the
macromolecules tending to limit molecular weights.

Marvel and Chambers obtained a little polymer from diallyl (DA) and hexamethylene dimercaptan by irradiation
with mercury light through glass in closed test tubes
(100). However, direct UV irradiation of cyclohexane
solutions in open beakers gave an immediate precipitate
of white polymer while polymer of lower molecular weight
remained in solution. Reaction times as long as 48 hr
were used. Chloroform was less satisfactory as a solvent,
since it gave yellowish polymers of low molecular weight.
The polymers of around 1400 average molecular weight were
waxy, whereas the sulfide polymers of 8000 to 14,000 mol.
wt. were solids less soluble in petroleum ether.

Marvel and Chambers proved that the polymers are linear
sulfides resulting from Posner-type addition by demonstrating that they are identical to the condensation
polymers prepared from reaction of the dimercaptan with
1,6-dibromohexane in presence of alkali. The x-ray
patterns and IR spectra were substantially the same. A
series of the linear polysulfides from different linear
dithiols and diallyl hydrocarbons gave capillary melting
points that fell in a rather narrow range. For example,
ethanedithiol + DA gave polymers melting at 82 to 86°C,
hexamethylenedithiol + DA at 71 to 76°C, and decamethylene-
dithiol + 1,10-undecadiene at 74 to 78°C. Inherent viscosities of the polymers in solution were 0.29 and lower,
but a crystalline x-ray pattern was obtained as in Fig.
3.3. Sulfide polymers containing p-phenylene groups in the
main chain melted at much higher temperatures. The polymer

from hexamethylenedithiol and $CH_2=CH\langle\bigcirc\rangle CH_2\langle\bigcirc\rangle CH=CH_2$ did

not melt even at 220°C. Copolymers from DA and branched
dienes such as 3,4-dimethyl-1,5-hexadiene reacting with a
dimercaptan were rubbery products of lower melting temperatures. Oxidation of some of the linear sulfide polymers
gave sulfone polymers soluble only in hot m-cresol.

Marvel and Aldrich (101) found that aqueous persulfate
emulsions similar to those used to prepare butadiene-
styrene synthetic rubber gave somewhat faster polymerizations of diallyl with hexamethylenedithiol and higher
viscosity products than obtained from photochemical reaction in solution. Certain emulsion reactions at 30°C
for 4 days with a slight excess of diallyl gave polysulfide
inherent viscosities of 0.9 to 1.4 (0.4% in benzene). In
the acidic persulfate, emulsion polymerizations at 30°C

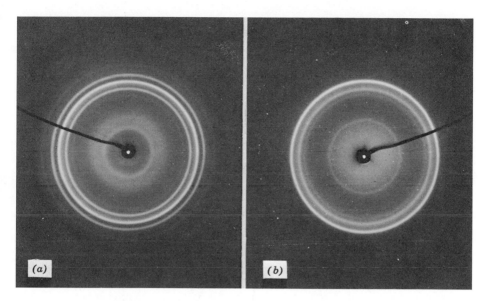

Fig. 3.3. X-Ray diffraction patterns of crystalline sulfide polymers:
(a) from ethanedithiol and diallyl, (b) from hexamethylenedithiol and
diallyl. From thesis of R. R. Chambers, Univ. of Illinois; cf. C. S.
Marvel and R. R. Chambers, J. Am. Chem. Soc. 70, 995 (1948)

about 10% of precoagulation occurred during 3 to 4 days of
reaction. Oxidation of the sulfide polymers by ozone or
permanganate gave sulfones containing some sulfoxide groups,
along with considerable degradation of the polymer chains.
The sulfide polymers of higher inherent viscosity than
about 0.36 could be cold drawn to oriented fibers.

In the persulfate emulsion polymerizations of diallyl
with hexamethylenedithiol, highest molecular weight polymers were formed by using pH 3.0 to 3.7 (102). Below 2
and above 7.0 polymer viscosities fell off sharply
(Fig. 3.4). Adding very small amounts of the vinyl polymerization inhibitor p-t-butylcatechol promoted formation
of high polymers. The following recipe in water, buffered
to initial pH 3.0 to 3.7, gave best results:

"Monomers"	4.3 g
Emulsifier MP 189	0.5 g
$(NH_4)_2S_2O_8$	0.0215 g
$NaHSO_3$	0.0108 g
$CuSO_4$	0.0011 g
p-t-Butyl catechol	0.007 g

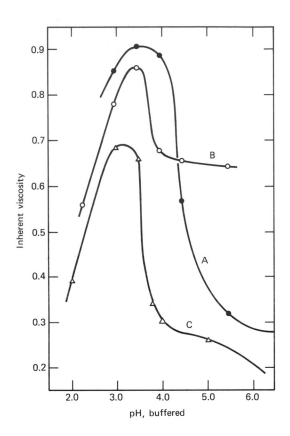

Fig. 3.4. Striking effects of pH in the persulfate emulsion polymerization of diallyl with hexamethylene dithiol upon inherent
viscosity and polymer molecular weight: curve A, acetate buffer;
curve B, phthalate buffer; curve C, citrate buffer. The maxima
may possibly result from optimum states of dispersion in contrast
to partial coagulation. C. S. Marvel and G. Nowlin, J. Am. Chem.
Soc. <u>72</u>, 5027 (1950).

The reactions were run under nitrogen in screw cap bottles
at 30°C for 5.5 days and then the latices were coagulated
by alum.

Marvel and Markhart discovered that in the emulsion
polymerization of diallyl and hexamethylenedithiol at 30°C
as much as 96% of the thiol groups react in the first 16
min, but the reaction was slow thereafter (103). The
polymer formed after a few minutes had inherent viscosities
of 0.4 to 0.7, but many more hours were required to reach
values of 1.0 and above. Polymerizations of other dithiols
with diolefins showed similar behavior. It was suggested
that slow reaction of terminal SH groups to form disulfide
linkages doubled the chain length (104). Hydrogenation
could cleave the high polymer which could then be oxidized
again by iodine to give inherent viscosity above 1.0 once
more. Azobis initiator could be employed for reactions of
18 hr at 50°C to give $\eta = 0.65$, which after oxidation by
iodine in chloroform gave polymers of $\eta = 1.0$.

Marvel and his students reacted diallyl and other diolefins with a variety of dithiols by emulsion to give novel
rubbers (105) and fibrous (106) polymers. The disadvantages of long reaction times and relatively low polymer
viscosities were not entirely overcome. In the reaction
of hexamethylenedithiol with the diolefinic compound
dibenzal acetone, Wexler found that polymerization in dry
benzene in the presence of piperidine gave relatively fast
rates (107). Highest inherent viscosity of 0.24 was
obtained with 1:1 initial molar ratios.

Mercaptans will add also to allylic double bonds in
preformed high polymers. For example, polybutadiene added
2.8 to 12.3% of 2-mercaptoethanol, mercaptoacetic acid, or
β-mercaptopropionitrile (108). Dioxane solutions with
azobis catalyst were heated 2 days at 50°C. The oil resistance of the polybutadiene rubbers was slightly improved
by reaction with such polar mercaptan compounds.

Marvel, Kotch, and Krainman studied H-polymerization of
diallyl with dithiodicarboxylic acids such as dithioadipic
acid (109):

$$CH_2{=}CHCH_2CH_2CH{=}CH_2 + HS\overset{O}{\overset{\|}{C}}(CH_2)_4\overset{O}{\overset{\|}{C}}SH \longrightarrow {-}{-}CH_2CH_2CH_2CH_2CH_2CH_2S\overset{O}{\overset{\|}{C}}(CH_2)_4\overset{O}{\overset{\|}{C}}S{-}{-}$$

Again the reactions had characteristics of slow stepwise
polymerization in which polymer growth seemed to stop when
polymer precipitated out. Benzene solutions of the reactants could be irradiated by UV light to give inherent
viscosities up to 0.21 (in chloroform). Benzene, cyclohexane, and other nontransferring solvents gave highest
inherent viscosities while chloroform gave poor results.
The solid polymers usually first deposited upon the vessel
walls. Infrared investigation showed no side methyl groups,

indicating that reaction occurred by Posner or reverse
addition. The thiol ester polymers melted higher than
the oxygen analogs. They were relatively stable against
aqueous hydrolysis at room temperature but were saponified
by 5% alcoholic sodium hydroxide. Hexamethylene thio-
terephthalate melted near 200°C. Diallyl and dithio-
sebacic acid gave polyester of inherent viscosity as high
as 2.0 and softening temperature near 250°C with decomposition. Emulsion with persulfate or hydrogen peroxide
gave higher viscosities than solution polymerization with
UV light. Aromatic dithiol dibasic acids seemed to undergo
some self-condensation rather than only adding to diallyl.

Self-addition of 4-β-mercaptoethyl-1-cyclohexene by
persulfate emulsion (4 days at 30°C) only formed polymers
of low inherent viscosity (110). Polymers of low molecular
weight were obtained also by reaction of vinylcyclohexene
with dimercaptans under UV light, e.g., for 18 days (111).
Viscous, liquid low copolymers have been prepared by addition of phenyl phosphine to diallyl under UV light for 20
hr (112). The patent also gives examples of polymers by
H-addition of phenylphosphine to diallyl ether and to
p-divinylbenzene.

References

1. M. Berthelot and S. de Luca, Ann. 100, 361 (1856); G. H. Jeffery
    and A. I. Vogel, J. Chem. Soc. 663 (1948); G. Hatta and A.
    Miyake, J. Org. Chem. 28, 3237 (1963).
2. A. Turk and H. Chanan, Org. Syn. 27, 7 (1947).
3. H. Dreyfus, U.S. 2,387,723 (Celanese); R. L. Hodgson and J. H.
    Raley, U.S. 3,052,735 (Shell); cf. F. N. Jones, U.S. 3,397,252
    and D. E. McCarthy, U.S. 3,484,502 (Du Pont).
4. B. Delarue, U.S. 3,565,966 (Rhône-Poulenc).
5. Fr. 1,323,328 (Huels); CA 59, 9784 (1965).
6. N. L. Bauld, Tetrahedron Lett. 859 (1962).
7. L. H. Slaugh and J. H. Raley, Tetrahedron 20, 1005 (1964);
    cf. U.S. 2,730,559; 2,755,322 (Shell).
8. T. M. O'Grady et al, U.S. 3,257,417 (Std. Oil Indiana).
9. V. P. Gol'mov, Zh. Obshch. Khim. 22, 2131 (1952); CA 48, 1240
    (1954).
10. J. M. Baker and H. Burton, J. Chem. Soc. 815 (1933).
11. W. E. Vaughn et al, J. Org. Chem. 7, 482 (1942).
12. M. S. Kharasch et al, J. Am. Chem. Soc. 69, 1100 (1947).
13. C. Harries and H. Tuerck, Ann. 343, 360 (1905).
14. R. Srinivasan, J. Phys. Chem. 67, 1367 (1963).
15. R. Montarnal and M. Brun, Fr. 1,313,710 (Institut Francais du
    Pétrole).
16. J. K. Stille, J. Polym. Sci. A 2, 405 (1964); cf. D. J. C. Wood
    and L. F. Wiggins, Nature 164, 402 (1949).

17. L. Bateman, J. Chem. Soc. 936 (1950); P. Karrer and S. Perl, Helv. Chim. Acta 33, 36 (1950).
18. C. L. Henne and A. Turk, J. Am. Chem. Soc. 64, 826 (1942); 66, 395 (1944).
19. K. Alder et al, Ber. 76, 35 and 52 (1943).
20. M. Munzer et al, Angew. Makromol. Chem. 11, 27 (1970).
21. Brit. 776,326 (Du Pont), appl. August 16 and 19, October 15 and November 22, 1954.
22. C. S. Marvel and R. D. Vest, J. Am. Chem. Soc. 79, 5771 (1957); cf. U.S. 2,719,182 (Sprague El.) and 2,457,306 (Celanese).
23. C. S. Marvel and J. K. Stille, J. Am. Chem. Soc. 80, 1740 (1958); cf. Brit. 776,326 (Du Pont).
24. A. Valvasori et al, Chim. Ind. (Milan) 44, 1095 (1962); CA 58 2554 (1962).
25. H. S. Makowski et al, J. Polym. Sci. A 2, 1549 (1964).
26. H. S. Makowski et al, U.S. 3,357,961 (Esso).
27. S. G. Olson, U.S. 3,435,020 (Hercules).
28. C. S. Marvel and W. E. Garrison, J. Am. Chem. Soc. 81, 4737 (1959).
29. J. A. Cornell et al, Rubber Division, ACS Meeting, May 1963.
30. W. F. Gresham and M. Hunt, U.S. 2,933,480 (Du Pont).
31. G. Natta et al, Ital. 638,953 (Monte); CA 60, 3179 (1964).
32. G. Natta et al, Belg. 626,106 (Monte), appl. 1961; CA 60, 16090 (1964); cf. CA 57, 10013 (1962).
33. Chem. Eng. News, 15 (Jan. 23, 1967); cf. H. K. Frensdorff and E. Karcher, U.S. 3,284,424 (Du Pont).
34. R. R. Souffie, U.S. 3,331,793 (Du Pont).
35. S. Suzuki, U.S. 3,320,168 (Chevron).
36. H. Hendricks and C. E. P. V. van den Berg, Ger. 1,142,701 (Stamicarbon), appl. 1960; CA 58, 10356 (1963); cf. Brit. 910,132 (Dow).
37. Brit. 940,714 (Huels) (German appl. 1961); CA 60, 749 (1964).
38. Brit. 930,985 (Shell); CA 59, 10259 (1963).
39. W. S. Anderson and J. Boor, U.S. 3,291,782 and 3,223,638 (Shell).
40. H. S. Makowski et al, J. Polym. Sci. A 2, 4973 (1964).
41. R. J. Angelo, Belg. 632,633 (Du Pont); CA 61, 799 (1964).
42. Brit. 827,365 (Goodrich-Gulf); CA 54, 19003 (1960).
43. K. Alder et al, Ber. 87, 447 (1954).
44. R. H. Rheinhard, U.S. 3,060,155 (Monsanto).
45. J. J. Drysdale and C. S. Marvel, J. Polym. Sci. 13, 513 (1954).
46. K. K. Georgieff et al, J. Appl. Polym. Sci. 8, 889 (1964); CA 61, 3199.
47. J. W. Breitenbach, J. Polym. Sci. 23, 949 (1957).
48. H. Corte and H. Heller, U.S. 3,544,488 (Bayer).
49. J. K. Stille and D. W. Thompson, J. Polym. Sci. 62, S 118 (1962).
50. W. E. Vaughn et al, U.S. 2,818,441 (Shell).
51. P. Pomerantz, J. Res. Natl. Bur. Std. (U.S.) 52, 59 (1954).
52. N. G. Gaylord, Rec. Trav. Chim. Pays-Bas 70, 1042 (1951).
53. H. E. de la Mare, U.S. 2,957,929 (Shell).
54. Ger. 886,907 (Badische); CA 52, 13799 (1958).

55. M. Marek et al, J. Polym. Sci. C 16, 971 (1967).
56. Brit. 827,365 (Goodrich-Gulf); CA 54, 19003 (1960).
57. C. E. Schildknecht and R. F. Williams, unpublished work;
    cf. U.S. 2,612,493 (Esso).
58. L. M. Welch, U.S. 2,729,626 and 2,671,073 (Esso).
59. P. O. Tawney, U.S. 3,452,700 (U. S. Rubber).
60. D. A. Guthrie and R. M. Thomas, U.S. 2,955,994 (Esso).
61. G. Natta et al, U.S. 3,300,467 (Monte).
62. R. Riemschneider et al, Monatsh. 92, 777 (1961); CA 56, 336
    (1960).
63. R. S. Mitchell et al, Macromolecules 1, 417 (1968).
64. R. G. R. Bacon and E. H. Farmer, J. Chem. Soc. 1072 (1937);
    cf. U. M. Slobodin, CZ II, 66 (1939).
65. R. W. Stackman, J. Macromol. Sci.-Chem. A 5, 251 (1971).
66. T. Alderson et al, U.S. 3,013,066; J. Am. Chem. Soc. 87, 5638
    (1965); R. Cramer, ibid. 89, 1633 (1967); U.S. 3,502,738
    (Du Pont).
67. R. J. Harder, U.S. 3,472,908 (Du Pont).
68. W. Schneider, U.S. 3,398,209 and 3,441,627 (Goodrich).
69. G. Hata, J. Am. Chem. Soc. 86, 3903 (1964); cf. A. C. L. Su
    and J. W. Collette, Polym. Preprints 12, 415 (September 1971).
70. J. W. Collette and A. C. L. Su, U.S. 3,565,967 (Du Pont).
71. C. Tosi et al, J. Appl. Polym. Sci. 16, 801 (1972).
72. A. J. Mital and J. F. Jones, U.S. 2,996,488 (Goodrich).
73. W. Reeve and R. J. Bianchi, J. Org. Chem. 34, 1921 (1969).
74. J. M. Wilbur and C. S. Marvel, J. Polym. Sci. A 2, 4415 (1964).
75. D. S. Trifan and J. J. Hoglen, J. Am. Chem. Soc. 83, 2021 (1961).
76. G. B. Butler and T. W. Brooks, J. Org. Chem. 28, 2699 (1963).
77. C. S. Marvel and E. J. Gall, J. Org. Chem. 25, 1784 (1960).
78. A. T. Blomquist et al, J. Am. Chem. Soc. 78, 6057 (1956).
79. G. C. Corfield and A. Crawshaw, J. Macromol. Sci.-Chim. 5, 1873
    (1971).
80. C. C. Wu and R. W. Lenz, Polym. Preprints 12, 209 (September
    1971); cf. D. E. Applequist and J. D. Roberts, J. Am. Chem.
    Soc. 78, 4012 (1956).
81. G. B. Butler et al, J. Polym. Sci. A 3, 723 (1965).
82. G. B. Butler and M. L. Miles, Polym. Eng. Sci. 6, 71 (1966).
83. G. B. Butler and M. A. Raymon, J. Macromol. Chem. 1, 201 (1966).
84. N. E. Boyer and M. P. Weaver, U.S. 3,526,672 (Borg-Warner).
85. G. B. Butler and M. L. Miles, J. Polym. Sci. A 3, 1609 (1965).
86. K. Toyoshima et al, CA 70, 58438 (1969).
87. Y. Choshi et al, J. Chem. Soc. Japan, Ind. Chem. Sec. 72, 999
    (1969).
88. L. P. Ellinger, Brit. 1,255,838 (BP Chemicals); CA 76, 86405
    (1972).
89. O. Ohara (Fukuoka, Japan), CA 76, 15003 (1972).
90. T. W. Campbell and A. C. Haven, J. Appl. Polym. Sci. 1, 73
    (1959); CA 53, 19437 (1959).

91. W. P. Baker, J. Polym. Sci. A 1, 655 (1963); cf. Brit. 776,326
     (Du Pont); S. Otsuka and A. Nakamura, U.S. 3,536,692 (Japan
     Syn. Rubber).
92. H. Tadokora et al, J. Polym. Sci. B 3, 697 (1965).
93. Ital. 792,570 (Monte); CA 70, 2054 (1969); cf. R. E. Rinehart,
     U.S. 3,534,014 (Uniroyal).
94. G. D. Shier, U.S. 3,442,883 (Dow); cf. R. J. Pasquale, CA 76,
     72849 (1972).
95. G. D. Jones and H. H. Roth, U.S. 3,541,058 (Dow).
96. A. Cozzone et al, Bull. Soc. Chim. (5), 32, 1656 (1966).
97. M. Bertrand, Compt. Rend. 244, 619 (1957), CA 51, 10360;
     Brit. 775,723 (Bayer), CA 51, 16511 (1957).
98. J. Braun and R. Murjahn, Ber. 59, 1207 (1926).
99. C. S. Marvel and R. R. Chambers, J. Am. Chem. Soc. 70, 993 (1948).
100. W. E. Vaughan and F. F. Rust, J. Org. Chem. 7, 472 (1942).
101. C. S. Marvel and P. H. Aldrich, J. Am. Chem. Soc. 72, 1978 (1950).
102. C. S. Marvel and G. Nowlin, J. Am. Chem. Soc. 72, 5026 (1950).
103. C. S. Marvel and A. H. Markhart, J. Polym. Sci. 6, 711 (1951).
104. C. S. Marvel, C. W. Hinman, and H. K. Inskip, J. Am. Chem. Soc.
     75, 1997 (1953).
105. C. S. Marvel and H. E. Baumgarten, J. Polym. Sci. 6, 127 (1951).
106. C. S. Marvel and P. D. Caesar, J. Am. Chem. Soc. 73, 1097 (1951).
107. C. S. Marvel and H. Wexler, J. Am. Chem. Soc. 75, 6318 (1953).
108. C. S. Marvel et al, Ind. Eng. Chem. 45, 2090 (1953).
109. C. S. Marvel and A. Kotch, J. Am. Chem. Soc. 73, 1100 (1951);
     Marvel and E. A. Kraiman, J. Org. Chem. 18, 707 (1953); cf.
     H. Behringen, Ann. 564, 219 (1949).
110. C. S. Marvel and L. E. Olson, J. Polym. Sci. 26, 23 (1957).
111. P. F. Warner, U.S. 3,484,355 (Phillips).
112. A. Y. Garner, U.S. 3,010,946 (Monsanto).

4. ALLYL HALIDES AND RELATED

Allyl chloride, first supplied by Shell about 1940, is the most important allyl halide. It is used in large volume for the manufacture of glycerol, epichlorohydrin, and epoxy resins. Smaller amounts go into allyl alcohol, the allyl monomers, allyl carbonates, and allyl phthalates; it is also an ingredient for allylamines, allyl isothiocyanate (mustard oil), allyl barbiturates (sedatives), cyclopropane (anesthetic), and allyl esters for perfumes. No large demand for methallyl chloride has developed comparable to that for allyl chloride. Homopolymers of allyl and methallyl halides are not used commercially.

Many applications have been limited through toxicity and relatively high cost of allyl chloride. Allyl halides have found relatively little use in direct applications although allyl chloride and methallyl chloride have been employed as fumigants and biocides (e.g., for soils and grains under specially controlled conditions). In animal tests allyl chloride has caused inflammation of lungs and skin, together with swelling of kidneys. Prolonged exposures have produced degenerative changes in the liver and kidneys. The irritating odor of allyl chloride and the accompanying headaches and dizziness may give warning against overexposure. Early work on toxicity and other properties has been reviewed (1).

Special precautions should be taken with volatile allyl chloride. The LD_{50} for allyl chloride in mice was found to be 500 mg/kg (2). The dead animals showed alterations in heart, liver, and kidneys. Decreases in catalase activity and in the number of erythrocytes were observed in the blood. When given orally in 20% solution in oil the LD_{50} for mice was 1.37 g/kg and for rats 1.0 g/kg (3). Doses of 0.25 g/kg caused hypertension and reduced body weight. Methallyl chloride did not show cumulative effects. Feeding grain containing 50 g/m^3 did not cause changes in rats. In Europe methallyl chloride has been recommended for disinfecting grains and other food crops stored in warehouses (4). The normal amount was 46 to 60 g/m^3 during 5 days.

Allyl and methallyl halides listed in Table 4.1 are colorless liquids (except gaseous allyl fluoride). They

TABLE 4.1

Allyl and Methallyl Halides, $CH_2=CHCH_2X$ and $CH_2=\underset{\underset{CH_3}{|}}{C}CH_2X$

	bp, °C	mp, °C	Density $\frac{20}{4}$	n_D^{20}	Other Properties and References
Allyl fluoride	about -3	--	--	--	Garlic odor; Hoffmann, J. Org. Chem. 14, 106 (1949); Meslans, Compt. Rend. 111, 882 (1890); Ann. Chim.[7], 1, 374 (1894); 3% soluble in water
Allyl chloride[a]	45	-135	0.9382 or 0.9397	1.4157	0.36% soluble in water 20°C; viscosity at 20°C = 0.0037; surface tension at 20°C = 23.06 dynes/cm
Allyl bromide	70	-119	1.3980	1.4697	Heat cap. at 25° = 0.26 cal/(°C)(g); surface tension 12°C = 28 dynes/cm; McCullough, JACS 51, 226 (1929); Tollens, Ann. 156, 152 (1870)
Allyl iodide	102	-98	1.7765	1.5437	Letsinger, JACS 70, 2818 (1948); Jeffrey, J.Chem.Soc. 663 (1948)
Methallyl chloride	72	--	0.918	1.4291	Smith, Org. Syn. 32, 91 (1952)
Methallyl bromide	95	--	1.339	1.4680	Jones, J. Chem. Soc. 1448 (1947)
Methallyl iodide	30(5)	--	--	1.4862	Tamele, Ind. Eng. Chem. 33, 116 (1941)

[a]Vapor pressure = 294 mm at 20°C.

are only slightly soluble in water when pure. The iodides
are quite unstable, becoming colored quickly on standing.
The bromides and chlorides may discolor to yellow or brown
on storage unless they contain water (5), propylene oxide
(6), or other stabilizers. It is possible that small
amounts of liberated halogen or HX, together with $FeCl_3$
from metal corrosion, may catalyze cationic polymerization
and that the low polymers formed lose HX to give colored
conjugated diene structures. Black polymer residues
usually remain after evaporation of stored commercial
monomer. The allyl and methallyl chlorides, bromides, and
iodides are miscible with a wide range of organic solvents.
The somewhat camphorlike odor of methallyl chloride is
much less unpleasant than the odor of allyl chloride. The
complex IR spectrum of allyl chloride is shown in Fig. 4.1.

Much early laboratory work was done with allyl bromide
before the availability of allyl chloride and, later, because of the high volatility of the latter. Gaseous allyl
fluoride is highly toxic and hydrolyzes rapidly with water,
forming hydrogen fluoride. Infrared spectra indicate that
allyl chloride, bromide, and iodide contain predominantly
the transoid forms (7). The proportion of the minor
cisoid rotational isomer decreases from allyl chloride to
allyl iodide.

PREPARATION OF ALLYL HALIDES

The first allyl halide prepared was the iodide obtained
by Berthelot and de Luca in 1854 by reaction of glycerol
with phosphorus and iodine (8). Soon after, Cahours and
Hofmann made the chloride, bromide, and iodide from allyl
alcohol and the phosphorus halides (9). Allyl fluoride
was prepared by reacting allyl iodide with silver fluoride
(10). Allyl and methallyl halides other than fluorides
can be prepared by reaction of the alcohols with hydrogen
halides (11):

$$CH_2=CHCH_2OH + HX \longrightarrow CH_2=CHCH_2X + H_2O$$

High temperatures and high acid concentrations produce
rearrangements. Methallyl alcohol with hydrogen iodide
even at 0°C gives rearrangement to t-butyl iodide and
iodine (12). Allylic halides can be prepared avoiding
rearrangement to secondary and tertiary isomers by reacting the alcohols with a reagent from triphenylphosphine
and carbon tetrachloride at 25°C for 48 hr (13).

The dehydrohalogenation of dihalides has not been
favorable for preparation of allyl and methallyl halides.
In Russia allyl chloride has been made from diallyl ether
and hydrogen chloride in the gaseous phase over transition
metal catalysts (14), and from allyl esters with HCl in

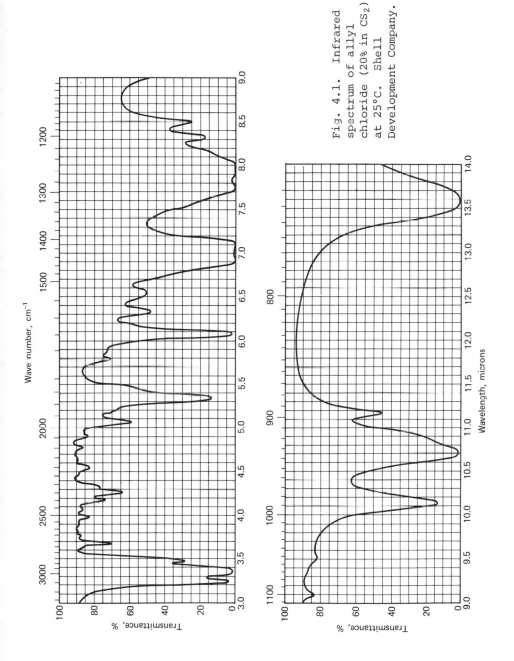

Fig. 4.1. Infrared spectrum of allyl chloride (20% in CS_2) at 25°C. Shell Development Company.

159

the liquid phase with CuCl catalyst (15).   Allyl bromide
was made by distilling with a stream of carbon dioxide a
mixture of glycerol, white phosphorus, and iodine (16).
Isopropyl iodide was a side product.   Addition of mercur
to a solution of allyl iodide (AI) in ethanol gave AHgI,
which was recrystallized.

The manufacture of allyl and methallyl chloride (AC an
MAC) is based on the replacement of allylic hydrogen ato
by chlorine under conditions minimizing simple additon o
chlorine to the double bond:

$$CH_2=\underset{\underset{CH_3}{|}}{C}-CH_3 + Cl_2 \longrightarrow CH_2=\underset{\underset{CH_3}{|}}{C}HCH_2Cl + HCl$$

In 1883 Sheshukov of St. Petersburg discovered that the
principal product from reaction of chlorine with isobute
at ordinary temperatures is not by addition but has the
structure of methallyl chloride or isobutenyl chloride
(17).   It reacted with dilute KOH to give an unsaturated
alcohol and an unsaturated ether.   Introducing Cl_2 into
either dry or wet gaseous isobutene at room temperature
or into liquid isobutene at -20°C gave largely MAC inste
of forming the dichloride by addition (18).   Methallyl
chloride was heated with a solution of 1.5 parts of KOH
in 20 parts of water to form methallyl alcohol or isopro
penyl carbinol.   Pogorshelski extended the researches of
Sheshukov.   Slow reaction at low temperatures between
isobutene and chlorine gave a mixture of $(CH_3)_3CCl$ and
methallyl chloride (19).   The MAC was reacted with $NaOC_2$
to form ethyl methallyl ether.   This added Br readily,
but the dihalide slowly evolved HBr on standing.   Heatin
MAC with potassium acetate formed methallyl acetate.

Burgin and co-workers of Shell found that the chlorina
tion of liquid isobutene at 0°C was extremely rapid,
whereas in the vapor phase at 70°C it occurred slowly,
even with light or catalysts (20).   There was little eff
of temperature on the ratio of chlorination products fou
Byproducts were 1,2-dichloroisobutane, dichloroisobutene
t-butyl chloride, and isocrotyl chloride.   Taft proposed
an ionic mechanism for low-temperature chlorinations (21
At high temperatures (e.g., 400-500°C), isobutylene and
chlorine have been reported to form largely t-butyl
chloride (22).   However, MAC has been made by passing
isobutene, HCl and O_2 rapidly over magnesium chloride or
pumice at 500°C (23).

Propylene behaves differently from isobutene in reacti
with chlorine.   It was necessary for Shell chemists to
develop precise conditions for reaction at higher temper
tures in the gas phase with controlled oxygen concentrat

for the commercial preparation of allyl chloride (24).
Both addition and substitution by chlorine were found to
be free radical chain reactions (25). Additions of tetraethyl lead accelerated the reactions and favored substitution over addition. Both reactions were retarded by
oxygen at least below 240°C. Effects of oxygen concentration are shown in Ref. 25. At 310°C controlled small
concentrations of oxygen promoted substitution and
retarded addition. Light accelerated the addition
reaction. Allyl chloride can be prepared by passing
propylene and chlorine through a spherical steel reactor
at 400 to 600°C with residence time 0.3-1.5 sec (26).
Principal byproducts are hydrogen chloride, coke, and
dichloropropanes.

Ziegler and co-workers found that allylic substitution
of halogen could be accomplished at moderate temperatures
in the laboratory by N-bromophthalimide or N-bromosuccinimide (27):

$$CH_2=CHCH_2R + (CH_2CO)_2NBr \longrightarrow CH_2=CHCHBrR, \text{ etc.}$$

Standing at room temperature or warming 40 min in carbon
tetrachloride solution was sufficient to induce allylic
brominations which occur by free-radical mechanism.

In the future it may become possible to manufacture
allyl chloride (AC) from hydrocarbons by oxyhalogenation
processes. Farbwerke Hoechst has made allyl halides by
reacting olefins and monohaloalkanes with oxygen and
hydrogen chloride over palladium-silica at 200°C (28).
Propylene, HCl, and O_2 (2:1:1) were passed at 225°C through
a tube containing tellurium dioxide on aluminum silicate
to produce small yields of AC (29). Mixed tellurium and
palladium catalysts may also be used. Propyl and isobutyl
chloride and olefins may be reacted with oxygen to obtain
AC and MAC at 200 to 300°C using catalysts made from
tellurium compounds and alkali hydroxides or alkali salts
(30). Allyl bromide was made by gradual addition of
sulfuric acid to a mixture of hydrogen bromide, water,
and allyl alcohol (31). Along with the allyl bromide was
distilled a fraction of dibromopropane. In the preparation of derivatives of allyl halides, especially allylic
bromides, a mixture of isomers was formed in the presence
of acid catalysts (32).

Methallyl bromide was prepared by heating 2 moles of
sodium bromide and 1.5 moles of MAC in 1 liter of acetone
at reflux for 5 hr (33). After the sodium chloride had
been filtered off, the methallyl bromide was recovered by
distillation. Bromine adds readily to isobutene to give
the relatively stable dibromide instead of methallyl
bromide. Methallyl iodide was made in low yield by heating

2 moles of sodium iodide and 1.5 moles MAC in 1 liter of
methyl ethyl ketone at reflux for 3 hr followed by distillation (33). The unstable methallyl iodide was
observed to decompose violently at elevated temperatures
and quite rapidly even on storage at room temperatures.
Cationic polymerization catalyzed by iodine and hydrogen
iodide may be involved.

Recently allyl halides were made from allyl esters of
organic acids. Allyl acetate or methallyl acetate was
reacted with dry HCl in presence of a cuprous compound and
a Lewis-acid such as $FeCl_3$ or $AlCl_3$ (34).

PROPERTIES OF ALLYL AND METHALLYL HALIDES

Physical properties of some allylic halides were given
in Table 4.1. Figure 4.1 shows the complex infrared
spectrum of allyl chloride (20% in carbon disulfide at
25°C). Two different cell depths were used in order to
obtain details of bands in regions both of high and low
absorption. References until 1948 on chemical properties
of allyl halides appear in the book "Allyl Chloride" by
Shell Chemical Company, New York and San Francisco (1949).
Many reactions of allyl chloride are shown in the syntheses
of allyl compounds in other chapters of the present book.
Here we emphasize reactions of allyl halides of special
interest to polymer chemists such as addition reactions.
Reaction with hydrogen in presence of active carbon or
nickel may not give saturated halides but instead alkene
and HX. However, allylic halogen compounds can be satisfactorily hydrogenated using rhodium on Al_2O catalyst (35).
Pyrolysis at 600°C may give benzene, HCl and H_2. Photoly-
sis of liquid allyl chloride under UV light of 2537 Å
formed nearly equal amounts of allene and propylene (36).

Allyl bromide and allyl chloride add Br , Cl , or BrCl
fairly rapidly. The principal reaction of allyl iodide
with chlorine or bromine involves both substitution and
addition (37):

$$CH_2=CHCH_2I + Cl_2 \longrightarrow CH_2ClCHClCH_2Cl + ICl$$

Iodine does not react readily with allyl iodide. At low
temperatures allyl iodide forms an unstable allyl iodide
dichloride in which Cl is attached to I (as in iodobenzene
dichloride) and the double bond is retained (38). Bromination of allyl chloride or allyl bromide in benzene in
the presence of oxygen gave, besides trihalopropane,
considerable amounts of halogenated dialkyl peroxides (39).
Since the halogens are weakly electron attracting, allyl
chloride and allyl bromide add bromine more slowly than
does ethylene. Swindale and co-workers found that the

relative rates of bromine addition in acetic acid at 25°C
were (40):

$C_2H_4 = 1$ $ACl = 0.019$ $ABr = 0.013$ $CH_2=CHCN = 0.0027$

The addition of bromine was accelerated by adding lithium
halides. Iodine did not add to allyl iodide in acetic
acid (41).

The normal or Markovnikov additions occur by ionic
mechanisms, whereas reverse addition of hydrogen bromide
to allyl halides may occur in the presence of light, peroxides, or metal catalysts in air (42). Reverse addition
of BrOH to allyl bromide occurred in the presence of water
and $HClO_4$ (43). Reverse additions of mercaptans (44) and
of olefins (45) to allyl chloride also were observed.
Reverse additions of HCl or HI to allyl halides are more
difficult. Dihalides obtained by reverse addition may be
reacted with zinc to give cyclopropane, a useful but
flammable anesthetic (46). That reverse additions do not
occur with HI nor with mercaptans has been attributed to
reduction of peroxides, but this is not so convincing.
Hydrogen iodide reacts with allyl iodide by successive
addition and reduction reactions to form isopropyl iodide
and propylene.

The addition of hypochlorous acid to allyl chloride (AC)
has been much studied in the manufacture of glycerol by
the Shell process. It largely follows normal addition
where positively charged Cl adds to the more negatively
charged carbon atom (47):

$$CH_2=CHCH_2Cl + \overset{+}{C}l(\overset{-}{O}H) \longrightarrow CH_2ClCH(OH)CH_2Cl$$

However, considerable proportions of 2,3-dichlorohydrin
are formed. Trichloropropane also forms in increasing
amounts at higher concentrations and higher temperatures
(48). Treatment of the mixture of dichlorohydrins with
alkali gives epichlorohydrin useful for synthesis of epoxy
resins and for addition copolymerization to form new
chlorohydrin rubbers. Cold treatment of AC with sulfuric
acid followed by dilution with water leads to normal
addition to form $CH_2CHOHCH_2Cl$ (49).

Normal addition of HBr occurs to allyl bromide in the
presence of $FeCl_3$ in the dark (50); also in acetic acid
(light or darkness) between room temperature and 100°C
(51). However, 1,3-dibromopropane is a principal product
by reaction with HBr under UV light at -78°C (52) and in
the presence of oxygen or peroxides (53).

Allyl halides $CH_2=CHCH_2X$ are very much more reactive
than alkyl halides RX in substitution of halogen and in
solvolysis reactions (54). The vinyl halides $CH_2=CHX$, in
contrast, are even less reactive in halogen substitution

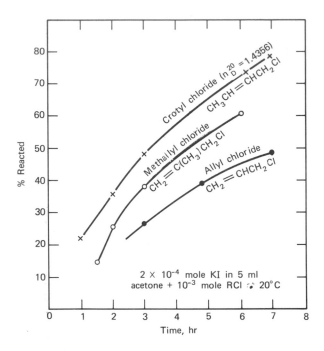

Fig. 4.2.   Relative reactivities of allylic chlorides in the reaction RCl + KI ⟶ RI + KCl.   Substitution of electron-releasing methyl groups into allyl chloride accelerated the substitution reaction with iodide ion.   However, allyl chloride and methallyl chloride reacted at nearly identical slow rates with $NaOC_2H_5$ at 35°C, whereas crotyl chloride was more reactive.   M. Tamele et al, Ind. Eng. Chem. <u>33</u>, 118 (1941).

and solvolysis reactions than alkyl halides. Rates of reaction of allyl bromide with KI in acetone at 25°C and 15°C were studied (55). Allyl bromide reacted much faster than AC, while AC reacted about 80 times as fast as n-propyl chloride (56). Methallyl chloride reacted faster than AC (Fig. 4.2). Using allyl bromide labeled with ^{14}C allylic rearrangement was demonstrated in the presence of $Ni(CO)_4$ and $NiBr_2$ (57). Such migrations of the double bond are believed to occur through π-bond complexing.
   Allyl bromide has been identified by forming the S-allyl-isothiourea-picrate [mp 155°C (58)] and by N-allyl saccharin [mp 98°C (59)]. Allyl bromide has found use in commercial organic syntheses and has been supplied by Freeman Industries, Inc., and by White Chemical Corporation.

Hughes found that allylic chlorides are capable of reacting with bases by two different mechanisms, unimolecular reaction with water or alcohol (SN_1) and bimolecular replacement of Cl by OH or OR (SN_2) (60). The second type of reaction is favored where the concentration of nucleophilic negative ion is high; diallyl ethers are byproducts. Both excess acid and alkali can be detrimental to the yield of allyl or methallyl alcohol. Some dimethallyl ether is formed even in hydrolysis of MAC in the presence of a large excess of water.

The high reactivity of allyl halides is shown by the rapid double replacement reactions such as allyl iodide + $HgCl_2$ and allyl chloride + CaI_2. Although addition of halogen occurs at lower temperatures, substitution of halogen into allyl halides can be carried out at 500°C (61):

$$CH_2=CHCH_2Cl + Cl_2 \longrightarrow CHCl=CHCH_2Cl + HCl$$

Allyl halides dispersed in water hydrolyze to allyl alcohol and HX much more rapidly than alkyl halides. Allyl iodide was hydrolyzed slowly in warm water (62).

The reaction of AC on heating with aqueous alkali to form allyl alcohol and some diallyl ether was accelerated by Cu_2Cl_2 (63). Rates of acid hydrolysis of AC have been studied (64). Allyl fluoride was very resistant to hydrolysis, but allyl chloride hydrolyzed slowly in water without catalyst. Allyl chloride was hydrolyzed 25 times faster than n-propyl chloride by a mixture of formic acid and water (65). An SN_1 mechanism was suggested. Vernon found that AC reacted with $NaOC_2H_5$ 3.7 times faster than n-propyl chloride. In many nucleophilic substitution reactions there are no sharp divisions between SN_1 and SN_2 (66).

Allyl chloride reacted with oxygen and bromine in benzene to give the explosive peroxide $CH_2ClCHBrCH_2OOH$, boiling 56°C at 0.25 mm (67). Allyl bromide was reacted with potassium t-butyl peroxide in isopropanol during 5 days to form allyl hydroperoxide (68). Lower yields were obtained with AC and with methallyl bromide at room temperature. Allyl chloride with phenyl lithium, at reflux for 5 hr in ether, gave allylbenzene and a little phenylcyclopropane (69). Allyl chloride in cyclohexane in the presence of di-t-butyl peroxide at 160°C gave allylcyclohexane and 3-chloropropylcyclohexane (70).

Commercial allyl chloride and allyl bromide were observed to hydrolyze in excess water at room temperature faster than allyl acetate (71). The initial pH in these tests was near 8.0 by addition of a little sodium carbonate. After 3 days AC gave pH 3.3 and AB 3.7. Starting at initial pH of 5.3 in distilled water under air, the

following values of pH were observed after two days:
AB = 3.1, AC = 3.5, MAC = 3.7. The pH values fell rapidly
during the first four hours and then leveled off. These
changes result largely from hydrolysis:

$$CH_2=CHCH_2Cl + HOH \longrightarrow CH_2=CHCH_2OH + HCl$$

Hydrolysis of AC was used for gradual precipitation of
AgCl from $AgNO_3$ solution in photographic research (72).
   Reactions of allyl halides with cyanides, cyanates, and
thiocyanates are complicated by the tendency for rearrangement to form organic isocyanides, isocyanates, and
isothiocyanates. Reaction of allyl iodide with silver
cyanide apparently gave allyl isocyanide ANC (73). Reaction of allyl halides with KCN gave largely crotonitrile
by isomerization to the conjugated structure (catalyzed by
the alkalinity of KCN):

$$CH_2=CHCH_2X + KCN \longrightarrow CH_3CH=CHCN + KX$$

However, allyl cyanide can be prepared by reaction of CuCN
with allyl halides (74). Allyl iodide reacts with silver
cyanate to form allyl isocyanate (73):

$$AI + AgOCN \longrightarrow ANCO + AgI$$

Reaction of allyl bromide with silver nitrate gave a mixture of 80% 3-nitro-1-propene and 20% allyl nitrite (75).
Allyl halides react with ammonium, potassium, or silver
thiocyanates under mild conditions to give at first the
thiocyanate which rearranges to allyl isothiocyanate
(mustard oil) (76). (See page 486.)
   Allyl halides can couple to form diallyl, a useful
allylic diene. Allyl chloride vapor passed over catalysts
of supported metallic copper at 100 to 250°C formed diallyl (77). Reaction of allyl bromide with ethyl magnesium chloride in cold ether gave 1-pentene, along with a
little diallyl or 1,5-hexadiene (78). Rapid addition of
allyl bromide to magnesium turnings in dry ether under
reflux also gave 1,5-hexadiene (79). Allyl halides react
with ketones and magnesium to give carbinols (80).
Table 4.2 presents some additional reactions of allyl
halides.
   Allyl halides can react with extraordinary speed to form
allyl ethers, esters, amines, and sulfides. These and
other allylations are discussed in separate chapters;
older references are given in the Shell book on allyl
chloride. In general, reaction conditions should be kept

TABLE 4.2

Some Other Reactions of Allyl Halides

Reactants	Product	References
$(EtO)_2SO + KOH$	ASO_3K	Arbuzov, CZ II 685 (1909); J. Chem. Soc. 452 (1909)
Ag_2SO_4	A_2SO_4 (explosion on distillation)	Braun, Ber. $\underline{50}$, 290 (1917); CA $\underline{11}$, 2796 (1917)
$Na_2S_2O_3$	$ASSO_3Na$ (yields ASH and ASSA)	Purgotti, CZ II, 316 (1892); Twiss, J. Chem. Soc. 36 (1914)
$AgNO_2$	ANO_2 (and some AONO)	Askenasy, Ber. $\underline{25}$, 1701 (1892); Henry, Rec. Trav. Chim. $\underline{17}$, 1 (1898); Reynolds, JACS $\underline{51}$, 279 (1929)
$(NaO)_3PO$	$AOP(ONa)_2$ $\overset{\text{\tiny II}}{O}$	Bailly, CZ III 93; 254; 752 (1919); cf. Zetzsche, CZ I, 1396 (1926)
$(AgO)_3PO$	$(AO)_3PO$ styrene copolymer	Cavalier, Compt. Rend. $\underline{124}$, 91 (1897); J. Chem. Soc. 310 (1897); Britton, U.S. 2,186,360 (now)
Na_3AsO_3, etc.	$AAsO(OH)_2$	Brit. 167,157 (LaRoche); CA $\underline{16}$, 614 (1922); Quick, JACS $\underline{44}$, 805 (1922)
KSnOOK, etc.	ASnOOH A_2SnO	Lesbre, CA $\underline{28}$, 4376 (1934); Jones J. Chem. Soc. 1446 (1947)
ACl coupled	1,5-hexadiene	Jones, J. Org. Chem. $\underline{32}$, 1667 (1967)

mild to prevent rearrangements. Methallyl halides also are very reactive, but a greater tendency toward rearrangement often reduces yields. Allyl bromide reacted 380 times as fast as butyl bromide in the alkylation of ketones in presence of sodium t-alkoxide (81). The 2,2-diallylcyclo-hexanone resulting boiled at 118°C (14 mm) and had $n_D^{14} = 1.4887$.

Double replacement reactions of methallyl chloride were discussed by Tamele and co-workers of Shell (82). Commercial methallyl chloride contained about 4% of $(CH_3)_2C=CHCl$, a less reactive compound. Methallyl chloride (MAC) reacted more rapidly with KI than did AC but less rapidly than crotyl chloride. Thus the introduction of a methyl group on the 2-position is apparently less activating upon

the chlorine reactivity than a methyl group in the 3-
position. With sodium alcoholate in anhydrous ethanol,
MAC showed about the same rates as AC at room temperature,
but slightly less at 50°C. At 116°C MAC was hydrolyzed
to alcohol and ether in less than 15 min by 10% aqueous
NaOH. If insufficient alkali is present to neutralize
acids, the product may be largely isobutyraldehyde by
rearrangement of the alcohol. Methallyl chloride and
steam passed at 320°C over solid Lewis acids such as
active alumina gave isobutyraldehyde (83).

Methallyl chloride reacted with paraformaldehyde in the
presence of sulfuric acid and acetic acid for 3 hr at
60°C (84). By neutralization of the water-insoluble layer
with sodium carbonate there was obtained 4-methyl-4-
(chloromethyl)-1,3-dioxane, and from this was made
2-(chloromethyl)-1,3-butadiene. The reaction of MAC with
chlorine in carbon tetrachloride at 20 to 30°C is complex
(85). The simple addition product $ClCH_2C(CH_3)ClCH_2Cl$ is
obtained in smaller yield than substitution products
$CH_2ClC(CH_3)=CHCl$ and $CH_2ClC(CH_2Cl)=CH_2$. A determination
of MAC content of air has been based on reaction of a
mole with one mole of a pyridine sulfate dibromide reagent (86). Methallyl chloride with HBr gave normal
addition with 12n HCl at 63°C, whereas reverse addition
occurred in presence of peroxides (87). Water added to
MAC in the presence of sulfuric acid at 10°C yielded
$CH_2ClC(OH)(CH_3)CH_3$. Heating MAC with methanol at 180°C
in a sealed tube gave isobutyraldehyde and the dimethyl-
acetal of the latter (88).

Victor Grignard reacted allyl iodide with magnesium
(89). Gradual addition of ethyl magnesium bromide to MAC
in ether gave Grignard coupling to 2-methyl-1-pentene
(90). Methallyl chloride can be reacted with magnesium
to give the Grignard reagent methallyl magnesium chloride
only when special precautions are taken to prevent the
coupling reaction MAMgCl + MAX ———→MA-MA. The coupling
can be prevented by adding the MAC very slowly to large
excesses of the Mg and anhydrous ether. The MAMgCl obtained can be reacted with carbonyl compounds to form
methallyl substituted alkanols. The Barbier modification
has been applied also in which acetone, MAC and magnesium
were reacted directly in anhydrous ether to give, for
example, 59 mole % $MAC(CH_3)OHCH_3$ and 37 mole % dimethallyl.

HOMOPOLYMERIZATIONS OF ALLYL CHLORIDE

Viscous liquid to soft solid allyl chloride polymers
have been prepared in many laboratories, but homopolymers
of high molecular weight and good color and utility have
not been available. Staudinger and Fleitman exposed allyl

chloride (AC) to sunlight for 4 months and to UV light for 4 weeks to obtain low yields of honeylike viscous liquid polymers of cryoscopic average molecular weight 624 to 698 (in benzene) (91). This polymer was observed to have $D_4^{20} = 1.258$ and $n_D^{20} = 1.5292$. A fraction of molecular weight 1886 was obtained by fractional precipitation. Long reduction of the polymer by hydrogen iodide and phosphorus gave an oily hydrocarbon. Heating the AC polymer 14 days with alcoholic KOH gave an oxygen-containing product, and reacting it with triethylamine formed a water-soluble substance. Allyl bromide also polymerized slowly in light to form a brown-black syrup containing polymer of low DP. Such colored low polymers from allyl halides are not pure allyl halide polymers because of loss of HX to form conjugated polyene chain segments --CH=CH-CH=CH-- .

Until recently only viscous, liquid low polymers of DP below 10 could be prepared. Bartlett and Altschul heated allyl chloride with 6.3% BP at 80°C obtaining a maximum of 83% polymerization (92). The ratio of monomer polymerized to peroxide used up was 36.4, somewhat higher than for allyl acetate, yet the molecular weight of the polymer cryoscopically was only 480, or DP near 6. The lower molecular weight compared with allyl acetate polymers formed under similar conditions was attributed to greater degradative chain transfer. In allyl chloride removal of both alpha-hydrogen and chlorine could give resonance-stabilized radicals of low capacity to propagate new polymer chains. Thus allyl chloride polymerizes under free radical initiator, but it acts as its own regulator to form only polymer of low molecular weight. Oligomers of allyl chloride prepared by heating with diacyl peroxide at 70°C for 27 hr were suggested as flame-retarding agents (93).

Kharasch and Buchi added diacetyl peroxide in allyl bromide dropwise to excess boiling allyl bromide (94). Among the products were the dimer 4-bromomethyl-5-bromo-1-pentene and a black polymer of the consistency of pitch.

Allyl chloride heated at 100°C for 11 hr in the presence of hydrogen chloride and BP gave a 45% yield of $H(C_3H_5Cl)_2Cl$, bp 147°C (100 mm), $n_D^{25} = 1.4790$, along with 25% $CH_3CCl = CHCl$ and 30% of products of higher molecular weight (95). Allyl chloride was polymerized to dark viscous oils and greases by treating with oxygen or a peroxide followed by reaction with hydrogen fluoride (96). In radical polymerizations of AC, the polymer molecular weight range of 600 was nearly independent of BP concentration and temperature below 100°C (97). Research directed toward raising the DP of allyl chloride polymers in radical-initiated systems has given only DP 7 to 8.

Bauer and Goetz (98) observed that on long standing in
light AC gives a honeylike viscous low-polymer solution.
Irradiation of commercial allyl chloride slowly formed a
dark viscous solution; methallyl chloride changed more
slowly in light. Under gamma radiation the yield of
polymer decreases in the order of ACl > ABr > AI (99).
Under a hot mercury arc lamp, redistilled liquid AC gave
a viscous, yellow polymer solution in only 2.5 hr (100).
There was a strong test for aldehyde by Schiff reagent.
By gamma radiation the yield of polymer from AC was substantially greater than from allyl acetate, ether, alcohol,
cyanide, or amine (101).

Homopolymerizations of allyl chloride by radical initiation in bulk over the wide temperature range of -136°C
(its melting point) and +100°C gave polymers of similar
low specific viscosities (102). For the low-temperature
polymerizations, alkyl boron-oxygen initiation was applied.
Similar results were obtained with allyl alcohol from its
melting point of -129 to +175°C. In copolymerizations
with ethyl acrylate, chain transfer to allyl chloride
relative to propagation was evident at -78°C as well as
at +30°C. Allyl chloride entered the copolymer chains
more readily at the higher temperature range.

Pastelike soluble AC homopolymers of DP 8 to 20 were
prepared in Japan, and these products containing double
bonds were copolymerized with vinyl chloride (103). When
azobis initiator was used, the reactivity ratios for
copolymerization of vinyl chloride with AC oligomer were
$r_1 = 0.63$ and $r_2 = 0.33$. Allyl chloride polymer of molecular
weight 647 was reacted with N-dodecylpiperidine, giving a
quaternary cationic surfactant (104). Allyl chloride
oligomers also were copolymerized with maleic anhydride by
heating with benzoyl peroxide 10 hr at 80°C (105).

Okuyama and Hirato of Osaka described interesting fila-
mentary growths of crosslinked polymers originating from
activated aluminum metal in contact with mixtures of allyl
chloride and toluene (106). For example, a mixture of
38 g AC and 26 g toluene did not react with the aluminum
foil until the metal surface was pricked with a pin, cut
with a knife, or rubbed. Red, gel-like polymer filaments
grew outward from points of scratching during an hour at
room temperature. Aluminum and/or aluminum trichloride
apparently catalyzed both polymerization and dehydrochlorination of the polymer. Bubbles of HCl gas were
evolved; the polymer was crosslinked and contained only
5 to 7% chlorine (AC polymer would contain 46% Cl). Wide
ratios of AC to toluene (0.1 to 10) produced such polymer
growths, and aluminum alloy such as Duraluminum also could
be used. The outer portions of the polymer filaments were
more transparent and yellowish. The sheathlike polymer

grew at about 3 cm/hr under one set of conditions.

Davidson polymerized AC with $AlBr_3$ catalyst at low temperatures (107). These cationic polymerizations gave low yields of discolored oils or soft solids. Polymerizations near room temperature formed yellow-brown polymers of DP about 30. Such low polymers were obtained in better yield by polymerization in refluxing heptane. Polymerizations in closed systems produced high-melting colored polyene structures from loss of HCl. Addition of zinc chloride as complexing agent accelerated the radical polymerization of AC as well as that of allyl acetate (108).

Bauer and Goetz refluxed methallyl chloride with 1% BP for 8 days obtaining only 33% yield of dimer. By irradiation with white light and 0.5% tetraethyl lead at 80°C for 5 days, a yield of 79% dimer, bp 75(10), $n_D^{20} = 1.4774$, was obtained along with 5% of higher polymers (109). That the DP and the rate were lower than obtained from allyl chloride was attributed to steric hindrance. Viscous liquid low polymers were also reported to form at 0 to 25°C by action of boron fluoride etherate (110). Allyl chloride and bromide reacted with metallic nickel to form π allyl nickel halide along with dark, oily polymers (111).

Recently solid low polymers were obtained by radical polymerization under high pressures (112). In one experiment 10 ml AC and 0.2 g isopropyl percarbonate were heated for 20 hr at 25 to 33°C under 100,800 psi. There was recovered 2.8 g of solid homopolymer as an amorphous elastomer. The relative viscosity was only 1.02 in solution, and the glass transition temperature was near 23°C. In another example, BP was used at 70°C and 74,000 psi to give a solid AC homopolymer.

Viscous brown to semisolid allyl chloride and methallyl chloride polymers have been reported by use of Lewis acids (113) or Ziegler-Natta and related transition metal heterogeneous catalyst systems. It was already known that BF_3 catalysts near -78°C could form viscous to resinous solids from allyl chloride. In one paper, $Al(C_2H_5)_3-TiCl_4$ catalyst in hexane was used at -70 to 20°C for up to 170 hr to give yellow polymers low in chlorine from dehydrohalogenation (114). These polymerizations showed characteristics of a cationic mechanism. In another case, 200 g AC, 3 ml $TiCl_4$, and 10 g aluminum powder were stirred in an autoclave at 25 to 45°C for an hour (115). The product was washed with water and dried, yielding 120 g of polymer.

Adelson and Dannenberg found that methallyl chloride responded better to polymerization with BF_3 at low temperatures than did AC. Methallyl chloride was treated with BF_3 at -75°C with agitation during 30 min and without agitation for 60 hr at -60 to -75°C. The mixture was then

allowed to stand at room temperature for 5 days. The
dried polymer, purified twice by dissolving in acetone
and precipitating by water, was a sticky syrup of DP
about 7 in 90% yield. With BF_3 activated by Raney nickel,
allyl chloride gave only polymers of lower DP including
dimer ($n_D^{20} = 1.4205$). As expected, methallyl chloride,
bearing the electron-donating methyl group, seems to
respond somewhat more to Ziegler-Natta type catalysts
than allyl chloride. In one example, 60 g of methallyl
chloride in isooctane was treated with $TiCl_4$ giving a
brown coloration (116). On adding triethyl aluminum, the
temperature rose from 20 to 50°C and a dark suspension of
sticky viscous polymer was formed. When $AlBr_3$ in heptane
was added dropwise at room temperature to methallyl
chloride-hexane (1:1), the mixture turned brown then black
(117). After refluxing 8 hr at 100°C and cooling, excess
methanol was added. A white solid polymer was reported
melting at 90°C.

Low polymers of allyl chloride terminated by trichloro-
silanes were prepared under gamma radiation from [60]Co
(118). For example, AC with $ClCH_2(CH_2)_2SiCl_3$ in 62 hr
gave 40% yield of telomer boiling 58 to 60°C at 6 mm with
$n_D^{25} = 1.4646$.

Allylic halogen compounds may be added as regulators in
polymerizations under conditions such that little if any
allyl compound may react except by chain transfer. Thus
AC has been added as a regulator in the cis-1,4-polymeri-
zation of butadiene using Et_2AlCl, cobalt octanoate, and
metallic aluminum as catalysts (119). Delayed introduction of allyl bromide was used to produce polybutadiene
having a bimodal molecular weight distribution (120).
Polymerization of dienes by organolithium catalysts may
give a narrow molecular weight range of rubber which is
difficult to process. By delayed addition of allyl
bromide, a polymer fraction of lower molecular weight
is formed, and it acts somewhat as a plasticizer. In one
example, there was mixed under nitrogen 180 ml of toluene,
12.75 g of butadiene, and 10.2 mg of butyl lithium. This
was rotated in a reactor at 55°C for 20 min and then 10.2
mg of allyl bromide was added and polymerization continued
for 2 hr. By elution, polybutadiene fractions of intrinsic
viscosity 0.95 and 1.50 were isolated from the product.
Allyl chloride was a potent regulator of molecular weight
in the polymerization of ethylene by Ziegler catalysts
(121). However, higher omega-haloalkenes, including
4-iodo-1-butene, have been homopolymerized by use of
organoaluminum-titanium halide catalysts (122).

Allyl bromide has been used to prepare allylated phenolic
resins (123). A mixture of formalin, phenol, water, allyl

bromide, and sodium hydroxide catalyst was stirred 8 hr
at room temperature under nitrogen. Vacuum distillation
left a yellow viscous liquid--allylated trimethylol phenol
resin intermediate--which was tested with epoxy resins in
coatings. Allylated phenol-formaldehyde resins, made from
phenol, formaldehyde, and AC with NaOH catalyst, were
reacted with toluene diisocyanate to form phenolic urethane polymer foams (124).

ALLYL HALIDES IN COPOLYMERIZATIONS

Allyl chloride (AC) reacts at moderate temperatures with
SO_2 in presence of peroxides to form solid sulfone copolymers (125). Allyl bromide would not so copolymerize, and
it inhibited reaction of AC with SO_2 (126). Terpolymers
of AC, isobutylene, and SO_2 were prepared where the proportion of AC units in the product was lower than in the
monomer mixture (127). The proportion of sulfur dioxide
units in these products was near 50 mole %, whereas the
proportion of isobutylene and allyl chloride can be varied
by the feed (128). In one example, a mixture of 80 mole %
AC and 20 mole % isobutylene was added to excess SO_2 at
-35°C in the presence of 1% dicumyl peroxide during 50 min.
The polysulfone formed was soluble in sulfuric acid and
melted near 190°C. Chemical instability to alkali and to
heat have discouraged practical applications. Japanese
copolymers of AC-1-butene-SO_2 were soluble in acetone.
Allyl chloride has been copolymerized with vinyl chloride
together with maleate or fumarate esters (129). It has
also been copolymerized with maleic anhydride by long
heating with lauroyl peroxide at 60°C (130). Additions
of AC or MAC greatly retard polymerizations of most vinyl
monomer, and the products have lower molecular weights
resulting from chain transfer. Figures 4.3 and 4.4 show
effects of allyl chloride on radical polymerization of
ethyl acrylate at 30°C and at -78°C.
Heating AC with sodium tetrasulfide gives hard polymers
(131), and with sodium sulfide and sulfur rubberlike
products have been observed (132). In the preparation of
Thiokol polysulfide polymers from reaction of dichloro
compounds, such as dichlorodiethyl formal and sodium polysulfide, AC may be added as a terminating agent; at the
same time, the addition provides reactive terminal olefin
groups to facilitate curing (133).
Allyl chloride is much more reactive with carbon tetrachloride in forming low molecular telomers than with $CHCl_3$
in presence of BP catalyst at 100°C (134). Yields of
telomers per mole of BP consumed are smaller than obtained
from isobutylene or propylene.
Alfrey and Harrison studied the retarding effects of

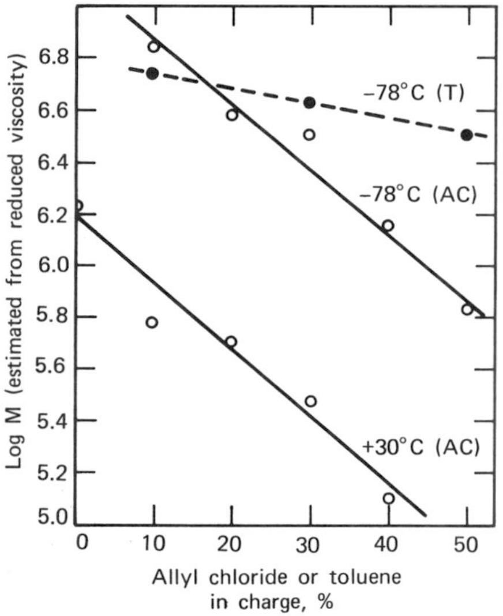

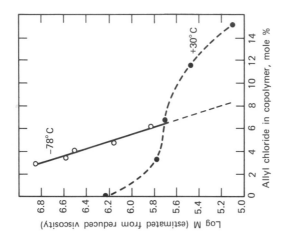

Fig. 4.4. Molecular weights of ethyl acrylate-allyl chloride copolymers as functions of allyl chloride units in the copolymers formed. Tributyl boron and oxygen were used for initiation. T. L. Dawson and R. D. Lundberg, J. Polym. Sci. A 3, 1803 (1965).

Fig. 4.3. Effects of allyl chloride (AC) addition upon molecular weights of ethyl acrylate polymers at 30°C and at -70°C using tributyl boron-oxygen initiation. TBB was 0.061 molar and O_2 was 0.024 to 0.029 molar. Addition of toluene (T) had much less effect. T. L. Dawson and R. D. Lundberg, J. Polym. Sci. A 3, 1803 (1965).

allyl chloride on polymerization of styrene (135). To a
styrene free radical chain end, AC monomer adds only about
0.032 times as fast as does styrene monomer. It was suggested that low rates of propagation contributed to the
low overall rates and the low DP of the products. Joshi
and Kapur, at 60°C with BP, found copolymerization reactivity ratios of vinyl monomers r_1 with allyl chloride
as r_2, respectively, as follows: alkyl methacrylates
about 50 and 0.1; alkyl acrylates about 8 and 0.1; vinyl
acetate 0.34 and 0.75; vinyl butyrate 0.59 and 1.62;
vinyl benzoate 0.46 and 0.88; styrene 37 and 0.3 (136).
Small amounts of copolymerized AC units in styrene or
methacrylate polymers were detected by quaternization with
pyridine and applying dyes (137).
    Tawney and co-workers added allyl chloride to improve
yields of soluble prepolymers of diallyl fumarate and
styrene, also of diallyl fumarate and vinylidene chloride
(138). Copolymerizations with 10% or more AC were run in
sealed tubes, for example, with BP at 60°C for over 100 hr.
Avoiding gelation by adding allyl chloride permitted
separation of fusible prepolymers which could be used with
additional peroxide and heating to give thermoset films
and coatings. Soluble copolymers of allyl acrylate and
methallyl chloride were prepared in bulk in sealed tubes
with BP by heating for 21 hr at 60°C (139). In one experiment, 10 parts of a copolymer containing 67% allyl acrylate
units and 36% methallyl chloride units were mixed with 4.5
parts of styrene and 0.09 part of BP. The mixture was
heated in a mold at 60°C followed by 120°C to give a
thermoset object. Fusible prepolymers of triallyl itaconate, styrene, and methallyl chloride were also molded
to thermoset shapes using peroxide catalyst. In contrast
to peroxide-initiated polymerizations, the addition of
10 mole % AC accelerated the photopolymerization of styrene
with 3660 and 3080 Å UV (140). The acceleration was less
than that produced by adding CCl_4.
    Vinylidene cyanide $CH_2=C(CN)_2$ with allyl chloride in
ratio 2:7 by weight and the presence of o,o-dichlorobenzoyl
peroxide at 50°C slowly formed copolymer of nearly 1:1
molar composition (141). Equal parts by weight of monomers
yielded copolymer containing 67 mole % vinylidene cyanide
units. Vinylidene cyanide and methallyl chloride in ratio
1:10 with dichlorobenzoyl peroxide at 40°C formed copolymer
containing 48 mole % vinylidene cyanide units. A monomer
mixture of about 4:1 gave copolymer containing 74 mole %
vinylidene cyanide units. Benzene was used as solvent for
the copolymerizations. Application of the copolymers as
fibers spun from solution in dimethyl formamide was considered.

Acrylonitrile has been copolymerized with minor proportions of allyl and methallyl halides. In one example, 82%
acrylonitrile and 18% AC required 24 hr at 50°C for copolymerization by aqueous persulfate-bisulfite-sodium
lauryl sulfate emulsion (142). The fibers therefrom were
heated with 10% N,N-dimethylhydrazine in water for quaternization and improved dye receptivity. Allyl chloride
copolymers were quaternized by pyridine and their compositions were estimated by dye partition techniques (143).
In copolymerizations with methyl methacrylate AC was more
effective in lowering intrinsic viscosity than was allyl
acetate (144). Boron alkyl-oxygen initiation was used
for mixtures of allyl chloride and ethyl acrylate at -78
to +100°C (145). It was confirmed that this is a free
radical system in which AC behaves both as comonomer and
as chain transfer agent. Chain transfer to AC relative
to propagation was as pronounced at -78°C as at 30°C.
Copolymer molecular weights were only slightly higher at
-78°C. Acrylic acid with AC gave reactivity ratios r_1 =
11.5 and r_2 = 0.01 (146).

Allyl chloride was evaluated as a regulating and chain-
terminating agent in the radical-initiated high-pressure
polymerization of ethylene (147). Also of interest has
been addition of allyl chloride in cationic polymerization
of olefins, especially with complex transition metal
halide catalysts. Block copolymers of diene and allyl
chloride units were made using Ziegler-type catalysts
(148). Flame-resistant copolymers of ethylene and allyl
halides were prepared using catalysts from alkyl aluminum
halides and $VOCl_3$ or $TiCl_4$ (149). Allyl halides have been
added in 1-olefin polymerization using Ziegler-type
catalysts in order to control molecular weight, raise
polymer melt index, and obtain better injection molding
of large articles such as waste baskets, garbage containers
and laundry hampers (150). In one example, propylene and
ethylene (95 to 5) were polymerized in a 1-gal stainless
steel pressure reactor. The solvent cyclohexane was added
first and after flushing with nitrogen, $TiCl_3$ was introduced followed first by triisobutyl aluminum in cyclohexane
solution and then by 0.938% allyl chloride based on the
monomers. The reactor was pressurized with the monomer
gases. The temperature was raised until polymerization
was initiated, after which the exothermic polymerization
maintained the temperature in the range of 105 to 125°C.
The melt index of the copolymer was 4.9, compared with 1.3
without added AC. Melt index was defined as the grams of
polymer extruded in 10 min through an orifice of 0.0825 in.
at 190°C when subjected to a load of 2160 g. Equimolar
amounts of either allyl chloride or allyl bromide to that

of titanium compound were sufficient to control high-density polyethylene by Ziegler catalysts to within the range of molecular weights useful in most commercial applications (151). However, methallyl chloride was found to have little influence upon molecular weight and was not incorporated into the polyethylene chain molecules. Allyl halide addition in polymerization of ethylene by complex transition metal catalysts not only controlled molecular weight but also prevented rapid deactivation of the catalysts (152).

Allyl halides with isobutylene in alkyl halides and conventional Lewis-acid catalysts have given copolymers of high molecular weight at low temperatures (153). For example, 5 to 25 parts of AC mixed with 75 to 95 parts of methyl chloride were treated with BF_3, $AlCl_3$, $AlBr_3$, or $SnCl_4$ to give strongly colored dispersions. When these were reacted with solutions of 10 to 35 parts of isobutylene and 65 to 90 parts of methyl chloride, copolymers of high molecular weight were formed. In one example, 2.25 g MACl in 15 g CH_3Cl at -50°F was treated with BF_3 until a yellow color persisted. Half of this solution was added to a mixture of 100 g CH_3Cl and 30 g isobutylene at -90°F. The copolymer, 92,000 mol. wt., contained 0.63% Cl.

Although allyl halides were reported to deactivate some Ziegler-Natta catalysts with production of hydrogen chloride, 6-chloro-1-hexene and other compounds with halogen removed from the allyl group formed copolymers with propylene (154). Copolymer yields were improved and loss of HCl was suppressed by adding to the catalyst small amounts of a Lewis base. In one example, in a stirred glass reactor the following were introduced in order: 100 ml n-heptane, 0.36 g $Al(C_2H_5)_2Cl$, 0.20 g $TiCl_3$, 0.079 g pyridine, and 1.42 g 6-chloro-1-hexene. The reactor then was pressurized with propylene to 50 psi for an hour. After releasing pressure and quenching the catalyst with 2 ml of methanol, the mixture was poured into 600 ml of methanol acidified with HCl. The copolymer was washed with methanol on a filter and was dried at 80°C. The reduced viscosity was 2.3 and the chlorine content indicated about 7% chloroalkene units. Using a highly stereospecific catalyst from $Al(C_2H_5)_2Cl$ and $TiCl_3$, copolymers of propylene with omega-halo-1-olefins were obtained showing crystalline fractions. The normally crystalline fractions displayed less halogen than the amorphous fractions.

The very low reactivity of AC (M_1) in ordinary free radical copolymerizations with vinyl, acrylic, and styrene monomers (M_2) is shown by r_1 values (155). The most favorable was $r_1 = 0.48$ and $r_2 = 2.3$, reported with ethyl

acrylate. The r_1 values for methallyl chloride (M_1) with
the same comonomers were essentially zero.

OTHER ALLYLIC HALIDES

   Examples of some other allylic halides are listed in
Table 4.2. Hatch and co-workers studied the relative
reactivities of a series of allylic chlorine compounds in
the SN_2 reaction with potassium iodide in acetone at 20°C
to form allylic iodine compounds (156). Vinyl chlorine
atoms did not react under these conditions. Bromallyl
chloride was slightly more reactive than allyl chloride.
Methallyl chloride was more reactive than allyl chloride.
Still more reactive were cis- and trans-1,3-dichloropro-
penes and cis- and trans-1,3-dichloro-2-methyl-1-propenes.
The rate of replacement of chlorine by iodine increased
as the substituent on the second carbon atom became less
electronegative (Cl > Br > H > CH).
   Many allylic halides (Table 4.3) are highly reactive,
unstable, and highly unpleasant to work with in the
laboratory. For example, $CH_2=CClCH_2Br$ (bp 120°C) strongly
attacks the mucous membranes. Chlorallyl chloride was
made by dehydrochlorination under various conditions (157).
A trichloropropane was heated with water in a sealed tube
(158):

$$CH_3CCl_2CH_2Cl \xrightarrow{H_2O} CH_2=CClCH_2Cl + HCl$$

Heating this reactive chloroolefin with KSCN gave chlorallyl isothiocyanate (159), possibly an even more reactive
compound than mustard oil. With triethylamine the chlorallyl chloride gave chlorallyl triethyl ammonium chloride
(160). The polymerization behavior of the chloroallyl
compounds may more closely resemble that of vinyl halides
than allyl halides. The instability and repulsive
physiological properties of these monomers discourage
research.
   Bromallyl bromide was made by heating 200 g of
$CH_2BrCHBrCH_2Br$ with 10 g of water and 50 g of NaOH (161).
There are numerous early references to reactions of bromallyl bromide. For example, cyclohexyl magnesium bromide
in ether gave allylic coupling with bromallyl bromide (162):

$$\text{⬡}MgBr + BrCH_2CBr=CH_2 \longrightarrow \text{⬡}CH_2CBr=CH_2 + MgBr_2$$

Photoisomerizations of allylic halides have been reviewed
(163). Because of electron attraction of the CCl_3 group,
$CH_2=\underset{CCl_3}{C}CH_3$ does not show typical allylic reactivity (164).
It is not attacked by Br_2 or by aqueous permanganate, but
rearranges in the presence of acids to form $Cl_2C=C(CH_3)CH_2Cl$

TABLE 4.3

Other Allylic Halogen Compounds

	bp, °C	n_D^{20}	References and Other Properties
$CH_2=CClCH_2Cl$	94	1.4603	Henne, JACS 63, 2692 (1941); Groll, Ind. Eng. Chem. 31, 1534 (1939); Coleman, U.S. 2,285,329 (Dow)
$CH_2=CBrCH_2Cl$	127	1.4968 (25°C)	Henry, Ber. 5, 453 (1872); Hatch, JACS 70, 1093 (1948)
$CH_2=CHCHCl_2$	83	1.4510	Andrews, JACS 70, 3458 (1948)
$CH_2=CHCCl_3$	115		Vernon, J. Chem. Soc. 3629 (1952); Vitoria, CZ I, 345 (1905); Henry, CZ I, 1697 (1905)
$CH_2=CHCHClCH_3$	--	--	Young, JACS 73, 1076 (1951) (reactions with amines)
$CH_2=CHCH_2CH_2Cl$	75	1.4233	Juvala, Ber. 63, 1993 (1930) (reactions with Br_2 and KI)
$CH_2=CHCH_2CCl_3$	129	1.4678	Kharasch, J. Org. Chem. 14, 82 (1949) (from $ABr + CBrCl_3$ in UV)
$CH_2=C(CH_3)CHCl_2$	112	$n_D^{24} =$ 1.4523	Tischtchenko, CA 33, 4190 (1939);
$CH_2=C(CH_3)CCl_3$	134	1.4770	McElvain, JACS 69, 2669 (1947); Kundiger, JACS 76, 615 (1954)
$CH_2=C(CH_3)CH_2Cl$	--	--	U.S. 2,067,392 (Shell)
$CH_2=C(CH_2Cl)_2$	138	1.4754	Mooradian, JACS 67, 942 (1945); Burgin, Ind. Eng. Chem. 31, 1414 (1939); Ropp, JACS 73, 3024 (1951) (dimer and trimer by BP); copolymers: Hulse, U.S. 2,571,883 (Hercules)
$CH_2=CClCH_2Br$	121	1.505	Henry, Bull. Soc. Chim. [2] 39, 526 (1883); Kremer, Bull. Soc. Chim. 166 (1948); Marvel, Org. Syn. 5, 49 (1925)
$CH_2=CBrCH_2Br$	141 or 70(60)	1.5157 or $n_D^{21} =$ 1.5352	Kopper, Monatsh. 62, 84 (1933); Peer, U.S. 3,197,514; Henry, Ann. 154, 371 (1870); Ber. 14, 404 (1881)
$CH_2=CHCHBr_2$	31(14)	1.532	Kirrmann, Compt. Rend. 202, 1934 (1936)
$CH_2=CHCH_2CH_2Br$	99	1.4622	Juvala
$CH_2=CBrCHBrCH_3$	75(20)	1.5464	Hurd, JACS 53, 293 (1931)
$CH_2=CCH_2Cl$ $\quad C_6H_5$	--	--	Keith, U.S. 3,361,835 (Sinclair) (from alpha-methylstyrene + Cl_2 at 500°C)
$CH_2=CBrCHBr_2$	190	--	Pinner, Ann. 179, 59 (1875); Ber. 8, 1562 (1875)
$CH_2=CClCH_2I$	95(40) with decomp.	--	Van Romburgh, Rec. Trav. Chim. I, 237 (1882)

179

There has been little polymerization research with many
of the possible other allylic halogen compounds. Most
will be expected to resist homopolymerization and to retard copolymerizations of active vinyl monomers. Attempts
to homopolymerize allylidene dichloride $CH_2=CHCHCl_2$
(prepared from acrolein and phosphorus pentachloride) by
heating with dicyclohexyl percarbonate at 50°C showed only
slow increase in molecular weight, even under pressure
(165). When 20% of the dichloride and 80% vinyl acetate
containing BP were heated at 50°C, 26 days was required
to reach a viscous solution (166). The copolymer showing
21.6% allylic dichloride units became brown on drying at
50°C. In free radical copolymerizations with methyl
methacrylate and with styrene under similar conditions,
only small proportions of allylic units entered into the
polymer products. Copolymerizations of chlorallyl halides
and alcohols with styrene and acrylic monomers also were
unpromising (167). In copolymerizations of chlorallyl
chloride with methacrylic acid, with methyl methacrylate,
and with styrene using radical initiator, the values of r_1
were near zero. However, fluorallyl chloride in copolymerizations with acrylonitrile and with methyl methacrylate
using isopropyl percarbonate catalyst at 25°C gave polymers
containing 32 and 46% fluoromonomer units, respectively
(168). Polymerizations of $CH_2=C(CH_3)CCl_3$ have been little
studied. This compound is insoluble in concentrated H_2SO_4
and is not attacked by Br_2 in CCl_4 or by aqueous potassium
permanganate. Apparently the CCl_3 group is sufficiently
electron attracting to prevent the CH_3 group from showing
allylic behavior. Heating with HCl or HF gives $CCl_2=$
$C(CH_3)CH_2Cl$ (169). Semisolid low polymers from
$(CH_3)_2CClCH=CH_2$ obtained in C_2H_5Cl with $AlCl_3$ catalyst
showed by NMR some structure of the type $--(CH_3)_2CCHCl-CH_2-$
resulting from chloride-shift polymerization (170).
   An interesting monomer containing 4 allylic hydrogen
atoms per molecule is $CH_2=C(CH_2Cl)_2$. This gave low polymers by heating with 5% BP for 6 days at 65°C (171). A
liquid polymer fraction boiling at 172 to 205°C at 1 mm
formed a white crystalline solid melting at 117°C which
was found to be a trimer. A higher boiling fraction was a
viscous brown polymer. Photopolymerization gave viscous
liquid products. Long heating at 65°C of mixtures of this
monomer with maleic anhydride and BP gave a viscous liquid
polymer of molecular weight 339 and a white powder of
molecular weight 836, which melted above 90°C. Of special
interest is the polymerization of $CH_2=CHCCl_3$ in UV light
at 20 to 50°C, which formed polymers melting at 150°C
(172). These polymers, believed to contain largely 1,3-
linkages $--CH_2CHCl-CCl_2--$, were insoluble in acetone and

in methylene chloride, but they dissolved in hot hexa-
chlorobutadiene. The polymers could be molded at 170°C
under pressure.

Compounds of the types $CH_2=CHCHClR$ and $CH_2=C(CH_3)CHClR$
have had few applications. Reaction of alpha-methallyl
chloride $CH_2=CHC(CH_3)Cl$ with dimethylamine in benzene at
70°C does not give normal SN_2 displacement; rather, the
principal product is trans-N,N-dimethyl-γ-methallylamine
(173). Reaction of α-methylallyl alcohol with $SOCl_2$ in
ether gave predominantly the crotyl chloride (174). However, in the presence of a tertiary amine, a mixture of
chlorides resulted. Inversion of configuration depended
upon reaction conditions.

Reaction of acrolein with PCl_5 gives the highly reactive
allylidene chloride $CH_2=CHCHCl_2$ (175). It is also formed
in small amounts by action of chlorine with allyl chloride
or propylene. Allylidene chloride copolymerized reluctantly with styrene, with vinyl acetate, and with methyl
methacrylate in the presence of radical catalysts (176).
Allylidene bromide was made by dehydration of $CH_3CH(OH)-$
$CHBr_2$ (177). It was stable for a week in a sealed tube
over a trace of KOH, but in the presence of hydrogen
bromide it rearranged to $CHBr=CHCH_2Br$.

FLUOROALKENES

More research is needed to determine to what extent
fluorinated allyl compounds behave as typical allyl compounds. However, some significant differences have been
observed already in line with the electronegativity of
fluorine. Fluorinated l-alkenes of more than two carbon
atoms polymerize less readily than vinyl fluoride and
vinylidene fluoride, and they generally require pressures
above 1000 atm for the formation of solid homopolymers of
moderate molecular weight.

Hexafluoropropylene (HFP) $CF_2=CFCF_3$ has been carefully
studied in free radical copolymerizations under high
pressure (178) and copolymers of HFP with larger proportions of $CF_2=CF_2$ and of $CH_2=CF_2$ are commercial products
(moldable Teflon X-100 resins and Viton A elastomers of
Du Pont) (179). Hexafluoropropylene or perfluoropropylene
and technical information about this normally gaseous
monomer have been supplied by the Pennwalt Corporation.
At lower pressures HFP gives only soft homopolymers of low
DP. Rigid, high softening polymers have been obtained by
use of fluorinated radical-forming initiators such as
$(CF_3C(O)O)_2$ or by gamma radiation at 3 to 15 katm and
temperatures above 200°C (180).

Lowry, Brown and Wall were able to make homopolymers of
hexafluoropropylene having intrinsic viscosities as high

as 2.0 dl/g. Both the rates and DP increased with rise
of pressure and temperature. From polymerizations at
5-15 atm at 100-230°C with gamma ray initiation they concluded that molecular weights were principally controlled
by transfer and by termination. Small amounts of
impurities produced pronounced differences in behavior of
different samples of this difficult monomer. Such difficulties are also encountered in manufacture of Viton
elastomers such as interpolymers of 60% $CH_2=CF_2$ and 40%
HFP units. These fluoropolymers are valued for their
resistance to space-age fuels and lubricants, permitting
continuous service over the temperature range of -65 to
400°F. The Viton elastomers were first introduced by Du
Pont in 1957 as sealing materials, but applications now
include coated fabrics, oil and fuel lines, molded
electrical connectors, cable coatings, gaskets, and caulks.
In combination with asbestos Viton copolymers can be used
even at 600 to 800°F.

Eleuterio, Sianesi, and Caporiccio revealed ionic polymerizations of hexafluoropropylene of unusual interest
(181). In one example, Eleuterio dissolved 0.2 g mercuric
trifluoromethyl mercaptide $Hg(SCF_3)_2$ in 75 ml perfluoro-
1,3-dimethylcyclobutane solvent in an autoclave at 225°C,
and this was pressured with HFP at 3000 atm for 14 hr.
An amorphous homopolymer softening at 225-250°C was recovered. The solid, clear polymer remained relatively
stiff at 90°C. Another catalyst type suggested was lead
tetrakis(perfluorocarboxylates). Interpolymers of HFP
with minor proportions of $CF_2=CF_2$ were made by mercuric
mercaptide or CoF_3 catalysts. These interpolymers could
be molded at 250°C; on cooling they were limp, transparent
films completely amorphous by x-ray studies. One may
conclude that these perfluoropropylene macromolecules were
largely atactic. The response of HFP to such nucleophilic
catalysts is not surprising in view of the electron-
attracting properties of the CF_3 group. Partially isotactic and normally-crystalline HFP polymers were reported
by Sianesi and Caporiccio by use of Ziegler-Natta type
catalysts. Polymer of crystal melting point above 115°C
formed very slowly. In one experiment a catalyst was
prepared from titanium isopropylate and aluminum isobutyl
in methylene chloride. This catalyst acting upon HFP for
15 days at 30°C and 9 days at 40°C gave a small yield of
white polymer. After extractions with CCl_4 and with
n-heptane the residual polymer fraction melted at 170°C.

Polymerizations of fluoroolefins at high pressures were
reviewed (182). A number of fluoropropylenes polymerize
under pressure more readily than either propylene of HFP.
Polymerization of $CH_2=CHCF_3$ by radical initiation was
reported in patents (183). Polymers of low DP were

reported by use of Ziegler-Natta type catalyst (184).
Brown at the U. S. National Bureau of Standards, using
high pressures and gamma radiation, found the polymer
molecular weights to increase with pressure but not with
temperature. Rates of polymerization of $CH_2=CHCF_3$ increased with the square root of radiation intensity. The
homopolymers were soluble in acetone and in hexafluorobenzene. Two solid polymers gave intrinsic viscosities
in acetone at 30°C of 0.15 and 3.5 dl/g; but in hexafluorobenzene solution the values were 0.25 and 8.5 dl/g,
respectively. Acetone was a poor solvent from which the
high polymer separated on cooling to 25°C (185).

Wall concluded that the importance of the transfer step
varied greatly with the reaction conditions (186). At
1800 atm and 99°C, about 98% of the growing radicals
terminated by transfer, whereas at 5000 atm and 25°C only
25% termination by transfer occurred. Thus the molecular
weight of the polymer from $CH_2=CHCF_3$ decreased with rising
temperature at constant pressure, and eventually the reaction became controlled largely by transfer. Differences
in temperatures of thermal decomposition of the polymers
were observed depending on polymerization conditions (187).
A polymer sample prepared at 10 katm and -80°C showed best
stability against thermal decomposition. As Fig. 4.5
illustrates, $CH_2=CHCF_3$ was quite reactive in copolymerization with $CF_2=CF_2$.

Electron withdrawal from the double bond of $CH_2=CFCF_3$
would be expected to give this monomer properties of a
vinyl halide monomer, and at high pressures it does in
fact form polymers of high molecular weight readily (188).
$CHF=CHCF_3$ gave only slow rates of polymerization from the
liquid state at 13 katm and 70°C, forming polymer of very
low DP (189).

Fluorinated higher 1-olefins such as $CH_2=CHCF_2CF_3$ and
$CH_2=CHCF_2CF_2CF_3$ polymerize more reluctantly than $CH_2=CHCF_3$.
With gamma-ray initiation, the two monomers required 10 to
15 katm to produce polymers with "appreciable intrinsic
viscosities" (190). The homopolymers were soluble in
hexafluorobenzene, but not in acetone. Maximum polymerization rates were near 30%/hr using the lowest dose rate
convenient.(about 2 krad/hr). Polymerization of the
liquid perfluoroethylethylene occurred most readily just
below its freezing point of 22°C at about 13 katm. In
the bulk polymerization of $CH_2=CHCF_2CF_2CF_3$ at 10 katm,
postpolymerization after irradiation was observed which
was attributed to a peroxide formed in situ.

Perfluoro-1-heptene required forcing conditions of 17
katm and high temperature for polymerization, even to
products of low DP (191). The compound $CF_2=CFCF_2CF=CF_2$
polymerized along with isomerization to the conjugated

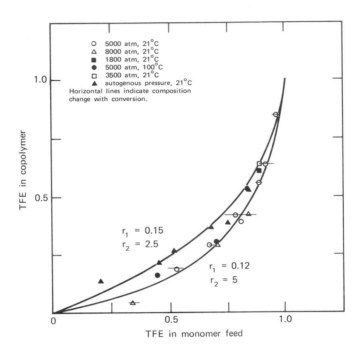

Fig. 4.5. In copolymerization with $CF_2=CF_2$ (TFE) initiated by
gamma-radiation, $CH_2=CHCF_3$ seems to be more reactive than typical
allyl halides. For all except highest proportions of TFE monomer
the copolymers were lower in TFE units than the monomer feed. The
reactivity ratios were little changed by pressure and by temperature. D. W. Brown and L. A. Wall, Polym. Preprints 7, 1122
(September 1966); cf. J. Polym. Sci. A1 6, 1367 (1968).

compound so that copolymers were actually formed (192).
Gamma radiation and high pressures were used. Polymers
high in units from the conjugated monomer were rubbery,
but those from largely unisomerized monomer were brittle
solids. Wall obtained polymers soluble in hexafluorobenzene by polymerization of a number of nonconjugated
perfluorodienes of the type $CF_2=CF(CF_2)_nCF=CF_2$ at high
pressures with gamma radiation (193). These homopolymers

were believed to contain cyclic structures $--CF_2 \diagdown \begin{smallmatrix} CF_2 \\ \\ CF_2 \end{smallmatrix} \diagup CF--$,

but high conversions and impurities led to crosslinked

TABLE 4.4

Some Fluoropropylenes and Related Compounds

	mp, °C	bp, °C	References
$CF_2=CHCH_3$	-161	-29	Henne, JACS 68, 496 (1946); Moss, Trans. Faraday Soc. 65, 415 (1969)
$CF_2=CHCF_3$	-153	-21	Henne; also McBee, JACS 70, 2024 (1948); Sianesi, Belg. 624,205; 626,289; 633,329 (Monte)
$CH_2=CHCF_3$	--	ca. -19	Henne; Robbins, CA 40, 3187 (1946)
$CH_2=CFCF_3$	-152	-28	Henne; Kung, CA 71, 10127 (1969)
$CHF=CFCF_3$	--	--	Sianesi, Belg. 642,942; 645,894 (Monte); CA 63, 4477 and 8617 (1965)
$CH_2=CFCF_2Cl$	-144	+12	Henne
$CH_2=CFCFCl_2$	-116	54	Henne
$CF_2=CFCF_3$	-156	-29	Park, Ind. Eng. Chem. 39, 356 (1947); Sianesi, Makromol. Chem. 86, 308 (1965) (photooxidation to ether polymers)
$CF_2=CClCF_3$	-130	+7	Henne, JACS 63, 3479 (1941); Miller, Ind. Eng. Chem. 39, 402 (1947)
$CF_2=CBrCF_3$	--	25	McBee, JACS 70, 2024 (1948)
$CF_2=CFCF_2Cl$	--	7.5 $n_D^{14}=1.3050$	Henne, JACS 70, 131 (1948)
$CF_2=CClCF_2Cl$	-121	45	Henne, JACS 63, 3479 (1941)
$CF_2=C=CF_2$	--	--	Banks, J. Chem. Soc. C, 1104 (1969)

polymers. Copolymerization of CF_2-CF_2 with $CH_2=CHCF_2CF_2CF_3$ at high pressures under radiation from ^{60}Co was studied (194). With Ziegler-Natta catalysts, $CH_2=CHCH_2CH_2CF_3$ polymerized more readily than $CH_2=CHCH_2CF_3$ (72 hr at 70°C) (195).

Allyl fluoride was made by heating allyl chloride and potassium fluoride in diethylene glycol (196, 197). Margulis and co-workers polymerized monomer of undisclosed properties by heating it with nitrogen in a sealed tube at 70°C for 100 hr with 2% azobis as initiator (197). The polymer sample contained 28.9% F and there was no evidence of evolution of HF. Infrared spectra showed conventional allyl halide polymer structure as well as nitrile groups. Allyl fluoride was copolymerized with vinyl acetate by heating at 80°C with free radical initiator to give a yellow, viscous resin, and it was copolymerized with vinyl chloride at 65°C.

Fluorocompounds of the type $CH_2=CHCF_2CF_2Y$ (where Y is COOR, $CONH_2$, or CN) were said to polymerize on heating with di-t-butyl peroxide at 125 to 150°C, but the properties of the products were not disclosed (197). Table 4.4 gives references and properties of some fluoropropylenes and related haloalkenes. Some of these such as $CH_2=CHCF_3$ showed anesthetic properties in tests with laboratory animals. Allyl fluoride was included in Table 4.1.

References

1. Allyl Chloride and Other Allyl Halides, Shell Chemical Corp., New York and San Francisco, 1949.
2. V. E, Karmazin et al, CA 66, 27414 (1967).
3. V. Druzhinina (Moscow), CA 60, 9811 (1964).
4. A. Y. Cherkovskaya et al, CA 60, 4718 (1964).
5. W. Engs, U.S. 2,341,140 (Shell).
6. Belg. 630,900 (Solvay); CA 60, 11895 (1964).
7. K. Radcliffe and J. L. Wood, Trans. Faraday Soc. 62, 2038 (1966).
8. M. Berthelot and S. de Luca, Ann. 92, 306 (1854).
9. A. Cahours and A. W. Hofmann, Ann. 100, 356 (1856); 102, 285 (1857); cf. B. Tollens, ibid. 156, 154 (1870).
10. H. Meslans, Compt. Rend. 111, 882 (1890); 114, 763 (1892).
11. O. Kamm and C. S. Marvel, Org. Syn. 1, 3 (1921); cf. J. F. Norris et al, J. Am. Chem. Soc. 38, 1076 (1916).
12. M. Sheshukov, J. Russ. Phys.-Chem. Soc. 16, 478 (1884).
13. E. W. Collington and A. I. Meyers, J. Org. Chem. 36, 3044 (1971); 37, 1466 (1972).
14. Y. A. Treger et al, USSR 162,120; CA 61, 8190 (1964).
15. Y. A. Treger et al, USSR 165,161; CA 62, 2707 (1965).
16. A. Zaitsev, Ann. 185, 191 (1877); G. Wagner, Ber. 9, 1810 (1875); cf. B. Tollens and A. Henninger, Ann. 156, 156 (1870); R. L. Datta and F. V. Fernandes, J. Am. Chem. Soc. 36, 1005 (1914).
17. M. Sheshukov, J. Russ. Phys.-Chem. Soc. 15, 355 (1883); Ber. 16, R 1869 (1883); cf. J. Kondakov's rule, Ber. 24, 929 (1891).
18. M. Sheshukov, J. Russ. Phys.-Chem. Soc. 16, 478-511 (1884); Ber. 17, R 412-15 (1884).
19. Z. A. Pogorshelski, J. Russ. Phys.-Chem. Soc. 36, 1129-84 (1904); CZ I, 668 (1905).
20. J. Burgin, W. Engs, H. P. A. Groll, and G. Hearne, Ind. Eng. Chem. 31, 1413 (1939); cf. M. L. Poutsma, J. Am. Chem. Soc. 87, 4285 (1965).
21. R. W. Taft, J. Am. Chem. Soc. 70, 3364 (1948).
22. N. T. Sultanov et al, CA 63, 11332 (1965); T. I. Akhmadiev et al, CA 74, 142396 (1971).
23. Brit. 963,031 (Monsanto); cf. Brit. 935,088 and U.S. 2,966,525.
24. H. P. A. Groll, G. Hearne, J. Burgin, and D. S. LaFrance, U.S. 2,130,084 (Shell); CA 32, 9096 (1938); Ind. Eng. Chem. 31, 1530 (1939).

25. F. F. Rust and W. E. Vaughn, J. Org. Chem. _5_, 472 (1940).

26. D. Brown, U.S. 3,120,568 (Halcon).

27. K. Ziegler, A. Spaeth, E. Schaaf, W. Schuman, and E. Winkelmann, Ann. _551_, 80 (1942).

28. L. Hoernig and G. Mau, Ger. 1,230,781; U.S. 3,489,816 and 3,513,207 (Hoechst); also Ger. 1,222,913 and 1,224,301-2 (Hoechst).

29. L. Hoernig and G. Mau, U.S. 3,462,502 (Hoechst); cf. CA _66_, 85450 (1967), _64_, 15740 and 17420 (1966).

30. L. Hoernig and G. Mau, U.S. 3,462,501 and 3,607,959 (Hoechst).

31. O. Kamm and C. S. Marvel, in Org. Syn. Coll. I, 27 Wiley, 1941.

32. M. S. Kharasch, J. Org. Chem. _2_, 489 (1937); E. A. Braude, Quart. Rev. (London) _4_, 419 (1950).

33. M. Tamele et al, Ind. Eng. Chem. _33_, 118 (1941).

34. H. Fernholz and H. Wendt, U.S. 3,627,849 (Hoechst).

35. G. E. Ham and W. P. Coker, J. Org. Chem. _29_, 194 (1964).

36. D. H. Volman and R. W. Phillips, Ber. Bunsenges, Phys. Chem. _72_, 242 (1968).

37. A. Oppenheim, Ann. _133_, 383 (1865).

38. J. Thiele and W. Peter, Ann. _369_, 149 (1909).

39. W. Bockemueller and L. Pfeuffer, Ann. _537_, 178 (1939).

40. L. D. Swindale et al, J. Chem. Soc. 812 (1950); cf. K. Nozaki and R. A. Ogg, J. Am. Chem. Soc. _64_, 704 (1942).

41. N. J. Bythell and P. W. Robertson, J. Chem. Soc. 182 (1938).

42. Recent examples: L. A. Nakhapetyan et al, CA _61_, 10577 (1964); G. N. Kovalev et al, CA _64_, 5989 (1966).

43. C. A. Clarke and D. L. H. Williams, J. Chem. Soc. B 1126 (1966).

44. P. F. Warner, U.S. 3,293,163 (Phillips).

45. F. F. Rust and H. S. Klein, U.S. 3,290,397 (Shell).

46. H. B. Hass et al, Ind. Eng. Chem. _28_, 1178 (1936).

47. H. Geyerfelt, Ann. _154_, 247 (1870); L. Henry, Ber. _3_, 347, 351 (1870); H. Essex and A. L. Ward, U.S. 1,477,113 (Du Pont).

48. H. Essex and A. L. Ward, U.S. 1,594,879 (Du Pont); I. I. Ioffe, CA _40_, 61 and 5771 (1946).

49. G. R. Bancroft, J. Am. Chem. Soc. _41_, 424 (1919); A. Michael, Ber. _39_, 2785 (1906); L. Henry and A. Dewael, Rec. Trav. Chim. Pays-Bas _22_, 319 (1903).

50. M. S. Kharasch et al, J. Org. Chem. _4_, 430 (1939).

51. J. P. Wibaut, Rec. Trav. Chim. Pays-Bas _50_, 329 (1931).

52. W. E. Vaughn et al, J. Org. Chem. _7_, 482 (1942).

53. M. S. Kharasch and F. R. Mayo, J. Am. Chem. Soc. _55_, 2495 (1943).

54. R. E. Robertson et al, Can. J. Chem. _37_, 803 (1959).

55. P. D. Bartlett and L. J. Rosen, J. Am. Chem. Soc. _64_, 544 (1942); A. Juvala, Ber. _63_, 1997 (1930).

56. J. B. Conant et al, J. Am. Chem. Soc. _47_, 488 (1925).

57. M. Dubini et al, Chim. Ind. (Milan) _45_, 1237 (1963).

58. E. L. Brown and N. Campbell, J. Chem. Soc. 1700 (1937).

59. L. L. Merrett et al, J. Am. Chem. Soc. _61_, 15 (1939).

60. E. D. Hughes, Trans. Faraday Soc. _34_, 185 (1938).

61. H. P. A. Groll and G. Hearne, Ind. Eng. Chem. _31_, 1530 (1939).

62. G. Niederist, Ann. 196, 350 (1879).
63. G. H. van de Griendt and L. M. Peters, U.S. 2,475,364 (Shell);
     J. G. Fife, Brit. 549,001 (Shell); CA 38, 549 (1944).
64. L. F. Hatch and R. R. Estes, J. Am. Chem. Soc. 70, 1093 (1948);
     cf. A. Kirrmann and H. Pourrat, Compt. Rend. 228, 1649 (1949).
65. C. A. Vernon, J. Chem. Soc. 423 and 4462 (1954).
66. R. H. DeWolfe and W. G. Young, Chem. Rev. 56, 764 (1956).
67. A. Reiche et al, Angew. Chem. 77, 219 (1965); Chem. Ber. 99,
     3244 (1966).
68. F. F. Rust and F. H. Dickey, U.S. 2,516,649 (Shell).
69. R. M. Maged and J. G. Welch, J. Am. Chem. Soc. 88, 5681 (1966).
70. L. K. Freidlin et al, CA 66, 55085 (1967).
71. L. Heavner and C. E. Schildknecht, unpublished; cf. S. A. El
     Khishen, Anal. Chem. 20, 1078 (1948).
72. J. J. Black et al, J. Photogr. Sci. 12, 86 (1964); CA 61, 5122
     (1964).
73. A. Cahours and A. W. Hofmann, Ann. 100, 356 (1856) and 102, 285
     (1857); E. Menne, Ber. 33, 657 (1900).
74. R. Breckpot, Bull. Soc. Chim. Belg. 39, 462 (1930); CA 25, 2412
     (1931); E. Rietz, Org. Syn. 24, 96 (1944).
75. R. B. Reynolds and H. Adkins, J. Am. Chem. Soc. 51, 284 (1929).
76. G. Gerlich, Ber. 8, 650 (1875); J. Houben, Ber. 36, 2897 (1903);
     W. Flemming and H. Buchholz, Ger. 705,651 (IG); CA 36, 1957
     (1942); Brit. 525,136 (IG); CA 35, 6604 (1941).
77. R. L. Hodgson and J. H. Raley, U.S. 3,052,735 (Shell).
78. J. F. Norris and J. M. Joubert, J. Am. Chem. Soc. 49, 885 (1927);
     B. T. Brooks, ibid. 56, 2000 (1934).
79. F. Cortese, J. Am. Chem. Soc. 51, 2266 (1929).
80. W. Jaworski, Ber. 42, 436 (1909).
81. J. M. Conia, Bull. Soc. Chim. France 533 (1950); CA 45, 1525
     (1951).
82. M. Tamele et al, Ind. Eng. Chem. 33, 115 (1941).
83. L. J. Hughes, U.S. 3,287,413 (Monsanto).
84. S. I. Sadykh-Zade, CA 66, 64975 (1967).
85. L. F. Hatch et al, J. Am. Chem. Soc. 69, 2615 (1947).
86. S. A. El Khishen, Anal. Chem. 20, 1078 (1948).
87. J. Burgin et al, Ind. Eng. Chem. 33, 385 (1941).
88. L. Schmerling and V. N. Ipatieff, U.S. 2,412,012 (Universal Oil).
89. V. Grignard, Compt. Rend. 132, 560 (1901).
90. F. C. Whitmore, J. Am. Chem. Soc. 62, 798 (1940); cf. Norris
     and Joubert, ibid. 49, 885 (1927).
91. H. Staudinger and T. Fleitmann, Ann. 480, 92 (1930).
92. P. D. Bartlett and R. Altschul, J. Am. Chem. Soc. 67, 821 (1945);
     cf. N. G. Gaylord and F. R. Eirich, J. Polym. Sci. 7, 575 (1951).
93. A. Suzui et al, Japan 71-28,772, CA 76, 46695 (1972).
94. M. S. Kharasch and G. Buchi, J. Org. Chem. 14, 84 (1949).
95. Brit. 582,663 (ICI); CA 41, 2274 (1947).
96. P. H. Carnell, U.S. 2,486,923-4 and 2,533,425 (Phillips).
97. I. Sakurada and G. Takahashi, CA 50, 601; T. L. Dawson and
     R. D. Lundberg, J. Polym. Sci. A 3, 1801 (1965).

98. W. Bauer and F. Goetz, Ger. 706,510; U.S. 2,338,893 (R & H).
99. D. Leaver et al, J. Chem. Soc. 3331 (1962); cf. Y. Tsuda, J.
     Polym. Sci. $\underline{49}$, 369 (1961).
100. Linda Heavner and C. E. Schildknecht, unpublished.
101. S. A. Dolmatov and L. S. Polak, CA $\underline{64}$, 3691 (1966).
102. T. L. Dawson and R. D. Lundberg, J. Polym. Sci. A $\underline{3}$, 1801 (1965).
103. T. Kagiya et al, J. Chem. Soc. Japan, Ind. Chem. $\underline{71}$ (5), 745
     (1968); CA $\underline{69}$, 97224 (1968); K. Fukui et al, ibid. $\underline{74}$, 64651
     (1971).
104. Y. Ishigami et al, CA $\underline{64}$, 5302 (1966).
105. I. Fugioka et al, Japan 71-29,495.
106. M. Okuyama and H. Hirata, Nippon Kagaku Zasshi $\underline{85}$, 342-3 (1964);
     Kolloid-Z. $\underline{212}$ (2), 162 (1966).
107. E. B. Davidson, IUPAC Symposium, Budapest, 1969; CA $\underline{75}$, 64312
     (1971).
108. V. P. Zubov et al, CA $\underline{69}$, 52516 (1968).
109. W. Bauer and F. Goetz, U.S. 2,338,893; Ger. 706,510 (Roehm und
     Haas).
110. Fr. 876,123 (Roehm und Haas); CZ II 576 (1943).
111. M. J. Piper and P. L. Timms, J. Chem. Soc. Chem. Com. 50 (1972).
112. F. P. Reding et al, U.S. 3,245,969 (Carbide).
113. D. Adelson and H. Dannenberg, U.S. 2,331,869 (Shell).
114. S. Murahashi et al, Bull. Chem. Soc. Japan $\underline{34}$, 631 (1961); cf.
     Japan 15,591 (1960) (Osaka Soda).
115. S. Nose and K. Satoshige, CA $\underline{55}$, 10961 (1961).
116. H. C. Evans and P. Kirby, U.S. 3,202,610 (Shell).
117. E. B. Davidson, U.S. 3,379,793 (Esso).
118. A. M. El-Abbady and L. C. Anderson, J. Am. Chem. Soc. $\underline{80}$, 1737
     (1958).
119. M. Gippin, U.S. 3,284,431 (Firestone); cf. Free radical
     copolymerization, J. B. Rivlin, U.S. 3,532,675 (Olin).
120. Netherlands appl. (Michelin) CA $\underline{66}$, 95999 (1967).
121. H. Weber and K. Kiepert, Makromol. Chem. $\underline{70}$, 54 (1964).
122. K. J. Clark and T. Powell, Polymer $\underline{6}$, 531 (1965).
123. N. G. Gaylord and A. M. Tringali, U.S. 3,291,770 (Interchemical).
124. J. E. Cantrill, Ger. Offen. 2,000,042 (GE), CA $\underline{73}$, 78117 (1970).
125. F. E. Frey et al, U.S. 2,114,292 (Phillips); C. S. Marvel and
     F. J. Glavis, J. Am. Chem. Soc. $\underline{60}$, 2622 (1938).
126. M. S. Kharasch and E. S. Steinfeld, J. Am. Chem. Soc. $\underline{62}$, 2559
     (1940).
127. W. W. Crouch and L. D. Jurrens, U.S. 2,531,403 (Phillips); cf.
     Ito et al, Makromol. Chem. $\underline{55}$, 15 (1962); CA $\underline{57}$, 12706 (1962).
128. S. E. Ross and H. Noether, U.S. 2,698,317 (Celanese); cf.
     J. Furukawa et al, CA $\underline{60}$, 1859 (1964).
129. G. P. Rowland and R. A. Piloni, U.S. 2,849,422-4 (Firestone);
     cf. S. Nose, CA $\underline{77}$, 6063 (1972).
130. Belg. 576,837 (Solvay).
131. A. G. Horney, U.S. 2,303,549 (Airco).
132. Brit. 509,796 (I.G.); H. Jacobi and W. Flemming, U.S. 2,259,470.
133. J. C. Patrick and H. R. Ferguson, U.S. 2,485,107 (Thiokol).

134. F. M. Lewis and F. R. Mayo, J. Am. Chem. Soc. 76, 457 (1954);
      cf. M. S. Kharasch et al, ibid. 69, 1105 (1947).
135. T. Alfrey and J. G. Harrison, J. Am. Chem. Soc. 68, 299 (1946).
136. R. M. Joshi and S. L. Kapur, J. Sci. Ind. Res. (India) 16B,
      441 (1957); CA 52, 5105 (1958); cf. CA 78, 17017 (1973).
137. M. K. Saha et al, J. Polym. Sci. A 2, 1365 (1964).
138. P. O. Tawney et al, U.S. 2,498,099 and 2,597,202;
      cf. U.S. 2,498,084 and 2,626,252 (U. S. Rubber).
139. P. O. Tawney, U.S. 2,643,991 and 2,553,430 (U. S. Rubber).
140. K. S. Bagdasar'yan and R. I. Milyutinskaya, CA 48, 13439 (1954).
141. H. Gilbert et al, U.S. 2,650,911 (Goodrich).
142. J. R. Caldwell and E. H. Hill, U.S. 3,194,797; cf. U.S.
      2,656,337 (Eastman); cf. M. Imoto et al, CA 62, 14831; A. I.
      Ageev et al, CA 69, 77781 (1968).
143. M. Kumarsata et al, J. Polym. Sci. A 2, 1365 (1964).
144. G. Smets et al, Bull. Soc. Chim. Belg. 57, 493 (1948); CA 43,
      8189 (1947).
145. T. L. Dawson and R. D. Lundberg, J. Polym. Sci. A 3, 1801 (1965).
146. G. Smets et al, J. Polym. Sci. A 2, 4835 (1964).
147. W. G. Tabar and R. A. Walther, Belg. 616,208 (Carbide).
148. Brit. 838,996 (Goodrich-Gulf); CA 54, 26016 (1960).
149. J. L. Nyce and R. S.-Y. Ro, Belg. 622,213 (Du Pont); CA 59,
      5335 (1963).
150. H. D. Lyons, U.S. 3,101,327 (Phillips); cf. Belg. 638,638
      (Huels); CA 62, 9258 (1965).
151. H. Weber and K. Kiepert, Makromol. Chem. 70, 54 (1964); cf.
      Belg. 626,642 (Monte), CA 60, 9378 (1964).
152. Brit. 953,958 (California Res.); CA 61, 1754 (1964).
153. E. A. Hunter and C. L. Aldridge, U.S. 3,299,020 (Esso); cf.
      J. P. Kennedy et al, J. Macromol. Sci. Chem. A 1, 977 (1967).
154. R. Bacskai, J. Polym. Sci. A 3, 2491 (1965).
155. L. J. Young, J. Polym. Sci. 54, 414 (1961).
156. L. F. Hatch, L. B. Gordon and J. J. Russ, J. Am. Chem. Soc. 70,
      1093 (1948).
157. M. Reboul, Ann. Chim. [3] 60, 38 (1860); W. Pfeffer and
      R. Fittig, Ann. 135, 359 (1865); Snirnow, CZ I, 576 (1904).
158. C. Friedel and R. D. Silva, J. Prakt. Chem. 113, 27 (1872).
159. L. Henry, Ber. 5, 188 (1872).
160. M. E. Reboul, Compt. Rend. 95, 995 (1882).
161. R. Lespieau and M. Bourguel, Org. Syn. 5, 49 (1925).
162. R. Lespieau and M. Bourguel, Org. Syn. 6, 20 (1926); cf.
      J. R. Johnson and W. L. McEwen, J. Am. Chem. Soc. 48, 473 (1926).
163. P. I. Abell and P. K. Adolf, Int. J. Chem. Kinetics I, No. 6
      (1970).
164. C. C. Price and H. D. Marshall, J. Org. Chem. 8, 532 (1943).
165. N. V. Klimentova et al, CA 69, 5248 (1968).
166. L. M. Minsk and W. O. Kenyon, U.S. 2,443,167 (Eastman).
167. W. O. Kenyon and J. H. VanCampen, U.S. 2,419,221 (Eastman);
      cf. P. O. Tawney, U.S. 2,567,304 (U.S. Rubber).
168. J. P. Henry and L. O. Moore, U.S. 3,215,746 (Carbide).

169. C. C. Price et al, J. Org. Chem. **8**, 532 (1943).
170. J. P. Kennedy et al, Makromol. Chem. **93**, 191 (1966).
171. G. E. Hulse, U.S. 2,571,883 (Hercules).
172. R. M. Murch, U.S. 3,240,690 (Dow); Brit. 888,730; CA **56**, 14471 (1939).
173. W. G. Young, J. Am. Chem. Soc. **79**, 4793 (1957).
174. W. G. Young et al, J. Am. Chem. Soc. **82**, 6163 (1960).
175. H. Huebner and A. Geuther, Ann. **114**, 36 (1860); L. J. Andrews and R. E. Kepner, J. Am. Chem. Soc. **70**, 345 (1948).
176. L. M. Minsk and W. O. Kenyon, U.S. 2,443,167 (Eastman).
177. A. Kirrmann, Compt. Rend. **202**, 1934 (1936).
178. M. I. Bro and B. W. Sandt, U.S. 2,946,763 and 2,943,080 (Du Pont).
179. R. A. Pasternak et al, Macromol. **4**, 470 (1971); J. S. Rugg and A. C. Stevenson, Rubber Age **82**, 102 (1958).
180. M. I. Bro et al, U.S. 2,988,542 and Brit. 840,080 (Du Pont); L. A. Wall et al, J. Polym. Sci. Al **4**, 2229 (1966).
181. H. S. Eleuterio, U.S. 2,958,685 and Brit. 884,161 (Du Pont); IUPAC Boston Meeting, July 1971); D. Sianesi and G. Caporiccio, Makromol. Chem. **60**, 213 (1963); Belg. 618,312 (Monte).
182. K. E. Weale, Quart. Rev. (London) **16**, 267 (1962).
183. G. H. Crawford, U.S. 3,084,144 (1963); E. M. Sullivan et al, U.S. 3,110,705 (1963).
184. D. Sianesi and G. Caporiccio, Makromol. Chem. **81**, 264 (1965).
185. D. W. Brown, Polym. Preprints **6** [2], 965 (1965).
186. L. A. Wall, Report on High Pressure Polymerization, U. S. National Bureau of Standards, Washington, D.C., July 1970.
187. S. Straus and D. W. Brown, Polym. Preprints **7**, 1128 (1966).
188. D. W. Brown, R. E. Lowry and L. A. Wall, J. Polym. Sci. A **1**, 1993 (1971); cf. W. K. Hu and W. A. Zisman, Macromolecules **4**, 688 (1971).
189. L. A. Wall, Polym. Preprints **7**, 1112 (1966).
190. D. W. Brown, Polym. Preprints **6** [2] 965 (1965).
191. D. W. Brown and L. A. Wall, SPE Trans. **3**, 300 (1963).
192. D. W. Brown et al, J. Polym. Sci. A **3**, 1641 (1965; **4**, 131 (1966).
193. L. A. Wall, Polym. Preprints **7**, 1112 (1966).
194. D. W. Brown et al, Polym. Preprints **10**, 1394 (September 1966).
195. C. G. Overberger and G. Khattab, J. Polym. Sci. Al **7**, 217 (1969).
196. F. W. Hoffman, J. Org. Chem. **14**, 105 (1949); S. Margulis et al, U.S. 3,524,839 (Israeli Mining).
197. Y. K. Kim, U.S. 3,505,945 and 3,503,945 (Dow).

5. ALLYLIC ALCOHOLS

Allyl alcohol is second only to propylene and allyl chloride among commercially important allyl compounds. Its high reactivity makes possible the synthesis of useful plastics, pesticides, perfumes, and intermediates. However, direct applications, e.g., as a biocide and monomer, are limited by high toxicity, frequent side reactions, and low molecular weights of the polymer products. Nevertheless, allyl alcohol has had some use as a biocide--for example, in fumigants for soil and grain. Further development awaits lower costs, which might result from oxidative synthesis without chlorination.

The instability of methallyl alcohol and difficulties in preparing its esters have contributed to its slow commerical progress. Methallyl compounds seldom have advantages over corresponding allyl compounds in polymers, perfumes, or pesticides.

There has been recent commercial interest in higher allylic alcohols such as dimethyl vinyl carbinol synthesized from acetylene and ketones without use of halogen. Such alcohols, $CH_2=CHC(R)(R)OH$, free of allylic hydrogen atoms, copolymerize favorably with some vinyl-type monomers such as ethylene. Some of the allylic alcohols and their derivatives have very pleasant odors and are useful as aroma chemicals in perfumes. In contrast allyl alcohol itself is most repulsive to the nose and eyes. Allyl alcohol and its oxidation product acrolein may be formed when other allyl compounds are decomposed by heating.

PREPARATION OF ALLYL ALCOHOL

In 1856 Cahours and Hofmann first prepared allyl alcohol by reacting allyl oxalate with ammonia (1). Later the liquid compound was made in the laboratory by heating glycerol with oxalic acid (2) or with formic acid (3). A small amount of allyl alcohol was available from wood distillation. For 30 years allyl alcohol has been available commercially from Shell by reacting allyl chloride with aqueous alkali (4):

$$CH_2=CHCH_2Cl + \overline{O}H \longrightarrow CH_2CHCH_2OH + \overline{C}l$$

Conditions may be 150°C, 200 psi with 5% sodium hydroxide in water. Higher concentrations of alkali may increase the undesired byproduct diallyl ether. In one continuous hydrolysis process, residence time was 10 to 15 min. The yield of purified allyl alcohol was 90%, the remainder being largely diallyl ether. The latter may be recycled because important industrial outlets for it have not been developed. Some variations in the preparation of allyl alcohol from allyl chloride have been patented (5). Hydrolysis by dilute HCl at 80°C may use a catalyst such as Cu_2Cl_2 (6). In a Russian process allyl chloride reacted with NaOH or Na_2CO_3 at 80 to 90°C and 7 atm in a column packed with oxides of copper and chromium (7). There has been interest also in preparing allyl alcohol by hydrolyzing diallyl ether (8) and allyl acetate (9).

A competitive process for allyl alcohol is the vapor phase isomerization of propylene oxide over a contact catalyst such as lithium phosphate (10). The exothermic reaction may be maintained at 230 to 270°C. Exit gases containing allyl alcohol, propionaldehyde, acetone, and unreacted propylene oxide are condensed and distilled. Enough water may be added in order to distill off the azeotrope of allyl alcohol-water (73-27%). By reacting 30% per pass, yields up to 80% may be attained. The expenses of chlorine, alkali, and byproduct diallyl ether are avoided in this process. Alumina, titania, thoria, or transition metal oxide catalysts also may be used. Improvements in lithium phosphate catalysts have been patented (11). In one example at 280°C a yield of 93% allyl alcohol with 42% conversion was reported (12).

Selective hydrogenation of acrolein to allyl alcohol has received recent attention and may become practical in the future. Literature until 1957 was reviewed, as well as some unpublished work of Shell (13).

$$CH_2=CHCHO + H_2 \xrightarrow{\quad cat. \quad} CH_2=CHCH_2OH$$

Yields up to 96% were reported using aluminum alcoholates such as the sec-butylate in reduction (14). Many reducing catalysts rearrange allyl alcohol to propionaldehyde. Cadmium metal with sulfuric acid at 80°C gave 82% allyl alcohol and 18% propionaldehyde in one example (15). Copper and cadmium salts were catalysts at 255°C and 55 atm H_2 (16). Reaction of acrolein-anthracene adduct (protected double bond) gave up to 95% allyl alcohol by Raney nickel at 100°C (17). The possible future manufacture of acrolein by oxidation of propylene for synthesis

of glycerol and acrylic monomers would favor this route to
allyl alcohol. Acrolein has been reduced by isopropanol
at 185°C over a catalyst obtained by calcining zinc
nitrate and magnesium oxide (18).

Other routes disclosed for preparation of allyl alcohol
(AA) include oxidation of propylene in acetic acid to give
allyl acetate (19). There are other methods for which
some references are given in Beilstein (20). The compositions of more than 50 azeotropes of allyl alcohol are
included in Beilstein. Free radical reactions of alcohols
with acetylene have given allylic alcohols (21). For
example, AA was obtained by reacting acetylene with
methanol in the presence of di-t-butyl peroxide for 5 hr
at 140°C and 300 psi. It is interesting that reaction of
cyclopropylamine with nitrous acid forms allyl alcohol,
apparently via rearrangement of the intermediate carbonium
ion to the more resonance-stabilized allyl carbonium ion.

PROPERTIES OF ALLYL ALCOHOL

When the nose is brought near liquid allyl alcohol,
there is a mild alcoholic odor for an instant, and then
the mucous membrane is attacked viciously. The painful,
penetrating effect is exceeded only by that accompanying
the acrylic compound acrolein. Allyl alcohol seems to be
oxidized quickly in animal tissues including nasal mucosa,
which then give tests for acrolein (22). Vapors of allyl
alcohol may cause severe irritation and injury to the eyes,
nose, throat, and lungs. The odor threshold is below 0.7
ppm. Contact with the skin causes irritation and effects
of the alcohol can be distributed through the body. Allyl
alcohol vapors are said to be more toxic than those of
allyl chloride. Precautions for safe handling are available (23). Allyl alcohol has strong bactericidal and
fungicidal action.

Allyl alcohol is miscible with lower alcohols, ethers,
acetone, aromatic hydrocarbons, and carbon tetrachloride.
It is not miscible with hexane or petroleum ether. Allyl
alcohol is more difficult to salt out than saturated
alkanols and it is soluble in solutions of many heavy
metal salts with complex formation (see Chapter 31). The
infrared spectrum of allyl alcohol is shown in Fig. 5.1.
References to 1945 on properties of allyl alcohol are
available (24). Important physical properties of allyl
and methallyl alcohols are given later in Table 5.1.

The high reactivity of allyl alcohol is shown in the
rapid formation of esters with organic acids or acid
anhydrides at moderate temperatures with little or no
mineral acid catalyst (25). This high reactivity is
favorable for preparing unsaturated polyesters terminated

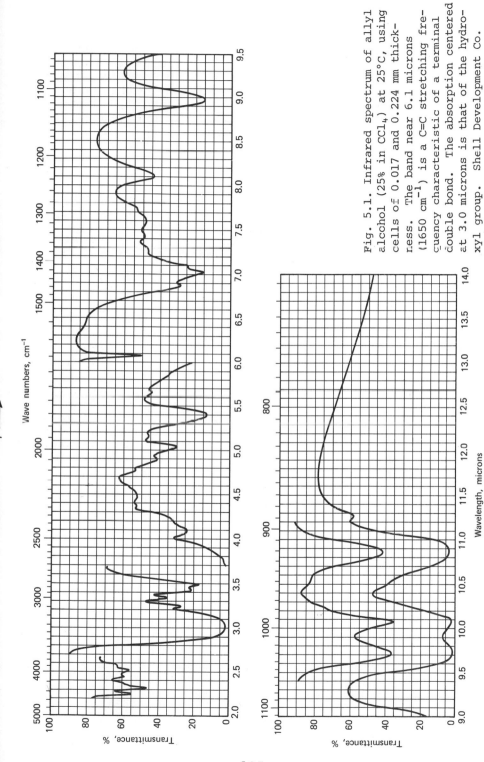

Fig. 5.1. Infrared spectrum of allyl alcohol (25% in CCl_4) at 25°C, using cells of 0.017 and 0.224 mm thickness. The band near 6.1 microns (1650 cm^{-1}) is a C=C stretching frequency characteristic of a terminal double bond. The absorption centered at 3.0 microns is that of the hydroxyl group. Shell Development Co.

with allyl groups for curing plastics and coatings (26).
Allyl alcohol (AA) shows extraordinary ease in forming
ethers. Hydrogen bromide adds to AA not only by simple,
normal addition, but also to give diallyl ether and its
addition products (as well as some 1,2 and 1,3-dibromo-
propane). When chlorine was bubbled into aqueous allyl
alcohol at 0 to 25°C and the products were neutralized by
sodium hydroxides ethers were recovered, especially
$(HOCH_2CHClCH_2)_2O$ (27). Allyl alcohol reacts rapidly with
concentrated hydrochloric acid containing $ZnCl_2$ (Lucas
reagent) to form allyl chloride.

Cahours and Hofmann observed violent reaction of AA
with sodium and with potassium, forming hydrogen and
gelatinous alkali alcoholate. Allyl alcohol reacted with
metallic sodium four times as fast as did n-propanol (28).
The compound is highly reactive and very difficult to
purify from peroxides, aldehydes, and liquid low polymers
which may form in the presence of air, moisture, and light.
Many of the early observations on the reactions and
stability of AA are of doubtful value because the initial
impurities were not known. The failure of allyl alcohol
to polymerize quickly to solid polymers is no reason to
consider it a stable compound. During distillation in
contact with the air, AA forms peroxides and liquid low
polymers that remain as high boiling residues. Land and
Waters were not able to eliminate peroxides from allyl
alcohol easily (29). Studies of autoxidation of AA in
air have shown the formation of acrolein, acrylic acid,
formic acid esters, ketones, and peroxides, as well as
low polymers (30). Heating AA with oxygen in the presence
of heavy metal catalysts gives largely acrolein with
smaller amounts of other aldehydes and acrylic acid:

$$CH_2=CHCH_2OH \xrightarrow[Cu]{O_2} CH_2=CHCHO \xrightarrow{O_2} CH_2=CHCOOH$$
$$+ H_2O$$

Hofmann and Cahours observed rapid reaction of AA with
potassium dichromate and sulfuric acid to give a mixture
of acrolein and acrylic acid. Oxidation of AA using
acidic permanganate solution gave formic acid as the
major product (31). Dichromate-sulfuric acid also
oxidized allyl alcohol to formic acid. This oxidation,
which is about six times as rapid as that of ethanol, has
been used as a method for estimating allyl alcohol (32).
The reaction of ozone with AA in acetic acid and the
reduction of the ozonide with zinc dust were investigated
(33). Oxidation of AA by MnO_2 under varied conditions
gave acrolein and other products (34). Allyl alcohol and
steam passed over $CuO-Cr_2O_3$ gave acrolein (45%), acrylic
acid (34%), and hydrogen (35). Allyl alcohol vapor and

air over active copper catalyst at 375°C formed principally acrolein (36). Allyl alcohol in ether was oxidized
by nickel peroxide even at 20°C for 6 hr giving acrolein
(37).

Allyl hydroperoxide $CH_2=CHCH_2OOH$ had long been postu-
lated, but it was made first in high concentration by
Dykstra and Mosher in 1957 (38). It was not made directly
from propylene or allyl alcohol but by reacting sodium
allylate in allyl alcohol-diethyl ether with methane
sulfonyl chloride at -5 to 25°C. The allyl methane sulfonate was then reacted with a solution of hydrogen peroxide, potassium hydroxide, and methanol near 0°C. The
hydroperoxide is extremely irritating to the membranes of
the nose. Contact with the skin caused blisters and a
burning sensation. It is unstable at 40 to 60°C, especially in actinic light; explosive polymeric residues may
form. Allyl hydroperoxide did not detonate on impact but
did detonate on heating or on contact with anhydrous BaO.
In contact with anhydrous K_2CO_3, delayed detonations were
observed. Substituted allyl hydroperoxides such as 2-
neopentylallyl hydroperoxide have been made by treating
the bromide with H_2O_2, KOH, and methanol at 25°C for
40 hr (39). Allyl hydroperoxide was also prepared by way
of reaction of $CH_2=CHCH_2ZnCl$ with H_2O_2 followed by
hydrolysis in dilute HCl (40). It could be detonated by
shock or by adding grains of sand.

Allyl alcohol may be detected in small amounts by
treatment with bromine water at 90°C and then heating the
product with 3-hydroxybenzoic acid in dilute sulfuric acid
to give green fluorescence (41). It can be identified as
the 2,4,6-trinitrobenzoate melting at 147°C and can be
estimated volumetrically by reaction with mercuric acetate
in methanol (42). Allyl alcohol vapors in air were
determined by absorption in water and titration with
bromide-bromate in dilute sulfuric acid (43). Allyl
alcohol adds Br_2 from bromide-bromate reagent more rapidly
than simple allyl esters. Only about a minute may be
required for titration. This is expected from the
electron-releasing OH group.

Irradiation of AA-water solutions in the presence of
oxygen gave acrolein, hydroxyacetaldehyde, formaldehyde,
and two types of hydroperoxides in proportions depending
on pH and oxygen pressure (44).

Heavy metal oxides and salts can catalyze the rearrangement of allyl alcohol (AA) to propionaldehyde and/or
dehydrogenation to acrolein, depending on conditions (45).
Complex formation of the π-allyl type, such as the red
rheniumIII-allyl alcohol found by Nicholson, may be
involved in such reactions (see Chapter 31). The isomerization of allyl alcohol to propionaldehyde may be brought

TABLE 5.1

Properties of Allyl and Methallyl Alcohols

Properties	Allyl	Methallyl
Boiling point, °C	97	114
Melting point, °C	-129 (glassy)	--
Density of liquid, 20/4	0.8520	0.8515
Refractive index, n_D^{20}	1.4135	1.4255
Vapor pressure, mm Hg at 20°C	17.3	--
Viscosity, cp 20°C	12	--
Surface tension, dynes/cm	25.7	--
Heat of vaporization at bp, cal/mole	9550	--
Heat of combustion (liq), kcal/mole	443	--
Solubility in water, % 20°C	miscible	19.4
Water solubility in % 20°C	miscible	33.8
Azeotropes with water (bp, % alcohol)	88.9°C, 72.3%	92.5°C, 59.7%

about by iron pentacarbonyl (45) or by cobalt hydro-carbonyl (46) at moderate temperatures. $CH_2=CHC(OH)(CH_3)_2$ did not isomerize, apparently because it contains no allylic hydrogen. The isomerization of AA to propionalde-hyde does not occur as readily on warming with mineral acids as in the case of methallyl alcohol. Over catalysts such as aluminum, zinc oxide, or copper, rearrangement of AA may be carried out at 300°C.

ADDITION REACTIONS OF ALLYL ALCOHOL

Because the electron-donating hydroxyl group makes the double bond of allyl alcohol relatively high in electron density, electrophylic reagents such as halogens add rapidly. For the same reason, unsymmetrical hydrogen compounds add usually in the normal or Markovnikov way, except in the case of SH compounds. Hydrogenation of AA in the presence of noble metal catalysts occurs readily (47), but few comparative rates with other olefinic com-pounds have been published. Electroreduction in acidic solution may lead to partial hydrogenolysis to propane, propylene, and other products (48). Allyl alcohol is hydrogenated readily to n-propanol using such catalysts as Cu, Co, Ni, or Pd at room temperature or above at moderate pressures.

Addition of bromine is rapid at room temperature and fairly fast at -20°C (49). Besides 2,3-dibromo-1-propanol a little 1,3-dibromo product was also formed at low tem-peratures (50). BrCl adds to allyl alcohol even faster than Br_2 (51). Reaction with I_2 was first order in halogen and also first order in allyl alcohol concentration (52). The violent addition of Cl_2 may be carried out by gradually introducing Cl_2 into AA and HCl at -15°C (53). The addition of chlorine occurs rapidly and with complications under most conditions. Even dry chlorine may give beside 2,3-dichloro-1-propanol some 1,3-dichloro-2-propanol (54), and chlorine may act as an oxidizing agent to form some acrolein. Chlorination of AA in saturated aqueous NaCl gave principally simple addition with no more than 1% of dichloroisopropanol (55). Ethers are also formed as byproducts in reactions with halogens. ICl added to allyl alcohol, yielding 70% 3-chloro-2-iodo-1-propanol and 30% 2-chloro-3-iodo-1-propanol (56). Different results were reported earlier (57). Rates of addition of I_2 to the double bond were studied in different solvents at different temperatures, with and without light (58).

Chlorine or bromine in the presence of water reacts with AA to form principally the halohydrin following Markovnikov addition, since Cl^+ is the attacking species (57, 59):

$$CH_2 = CHCH_2OH \xrightarrow{\text{ClOH}} CH_2ClCHOHCH_2OH$$

Besides the HCl from the reaction of Cl_2 with water, additional HCl may be added initially as a catalyst. In the presence of concentrated HCl not only does addition occur, but diallyl ether forms (60); allyl chloride is a product in the presence of catalysts such as Cu_2Cl_2 (61). Winstein and Goodman studied reactions of aqueous halogen with methallyl alcohol (62).

The additions of mercaptans to allyl alcohol occur largely by Posner or reverse addition, the influence of sulfur prevailing over the electron-donating OH group of the alcohol. Hydrogen sulfide adds to give $HSCH_2CH_2CH_2OH$ in the liquid phase under UV light (63). With catalysts such as cobalt sulfide or azobis at 80°C, good yields of the reverse addition products were obtained (64). Methyl mercaptan added to AA under UV radiation to give 98% yield of 3-mercaptomethyl propanol (65). Reaction occurred more rapidly in the presence of oxygen than under nitrogen (66). Benzyl mercaptan added to AA as we would expect from Posner's rule, giving largely $C_6H_5CH_2S(CH_2)_3OH$ (67). Also reported to occur in the reverse manner to allyl alcohol were the additions of thiolacetic acid (catalyzed by peroxides) (68), 2-mercapto-3-pentanone (69), primary and secondary amines (70), and mono(2-cyanoethyl) phosphine (71). However, Markovnikov or normal addition was reported from the reaction of allyl alcohol with thiophenol in the presence of free sulfur (72). Bisulfites such as $NaHSO_3$ also add in the reverse way to AA with oxygen or peroxides as catalysts (73). By

dehydration of the product Shell prepared the sultone
$\underset{\underset{O}{O_2 S}}{\overset{CH_2 - CH_2}{\diagup}} CH_2$. By adding concentrated sulfuric acid slowly
to allyl alcohol in acetic anhydride-acetic acid solution
and then warming at 60°C for 2 hr, there is formed
$HSO_3 CH_2 CHOHCH_2 OH$ (74). Allyl alcohol adds to propylene
oxide in the presence of basic catalysts to give the
allyloxypropanol isomer expected from the influence of
the electron-releasing methyl group (75):

$$CH_3\underset{O}{\overset{\diagup}{CH-CH_2}} + HOCH_2 CH=CH_2 \longrightarrow CH_3\underset{OH}{\overset{|}{C}}HCH_2 OCH_2 CH=CH_2$$

   Allyl alcohol reacts with benzene in the presence of
Lewis-acid catalysts such as $H_2 F_2$, BF_3, and $H_2 SO_4$ to form
allylbenzene and/or 1,2-diphenyl propanol (76). Allyl
alcohol was reported to react with m-cresol in the
presence of sulfuric acid to form an isopropenyl ether,
which rearranged to a polymer of an isopropenyl m-cresol
(77). This reaction required several days and may have
involved rearrangement of AA to propionaldehyde, which
condensed with the cresol to form polymers. On heating AA
with trimethylhydroquinone in benzene in presence of $ZnCl_2$
at 200°C, there was formed 5-hydroxy-2,4,6,7-tetramethyl
coumarane (78). Carbon tetrachloride added to AA to give
$CCl_3 CH_2 CHClCH_2 OH$ (79). The reaction was best carried out
with $FeCl_3 \cdot 6H_2 O$ as catalyst in propanol at 110°C during
7 hr. The iron chloride had a remarkable activating
effect on CCl_4 giving a 55% yield with very little product
of higher molecular weight. Diethylammonium chloride was
added to prevent precipitation of $Fe(OH)_3$. A free radical
mechanism as proposed by Kharasch for such reactions was
supported. When $CF_3 I$ was added to AA under the influence
of peroxides, $CF_3 CH_2 CHICH_2 OH$ was produced; better yields
were obtained by irradiation with UV light (80).
   Hydroxylation of AA by $H_2 O_2$ in the presence of catalysts
may form glycerol:

$$CH_2 = CHCH_2 OH + H_2 O_2 \longrightarrow CH_2 OHCHOHCH_2 OH$$

Milas and Sussman carried out the reaction in t-butanol
with osmium tetroxide catalyst at room temperature or
below (81). Besides OsO_4, tungsten trioxide was effective as a catalyst at 50 to 70°C using acetic acid as
solvent (82). An ionic mechanism was proposed in contrast to the OH radical mechanism of Milas. It was not
clear whether peracetic acid was the active oxidizing
agent.
   More than 60 years ago, Prileschajew obtained glycidol
by epoxidation of allyl alcohol by benzoyl hydroperoxide

(83). More recently, Raciszewski found that glycidol is the primary product on treating allyl alcohol with aqueous H_2O_2 in the presence of H_2WO_4 catalyst (84). Effects of concentrations of the reactants and temperature were investigated and two mechanisms were considered. Allyl alcohol has been heated with peracetic acid in water and organic solvent to form glycerol (85). Epoxidation of AA with peracetic acid was carried out in a distillation column with a high boiling solvent such as diisobutyl ketone (86). Other conditions and catalysts for hydroxylation of allyl alcohol using peroxides have been disclosed (87). The action of H_2O_2 on allyl alcohol under radiation has been studied. The mercury resonance light 2537 Å with allyl alcohol and H_2O_2 in 3M aqueous solution gave little glycerol, but instead several polyols of six carbon atoms such as 1,2,5,6-hexanetetraol (88). Electron-spin resonance (ESR) was used to study the action of H_2O_2 on allyl alcohol under UV radiation (89). Allyl alcohol adds to acrylonitrile in the presence of NaOH to form $AOCH_2CH_2CN$ (5 hr, 50-60°C) (90). Many other reactions of AA not discussed here are typical of reactive alcohols.

Adducts or complexes between AA and conjugated dienes such as butadiene and cyclopentadiene have been studied (91). A one-step reductive coupling of allyl alcohols has formed 1,5-dienes (92). Allyl alcohol reacted on heating with CO and H_2 in presence of a cobalt carbonyl to give 4-hydroxybutyraldehyde (93), with CO and water in the presence of $Ni(CO)_4$ to form crotonic acid (94), and with CO, methanol, and acetylene in presence of nickel compounds to give isomers of methyl hexadienoate (95). Allyl alcohol can yield cyclic compounds; for example, reaction with ammonia yields 3-methyl pyridine (96).

The literature contains many examples of reactions of allyl alcohols with polymers. For example, alcoholysis of acrylate ester polymers by AA has given fusible allyl acrylate polymers (97). In order to improve "air-drying" properties in films, unsaturated polyesters have been reacted with mixtures of AA and formaldehyde (98). These "formalized polyesters" were copolymerized with styrene or other ethylenic monomers. Organic polyisocyanates were reacted with allyl or methallyl alcohol to form polyfunctional allyl urethanes (99). Heating these products with free radical catalysts may give thermoset moldings.

POLYMERIZATIONS OF ALLYL ALCOHOLS

Well-defined homopolymers of high molecular weight never have been prepared from allyl alcohol; nor have the structures of the viscous liquid low polymers resulting

from allyl alcohol under different conditions been studied
very closely. Samples of commercial allyl alcohol were
observed to become viscous on long storage, and gummy
polymers insoluble in water could be separated (100).
According to Shell patents, heating allyl alcohol with air
or peroxides gave viscous liquid polymers of only 5 to 10
monomer units (101). These products also contained some
acid and ester compounds. When allyl alcohol was heated
with H_2O_2 at 100°C for 116 hr and the volatiles distilled
off in vacuum, a viscous residue of $n_D^{20}=1.5143$ and molecular weight about 300 remained (102). Such liquid low
polymers were reacted with drying oil acids, also with
maleic anhydride and allyl glycidyl ether for the purpose
of preparing polyesters (103). Allyl alcohol low polymers
were obtained also by saponification of allyl acetate low
polymers, for example, in methanol under reflux (104).

Degradative chain transfer as elucidated by Bartlett
and discussed in other chapters applies also to allyl
alcohols. It was reported that activation energies for
addition of ethyl radicals to the double bond of AA and
for abstraction of allylic hydrogen are about equal (105).
Massive amounts of peroxide, e.g., 25 g BP with 51 g AA
and 10 ml benzene, and heating 18 hr at 60°C, produced low
yields of soft solid polymers which dissolved in pyridine,
acetic acid, or alcohols (106). The highest molecular
weight fraction had intrinsic viscosity of only 0.037 in
ethanol. Heating AA with 2% BP at 80°C gave only 3%
polymer in 24 hr (107).

Because of partial oxidation of allyl alcohol and methallyl alcohol to acrolein and methacrolein (and these to
acrylic and methacrylic acids), the polymers formed under
oxidizing conditions may have very complex structures.
Solid homopolymers of AA having molecular weights above
2000 and showing some evidence of crystallinity were
reported by slow polymerization at low temperatures with
excess boron fluoride (108). Experimental details were
not disclosed completely, and the process and polymers
have not been regarded as promising commercially.

Recent researches in Russia have given promise of polymers from AA of high molecular weight by use of high
energy radiation, but the structures of the products
remain uncertain. Dolmatov and Polak reported solid
polymers from allyl alcohol by use of radiation from [60]Co
in bulk and more rapidly in water solution (109). Kargin,
Polak, Kabanov, Zubov, and Pankova of the Topchiev Institute, Moscow, prepared AA polymers of high molecular
weight by means of high energy radiation in presence of
inorganic complexing agents that formed coordination bonds
with functional groups of the monomer (110). Salts of
metals of Groups I and II as well as inorganic acids with

[60]Co or x-ray radiation promoted formation of solid, amorphous, white polymers from AA. In the first patent example a solution of 27% $CaCl_2$ in allyl alcohol was irradiated with ^{60}Co at 35°C at a dose rate of 300 r/sec during 10 hr. A rate of 1.5% conversion to polymer per hour was obtained. After adding NaOH to precipitate $Ca(OH)_2$, a solid polymer of molecular weight 5×10^5 was recovered by evaporating the solution. Without the use of $CaCl_2$ the rate was only 0.15% per hour and the polymer had a molecular weight below 1000.

In another patent example of Kargin and co-workers, a solution of $MgCl_2$ in allyl alcohol (molar ratio 1.0 to 0.5) was irradiated one hour by fast electrons from an accelerator of 1.6 Mev at 20°C and at a dose rate of 1200 r/sec. The rate of conversion to polymer was 18% per hour or 30 times as fast as without $MgCl_2$. The AA polymer was precipitated from a methanol solution of the reaction products by addition of ethyl acetate. Less effective complexing agents in accelerating radiation polymerization of AA included LiCl, $ZnBr_2$, and HCl. The polymer properties and methods for determining molecular weights were not discussed. However, the new AA polymers were reported to be superior to polyvinyl alcohols in heat and chemical resistance. Dehydration of the polymers from AA to form colored polyene chains did not occur below 300°C. Benzoyl peroxide or other radical initiators could be used instead of radiation with complexing agents for preparing allyl alcohol polymers of molecular weights above 40,000 (111). Electrophilic complexing agents such as $ZnCl_2$ were believed by the Moscow group to lower the frequency of degradative chain transfer as well as increase initiation. Treatment of AA vapors with a radio-frequency glow discharge gave partially soluble polymer films of uncertain composition (112).

Since high polymers of allyl alcohol cannot be made readily by direct homopolymerization, numerous efforts have been made to make them by chemical reactions from other preformed polymers. Marvel and co-workers reduced copolymers of butadiene and methyl acrylate in tetrahydrofuran (THF) solution by $LiAlH_4$ (113). Apparently the $COOCH_3$ side groups were transformed partially to CH_2OH groups to form butadiene-allyl alcohol copolymer. Polyacrylyl chloride, polymethyl acrylate, and polyacrolein were reduced with $LiAlH_4$ in THF in order to prepare allyl alcohol polymers (114). By gradual addition of methyl acrylate polymer in N-methylmorpholine to a suspension of $LiAlH_4$ in N-methylmorpholine, and by refluxing for 3 hr, Cohen isolated a "resinous polyallyl alcohol" soluble in methanol-water (90 to 10 by volume)(115). Examination of

the IR bands supported a simple polyallyl alcohol structure. Vinyl acetate-allyl acetate copolymers were saponified to form vinyl alcohol-allyl alcohol copolymers (2 to
10 mole % AA groups) (116). Water solutions were spun to
fibers.

Copolymers of maleic anhydride-vinyl acetate-styrene
(2:1:1 by weight) were esterified by heating in toluene
with allyl alcohol and a little NaHSO₄ (117). Thirty
parts of the resulting soft solid polymer were dissolved
in 45 parts of styrene along with 3 parts of hexanol and
3.5 parts of BP, and this solution was used for making
thermoset glass fiber laminates. Later high styrene-
maleic anhydride copolymers were partially esterified by
allyl alcohol, and the ammonium salts were suggested as
copolymerizable dispersing agents in emulsion copolymerization of styrene with ethyl acrylate (118).

Allyl alcohol was used to prepare allyl-terminated
unsaturated polyesters (119). In one example, a mixture
of 500 parts phthalic anhydride, 103 parts ethylene glycol,
225 parts AA, 225 parts toluene and 3.4 parts toluene
sulfonic acid was raised to 180°C during 2.5 hr. A viscous
liquid resin of acid number 38 was obtained. Equal parts
of this resin and propylene glycol maleate polyester along
with BP were cured at 120°C.

Allyl alcohol gave telomers with CCl₄ under Lewis acid
catalysis with anhydrous FeCl₃ upon CaCO₃-MgSO₄ (120).
The 2,4,4,4-tetrachlorobutanol obtained was reacted with
lime to form 4,4,4-trichlorobutylene oxide. Because of
the low reactivity of AA in entering into growing high
polymer chains, it may be added as a regulator where little
actual copolymerization occurs. Allyl alcohol was a
favorable diluent in the rhodium-nitrate-catalyzed polymerization of butadiene (121). Ethanol gave much lower
conversions at 50°C. Badische has added AA in emulsion
polymerization of vinyl propionate to improve application
and adhesive properties in coatings.

ALLYL ALCOHOL-STYRENE COPOLYMERIZATIONS

Addition of allyl alcohol to free radical polymerization
of vinyl monomers can be used to give polymers of controlled
low molecular weight and in order to form soluble partial
polymers from polyfunctional monomers. D'Alelio added AA
in the solution polymerization of divinyl benzene in order
to increase the yield of fusible partial polymer (122). In
early work, the extents to which allyl alcohol entered the
products as comonomer units or telomerization occurred were
uncertain. Tawney and co-workers of U.S. Rubber added 7%
or more AA to diallyl fumarate or maleate to obtain soluble
interpolymers by heating with peroxide initiators (123).

Soluble prepolymers of styrene, allyl acrylate, and an
AA could be cured in films by second-stage heating with
radical catalysts (124). Copolymers of styrene, diallyl
fumarate, and AA were found to liberate irritating vapors
of allyl alcohol on storage (125). Soluble interpolymers
of styrene, diallyl fumarate, and allyl alcohol could be
made by heating at 100°C for 10 hr with BP (126). Soluble
partial copolymers from equal parts of triallyl aconitate
and AA with 2% BP were made by heating for 336 hr at 60°C
(127). Additional peroxide was added at 24-hr intervals.
The precipitated copolymer could be cured by heating at
120 to 150°C.

In Shell laboratories polymers were made from mixtures
of allyl alcohol and styrene heated for 4 hr at 125°C and
260 psi while air was supplied continuously (128). The
complex oxidation products were not identified, but crude
products of low molecular weight having 3 to 4 OH groups
per chain molecule were esterified with drying oil unsaturated acids for application as synthetic drying oils
in coatings. Later peroxide catalysts such as t-butyl
peroxide were used in the absence of air. In order to
obtain more allyl alcohol units in the copolymer, the
styrene could be added in portions during 5 hr at 135°C,
giving in one case a 23% yield of resin of molecular
weight 1300 (129). This was reacted with dehydrated
castor oil acid in xylene solution for 3 hr at 230°C.
Styrene-allyl alcohol low polymers were reacted also with
ethylene oxide (130) and with polyisocyanates (131), and
were tested with shellac in coatings (132).

Favorable results were obtained by blending styrene-
allyl alcohol copolymer of DP 4 to 10 with commercial
acrylic baking enamels such as methylolated styrene-ethyl
acrylate-acrylamide interpolymers (133). Addition of 20%
of styrene-allyl alcohol having about 5 OH groups per
molecule to N-methylolated acrylamide interpolymers
lowered the baking temperature required for insolubilization of the films by crosslinking (134). An example of
such a reactive interpolymer was prepared by refluxing the
following mixture:

Compound	Parts by weight
Styrene	39
Ethyl acrylate	44
Acrylamide	15
Acrylic acid	2
n-Butanol	100
Cumene hydroperoxide	1
t-Dodecyl mercaptan	1

Increments of 0.5 part hydroperoxide were added at 2 and
4 hr after beginning of reflux. Then there was added a
solution of 0.4 mole formaldehyde (40% in butanol) together with 0.33 part maleic anhydride. Refluxing was
continued for 3 hr, then half of the butanol was removed
from the interpolymer solution by distillation and replaced by xylene before use in coating formulations. When
cured by heating for a half hour at 300°F, films modified
by 20% styrene-allyl alcohol copolymer as crosslinking
agent were harder than those without modification.

Monsanto studied styrene-allyl alcohol copolymers and
terpolymers leading to their commercial "resinous polyols".
In one example, 70 parts styrene, 30 parts AA, and 1 part
di-t-butyl peroxide were heated at 200°C for 10 min (135).
A colorless syrup was obtained which, after the volatiles
had been distilled off, left 53% of a clear, brittle resin
containing 6% allyl alcohol units. The following ter-
polymerizations of styrene and allyl alcohol were disclosed
by Chapin and co-workers of Monsanto:

Terpolymers	Patent
Acrylic esters	U.S. 2,897,174
Acrylic acid	U.S. 2,899,404
Acrylonitrile	U.S. 2,900,359
Conjugated dienes	U.S. 2,951,831
	and 2,950,270
Unsaturated dicarboxylic acids	U.S. 2,961,423
Divinyl benzenes	U.S. 2,962,462
Vinyl benzyl alcohol	U.S. 3,069,399
Acrylic acid	Brit. 787,372

Terpolymers made at lower temperatures were found to contain more styrene units. One suggested use of aqueous
emulsions of drying oil esters of the styrene-allyl
alcohol copolymers was with butylated melamine formaldehyde
for baking enamels (136). Allyl alcohol, styrene, and
maleic anhydride with t-butyl peroxides were reacted under
pressure at 165°C to give yellow syrupy terpolymer solutions from which soluble, partially esterified brittle
solids were obtained (137). One contained 6% free hydroxyl
groups and 0.55% free carboxyl groups.

Monsanto's RJ-100 resinous polyol based on styrene and
allyl alcohol (S-AA) had the approximate composition (138):

$$\left[\underset{\underset{C_6H_5}{|}}{C}HCH_2\overset{\overset{CH_2OH}{|}}{C}HCH_2\underset{\underset{C_6H_5}{|}}{C}HCH_2 \right]_{4\ to\ 7} \qquad \text{S-AA copolymer}$$

It was reacted with phthalic anhydride to give ammonia-
water dispersible paints for high-gloss coatings and
polishes and has been used with polyisocyanates for polyurethane coatings. It could be blended with alkyds (139)
to reduce cost and improve film flexibility (140). S-AA
copolymers were esterified with soya fatty acids by heating
in xylene up to 265°C during 438 min while distilling off
water (141). Ammonium salts of the products could be dispersed in water. Blends of S-AA copolymer, epoxy resin,
and H_3PO_4 curing agent were applied as priming coatings on
steel from methyl isobutyl ketone solution (142). Molecular weights of the resins were 1100 and 950, respectively.
Favorable properties of styrene-allyl alcohol copolymers
and their reaction products with drying oil acids and with
maleic acid were discussed (143). S-AA copolymer of
molecular weight 1150 and 5.2 OH groups per molecule was
evaluated with butylated melamine-formaldehyde in coatings
(144). Crosslinking of S-AA copolymers has been studied
(145), and the copolymers have been used in printing inks
for plastics (146).
    Styrene-allyl alcohol low copolymer esters of drying
oils were said to form stable emulsions (147) that gave
improved penetration and adhesion in latex paints on bare
wood, and on chalked and corroded old paint surfaces.
S-AA copolymer containing 6% OH groups was partially
esterified by tall oil fatty acid and the product reacted
with maleic anhydride to form a polyester (148). This was
added to latex paints as an "internal primer."
    Although the principal interest in styrene-allyl alcohol
polymers has been coatings, they have been reacted with
ethylene oxide to form surface active products (149). A
copolymer of allyl alcohol (3.6%) - butadiene (79%) -
styrene (17.4%) was suggested as a food can coating (150).
Mixtures with aminoplasts were used as coatings for
thermoplastics (151). Blends with epoxy resin and phenol-
formaldehyde (with phosphoric acid catalyst) were cured on
metals at 250°C (152). S-AA copolymers such as Monsanto's
RJ-101 were blended with terpolymers of styrene-hydroxy-
ethyl acrylate and alkyl acrylates for can coatings cured
by high energy radiation (153).
    Terpolymers of styrene-allyl alcohol-maleic anhydride,
or with unsaturated dicarboxylic acids, were studied by
Monsanto (154). Above 15% acid groups they gave thermosetting polymers. By copolymerization in solvents, the
reactions could be interrupted to yield soluble intermediate polymers for coatings.

OTHER ALLYL ALCOHOL COPOLYMERIZATIONS

    Vinyl acetate-allyl alcohol copolymers were evaluated
in finishes (155) and they were used as suspending agents

in aqueous-suspension polymerizations of vinyl chloride
to produce porous, nonglassy particles of polyvinyl
chloride (156). Allyl alcohol has been interesting as a
copolymerizing regulator for preparation of dyeable
acrylonitrile polymers for fibers. Acrylonitrile copolymers containing 0.6 to 0.7% combined AA units were
patented (157). In one example, copolymers of average
molecular weight 64,000 were used for spinning. A continuous aqueous copolymerization process at 30 to 40°C
used $NaClO_3$-Na_2SO_3 acidified by H_2SO_4 and up to 15% AA in
the monomer feed (158). Copolymers of acrylonitrile-allyl
alcohol of relatively high molecular weight were prepared
by dye-sensitized photopolymerization (159). Dyeable
copolymer fibers were made by Montecatini from 92% acrylonitrile, 3% 2-vinylpyridine and 5% AA using persulfate-
bisulfite aqueous emulsion with sodium dodecane sulfonate
(160). Reaction required 2 hr at 50°C. In an alternate
process a copolymer of AA and vinylpyridine was blended
with acrylonitrile homopolymer. Terpolymers of acrylo-
nitrile-methyl acrylate-allyl alcohol were prepared in
concentrated $ZnCl_2$ aqueous solution with $NaClO_3$ catalyst
at 60°C for use as dyeable fibers (161).

Allyl alcohol has been added to control molecular weights
of polyethylenes. Thus ethylene-allyl alcohol copolymer
waxes of 1500 mol. wt. and 2.5% OH (suitable for shoe
polish, crayons, and carbon paper) have been made at high
pressures with azo catalyst (162). Smaller proportions of
AA may be used to obtain polyethylene plastics of improved
processing characteristics. In one example, ethylene with
0.2% allyl alcohol and 0.017% t-butyl perbenzoate was fed
into an autoclave at 435°F and 20,000 psi (163). The
copolymer obtained had high tensile strength and modulus,
but lower softening temperature than ethylene homopolymer
made under similar conditions. Allyl alcohol-ethylene
copolymers were used as auxiliary suspending agents in the
suspension polymerization of vinyl chloride (164). Terpolymers of butadiene, allyl alcohol, and mesityl oxide
were made by heating with t-butyl peroxide at 140°C (165).
One product had a molecular weight of 1260 by boiling
point rise; it contained 7.7% AA, 12.8% mesityl oxide, and
79.5% butadiene units. Films could be heated with maleic
anhydride at 235°C for curing.

Copolymerization of AA with sulfur dioxide occurs
readily, yielding approximately 1:1 molar copolymers,
generally believed to have polysulfone structure
$--CH_2CH-SO_2--$. It was once suggested, however, that poly-
CH_2OH
mers of allyl sulfonic acid were formed (166). In one
experiment a liquid mixture of 650 g SO_2 and 580 g AA was

held under pressure at 60°C for 16 hr to give an amorphous polymer mass. With added BP, 10 hr at 60°C gave equivalent yield. The polymers are insoluble in common organic solvents, but they dissolve in dilute aqueous alkali and in pyridine. Salts such as $LiNO_3$ and $AgNO_3$ catalyze the copolymerization of allyl alcohol and SO_2 (167). Polymers made at 0 to 20°C using silver nitrate catalyst were swollen in water and slowly hydrolyzed in boiling water (168). Copolymers of AA, acrylic acid, and SO_2 were prepared using persulfate catalyst at 40°C for 16 hr (169). Films of the copolymer cast from water solution became insoluble on heating. Apparently allyl acrylate was not an intermediate. In Japanese work AA and SO_2 were reacted under gamma radiation both in the liquid and solid phases to give 1:1 molar copolymers regardless of the temperature or feed mixture (170). Since the reaction was inhibited by diphenyl picryl hydrazyl and the copolymer yield was proportional to the square root of radiation dose, the copolymerization was believed to have a radical mechanism.

Sulfone copolymers were tested as flocculating agents but were less effective than acrylamide and styrene sulfonate polymers (171). In one example the following were reacted:

Reactant	parts
Monoallyl ether of a polyethylene glycol	100
Allyl alcohol	15
Na sulfoethyl acrylate	20
$LiNO_3$	2, in 8 p water

To this mixture at -5°C was added 48 parts of liquid SO_2. After 20 hr at 38°C under pressure, a sticky copolymer was recovered; a 1% solution of the product in water had a viscosity similar to glycerin. By adding AA in this copolymerization, more sulfone units could be incorporated into the polymer. Higher proportions of AA resulted in less soluble polymers.

Copolymerization of maleic anhydride with allyl or methallyl alcohol with suitable radical initiators gave soluble 1:1 molar copolymers (172). These contained carboxyl groups and lactone rings under the conditions of polymerization at -10 to +70°C.

There is much interest in modifying monomer reactivities by complexing with inorganic or organic compounds. Complexes of AA with $ZnCl_2$ were reported to polymerize as much as 22 times the normal rate from allyl alcohol under high-energy radiation (173). Electronic displacement from

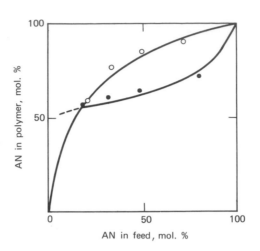

Fig. 5.2.   Photocopolymerization of acrylonitrile (AN) with allyl
alcohol in ethyl acetate at 0°C in the presence of azobis.   Adding
an equimolar amount of zinc chloride (based on AN) greatly accelerated the rates and reduced the AN content of the copolymers:
open circles, no $ZnCl_2$; solid circles, with $ZnCl_2$.   M. Imoto, T.
Otsu, and B. Yamada, Kogyo Kagaku Zasshi 68, 113 (1965).

the double bond through interaction with electrophilic
zinc chloride was suggested as causing less chain transfer
at the allylic hydrogen atoms.   However, the polymer intrinsic viscosities were under 0.1 and only slightly
increased by adding zinc chloride.   The effects of zinc
chloride on compositions of copolymers of allyl alcohol
with acrylonitrile are shown in Fig. 5.2 from research in
Japan.
     The following additional copolymers of allyl alcohol by
radical initiation were studied:

Copolymer	References
Cyclopentadiene-AA + acrylates, etc.	Yuska, U.S. 2,677,671 (Inter-chemical); Nichols, U.S. 2,683,126 (Interchemical)
Cyclopentadiene for coatings	Gerhart, U.S. 2,689,232 (PPG)
Acrylate or methacrylate esters (with ethyl acrylate gave 4.4% OH, soft solid)	Chapin, U.S. 2,817,651 (Monsanto)
Methacrylates (e.g., lauryl) (using azobis at 80-100°C obtained copolymer of 10,000 mol. wt.	Nelson, U.S. 3,033,790 (Shell) Nelson, U.S. 3,033,828 (Shell)

Methyl acrylate   Daumiller, CA <u>54</u>, 20311
  (with toluene diisocyanate gave foam)   (Badische)
Vinyl chloride and maleate esters   Chapin, U.S. 3,036,029 (Monsanto)
Vinyl chloride and acrylate esters   Chapin, U.S. 3,084,136 (Monsanto)
N-Methylol acrylamide, etc.   Nagata, U.S. 3,347,836 (Nippon
  (thermosetting copolymers)   Paint)

METHALLYL ALCOHOL

That methallyl alcohol or isopropenyl carbinol has found
less use than allyl alcohol can be attributed in part to
its easy rearrangement to isobutyraldehyde in presence of
acid catalysts (equation on p. 41). This isomerization
promoted by the electron-releasing methyl group is in
harmony with the nonexistence of vinyl and isopropenyl
alcohol and the greater stability of enols bearing electron-
attracting groups. Homopolymers of high molecular weight
have not been obtained from methallyl nor from the large
number of other substituted allyl alcohols which are known.
An advantage of methallyl alcohol is its camphorlike odor
which is not unpleasant in low concentrations. In higher
concentrations it gives much less delayed sting in the nose
than does allyl alcohol. Methallyl alcohol is about 19%
soluble in water, while about 34% water dissolves in the
unsaturated alcohol at room temperature.

Methallyl alcohol was first made by Sheshukov in Russia
before 1884 by heating methallyl chloride with aqueous
potassium carbonate (174). He found that addition of Br_2
was rapid. The isomerization of the alcohol to isobutyraldehyde occurred readily with mineral acid but only slowly
with acetic acid. Tamele and Groll of Shell used 10% NaOH
at 120°C under pressure to obtain 95% yields of methallyl
alcohol from methallyl chloride (175). Dimethallyl ether
is a troublesome byproduct, since it forms with methallyl
alcohol an azeotrope of boiling point near that of the pure
alcohol. Hydrolysis of methallyl chloride with insufficient
alkali can lead to losses by isomerization of the alcohol
to isobutyraldehyde. Modifications for improving yields of
methallyl alcohol have included adding Zn, Co or Cr compounds (176), or surfactants (177). Methallyl alcohol can
be obtained also from cleaving the ether at 225°C over
contact catalysts such as cupric sulfate on alumina (178).
Of course the alcohol also can be made by saponification or
reduction of the acetate or other ester.

Manufacture of methallyl alcohol by reduction of methacrolein may become feasible if methacrolein becomes an
intermediate in large-scale synthesis of methacrylates from
isobutylene. Most reducing agents disclosed in patents,
such as aluminum alcoholates and $LiAlH_4$, are not very
practical commercially (179). Methacrolein dimer may be

hydrogenated to a pyran derivative, which may be pyrolyzed to give methallyl alcohol and methacrolein (180). Isomerization of isobutene oxide at 240°C with lithium phosphate in biphenyl-diphenyl oxide solvent gave methallyl alcohol (181).

A number of allylic alcohol derivatives have been found in nature. Methallyl alcohol was reported among the volatile substances in honey. Vinyl ethyl carbinol or 1-ethylallyl alcohol is one of the volatiles in raspberries and strawberries. Levo-1-pentylallyl alcohol occurs in certain mushrooms, in cypress leaves, and in oil of pennyroyal. Allyl carbinol was found in oil of rape seed.

The isomerization of methallyl alcohol to isobutyraldehyde discovered by Sheshukov is an interesting reaction (182). Hearne and co-workers found that, by heating methallyl alcohol in 12% sulfuric acid, an azeotrope of the aldehyde and water (bp 60.5°C) can be distilled off continuously (183). Under the same conditions dimethallyl ether is hydrolyzed to methallyl alcohol and rearranged to the aldehyde. Concentrations of H_2SO_4 in water above 20% reduced the yields of isobutyraldehyde because of formation of polymers and other byproducts. At higher temperatures and with longer contact times, another byproduct becomes evident, the isobutyracetal of isobutylene glycol. Methallyl alcohol can be rearranged in organic solvents, e.g., 1 g H_2SO_4 in 150 g isobutyric acid (184). It was also rearranged by passing the vapors over porous catalysts such as pumice or activated charcoal at 300°C (185). On heating with aqueous H_2SO_4, the unsaturated alcohol $CH_2=C(CH_3)CHOHCH_3$ yielded a mixture of 88% 3-methyl-2-butanone and 12% 2-methyl-1-butanol.

In order to test the mechanism of rearrangement, Currell and Fry isomerized methallyl-1-C^{14} alcohol made by the action of $LiAlH_4$ on labeled ethyl methacrylate (186). Eight grams of the radioactive methallyl alcohol was added to 30 ml of 12% sulfuric acid, and the resulting two-phase system was refluxed for 2 hr. Isobutyraldehyde was recovered by distillation in 94% yield. The distribution of radioactive hydrogen in the oxidation products of the aldehyde suggested that addition of hydrogen ion to the double bond of methallyl alcohol is the rate-determining step. A number of compounds of the type $CH_2=CRCH_2OH$ were isomerized to aldehydes by heating for 48 hr under reflux with 0.4N sulfuric acid (187). Methallyl alcohol was rearranged using rhodium chloride as catalyst (188). Small amounts of isobutylene, propylene, and propane were also formed. A mechanism involving π-allyl bonding to the heavy metal ion was suggested. Boiling methallyl alcohol-water (1:1) was treated with dilute ruthenium trichloride to give isobutyraldehyde, along with some methacrolein and isobutylene (189). By heating methallyl alcohol with isobutyraldehyde, the derivative of 1,3-dioxolane can be formed (190).

Methallyl alcohol may be oxidized to methacrolein in animal tissues but apparently less rapidly than in the case of allyl alcohol. By oxidation with air in the presence of steam over silver contact catalysts at 500°C, methacrolein was obtained in 90% yield, along with a little isobutyraldehyde (191). The majority of members of a panel found that the odor of methallyl alcohol was first objectionable at 25 ppm in air (192). Some physical properties of methallyl alcohol are given in Table 5.1 (p. 198). Infrared spectra of methallyl alcohol and its isomers were compared (193).

ADDITION REACTIONS OF METHALLYL ALCOHOL

Under the influence of the electron-donating CH_3 and OH groups, halogens add rapidly to methallyl alcohol even at -20°C (194). Methallyl alcohol reacted with chlorine water to give a good yield of the ether by Markovnikov addition (195):

$$2CH_2=C(CH_3)CH_2OH + 2ClOH \longrightarrow 2CH_2ClC(CH_3)(OH)CH_2OH \xrightarrow{H^+}$$

$$[CH_2ClC(CH_3)CH_2OH]_2O$$

Aqueous acidic solutions at higher temperatures would be expected to give some isobutyraldehyde and other bypro- ducts. Sheshukov found that methallyl alcohol, or iso- propenyl carbinol as he called it, reacted with HI to give largely t-butyl iodide. Hydrogenation to isobutanol occurs readily over the usual catalysts such as finely divided nickel (196). Methallyl alcohol was reacted with 30% H_2O_2 in the presence of tungstic oxide for 7 hr at 55°C to give an 83% yield of 2-methyl glycerol (197). A slower analogous reaction occurred with 1-methylallyl alcohol. Similarly several 1-alkylallyl alcohols and 1-phenylallyl alcohol were hydroxylated by reaction with hydrogen peroxide in formic acid to give substituted glycerols (198).

Molasseslike viscous liquid polymers ($n_D^{30} = 1.507$) were made by irradiating methallyl alcohol with UV for 18 days (199). Refluxing methallyl alcohol with peroxide cata- lysts or inorganic acid catalysts gave slow reaction to polymers of low DP. Refluxing methallyl alcohol in con- tact with pure oxygen for 3 days gave liquid low polymers that showed evidence of acid peroxide and ester groups (200). Adding a small amount of diallyl ether accelerated the polymerization.

Additions of methallyl alcohol strongly retard free radical polymerizations of styrene, acrylic, and vinyl monomers. However, small amounts of allylic carbinols may be useful regulators in some radical copolymerization systems. Copolymers with styrene and with MMA by radical initiation were evaluated in coatings crosslinked through OH groups (201).

Lewis-acid catalysts such as BF_3, BF_3 etherate, and H_2SO_4 acting on methallyl alcohol at -78 to 0°C may form soft yellow, solid polymers (202). These did not contain OH groups and they probably result from isobutyraldehyde polymerization.

Methallyl alcohol-sulfone polymers were prepared using excess SO_2 (203). Mixtures of methallyl alcohol isobutene, and SO_2 also were copolymerized. Copolymers of acrylonitrile with 15 to 30% methallyl alcohol units were soluble in acetone-water (90:10) (204). Copolymers of minor proportions of methallyl alcohol with acrylonitrile, with diallyl itaconate, or with methyl acrylate were crosslinked by reaction with diisocyanates, dibasic acids, or their anhydrides (205).

Copolymers containing alternating sodium methacrylate and methallyl alcohol units were prepared from methacrolein polymers (206). The methacrolein polymer was prepared by free radical emulsion polymerization. The acetal derivative of the polymethacrolein was reacted with NaOH to form alcohol and acid groups by a Cannizzaro reaction. The free acid form of the copolymer formed a soluble polylactone containing some residual OH and COOH groups. Coatings were crosslinked by heating above 200°C. Methacrolein polymers were reduced by $LiAlH_4$ in tetrahydrofuran to form alleged methallyl alcohol polymers softening near 120°C (207). Cyclization occurred under some conditions of reduction.

Properties and references to other 2-substituted allyl alcohols are given in Table 5.2. Polymers of high molecular weight have not been prepared from these. The compounds are normally liquids. However, 2-phenylallyl alcohol melts at 19°C.

OTHER MONOALLYL ALKANOLS

Little research has been done on the polymerization and copolymerization of the many known monoallyl alkanols. A number of the allylic alkanols have had interest for theoretical research, synthesis of dienes, pharmaceuticals, and terpene alcohols. In minor porportions they might be used as regulating comonomers with the more reactive vinyl-type monomers. Allyl carbinol $CH_2=CHCH_2CH_2OH$ and its alkyl derivatives were made from allyl magnesium

TABLE 5.2

2-Substituted Allyl Alcohols, $CH_2=CRCH_2OH$

R group	bp, °C	d_4, g/ml	n_D^{20}	References
CH_3	116	0.8515	1.4252	Ryan, JACS **62**, 3469 (1940)
C_2H_5	135	0.8682	--	Kondakov, Ber. **20**, 148 (1888)
$CH_2CH_2CH_3$	152	--	1.4318	Green, J. Chem. Soc. 3262 (1957)
$CH(CH_3)_2$	147	--	--	Green
$(CH_2)_8CH_3$	137(14.5)	0.851	1.4323	Normant, Compt. Rend. **240**, 314 (1955)
$CH_2C_6H_5$	82(0.5)	--	1.5359	
C_6H_5	105(4)	1.0509 (mp 19)	1.5675	Butler, U.S. 2,537,622 (Monsanto); Searles, J. Org. Chem. **24**, 1839 (1959)

bromide and carbonyl compounds (208). Fair yields of allyl carbinol were obtained using monomeric formaldehyde and the Grignard reagent. Allyl carbinol also has been made from vinyl magnesium bromide and ethylene oxide (209), from propylene and paraformaldehyde (210), and from allyl alcohol and paraformaldehyde (211). Allyl carbinol is one of the products obtained from passing vapor of 1,3-butanediol over activated bentonite at 350°C (212).

Allyl carbinol, a pleasant smelling liquid, is rearranged to ill-smelling n-butyraldehyde by passing over copper at 300°C. On heating with HCl in a sealed tube at 100°C it formed 3-chloro-1-butanol and a little allyl chloride (213). Warming allyl carbinol with HBr and H_2SO_4 gave 1-bromo-2-butene and 4-bromo-1-butene in ratio 3:1 (214). Butadiene can be made by dehydration (215). Allyl carbinol has been reacted with CO and water in the presence of $Ni(CO)_4$ to give a mixture of alpha-methylbutyrolactone and delta-valerolactone (216). Oxidation with permanganate gave 3-hydroxy tetrahydrofuran (217).

Methallyl carbinol $CH_2=C(CH_3)CH_2CH_2OH$ was obtained by reaction of isobutylene with paraformaldehyde in acetic acid-acetic anhydride at 190°C (218). The one-step Barbier-Grignard reaction was used to make $CH_2=CHCH(CH_3)C(OH)(CH_3)-(C_2H_5)$ from allyl magnesium bromide and methyl isopropyl ketone (219). Previously $CH_2=C(CH_3)CH_2CH(OH)CH_3$ had been made by dehydration (220). Pyrolysis of methallyl carbinol at 400°C gave a mixture of 1,3- and 1,4-pentadienes (221). Minor proportions of methallyl carbinol have been copolymerized with ethylene by a free radical-high pressure process (222). The copolymers contained more alkenol groups than expected.

Methyl vinyl carbinol or l-methyl allyl alcohol
(CH_2=CH_2CH(OH)CH_3 was obtained along with crotyl alcohol
CH_3CH=$CHCH_2OH$ by reaction of 3-chloro-1-butene or trans-
1-chloro-2-butene with aqueous alkali (223). It slowly
forms by rearrangement of crotyl alcohol in presence of
dilute sulfuric acid (224). It also can be prepared from
hydrogenation of HC≡C-CH(OH)CH_3 (225), from acrolein and
CH_3MgBr (226), and from the reduction of vinylethylene
oxide (227). Saponification of levo-methyl vinyl carbinyl
acetate gave partially racematized alcohol (228). Reduction of vinyl methyl ketone by $LiAlH_4$ was used to prepare
the alcohol (229). It was isomerized in dioxane-water
with sulfuric acid catalyst (carbonium ion mechanism) to
form crotyl alcohol along with byproduct ethers. Partial
allylic rearrangement of crotyl alcohol with acid catalyst
forms an equilibrium mixture containing 70% of l-methyl-
allyl alcohol. Methyl vinyl carbinol is miscible with
water at room temperatures, and with 26% water it forms
an azeotrope that boils at 86°C. The name is convenient,
but it must be remembered that it behaves as an allyl
compound rather than as a vinyl compound.
Methyl vinyl carbinol was oxidized by air in the
presence of ZnO + CuO catalysts at 300°C producing methyl
vinyl ketone (230). Oxidation at 30 to 40°C by H_2O_2 in
the presence of WO_3 or OsO_4 formed 1,2,3-butanetriol (231).
On standing with 10 to 30% H_2SO_4 or on heating with dilute
acid, partial isomerization to crotyl alcohol occurred and
ethers of both alcohols were formed (232). An optically
active methyl vinyl carbinol polymer was obtained by
asymmetric reduction of methyl vinyl ketone polymer using
$LiAlH_4$ and d-camphor (233). Polymers of this structure
cannot be made directly from the compound.
The literature contains other syntheses for l-substi-
tuted allyl alcohols CH_2=CHCH(OH)R, including the following:

Reactants	References
Acrolein + RMgBr	Prevost, Ann. Chim. [10] 10, 113 and 147 (1928); Lauer, J. Org. Chem. 26, 4785 (1961)
Aldehyde + AMgBr	Ginnings, JACS 61, 807 (1939)
Dehydration of CH_2OHCH_2CH(OH)CH_3	Biedermann, Ger. 964,044; CA 53, 11223 (1959)
Hydrogenation of CH_2=CHC(O)CH_3	Brendlein, Ger. 858,247 (Degussa); CA 50, 1891 (1956)
Hydrolysis of acetylated butadiene	Kamlet, U.S. 2,423,599 (Publicker)
Radical addition of ethanol to acetylene	Cywinski, J. Org. Chem. 30, 3814 (1965)

Allyl alcohols substituted in the 1-position by a halomethyl group such as $CH_2=CHCH(OH)CH_2Cl$ can be made by action of hypohalous acid or N-bromosuccinimide on butadiene (234). The 1-(chloromethyl)allyl alcohol hydrolyzes in water and in dilute acids; with NaOH it is converted rapidly into 3,4-epoxy-1-butene (235).

1-Alkyl-substituted allyl alcohols can rearrange to saturated ketones on heating with 12% H_2SO_4. Over Cu or Ni at 210 to 325°C, they also give ketones (236). More dilute sulfuric acid (1%) with heating or strong acid at room temperature gives rearrangement forming crotyl alcohols (237). The extensive literature on rearrangement of 1-substituted allyl alcohols was reviewed by de la Mare (238). The products obtained from reactions of substituted allyl alcohols with hydrogen halides, phosphorus halides, and thionyl halides are complex because of the possible rearrangements (239). On standing with perbenzoic acid in chloroform for 24 hr at 25°C, racemic 1-methylallyl alcohol gave a 50% yield of the epoxy alkanol (240). Monoperphthalic acid has been used to epoxidize a series of 1-alkyl allyl alcohols (241). Additional references to a number of 1-substituted allyl alcohols are given in Table 5.3.

There is extensive literature on vinyl dimethyl carbinol and other compounds of structure $CH_2=CRCR_2OH$. These do not contain allylic hydrogen atoms for degradative chain transfer. That they nevertheless do not polymerize very readily by free radical methods confirms that low activation of the ethylenic group is an important factor in the reluctance of allyl compounds to polymerize. Beginning in 1969, Hoffmann-LaRoche has supplied dimethyl vinyl carbinol $CH_2=CHC(OH)(CH_3)_2$ and a number of allylic alcohols of the type $CH_2=CHCRR'(OH)$. These compounds are obtained from partial hydrogenation of acetylenic alcohols derived from reaction of ketones with acetylene. Acetylenic and allylic alcohols are intermediates in the synthesis of vitamins A and E, as well as "aroma" or perfume chemicals such as linalool and ethyl linalool. The vinyl dialkyl carbinols have mild pleasant odors, suggesting camphor slightly, and they produce no delayed sting in the nose.

Minor proportions of vinyl dimethyl carbinol have been copolymerized with ethylene at 209°C and 18,000 psi in alkane solvent (242). Films and moldings and good impact strength and moisture barrier properties. Vinyl acetate-dimethyl vinyl carbinol (9 to 1) were copolymerized by heating at reflux with azobis for 2.5 hr (243), and vinyl alcohol copolymers were prepared from the product. Two other allylic tertiary alcohols having mild camphorlike odors have been supplied by LaRoche. These liquid alcohols are vinyl methyl ethyl carbinol or 3-hydroxy-3-methyl-1-

TABLE 5.3

1-Substituted Allyl Alcohols, $CH_2=CHCH(OH)R$

R group	bp, °C	$n_D(°C)$	References
CH_3	98	1.4127(20)	Wohl, Ber. **41**, 3621 (1908)
C_2H_5	116	1.4223(25)	Kohler, Am. Chem. J. **38**, 525 (1908)
			Kepner, JACS **71**, 118 (1949)
$CH_2CH_2CH_3$	134	1.4269(16)	Niemann, J. Org. Chem. **8**, 399 (1943)
$CH(CH_3)_2$	126	1.4316(16)	Bacon, J. Chem. Soc. 1076 (1937);
			Ginnings, JACS **61**, 807 (1939)
$CH_2CH_2CH_2CH_3$	155	1.4336(20)	Smets, CA 8315 (1950); Hills,
			J. Chem. Soc. 580 (1936)
$CH_2CH(CH_3)_2$	146	1.4285(21)	Delaby, Compt.Rend. 203, 1521 (1936)
$(CH_2)_4CH_3$	174	1.4378(20)	Gredy, Bull. Soc. Chim. [5], **3**,
			1094 (1936) (mushroom odor)
C_6H_5	216	1.5450(21)	Klager, Ber. **39**, 2553 (1906)
			(pleasant ketonelike odor)

pentene and 1-vinylcyclohexanol or 1-hydroxy-1-vinylcyclo-
hexane. The compound isophytol $CH_2=CHC(OH)(CH_2)C_{11}H_{23}$ is
supplied by BASF and might form interesting copolymers of
vinyl acetate and of vinyl alcohol.

Vinyl ethyl carbinol is a liquid smelling somewhat like
allyl alcohol but not as strong. It has been made from
acrolein and ethyl magnesium bromide (244, 247). It boils
at 115°C, $n_D^{25} = 1.4223$, and it is incompletely miscible
with water. Slow distillation with Raney nickel gives
diethyl ketone (245). Repeated distillation at atmospheric
pressure gave the ether (246). Reaction of vinyl ethyl
carbinol with HCl or PCl_3 gave 3-chloro-1-pentene and
1-chloro-2-pentenes (largely trans). Reaction with HBr
and H_2SO_4 during 2 days gave 1-bromo-2-pentene and 3-bromo-
1-pentene (247). The sodium salt of vinyl ethyl carbinol
reacts with diethyl amine on heating to give some
1-diethylamino-3-pentanol (248). Heating 1-ethyl allyl
alcohol in benzene with trimethyl hydroquinone and $ZnCl_2$
in a sealed tube at 200°C gave an hydroxycoumaran derivative (249). Optically active forms of vinyl ethyl carbinol
have been separated (250).

A useful method to prepare allyl-substituted methanols
(allyl carbinols) is the reaction of the Grignard reagent
with carbonyl compounds:

$$AMgBr + {\displaystyle \mathop{>}^{}C=O} \xrightarrow{ether} {\displaystyle \mathop{>}^{A}}C-OMgBr \xrightarrow{HCl} {\displaystyle \mathop{>}^{A}}C_{OH} + MgClBr$$

Many references until 1948 on preparation of allyl carbinols using allyl Grignard reagents have been listed (251). Before the work of Grignard, allyl-substituted alcohols were made using allyl halides and zinc, for example:

$$ABr + Zn + HCHO \longrightarrow ACH_2OZnX \xrightarrow{H_2O} ACH_2OH + Zn(OH)X$$

References to this old literature using allyl bromide and iodide have been given (251). Diallyl carbinols have been prepared from reaction of esters with allyl magnesium bromide or with zinc and allyl iodide (251). Some additional allylic alcohols are listed in Table 5.4 where A and MA represent allyl and methallyl groups. Apparently homopolymers of high molecular weight have not been prepared from any of these.

An example of a diallylic alcohol is allyl vinyl carbinol which can be made by reaction of allyl magnesium bromide with acrolein (252).

CYCLIC ALLYL ALCOHOLS

Reactions of cyclic ketones with Grignard reagents give allyl cycloalkanols such as 1-allylcyclohexanol which has a fine odor suggestive of menthol, camphor, and peppermint (253):

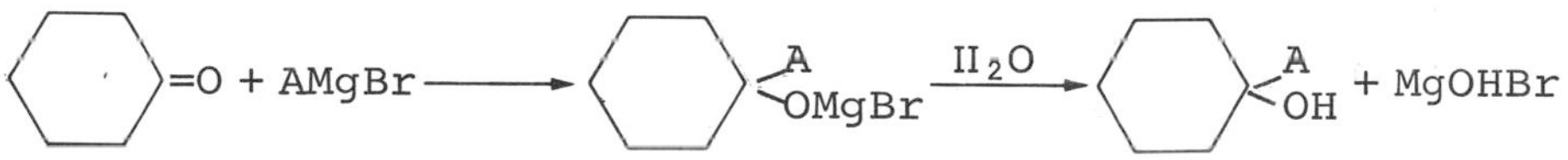

Early investigators reacted magnesium with mixtures of a ketone, allyl halide, and ether to obtain the cycloalkyl allyl alcohol derivatives. The allyl compound 1-vinyl-1-cyclohexanol (mp, 4°C; bp, 170°C; $n_D^{20}-1.4755$) has received considerable attention (254). It has a camphor-like odor. 1-Allyl-1,2,3,4-tetrahydro-1-phenanthrol (mp, 104°C) reacted with concentrated sulfuric acid to give a red to green fluorescent product (255). References to some other cyclic allyl alcohols are given in Table 5.5. Polymerizations and copolymerizations of these compounds have received little attention.

A number of allyloxybenzyl and related alcohols have been prepared. For example, saligenin and allyl bromide reacted in ethanol in the presence of potassium carbonate formed 2-allyloxybenzyl alcohol (256). This compound, boiling at 150°C, resinified on heating with evolution of HCHO and water. Beta-santalol, a constituent of sandalwood oil, is an allylic bicyclic primary alcohol (257).

Vinyl aryl carbinols can be prepared by reacting aryl magnesium bromide and acrolein in ether with cooling. Thus 1-naphthylallyl alcohol [bp 187°C(19)] was prepared

TABLE 5.4

Monoallyl and Monomethallyl Carbinols

Carbinol	bp, °C	n_D^{20}	References and Other Properties
ACH_2OH	113	1.4224	Juvala, Ber. 63, 1992 (1930); Newman, J. Chem. Soc. 262 (1936) (3,5-dinitrobenzoate, mp 59°C);
$ACH(OH)CH_3$	116	1.4225	Ou, Ann. Chim. Phys. [11], 13, 182 (1940); Wagner, Ber. 27, 2434 (1895)
$AC(OH)(CH_3)_2$	118	1.4263	Fischer, Ber. 76, 735 (1943); Colonge, Bull. Soc. Chim. 433 (1948); Henze, J. Org. Chem. 7, 329 (1942)
$ACH(OH)C_2H_5$	132	1.4330	Henze, J. Org. Chem. 7, 328 (1942); Fournier, Bull. Soc. Chim. [3] 11, 124
$ACH(OH)CH(CH_3)_2$	142	1.4365	Fournier, Bull. Soc. Chim. [3] 11, 359; Karassew, CZ I, 28 (1942)
$ACH(OH)CH_2CH(CH_3)_2$	164	1.4380	Fournier
$ACH(OH)R$, e.g., allyl citronellols	124	--	Rupe, Annalen 402, 173 (1913) (some with roselike odor)
$MACH_2OH$	133	1.4225	Ou
$MACH(OH)CH_2CH_2CH_3$	151	1.4342	Henze, J. Org. Chem. 7, 329 (1942)
$AC(OH)(CH_3)_2$	117	1.4263	Henze, J. Org. Chem. 7, 329 (1942); camphorlike odor
$AC(OH)CH_3(C_2H_5)$	139	--	Zaitsev*, J. Russ. Phys. Chem. Soc. 24, 469 (1892); CZ I, 635 (1893)
$AC(OH)(CH_3)CH(CH_3)_2$	160	--	Schryver, J. Chem. Soc. 63, 1336; turpentine odor
$AC(OH)(C_2H_5)_2$	156(727)	--	Zaitsev, Ann. 196, 113 (1879); camphor odor
$MAC(OH)(CH_3)_2$	132	--	Lemaire, CZ I, 1982 (1909)
ACH_2CH_2OH	141	1.4299	Brooks, Org. Syn. Coll. Vol. III, 698 (1955); Schniepp, JACS 67, 56 (1945)

ACH$_2$CH(OH)CH$_3$	140	--	Crow, Ann. 201, 42 (1880); Gardner, J. Chem. Soc. 91, 851 (1907); Fischer, Ann. 520, 59 and 69 (1935)
AC(CH$_3$)$_2$OH	119	--	Zaitsev*, Ann. 185, 151 and 175 (1877); camphor odor
MACH$_2$OH	130	1.4347	U.S. 2,308,192 (Shell); U.S. 2,335,027 (Esso) Laforque, Compt. Rend. 227, 353 (1948)
MACH$_2$CH(OH)CH$_3$	131	1.4364	Young, J. Chem. Soc. 1452 (1938); Tamele, Ind. Eng. Chem. 33, 120 (1941)
CH$_2$=CHCH(CH$_3$)CH$_2$OH	121	1.4262	Roberts, JACS 67, 149 (1945); Imhoffen, Ber. 82, 315 (1949); (3,5-dinitrobenzoate, mp 58°C)
CH$_2$=CH(CH$_2$)$_4$CH$_2$OH	105	1.4403	Linstead, J. Chem. Soc. 1973 (1937)
CH$_2$=CHCH(CH$_3$)CH(OH)CH$_3$	128	1.4315	Roberts, JACS 67, 149 (1945); Ou, Compt. Rend. 208, 529 (1939)
CH$_2$=CHCH(CH$_3$)CH(OH)CH$_2$CH$_3$	141	n_D^{22} = 1.5365	Ou, Compt. Rend. 208, 529 (1939)
CH$_2$=CHCH$_2$CH$_2$(CH$_3$)$_2$OH	143	--	Linstead, J. Chem. Soc. 473 (1936); Colonge, Compt. Rend. 225, 1161 (1947); (mp -23°C)
CH$_2$=CHCH$_2$C(CH$_3$)OHCH$_2$CH$_3$	138	1.4370	Milas, JACS 57, 581 (1935); Henze, J. Org. Chem. 7, 328 (1942)
CH$_2$=CHCH(CH$_3$)C(OH)(CH$_3$)$_2$	95(200)	1.4367	Roberts
CH$_2$=C(CH$_3$)CHOH(CH$_3$)	117	n_D^{17} = 1.4288	Kondakov, J. Russ. Phys. Chem. Soc. 17, 276; Keefer, JACS 71, 3907 (1949)
CH$_2$=C(CH$_3$)CH(OH)C$_2$H$_5$	134	1.4336	Bruson, JACS 59, 2013 (1937); U.S. 2,200,538 and U.S. 2,290,274 (Parke-Davis)
CH$_2$=C(CH$_3$)CH(OH)CH$_2$CH$_2$CH$_3$	154	1.4370	Bruson and Parke-Davis

*Formerly spelled Saytzeff or Saizev

TABLE 5.5

Some Allyl-Substituted Cycloalkanols

	bp, °C	n_D^{20}	References
1-Allyl-cyclohexanol	192 77(11)	1.4787	Available from Pfaltz & Bauer, Aldrich; Aldersley J.Chem.Soc. 10 & 14 (1940
1-Allyl-cyclopentanol	63(10)	1.4683	Crane, JACS **67**, 1237 (1945)
2,2-Diallyl cyclohexanol	116(6)	1.4860	Newman, JACS **68**, 2114 (1946
1-Methallyl-cyclopentanol	98.5(40)	1.4720	Crane
1-Allylmenthol	250 130(22)	--	Zaitsev, CZ I, 1402 (1909); Jaworski, Ber. **42**, 437 (1909); CZ I 856 (1909); peppermint-camphor odor
Allylpulegone	135(27)	--	Jaworski
Allylborneol or 2-allylcamphanol	130(20)	1.4894(22)	Jaworski; Gordon, Bull.Soc. Chim. [3] **28**, 31 (1902)
Allylfenchyl alcohol	--	ca. 1.491	Zaitsev, CZ I, 783 (1914) (from fenchone, AI and Zn
1-Methyl-3-allyl-cyclohexanol	108(17)	1.4721(25.4)	Cornubert, Ann. Chim. [9] **16**, 185 (1921)
1-Methyl-4-allyl-cyclohexanol	112(19)	1.4666(24.4)	Cornubert, Compt. Rend. **159** 76 (1914); menthol-like odor
1-Cyclohexyl-1-allyl alcohol	91(12)	1.4770	Danilov, CZ I, 2588 (1937)

(258). Optical isomerism in 1-phenylallyl alcohol has been studied (259). The compound, which boils at 106°C at 16 mm and has n_D^{19} = 1.5390, was largely converted to cinnamyl alcohol on treatment with dilute sulfuric acid (260). In contrast, 1-cyclohexylallyl alcohol remained unchanged. The 1-o-tolylallyl alcohol (261) and 1-phenylmethallyl alcohol (262) also have been prepared. Solid 1:1 molar copolymers of 2-phenylallyl alcohol and maleic anhydride were made at 50 to 60°C using azobis as initiator (263).

Cyclic aldehydes react with allyl Grignard reagents to form allyl aryl carbinols and allyl cycloalkyl carbinols. It is also possible to react an aldehyde with allyl iodide and zinc in dry ether (264):

The product slowly decomposed during distillation at 229°C.
Allyl diphenyl carbinol, bp 183°C(32 mm), also decomposed
when distillation was attempted at atmospheric pressure
(265).

Aryl-substituted allyl alcohols do not homopolymerize
readily, but 2-phenylallyl alcohol was copolymerized with
acrylonitrile by heating with BP in solvents such as
dimethyl formamide (266). Crosslinked products were
obtained under some conditions. Minor proportions of
2-phenylallyl alcohol could be copolymerized with styrene
in bulk or emulsion using peroxide catalysts (267). Solid
homopolymers of 2-phenylallyl alcohol, but of molecular
weight only about 1100, were made using azo catalyst at
50 to 65°C (268).

POLYFUNCTIONAL ALLYLIC ALCOHOLS

Allyl-substituted polyols have been used in the preparation of polyesters bearing allyl groups (269). Vinyl
ethylene glycol $CH_2=CHCHOHCH_2OH$ is an allyl compound
prepared long ago (270) and recently suggested in copolymerizations with acrylonitrile and acrylate esters in
persulfate emulsion (271). It can be isomerized to
α-hydroxybutyraldehyde at 130°C over palladium on charcoal
(272) or by heating with 10% HCl at 100°C (273). Kharasch
and Buechi (274) made 2-allyl-1,3-propanediol; with
related diols, this compound has been studied in the
synthesis of reactive polyesters and polyurethanes (275).
The diol $CH_2=C(CH_2CH_2OH)_2$ was prepared from 3-methallyl
carbinol and formaldehyde (276). Polyurethanes were made
by reaction of aromatic diisocyanates with 1,2-divinyl
ethylene glycol (277). Tough films were obtained. The
reaction of $CH_2=CHCH_2CHO$ with formaldehyde in the presence
of sodium hydroxide at 90°C gave $CH_2=CHCH_2C(CH_2OH)_3$,
melting at 91°C (278).

The allylic polyol $CH_2=C(CH_2OH)_2$ can be made from pentaerythritol, e.g., by heating the monobromide of the latter
with $Ba(OH)_2$ or Ag_2O (279). Several methods have been
employed to make 2-hydroxyallyl carbinol, for example,
alkaline hydrolysis of 1,4-dihalo butadienes (280) or of
4-chloro-1-buten-3-ol (281). The diol $CH_2=CHCH(OH)CH_2CH_2OH$
was obtained from a Prins reaction of butadiene with
formaldehyde (282). The Millmaster Chemical Company has
supplied 2-allyl-2-ethyl-1,3-propanediol.

The most important doubly unsaturated allylic alcohol is
the tertiary alcohol, linalool, a C_{10} derivative of
1-methylallyl alcohol. It is 3,7-dimethyl-1,6-octadien-
3-ol of formula $(CH_3)_2C=CHCH_2CH_2\underset{OH}{C}(CH_3)CH=CH_2$. So widely
does l-linalool occur in flower and tree oils that it may

be a common precursor of terpenes in plants (perhaps via
alpha-terpineol). This essential oil, which has a fine
rose odor, is the chief constituent of linaloe oil from
Mexican rosewood and it occurs in oils of sassafras,
Ceylon cinnamon, lavender, orange, grapefruit, lemon,
bergamot, thyme, sage, and in many flower oils. Dextro-
rotary d-linalool is known as coriandrol. Linalools are
widely used in formulating perfumes and for flavoring
beverages, fruit juices, wines, and teas. The acetate
ester also occurs widely in plant oils and finds use in
perfumery.

Racemic dl-linalool can be prepared by slow addition of
6-methyl-5-hepten-2-one in dry ether to sodium acetylide
dispersed in liquid ammonia (283). The acetylenic alcohol
recovered was partially hydrogenated and linalool was
recovered by distillation (bp, 92°C at 17 mm, $n_D^{25} = 1.4600$).
Another synthetic derivative of 1-methylallyl alcohol is
3,7,9-trimethyl-1-decen-3-ol which has a pleasant "fresh
leafy fragrance" (284). Some of the allylic tertiary
alcohols which have valuable aromas can be prepared by
rearrangement of esters followed by hydrolysis (285). By
heating linalool with a catalyst such as acetic anhydride,
it is partially isomerized to the primary alcohols geraniol
and nerol.

There are very few records of polymerization or copolymerization of polyfunctional allyl alcohols even though
some have been distilled at high temperatures. Unstable
copolymers which might slowly liberate monomeric perfume
components could be worth investigation. Additional
references to polyfunctional allylic alcohols are listed
in Table 5.6.

HALOALLYL ALCOHOLS

The 2-chloro and 2-bromoallyl alcohols were prepared
from the 2-haloallyl chlorides and aqueous alkali carbonate
or by treating $CH_2ClCHXCH_2Cl$ with alkali (286). Chloroallyl alcohol also has been made by reduction of chloroacrolein (287) and by adding HCl to propargyl alcohol at
60°C in the presence of $HgCl_2$ (288). These monomers are
unstable and most unpleasant to handle. Properties of
some haloallyl alcohols are given in Table 5.7.

Haloallyl compounds $CH_2=C-\overset{\frown}{C}H$, being derivatives of vinyl
X
halides, may be expected to polymerize in radical systems
somewhat more readily than other allyl compounds. This
should be true especially when electron-attracting groups
are attached to the third carbon (instead of electron-
releasing OH group). Chlorallyl alcohol, acetate, and
chloride have been copolymerized with styrene and acrylic
monomers (289).

TABLE 5.6

Some Polyfunctional Allylic Alkanols

Compound	bp, °C	Refractive index	References, etc.
Diallyl carbinol	151	--	Everett, J. Chem. Soc. 3131 (1950); Zaitsev, Ann. 185, 129 (1876)
Sym-diallylisopropanol	185	--	Oberreit, Ber. 29, 2002 (1896)
Divinyl carbinol	63 (94)	$n_D^{20} = 1.4450$	Available from Lithium Corp.
Triallyl carbinol	68 (9)	--	Stetter, Chem. Ber. 86, 589 (1953) (Terpene-like odor)
2,2-Diallylethanol	192	$n_D^{21} = 1.468$	Reformatski, Ber. 41, 4086 (1908)
Diallyl propyl carbinol	173	--	Oberreit, Ber. 29, 2007 (1896)
	194		Zaitsev, Ann. 193, 362 (1878) (Dehydrated on boiling)
3,4-Dihydroxydiallyl (or divinyl ethylene glycol)	55 (0.3)	$n_D^{25} = 1.4739$	Braun, J. Org. Chem. 28, 1383 (1963)
Diallyl methyl carbinol	159	$[d_{20}^{20} = 0.8626]$	Zaitsev, J. Prakt. Chem. 76, 100 (1907)
2-Allyl-2-methallyl-1,3-propanediol, etc.	--	--	Wasson, Can. J. Chem. 41, 3070 (1963)
2,5-Divinyl-2,5-hexanediol	--	--	Metalova, CA 75, 5601 (1971)
1-Linalool	198	$n_D^{20} = 1.4652$	Cornforth, Tetrahedron 18, 1351 (1962); Ohloff, Ibid. 18, 37 (1962)
d-Linalool or coriandrol	198	--	Ohloff (In orange and coriander oils)
3,3-Diallylborneol	[mp, 47°C]	--	Haller, Ann. Chim. [9] 9, 212 (1918)
Allylmethyl diallyl carbinol	217	--	Reformatski, Ber. 41, 4091 (1908) (Peppermint odor)
2-Methyl-1,1,5,5-tetraallyl-6-cyclohexanol	193 (21)	$n_D^{25} = 1.5070$	Cornubert, Ann. Chim. [9] 16, 185 (Partially polymerized)
Tetraallylcyclohexanol	173 (15)	--	Haller, Compt. Rend. 156, 1202 (1914) (Piercing odor)

TABLE 5.7

Some Halogen-Substituted Allyl Alcohols

	bp, °C	Refractive index	References
$CH_2=CBrCH_2OH$	154	$n_D^{18}=1.5000$	Henry, Ber. 5, 453 (1872); Henry, Ber. 14, 404 (1881)
$CH_2=CClCH_2OH$	140	$n_D^{20}=1.4588$	Henry, Compt. Rend. 95, 849 (1882); van Romburgh, Rec. Trav. Chim. 1, 233 (1882) Hinman, CA 49, 4925 (1955) (Killed fruit fly larvae)
$CH_2=CFCH_2OH$	55(100)	$n_D^{25}=1.5515$	Henry, U.S. 3,215,746 (Carbide a toxic compound; for synthesis of antinarcotics
$CH_2=CHCH(OH)CH_2Cl$	147	$n_D^{20}=1.4649$	Evans, J.Chem.Soc. 239 (1949); Kadesch, JACS 68, 44 (1946) (Odor similar to that of allyl chloride)
$CH_2=CHCH(OH)CH_2Br$	162	$n_D^{20}=1.5000$	Petrov, CA 32, 5370 (1938); Bottini, J.Org.Chem. 27, 271 (1962) (Odor similar to that of allyl bromide)

There is very little information on the 2-halomethallyl alcohols $CH_2=C(CH_2X)CH_2OH$, which seem to be difficult to prepare and purify. Chloromethallyl alcohol was reported from the reduction of chloromethacrolein by an aluminum alcoholate (287).

There has been some research on 1-(chloromethyl)allyl alcohols. Carbon tetrachloride gave addition of chlorine to the double bond, but a chlorinated polyether byproduct was also formed (290). In order to add ClOH to 1-(chloromethyl)allyl alcohol, t-butyl hypochlorite was used (291).

References

1. A. Cahours and A. W. Hofmann, Ann. 100, 356 (1856); 102, 285 (1857); cf. N. Zinin, Ann. 96, 361 (1855).
2. B. Tollens and A. Henninger, Ann. 156, 134 (1870).
3. O. Kamm and C. S. Marvel, Org. Syn. Coll. Vol. I, 42 (1941).
4. H. P. A. Groll and G. Hearne, Ind. Eng. Chem. 31, 1530 (1939); M. W. Tamele and H. P. A. Groll, U.S. 2,072,015-6 (Shell); E. C. Williams, Trans. Am. Inst. Chem. Eng. 37, 157 (1941); Chem. Met. Eng. 47, 834 (1940); A. W. Fairbairn et al, Chem. Eng. Progr. 43, 280 (1947).

5. E. C. Britton and G. H. Coleman, U.S. 2,176,055 (Dow); F. Koehler, U.S. 2,323,781 (R & H); M. A. Pollack and A. G. Chenicek,
U.S. 2,313,767(PPG); Belg. 577,345 (Solvay), CA 54, 4386 (1960).
6. G. H. van de Griendt and L. M. Peters, U.S. 2,475,364 (Shell).
7. A. I. Chernyshev et al, USSR 173,225 and 174,179 (1965);
CA 64, 593 (1966).
8. H. A. Cheney et al, U.S. 2,434,394 (Shell).
9. A. B. Ash et al, U.S. 2,441,540; cf. U.S. 2,485,694 (Wyandotte).
10. C. O. Young and G. H. Law, U.S. 1,917,179; H. H. Law and R. W.
McNamee, U.S. 2,159,507; G. W. Fowler and J. T. Fitzpatrick,
U.S. 2,426,264 (all to Carbide).
11. A. Thizy et al, Fr. 1,271,563 (Progil); W. I. Denton, U.S.
2,986,585; 3,090,815; 3,092,668 (Olin).
12. G. Schreyer et al, Ger. Offen. 1,810,210 (Degussa), CA 72, 132033.
13. H. D. Finch, Chapter 5 in Acrolein, C. W. Smith, Ed., Wiley, 1962.
14. H. D. Finch and A. DeBenedictus, U.S. 2,779,801 and S. A. Ballard
and E. A. Youngman, U.S. 2,991,305 (Shell).
15. R. W. Foreman, U.S. 3,109,865 (Std. Oil Ohio); cf. U.S. 3,227,640.
16. C. J. Duyverman, U.S. 3,466,339 (Stamicarbon); cf. Belg. 667,461
(Kyowa)
17. C. W. Smith and R. T. Holm, U.S. 2,761,883 (Shell).
18. S. Kunichika et al, CA 66, 28335 (1967).
19. E. C. Shokal and T. W. Evans, U.S. 2,428,590 (Shell); G. Stein,
Angew. Chem. 54, 148 (1941).
20. Beilstein's Handbuch der Organischen Chemie, 4th ed., 3rd supple-
ment lit. to 1949, F. Richter, Ed., Springer, 1958, Vol. I,
p. 1873.
21. N. F. Cywinski and H. J. Hepp, J. Org. Chem. 30, 3814 (1965).
22. M. Legator and D. Racusen, J. Bacteriol. 77, 120 (1959); also
unpublished work by G. E. Carvell.
23. N. I. Sax, Dangerous Properties of Industrial Materials,
Reinhold, 1957; Data sheets of Borden, Dow, and Shell.
24. Booklet, Allyl Alcohol, Shell Chemical Company (1946).
25. A. Kailan and F. Adler, Monatsh. Chem. 63, 164 (1933);
S. Schwebel, ibid. 63, 64 (1933).
26. Y. Jen and J. A. Seneker, U.S. 3,176,050 (Cal. Res.).
27. K. B. Cofer and R. W. Fourie, U.S. 3,093,689 (Shell).
28. D. Swern et al, J. Am. Chem. Soc. 71, 1153 (1949).
29. H. Land and W. A. Waters, J. Chem. Soc. 2129 (1958).
30. K. A. J. Singer, Peintures, Pigments, Vernis 34, 547 (1958);
35, 8 (1959).
31. W. H. Hatcher and C. T. Mason, Can. J. Res. 10, 318 (1934).
32. P. Jaulmes and R. Mestres, Chim. Anal. 40, 413 (1958).
33. A. van Dormael, Bull. Soc. Chim. Belg. 52, 109 (1943).
34. R. J. Gritter et al, J. Org. Chem. 24, 1051 (1959).
35. S. Goldschmidt et al, Ber. 67B, 208 (1934).
36. H. P. A. Groll and H. W. de Jong, U.S. 2,042,220 (Shell); cf.
R. Delaby, Bull. Soc. Chim. France [4] 53, 303 (1933).
37. K. Nakagawa et al, J. Org. Chem. 27, 1601 (1962).

38. S. Dykstra and H. S. Mosher, J. Am. Chem. Soc. 79, 3474 (1957).
39. J. Hoffman, J. Org. Chem. 22, 1747 (1957).
40. H. E. Seyfarth et al, Angew. Chem. Int. Ed. 4, 1074 (1965).
41. E. Eegriwe, Z. Anal. Chem. 100, 33 (1935).
42. R. W. Martin, Anal. Chem. 21, 922 (1949).
43. N. B. Baranow, CA 34, 1591 (1940); E. A. Peregud, CA 38, 699 (1944); V. Hamann and A. Herman, Microchim. Acta 105 (1961).
44. P. G. Clay et al, Proc. Chem. Soc. 125 (1959) and 22 (1962).
45. G. F. Emerson and R. Pettit, J. Am. Chem. Soc. 84, 4591 (1962); R. W. Goetz and M. Orchin, ibid. 85, 1549 (1963); J. K. Nicholson et al, Proc. Chem. Soc. 282 (1963).
46. R. W. Goetz and M. Orchin, J. Am. Chem. Soc. 85, 1549 (1963).
47. G. W. Watt et al, J. Am. Chem. Soc. 76, 5958 (1954).
48. M-E. Manzhelei and A. F. Sholin, CA 56, 13,955 (1962).
49. M. L. Wolfram et al, J. Org. Chem. 25, 1079 (1960); W. Walisch and J. E. Dubois, Chem. Ber. 92, 1028 (1959).
50. A. Bigot, Ann. Chim. Phys. [6] 22, 433 (1891); CZ I, 866 (1891).
51. N. J. Bythell and P. W. Robertson, J. Chem. Soc. 179 (1938).
52. A. Berthoud and M. Mosset, J. Chim. Phys. 33, 272 (1936).
53. G. Bohm and W. Dietrich, U.S. 3,290,395 (Huels).
54. H. Huebner and K. Mueller, Ann. 159, 179 (1871); H. Tornöe, Ber. 24, 2670 (1891); J. Chem. Soc. A II 1442 (1891); H. King and F. L. Pyman, J. Chem. Soc. 1257 (1914).
55. P. B. D. delaMare and J. G. Pritchard, J. Chem. Soc. 3990 (1954).
56. P. B. delaMare et al, J. Chem. Soc. 3429 (1963).
57. Allyl Alcohol booklet, Shell, 1946.
58. P. W. Robertson et al, J. Chem. Soc. 1324 (1933); 179 (1938); cf. 1628 (1950); K. Nozaki and R. A. Ogg, J. Am. Chem. Soc. 64, 713 (1942).
59. L. Smith and S. Skyle, Acta Chim. Scand. 4, 39 (1950); 5, 1415 (1951); H. Hibbert et al, Can. J. Res. 4, 122.
60. E. Moffett et al, J. Am. Chem. Soc. 56, 2009 (1934).
61. J. Jacques, Bull. Soc. Chim. France [5], 12, 844 (1945).
62. S. Winstein and L. Goodman, J. Am. Chem. Soc. 76, 4368 (1954).
63. W. E. Vaughan and F. F. Rust, U.S. 2,398,479 (Shell).
64. A. M. Alvarado, U.S. 2,402,586 (Du Pont).
65. P. S. Pinkney, U.S. 2,551,813 (Du Pont); cf. Kaneko, CA 33, 2105 (1939).
66. T. Hoshino et al, CA 48, 5789.
67. R. Brown et al, J. Chem. Soc. 3315 (1951).
68. R. Brown et al, J. Chem. Soc. 2123 (1951).
69. K. Ruehlmann et al, J. Prakt. Chem. 10, 325 (1960).
70. Allyl Alcohol Booklet, Shell, p. 39.
71. M. M. Rauhut et al, J. Org. Chem. 26, 5138 (1961).
72. R. C. Fuson and J. H. Koehneke, J. Org. Chem. 14, 707 (1949).
73. M. S. Kharasch et al, J. Org. Chem. 3, 175 (1938); cf. J. H. Helberger, Ann. 588, 71 (1954).
74. H. Friese, Ber. 71B, 1303 (1938).
75. D. Swern et al, J. Am. Chem. Soc. 71, 1152 (1949).
76. W. S. Calcott et al, J. Am. Chem. Soc. 61, 1012 (1939);

J. H. Simon and S. Archer, ibid. 61, 1521 (1939); J. F. McKenna and F. I. Sowa, ibid. 59, 471 (1937); cf. Nasarova et al, CA 39, 916 (1945).

77. J. B. Niederl et al, J. Am. Chem. Soc. 53, 3390 (1931); 55, 292 (1933); cf. W. Baker et al, J. Chem. Soc. 76 (1951); 1778 (1952).
78. L. I. Smith et al, J. Am. Chem. Soc. 61, 2617 (1939).
79. M. Asscher and D. Volsi, J. Chem. Soc. 1887 (1963); cf. J. S. Rose and A. A. Shuragian, Belg. 632,171 (Olin).
80. J. D. Park et al, J. Org. Chem. 26, 2089 (1961).
81. N. A. Milas and S. Sussman, J. Am. Chem. Soc. 58, 1302 (1936); 59, 543 (1937); U.S. 2,437,648 (Res. Corp.); cf. E. Bauer, Ger. 890,943; cf. W. D. Lloyd et al, CA 78, 34336 (1973).
82. M. Mugdan and D. P. Young, J. Chem. Soc. 2988 (1949); cf. I. Bergsteinsson, U.S. 2,373,942 (Shell).
83. N. Prileschajew, Ber. 42, 4813 (1909); CZ II, 268 (1911).
84. Z. Raciszewski, J. Am. Chem. Soc. 82, 1267 (1960); cf. Netherlands Appl. (Hoechst), CA 63, 2898 (1965).
85. H-P. Liao et al, U.S. 3,454,655 (FMC).
86. C. J. Wenzke and S. A. Mednick, U.S. 3,509,183 (FMC).
87. J. J. Tjepkema, U.S. 2,813,910 (Shell); C. W. Smith, U.S. 2,731,502 (Shell).
88. D. H. Volman et al, J. Am. Chem. Soc. 81, 756 and 4141 (1959).
89. K. A. Mass and D. H. Volman, Trans. Faraday Soc. 60, 1202 (1964).
90. Brit. 544,421 (Am. Cyan.); CA 36, 65848 (1942).
91. J. Nichols, U.S. 2,557,136 and 2,596,279 (Interchemical); S. W. Tinsley et al, U.S. 3,014,048; Brit. 932,144 (Carbide); E. K. Fields, J. Am. Chem. Soc. 76, 2709 (1954).
92. K. B. Sharpless et al, J. Am. Chem. Soc. 90, 209 (1968).
93. H. Adkins and G. Krsek, J. Am. Chem. Soc. 70, 383 (1948).
94. W. Reppe and Kroeper, Ann. 582, 50 and 65 (1953).
95. G. P. Chiusoli and S. Merzoni, U.S. 3,238,246 (Monte).
96. H. Hoog and W. F. Engel, U.S. 2,603,645 and 2,605,264 (Shell).
97. Brit. 578,266 (PPG); cf. Brit. 578,267; CA 41, 2275 (1947).
98. H. Tanaka and I. Neshikawa, U.S. 3,441,632 (Hitachi).
99. B. F. Dannels and A. F. Shephard, U.S. 3,557,249 (Hooker).
100. F. F. Blicke, J. Am. Chem. Soc. 45, 1562 (1923).
101. H. Dannenberg and D. E. Adelson, Brit. 566,344; U.S. 2,541,155 (Shell).
102. D. E. Adelson and H. F. Grey, U.S. 2,555,775 (Shell).
103. Brit. 779,838 (Shell).
104. D. E. Adelson et al, U.S. 2,467,105 and 2,473,124 (Shell).
105. A. C. R. Brown and D. G. L. James, Can. J. Chem. 40, 796 (1962).
106. Fanny Boyer-Kawenoki, Bull. Soc. Chim. France 624 (1959); CA 53, 15631 (1959); cf. CA 54, 25951 (1960).
107. N. G. Gaylord and F. R. Eirich, J. Am. Chem. Soc. 73, 4981 (1951).
108. I. Goodman and J. Mather, Brit. 854,207 (ICI); CA 55, 11920 (1961); Belg. 571,019 (ICI).
109. S. A. Dolmatov and L. S. Polak, CA 60, 3102; CA 64, 3691 (1966); cf. W. D. McHenry (NASA), CA 60, 1845 (1964); cf. U.S. 3,285,897 (high pressures).

110. V. A. Kargin, L. S. Polak, V. A. Kabanov, V. P. Zubov and V. F. Pankova, U.S. 3,666,740, appl. Aug. 14, 1969.

111. V. A. Kargin et al, Ger. appl. (Topchiev Inst.), CA 74, 142661 and 75, 152175 (1971).

112. A. R. Denaro et al, Eur. Polym. J. 6, 487 (1970).

113. C. S. Marvel et al, J. Am. Chem. Soc. 77, 177 (1955); J. Org. Chem. 24, 599 (1959).

114. R. C. Schulz et al, CA 54, 1281; Makromol. Chem. 42, 205 (1961); 54, 146 (1962).

115. H. L. Cohen et al, J. Org. Chem. 26, 1274 (1961); cf. B. Houel, CA 56, 227 (1961); D. A. Levis and P. A. Small, CA 56, 13097 (1961).

116. M. Matsumoto et al, U.S. 2,909,502 (Kurashiki and Airco); cf. R. Oda et al, U.S. 3,079,356 (Sumitomo).

117. R. A. Jacobson, U.S. 2,519,764 (Du Pont).

118. J. A. Verdol et al, U.S. 3,363,029 (Sinclair).

119. E. L. Kropa, U.S. 2,443,740 (Am. Cyan.).

120. J. A. Zaslowsky, U.S. 3,399,217; cf. U.S. 3,399,241 (Olin).

121. J. E. Burleigh and S. R. Collins, U.S. 3,296,277 (Phillips).

122. G. F. D'Alelio, U.S. 2,378,196 (G.E.); cf. U.S. 2,363,836.

123. R. H. Snyder, U.S. 2,504,052; P. O. Tawney, U.S. 2,546,798 (U.S. Rubber).

124. R. H. Snyder, U.S. 2,441,515-6 (U.S. Rubber).

125. C. A. Heiberger, U.S. 2,461,735 (U.S. Rubber).

126. A. W. Meyer, U.S. 2,521,078 (U.S. Rubber).

127. P. O. Tawney, U.S. 2,599,027 (U.S. Rubber).

128. E. C. Skokal and P. A. Devlin, U.S. 2,588,890 and 2,630,430 (Shell); cf. U.S. 3,013,999 (Shell).

129. E. C. Skokal and P. A. Devlin, U.S. 2,940,946 (Shell).

130. A. C. Mueller and T. F. Bradley, U.S. 2,961,424 (Shell).

131. R. W. H. Tess, U.S. 2,965,615 (Shell).

132. A. C. Mueller and R. W. H. Tess, U.S. 3,095,389 (Shell).

133. R. M. Christenson et al, U.S. 3,037,963 (PPG).

134. R. M. Christenson and F. S. Shahade, U.S. 3,118,852 (PPG); cf. U.S. 3,293,201 (PPG).

135. Brit. 787,420 (Monsanto); E. C. Chapin and R. F. Smith, U.S. 2,894,938 (Monsanto).

136. R. J. Carney and F. J. Hahn, U.S. 3,069,368 (Monsanto).

137. J. M. Gethins and R. I. Longley, U.S. 2,995,535 (Monsanto).

138. M. R. Sullivan and F. J. Hahn, ACS Org. Coatings and Plastics Chem. Preprint 25, 1 (April 1965). (Some aldehyde groups are also present in commercial S-AA.)

139. D. J. Kay, U.S. 3,218,282 (Hooker).

140. F. J. Hahn, U.S. 3,287,295 (Monsanto).

141. F. J. Hahn, U.S. 3,355,403, cf. U.S. 3,434,989 (Monsanto).

142. W. A. Higgins, U.S. 3,133,838; cf. U.S. 3,055,865 (Lubrizol).

143. F. J. Hahn et al, Off. Digest Fed. Soc. Paint Technol. 37, 1251 and 1279 (1965).

144. R. H. Reiter, U.S. 3,211,579 (Am. Cyan.); cf. U.S. 3,576,775 (Celanese).

145. R. F. Bates and G. J. Howard, J. Polym. Sci. C 16, 921 (1967).
146. B. V. Burachinsky and Y. P. Jacob, U.S. 3,397,074 (Interchemical).
147. R. J. Carney and F. J. Hahn, U.S. 3,069,368 (Monsanto).
148. K. Sekmakas, U.S. 3,558,536 (Desoto); cf. A. P. Sahni,
      U.S. 3,551,368 (Monsanto).
149. J. F. Gerecht et al, U.S. 2,806,844 (Colgate).
150. Brit. 850,894 (Shell).
151. R. W. Hill and F. R. Galiano, U.S. 3,402,219 and 3,480,693
      (Gulf Oil); cf. U.S. 3,480,574 (Monsanto); W. J. Brinton,
      U.S. 3,557,033.
152. J. W. Forsberg, U.S. 3,454,418; E. R. Farone, U.S. 3,519,493
      (Lubrizol).
153. S. B. Radlove and A. Ravve, U.S. 3,546,002 (Continental Can);
      cf. CA 73, 16459 (1970).
154. Brit. 911,763; CA 58, 14238 (1963).
155. E. C. Chapin and R. F. Smith, U.S. 2,945,835 and 2,962,460
      (Monsanto).
156. G. Natta and G. Benetta, U.S. 3,228,919 (Edison, Milan); cf.
      R. H. Martin, U.S. 2,957,857 (Monsanto).
157. E. L. Kropa and A. S. Nyquist, U.S. 2,624,722 (Am. Cyan.);
      Brit. 695,449; CA 48, 3701 (1954); cf. K. Jost, CA 55, 9952
      and 9963 (Badische).
158. W. C. Mollison, U.S. 2,777,832 (Am. Cyan.).
159. G. Oster and Y. Mizutani, J. Polym. Sci. 22, 173 (1956).
160. D. Maragliano and E. Cernia, Ital. 544,920; CA 53, 1766 (1959)
      and Ital. 521,884; CA 54, 3906 (1960).
161. M. Taniyama et al, U.S. 3,287,307 (Toho Rayon).
162. G. D. Buckley et al, Brit. 669,771 (ICI).
163. R. P. Shouse and C. R. Donaldson, U.S. 3,267,085 (Nat. Distillers).
164. R. H. Martin, U.S. 2,979,487; cf. U.S. 2,977,334 (Monsanto).
165. E. Bergman and P. A. Devlin, U.S. 3,245,934 (Shell).
166. F. Hoelscher, Ger. 842,048 (Badische).
167. F. E. Frey, U.S. 2,280,818 and 2,114,292 (Phillips).
168. S. N. Ushakov et al, CA 44, 1746 (1950).
169. R. R. Dreisbach and J. F. Mulloy, U.S. 2,794,014 (Dow).
170. M. Ito and Z. Kuri, J. Chem. Soc. Japan (Ind. Chem.) 69, 531
      (1966); CA 65, 12292 (1966).
171. B. W. Wilson, U.S. 3,308,102 (Dow).
172. G. Sackman and G. Kolb, Makromol. Chem. 149, 51 (1971) and
      Colloqium, Freiburg, March 1972.
173. V. A. Kargin et al, Polym. Sci. (USSR) (in English) 9, 329 (1967);
      V. P. Zubov et al, J. Polym. Sci. C 23, 147 (1968);
      V. F. Kulikova et al, CA 66, 76339 (1967).
174. M. Sheshukov, J. Russ. Phys.-Chem. Soc. 16, 478 (1884); cf.
      Ber. 17, 414 ref. (1884); cf. Pogorshelski, CZ I, 668 (1905).
175. M. W. Tamele and H. P. A. Groll, U.S. 2,072,015 (Shell);
      Ind. Eng. Chem. 33, 115 (1941).
176. F. Koehler, U.S. 2,323,781 (R & H).
177. M. A. Pollack and A. G. Chenicek, U.S. 2,313,767 (PPG).
178. H. A. Chenly et al, U.S. 2,434,394 (Shell).

179. U.S. 2,779,801; Brit. 755,600; U.S. 2,761,883 (Shell); U.S.
      3,109,865 (Std. Oil Ohio); U.S. 2,710,884 (Carbide' " ".
      Green and W. J. Hickinbottom, J. Chem. Soc. 3262 (1957).
180. H. R. Guest and B. W. Kiff, U.S. 2,710,884 (Carbide); cf.
      C. W. Smith et al, J. Am. Chem. Soc. 73, 5270 (1951).
181. R. L. Rowton, U.S. 3,238,264 (Jefferson) and U.S. 3,325,245.
182. M. Sheshukov, J. Russ. Phys.-Chem. Soc. 16, 478 (1884); cf.
      Z. Pogorshelski, ibid. 36, 1129 (1904).
183. G. Hearne, M. Tamele, and W. Converse, Ind. Eng. Chem. 33, 805
      (1941); U.S. 2,078,534 and 2,010,076 (Shell).
184. H. P. A. Groll and G. Hearne, U.S. 2,046,556 (Shell).
185. H. P. A. Groll and C. J. Ott, U.S. 2,097,154 (Shell).
186. D. Currell and A. Fry, J. Am. Chem. Soc. 78, 4377 (1956).
187. M. B. Green and W. J. Hickinbottom, J. Chem. Soc. 3262 (1957).
188. R. E. Rinehart and R. W. Fuest, Chem. Eng. News 43, No. 7 40
      (1965).
189. J. K. Nicholson and B. L. Shaw, Proc. Chem. Soc. 282 (Sept. 1963).
190. G. Hearne et al, Ind. Eng. Chem. 33, 806 (1941).
191. G. Hearne et al, Ind. Eng. Chem. 33, 808 (1941); U.S. 2,042,220
      (Shell).
192. L. Silverman et al, J. Ind. Hyg. Toxicol. 28, 262 (1946).
193. J. Canceill et al, Compt. Rend. 265B, 918 (1967).
194. M. L. Wolfrom et al, J. Org. Chem. 25, 1079 (1960).
195. K. B. Cofer and R. W. Fourie, U.S. 3,093,689 (Shell).
196. H. P. A. Groll and J. Burgin, U.S. 2,055,437 (Shell).
197. I. Bergsteinsson, U.S. 2,373,942 (Shell).
198. J. Wiemann and J. Gardan, Bull. Soc. Chim. France 433 (1958).
199. J. D. Ryan and F. B. Shaw, J. Am. Chem. Soc. 62, 3469 (1940).
200. H. F. Pfann and E. L. Kropa, U.S. 2,401,959 (Am. Cyan.).
201. H. Maeda et al, CA 70, 69314 (1969).
202. R. F. Williams and C. E. Schildknecht, unpublished.
203. S. E. Ross and H. D. Noether, U.S. 2,698,317 (Celanese).
204. J. E. Caldwell, U.S. 2,525,521 and 2,591,670 (Eastman).
205. Brit. 590,373 (U.S. Rubber); CA 42, 416 (1948).
206. H. D. Anspon et al, J. Polym. Sci. Al 6, 2001 (1968); W. E.
      Smith et al, U.S. 3,420,798 (Gulf Oil); I. V. Andreva et al,
      CA 72, 3823 (1970).
207. I. V. Andreva et al, CA 68, 30340 (1968).
208. K. Ziegler, Ber. 54, 737 (1921); H. Gilman et al, J. Am. Chem.
      Soc. 55, 4691 (1933); Org. Syn. Coll., Vol. I, Wiley, 1941;
      M. G. Ettlinger and J. E. Hodgkins, J. Am. Chem. Soc. 77, 1831
      (1955).
209. D. J. Foster and E. Tobler, J. Org. Chem. 27, 834 (1962).
210. N. O. Brace, J. Am. Chem. Soc. 77, 4666 (1955).
211. S. Olsen, Acta Chem. Scand. 4, 901 (1950); CA 45, 2403 (1951).
212. A. N. Bourns and R. V. V. Nicholls, Can. J. Res. B 26, 83 (1948).
213. J. Verhulst, Bull. Soc. Chim. Belg. 40, 86 (1931).
214. R. Linstead, J. Chem. Soc. 1996 (1934); cf. Pariselle, Ann,
      Chim. Phys. [8] 24, 319.

215. R. G. R. Bacon and E. H. Farmer, J. Chem. Soc. 1065 (1937);
     A. T. Blomquist and J. A. Verdol, J. Am. Chem. Soc. 77, 78
     (1955); C. E. Boord, U.S. 2,414,012 (Wingfoot).
216. W. Reppe and H. Kroeper, Ann. 582, 50 and 65 (1953).
217. H. Pariselle, Ann. Chim. Phys. [8] 24, 347 and 367 (1911).
218. A. T. Blomquist and J. A. Verdol, J. Am. Chem. Soc. 77, 78 (1955).
219. N. P. Dreyfus, J. Org. Chem. 28, 3269 (1963).
220. B. J. Hudson et al, Tetrahedron 1, 284 (1957).
221. J. R. Long, U.S. 2,398,103 (Wingfoot); CA 40, 4075 (1946).
222. H.-G. Trieschmann et al, U.S. 3,664,988 (BASF).
223. W. G. Young and L. J. Andrews, J. Am. Chem. Soc. 66, 422 (1944).
224. K. Nozaki et al, J. Am. Chem. Soc. 61, 2564 (1939).
225. W. Reppe et al, Ann. 596, 58 (1955); cf. U.S. 2,418,441 (Du Pont).
226. H. van Risseghem, Bull. Soc. Chim. Belg. 39, 350 (1930);
     R. Delaby and J. Lecomte, Bull. Soc. Chim. France [5] 4, 741 (1937).
227. L. W. Trevoy and W. G. Brown, J. Am. Chem. Soc. 71, 1675 (1949).
228. S. A. Morell and A. H. Auerheimer, J. Am. Chem. Soc. 66, 793 (1944).
229. W. G. Young and J. S. Franklin, J. Am. Chem. Soc. 88, 785 (1966).
230. J. J. Kolfenbach et al, Ind. Eng. Chem. 37, 1179 (1945).
231. I. Bergsteinsson, U.S. 2,373,942 (Shell); W. Reppe et al,
     Ann. 596, 99 and 137 (1955).
232. W. G. Young et al, J. Am. Chem. Soc. 61, 2564 (1939); cf.
     U.S. 2,435,078 (Shell).
233. Y. Minoura and H. Yamaguchi, J. Polym. Sci. A 1 6, 2013 (1968).
234. R. M. Evans and L. N. Owen, J. Chem. Soc. 239 (1949);
     A. A. Petrov, CA 32, 5369 (1938), A. T. Dottini and V. Dev,
     J. Org. Chem. 27, 971 (1962).
235. R. G. Kadesch, J. Am. Chem. Soc. 68, 44 (1946).
236. R. Delaby and J-M. Dumoulin, Compt. Rend. 180, 1279 (1925);
     CA 19, 2185 (1925).
237. G. W. Hearne and D. S. LaFrance, U.S. 2,435,078 (Shell); W. G.
     Young et al, J. Am. Chem. Soc. 61, 2564 (1939); ibid. 88,
     785 (1966).
238. P. B. D. de la Mare, chapter on Rearrangements in the Chemistry
     of Allylic Compounds in Molecular Rearrangements, P. de Mayo,
     Ed. Vol I, Interscience, 1963.
239. E. A. Braude, Quart. Rev. (London) 4, 404 (1950).
240. C. Niemann et al, J. Org. Chem. 8, 397 (1943).
241. F. L. Sassiver and J. English, J. Am. Chem. Soc. 82, 4891 (1960);
     cf. CA 57, 16523 (1962).
242. G. E. Waples, U.S. 3,471,459 (Dow).
243. M. K. Lindemann, U.S. 3,441,547 (Airco).
244. E. P. Kohler, Am. Chem. J. 38, 525; cf. Hurd, J. Am. Chem. Soc.
     59, 104 (1937).
245. R. Paul, Compt. Rend. 208, 1319 (1939); CA 33, 5803 (1939).
246. G. Smets, CA 8315 (1950).
247. C. D. Hurd and R. W. McNamee, J. Am. Chem. Soc. 59, 105 (1937);
     W. G. Young et al, ibid. 61, 3071 (1939).
248. O. Hromatka, Ber. 75, 382 (1942).
249. L. I. Smith et al, J. Am. Chem. Soc. 61, 2617 (1939).

250. G. K. Kamai, CZ I, 3047 (1932); CA 26, 4300 (1931).
251. Allyl Chloride, Shell Chemical Corp. 1949, pp. 64-71. Also
      see Beilstein. For improved Barbier and Grignard procedure,
      see M. P. Dreyfuss, J. Org. Chem. 28, 3269 (1963).
252. J. C. H. Hwa and H. Sims, Org. Syn. 41, 49 (1961).
253. V. Mazurewitsch, CZ II, 1922 (1911); M. Zaitsev, ibid. I, 23
      (1913); H. J. Backer et al, Rec. Trav. Chim. Pays-Bas 60, 391
      (1941).
254. P. Karrer, Helv. Chim. Acta 25, 29 (1942); Nasarov, CA 42,
      7730 (1948); W. Reppe et al, Ann. 596, 1 and 58 (1955);
      supplied by Roche.
255. W. Schlenk, J. Soc. Chem. Ind. 52, 209 (1933).
256. L. Claisen and O. Eisleb, Ann. 401, 106 (1913).
257. W. Treibs, Chem. Ber. 84, 47 (1951); G. Brieger, Tetrahedron
      Lett. 2123 (1963).
258. H. Burton, J. Chem. Soc. 759 (1931).
259. J. Meisenheimer, Ann. 479, 211 (1930); Duveen, J. Chem. Soc.
      1697 (1939); Coppock, ibid. 1069 (1938).
260. A. Valeur and E. Luce, Bull. Soc. Chim. France 27, 611 (1920);
      cf. E. A. Braude, J. Chem. Soc. 1971 and 1982 (1948).
261. R. Delaby, Compt. Rend. 194, 1248 (1932).
262. P. G. Stevens et al, J. Am. Chem. Soc. 62, 1424 (1940);
      E. A. Braude and E. A. Evans, J. Chem. Soc. 3333 (1956).
263. J. A. Verdol and M. O. Thienot, U.S. 3,511,820 (Sinclair).
264. M. H. Fournier, Bull. Soc. Chim. France [3] 9, 600 (1893).
265. V. Yaworskii, Ber. 42, 437 (1909); CA 3, 1019 and 1760 (1909).
266. W. C. Keith, U.S. 3,230,205 (Sinclair).
267. C. L. Mills and J. M. Butler, U.S. 2,563,611 (Monsanto);
      J. A. Verdol and M. O. Thienot, U.S. 3,536,684 (Sinclair).
268. M. O. Thienot and J. A. Verdol, U.S. 3,497,486 (Sinclair).
269. T. W. Evans and D. E. Adelson, U.S. 2,435,429 (Shell).
270. A. Henninger, Ann. Chim. (Paris) [6] 7, 213 (1886).
271. H. A. Braun, U.S. 3,157,623 (Du Pont).
272. C. M. Himel and L. O. Edmunds, U.S. 2,683,175 (Phillips).
273. L. E. Craig et al, J. Am. Chem. Soc. 72, 3277 (1950).
274. M. S. Kharasch and G. Buechi, J. Org. Chem. 14, 84 (1949);
      cf. B. K. Wasson and J. M. Parker, U.S. 2,926,190.
275. L. F. Theiling and R. J. Knopf, U.S. 2,982,790 (Carbide);
      B. K. Wasson et al, Can. J. Chem. 39, 923 (1961); Data sheet,
      Millmaster Chem. Corp., New York, 1967.
276. R. C. Schreyer, U.S. 2,789,996 (Du Pont); cf. A. T. Blomquist
      and J. A. Verdol, J. Am. Chem. Soc. 77, 78 (1955).
277. E. F. Hoegger and J. H. Werntz, U.S. 3,376,266 (Du Pont).
278. L. F. Theiling and R. J. Knopf, U.S. 2,982,790 (Carbide).
279. R. Lukes and J. Plesek, CA 50, 9288 (1956); cf. A. Mooradian
      and J. B. Cloke, J. Am. Chem. Soc. 67, 942 (1945); F. Nerdel
      et al, Chem. Ber. 91, 938 (1958).
280. C. Prevost, Bull. Soc. Chim. France [5] 11, 223 (1944).
281. W. E. Bissinger et al, J. Am. Chem. Soc. 69, 2955 (1947).

282. E. Hanschke, Chem. Ber. 88, 1043 (1955); S. Olsen et al,
     Acta Chem. Scand. 6, 859 (1952); CA 47, 5873 (1953).
283. J. D. Surmatis, U.S. 2,848,502 (LaRoche)
284. J. D. Surmatis, U.S. 2,824,896 (LaRoche).
285. J. D. Surmatis, Swiss 361,568 (LaRoche), CA 59, 8593 (1963);
     cf. Brit. 814,636 (LaRoche).
286. G. H. Coleman and R. W. Sapp, U.S. 2,285,329 (Dow); G. Kremer,
     Bull. Soc. Chim. [5] 15, 165 (1948); W. E. Noland and B. N.
     Bastian, J. Am. Chem. Soc. 77, 3396 (1955).
287. H. D. Finch et al, U.S. 2,779,801 (Shell).
288. J. W. Copenhaver et al, Acetylene and Carbon Monoxide Chemistry,
     Reinhold, 1949, p. 125.
289. W. O. Kenyon and J. H. Van Campen, Brit. 576,022-3; U.S.
     2,419,221 (Eastman).
290. W. E. Bissinger et al, J. Am. Chem. Soc. 69, 2955 (1947).
291. R. M. Evans and L. N. Owen, J. Chem. Soc. 239 (1949).

6. ALLYL ACIDS AND RELATED

Allyl- and allyloxyacids and their esters are of interest as drugs, antiseptics, and herbicides. Homopolymers of high molecular weight have not been prepared from monoallyl-substituted aliphatic or aromatic acids. The lower acids, especially vinyl- and allylacetic acid, isomerize readily. The higher acids such as undecylenic acid may be oligomerized by Lewis-acid catalysts to give polyfunctional acids. Esters of allyl-substituted acids such as allylmalonate esters have been little studied in polymerization although they are available as drug intermediates. Both the allyl acids and their esters have interesting possibilities in copolymerizations as regulators and modifiers of copolymer properties. Their little application in polymers is in contrast to acrylic acid, methacrylic acid, and their esters, important commercial monomers in which the COOH or COOR group directly activates the ethylenic nucleus (see Chapter 1).

The simplest allyl acid is 3-butenoic acid or vinylacetic acid. It can be prepared by hydrolyzing allyl cyanide (1):

$$CH_2=CHCH_2CN + H_2O + HCl \longrightarrow CH_2=CHCH_2COOH + NH_4Cl$$

The boiling and melting points of the unsaturated acid are 163°C and -36°C, respectively.

Cold aqueous NaOH solution may be used in purifying the allyl carboxylic acid, but at higher temperatures alkalies catalyze isomerization to crotonate salts. Vinylacetic acid also may be made by carbonation of allyl magnesium bromide (2), and by malonic ester synthesis (3).

Acid hydrolysis of sinigrin gave vinylacetic acid (4). The acid can be made also by metalation, for example, sodium metal reacted with butadiene and propylene followed by solid carbon dioxide (5). Allyllithium or diallylzinc may be reacted with solid CO_2 granules in ether in the presence of a little hydroquinone (6). The mixture was treated with diluted sulfuric acid, and vinylacetic acid was distilled off (bp, 70°C at 10 mm and n_D^{20} = 1.4223). The complex allylpalladium chloride may be reacted in

benzene with CO for 5 hr at 200 atm and 50°C to give the
unsaturated acid (7). Allyl chloride was reacted with CO
in presence of Ni(CO)₄ to give vinylacetic acid along
with crotonic acids (8).

Hydrogen bromide adds to vinylacetic acid to give
largely 3-bromobutyric acid. Hydrogenation of vinylacetic
acid using noble metal catalysts is slower than that of
olefinic acids having the carboxyl group further removed
from the double bond (9).

Some viscous liquid polymeric products have been observed following heating and irradiation of systems to which
vinylacetic had been added (10). We do not know the
nature of these low polymers or whether rearrangements
were involved. Reaction of vinylacetic acid with diacetyl
peroxide gave products of allylic shift, $(CH_2CH=CHCOOH)_2$
and $CH_3CH=CHCOOH$, as well as a viscous water-soluble fraction and an unsaturated solid $(C_5H_8O_2)_5$ (11). Other
references to reactions of vinylacetic acid are given in
Beilstein. Vinylchloroacetic acid (bp 104°C at 19 mm and
$n_D^{22} = 1.459$) isomerizes in the presence of alkali to give
salts of trans-chlorocrotonic acid (12).

Esters of vinylacetic acid were made by Chiusoli directly by carboxyalkylation of allyl halides by CO and
alcohols in presence of nickel or cobalt carbonyl catalysts (13). Some isomerization to crotonate esters also
occurred.

$$CH_2=CHCH_2X + CO + ROH \xrightarrow{Ni(CO)_4} CH_2=CHCH_2COOR + HX$$

$$\text{also } CH_3CH=CHCOOR$$

With sodium methoxide and sodium cobalt carbonylate
NaCo(CO)₄ as catalysts at 25°C for 16 hr, methyl 3-bute-
noate was obtained from allyl bromide without evidence of
isomerization (14). Carbonylation of π allylpalladium
chloride complex in ethanol or benzene at 70°C with CO
under 100 kg/cm² at 70°C gave ethyl vinylacetate along
with some methyl crotonate and vinyl acetyl chloride (15).
Carbonylation of allyl alcohol in presence of a fluoroaryl
phosphine-platinum complex at 200°C and 1000 atm of CO
formed the allyl ester of vinylacetic acid (bp 146°C and
$n_D^{25} = 1.4311$) (16). Running the synthesis at 250°C produced some diallyl ether and allyl crotonate as byproducts.
A Japanese patent discloses synthesis of esters of vinylacetic acid by reaction of carbon monoxide and alcohol
with allyl alcohols, halides, ethers, or esters in presence
of noble metal catalysts (17). In one example, 46 g
ethanol, 5 g allyl chloride, 5 g palladium dichloride and
CO were reacted 4 hr at 75 kg/cm² at 65°C. There was
recovered 94% ethyl vinylacetate (bp 50°C at 43 mm) and
we trust the expensive catalyst also could be recovered.

A 4-chlorobutyl ester of vinylacetic acid (bp 112°C at 13 mm) also was synthesized. Some methyl ester of vinylacetic acid was obtained during pyrolysis of cis-methyl crotonate at 400 to 560°C (18).

Allylacetic acid or 4-pentenoic acid (bp 93°C at 20 mm and $n_D^{20} = 1.4281$) is relatively unstable (19). It can be made by heating allylmalonic acid at 180°C (20). When heated with 50% H_2SO_4 or I_2 it rapidly formed gamma-valerolactone (21). On treatment with HCl it formed 4-chlorovaleric acid; the direction of addition was little influenced by the type of solvent, by addition of peroxides or by antioxidants (22). However, the 5-bromovaleric acid was obtained by addition of HBr in hexane or toluene in the presence of antioxidants (19). Kharasch and McNab obtained in hexane the expected 4-bromovaleric acid in presence of diphenylamine, and the 5-bromoacid in presence of the peroxide ascaridole (23). In rats 4-pentenoic acid caused ketonuria and hypoglycemia in contrast to pentanoic and 2-pentenoic acids (24).

Allylacetic acid has not given homopolymers of high molecular weight. It reacted with SO_2 in the presence of ascaridole (peroxide) at room temperature to form white, rubberlike sulfone copolymers (25). The copolymer dissolved in ethanol and in aqueous alkali. Esters of allyl acetic acid (Table 6.1) have not given high polymers.

TABLE 6.1

Some Derivatives of Allylacetic Acid

Compound	bp,°C	Refractive index	References
Ethyl ester	144	$n_D^{20} = 1.4142$	Schjånberg, Z. Phys. Chem. A 178, 276 (1936); Linstead, J. Chem. Soc. 575 (1933)
n-Propyl ester	166	$n_D^{20} = 1.4199$	Schjånberg (Studied saponification)
n-Butyl ester	186	$n_D^{20} = 1.4241$	Schjånberg
Perfluoroalkyl-esters	-	--	Katsushima, U.S. 3,457,247 (Daiki
Allyl ester	162	$n_D^{25} = 1.4198$	McElvain, JACS 64, 2529 (1942) (Made from acetals)
Methallyl ester	167	--	Ritter, J. Org. Chem. 27, 622 (1962)
Vinyl ester	65(40mm)	--	Fang, Belg. 631,528 (Rohm & Haas)

Ethyl esters of allylacetic acid and of methallyl acetic
acid did not polymerize when distilled at 144 and 167°C,
respectively (26). Vinyldiallylacetic acid (bp 110°C at
2.5 mm and $n_D^{25} = 1.4743$) rearranged to 2,4-diallylcrotonic
acid on heating at 185°C (27). Diallylacetic acid was
only slightly more toxic than dipropylacetic acid in tests
with mice. Exploratory polymerizations of the esters were
reported (28).

Isopropenylacetic acid or methallylformic acid (bp 70°C
at 5 mm and $n_D^{20} = 1.4308$) has been obtained from reaction
of ketene with acetone in the presence of $AlCl_3$ below
20°C (29), and from the reaction of methallyl magnesium
bromide in ether with CO_2 (30). On standing with H_2SO_4
in ether or on heating with KOH solution it gave 3,3-
dimethylacrylic acid (31). The methyl ester of isopropenylacetic acid (bp 41°C at 27 mm and $n_D^{20} = 1.4168$) was
prepared by heating the silver salt with methyl iodide
(30). The ethyl ester (55°C at 20 mm and $n_D^{20} = 1.4400$) was
prepared starting from ketone and acetoacetic ester (32).

Copolymers of ethyl acrylate and vinyllactic acid were
prepared by heating with azobis (33). Solutions of copolymer in toluene gave solvent-resistant coatings on
steel. Polymers of low molecular weight resulted from
heating beta-vinylpropionate alkyl esters with t-butyl
peracetate at 115°C (34).

The allyl long-chain acid undecylenic acid $A(CH_2)_7COOH$,
obtained by heating castor oil at 300°C or above with acid
catalysts, has been much studied. Addition of hydrogen
halide ordinarily gives the 10-haloacid. Pure undecylenic
acid in the absence of oxygen gave with HBr the 10-bromo-
decanoic acid, but in the presence of oxygen and peroxides,
the 11-bromoacid was the chief product (35). In France,
reverse addition of HCl to undecylenic acid has been used
to prepare 11-aminoundecanoic acid and 11-nylon. Some
undecylenic acid derivatives such as the zinc salt have
useful bacteriocidal and fungicidal action.

Undecylenic acid has not given well-characterized homopolymers of high molecular weight, but evidence of polymerization and copolymerization has been recorded. Heating
the anhydrous sodium salt with excess alkali was suggested
to give salts of polymeric carboxylic acids (36). When BF_3
was passed into melted undecylenic acid, an exothermic
complex reaction occurred giving brown viscous liquid products (37). After freeing from catalyst residues, the
oily polymer showed properties of an ester and of an acid.
Vinyl-type homopolymers of high DP made under controlled
conditions from undecylenic acid might develop considerable
technical interest. Undecylenic acid and propylene were
copolymerized with SO_2 by addition of a solution of lithium

nitrate in alcohol as catalyst (38). Terpolymers of
ethylene, propylene, and undecylenic acid were prepared
using catalysts from reaction of dialkylaluminum chloride
and $VOCl_3$ (39). Solutions of these polymers in xylene
could be gelled by addition of a little tetraethylene
pentamine and aluminum isopropylate.

Undecylenic acid and liquid SO_2 reacted in ethanol in
the presence of ascaridole to form rubbery polymers (40).
The 1:1 molar copolymers melted above 200°C, were insoluble
in organic solvents, but dissolved with reaction in liquid
ammonia to give cyclic sulfones. Terpolymers with 1-
pentene also were prepared. Methyl undecylenate formed
glassy sulfone polymers which became viscous liquids on
heating and decomposed above 200°C. The copolymer was
soluble in acetone and in warm ethanol.

Alkyl esters of undecylenic acid have been studied, but
soluble homopolymers of high molecular weight have not
been reported (41). Vinylundecylenate (bp 124°C and
$n_D^{30} = 1.4442$) has been made by reacting the organic acid
with excess vinyl acetate in the presence of mercuric
acetate and sulfuric acid (42). The vinyl and chloroallyl
esters homopolymerized to solid gels on heating with 1%
benzoyl peroxide (e.g., for 16 hr at 80°C). Allyl and
methallylundecylenate showed little evidence of polymerization under similar conditions. Copolymerization of the
doubly unsaturated esters with vinyl acetate gave crosslinked products. Tertiary butyl perundecylenate (bp 76°C
at 0.07 mm and $n_D^{20} = 1.4453$) quickly polymerized on heating
at 130°C to give a white polymer that was insoluble in
organic solvents (43). Ethylene glycol diundecylenate and
glycerol triundecylenate have been prepared (44).

Allylmalonic acid HOOCC(A)HCOOH (mp 105°C) and its derivatives have been known for a long time because of their
relation to synthesis of barbiturates. Conrad and Bischoff
reacted sodium malonic ester with allyl iodide (45). The
acid was obtained by saponification or hydrolysis (46).
On heating allylmalonic acid at 100 to 180°C, decarboxylation occurs to CO_2 and allylacetic acid (47). Allylmalonic
acid decolorizes permanganate solution rapidly. The
dimethyl, diethyl, dipropyl, and dibutyl esters of allylmalonic acid were prepared by conventional esterification
with sulfuric acid catalyst (48).

Diallylmalonic acid (mp 137°C) was prepared as well as
the diethyl ester (bp 244°C and $n_D^{15} = 1.4477$) and the di-
amide (mp 202°C) (49). The diethyl ester of 2,2-diallyl-
malonic acid was copolymerized with ethyl acrylate by
heating with azobis in benzene at 50°C for 16 hr (50).
(Also made were diethyl and diallyl esters.)

A series of esters of allylmalonic acid were prepared
(51). Esters of allyl-substituted malonic acids are com-

mercially available for the synthesis of barbiturates.
For example, Benzol Products Company supplies monoallyl,
diallyl, allyl methallyl, and allyl 1-methylbutyl diethyl
malonates, $R\dot{R}C(COOC_2H_5)_2$. Pyrolysis of methallylmalonic
ester (bp 113 at 7 mm and $n_D^{25} = 1.4381$) was studied by
Kimel and Cope (52). Allyl- and diallylmalonate esters
with thiourea do not give the expected thiobarbituric
acids but form allylpyrimidine derivatives instead (53).
Diallylmalonic acid was prepared, but physical properties
were not reported (54). Allylsuccinic acid (mp 100°C) was
made (55). With permanganate it was oxidized to tricarballylic acid. The diethyl ester of 2-allyladipic acid
(bp 149°C at 13 mm) was made but not the free acid (56).
The diallylcrotonic acid $ACH_2CH=CHCOOH$ (bp 118°C at 1.5 mm)
was obtained by heating vinyldiallylacetic acid at 185°C
under nitrogen (57). Allyl acids prepared include 3-methy-
leneglutaric acid $CH_2=C(CH_2COOH)_2$ (58) and its homologs.
References to other allyl-substituted fatty acids are given
in Table 6.2.

Heating propylene with maleic anhydride at 250°C gave
allylsuccinic anhydride (bp 140°C at 16 mm) (59). Meth-
allylsuccinic anhydride (138°C at 9 mm) has been prepared
from reaction of isobutylene with maleic anhydride in
benzene (60). Alpha-methallyl-gamma-butyrolactone also
was made. Methallyl succinic anhydride was reacted with
ethylene glycol under nitrogen at 200°C for 5 hr giving a
polyester bearing methallyl groups (61). This was incorporated into cellulose acetate spinning solutions in order
to protect anthraquinone dyes against gas fading. The
anhydride of 3-allylphthalic acid was reported to melt at
160°C (62). Carbonylation of allyl acetate by CO in the
presence of PdCl gave 3-butenoic-acetic mixed anhydride
(63).

Few allyl derivatives of acyl chlorides have been
studied. However, allylacetyl chloride, a sharp-smelling
liquid boiling at 128°C, was prepared by reacting the acid
with PCl_3 (64).

Allyloxyacetic acid (126°C at 22 mm) was made by reacting
sodium allylate with chloroacetic acid (65). It was homopolymerized to an orange viscous liquid by heating for
7 hr with 0.5% BP at 88 to 108°C (66). After residual
monomer had been removed under vacuum, the viscous liquid
polymer had a viscosity of 72 poises at 25°C. Equal
weights of allyloxyacetic acid and vinyl acetate containing
5% BP were heated for 2 hr at 67 to 87°C to give a viscous
liquid copolymer from which residual monomer was removed
at 100°C under vacuum. Conversion was 68% to copolymer
containing 47% units of allyloxyacetic acid. Examples
were given of copolymerizations with butyl methacrylate,
vinyl chloride, acrylonitrile, and allyl glycidyl ether.

TABLE 6.2

Some Allyl-Substituted Acids

Acid	bp, °C	mp, °C	n_D^{20}	References
Vinylacetic	163 or 70(12)	-35	1.4220	Morton, JACS 67, 2227 (1945); Linstead, J. Chem. Soc. 559 (1933)
Allylacetic	187	-22	1.4281	Linstead; Michael, JACS 65, 684 (1943)
Methylvinylacetic	102(50)	--	1.4231	Lane, JACS 66, 543 (1944)
Isopropenylacetic (methallylformic)	70(5)	-21	1.4308	Wagner, JACS 71, 3216 (1949)
3-Allylpropionic	105(13) 203	--	1.4343	LaForge, JACS 70, 3709 (1948); Wallach, Ann. 343, 48 (1905)
4-Allylbutyric	125(15)	--	1.4404	Linstead, J. Chem. Soc. 1973 (1937)
7-Allylheptanoic	158(14)	--	1.4468	Ellis, J. Biol. Chem. 113, 222 (1936)
Undecylenic acid	180(26.5)	24.5	1.4457	Jordan, JACS 71, 2378 (1949)
Allylsuccinic	--	93	--	Hjelt, Ber. 16, 334 (1883)
Allylaspartic	--	--	--	Laliberte, Can. J. Chem. 40, 163 (1962)
Allylmalonic	--	103	--	Conrad, Ann. 204, 168 (1880)
Diallylhydroxyacetic	--	48	--	Zaitsev, Ann. 185, 183 (1877)
Allylisopropylmalonic	--	112	--	Hjelt, Ber. 29, 1856 (1896)
Diallylmalonic	--	133[a]	--	Bernouille, Helv. Chim. Acta 2, 511 (1919)
3-Vinyladipic	--	--	--	Perry, U.S. 3,359,311 (Esso)
2-Allylpimelic	--	--	--	Greive, Ber. 76, 1075 (1943); (also diallyl ether)
2-S-Diallyladipic	155(0.05)	103	--	Marvel, J. Org. Chem. 25, 2207 (1960)
Diallylcyanoacetic	--	--	--	Ger. 473,329 (IG)

[a] With decarboxylation

242

Allyl-substituted lactones can be hydrolyzed to allyl-substituted hydroxyacids. For example, a diallyl butyro-lactone was hydrolyzed to $HOCA_2CH_2CH_2COOH$ (67).

Allylbenzoic acids have interest as modifiers of eugenol in chelate cements for dentistry (68). Tests were made with m-allylbenzoic acid (mp 62°C). Some allyl-substituted aromatic acids are listed in Table 6.3. The esters of a few of these polyfunctional allyl compounds have been observed to polymerize, but not the free acids. Thus the methyl ester of 2-allyloxy-3,5-diallyl benzoic acid [bp 182(10)°C] resinified on heating (69). Likewise, ethyl 4-allyloxy-3,5-diallyl benzoate, an oil boiling 190(10)°C, polymerized to a solid resin on heating. The Upjohn Company evaluated a number of allyloxybenzoate esters as herbicides and reported boiling points and refractive indices (70). A number of these such as n-butyl-4-allyl-oxybenzoate killed crabgrass and chickweed but not lawn grasses, corn, or beans. Derivatives of 2- and 4-allyl-oxybenzoic acids were depressants of the central nervous system, judging from sleeping times of mice (71).

Claisen and co-workers studied the rearrangement of the allyloxy-substituted acids on heating to form allyl phenols but showed little interest in polymerizations. Allyl phenols are acidic compounds, discussed in Chapter 17.

TABLE 6.3

Some Allyl-Substituted Cyclic Acids and Derivatives

Compound	bp,°C	mp,°C	n_D^{25}	References
Allylphenyl acetic acid	260 or 160(25)	34	--	Wislicenus, Ber. 29, 2601 (1896) (from A phenyl malonic acid
Allylbenzylacetic acid; (allylben-zyl malonic acid)	146(1.7)	115	1.5180	Arnold, JACS 75, 1044 (1953); U.S. 2,526,108
Allylphenylmalonic acid	--	145	--	Pickard & Yates, J.Chem. Soc. 95, 1015 (1909); Wislicenus; (also diethyl esters)
3,5-Diallyl salicylic acid	--	99	--	Claisen, Ann. 401, 77 (1913) (Melted with decarboxylation)
2-Allyloxy-3,5-diallyl benzoic acid	--	55	--	Claisen; (heating at 100°C gave 2,4,6-tri-allyl phenol)
Diallyl phthalid (from phthalic anhydride and AMgBr)	185(0.4)	--	1.5361	Orlow, CZ I, 1417 (1913)

References

1. E. Rietz, Org. Syn. 24, 96 (1944); cf. U.S. 3,539,624 (Dow).

2. J. Houben, Ber. 36, 2897 (1903).

3. R. P. Linstead et al, J. Chem. Soc. 560 (1933); cf. A. L. Naum-
    chuk, CA 51, 8651 (1957).

4. M. G. Ettlinger and A. J. Lundeen, J. Am. Chem. Soc. 78, 4172
    (1956).

5. C. E. Frank et al, U.S. 2,954,410 and 2,966,526 (Nat. Distillers).

6. D. Seyferth and M. A. Weiner, J. Org. Chem. 26, 4797 (1961);
    cf. M. Gaudemar, CA 57, 5939 (1962).

7. R. Long and G. H. Whitfield, J. Chem. Soc. 1852 (1964);
    Brit. 1,007,707 (ICI).

8. G. Chiusoli, Angew. Chem. 72, 74 (1960).

9. W. P. Dunworth and F. F. Nord, J. Am. Chem. Soc. 74, 1457 (1952).

10. H. I. Waterman et al, CA 44, 3884 (1950); G. Nagashima, CA 48,
    9107 (1954); S. A. Dolmatov and L. S. Polak, CA 60, 3102 (1964)
    64, 3691 (1966).

11. K. B. L. Mathur and R. S. Thakur, J. Chem. Soc. 3231 (1956).

12. Rambaud, Compt. Rend. 197, 769 (1933).

13. G. P. Chiusoli, Gazz. Chim. Ital. 89, 2779 (1959); cf. CA 61,
    10585 (1964).

14. R. F. Heck and D. S. Breslow, J. Am. Chem. Soc. 85, 2779 (1963).

15. J. Tsugi et al, Tetrahedron Letters 1811 (1963) and J. Am. Chem.
    Soc. 86, 4491 (1964).

16. G. W. Parshall (Du Pont), Z. Naturforsch. 18b, 772 (1963);
    CA 59, 15175 (1963).

17. Fr. 1,389,856 (Toyo Rayon), CA 63, 501 (1965).

18. J. N. Butler and G. J. Small, Can. J. Chem. 41, 2492 (1963).

19. R. P. Linstead et al, J. Chem. Soc. 583 (1933); cf. A. Michael
    and H. S. Mason, J. Am. Chem. Soc. 65, 684 (1943).

20. M. Conrad and C. A. Bischoff, Ann. 204, 168 (1880); CZ 739
    (1880); cf. G. P. Shulman and J. Osteraas, Can. J. Chem. 41,
    2718 (1963).

21. M. F. Ansell and M. H. Palmer, J. Chem. Soc. 2640 (1963);
    M. M. Campos and L. Amaral, CA 63, 4159 (1965).

22. E. Schjänberg, Ber. 70, 2389 (1937).

23. M. S. Kharasch and J. B. McNab, Chem. & Ind. (London) 989 (1935).

24. A. E. Senior and H. S. A. Sherratt, Biochem. J. 100, 71P (1966).

25. L. L. Ryden et al, J. Am. Chem. Soc. 59, 1014 (1937); cf. Snow
    and Frey, Ind. Eng. Chem. 30, 179 (1938).

26. J. J. Ritter and T. J. Kaniecki, J. Org. Chem. 27, 622 (1962).

27. D. E. Whyte and A. C. Cope, J. Am. Chem. Soc. 65, 2004 (1943).

28. S. G. Matsoyan et al, CA 59, 7656 (1963).

29. A. B. Boese, U.S. 2,382,464 (Carbide).

30. R. B. Wagner, J. Am. Chem. Soc. 71, 3216 (1949).

31. A. Mooradian and J. B. Cloke, ibid. 68, 788 (1946).

32. H. J. Hagemeyer, Ind. Eng. Chem. 41, 769 (1949); also patents,
    CA 43, 1055 (1949).

33. French appl. (Badische) CA 74, 142682 (1971).

34. J. E. Masterson, Belg. 633,935 (Rohm and Haas), CA 61, 5813 (1964).
35. R. Ashton and J. C. Smith, J. Chem. Soc. 437 (1934); P. L. Harris et al, ibid. 1109 and 1575 (1935); G. Champetier, Compt. Rend. 225, 632 (1947).
36. J. H. Percy and J. Ross, U.S. 2,341,239 (Colgate).
37. J. R. Cann and E. D. Amstutz, J. Am. Chem. Soc. 66, 839 (1944).
38. F. E. Frey et al, U.S. 2,192,467 (Phillips).
39. C. A. Stewart, U.S. 3,250,754 (Du Pont).
40. L. L. Ryden, F. J. Glavis, and C. S. Marvel, J. Am. Chem. Soc. 59, 1014 (1937); cf. ibid. 76, 61 (1954).
41. Beilstein, E III 2-1364.
42. E. F. Jordan and D. Swern, J. Am. Chem. Soc. 71, 2378 (1949); Org. Syn. 30, 108 (1950).
43. N. A. Milas and D. M. Surgenor, J. Am. Chem. Soc. 68, 642 (1946).
44. B. Flaschenträger et al, Ann. 552, 109 (1942).
45. M. Conrad and C. A. Bischoff, Ann. 204, 168 (1880).
46. W. H. Perkin and J. L. Simonsen, J. Chem. Soc. 91, 822 (1907); R. P. Linstead and H. N. Rydon, J. Chem. Soc. 582 (1933).
47. J. F. Norris and H. F. Tucker, J. Am. Chem. Soc. 55, 4700 (1933); R. P. Linstead et al, J. Chem. Soc. 582 (1933).
48. G. H. Jeffery and A. I. Vogel, J. Chem. Soc. 663 and 668 (1948).
49. H. Leuchs, Ber. 47, 2584 (1914).
50. C. D. Wright, U.S. 3,247,170 (3M).
51. G. H. Jeffrey and A. I. Vogel, J. Chem. Soc. 663 (1948).
52. W. Kimel and A. C. Cope, J. Am. Chem. Soc. 66, 1613 (1944).
53. T. R. Johnson and A. J. Hill, J. Am. Chem. Soc. 36, 364 (1914).
54. G. Schwarzenbach, Helv. Chim. Acta 16, 530 (1933); A. Dinglinger and E. Schroer, Z. Physik. Chem. A 179, 425 (1937).
55. K. Alder et al, Ber. 76, 44 (1943).
56. H. Staudinger and Ruzicka, Helv. Chim. Acta 7, 446 (1924).
57. D. E. Whyte and A. C. Cope, J. Am. Chem. Soc. 65, 2004 (1943).
58. A. A. Morton et al, J. Am. Chem. Soc. 67, 2227 (1945); 69, 160 (1947).
59. K. Alder et al, Ber. 76B, 27 (1943); Org. Syn. 31, 85 (1951); J. Org. Chem. 26, 2594 (1961); R. M. Anderson and J. C. Wygant, U.S. 3,256,506 (Monsanto).
60. D. D. Philips et al, J. Am. Chem. Soc. 77, 5977 (1955); 80, 3663 (1958).
61. A. B. Conciatori, U.S. 3,021,188 (Celanese).
62. E. Zbiral et al, Monatsh. 92, 654 (1961); CA 56, 12784 (1962).
63. J. Tsuji et al, J. Am. Chem. Soc. 86, 4350 (1964).
64. L. Henry, CZ II 663 (1898); Buu-Hoi, Compt. Rend. 217, 28 (1943).
65. R. S. Barker and L. N. Whitehill, U.S. 2,448,246 (Shell).
66. H. S. Rothrock, U.S. 2,607,760 (Du Pont); cf. W. R. Saner, U.S. 2,755,186 (Du Pont).
67. A. Kasansky, CZ I 1330 (1904).
68. G. M. Brauer et al, J. Res. Natl. Bureau Stds. 68A, 619 (1964).
69. L. Claisen, Ann. 401, 79 (1913); cf. L. Claisen and O. Eisleb, Ber. 45, 3163 (1912).
70. Brit. 883,234 (Upjohn); cf. CA 60, 14442 (1964) and 62, 14582 (1965).
71. R. B. Moffett and P. H. Seay, J. Med. Pharm. Chem. 2, 213 (1960).

7. ALLYL ALDEHYDES AND KETONES

Allyl aldehydes are relatively unstable and they do not polymerize readily by vinyl-type addition reactions. There is little literature on these compounds in contrast to that of the highly reactive, readily polymerizing acrylic aldehydes, acrolein, and methacrolein (1) which bear activating substituents directly attached to the ethylenic group.

Allyl ketones find application in perfumes and drugs. Allyl derivatives of aromatic ketones and of hydroxyaromatic ketones are of interest in stabilization of plastics, coatings, and fibers against ultraviolet. They may be added as conventional light stabilizers or they can be used as minor comonomers in copolymerizations with propylene, vinyl chloride, methyl methacrylate, or other monomers. The monofunctional allyl ketones do not homopolymerize readily by use of conventional catalysts in contrast to vinyl ketones (2).

The contrasting formulas of the typical allyl compounds and the extremely pungent and reactive vinyl carbonyl compounds are shown:

Allylacetaldehyde	$CH_2=CHCH_2CH_2CHO$
Acrolein	$CH_2=CHCHO$
Allylacetone	$CH_2=CHCH_2CH_2COCH_3$
Vinyl methyl ketone	$CH_2=CHCOCH_3$

Allyl formaldehyde or 3-butenal is not well known. It undergoes rapid isomerization to the conjugated propenyl compound crotonaldehyde:

$$CH_2=CHCH_2CHO \longrightarrow CH_3CH=CHCHO$$

Crotonaldehydes, of which the common commercial compound is largely the trans isomer, are not discussed in this book. Allylacetaldehyde and allyl-substituted higher aldehydes, as well as monofunctional allyl ketones, have not been polymerized to products of high molecular weight and known structure.

Allylacetaldehyde or 4-pentenal is an odorous liquid that can be prepared by rearrangement of allyl vinyl ether on pyrolysis (3). The reaction can be carried out in vapor phase at 255°C:

246

$$CH_2=CHCH_2OCH=CH_2 \longrightarrow CH_2=CHCH_2CH_2CHO$$

The 2,4-dinitrophenyl hydrazone from the aldehyde melted at 120°C. The large negative entropy of activation of the first-order isomerization reaction was believed to support a cyclic mechanism (4).

Treating 2-allylbutyraldehyde in hexane with BF_3 etherate at room temperature formed only low yields of brown, tar-like polymers (5). These could be drawn out to short fibers. Heating the allylbutyraldehyde with lauroyl per-oxide for 24 hr at 60°C gave low yields of semisolid, yellow polymers. The compound $CH_2=CHC(CH_3)_2CHO$ (bp 88°C at 70 mm) is of special interest because it does not con-tain allylic hydrogen atoms. Semicarbazone derivatives were reported to melt at 160°C and 175°C (6).

Methallylformaldehyde was made by dehydrogenation of the alcohol over copper at 300°C (7). The acetal was prepared by reaction of methallyl magnesium bromide with ethyl orthoformate (8). The 2,4-DNPH derivative of the aldehyde and the semicarbazone were reported to melt at 191 and 205°C, respectively. The diethyl acetal of $CH_2=CHCH(CH_3)CHO$ was prepared (9). Physical properties of some allylic aldehydes are given in Table 7.1. The allyl radical is abbreviated A and the methallyl radical MA.

A number of liquid allyloxybenzaldehydes, allyloxy allylbenzaldehydes, and allyl hydroxybenzaldehydes were prepared by Claisen who studied their rearrangement to allylphenol derivatives and their oxidation to benzoic acid derivatives (10). For example, heating 4-allyloxy-3,5-diallylbenzaldehyde above 170°C gave largely 2,4,6-triallylphenol.

TABLE 7.1

Some Allyl and Allyloxy Aliphatic Aldehydes

Aldehyde	bp, °C	n_D^{20}	References
ACH_2CHO	104	1.4191	Hurd, JACS 60, 1910 (1938)
HCA_2CHO	67(7)	1.4486	Zifferero, CA 50, 12053 (1956)
MACHO	118	--	Groll, U.S. 2,042,220 (Shell)
			CZ I, 1793 (1937)
$ACH(CHO)_2$ etc.	--	--	Van Hook, U.S. 2,582,212 (R & H)
$AOCH_2CH_2CHO$	54(14)	1.4291	Hall, J. Chem. Soc. 3388 (1954)
$MAOCH_2CH_2CHO$	62(9)	1.4352	Hall

ALLYL KETONES

Some simple allyl ketones are not very stable. Methyl allyl ketone, for example, rearranges in the presence of mineral acid or basic catalysts to form methyl propenyl ketone (11):

$$CH_3CCH_2CH=CH_2 \xrightarrow{H^+} CH_3CCH=CHCH_3$$
$$\underset{O}{\|} \qquad\qquad\qquad \underset{O}{\|}$$

Methyl allyl ketone, having an unpleasant odor, was prepared by the slow addition of allyl magnesium bromide to acetic anhydride in ether at -70°C (12). Allylacetone or 5-hexen-2-one was made by Hurd and Pollack by pyrolysis of allyl isopropenyl ether vapor at 255°C (13). It has been made frequently by way of allylacetoacetate (14). Oxidation of allylacetone by dilute aqueous permanganate gives levulinic acid (15). The oxime of allylacetone is a colorless liquid of piercing odor and boiling point of 90°C. Posner studied the reverse addition of ethylmercaptan to allylacetone and oxidation of the product to a sulfone (16). Reactions of allylacetone with aromatic ketones in the presence of $AlCl_3$ were studied (17). Viscous liquid polymers were obtained by treating allylacetone with small amounts of sulfuric acid below room temperatures and also by heating with benzoyl peroxide for 24 hr at 60°C (18).

Allylpinacolone $ACH_2COC(CH_3)_3$ (bp 64°C at 14 mm) was made by reacting the ketone with sodamide and allyl iodide in ether (19). Allylacetylacetone or diacetyl allyl methane was found to boil at 200°C and showed $n_D^{14} = 1.4698$ (20). Alpha-allyl diethyl ketone,bp 156°C,was reported (21). An allyl isopropyl acetone, 2-methyl-7-octen-4-one, was prepared (bp 63°C at 14 mm and $n_D^{13} = 1.4288$) (22).

Methallylacetone was made by passing ketene into aceto-nylacetone in presence of BF_3 etherate at 0°C and then heating the reaction mixture to 108°C (23). It was also made by way of methallyl-substituted acetoacetic ester (14). The semicarbazone of methallyl acetone melted at 137°C. A number of ketones were alpha-alkylated in basic medium (24). For example, a mole of ketone, 1.3 mole of alkenyl halide, 2 g of triethanolamine and 3 mole of potassium chips were heated in an autoclave for 7 hr at 100°C.

Methallyl acetone largely rearranges on standing with p-toluene sulfonic acid or with KOH to form an equilibrium mixture containing about 90% mesityl oxide (methyliso-butenyl ketone):

$$CH_2=C(CH_3)CH_2CCH_3 \longrightarrow (CH_3)_2C=CHCCH_3$$
$$\underset{O}{\|} \qquad\qquad\qquad\qquad \underset{O}{\|}$$

The greater stability of mesityl oxide is reflected in a
larger volume of literature on this compound collected in
Beilstein. An allyloxy derivative of 1-(p-hydroxyphenyl)-
3-butanone has a raspberry flavor (25).

Unsymmetrical diallyl acetone $A_2CHCOCH_3$ was prepared by
the acetoacetic ester synthesis (26). Butler polymerized
diallyl ketone by heating with 5% di-t-butyl peroxide for
12 hr near 50°C (27). A surprisingly high-melting polymer
was formed which was soluble in chloroform. A cyclic
structure was suggested -- ⬡CH_2 -- . Chloroform-soluble
homopolymers were also obtained by Butler from free radical
polymerization of the symmetrical bifunctional monomers
bis(2-chloroallyl)ketone, bis(2-cyanoallyl)ketone, and
bis(2-phenylallyl)ketone.

Allyl-substituted acetoacetic ester $CH_3C(O)C(A)HCOOC_2H_5$
or ethyl alpha-allylacetoacetate is a commercially important intermediate for synthesis of pharmaceuticals, perfumes, and biocides. The allyl derivatives of the free
keto acid are unstable, giving substituted ketones when
heated. The monoallyl derivative of ethyl acetoacetate
boils around 200°C with decarboxylation. The technical
compound has a sharp ester odor, refractive index n_D^{25}
about 1.435, and turns brown on prolonged exposure to
light and air. Mild alkaline hydrolysis gives monoallyl-
acetone. Stronger conditions of hydrolysis produce allylacetic acid or its salts. The carbonyl oxygen atom of
allyl acetoacetic ester also is reactive. This allyl compound is prepared by the well-known acetoacetic ester
synthesis (28).

Diallylketene was observed by Staudinger and co-workers
to polymerize to about 75% conversion after 5 days at 25°C
(29). One sample of $A_2C=CO$ polymerized more slowly than a
sample of allylmethylketene.

The allyl group in allethrin, a commercial insecticide,
is attached to a cyclopentenone ring. It is the allyl
analog of cinerin I, one of the active components of pyrethrins. Allethrin is about as toxic as natural pyrethrins
to houseflies, but less toxic to many field insects. Many
allyl-substituted esters of chrysanthemum carboxylic acids
evaluated in recent years as insecticides are listed in
Chemical Abstracts. Research on such derivatives of vinylcyclopropane carboxylic acids has been active in Japan (30).

A few allyl derivatives of open-chain diketones have been
prepared, for example, 3-allyl and 3,3-diallyl derivatives
of 2,4-pentanedione, which were reported to have the
following boiling points and refractive indices, respec-

TABLE 7.2

Some Acyclic Allyl Ketones

Ketone	bp, °C	n_D^{20} or n_D^t	References, etc.
$ACOCH_3$	108	--	Newman, JACS 67, 154 (1945); semicarbazone 145°C
$ACOC_2H_5$	124-127	1.4244	Blaise, Bull. Soc. Chim. [3] 33, 40 (1905); Coppens, Bull Soc. Chim. Belg. 38, 310 (19
ACH_2COCH_3	130	1.4199	Hurd, JACS 60, 1905 (1938) and 77, 3284 (1955); semicarbazone 103°C
$A(CH_2)_4COCH_3$	147	--	Blaise.
$ACH_2COC(CH_3)_3$	65(18)	1.4290	Colonge, Bull. Soc. Chim. 16 (1949), oxime 64.5°C
$AC(CH_3)HCOCH_3$	138	n_D^{25}= 1.4215	Cope, JACS 62, 444 (1940)
$AC(C_3H_7)HCOCH_3$	169	n_D^{25}= 1.4305	Cope, JACS 63, 1849 (1941)
$ACH_2CO(CH_2)_4CH_3$	88(15)	n_D^{22}= 1.4365	Killian, JACS 58, 893 (1936)
$MACOCH_3$	121	1.4213	Stross, JACS 69, 1627 (1947)
$MACH_2COCH_3$	149	1.4279	Hagemeyer, Ind. Eng. Chem. 41, 769 (1949)
$A_2CHCOCH_3$	175	--	Wolff, Ann. 201, 48 (1880)
ACH_2COCH_2A	186	--	Volhard, Ann. 267, 87 (1892)
$ACOC(CH_3)=CH_2$	47(11)	n_D^{19}= 1.4712	Nasarow, CZ I, 1244 (1942)
A_3CCOCH_3	100(13)	--	Conia, CA 53, 13199 (1959)
$ACH(COCH_3)_2$	92(16)	$n_D^{13.5}$= 1.4698	Perkin, J. Chem. Soc. 65, 825 (1894)
$ACH_2COCH_2COCH_3$	89(16)	1.4791	Leser, Bull. Soc. Chim. [3] 27, 65 (1902

tively (31): bp 83 at 15 mm, $n_D^{21} = 1.4642$ and bp 113 at 16 mm, $n_D^{21} = 1.4678$. Physical properties of some acyclic allyl and methallyl ketones are listed in Table 7.2.

Allyl derivatives of camphor and menthone were prepared early in this century by Haller and co-workers in France for evaluation in perfumery. They developed a practical synthesis for alkyl and alkylene derivatives of cyclic ketones (32). A dry ether solution of the ketone was mixed with finely powdered sodamide and heated until ammonia was no longer evolved. A slight excess of allyl halide was added and the mixture warmed until precipitation of sodium halide was completed. The allylmenthone or allylcamphor was recovered from the washed and dried ether solution by distillation. Optical rotation was inverted in passing from menthone to allyl- or alkylmenthones. The odors were of interest rather than polymerization behavior, and not even triallyl menthone was observed to polymerize. Lorette

TABLE 7.3

Some Allyl Cyclic Ketones

Ketone	bp, °C at mm	n_D	References
ACH₂COC₆H₅	238(710) 126(17)	n_D^{21} = 1.5279	Baeyer, Ber. 16, 2132 (1883); Perkin, J. Chem. Soc. 45, 187 (1884)
A₂CHCOC₆H₅	146(18)	n_D^{21} = 1.5278	Haller, Compt. Rend. 156, 828 and 1205 (1913); Cuvigny, Bull. Soc. Chim. 1872 (1965)
A(C₆H₅)CHCOC₆H₅	337	--	Buddeberg, Ber. 23, 2067 (1890)
A₃CCOC₆H₅	170(18)	n_D^{21} = 1.5327	Haller, Compt. Rend. 158, 828 (1914) and Cuvigny
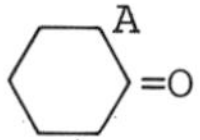 (A =O cyclohexanone)	87(16 or 94(16)	n_D^{22} = 1.4690	Howard, Org. Syn. 42, 14 (1962); Payne, CA 57, 11133 (1962); Cope JACS 63, 1848 (1941)
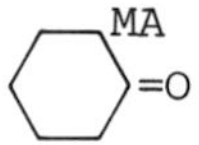 (MA =O cyclohexanone)	118(33)	n_D^{26} = 1.4669	Cantor, JACS 86, 2902 and 2907 (1964)
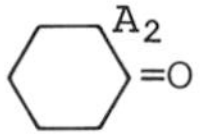 (A₂ =O cyclohexanone)	118(14)	n_D^{14} = 1.4887	Conia, Bull. Soc. Chim. 537 (1950)
Other diallyl cyclohexanones	--	--	Cornubert, Compt. Rend. 158, 1901 (1914) (Good odors)
2-Allyl-2-isopropenylcyclopentanone	--	--	Conia, CA 65, 8774 (1966)
2-Allylcycloheptanone	90(10)	--	Opitz, Ann. 649, 26 (1961)
2-Allylcamphor	130(20)	--	Haller, Compt. Rend. 136, 791 (1903)
3,3-Diallyl camphor	155(16)	--	Haller, ibid. 156, 828; 1140 & 1205 (1913)
2-Allylmethone	137(20)	--	Haller, ibid. 138, 1140 (1904) (Application in perfumes)
Triallylmenthone	167(14)	--	Haller, Compt. Rerd. 156, 1205 (1913)
Allylthujone	110(15)	--	Haller, ibid. 140, 1628 (1905)
2-Allylionone	--	--	Givaudan index (Pineapple odor)

and Howard prepared alpha-allyl-substituted cyclopentanones and cyclohexanones by pyrolysis of diallyl ketals (33).

Table 7.3 gives some references to allyl derivatives of aromatic ketones, cyclohexanone, and other cyclic ketones. It is very surprising that allyl-2-methyl cyclohexyl ketone was reported by Nazarov and co-workers to polymerize on standing for 2 months (34). Slow addition of allyl bromide to the reaction product of cyclohexanone and sodamide near 0°C, followed by heating and distillation, gave principally 2-allylcyclohexanone (bp 92°C at 17 mm) having an odor suggesting menthol, along with some diallylcyclohexanone (bp 124°C at 17 mm) (35).

Very little seems to be known about allyl phenyl ketone, a compound which one might expect to isomerize readily to crotonophenone.

Copolymers of l-alkenes and of vinyl monomers with minor proportions of allyl derivatives of aromatic ketones or of hydroxyaryl ketones have interesting possibilities for plastics and fibers "internally stabilized" against degradation by ultraviolet light. For example, ethylene has been copolymerized by Tocker of Du Pont with small amounts of 3-allyl-2-hydroxybenzophenone using catalysis by vanadium oxychloride (VOCl$_3$) and lithium butyl (36). Allyl derivatives of aromatic ketones seem to merit more consideration also as UV and antioxidant stabilizers for plastic without copolymerization.

A number of allyloxy derivatives of aromatic ketones have been investigated in synthesis of drugs and stabilizers for plastics. Allyloxy and methallyloxacetophenones were prepared and by Claisen rearrangement the alkenyl hydroxyaryl ketones were made (37). Physical properties were reported for these as well as corresponding derivatives of propiophenone and butyrophenone. Depressing action upon the central nervous system was observed from 4-(allyloxy)-3,5-dimethoxyacetophenone (38).

For stabilization of isotactic polypropylene in films, Geigy researchers added 0.6% of 2-allyloxy-4-butoxybenzophenone (39). Similar derivatives were suggested for stabilizing polyesters, polyamides, polyvinyl chloride, and acrylic polymers. For stabilization against UV light, styrene was copolymerized with minor proportions of 2-hydroxy-4-allyloxybenzophenone by Sharetts and Melchore of American Cyanamide (40). The monomers containing 0.2% t-butyl hydroperoxide and 0.005% t-butylcatechol were heated in a sealed tube for 4 days at 100°C and 3 days at 130°C for copolymerization.

References

1. C. W. Smith, Acrolein, Wiley, 1962; R. C. Schulz, Angew. Chem. 76, 357 (1964).
2. C. E. Schildknecht, Vinyl and Related Polymers, Wiley, 1952.
3. C. D. Hurd and M. A. Pollack, J. Am. Chem. Soc. 60, 1910 (1938).
4. F. W. Schuler and G. W. Murphy, J. Am. Chem. Soc. 72, 3155 (1950).
5. David Walborn and C. E. Schildknecht, unpublished.
6. Y. Deux, Compt. Rend. 216, 415 (1943); H. Adkins and D. Folkers, J. Am. Chem. Soc. 53, 1423 (1931).
7. H. P. A. Groll and H. W. de Jong, U.S. 2,042,220 (Shell).
8. D. Kritchevsky, J. Am. Chem. Soc. 65, 487 (1943).
9. H. H. Inhoffen et al, Ber. 82, 315 (1949).
10. L. Claisen, Ann. 401, 95 (1913).
11. E. E. Blaise, Bull. Soc. Chim. France [3] 33, 44 (1905).
12. M. S. Newman and W. T. Booth, J. Am. Chem. Soc. 67, 154 (1945).
13. C. D. Hurd and M. A. Pollack, J. Am. Chem. Soc. 60, 1910 (1938); cf. L. Stein and G. W. Murphy, ibid. 74, 1041 (1952).
14. F. Zeidler, Ann. 187, 35 (1877); W. Kimel and A. C. Cope, J. Am. Chem. Soc. 65, 1996 (1943); cf. M. S. Schlecter, ibid. 71, 3168 (1949).
15. J. Braun and F. Stechele, Ber. 33, 1472 (1900).
16. T. Posner, Ber. 37, 507 (1904).
17. J. Colonge and L. Pichat, Compt. Rend. 226, 674 (1948).
18. Erick Hagmann and C. E. Schildknecht, unpublished.
19. A. Haller and E. Bauer, Compt. Rend. 158, 825; CA 8, 2672 (1914).
20. K. Auwers, Ann. 462, 129 (1928); cf. L. Claisen, Ber. 45, 3157 (1912).
21. A. C. Cope, J. Am. Chem. Soc. 63, 1850 (1941).
22. B. Helferich, Ber. 57, 1619 (1924).
23. H. J. Hagemeyer, Ind. Eng. Chem. 41, 769 (1949); U.S. 2,450,132 (Eastman).
24. W. C. Menly and P. Gradeff, Belg. 640,555 (Rhône-Poulenc).
25. M. Winter, Helv. Chim. Acta 44, 2110 (1961); cf. Belg. 631,124; CA 60, 13144; U.S. 2,799,706.
26. M. Kijewska et al, CA 34, 715 (1940).
27. G. B. Butler, U.S. 3,044,986 (Peninsular Chem.).
28. C. R. Hauser and B. E. Hudson in Organic Reactions, R. Adams et al., Eds., Wiley, 1942; also reference 26.
29. H. Staudinger et al, Helv. Chim. Acta 6, 291 (1923).
30. For example, T. Sasaki et al, J. Org. Chem. 37, 466 (1972).
31. T. Cuvigny and H. Norman, Bull. Soc. Chim. France 1872 (1965); CA 63, 13136 (1965).
32. Albin Haller, Compt. Rend. 138, 1139-1142 (1904).
33. N. B. Lorette and W. L. Howard, J. Org. Chem. 26, 3112 (1961); U.S. 3,114,772 (Dow).
34. I. N. Nazarov et al, CA 49, 6846 (1955).

35.   C. A. Vanderwerf and L. V. Lemmerman, Org. Syn. Coll. Vol. 3,
      44, Wiley, 1955; cf. G. Stork and S. R. Dowd, J. Am. Chem. Soc.
      85, 2178 (1963).
36.   S. Tocker, Brit. 893,507 (Du Pont), CA 57, 13995 (1962).
37.   N. P. Buu-Hoi et al, Bull. Soc. Chim. France 11, 23 (1964),
      CA 60, 10581 (1964).
38.   R. B. Moffet et al, J. Med. Chem. 7 [2], 178 (1964), CA 60,
      11930 (1964).
39.   Brit. 958,167 (Geigy), CA 65, 7387 (1966).
40.   R. R. Sharetts and J. A. Melchore, U.S. 3,215,665 (Am. Cyan.);
      cf. J. Fertig et al, U.S. 3,173,893 (National Starch).

8. MONOALLYL ESTERS

The allyl esters of fatty acids $CH_2=CHCH_2OC(O)R$ generally have slightly pungent yet pleasant odors which are quite different from those of esters of saturated alcohols. In the United States large amounts of allyl esters of the acids of 3 to 7 carbon atoms have been used in perfumery-- a development that followed naturally from allylic terpene esters. Some perfume chemists say that certain monoallyl esters impart sexy odors. Many have fruity odors along with subtle, mild pungence. The U.S. Food and Drug Administration has found a number of the allyl esters to be much less toxic than allyl alcohol but more toxic than the corresponding n-propyl esters. The following values of LD_{50} were reported by Jenner and co-workers for rats (unless otherwise designated) (1):

Compound	LD_{50}, mg/kg		
Amyl alcohol	3030		
Allyl alcohol	70		
Allyl formate	124;	136	(mouse)
Allyl acetate	142;	170	(mouse)
Allyl acetic acid	470;	610	(mouse)
Allyl butyrate	250		
Propyl butyrate	1500		
Allyl caproate	218;	280	(guinea pig)
Allyl cinnamate	1520		
Allyl heptanoate	500;	630	(mouse)
Allyl isothiocyanate	339		
Anisole	3700		
p-Allyl anisole	1820;	1250	(mouse)
Safrole	1950		
Allyl cyclohexane	585;	380	(guinea pig)

The FDA has cleared more than 20 allyl esters as additives for perfumes and foods. The lower esters such as allyl formate and acetate are more toxic and they produce more sting in the nose. This may result from hydrolysis followed by oxidation in the nasal tissues (cf. page 194). Allyl formate has a sharp, somewhat onionlike odor. Allyl

255

acetate, allyl caproate, allyl cyanide, and triallyl
phosphite in saturated water solution were not very toxic
to seven types of bacteria (2). Methallyl esters have
found few uses and are more difficult to prepare than
simple allyl esters. Monoallyl and monomethallyl esters
have not found much direct application in polymer products
in contrast to polyfunctional allyl esters.

PREPARATION OF ALLYL AND METHALLYL ESTERS

Allyl acetate was made in 1855 from reaction of allyl
iodide and silver acetate (3), and much later by esterification from allyl alcohol, acetic anhydride, and a
little sulfuric acid (4). The use of acyl chlorides or
common esterification conditions of heating with strong
mineral acid catalysts often gives poor yields of allyl
esters, along with byproducts through rearrangement of the
alcohol. One of the best laboratory methods for preparing
allyl and methallyl esters is to warm the alcohol with the
acid anhydride.

Allyl acetate was prepared by Bartlett and Altschul by
heating under reflux a mixture of 1.5 moles allyl alcohol,
0.75 mole acetic anhydride, 1.5 moles acetic acid, and
5 ml sulfuric acid (5). The resulting mixture was poured
slowly with stirring into ice water and extracted with
cold sodium sulfite and cold saturated calcium chloride
solutions. After drying with solid potassium carbonate,
the ester was distilled twice in a stream of pure nitrogen.
The mid fraction boiled at 53.2 to 54°C with $n_D^{24} = 1.3985$.
This material showed no active hydrogen by CH_3MgI.

Allyl esters of simple monocarboxylic acids can be made
by mild esterification of allyl alcohol with the organic
acid, either with or without small amounts of mineral acid
catalyst (6). Excess mineral acid catalyst and high
temperatures must be avoided in order to minimize rearrangements to mixtures of esters and aldehydes (7).
Esterification of allyl alcohol by acetic acid with an
acidic ion exchange catalyst (Amberlite IR 120) at 18 to
60°C was found to be a second-order reaction with activation energy of 12.3 kcal/mole (8). Reaction of allene
with acetic acid in the presence of palladium acetate,
and benzonitrile at 90°C gave allyl acetate and isopropenyl acetate (4:1) (9).

Preparations of allyl esters by esterification were
discussed by DeWolfe and Young, who presented several
possible ionic mechanisms (via protonation of the acid
and via allylic carbonium ion) (10). When alpha- and
gamma-methyl allyl alcohols are esterified by strong acids
such as trichloroacetic acid, mixtures of isomeric esters
are obtained by rearrangement. Primary, secondary, and

tertiary allylic alcohols give esters without rearrangement when reacted with anhydrides in the presence of one
or more equivalents of a t-amine such as pyridine:

$$CH_2=CH\overset{|}{C}OH + (RC=O)_2O \xrightarrow{\text{pyridine}} CH_2=CH\overset{|}{C}O\underset{O}{\overset{||}{C}}R + RCOOH$$

Allylic esters can be prepared also by alcoholysis or
transesterification. The basic catalysts such as alkali
alkoxides do not cause rearrangements. For example, the
ethyl ester of the desired acid may be reacted with the
allylic alcohol (11):

$$CH_2=CHCH_2OH + C_2H_5O\underset{O}{\overset{||}{C}}R \xrightarrow{\text{NaOR}} CH_2=CHCH_2O\underset{O}{\overset{||}{C}}R + C_2H_5OH$$

Alcoholysis with alkaline catalyst under mild conditions
does not attack epoxy or oxirane groups. Thus the methyl
ester of β-methyl acrylic acid was epoxidized (12). The
epoxy ester was warmed with excess allyl alcohol in the
presence of alkali alkoxide to form the allyl epoxybuty-
rate. By way of the cyanohydrin of acrolein, Phillips
and Frostick made allyl 3,4-epoxy-2-hydroxybutyrate (bp
90°C at 1 mm, $n_D^{30} = 1.4590$) and related derivatives (13).
Recently there has been interest in direct industrial
synthesis of allyl acetate and methallyl acetate by oxidation of olefins in the presence of acetic acid using
noble metal catalysts. For example, a mixture of propylene and oxygen (ratio 18.5 to 2.75 by vol.) was bubbled
at 100°C through a solution of 0.04 mole of palladium
chloride, 0.04 mole of lithium chloride, 1 mole of lithium
acetate, and 0.3 mole of cupric acetate in 1 liter of
acetic acid (14). There was recovered allyl acetate,
isopropenyl acetate, and a little acrolein. One process
employed $PdCl_2$ on alumina-bentonite as catalyst at 10 to
50 atm and 100°C (15); bismuth may promote the noble
metal catalysts (16). If these processes succeed, allyl
acetate may later replace allyl chloride and allyl alcohol
as intermediates for synthesis of polyfunctional allyl
monomers. Other references to monoallyl esters are given
in Table 8.1. A number of these allyl esters are supplied
by Polysciences Inc.
Methallyl acetate and some butyl acetate were obtained
by passing a mixture of 20 parts of isobutene and 15 parts
of oxygen (diluted with 65% N_2) through a solution of
palladium chloride, lithium chloride, lithium acetate,
and cupric acetate in acetic acid at 150°C and 5.2 kg/cm^2
during 4 hr (17). In another patent example, isobutylene
was oxidized in acetic acid with $PdCl_2$, sodium acetate,
and $CuCl_2$ in a Parr bomb maintained for 12 hr at 60°C (18).
Methallyl esters cannot be prepared in good yields by

TABLE 8.1

Properties of Some Allyl Esters, $CH_2=CHCH_2OOCR$

Ester	bp, °C		References
AOOCH	83	$d_4^{18}=0.848$	Toxicity: Koga, CA 799 (1941); Palomaa, Ber. _61_, 1771 (1928)
AOOCCH$_3$	104	$n_D^{20}=1.4040$	Jeffery, J.Chem.Soc. 669 (1948); toxicity: Smyth, J. Ind. Hyg. _31_, 61
AOOCC$_2$H$_5$	123	$n_D^{20}=1.4158$	Jeffery
AOOCCH$_2$CH$_2$CH$_3$	142 80(70)	$n_D^{20}=1.4158$	Jeffery; Polysciences lit.
AOOCCH(CH$_3$)$_2$	129	$n_D^{20}=1.4125$	U.S. 3,079,429 (Texaco)
AOOC(CH$_2$)$_3$CH$_3$	162	--	Hofmann, Ann. _102_, 296 (1857)
AOOCCH$_2$CH(CH$_3$)$_2$	88(90	--	Polysciences lit.
AOOC(CH$_2$)$_4$CH$_3$	93(30) or 185	$n_D^{20}=1.4243$	Swern, JACS _70_, 2336 (1948); U.S. 2,160,941 (Dow); U.S. 2,356,871 (PPG)
AOOC(CH$_2$)$_6$CH$_3$	88(5.5) or 228	$n_D^{30}=1.4271$	Swern; (fruity odor)
AOOC(CH$_2$)$_7$CH$_3$	151(50)	$n_D^{30}=1.4302$	Swern; (fruity odor)
AOOC(CH$_2$)$_8$CH$_3$	118(5.2)	$n_D^{30}=1.4320$	Swern
AOOCCF$_3$	79	--	Polysciences lit.; Coover, U.S. 2,759,912 (Eastman)
AOOCCH$_2$Cl	163	--	Polysciences lit.
AOOCCH$_2$Br	73(10)	--	Polysciences lit.
AOOCCH=CHCH$_3$	83(5)	$n_D^{30}=1.4416$	Frostick, JACS _81_, 3350 (1959)
AOOC(CH$_2$)$_5$CN	115(0.55)	$n_D^{30}=1.4457$	Freure, U.S. 3,021,356 (Carbide)
AOOCCHBrCH$_2$CH$_3$	80(70)	--	Polysciences lit.
AOOCC$_6$H$_5$	228	--	Hofmann; Sabatier, Compt. Rend. _152_, 360 (1911); Perkin, J. Chem. Soc. _69_, 1227 (1896)
AOOCCH$_2$C$_6$H$_5$	92(3)	--	Polysciences lit.
AOOCCH(OH)CH$_3$	176(740)	--	Polysciences lit.

conventional esterification in the presence of mineral acids, because methallyl alcohol so readily rearranges to isobutyraldehyde. Methallyl acetate was made by warming the alcohol with acetic anhydride (19).

Allyl and methallyl isobutyrate have been made by pyrolysis of acetals (20). Thus isobutyraldehyde and allyl alcohol in CH_2Cl_2 together with $CaCl_2$ were heated at 60°C until 58% of the theoretical water had been removed. The diallyl acetal of isobutyraldehyde recovered was passed through a Pyrex tube at 346°C to form allyl isobutyrate. Allyl diazoacetate $AOC(O)CHN_2$ was prepared (21).

Allyl and methallyl esters of fatty acids are liquids

TABLE 8.2

Properties of Some Methallyl Esters, $CH_2=C(CH_3)CH_2OOCR$

Ester	bp, °C	Refractive index n_D^{20}	References
MAOOCCH	103	1.4135	Ryan, JACS <u>62</u>, 3469 (1940)
MAOOCCH$_3$	124	1.4129	Ryan (liquid polymers); Lewis, JACS <u>70</u>, 1528 (1948); Hearne, Ind. Eng. Chem. <u>33</u>, 807 (1941)
MAOOCC$_2$H$_5$	142	1.4170	Ryan; U.S. 2,207,611 & 613 (Dow)
MAOOC(CH$_2$)$_2$CH$_3$	161	1.4230	Ryan; Dow
MAOOCCH(CH$_3$)$_2$	152	1.417	Weis, Fr. 1,384,693 (Ugine); Chafetz, U.S. 3,079,429 (Texaco)
		n_D^{30}	
MAOOC(CH$_2$)$_4$CH$_3$	135(100)	1.4250	Swern, JACS <u>70</u>, 2336 (1948)
MAOOC(CH$_2$)$_6$CH$_3$	148(50)	1.4308	Swern; U.S. 2,341,060 (National Oil Products)
MAOOC(CH$_2$)$_7$CH$_3$	164(50)	1.4335	Swern
MAOOC(CH$_2$)$_8$CH$_3$	175(50)	1.4354	Swern; National Oil Products
MAOOC(CH$_2$)$_{10}$CH$_3$	165(10)	1.4390	Swern, National Oil Products
MAOOC(CH$_2$)$_{14}$CH$_3$	186(4)	1.4450	Swern
MAOOC(CH$_2$)$_{16}$CH$_3$	205(4.6)	1.4471	Swern; U.S. 2,207,611 (Dow)
MAOOCCF$_3$	85	--	Coover, U.S. 2,759,912

widely miscible with organic solvents, except for the esters of long-chain acids. Allyl palmitate and allyl stearate as supplied by Polysciences melt at 26 and 38°C, respectively. Boiling points and refractive indices of a number of methallyl esters are given in Table 8.2.

Hanna and Siggia studied rates of addition of Br$_2$ to allyl propionate compared with other olefinic compounds in solvents of different polarity (22). The promotion by solvents of high dielectric constant was attributed to stabilization of the intermediate carbonium ions and to possible enhancement of separation of charges in bromine molecules by solvation. Relative rates of bromine addition were as follows:

Solvent	Dielectric constant of solvent	Relative rates of Br$_2$ addition	
		Vinyl acetate	Allyl propionate
Formic acid	57	--	230
Methanol	33	>2000	43
Acetic anhydride	20	>2000	27
Acetic acid	6	73	3
CCl$_4$	2	0.002	<0.002

Under these conditions isopropenyl acetate and 1-octene
reacted faster than vinyl acetate. Addition of bromine
to diethyl fumarate and to tetrachloroethylene was too
slow to measure. A linear relation was found between the
Hammett sigma values of substituents in vinyl acetate,
allyl propionate, and a butenenitrile with dielectric
constants of solvents producing a constant rate. Bromine
was added to allyl acetate in CCl_4 with $CaBr_2$ as catalyst
at 0°C, giving 93% 2,3-dibromopropyl acetate (23). Without catalyst, much tribromopropane was formed.

Hydroboration of different allyl derivatives showed
increasing Markovnikov addition to the double bond as
electronegativity of substituents attached to the allyl
group decreased (24). Thus the percentage addition of
boron to the second carbon atom decreased as follows: A
tosylate 45%, A chloride 40%, A acetate 35%, A benzoate
25%, A borate 18%, A phenyl ether 32%, A phenyl sulfide
22%, A ethyl ether 19%, A alcohol 24%. Reactions of the
allyl tosylate, chloride, benzoate, and acetate eliminated
some propylene during hydroboration. Allyl acetate pyrolysis at 400 to 500°C was believed to proceed by a free
radical mechanism to form mainly CO_2, 1-butene, CO, and
CH_4 (25). Allyl acetate hydrolyzes faster in water and in
dilute mineral acid solution than do methyl and ethyl
acetate.

Cyclization of methallyl acetate in concentrated H_2SO_4
was observed to form 2,4-dimethyl dioxolenium cations (26).
Cycloaddition of methallyl cations to cyclopentadiene or
other conjugated dienes gave bridged olefinic seven-
membered rings (27). Cations were generated also by reaction of methallyl iodide with silver trichloroacetate to
form methallyl trichloroacetate.

ALLYL ACETATE POLYMERIZATION

Bartlett and Altschul made pioneer studies of the free-
radical-initiated polymerization of allyl acetate (28).
When allyl acetate was heated at 80°C with 6% BP, the
peroxide was only 90% decomposed after 13 hours and required about 48 hours for essentially complete reaction;
at that time, only about half of the initial allyl acetate
had polymerized. After removing residual volatiles by
distillation, a soft transparent polymer remained. The
polymer was soluble in common organic solvents and had an
average DP of 13. Different samples of allyl acetate,
prepared with care to remove free alcohol and acid, and
distilled in the absence of air, gave reproducible results.
The course of polymerization was followed by bromine addition to residual monomer (bromate-bromide reagent), by
dilatometry, and by weighing the polymer formed. Small

amounts of water, hydrogen chloride, or pyridine did not affect the rates. The rate of polymerization was approximately proportional to the first power of peroxide concentration.

Oxygen was capable of retarding polymerization of allyl acetate, but the effect of brief exposure to air was not great. Bartlett and Altschul found several evacuations and flushings with nitrogen to be effective in eliminating inhibition by oxygen. The result of these precautions was about 10% increase in rate over that observed under air without agitation. Study of the polymerization of allyl acetate using 5.9% p-chlorobenzoyl peroxide at 80°C showed that about 77% of the polymer molecules contained terminal fragments from the peroxide. Assuming only one Cl atom per polymer chain, 23% of the polymer may have been initiated by fruitful chain transfer. The consumption of peroxide and the CO_2 evolved were increased by passing O_2 through the reaction mixture. In different polymerization experiments, however, the CO_2 evolved was not proportional to the polymer yield. In the course of this work, allyl acetate low polymers were saponified by $NaOCH_3$ to obtain allyl alcohol polymers of low DP which then were reacetylated.

Bartlett and Altschul found the decomposition of benzoyl peroxide in allyl acetate to be nearly unimolecular at 80°C over a tenfold range of initial concentration (29) (Figs. 8.1 and 8.2). The ratio of remaining monomer to remaining peroxide (dM/dP) was fairly constant throughout. However, it changed from 28.8 to 15.6 as initial peroxide was increased from 1.01 to 9.95%; the polymer DP was remarkably constant at 13.7 ± 0.4 in all cases. Thus the DP was determined by the degradative chain transfer process. RH represents a terminated polymer molecule in the following equation:

$$\overset{|}{R} + CH_2{=}CHCH_2OOCCH_3 \longrightarrow RH + CH_2{=}CH\overset{|}{C}HOOCCH_3$$

Only a fraction of the monomer radicals can continue polymerization. Persulfate emulsion polymerization of allyl acetate gave similar results (30). For an allyl acetate polymer of η_{sp}/C (benzene) of 0.37, the Staudinger constant Km was found to be 3.3×10^{-4} by means of freezing point depression. Properties of the polymers in bulk were not reported by Bartlett and Altschul. In Japan, polymerizations of allyl acetate with azobisisobutyronitrile as initiator showed similar characteristics to those made with BP (31). After solution polymerization in CCl_4 from 2 to 3 chlorine atoms per polymer molecule were found chemically bonded. From initial concentrations of 0.5, 1.0, and 5% BP, the allyl acetate polymers had cryoscopic molecular weights of 1340, 1295, and 1120.

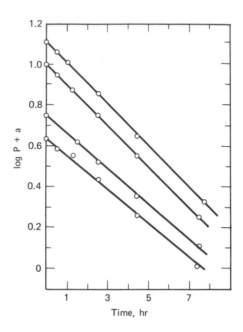

Fig. 8.1. The decomposition of benzoyl peroxide in allyl acetate at 80°C is nearly unimolecular. The data from four polymerizations at concentrations from 1.01 to 9.95% benzoyl peroxide were displaced vertically by a distance a for convenience. P. D. Bartlett and R. Altschul, J. Am. Chem. Soc. 67, 817 (1945).

Gaylord and Eirich found that rates of BP decomposition in different esters at 80°C decreased in the order: A trimethyl acetate, A propionate, A acetate, A ethyl carbonate (32). This suggested that radicals attack not only the allylic hydrogen atoms but also the acyl groups of the esters. The above order was reversed in DP of the polymers formed, all of which had η_{sp}/C below 0.1. Isopropenyl acetate behaved similarly. Litt and Eirich proposed a more detailed mechanism for allyl acetate homopolymerization (33).

In the free radical homopolymerization of allyl acetate and of vinyl acetate, benzoquinone, chloranil, and other agents were added in small amounts as inhibitors (34). Spectra suggested formation by C- and O- alkylation of colored products. Quantitative expressions for inhibitor activities were derived. Isomeric dinitrobenzenes and syn-trinitrobenzene were added in small concentrations as

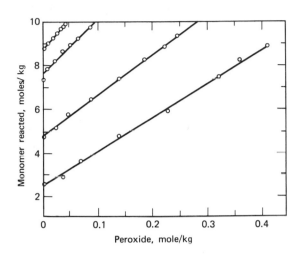

Fig. 8.2. From the experiments of Fig. 8.1 there was a linear relationship between the amount of peroxide reacted and the amount of allyl acetate used up in polymerization. The ratio dM/dP was constant in any run and varied from 15.6 to 28.8 over a tenfold range of peroxide concentration. P. D. Bartlett and R. Altschul, J. Am. Chem. Soc. **67**, 817 (1945).

inhibitors of the benzoyl-peroxide-initiated polymerization of allyl acetate (35). The trinitrobenzene was the most effective, stopping about six chain reactions per molecule. At 80°C with BP under high pressures (up to 8500 kg/cm^2), Walling and Pellon found that the DP of allyl acetate polymers formed was increased only slightly (36). The overall rate increased exponentially with pressure and became proportional to the square root of the peroxide concentration.

Bartlett and Tate studied the radical polymerization of allyl-1,1-dideutero acetate (37). This was found to give degradative transfer only about one-third as fast as the hydrogen compound, and molecular weights of polymer were increased by a factor of 2.4. The rates of polymerization still were linear functions of peroxide decomposed, in contrast to the square-root dependence more often encountered in the absence of degradative chain transfer. Simpson used free radical conditions similar to those of Bartlett and Altschul to obtain allyl acetate homopolymer of molecular weight 1400, which could be saponified by half-normal alcohol KOH (38). Heating allyl benzoate with

6% BP at 80°C for 48 hr gave polymer of DP about 11.   The
allyl alcohol polymer derived therefrom could not be ben-
zoylated again.   Also investigated were BP-initiated
telomerizations of allyl acetate with $CHCl_3$ and with CCl_4
(39).

Complexing allyl acetate with zinc chloride was reported
by Kargin, Zubov and co-workers to increase rates of conversion to polymer under high energy radiation (40).   With
increasing $ZnCl_2$ content beyond 1:1 molar, the polymer
products passed through viscous liquids to nonfluid
rubbery masses.   The higher polymers were thought to have
a branched microgel structure.   Displacement of electron
density in the monomer (electron withdrawal) was believed
to reduce chain transfer at the underlined allylic hydro-
gen atoms: $CH_2=CHC\underline{H}_2O\underset{\underset{CH_3}{|}}{C}=O:ZnCl_2$.   Intrinsic viscosities
of the soluble polymers were still very low, below 0.1.

Simple allyl esters do not polymerize readily by oxidative processes or "drying" of films as do certain allyl
ethers and drying oils.   Oxygen absorption by allyl
butyrate was only about one-thirtieth that of allyl butyl
ether (41).   Complex low polymers from allyl acetate prepared slowly by heating in the presence of oxygen were
partially saponified to yield products soluble in water
and in methanol (42).   Little has been reported on the
polymerizations of allyl formate, but the monomer has been
copolymerized with maleic anhydride, vinyl stearate, and
vinyl triethoxysilane to prepare interesting protective
coatings for glass (43).

Viscous liquid low polymers from allyl acetate and some
other monoallyl esters were obtained in Shell laboratories
by long heating with halides and other salts of Ba, Cd,
Hg, and Pb (44).   Catalysts were preferably added inter-
mittently, and polymerizations were terminated at low
yields.   Most of these catalysts are very weak Lewis acids,
but BF_3 was also included in one example.   Structures of
the polymers apparently received little attention.   When
1% nickel acetate was used as catalyst at 210°C under
pressure, 79% conversion to liquid, low polymer was
observed (45).

Small amounts of brown, viscous liquid or semisolid low
polymers are encountered by action of acids, bases, or
alkali metals upon allyl esters, but caution should be
taken before accepting these as simple addition polymers.
Especially in the presence of moisture allyl or methallyl
alcohol and their rearrangement products, propionaldehyde
and isobutyraldehyde, may lead to complex products.   Butyl
magnesium halides and their oxidation products at room
temperature gave yellow rubbery to harder polymers from
allyl acetate in low yields (46).   Addition of boron

fluoride-etherate to allyl n-caproate or n-heptanoate in
hexane (2 to 3) at room temperature gives no immediate
visible change and only slight temperature rise (47).
However, on long standing, small amounts of brown to black
low polymers were formed.

Goodman and Mather of ICI reported solid polymers from
monoallyl esters by cationic conditions (48). Although
the products have not been regarded as commercially promising, these systems should be further investigated.
Allyl acetate or allyl formate with 6% benzene diazonium
fluoroborate and benzene were heated in a sealed tube under
nitrogen at 200°C to form brownish solids that were soluble
in benzene and could be precipitated by methanol. The
solids melted at about 190°C with decomposition. A 10%
solution of allyl acetate in dry petroleum ether was cooled
to -78°C. Over the cold stirred liquid was passed a stream
of BF$_3$ gas for 2 hr after which stirring was continued for
4 hr. Under those conditions both allyl alcohol and allyl
acetate gave solid polymers exhibiting x-ray patterns
indicating some polymer crystallinity. The allyl acetate
polymer crystals melted at about 145°C. The x-ray cell
dimensions were 3.3, 4.3, and 6.1 Å. Under similar conditions, polymers were obtained from allyl trichloroacetate; these decomposed without melting above 330°C. The
mechanisms and structures of these polymeric products
remain to be elucidated.

ALLYL ACETATE IN COPOLYMERIZATIONS

Bartlett and Nozaki found that the separate homopolymerizations of allyl acetate and of maleic anhydride slowly
gave products of similar low DP but that the copolymerizations occurred with remarkable speed (49). Their work
suggested the importance of alternating polarity from
inductive effects in radical copolymerization of substituted ethylenic compounds. With 5% BP under air at 80°C,
the copolymerization of equimolar mixtures of allyl acetate
and maleic anhydride occurred violently with spontaneous
heating and consumption of 97% of the peroxide. With
purified monomers at 60°C under air, copolymerization
occurred without temperature rise; when repeated under
nitrogen, the copolymerization occurred more rapidly, with
temperature rise to 71°C in 18 min. Even by the most
careful degassing the retarding effects of oxygen could
not be eliminated completely. Separate bulk polymerization
of maleic anhydride at 55°C with BP showed that its reactivity was sensitive to molecular oxygen. Not only were
the copolymerizations about 100 times as fast as the homopolymerizations, but also the decomposition of peroxide was
correspondingly rapid. Pink colors were observed which

recently have been attributed to free radical formation
of ion-radical complexes between maleic anhydride and
donor molecules.

Monomer selectivity was found to be very high in giving
1:1 molar copolymers (49). Even when the molar ratio of
allyl acetate to maleic anhydride was 58:1, the monomers
entered the copolymer alternately within an experimental
error of 10%. The copolymers always precipitated as
formed, and this discouraged kinetic studies. Since none
of the 1:1 copolymers dissolved completely in organic
solvents, some crosslinking was suspected. A few copolymer samples dissolved almost completely in acetone. However, all the copolymers dissolved completely in water
with slow hydrolysis to allyl acetate-maleic acid copolymers. Apparent copolymer molecular weights in acetone
were larger than those obtained in water by osmometry.
Allyl acetate was copolymerized in bulk at 80°C with
diethyl maleate, diethyl fumarate, 1,2-dichloroethylene,
tetrachloroethylene, crotonaldehyde, and with styrene
(BP at 80°C), but the rates were much slower than with
maleic anhydride. Bartlett and Nozaki suggested that
alternating copolymerization might be explained by inductive effects. Applications of allyl acetate-maleic
anhydride and allyl acetate-maleate ester copolymers have
been investigated (50).

Allyl acetate was copolymerized with diethyl maleate by
long heating with 5% BP at temperatures up to 100°C (51).
A 75% yield of tacky resin was recovered. Allyl acetate
and styrene copolymerized reluctantly even at relatively
high temperatures to give polymers of low molecular weight
(52). In contrast, sulfur dioxide formed copolymers
readily, for example, under high energy radiation (53).
Copolymers of allyl esters with vinyl esters have had
interest for preparation of modified vinyl alcohol copolymers of controlled molecular weight (54). With most vinyl
and acrylic monomers, however, copolymerization reactivity
ratios of allyl and of methallyl acetate are unfavorable
(55). The e values for polarity of these two monomers are
very low, -1.13 and -1.33, while the Q values of reactivity are only 0.028 and 0.037. Small additions of allyl
esters can produce large differences in reaction rates and
in structures of vinyl and acrylic polymers. These may be
useful in reducing polymer molecular weights, branching,
and crosslinking.

Delayed addition of allyl acetate by Kuehne led to
relatively high softening terpolymers of vinyl chloride-
vinyl laurate-allyl acetate (56). Suspension copolymerization of the two monomers was run with methyl cellulose
and lauroyl peroxide at 53°C until the pressure from vinyl
chloride had fallen to 6.8 atm. Then the allyl acetate

was added and the temperature was raised gradually to 70°C
during 3 hr for completion of the copolymerization. Allyl
acetate (M_1) copolymerized readily with vinyl acetate (M_2)
using radical catalysts that gave reactivity ratios
$r_1 = 0.45$ to 0.7 and $r_2 = 0.6$ to 1.0 (57). Vinyl acetate
was copolymerized with 3 mole % allyl acetate; then the
copolymer was saponified by sodium hydroxide and fibers
were spun from 13% solutions in water (58). The spun
fibers were soluble in water at 20°C, in comparison to
vinyl alcohol homopolymer fibers which dissolved only at
70°C or above. Liquid low copolymers of 1-octene with 10%
allyl acetate (mol. wt. 350-750 and saponification value
47%) were observed to kill insect eggs, and they were suggested as insecticides nontoxic to plants (59).

The following reactivity ratios have been reported for
radical-initiated copolymerizations of allyl acetate (M_1)
with other monomers (M_2) (60):

r_1	M_2	r_2
0.007	Maleic anhydride	0.13
0	Methyl acrylate	5.0
0	MMA*	23
0.01	Styrene*	90
0	Vinylidene chloride*	6.6

Allyl acetate and organic acids formed telomers on
heating for 5 hr at 155°C (61). From 37 to 52% of the
products were 1:1 molar adducts of the type

$$CH_3COCH_2CH_2CH_2CHR$$
$$\underset{O}{\|} \qquad\qquad COOH$$

Allyl acetate and other allyl compounds were added in the
polymerization of trioxane to modified acetal polymers
(62). In one example, 600 g formaldehyde trimer, 3 g allyl
acetate, and 400 g cyclohexane were refluxed under N_2 at
60°C and 1.0 ml BF_3 etherate catalyst was added. After
polymerization, the polymer was precipitated by adding
acetone and a little triethylamine.

In the absence of water or other Lewis-basic impurities,
allyl acetate as well as allyl formate copolymerized
slowly with vinylidene cyanide $CH_2=C(CN)_2$ in the presence
of o-chlorobenzoyl peroxide at 50°C (63). With high proportions of the allyl esters, the copolymers contained
near 1:1 molar components; otherwise the copolymer products contained more than molar proportions of vinylidene
cyanide units. Monoallyl monomers have possibilities for
modifying preformed polymers by grafting on short side

_____
*With methallyl acetate M_1 the r_1 values were zero also.

chains. Attempts to prepare interesting graft copolymers
of allyl acetate have been made upon polystyrene (64) and
on wool (using ^{60}Co radiation (65). Chelates of allyl
esters with metal compounds have been investigated for
grafting to polyesters and polyamides under high-energy
radiation (66). Allyl formate was copolymerized with
methyl acrylate by long heating with 3% BP at 80°C (67).
A clear acetone-insoluble solid was obtained. In free
radical copolymerizations, simple allyl esters in small
proportions may have valuable regulating effects; in addition, they may serve to introduce useful ester or hydroxyl
groups. For example, allyl acetate was copolymerized with
$CF_2=CF_2$ in water using azobis for 14 hr at 75 to 80°C and
2000 psi (68). The white copllymer dispersed in dioxane
and methanol was saponified by heating with $NaOCH_3$. The
TFE-allyl alcohol copolymer obtained was complexed with
polysilicic acid in coatings.
  Other radical copolymerizations with monoallyl esters of
lower fatty acids were reported:

Comonomers with AOC(O)R	References
Acrylonitrile, methyl chloroacrylate, etc.	Nozaki, J. Polym. Sci. 1, 455 (1946)
MMA	Sengupta, J. Macromol. Chem. 1, 481 (1966)
Acrylonitrile, vinyl imidazole, etc.	Brit. 723,558 (Chemstrand); Lodhi, CA 75, 21185 (1971)
Methyl cyanoacrylate	Kinsinger et al, J. Appl. Polym. Sci. 9, 429 (1965)
1-Octene	Dierick, U.S. 2,784,136 (Shell)
Ethylene	Steinberg, CA 58, 10308 (1963); CA 59, 14116 (1963); CA 60, 14615 (1964) (^{60}Co radiation)
Vinyl chloride, etc.	Christenson, U.S. 2,882,251 (PPG); Burnett, Proc. Roy. Soc. A221, 41 (1954)
Unsaturated polyesters	Lundberg, U.S. 2,856,378 (Am. Cyan.)
Vinylidene cyanide	U.S. 2,650,911

ALLYL ESTERS OF HIGHER ACIDS

  The monoallyl esters of propionic acid and higher acids
have not been transformed into high-molecular-weight
linear homopolymers. Viscous liquid low polymers result
from heating with peroxide catalysts; for example, allyl
propionate and caprylate formed polymers of DP 4 to 5 (69).
Such low polymers may gradually become crosslinked on
heating in air at 200°C. A waxlike polyallyl stearate-

acetate could be made by heating polyallyl acetate of DP 9 with stearic acid.   Liquid allyl caprylate polymers were suggested as synthetic lubricants (70).   In free radical polymerizations, decreasing chain transfer was observed in the series A acetate > A propionate > A laurate (71).   Chain transfer was even more effective with allyl benzoate, allyl chloroacetate, and allyl chloride.

Little has been reported on copolymerizations with allyl esters of acids of 3 to 10 carbon atoms.   Copolymers of allyl propionate as well as allyl butyrate with 1-octene (mol. wt. 400-800) were suggested as plasticizers (72). For example, the monomers in methyl ethyl ketone with 0.05% di-t-butyl peroxide were heated in a sealed glass tube for 4 hr at 200°C.   Allyl caproate was copolymerized with m-vinyl phenol in bulk and in emulsion (73).   Reaction of allyl alcohol with lactic acid or its polymers in the presence of a little sulfuric acid gave allyl lactate (bp 60°C at 8 mm and n_D^{20} = 1.4369 (74).

A number of allyl esters of aromatic acids have attractive odors.   Allyl benzoate can be prepared by reacting propylene with t-butyl perbenzoate (75).   Methallyl benzoate (bp 227°C or 123°C at 24 mm and n_D^{20} = 1.5174) can be prepared by heating a benzoate salt with methallyl chloride or alcohol (76).   Early references to allyl benzoate were given in Table 8.1.

Heating allyl benzoate with BP at 60°C only increased viscosity slightly after 16 hr (77).   Allyl benzoate and diethyl maleate gave soft solid copolymers under similar conditions.   Allyl benzoate and sulfur dioxide gave sulfone polymers by radiation with gamma rays from ^{60}CO at 15°C (78).   These could be molded to white, translucent, somewhat flexible films melting near 170°C with decomposition.   Allyl o-toluate (103°C at 5 mm and n_D^{20} = 1.520) formed o-allyltoluene on pyrolysis and methallyl benzoate formed methallylbenzene (79).   Copolymers of allyl benzoate with maleic anhydride were esterified with long-chain alcohols for testing as oil additives (80).   Metal chelates of allyl salicylate polymers, particularly copolymers with ethylene, were studied by Du Pont as stabilizers of polymers against UV light (81).   When allyl benzoate was heated at 60°C with increasing o-chlorobenzoyl peroxide concentrations (1 to 7%), the polymer molecular weight fell from 1430 to 1010 and the combined chloride in the polymer increased from 0.54 to 0.95% (82).   Copolymerizations with vinyl acetate were also studied.   Allyl benzoate was copolymerized with maleic anhydride at 50°C or above (83). The products were esterified by long-chain alcohols for evaluation as oil additives.

Allyl salicylate (bp 249°C), available from Pfaltz and Bauer, has a pungent wintergreen odor.   Allyl-β-phenyl

propionate is an odorant for perfumes (84). Allyl 4-
aminobenzoate has been suggested as a UV-absorbing stabilizer and comonomer (85).

Some monoallyl esters of long-chain and unsaturated
acids (Table 8.3) have been evaluated in copolymerizations,
but their low reactivity has led to little commercial use.
However, a number of the monoallyl esters of higher acids
have found application in flavors and perfumes. The following characteristics have been recorded (86):

	bp, °C	Refractive index	Odor
Allyl crotonate	65(20)	--	--
Allyl hexadienoate	68(1)	n_D^{24}=1.5066	Pineapple, anise, honeydew
Allyl nonylenate	105(1)	n_D^{24}=1.4535	Pineapple
Allyl 9-undecylenate	118(1)	n_D^{23}=1.4479	Pineapple, coconut
Allyl cinnamate	108(1) 125(2)	n_D^{23}=1.5661	Peach, apricot
Allyl anthranilate	106(1)	n_D^{24}=1.5720	Green leaf

Allyl undecylenate of quince odor on heating with a
little sulfuric acid gives gamma-undecalactone with peach
odor (75). The allyl esters used in perfumes and as food
additives include simple allyl esters of fatty acids
(propionate to undecanoate), allylhexyl esters of acetic
to valeric acid, allyl sorbate, allyl phenylacetate,
allylcyclohexyl propionate, allyl 2,4-hexadienoate, allyl
tiglate, and allyl furoate. Allyl esters of perfumery
grade are supplied in America by Dodge and Olcott, Elan
Chemical Company, Florasynth Inc., Pfaltz and Bauer,
Ritter & Company, and others.

Allyl esters of long-chain acids can be prepared conveniently by direct esterification or by alcoholysis. For
example, allyl 10-hendecenoate was prepared by refluxing
for 5 hr a mixture of 0.4 mole of the long-chain acid,
0.8 mole of allyl alcohol, 1.5 g of naphthalene-2-sulfonic
acid, and 250 ml of benzene (87). The water was removed
azeotropically and the benzene returned continuously. The
acid catalyst was then neutralized by $NaHCO_3$, and the
benzene and remaining alcohol were distilled off. The
allyl ester distilled at 50 mm near 180°C to give about
90% yield. Methallyl hendecenoate was prepared by alcoholysis of methyl 10-hendecenoate. A piece of metallic
sodium (0.35 g) was dissolved in 1.75 moles of methallyl
alcohol and 0.35 mole of methyl ester was added (87). The
solution was heated at 95°C for 5 hr under nitrogen and

TABLE 8.3

Some Allyl Esters of Long-Chain Acids

Allyl Ester	bp, °C at mm	n_D^{30}	References
Undecylenate	112(2)	1.4449	Frostick, JACS 81, 3350 (1959)
Laurate (mp = 6°C)	164(20)	1.4370	Swern, JACS 70, 2334 (1948); Schwann, J. Polym. Sci. 40, 457 (1959); U.S. 2,616,852
Myristate	157(4)	1.4404	Swern
Palmitate (mp = 27°C)	169(4)	--	Swern
Stearate (mp = 35°C)	188(4)	1.4457	Swern; Cheng, CA 49, 7287
18-Hydroxystearate	--	--	Kulkarni, CA 76, 14968 (1972) (Kamala seed)
Oleate	148(4)	1.4539	Harrison, JACS 73, 839 (1951)
Linoleate	150(0.75)	1.4632	Harrison; also Frostick, JACS 81, 3350 (1959)
Chlorallyl laurate	152(4.2)	1.4484	Swern
1-Methylallyl laurate	157(10)	1.4350	Swern
Chlorallyl palmitate (mp = 29°C)	--	1.4524	Swern
1-Methylallyl palmitate	180(4.3)	1.4413	Swern
Chlorallyl stearate (mp = 37°C)	202(2)	1.4497 (at 40°C)	U.S. 2,208,960 (Dow); Swern
1-Methylallyl stearate	199(4.2	1.4440	Swern

then excess methallyl alcohol was distilled off at 100 mm. The methallyl ester was distilled at 10 mm and 152°C in 70% yield ($n_D^{30} = 1.4467$). Copolymers of the allyl and methallyl unsaturated long chain esters were prepared with vinyl acetate by heating with peroxide catalyst in bulk. Copolymers from 1 to 40% allyl compound were insoluble in solvents because of crosslinking through attack of hydrogen atoms of the pendant hydrocarbon branches by chain transfer.

Allyl palmitate and allyl stearate were studied by Swern and Jordan of the Eastern Research Laboratory, U.S. Department of Agriculture (88). These low-melting solid allyl esters can be prepared directly by esterification with toluene sulfonic acid catalyst. Methallyl esters of long-chain acids were made by alcoholysis of methyl esters by methallyl alcohol in the presence of a little sodium methallylate. On heating with benzoyl peroxide, allyl and methallyl palmitate and stearate polymerized very slowly, yielding products of very low DP. A more reactive monomer was chloroallyl oleate. Copolymerization of vinyl acetate with chloroallyl, methallyl and allyl oleate by heating

with 0.5% BP gave insoluble crosslinked copolymers apparently by chain transfer with the long hydrocarbon segments. Chloroallyl esters of the long-chain acids slowly
liberated HCl at room temperature and discolored.

Heating allyl stearate with 2% BP at 80°C gave only 40%
conversion to low polymer in 12 hr (89). Polymerization
rates were slower than with corresponding vinyl esters.
Reaction rates of allyl esters of multiple unsaturated
acids were little different whether the double bonds were
conjugated or separated by CH_2 groups. Bulk peroxide-
initiated polymerizations of the allyl esters of oleic,
linoleic, and 10,12-octadecadienoic acids stopped far
short of completion. When high concentrations of peroxide
initiator were added, crosslinked polymers resulted even
at low conversion. The soluble homopolymers prepared from
the polyallylic monomers had very low intrinsic viscosities (0.04-0.06 in benzene).

Allyl n-caproate copolymerized with maleic anhydride on
heating with 0.5% acetyl peroxide at 70°C for 24 hr,
giving a hard, fusible copolymer (90). Higher esters such
as allyl stearate gave softer copolymers. The copolymers
were esterified by methanol and by glycols, and inorganic
salts were made. Allyl stearate, also allyl laurate, were
copolymerized with maleic anhydride or maleate esters of
long-chain alcohols to give polymers in the 1000-2000
molecular weight range, and these were suggested as modifiers of lubricating oils (91). From 0.5 to 2% may be
added to hydrocarbon oils as pour point depressants and
viscosity index improvers. Allyl, methallyl, and chlorallyl esters of 9,10-dihydroxystearate as well as allyloxy
hydroxystearates were examined by Swern and co-workers for
compatibility with film-forming polymers and other properties (92). Allyl 9,10-epoxystearate and related allyl
epoxy compounds were copolymerized with vinyl chloride in
solution, in emulsion, and in suspension (93). Chloroallyl linoleate was copolymerized with styrene by radical
and by cationic initiation (94). Viscous liquid low
polymers of allyl n-octanoate and of allyl laurate and
their copolymers were also evaluated as additives for
lubricating oils (95). Copolymerization reactivity ratios
for allyl laurate with vinyl acetate were determined to be
$r_1 = 0.71$ and $r_2 = 0.08$ for o-chlorobenzoyl peroxide-
initiated reactions in bulk at 60°C (96). Low polymers of
allyl myristate and of allyl palmitate showed value as
pour point depressants for certain types of oils (97).
The relative ease of esterifying long-chain acids by allyl
alcohol has led to a technique for identifying acids from
natural oils, fats, and waxes by paper or layer chromatography of the allyl esters (98). The esters were reacted
with mercuric acetate and then spotted on paper that was

treated with undecane-acetonitrile as the mobile liquid
phase. The chromagraphs were sprayed with diphenyl car-
bazone and developed with ammonia, or else mercury was
determined quantitatively. A number of allyl esters of
long-chain acids (including undecanoic, 10,11-dibromo-
undecanoic, and 12-hydroxstearic acids) did not homopolymerize or copolymerize readily with styrene or methyl
methacrylate (99). However, they did copolymerize with
vinyl chloride by aqueous persultate emulsion. For
example, allyl 11-iodoundecanoate (M_1) and vinyl chloride
(M_2) gave $r_1 = 0.42$ and $r_2 = 1.64$ (azobis in benzene at
60°C). The pressure of 8 atm was suggested as contributing
to the successful copolymerizations with vinyl chloride.
   Neher and co-workers prepared copolymers of allyl esters
of long organic acids by use of massive amounts of peroxide catalysts (100). The copolymers gave promise as
pour point depressants for lubricating oils. For example,
a mixture of 162 g allyl stearate, 120 g laurate, and 10 g
lauroyl peroxide was added in portions to a reaction flask
swept by nitrogen and heated at 110°C. At 10-min intervals,
lots of about 17.5 g of monomer-catalyst mixture were added.
After 3 hr, additions were made hourly along with 0.5-g
portions of additional peroxide. After 7.75 hr total
polymerization time, the product was dissolved in 250 g of
toluene. Conversion to copolymer of 93% was obtained. A
30% copolymer solution in toluene was observed to have a
viscosity of 8 centistokes at 100°F. Lubricating oil was
added and the toluene distilled off. A concentrate of 37%
copolymer in oil was added in small amounts to motor oils.
The following copolymers also were tested as pour point
depressants:

   1. A stearate - A butyrate (1:3.09 molar, mol. wt. about 1000)
at 0.5% in oil gave pour point of -5°F.
   2. A stearate - A laurate (1:6 molar) gave at 2% in oil a pour
point of -20°F.
   3. A stearate - A myristate (1:1.5 molar) showed at 0.5% in oil
a pour point of -5°F.

Allyl esters of epoxidized long-chain acids have been
prepared and tested as comonomers for "built-in" stabilization of vinyl chloride and other halogen-containing
copolymers. For example, allyl-9,10,12,13-diepoxystearate
was prepared by warming allyl alcohol, the acid, and peracetic acid in toluene (101). An allyl dimethyl epoxy-
hexanoate had a pungent fruity odor (102). An allyl
trimethyl-5-hexenoate was epoxidized in ethyl acetate
solution during 8 hr at 40°C (103). Sodium salts of allyl
2-sulfopalmitate and allyl 2-sulfostearate were prepared
by esterification with allyl alcohol (104). These surface-
active monomers were relatively stable against hydrolysis.

Polymers of about DP 10 were made in persulfate aqueous
systems heated for 8 hr at 70°C. These low polymers
formed stable aqueous dispersions.

MONOALLYL ESTERS OF OTHER ACIDS

Allyl glycolate was polymerized by Adelson and Dannenberg to a viscous oil by long heating in air at 130°C
(105). The polymer was somewhat soluble in hot water and
gave turbidity on cooling aqueous solutions. Allyl glycolate containing 1.6% di-t-butyl peroxide was heated at
180°C for 1 hr, giving 76% conversion to polymer of
viscosity greater than 10 poises. These polymers were
said to have DP 4 to 20. An allyl glycolate polymer was
hydrolyzed in boiling water for 5.5 hr to form a yellow,
viscous liquid allyl alcohol polymer that was soluble in
water. An allyl acetate low polymer did not hydrolyze
readily by this treatment. A viscous semisolid polymer
was formed by heating allyl salicylate for 75 hr at 130°C.
Allyl acetoxy glycolate (102°C at 21 mm and n_D^{20} = 1.4318)
was prepared by portionwise addition of the acyl chloride
to allyl alcohol (105):

$$\begin{matrix} CH_2OCOCH_3 \\ | \\ CH_2COCl \end{matrix} + AOH \longrightarrow \begin{matrix} CH_2OCOCH_3 \\ | \\ CH_2COOA \end{matrix} + HCl$$

Viscous liquid polymers were obtained by long heating at
80°C with 3% BP. Further heating of films with peroxide
followed by cooling to room temperature gave a nonflowing
solid polymer. A polymer of allyl acetoxy glycolate made
by heating at 206°C with 1.0% di-t-butyl peroxide for
105 min. was found to have DP 7 by boiling point rise in
toluene. Viscous liquid polymers of allyl lactate also
were prepared. Allyl esters of naphthenic acids (acid
number ca. 165) were polymerized to viscous liquids of
DP about 4 by prolonged heating with peroxides at high
temperatures (106).

Allyl acetoacetate and derivatives can be prepared from
reaction of allylic alcohols with diketene (75). The
allylic esters of acetoacetic ester $CH_3C(O)CH_2COOCHRCR=CHR$
can undergo Cope rearrangement to form gamma, delta-
unsaturated ketones which are intermediates for synthesis
of perfume components, such as linalool, as well as
vitamin A. Diketene, for which $\begin{matrix} CH_2=C-CH_2 \\ | \quad | \\ O-C=O \end{matrix}$ and other

formulas have been proposed, does not behave as an allyl
compound and may be stored as a solid below its melting
point of 6.5°C in order to prevent spontaneous exothermic
polymerizations.

Allyl acetoacetate polymer of DP 8 to 10 containing a
little ferric acetylacetonate gave clear hard films by

baking at 120°C (107).   Jones prepared interpolymers from
4 parts of acrylamide with 2 parts of allyl acetoacetate,
some derivatives of which showed thermally reversible
gelation in water (108).   Copolymerization was carried out
in methanol solution under nitrogen by exposing to UV
light for 20 hr.   The interpolymer precipitated during the
reaction and could be recovered by filtration.   A derivative was formed by reaction of 25 parts of a 10% solution
of the polymer in water with 2 parts of a 2% solution of
adipohydrazide, acidified by acetic acid.   There was formed
a gel having melting point of 55°C and setting temperature
of 42°C.   Allyl acetoacetate acted as a telomerizing agent
in vinyl acetate polymerization at 180°C with t-butyl
peroxide in cumene (109).   Polymer of molecular weight 1460
contained approximately one telogen group per macromolecule.
Crosslinking of copolymers of allyl esters of β-ketoacids
and vinyl monomers was accomplished by reaction with formaldehyde and a polyamine (110).   Thus 2.8 g of a solution
of 30% HCHO in methanol and 1.6 g of hexamethylene diamine
were added with vigorous stirring to a 40% solution in
methanol of a copolymer of vinyl acetate-allyl acetyl
acetate (10:1).   Crosslinking occurred at room temperature
to a solid yellowish resin said to be stable at 150°C.

Allyl chloroacetate gave homopolymers of about 990 mol.
wt. using 0.4 mole % peroxide as initiator (111).   Copolymers of acrylonitrile with minor proportions of allyl
chloroacetate in dimethyl formamide solution were quaternized by reaction with trimethylamine; then they were
wet-spun to form dyeable fibers (112).   Allyl chloroacetate copolymerizations have been studied with vinyl
acetate (113), methacrylic anhydride (114), and butadiene
(115).   Allyl chloroacetate was a comonomer in the synthesis of acrylic-vulcanizable elastomers having a good
balance of properties including low temperature and oil
resistance (116).   In one example of aqueous emulsion copolymerization, para-menthane hydroperoxide, sodium
formaldehyde sulfoxylate, and an Igepal phosphate ester
were used at 30°C.   The monomer mixture consisted of 50%
n-butyl acrylate, 45% methoxyethyl acrylate, and 3% allyl
chloroacetate.   Allyl trifluoroacetate, boiling at 86°C,
is available from Pierce Chemical Company.

The literature contains scattered references to monoallyl esters of polyfunctional organic acids but none of
these has become important.   For example, monoallyl
chlorendate (mp 152°C) was obtained by reacting chlorendic
anhydride with a molar proportion of allyl alcohol at room
temperature (117).   The monoallyl tetrachlorophthalate
melted at 104°C.   Monoallyl trimellitate (mp 215°C) was
made by heating trimellitic anhydride with allyl alcohol
in methyl isobutyl ketone solution.   A copolymerization

was carried out by gradually adding a solution of 80 parts
of vinyl acetate, 20 parts of monoallyl succinate, and 2
parts of BP during 2 hr to 100 parts of isopropanol maintained at reflux temperature. Refluxing was continued for
10 hr, and then isopropanol was distilled off. Ammonium
hydroxide was added to give a 50% solution of the copolymer salt in water, and this dispersion with added hexa-
methylol melamine was tested as a water paint. The
coatings could be cured by heating at 300°F for 30 min.

   Allyl alkyl esters of o-phthalic acid have been made by
reacting phthalic anhydride with allyl alcohol or with
allyl chloride (118). Monoallyl mixed esters have been
formed by reaction of chloroformates with esters of lactic
acid (119). Monoallyl malonates have been used to prepare
polyesters bearing free allyl groups (120). Monoallyl
succinate was copolymerized with acrylate or vinyl esters
(121). The ammonium salts of the copolymers were used
with crosslinking resins such as hexamethylolmelamine for
coatings. Monoallyl epoxyalkyl esters of polyfunctional
acids have been prepared but little has been reported on
their polymerizations. However, allyl 2,3-epoxypropyl
maleate and phthalate have been prepared and polymerized
(122). References to a number of other monoallyl esters
are given in Table 8.4.

   Natta and co-workers reported that stereoregular crystalline polymers could be prepared from allyl esters of
unsaturated acids by use of catalysts such as butyl
lithium, phenyl magnesium bromide, or $LiN(C_2H_5)_2$ (123).
Allyl sorbate was one of the compounds studied. Allyl
sorbate as well as allyl-3,5-hexadienoate had been polymerized earlier to fusible low polymers by heating with
peroxide catalyst followed by crosslinking on heating in
air with cobalt resinate catalyst (124).

   Warson and co-workers (125) made allyl esters of epoxy
acids by the Darzens reaction, the first method used to
make epoxy compounds in 1904 (126). In one example, allyl
alcohol and monochloracetic acid were esterified using
p-toluene sulfonic acid as catalyst. The redistilled
allyl monochloroacetate was condensed with acetone using a
dispersion of NaH in benzene as catalyst at 5 to 10°C.
There was recovered allyl β,β-dimethyl glycidate (bp 77°C
at 7 mm and n_D^{25} = 1.4378). This monomer was heated at
95°C for 6 hr with addition of 5% BP in 5 increments,
giving 56% conversion to solid yellow polymer. The
polymer retained epoxy groups through which it could be
crosslinked by heating at 150°C in the presence of 1%
H_3PO_4 or 2% maleic anhydride. Liquid low polymers were
converted to amber sticky solids by reaction with diethylene triamine.

TABLE 8.4

Miscellaneous Monoallyl Esters

Ester	Remarks	References
Allyl lactate	Ethyl acrylate copolymers	Fisher, CA <u>43</u>, 5988 (1949); Rehberg, Org. Syn. Coll. III, 46
Allyl levulinate	Low homopolymers	Adelson, U.S. 2,475,273 (Shell)
Allyl sulfosucci- nates	Copolymerizable emulsifiers	Ingleby, U.S. 3,219,608
Allyl ethyl carbonate	Acrylonitrile copolymers	Bier, CA <u>53</u>, 1849 (1959)
Allyl rosinates	And derivatives; maleic anhydride copolymers	Gould, U.S. 2,639,273 (Hercules)
Allyl bicyclo- heptene car- boxylates	Fusible polymers curing above 250°C	Schweiker, U.S. 3,164,573 (Velsicol)
Allyl 2,3-epoxy- butyrate	--	Stevens, U.S. 2,680,109 (PPG)
Allyl 3,4-epoxy-2- hydroxy butyrate	--	Kilsheimer, Brit. 856,643; CA <u>55</u>, 13922 (1961)
Allylcyclohexyl propionate	Commercial perfume	--
Allyl cinerin or allethrin	Allylcyclopente none ester of chrysanthemum acid used as insecticide	Stansbury, U.S. 2,768,965; Sanders, Ind. Eng. Chem. <u>46</u>, 414 (1954)

Allyl esters that bear a second reactive allyl group in the molecule can form solid crosslinked polymers by heating with peroxide or azo catalysts. An example is allyl β-allyloxypropionate, made by reacting allyl alcohol and sec-butyl acrylate in presence of sodium alcoholate (127). This reaction required 16 hr at 27°C or 4 hr at reflux in the presence of an inhibitor of radical polymerization. The $AOOCCH(OA)CH_3$ polymerized on heating with BP, giving insoluble, hard, colorless coatings. The monomer was less effective than allyl methacrylate as crosslinking comonomer with acrylate esters.

ESTERS OF SUBSTITUTED ALLYLIC ALCOHOLS

Copolymerization reactivity ratios of 2-chloroallyl esters by free radical initiation are unfavorable with most ethylenic monomers. With chlorallyl acetate M_1 the

reluctance to copolymerize is shown by the following reactivity ratios (128):

r_1		r_2
0	Maleic anhydride	0
0	Methyl acrylate	0.7
0	MMA	1.0
0	Styrene	4.1
0	Vinyl chloride	1.16

However, copolymerizations have been made with vinyl
esters (129), with conjugated dienes (130), and with
other monomers (131). Tribromoallyl esters of different
organic acids have been copolymerized with styrene in bulk
at 120°C for evaluation as flame-resistant molding plastics. Some of the esters were too unstable for conventional aqueous polymerizations. Polymers from such
chlorine- and bromine-containing monomers generally discolor rapidly in sunlight.
  Besides the esters of allyl and methallyl alcohol with
organic acids, esters have been made with many of the
substituted alcohols containing allylic double bonds discussed in Chapter 5. Few of these have been studied in
polymerization or copolymerization. Some examples of
such allylic compounds are listed in Table 8.5.
  Esters of linalool and of eugenol occur widely in
natural and synthetic fragrances. They have very low
toxicities as shown in the following values of LD_{50} reported for rats by Jenner and co-workers of the Food and
Drug Administration (132):

Linalool (allylic alcohol)	2,790
Linalyl acetate	14,550
Linalyl isobutyrate	>36,300
Linalyl cinnamate	9,960
Eugenol (allylic phenol)	2,680
Isoeugenol	1,560
Eugenyl acetate	1,670
Eugenyl methyl ether	1,560

Linalyl acetate (bp 220°C and n_D^{20} = 1.4460) is the most
valuable constituent of bergamot and lavender oils.
Theories concerning the structure of linalool derivatives
and related terpenes have been reviewed recently (133).
In preparation of esters from terpene alcohols precautions
must be taken to avoid undesired rearrangements. For
example, linalyl acetate was prepared from reaction of
linalool, acetic anhydride, and pyridine (134). Without
pyridine only geranyl acetate could be recovered.

TABLE 8.5

Esters of Some Other Allylic Alcohols

Ester	bp, °C at mm	Refractive index	References
$CH_2=CClCH_2OCOCH_3$	68(33)	$n_D^{20}=1.4285$	Fr. 1,378,346 (Ciba); CA 62, 7645 (1965)
$CH_2=CClCH_2OCOC_2H_5$	62(14)	--	Fr. 1,378,346 (Ciba); CA 62, 7645 (1965); also made succinates and phthalates
$CH_2=CHCH(C_2H_5)OCOCH_3$	133	$n_D^{20}=1.4138$	Meisenheimer, Ann. 479, 261 (1930); Smets, CA 8316 (1950)
$CH_2=CHCH_2CH_2OCOCH_3$	151	$n_D^{19}=1.4176$	Paul, Compt. Rend. 193, 599 (1931); Schniepp, JACS 67, 55 (1945)
$CH_2=C(CH_3)CH_2CH(CH_3)OCOCH_3$	50(15)	$n_D^{20}=1.4201$	Duveen, J. Chem. Soc. 1453 (1936); Young, Ibid. 1452 (1938)
$CH_2=CHCH(CH_3)OCOC_6H_5$	191(13)	$n_D^{14}=1.5115$	Vavon, Compt. Rend. 154, 1706 (1912)
$CH_2=CHCH_2CH(C_2H_5)OCOC_6H_5$	261 or 192(110)	--	Fournier, Bull. Soc. Chim. [3] 15, 885 (1896

Normal esters can be obtained by reaction of allylic
alcohols with acyl chlorides in presence of pyridine or
other t-amine. Any liberated excess HCl may cause isomerization. Labile allylic alcohols may dehydrate in the
presence of pyridinium ions (135). Improved yields were
obtained by carrying out acetylations in nonpolar solvents
such as pentane or benzene, whereby pyridine hydrochloride
precipitated as reaction progressed. Optically active
substituted-allyl alcohols have been acetylated by acyl
chlorides in pyridine without rearrangement so that the
configurations were retained (136).

ALLYLIDENE DIESTERS

Allylidene diacetate can be prepared by reaction of
acrolein with acetic anhydride in the presence of a catalytic amount of sulfuric acid (137). Allylidene diacetate
and methallylidene diacetate are used commercially as
sources of acrolein and methacrolein in presence of water:

$$CH_2=CHCH(OOCCH_3)_2 + H_2O \longrightarrow CH_2=CHCHO + 2CH_3COOH$$

Because of the extremely unpleasant physiological effects

of handling and storing the acrylic compounds acrolein and methacrolein, the application of the allylidene diesters is more favorable. These diesters are toxic, however. Allylidene diacetate is a potent fungicide with relatively low phytotoxicity. Properties of the compounds supplied by Union Carbide are as follows:

Properties	Allylidene diacetate	Methallylidene diacetate
Boiling point, °C	77(12)	191 or 82(10)
Freezing point, °C	-37	-15
Solubility in water, %	1.8	0.7
Water solubility in, %	1.4	1.2
n_D	1.4170(30.5)	1.4250(20)
Specific gravity 20/20	1.0749	1.510
Viscosity at 20°C, cp	--	3.7

Allylidene diacetate may also be used as an acetylating agent. Starches were partially acetylated in alkaline medium (138).

Allylidene diacetate does not homopolymerize very readily by radical initiation, but the addition of potassium azodisulfonate in the presence of water is said to produce polymer at 0 to 30°C (139). This may have been a polymer of acrolein in part. Allylidene diacetate (ADA) was homopolymerized by heating with azobis (140). Copolymerization ratios for ADA-vinyl chloride were $r_1 = 0.58$, $r_2 = 1.75$. The Q and e values of ADA were 0.26 and 0.20, respectively.

Allylidene diacetate deaerated in sealed tubes at 21°C was exposed to gamma rays from [60]Co (141). Conversion of 63% to polymer of molecular weight 1800 to 2700 was obtained. The polymer was soluble in acetone and had a melting point near 140°C. It was converted to an aldehyde polymer. Izard studied copolymers of vinyl acetate with minor proportions of ADA made by free radical suspension at 90°C for 8 hr (142). Alkaline alcoholysis of these gave water-woluble vinyl alcohol-acrolein copolymers, which could be crosslinked in place by heating alone or with small amounts of acid catalysts. The dried polymers did not dissolve in cold water but did dissolve in hot water. These modified polyvinyl alcohols may be useful in finishing fabrics (143) and for wet-strength paper. Polymeric pyrazolone dyestuffs were made using ADA copolymers (144). The vinyl alcohol copolymers were reacted with formaldehyde to give modified polyvinyl formal sponges (145). Saponified copolymers of styrene-allylidene diacetate (146) as well as vinyl chloride copolymers were made in Japan. Other copolymers of allylidene di-

esters have been studied (148).   Chloroallylidene diacetate has been examined as minor comonomer with vinyl
chloride (149).   The copolymerizations were carried out
in acetone with diacetyl peroxide as initiator at 40°C
for 66 hr under nitrogen atmosphere.

References

1.   P. M. Jenner et al (Food & Drug Admin.), Food Cosmetic Toxicol.
     2, (3), 327 (1964); CA 62, 4516 (1964); cf. ibid. 62, 5789 (1964).
2.   G. M. Koons, Gettysburg College, unpublished.
3.   N. Zinin, Ann. 96, 361 (1855).
4.   V. Deulofeu, CA 22, 4104 (1928).
5.   P. D. Bartlett and R. Altschul, J. Am. Chem. Soc. 67, 815 (1945).
6.   E. Charon, Ann. Chim. [7] 197 (1899); L. W. J. Newman and H. N.
     Rydon, J. Chem. Soc. 261 (1936).
7.   E. A. Braude, Quart.(London) Rev. 4, 404 (1950) (rearrangements
     reviewed).
8.   M-D. Lee et al (Taiwan), CA 63, 11285 (1965).
9.   J. E. Lloyd, U.S. 3,574,717 (ICI).
10. R. H. DeWolfe and W. G. Young, Chem. Rev. 56, 819 (1956).
11. cf. W. Kimmel and A. C. Cope, J. Am. Chem. Soc. 66, 1613 (1944).
12. F. C. Frostick and B. Phillips, Brit. 863,445 (Carbide); J. Org.
     Chem. 25, 1420 (1960); U.S. 2,927,931 (Carbide).
13. F. C. Frostick and B. Phillips, U.S. 2,786,068 (Carbide).
14. Netherlands Appl. (ICI), CA 64, 4949 and 17430 (1966); cf.
     Brit. 969,162.
15. H. Holzrichter et al, Fr. 1,346,219 (Bayer); cf. Ger. Offen.
     2,038,120 (Hoechst), CA 76, 112730 (1972).
16. H. Fernholz et al, U.S. 3,670,014 (Hoechst); cf. C. N. Winnick,
     Ger. Offen. 2,120,715 (Halcon); CA 76, 100313 (1972).
17. Fr. 1,367,291 (ICI); CA 62, 1572 (1965).
18. W. C. Baird, Fr. 1,443,087 (Esso); CA 66, 37451 (1967).
19. M. Sheshukov, J. Russ. Phys. Chem. Soc. 16, 478 (1884).
20. H. Chafetz, U.S. 3,079,429 (Texaco).
21. W. Kirmse and H. Dietrich, Ber. 98, 4027 (1965).
22. J. G. Hanna and S. Siggia, Anal. Chem. 37, 690 (1965).
23. R. W. Rimmer, Belg. 633,648 (Du Pont); CA 61, 575 (1964).
24. H. C. Brown and O. J. Cope, J. Am. Chem. Soc. 86, 1791 (1964).
25. R. Louw and E. C. Kooyman, Rec. Trav. Chim. Pays-Bas 84, 1511
     (1965).
26. C. U. Pittman and S. P. McManus, Tetrahedron Lett. 339 (1969);
     J. W. Larsen et al, ibid. 539 (1970).
27. H. M. R. Hoffmann et al, J. Chem. Soc. B 57 (1968).
28. P. D. Bartlett and R. Altschul, J. Am. Chem. Soc. 67, 812 (1945);
     cf. D. E. Adelson and H. F. Gray, U.S. 2,512,410 (Shell).
29. P. D. Bartlett and R. Altschul, J. Am. Chem. Soc. 67, 816 (1945).
30. R. D. Bartlett and K. Nozaki, J. Polym. Sci. 3, 216 (1948).

31. I. Sakurada and G. Takahashi, CA 51, 3255 (1957); CA 52, 1670
     (1950), cf. CA 50, 601 (1956).
32. N. G. Gaylord and F. R. Eirich, J. Am. Chem. Soc. 74, 334 (1952);
     cf. Gaylord et al, J. Polym. Sci. B 2, 151 (1964).
33. M. Litt and F. R. Eirich, J. Polym. Sci. 45, 379 (1960).
34. P. D. Bartlett, G. S. Hammond, and H. Kwart, Discussions Faraday
     Soc. No. 2, 342-352 (1947); CA 43, 5740 (1949).
35. G. S. Hammond and P. D. Bartlett, J. Polym. Sci. 6, 617 (1951);
     CA 45, 6867 (1951).
36. C. Walling and J. Pellon, J. Am. Chem. Soc. 79, 4782 (1957);
     J. Polym. Sci. 48, 335 (1960).
37. P. D. Bartlett and F. A. Tate, J. Am. Chem. Soc. 75, 91 (1953);
     cf. D. G. F. James et al, J. Polym. Sci. A 3, 75 (1965).
38. W. Simpson et al, J. Polym. Sci. 10, 492 (1953).
39. F. M. Lewis and F. R. Mayo, J. Am. Chem. Soc. 76, 457 (1954).
40. V. A. Kargin et al, Polym. Sci. USSR 9, 327 (1967); cf. V. F.
     Kulikova et al, CA 66, 76339 (1967); V. P. Zubov et al, J.
     Polym. Sci. C No. 23, 147 (1968).
41. J. Mleziva et al, Dtsch. Farben Z. 21, 119 (1967).
42. D. E. Adelson and T. W. Evans, U.S. 2,473,124 (Shell).
43. P. Lagally, U.S. 3,108,920 (Brockway Glass).
44. D. E. Adelson et al, U.S. 2,402,481-6 (Shell).
45. R. R. Whetstone, U.S. 2,476,936 (Shell).
46. J. W. Hafey and C. E. Schildknecht, unpublished.
47. C. E. Schildknecht, D. Walborn, and E. Hagamann, unpublished.
48. I. Goodman and J. Mather, Brit. 854,207 (ICI); CA 55, 11920
     (1961).
49. P. D. Bartlett and K. Nozaki, J. Am. Chem. Soc. 68, 1495 (1946).
50. G. F. D'Alelio, U.S. 2,308,495 (GE); Brit. 727,476.
51. E. P. Irany, U.S. 2,557,189 (Celanese).
52. cf. G. S. Ronay and J. R. Vinograd, U.S. 2,663,701 (Shell).
53. S. Fukioka et al, J. Chem. Soc. Japan (Ind. Chem.) 69, 334 (1966).
54. E. W. Moffett and R. E. Smith, U.S. 2,424,838 (PPG); H. Rudolf,
     U.S. 2,611,754 (Shawinigan); S. Matsumoto et al (Kurashiki),
     CA 52, 9669.
55. L. J. Young, J. Polym. Sci. 54, 411 (1961); cf. G. M. Burnett
     and W. W. Wright, Proc. Roy. Soc. (London) A221, 41 (1954).
56. S. Kuehne, U.S. 3,230,203 (Hoechst).
57. L. J. Young, J. Polym. Sci. 54, 411 (1961).
58. M. Matsumoto et al, Japan. 6485 (1955); CA 52, 5848 (1958).
59. G. F. E. M. Dierick et al, U.S. 2,784,136 (Shell).
60. L. J. Young, J. Polym. Sci. 54, 411 (1961).
61. G. I. Nikishin (Moscow), CA 55, 22120 (1961).
62. W. Wilson and H. May, Brit. 1,022,563 (Brit. Ind. Plastics).
63. H. Gilbert et al, U.S. 2,650,911 (Goodrich).
64. P. Weiss et al, J. Polym. Sci. 35, 343 (1959).
65. V. Stannett et al, J. Polym. Sci. A 3, 3763 (1965); CA 66,
     38733 (1967).
66. E. T. Cline and D. Tanner, U.S. 3,068,122 (Du Pont).
67. E. P. Irany and I. Skeist, U.S. 2,557,189 (Celanese).

68. Brit. 1,134,368 (Du Pont); CA 70, 30194 (1969).
69. F. A. Bent, U.S. 2,541,148 (Shell).
70. T. W. Evans, U.S. 2,524,563 (Shell).
71. I. Sakurada and G. Takahashi (Kyoto), CA 50, 6892 (1956).
72. F. J. F. van der Plas, Ger. 967,485 (Shell); CA 54, 13746 (1960).
73. E. M. Evans and J. E. S. Whitney, Brit. 691,038 (Brit. Resin
    Prods.); CA 47, 10899 (1953).
74. C. E. Rehberg, Org. Syn. Coll. 3, 46, Wiley, 1955.
75. C. R. Noller, Chemistry of Organic Compounds, Saunders, 1965.
76. C. D. Hurd, J. Am. Chem. Soc. 61, 1156 (1939); cf. Swedlund, J.
    Chem. Soc. 131 (1945) and U.S. 2,207,611 (Dow).
77. E. P. Irany, I. Skeist, and V. F. Maturi, U.S. 2,557,189
    (Celanese).
78. J. B. Gardner et al, U.S. 3,386,971 (Dow).
79. F. Weiss et al, CA 63, 14748 (1965).
80. J. J. Giammaria, U.S. 2,615,864 (Socony).
81. C. T. Handy and H. S. Rothrock, U.S. 2,883,361 and 2,933,474
    (Du Pont).
82. I. Sakurada and G. Takahashi, CA 50, 601 (1956).
83. J. J. Giammaria, U.S. 2,615,864 (Socony).
84. M. Dunkel, U.S. 3,663,601 (Universal Oil Products).
85. M. Skoultchi and E. A. Meier, U.S. 3,666,732 (Nat. Starch).
86. A. Seldner, Am. Perfumer 54, 295 (1949); CA 44, 13675 (1950);
    cf. Polysciences Inc. lit., 1971.
87. D. Swern and E. F. Jordan, U.S. 2,541,126 (USA).
88. D. Swern and E. F. Jordan, J. Am. Chem. Soc. 69, 2440 (1947);
    70, 2334 (1948); cf. Price and Kapp, U.S. 2,341,060; Coleman
    and Hadler, U.S. 2,208,960.
89. E. A. Harrison and D. H. Wheeler, J. Am. Chem. Soc. 73, 839
    (1951).
90. J. L. Jones, U.S. 2,533,376 (L-O-F); cf. U.S. 2,698,298.
91. J. J. Giammaria, U.S. 2,616,852 and 2,698,298 (Socony); cf.
    Brit. 816,580 (Cal. Res.); CA 54, 2729 (1960).
92. D. Swern et al, J. Am. Oil Chemists Soc. 27, 281 (1950);
    CA 44, 8162 (1950).
93. J. R. Kilsheimer and B. R. Thompson, Ger. 1,025,148 (Carbide);
    CA 54, 23445 (1960).
94. E. Dyer and W. C. Meisenhelder, J. Am. Chem. Soc. 73, 1434 (1951).
95. T. W. Evans and R. Whetstone, U.S. 2,524,563; R. G. Larsen and
    K. E. Marples, U.S. 2,441,023 and 2,541,590 (Shell); H. T.
    Neher et al, U.S. 2,600,419-21; cf. U.S. 2,600,446-8 and
    2,600,382-3 (R & H).
96. I. Sakurada et al, CA 51, 2652 (1957).
97. J. M. Butler, U.S. 2,541,686-7 (Monsanto).
98. H. P. Kaufman and J. Pollerberg, Lab. Sci. (Milan) 11, 21-33
    (1963).
99. Roberta C. L. Chow and C. S. Marvel, J. Polym. Sci. A1 6,
    1515 (1968).
100. H. T. Neher, L. N. Bauer, and W. L. VanHorne, U.S. 2,600,619)
    (Röhm & Haas).

101. B. Phillips and F. C. Frostick, U.S. 2,779,771 (Carbide); cf.
      Brit. 789,034 (Carbide).
102. J. Levy and R. M. Lusskin, U.S. 2,889,340 (Trubek).
103. J. W. Lynn, J. Org. Chem. 26, 1284 (1961).
104. R. G. Bistline et al, U.S. 2,844,606 (USA); J. Am. Oil Chemists
      Soc. 33, 44 (1956); CA 50, 3782 (1956).
105. D. E. Adelson and H. Dannenberg, U.S. 2,464,741 (Shell); cf.
      U.S. 2,503,699 (Shell).
106. D. E. Adelson and H. Dannenberg, U.S. 2,482,606 (Shell).
107. J. W. Coates and W. C. Crawford, Brit. 582,899-900 (ICI).
108. G. D. Jones, U.S. 2,480,810 (GAF); cf. Hoover, U.S. 2,933,475
      (Du Pont) (chelates with beryllium or aluminum).
109. A. Kuehlkamp and G. Werner, U.S. 3,888,110 (Hoechst).
110. Neth. Appl. (Hoechst), CA 63, 3131 (1965).
111. I. Sakurada and G. Takahashi, CA 50, 601 (1956).
112. G. E. Ham et al, Ind. Eng. Chem. 45, 2323 (1953); U.S.
      2,656,326; 2,696,483; 2,719,834; 2,775,573 (Chemstrand); cf.
      Ital. 557,963 (Monte), CA 53, 1766 (1959).
113. W. Kunze and K. Roehrich, Ger. 1,072,082 (Cassella); CA 55,
      11847 (1961).
114. J. C. H. Hwa and L. Miller, J. Polym. Sci. 55, 197 (1961).
115. A. M. Clifford, U.S. 2,448,703 (Wingfoot).
116. A. H. Jorgensen, U.S. 3,488,331 and 3,624,058 (Goodrich).
117. I. H. Evans, U.S. 3,293,325 (CIL).
118. Fr. 1,396,837 (Ciba); CA 63, 827 (1965).
119. C. E. Rehberg et al, J. Org. Chem. 13, 254 (1948).
120. A. I. Tarasov et al, CA 68, 6575 (1968).
121. I. H. McEwan, Can. 729,729; CA 65, 4110 (1966).
122. Brit. 638,516 (Shell); CA 44, 8165 and 1532 (1950).
123. G. Natta, M. Donati, and M. Farina, Belg. 620,901 and
      621,007 (Monte).
124. M. L. A. Fluchaire and G. Collardeau, Fr. 965,773-4 (Rhône-
      Poulenc); CA 46, 2847 and 2850 (1952).
125. H. Warson, R. J. Parsons, and L. A. Simmonds, Brit. 995,726
      (Vinyl Products).
126. G. Darzens, Compt. Rend. 139, 1214 (1904).
127. C. E. Rehberg and C. H. Fisher, U.S. 2,504,151 (USA); A.C.S.
      Preprint, Paint, Varnish, & Plastics Chem., September 1945.
128. J. Brandrup and E. H. Immergut, Polymer Handbook, Interscience,
      1966.
129. W. Kunze, Ger. 965,966 (Cassella); CA 53, 10790 (1959).
130. A. M. Clifford, U.S. 2,448,703 (Wingfoot).
131. P. O. Tawney, U.S. 2,560,495 (U.S. Rubber).
132. P. M. Jenner et al, Food Cosmetic Toxicol. 2 (3), 327-43 (1964);
      CA 62, 4516 (1965).
133. D. V. Banthorpe et al, Chem. Revs. 72, #2, 121 (1972).
134. W. D. Young and I. D. Webb, J. Am. Chem. Soc. 73, 780 (1951).
135. J. J. Kenyon et al, J. Chem. Soc. 312 (1938); J. A. Mills,
      J. Chem. Soc. 2332 (1951).
136. H. L. Goering et al, J. Am. Chem. Soc. 77, 4042 and 6249 (1955).

137. H. Huebner and A. Genther, Ann. <u>114</u>, 47 (1860); A. Wohl and
      R. Maag, Ber. <u>43</u>, 3293 (1910); C. W. Smith et al, U.S. 2,575,896
      (Shell); A. Kirrman, Bull. Soc. Chim. [5] <u>5</u>, 256 and 915 (1938).
138. Brit. 912,387 (Staley); CA <u>58</u>, 5876 (1963).
139. J. A. Robertson, U.S. 2,468,111 (Du Pont).
140. T. Oota et al, Kogyo Kagaku Zasshi <u>71</u>, 736 (1968); CA <u>69</u>,
      67781 (1968).
141. Y. Toi and Y. Hachihama, CA <u>61</u>, 7108 (1964).
142. E. F. Izard, Ind. Eng. Chem. <u>42</u>, 2108 (1950); U.S. 2,485,239
      and 2,569,932 (Du Pont).
143. M. A. Dahlen and O. Shaus, Brit. 620,791.
144. D. M. McQueen, U.S. 2,477,462 (Du Pont).
145. C. L. Wilson, Ger. 1,117,304; CA <u>56</u>, 14484 (1962).
146. R. Oda and T. Saigusa (Kyoto); CA <u>51</u>, 4753 (1957).
147. T. Oota et al, J. Chem. Soc. Japan Ind. Chem. <u>71</u>, 736 (1968).
148. L. M. Minsk and C. C. Unruh, U.S. 2,471,404 and 2,443,167
      (Eastman).
149. Brit. 795,279 (Carbide); CA <u>53</u>, 771 (1959).

9. ALLYLIC ISOPROPENYL COMPOUNDS

A chapter is devoted to these isopropenyl or α-methyl-vinyl compounds which, in reluctance to polymerize as well as in structure, belong with the allyl compounds. They have special interest in demonstrating the relationships of reactivity to polarity of substituents. The isopropenyl compounds of the type $CH_2=C(CH_3)Y$, where Y is an ortho and para directing group in benzene substitution (electron repelling or halogen) and without conjugation, are the ones classified here as allyl compounds. Propylene and other alkenes which are allylic isopropenyl compounds have been discussed in Chapters 2 and 3. Isopropenyl esters, halides, and ethers are the most important other allylic isopropenyl compounds.

The chemistry of these isopropenyl compounds is in sharp contrast to that of the methacrylic monomers, in which the hydrogen atoms of the methyl group do not show allylic reactivity and the double bond is activated by electron-withdrawing groups such as CN, COOH, COOR, $CONH_2$, and CHO, which also have multiple bonds in conjugation with the ethylenic double bond. The methacrylic monomers, not discussed in this book, polymerize readily by radical and anionic methods; a large literature describes their technology (1). Besides the methacrylic compounds, some additional isopropenyl compounds, bearing other strongly negative groups such as $CH_2=C(CH_3)NO_2$, homopolymerize readily and do not behave as allyl compounds.

The reactivity of isobutylene or isopropenylmethane resembles somewhat that of allyl compounds in free radical systems. However, in contrast to monoallyl compounds, it forms homopolymers of high molecular weight readily by Lewis-acid initiators at low temperature, and its large literature (1) is not discussed here. The same may be said for alpha-methylstyrene and perhaps for isopropenyl-naphthalenes. Diisopropenyl or 2,3-dimethylbutadiene has resonance to enhance polymerization, and of course it does not behave as a diallyl compound.

Isopropenyl alcohol (the enol form of acetone) and iso-propenylamine, like the corresponding vinyl compounds,

TABLE 9.1

Hammett Sigma Values of Substituents
of Isopropenyl Compounds [a]

Compound	Substituent	Sigma Value
Isopropenylamine (unstable)	NH_2	-0.66
Isopropenyl alcohol (unstable)	OH	-0.37
Isopropenyl phenyl ether	OC_6H_5	-0.32
Isopropenyl methyl ether	OCH_3	-0.27
Isopropenyl ethyl ether	OC_2H_5	-0.24
Isobutene	CH_3	-0.17
2-Methyl-1-butene	CH_2CH_3	-0.15
α-Methylstyrene	C_6H_5	-0.01
Propylene	H	0
Isopropenyl methyl sulfide	SCH_3	0
N-isopropenyl acetamide	$NHCOCH_3$	0
Isopropenyl fluoride	F	+0.06
Methallyl chloride	CH_2Cl	+0.18
Isopropenyl mercaptan	SH	+0.15
Isopropenyl chloride	Cl	+0.22
Isopropenyl bromide	Br	+0.23
Isopropenyl iodide	I	+0.28
Isopropenyl acetate	$OCOCH_3$	+0.31
Isopropenyl thioacetate	$SCOCH_3$	+0.44
Methyl methacrylate	$COOCH_3$	+0.38
Methacrylic acid	COOH	+0.41
Isopropenyl phenyl ketone	COC_6H_5	+0.46
Isopropenyl methyl ketone	$COCH_3$	+0.50
Isopropenyl methyl sulfone	SO_2CH_3	+0.49
$CH_2=C(CH_3)CF_3$	CF_3	+0.54
Methacrylonitrile	CN	+0.66
$CH_2=C(CH_3)NO_2$	NO_2	+0.78
Isopropenyl trimethyl ammonium salts	$N(CH_3)_3^+$	+0.82

[a] J. Hine, Physics of Organic Chemistry, McGraw-Hill, 1962, and other references. (Based, for example, upon ionization constants of para-substituted benzoic acids.)

are not sufficiently stable to exist. In these cases the OH and NH_2 substituents are strongly electron repelling (i.e., have low Hammett sigma values, Table 9.1). Excessive electron donation apparently makes such ethylenic compounds unstable. Of the isopropenyl compounds showing allylic behavior, the most studied have been monoisopropenyl esters, isopropenyl halides, and monoisopropenyl ethers. None of these isopropenyl compounds forms high polymers very readily, and all copolymerize reluctantly with most vinyl-type monomers by conventional methods. Compounds having substituents with Hammett sigma values above about 0.4 and conjugation, such as methyl methacrylate (MMA), polymerize much more readily than typical monoallyl compounds.

Table 9.1 shows that the isopropenyl compounds behaving as allylic compounds generally have substituent groups of Hammett sigma values -0.30 to +0.31. Resonance explains why the polymerization reactivity of the methacrylic monomers is higher than can be attributed to sigma values alone. Other isopropenyl compounds expected to behave as allyl compounds include N-isopropenyl amides and isopropenyl sulfides. Just as polyvinyl alcohols can be made by saponification of polyvinyl acetate but not from vinyl alcohol, polymers approaching those from the nonexistent isopropenyl alcohol and isopropenyl amine can be made indirectly in some cases. Thus N-isopropenylamine polymer has been made by acid-catalyzed hydrolysis of ethyl-N-isopropenyl carbamate polymer (2).

ISOPROPENYL ACETATE

Isopropenyl acetate or α-methylvinyl acetate was made available first in 1949 by Tennessee Eastman and later by Union Carbide. It can be prepared from acetone and ketene in the presence of a little sulfuric or chlorosulfonic acid (3):

$$CH_2=\underset{\underset{OH}{|}}{C}-CH_3 \;+\; CH_2=C=O \;\xrightarrow{80°C}\; CH_2=\underset{\underset{CH_3}{|}}{C}OOCCH_3$$

Since the reaction is reversible, the monomer is unstable in the presence of acid catalysts.

Isopropenyl acetate is a pleasant smelling liquid of low toxicity. The following properties have been listed in a data bulletin of Union Carbide (June 1967):

Boiling point, °C	97.7 or 70.5 at 300 mm
Vapor pressure at 20°C, mm	30
Freezing point, °C	-93
Heat of vaporization, Btu/lb.	146
Coefficient of expansion	0.00128 per °C
n_D^{20}	1.4002
Viscosity at 20°C, cps	0.59
Solubility in water at 20°C	3.3%
Solubility of water in, 20°C	1.2%

Isopropenyl acetate is miscible with acetone, benzene, ether, heptane, methanol, and CCl₄. Isopropenyl acetate (IA) is useful as an acetylating agent where mild conditions are needed and when the presence of acetic anhydride would cause dehydration or other side reactions. Thus IA may be used to prepare acetate esters from t-alcohols, hydroxyesters, or hydroxynitriles where the coproduct acetone causes no difficulties. Isopropenyl acetate adds bromine or chlorine readily. It reacts with higher acids in the presence of mercuric acetate and/or sulfuric acid catalyst to give isopropenyl esters of higher acids (acidolysis) (4):

$$CH_2=C(CH_3)OOCCH_3 + HOOCR \longrightarrow CH_2=C(CH_3)OOCR + HOOCCH_3$$

Among the isopropenyl esters that have been available are the benzoate, n-butyrate, 2-ethylhexanoate, and 2-ethyl-butyrate. Isopropenyl benzoate has had some application as a benzoylating agent.

For preparing isopropenyl propionate, a mixture of 100 g IA, 38 g propionic acid, 1.0 g mercuric acetate and 0.275 g H_2SO_4 was heated for 1.5 hr on a steam bath. The mixture was filtered and washed with cold saturated Na_2CO_3. The dried organic liquid layer was then distilled giving iso-propenyl propionate. Phillips heated for 2 hr at 120°C a solution of 1080 parts of anhydride of 2-ethylhexanoic acid, 400 parts of IA, and 3 parts of sulfuric acid (4). After 5 parts of sodium acetate had been added, the mix-ture was distilled, giving 47% of isopropenyl 2-ethylhexa-noate. Relatively low yields of isopropenyl benzoate were reported. Isopropenyl stearate was made by reacting methylacetylene with stearic acid at 160°C and 500 psi with suitable precautions (5).

In contrast to vinyl acetate, which forms solid polymers of high molecular weight readily on heating with peroxides or by exposure to UV light, isopropenyl acetate forms only homopolymers of low DP. The methyl group participates strongly in chain transfer and apparently donates electron density, opposing the weak electron attraction of the acetate group. Under free radical conditions, the kinetics of formation of liquid low polymers (6) are similar to those described by Bartlett and co-workers for allyl acetate. Degradative chain transfer is dominant with a constant ratio of dM/dP, the ratio of change in monomer concentration to change of peroxide concentration (Fig. 9.1). Massive proportions of benzoyl peroxide were required to make solid polymers from IA (7). To 2000 g of monomer was added 1000 g of BP, and the mixture was heated under nitrogen at 60°C for 10 days. There was recovered 1225 g of brittle yellowish resin. It softened about 82°C and was soluble in alcohol. Heating the polymer for an hour at 210°C gave 33 g of liquid distillate consisting largely of acetic acid. Polymers from isopropenyl propio-

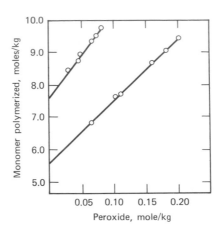

Fig. 9.1. Monomer polymerized plotted against peroxide concentration
in the bulk polymerization of isopropenyl acetate at 80°C: upper
curve, 2% BP; lower curve, 5% BP. N. G. Gaylord and F. R. Eirich, J.
Am. Chem. Soc. 74, 341 (1952).

nate, n-butyrate, and benzoate behaved similarly on heating. Heating isopropenyl acetate with t-butyl perbenzoate
at 96°C for 16 hr gave a low yield of polymer of number-
average molecular weight about 1500 softening at 81°C (8).
  Reding and co-workers prepared solid homopolymers from
isopropenyl acetate by operating at high pressures, but
the specific viscosity of the polymer was low (9). Polymerization was carried out in a tube of 5/16 in. diameter
containing 10 ml of IA and 0.05 g of BP. After 14 hr at
76,800 psi and 68 to 78°C, there was recovered 1.23 g of
solid, "glasslike" polymer. Its relative viscosity was
1.07 and the glass transition temperature was 45°C.
Chemists of the National Research Council of Canada made
polymers of IA of intrinsic viscosity of 0.1 (in acetone
at 25°C), and from these they prepared by saponification
methyl vinyl alcohol polymers having molecular weights
above 10,000 and melting above 100°C (10). Attempts to
form high polymers from IA by Lewis-acid catalysts at room
temperature and below gave only discolored liquid products
(11).
  Trifluoromethyl vinyl acetate (bp 85°C at 130 mm and
n_D^{25} = 1.3877) did not homopolymerize readily by radical
conditions, but it copolymerized with styrene, vinylpyridine, and vinyl acetate (12). Vinyl alcohol copolymers were prepared.

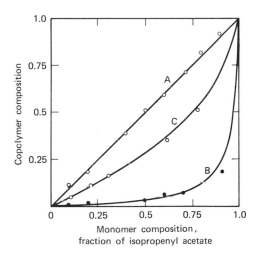

Fig. 9.2. Copolymerization of vinyl acetate with isopropenyl acetate at 65°C with 0.2% BP was close to the ideal case, where $r_1 = r_2 = 1$ and the proportion of monomer units in the copolymer was the same as that in the feed monomer (curve A). The copolymers of vinyl chloride with isopropenyl acetate were low in isopropenyl acetate units (curve C). Isopropenyl acetate and MMA gave products very low in isopropenyl acetate units (curve B). R. Hart and G. Smets, J. Polym. Sci. 5, 55 (1950).

Isopropenyl acetate copolymerizes readily with vinyl acetate but less readily with vinyl chloride and with methyl methacrylate (Fig. 9.2). Isopropenyl acetate co-polymerized in small proportions with ethylene, styrene, acrylonitrile, vinylidene chloride, vinyl chloride (13), vinyl acetate, maleate, and fumarate esters (14). With maleic anhydride, nearly equimolar interpolymers form readily (15). In reluctance to participate in free radical copolymerization with butadiene and other conjugated vinyl monomers, isopropenyl acetate resembles vinyl acetate. Isopropenyl acetate copolymerizes with vinyl esters using radical initiators, and the vinyl alcohol-isopropenyl alcohol copolymers obtained by saponification have been studied in Japan (16).

Properties of some isopropenyl esters are listed in Table 9.2.

ISOPROPENYL CHLORIDE

Isopropenyl chloride (IC) or 2-chloropropene, a liquid boiling near room temperature, is one of the products from

Table 9.2

Isopropenyl Esters

Ester	bp, °C	Refractive index	Reference
Propionate	114	n_D^{25} = 1.3990	Dickey, U.S. 2,646,437 (Eastman)
n-Butyrate	133	n_D^{26} = 1.4078	Dickey
Caproate	111(90)	n_D^{26} = 1.4173	Dickey
2-Ethylhexanoate	69(5)	n_D^{30} = 1.4213	Phillips, U.S. 2,466,738 (Carbide)
Crotonate	137(738)	--	Dickey
Sorbate	75(5)	n_D^{30} = 1.5021	Phillips
Benzoate	84(5)	n_D^{20} = 1.5165	Phillips

high-temperature chlorination of propylene (17). It has
been identified by an IR band at 11.38 microns. It can be
prepared from acetone and PCl_5 (18, 19) or from 1,2-di-
chloropropane and sodium ethylate (20). Like many halogen
compounds, IC has anesthetic action. Ionic addition of
HCl to allene gave not only normal addition to form IC but
also cis- and trans-1,3-dichloro-1,3-dimethylcyclobutanes
(21). Isopropenyl chloride added HCl in the dark with
$FeCl_3$ catalyst at 0°C, forming 2,2-dichloropropane (20);
with HBr it also gave normal addition (19). In the
presence of ascaridole as peroxide catalyst, HBr formed
largely $CH_2BrCH_2ClCH_3$ (20). Isopropenyl bromide adds HBr
easily (23). When heated with potassium hydroxide or triethylamine, it forms allene. Treatment of isopropenyl
bromide with mercuric acetate gives acetone. Isopropenyl
chloride and bromide are more potent than cyclopropane and
diethyl ether as anesthetics (24).

   Properties of some isopropenyl halides are given with
references in Table 9.3.

   Isopropenyl chloride as well as vinyl chloride formed
liquid peroxides when exposed to UV light under oxygen at
-15 to -20°C (25). Elimination reactions of IC and of
vinyl halides with alkali alkoxides were reviewed (26).
Isopropenyl chloride in small proportions has been copolymerized with vinylidene chloride (27).

   Isopropenyl bromide can be prepared by methods similar
to those used for IC, and it was also prepared by addition
of HBr to allene under free radical conditions (28). Thus
UV light promoted gas-phase reaction of equimolar quantities of allene and HBr, giving 2-bromopropene as the major
product. However, in the liquid state a range of products
was formed. Attack of the terminal position by bromine

Table 9.3

Isopropenyl Halides

	mp, °C	bp, °C	n_D^{20}	References
$CH_2=C(CH_3)Cl$	-138	23	1.3969	Ingold, J. Chem. Soc. 2744 (1931); Rogers, JACS 69, 1243 (1947)
$CH_2=C(CH_3)Br$	-126	49	1.4440	Kharasch, J. Org. Chem. 2, 300 (1937); ibid. 4, 431 (1939); Farrell, JACS 57, 1282 (1935); also Rogers; Linnemann, Ann. 138, 125; ibid. 143, 348
$CH_2=C(CH_3)I$	--	103	--	Oppenheim, Z. Chem. 719 (1865); Kahovec, Z. Phys. Chem. B, 47, 179 (1940) (Raman spectrum)

increased from 7% at room temperature to 24% at -40°C and 36% at -70°C. The proportion of terminal attack decreased with increasing concentration of initial allene. Isopropenyl bromide was made by the dehydrohalogenation of $CH_3CHBrCH_2Br$ by sodium phenate in ethanol (29). Also formed were 38% cis- and 18% trans- $CH_3CH=CHBr$. Kharasch and Fuchs observed the formation of alpha-methylstyrene and hydrocarbon polymer when phenyl magnesium bromide reacted with isopropenyl bromide in ether in the presence of $CoCl_2$ (30).

Aeltermann and Smets observed chain transfer in radical polymerizations of vinyl monomers with isorpopenyl chloride (31):

$$\dot{R} + CH_2=CClCH_3 \longrightarrow RH + C\dot{H}_2 \cdots \dot{C}Cl \cdots CH_2$$

The resonance-stabilized radical has low reactivity in continuing polymerization. For example, addition of IC markedly reduces the rates of radical polymerization of vinyl chloride (32).

Isopropenyl chloride and bromide have formed only homopolymers of comparatively low molecular weight. Otsu obtained very low molecular weight brown polymers from free radical and ionic homopolymerizations, but copolymerizations occurred on heating with peroxides in the case of the following comonomers (with r_1 and r_2 values, respectively) (33): vinyl acetate (1.84 and 0.22), vinyl chloride (5.21 and 0.18), acrylonitrile (0.01 and 1.40), and maleic anhydride (0.06 and 0.06). Isopropenyl chloride was more reactive in radical copolymerizations than

cis- and trans-propenyl chlorides (Figs. 9.3 and 9.4).
Complexes of isopropenyl chloride with thiourea only gave
low yields of low polymer under high energy radiation (34).
By Ziegler-type catalyst, a yellow low polymer was made
from isopropenyl chloride (in heptane at -20°C for 5 hr)
(35).

ISOPROPENYL ALKYL ETHERS

Just as isopropenyl esters differ from vinyl esters in
chemical behavior, the isopropenyl alkyl ethers differ
substantially from vinyl alkyl ethers. They have not been
generally classified as allylic compounds; nevertheless,
the reluctance of the isopropenyl alkyl ethers to form
high polymers suggests that they belong to the allylic
family.

Favorskii in 1888 described the preparation of allene
gas by refluxing propylene dibromide with alcoholic KOH
(36). The isomeric dibromo and dichloropropanes are
believed to give mixtures of allene and methylacetylene
(37).

$$CH_3CHBrCH_2Br \qquad\qquad CH_2=C=CH_2 \;+$$
$$\xrightarrow{\quad KOH \quad}$$
$$or \quad CH_3CCl_2CH_3 \qquad\qquad CH_3-C\equiv CH + KBr + H_2O$$

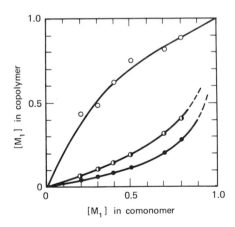

Fig. 9.3. Monomer-copolymer composition curves for copolymerizations
of propenyl chlorides (M_1) and vinyl acetate (M_2) initiated by azobis
at 60°C: open circles, isopropenyl chloride-vinyl acetate; half-open
circles, trans-propenyl chloride-vinyl acetate; solid circles, cis-
propenyl chloride-vinyl acetate. T. Otsu, A. Shimizu, and M. Imoto,
J. Polym. Sci. A 3, 619 (1965).

Favorskii absorbed the gas in cold absolute ethanol.
Portions of the solution were heated with KOH and alcohol
in sealed tubes for 12 hr at 170°C. After cooling and
adding water, an oily layer of ethyl isopropenyl ether was
obtained. The same ether could be made directly by reacting alcoholic KOH with propylene dibromide at 170°C.
Isopropenyl ether was quickly hydrolyzed by 1% H_2SO_4:

$$CH_2=C(CH_3)OC_2H_5 + H_2O \xrightarrow{(H_2SO_4)} CH_3COCH_3 + C_2H_5OH$$

Favorskii observed that ethyl isopropenyl ether on treatment with HI was resinified.

Shostakovskii and Gracheva obtained isopropenyl alkyl
ethers by reaction of methylacetylene with primary or
secondary alcohols and powdered KOH for 18 hr under pressure at 220°C (38). Isopropanol was reacted at even
higher temperatures.

Claisen made methyl and ethyl isopropenyl ethers by
heating acetals of acetone (39):

$$(CH_3)_2C(OR)_2 \longrightarrow CH_2=C(CH_3)OR + ROH$$

The reactions could be accelerated by heating in the
presence of P_2O_5 and quinoline. Low yields of isopropenyl

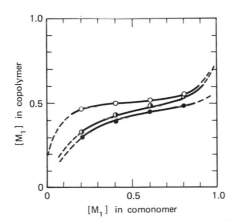

Fig. 9.4. Monomer-copolymer composition curves for copolymerization
of propenyl chlorides (M_1) with maleic anhydride (M_2) initiated by
azobis at 60°C: open circles, isopropenyl chloride-maleic anhydride;
half-open circles, trans-propenyl chloride-maleic anhydride; solid
circles, cis-propenyl chloride-maleic anhydride. T. Otsu, A. Shimizu,
and M. Imoto, J. Polym. Sci. A 3, 623 (1965).

ethers were obtained by heating a mixture of a mole of
ketal, 1.1 moles of P_2O_5 and 1.1 moles of quinoline at
140 to 180°C (40). Isopropenyl ethyl ether was made by
passing vapors of the ketal over silver on asbestos at
280°C. Isopropenyl alkyl ethers were prepared with difficulty by reacting IC or 1,2-dichloropropane with sodium
alkoxide (41). For example, the mixtures were heated in
an autoclave at 180°C and 20 atm. Isopropenyl alkyl
ethers also can be made starting from reaction of acetoacetic ester with ethyl orthoformate (42) and from α-bromo
ethers (43). Heating 1-chloro-2-isopropoxy propane or
2-chloro-2-isopropoxy propane with $KOCH_3$ gave isopropenyl
isopropyl ether and by-products (44).

Lower alkyl isopropenyl ethers hydrolyze readily in
water or dilute acids to give alcohols and acetone. Krantz
showed that methyl isopropenyl ether hydrolyzes in the
blood. He also evaluated vinyl isopropenyl ether as an
anesthetic (45). Rates of hydrolysis and colorimetric
determination in the blood were studied.

Only viscous liquid, balsamlike, and soft solid polymers
of relatively low molecular weight have been prepared from
isopropenyl alkyl ethers. After the observation of
Favorskii, chemists of I.G. Farben reported sticky, balsam-
like polymers from ethyl isopropenyl ether using $AlCl_3$
catalyst in benzene at up to boiling temperature (46).
Shostakovskii and Gracheva also obtained viscous liquid
polymers from ethyl, propyl, and n-butyl isopropenyl
ethers by cationic polymerizations (e.g., with $FeCl_3$ catalyst in dioxane at -15°C) (47). Isobutyl isopropenyl
ether gave the most solid polymers of the monomers examined
in Russia. Even t-butyl isopropenyl ether gave only viscous liquid polymers. Isopropenyl n-butyl ether did not
copolymerize with vinyl n-butyl ether in dioxane at -17°C
with $FeCl_3$ or BF_3 catalyst. The lower alkyl isopropenyl
ethers did not homopolymerize on heating with azobis at
60°C for 150 hr. By use of BF_3 etherate with mercuric
acetate as catalyst, dimers to tetramers (terminated by
acetal groups) were obtained at 32 to 37°C from isopropenyl alkyl ethers (48). Only viscous liquid and soft
solid homopolymers of relatively low specific viscosity
were obtained by Schildknecht and Farber from isopropenyl
alkyl ethers, even with special precautions at low temperatures with different diluents and boron fluoride-
etherate catalysts (49).

The preparation of syndiotactic polymers from methyl
isopropenyl ether by cationic polymerization was reported
(50), but has been disputed (51). The monomer (bp 32.5-
33.5°C; n_D^{25} = 1.3768) in toluene was polymerized by $FeCl_3$
and by I_2 at -78°C. Little or no reaction was obtained
using BF_3 etherate, $AlC_4H_9Cl_2$, and $Al(C_4H_9)_2Cl$ at -78, or

FeCl$_3$ at 0 or 50°C. The methyl isopropenyl ether polymers
obtained having intrinsic viscosities in benzene of 0.2 to
0.3 were transparent solids melting at 90 to 100°C, insoluble in water and in methanol. The alleged syndiotactic
crystalline polymer by x-ray diffraction showed an identity
period of 16.4 Å, which would require a spiraled main
chain (e.g., an eightfold helix with 3 turns).

Free radical copolymerizations with isopropenyl ethers
have been attempted with acrylonitrile (52). Isopropenyl
phenyl ether rearranges on heating with sulfuric acid and
acetic acid to form o-isopropenyl phenol (53). References
to other isopropenyl ethers are given in Table 9.4.

Diisopropenyl ether was prepared by portionwise addition
of 171 g of 2,2'-dichlorodiisopropyl ether to 200 g of
powdered KOH with warming (54). The dialkenyl ether was
distilled off as formed (bp 86°C; n_D^{25} = 1.4110). The
ether was stored over KOH to prevent hydrolysis to acetone.
Isopropenyl trimethylsilyl ether was polymerized near
-78°C using Lewis-acid catalyst (55). The polymers were
hydrolyzed to give isopropenyl alcohol polymers of low
molecular weight (e.g., $\eta_{sp}/C = 0.04$) in methanol solution.
The polymers were unstable in acid solutions.

OTHER ISOPROPENYL COMPOUNDS

Isopropenyl allyl ether rearranged in the vapor phase at
143 to 193°C giving allylacetone (56).

$$CH_2=C(CH_3)OCH_2CH=CH_2 \longrightarrow CH_3COCH_2CH_2CH=CH_2$$

Very few nonconjugated diisopropenyl compounds have been
prepared and investigated. Diisopropenyl ether was re-

TABLE 9.4

Properties of Some Isopropenyl Ethers, $CH_2=C(CH_3)OR$

R group	bp, °C	n_D^{20}	References
Methyl	35	1.375	Krantz, CA 37, 4799 (1943); Shos-takovskii, CA 47, 12217 (1953)
Ethyl	62	1.3927	Dolliver, JACS 60, 441 (1938)
n-Propyl	78	1.3990	Shostakovskii
Isopropyl	78	1.3932	Shostakovskii, CA 51, 15496 (1957)
n-Butyl	112	1.4111	Lutsenko, CA 42, 4148 (1948)
Isobutyl	99	1.4050	Shostakovskii
Phenyl	169	n_D^{23} = 1.5172	Niederl, JACS 55, 284 (1933)
Benzyl	192	--	Autenrieth, Ber. 29, 1647 (1896)
Vinyl	55	--	Krantz, CA 38, 162 (1944)
Allyl	88	1.4190	Stein, JACS 74, 1041 (1952)

ported as a liquid, boiling at about 135°C and fairly stable to aqueous acids (57); this needs confirmation, however. Diisopropenyl esters of dicarboxylic acids, acetals, ketals, and mercaptals have been little studied. Diisopropenyl acetylene polymerizes readily on standing in air and light (58). Brown solid masses of crosslinked polymer grow in the monomer. These polymers exploded on heating at 50°C or above. The polymerization was inhibited by Lewis acids (59).

References to some isopropenyl sulfide compounds are given in Table 9.5. The isopropenyl alkyl sulfides have repulsive odors. Isopropenyl heterocyclic compounds such as oxazoles, oxadiazoles, and thiazoles did not polymerize readily under free radical conditions (60).

TABLE 9.5

Some Other Isopropenyl Compounds

Compound	bp, °C	References
Isopropenyl ethyl sulfide	115	U.S. 2,066,191 (Esso); cf. Autenrieth, Ann. 254, 239 (1889)
Isopropenyl n-butyl sulfide	161	Sporzynski, CZ II, 1704 (1936) (refractive index n_D^{15} = 1.4753)
Isopropenyl benzyl sulfide	ca. 225	Autenrieth, Ber. 29, 1652 (1896)
CH_2=$C(CH_3)(CH_2)_mOR$	--	German Appl. Rohm & Haas 1958

References

1. C. E. Schildknecht, Vinyl and Related Polymers, Wiley, 1952; Polymer Processes, Interscience, 1956.
2. Belg. 540,976 (Gevaert); CA 52, 5041 (1958).
3. B. H. Gwynn and E. F. Degering, J. Am. Chem. Soc. 64, 2216 (1942); U.S. 2,383,965; H. J. Hagemeyer, Ind. Eng. Chem. 41, 765 (1949); cf. U.S. 2,487,849 and 3,142,700 (Roche).
4. B. Phillips, U.S. 2,466,738 (Carbide); J. B. Dickey and T. E. Stanin, U.S. 2,646,437 (Eastman).
5. S. Serota and E. S. Rothman, U.S. 3,666,781 (USA).
6. R. Hart and G. Smets, J. Polym. Sci. 5, 55 (1950); N. G. Gaylord and F. R. Eirich, ibid. 5, 743 (1950).
7. E. W. Taylor and C. C. Unruh, U.S. 2,751,372 (Eastman).
8. R. Lanthier, Brit. 927,975 (Shawinigan).
9. F. P. Reding et al, U.S. 3,328,369 (Carbide).
10. S. Bywater and E. Whalley, U.S. 3,349,068 (Natl. Res. Council Canada); Polym. Preprints 1, 143 (April 1960).
11. C. E. Schildknecht and David Walborn, unpublished.
12. H. C. Haas and N. W. Schuler, J. Polym. Sci. A 2, 1641 (1964).

13. R. L. Hasche and E. M. McMahon, U.S. 2,453,317 (Eastman); H. J. Hagemeyer and D. C. Hull, Ind. Eng. Chem. 41, 2920 (1949).

14. C. C. Unruh and W. O. Kenyon, U.S. 2,452,165; cf. U.S. 2,507,153 (Eastman).

15. L. K. J. Tong and W. O. Kenyon, J. Am. Chem. Soc. 71, 1925 (1949).

16. J. Nishino et al, CA 74, 142574 (1971).

17. H. P. A. Groll and G. Hearne, Ind. Eng. Chem. 31, 1534 (1939); V. I. Kolbasov et al, CA 56, 5404 (1962).

18. C. Friedel, Ann. 134, 263 (1865).

19. A. L. Henne and M. W. Renoll, J. Am. Chem. Soc. 59, 2435 (1937).

20. M. S. Kharasch et al, J. Org. Chem. 2, 300 (1937).

21. K. Griesbaum et al, J. Am. Chem. Soc. 87, 3151 (1965); cf. Angew. Chem. 76, 782 (1964).

22. M. S. Kharasch et al, J. Org. Chem. 4, 431 (1939).

23. R. Reboul, Ann. Chim. [5] 14, 477 (1878); E. Linnemann, Ann. 138, 125 (1866).

24. B. E. Abreu et al, Anesthesiology 2, 535 (1941); CA 35, 8094 (1941).

25. M. Lederer, Angew. Chem. 71, 162 (1959).

26. S. I. Miller, J. Org. Chem. 26, 2619 (1961).

27. R. C. Reinhardt, U.S. 2,160,947 (Dow).

28. K. Griesbaum et al, J. Org. Chem. 29, 2404 (1964).

29. P. S. Skell and R. G. Allen, J. Am. Chem. Soc. 80, 5997 (1958).

30. M. S. Kharasch and C. F. Fuchs, J. Am. Chem. Soc. 65, 505 (1943).

31. M. Aeltermann and G. Smets, Bull. Soc. Chim. Belg. 60, 459 (1951).

32. D. Braun and F. Weiss, Angew. Makromol. Chem. 13, 55 (1970).

33. T. Otsu et al, J. Polym. Sci. A 3, 615 (1965); cf. CA 62, 644 (1965).

34. J. F. Brown and D. M. White, J. Am. Chem. Soc. 82, 5671 (1960).

35. S. Nose and K. Satomo, Japan 18,992 (1963); CA 60, 3122 (1964).

36. A. Favorskii, J. Russ. Chem. Soc. 20, 518 (1888); J. Chem. Soc. Abstr. 360 (1899); J. Prakt. Chem. [2] 37, 538 (1888).

37. F. C. Whitmore, Organic Chemistry, Van Nostrand, 1937.

38. M. F. Shostakovskii and E. P. Gracheva (Moscow), CA 47, 12217 (1953); CA 51, 15496 (1957).

39. L. Claisen, Ber. 31, 1021 (1898).

40. H. P. Crocker and R. H. Hall, J. Chem. Soc. 2052 (1955).

41. W. Reppe, Ann. 601, 81 (1956); cf. Brit. 332,605 (1929).

42. M. A. Dolliver et al, J. Am. Chem. Soc. 60, 440 (1938).

43. W. M. Lauer and M. A. Spielman, J. Am. Chem. Soc. 53, 1533 (1931); cf. M. L. Sherill and G. F. Walter, ibid. 58, 742 (1936); J. M. Wilkinson et al, U.S. 2,619,505 (GAF).

44. J. A. Flint and G. T. Merrall, Brit. 1,004,809 (Shell); CA 63, 17902 (1965).

45. J. C. Krantz et al, Anesthesiol. 5, 160, 498 (1944); CA 39, 2330 (1945); J. Pharmacol. Exp. Ther. 83, 40 (1945); CA 39, 2135 (1945).

46. Ger. 524,189; Brit. 378,544 (I.G. Farben); CZ I 1030 and II 3483 (1933).

47. M. F. Shostakovskii and E. P. Gracheva (Moscow), CA 51, 1895 and 15496 (1957).

48. R. I. Hoaglin et al, J. Am. Chem. Soc. 80, 5460 (1958).

49. C. E. Schildknecht and L. Farber, Stevens Institute of Technology, unpublished.

50. M. Goodman and Y.-L. Fan, J. Am. Chem. Soc. 86, 4922 (1964); Macromol. Chem. 1, 163 (1968).

51. K.-J. Liu and S. J. Lignowski, Polym. Lett. 6, 191 (1968); K. Matsuzaki et al, J. Polym. Sci. B 6, 195 (1968).

52. G. E. Ham and A. B. Craig, U.S. 2,643,986 (Chemstrand).

53. J. B. Niederl and E. A. Storch, J. Am. Chem. Soc. 55, 284 (1933); cf. 55, 4151 (1933).

54. B. T. Gillis and K. F. Schimmel, J. Org. Chem. 25, 2187 (1960).

55. Y. Wakatsuki et al, Chem. High Polym., Japan, 25, 673 (1968).

56. L. Stein and G. W. Murphy, J. Am. Chem. Soc. 74, 1041 (1952); cf. C. D. Hurd and M. A. Pollack, J. Am. Chem. Soc. 60, 1905 (1938).

57. L. Knorr and P. Roth, Ber. 39, 1426 (1906); cf. Ber. 39, 2881 and 42, 547.

58. C. E. Schildknecht, unpublished.

59. R. J. Burch and C. E. Schildknecht, U.S. 3,117,167 (Airco).

60. Y. Iwakura et al, J. Polym. Sci. A 10, 1133 (1972).

10. DIALLYL CARBONATES

Chapters 10, 11 and 12 discuss polyfunctional allyl
esters which polymerize to solid high polymers by free
radical initiation. The most important diallyl carbonate
monomer is not A_2CO_3, but $ACO_3CH_2CH_2OCH_2CH_2CO_3A$. It is
known as diallyl diglycol carbonate or diethylene glycol
bis(allyl carbonate) with tradename CR-39 monomer. Unlike
most commercial polymer developments, CR-39 diallyl carbonate polymers--from monomer discovery to final success
in optical applications--must be largely credited to one
research group and one company: Pittsburgh Plate Glass
Company, later PPG Industries. The objectives of better
scratch resistance and optical properties than those of
Plexiglas MMA polymers were accomplished in clear cast
sheets after innumerable difficulties. Safety glasses,
prescription optical lenses, safety shields, and optical
cements are major products. Custom-cast copolymers with
MMA and other comonomers are available for specific applications. Fiber-reinforced and pigmented formulations are
seldom used, and thermosetting molding materials have not
become important. Articles are made by machining or cast
directly from monomer into crosslinked, cured sheets,
rods, tubes, lenses, or other shapes.

GLYCOL BIS(ALLYL CARBONATE) MONOMERS AND POLYMERS

Strain and co-workers of Pittsburgh Plate Glass Company
first explored methods of preparing organic glasslike
transparent sheets by cast polymerization of such polyfunctional ethylenic monomers as allyl acrylate, allyl
maleates, and methacrylates (1). For example, ethylene
dimethacrylate gave harder, more scratch-resistant, clear
polymer sheets than Plexiglas, but the products were very
brittle. In this work they were able to isolate fusible,
soluble prepolymers from a number of polyfunctional monomers by mild polymerization conditions. These prepolymers
could be molded to crosslinked thermoset shapes as a
separate step. Rothrock earlier had cured dimethallyl
fumarate prepolymer as a surface coating (2). Perhaps
thickened linseed oil should be regarded as the first

useful prepolymer. For the development of cast sheets
superior to methyl methacrylate polymer in scratch resistance, Strain and co-workers before 1940 turned to the
study of allyl carbonate esters. The monomers below have
the two allyl groups separated by 8 and 11 atoms, respectively. These gave more scratch-resistant, more shock-
resistant, and less brittle cast polymers than attainable
by polymerizing MMA. CR-39 (Columbia Resin-39) or Ally-
mer-39 was selected for commercialization.

$$CH_2OCO-CH_2-CH=CH_2$$
$$CH_2OCO-CH_2-CH=CH_2$$

$$O \begin{cases} CH_2CH_2OCO-CH_2-CH=CH_2 \\ CH_2CH_2OCO-CH_2-CH=CH_2 \end{cases}$$

Ethylene glycol
bis(allyl carbonate)

Diethylene glycol bis-
(allyl carbonate) CR-39

The first monomer above may be prepared from allyl
chloroformate, which can be made by bubbling phosgene into
allyl alcohol while cooling below 20°C (3). For example,
after 0.9 mole of phosgene was absorbed per mole of allyl
alcohol, the liquid was allowed to stand for an hour; then
it was washed with water to remove unreacted allyl alcohol
and dried over calcium chloride. To a solution of a mole
of ethylene glycol in excess pyridine cooled below 15°C,
there was added dropwise 2.2 moles of the allyl chloroformate. After an hour at room temperature the liquid product
(CR-38) was washed by dilute HCl followed by aqueous salt
solution until neutral and dried over calcium chloride.

$$2CH_2=CHCH_2-OCCl \ + \ \begin{matrix} CH_2-CH_2 \\ OH \quad OH \end{matrix} \xrightarrow{\text{pyridine}} CH_2=CHCH_2-OC-OCH_2 \ + \ C_5H_5N\cdot HCl$$

$$CH_2=CHCH_2OCOCH_2$$

allyl chloroformate
bp 180°C at 717 mm

By a similar procedure, diethylene glycol bis(allyl carbonate) could be prepared using diethylene glycol. The monomer from diethylene glycol proved to be the more favorable
for commercial development. Later CR-39 monomer was made
by reacting diethylene glycol bis(chloroformate) with allyl
alcohol in the presence of sodium hydroxide (4):

$$O(CH_2CH_2OCCl)_2 \ + \ 2CH_2=CHCH_2OH \ + \ 2NaOH \longrightarrow$$

$$O(CH_2CH_2OCOCH_2CH=CH_2)_2 \ + \ 2NaCl \ + \ 2H_2O$$

Properties of some diallyl glycol carbonate monomers are
as follows:

	bp, °C	n_D^{20}	d_4^{20}
Ethylene glycol bis(allyl carbonate	122(1)	1.4443	1.114
Diethylene glycol bis(allyl carbonate)			
(CR-39)	160(2)	1.4503	1.143
Triethylene glycol bis(allyl carbonate)	polymerized	1.4520	1.135
Ethylene glycol bis(methallyl carbonate)	142(2)	1.4490	1.110

Infrared spectra of CR-39 monomer and polymer are shown
in Figs. 10.1 and 10.2. The diallyl glycol carbonates
are relatively nontoxic, although they can irritate the
skin of some persons. Pure CR-39 monomer is a colorless
liquid of mild odor and low viscosity of 9 cp at 25°C.
It is sometimes modified by addition of other monomers to
accelerate initial polymerization. CR-39 monomer is
soluble in pentanols, carbon disulfide, gasoline, and
petroleum ether, and is insoluble in ethylene glycol and
glycerol. The melting point is about -4°C and surface
tension is 35 dynes/cm. CR-39 monomer is stable against
saponification by dilute aqueous alkali. The monomer is
compatible with dibutyl phthalate, triacetin, polyvinyl
acetate, polymethyl methacrylate, and some alkyd resins.
It has limited compatibility with Aroclors, methyl abie-
tate, cellulose nitrate, and cellulose acetate; it is also
incompatible with hydrocarbon polymers such as polystyrene
and polyisobutylene.

Unlike styrene and most acrylic monomers, diallyl esters
such as the diallyl glycol carbonates do not undergo thermal polymerization readily on heating without radical
catalysts. Diethylene glycol bis(allyl carbonate) monomer
has relatively good stability on storage, but some increase in viscosity of uninhibited monomer may be observed
after several months. Because of the relatively low reactivity of this diallyl ester, rather high concentrations
of peroxide or azo catalysts (e.g., 3-5%) are needed to
obtain polymerization in reasonable time. Monomer catalyzed by about 3% benzoyl peroxide can be stored at 10°C or
lower, whereas -5°C or lower is needed to prevent polymerization on storage when isopropyl percarbonate is present.
The reactive isopropyl percarbonate is a preferred catalyst for commercial cast polymerizations since it permits
polymerization at temperatures 15 to 25°C lower than with
benzoyl peroxide; also, the resulting castings have better
color and better color stability. In bulk or cast polymerizations of CR-39 a gel stage with crosslinking may
occur rather quickly. However, with precautions a uniform
syrup can be obtained which is useful as a lens cement.
Air and oxygen act as inhibitors of the polymerizations.
Ionic methods of polymerization have not succeeded in
forming homopolymers of high molecular weight.

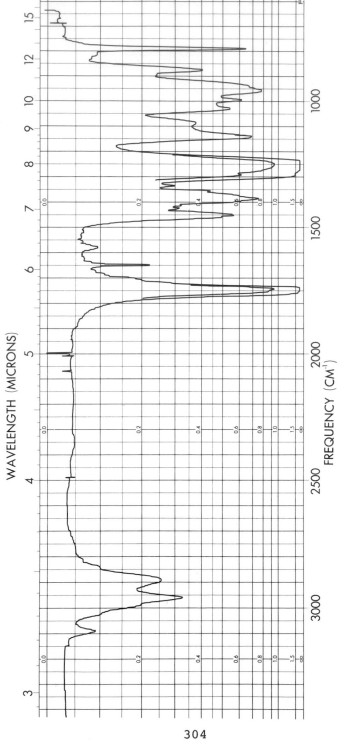

Fig. 10.1. Infrared spectrum of liquid CR-39 monomer, diethylene glycol bis(allyl-carbonate). Prominent are absorptions near 3.4 microns associated with carbon-hydrogen bonds, a strong band at 5.7 microns from the carbonyl group, and a weaker absorption at 6.1 microns from the allyl double bond. Courtesy of PPG Industries.

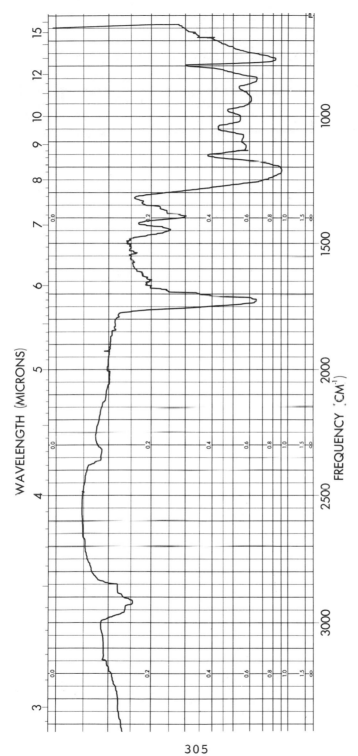

Fig. 10.2. Infrared spectrum of solid CR-39 cast polymer by reflectance at 45°C. As in the monomer, bands appear at 5.7, 6.9, 7.2, 8.8, and 13 microns. The 8.0-micron band of the monomer is shifted to about 8.2 microns in the polymer. Double bond absorptions are no longer present. Courtesy of PPG Industries.

CAST POLYMERIZATIONS OF CR-39 MONOMER

The diethylene glycol bis(allyl carbonate) and related monomers with peroxide catalyst can be cast directly in plate glass cells having flexible separating gaskets. Less often a syrup of prepolymer solution in monomer is prepared before the cell is filled. Thus Muskat and Strain described heating 1,2-propylene glycol bis(methallyl carbonate) containing 3% benzoyl peroxide at 60°C until a viscous solution had formed. This liquid was cast in a cell consisting of two polished glass plates provided with flexible edge seals.

The course of CR-39 polymerization in bulk may be followed by density increase (Fig. 10.3), by iodimetric determination of residual double bonds, or by heat evolution (5). Dial and Gould described precautions in casting monomer catalyzed by CR-39 to obtain perfect sheets, free of cracks or other imperfections (6). The shrinkage of 14% for complete polymerization normally does not occur

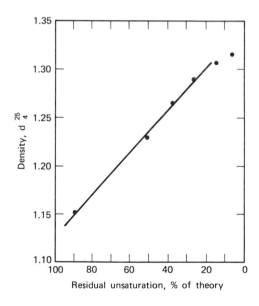

Fig. 10.3. Bulk polymerization of CR-39 monomer using peroxide initiators can be followed by increase in density which is linearly related to consumption of double bonds (at least to 80% conversion). W. R. Dial, W. E. Bissinger, B. J. DeWitt, and F. Strain, Ind. Eng. Chem. 47, 2448 (1955).

uniformly in three dimensions in the cell. Polished
metal plates or glass plates were used, with a peripheral
wall or gasket of vinyl chloride polymer elastomer (tubing
or extruded rectangular profile) held by spring clamps. A
small pressure, e.g., 4 psi, gauge may be applied to the
plates to promote contraction in thickness. The pressure
may be released at 85 to 90% conversion and polymerization
completed outside of the glass cell. Alternatively, non-
elastic separators of the two glass plates can be used
(as in the old German Plexiglas casting process), and the
separators may be removed after gelation. The gel state
may contain 30% or more polymer insoluble in acetone. The
horizontal cells may be heated at 60°C to obtain the gel
state and then up to 125°C for completion of polymerization. Dial and co-workers recommended different casting
cycles using 3.0% diisopropyl percarbonate, e.g., for 1/8
in. sheets 15 hrs at 46-105°C and for 0.5 in. sheets 24
hr at 39-105°C (with temperatures rising more rapidly near
the end).
   In CR-39 polymerization the temperatures must be adjusted to the peroxide added. Alkyl percarbonates, though
hazardous, may be used at 20 to 50°C (Fig. 10.4), whereas
ketone peroxides are active at 100 to 150°C. BP is useful

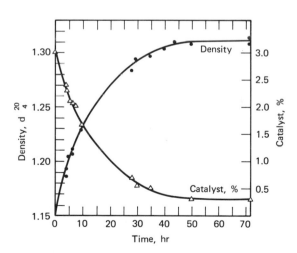

Fig. 10.4. Polymerization of diethylene glycol bis(allylcarbonate)
at 45°C with initial addition of 3.0% diisopropyl percarbonate.
Rates of polymerization and catalyst consumption decrease with time
at a given temperature. W. R. Dial et al., Ind. Eng. Chem. _47_,
2448 (1955).

at intermediate temperatures. Below 60°C and at lower
conversions, first order kinetics applied, e.g., the rate
of monomer consumption was proportional to the first
power of catalyst concentration.

CR-39 monomer can be cast in cells that are water heated
or air oven heated. Sheets of 1/8 to 1/4 in. thickness
may require 12 hr at 70 to 90°C for polymerization to
solid sheets, or 23 hr at 66 to 85°C for 1/4 to 1/2 in.
sheets (7). After the cycle is completed and the mold is
opened, the polymer should be allowed to postcure for 2 hr
at 115°C. Diethylene glycol bis(allyl carbonate) shrinks
about 14% during complete polymerization. The casting
processes resemble those used for methyl methacrylate.
However, after about 4% shrinkage CR-39 generally gives a
weak gel stage that is easily fractured.

Helpful precautions and instructions for casting CR-39
to form transparent sheets, rods, tubes, lenses, and other
useful shapes are given in data sheets of Pittsburgh Plate
Glass. For experimental work, addition of 3% benzoyl peroxide by stirring at 50°C is recommended; 3% isopropyl
percarbonate (IPP) is used for large-scale operations that
justify the special treatment required in handling this
unstable catalyst. IPP has been transported by truck
packed with Dry Ice. Since warming even at 90°F can
initiate polymerization with increasing viscosity, IPP
should be dissolved at room temperature. The catalyzed
monomer should be filtered through a fine filter paper or
an asbestos mat filter in order to give castings of high
clarity. CR-39 monomer catalyzed by benzoyl peroxide may
be stored at 50°F or lower. Any catalyst that crystallizes out should be redissolved on warming. At room
temperature or 25°C IPP-catalyzed monomer will remain
nearly constant in viscosity for about 24 hr, but it can
be kept only about an hour at 100°F before casting.

A Pittsburgh Plate Glass bulletin describes three
methods of casting tubes from CR-39. A metal rod insert
may be used in Pyrex glass tubing. The mold may be coated
with a solution of polyisobutylene in toluene. Centri-
fugal casting can be used for larger tubes of 2 to 3 in.
diameter. A relatively low-melting center insert may be
used in a third method. In tube casting involving a
sizable gas space, retardation of polymerization by oxygen
of air may be troublesome, and in such cases, replacement
by nitrogen or carbon dioxide is recommended. When isopropyl percarbonate was used and relatively low temperatures were required for polymerization, a little paraffin
wax (mp 70°C) was added to reduce contact with inhibiting
oxygen.

Practical methods for casting CR-39 as well as methyl
methacrylate were disclosed by Beattie of the Polycast

Corporation (8). CR-39 behaved somewhat like styrene in
fairly uniform rate of free radical bulk polymerization
with gradual evolution of heat of polymerization. Methyl
methacrylate casting is more difficult because greater
Trommsdorff effect causes the rate to become suddenly very
rapid (9). However, in all three cases polymerization
rates in the last stages at high conversion are slow, so
that higher temperatures must be used for full curing.
   CR-39 and related crosslinking monomers do not easily
give smooth syrups by prepolymerization but go to a gel
state at relatively low conversion. Although somewhat
elastic these gels have little strength. If the gel is
ruptured it never rejoins and the casting is spoiled. At
about 25% cure the CR-39 gels have mechanical properties
resembling somewhat those of the white of hard-boiled egg.
Precautions for overcoming these difficulties have been
developed.
   One method suggested for avoiding cracking in cast sheets
was based on minimizing polymerization on one side of the
sheet by contact with oxygen (10). Later the tacky soft
side could be coated with catalyzed monomer or syrup, and
polymerization could be completed in contact with a smooth
plate. Soft, limp sheets of partially cured CR-39 may be
shaped in or over a mold followed by curing with heat
radiation.
   During World War II, glass-cloth-reinforced CR-39 polymers found first application as supports for self-sealing
gasoline tanks of B-17 airplanes. Later CR-39 was largely
replaced in glass-fiber-reinforced plastics by less expensive polyester-styrene resins.

CAST COPOLYMERIZATIONS OF CR-39 MONOMER

   Data sheets of Pittsburgh Plate Glass Company describe
modifications of the above procedures for casting CR-39-
methyl methacrylate in ratio 70:30 by weight. There is
more difficulty from extraction of plasticizer by MMA from
the Tygon polyvinyl chloride gasket. The gasket can be
covered or coated by cellophane, polyvinyl alcohol, or
polyisobutylene. For the 70:30 mixture, addition of 1.0%
benzoyl peroxide was suggested. When more than 10% MMA is
copolymerized, the reaction should be started below 70°C
in order to prevent premature rapid polymerization accompanied by boiling and bubble formation. For example, the
mixture may be heated for 2 hr at 65°C to give a viscous
syrup of copolymer dissolved in the monomers for casting.
The temperatures for the filled mold may rise slowly to
70°C during 72 hr. The same postcure is recommended as
for unmodified CR-39. When 1.5% IPP is the catalyst, the
successful copolymerization of sheets requires much lower

temperatures, e.g., 25-35°C, to make a syrup and then heating the cell at 35°C for 16 hr followed by 72 hr at 45°C. With this very active initiator the postcure may be 3 hr at 90°C.

There has been less experience with copolymerization of 70% MMA with 30% CR-39. It is necessary to use flexible gaskets that are hollow or have a high degree of resiliency because of the greater shrinkage during MMA polymerization. Rectangular soft Tygon 3603 gasket with circular bore is suitable. It is suggested that 0.3% benzoyl peroxide or 0.5% IPP be added as catalyst. If a syrup is not made before casting, the cell must be tight to prevent leaks from the low viscosity MMA monomer. The copolymerization may start at 45°C and be completed at 100°C.

Properties of cast CR-39 copolymer sheets in comparison to homopolymer sheets given in Table 10.1 are based on data of Pittsburgh Plate Glass Company. Comonomers were methyl methacrylate, vinyl acetate (VAC) and maleic anhydride (MA).

TABLE 10.1

Properties of Cast CR-39 Copolymers

	CR-39	Comonomer MMA	Comonomer MMA	Comonomer VAC	Comonomer MA
Composition					
CR-39 %	100	70	30	70	70
Comonomer %	0	30	70	30	30
Specific gravity, 25°C	1.32	1.28	1.23	1.26	1.41
Hardness, Rockwell	M100	M102	M101	M93	M120
Tensile strength, psi	5800	7900	8700	8600	13500
Flexural strength, psi	9000	11600	14700	15200	19500
Modulus, 10^5 psi	2.9	3.0	3.3	3.7	5.1
Compressive strength, psi	23000	--	--	24700	30000
Modulus, 10^5 psi	2.7	--	--	4.6	4.4
Impact strength, ft-lb/in.					
Notched Izod	0.3	0.3	0.3	0.3	0.2
Unnotched Izod	1.9	2.4	5.7	2.6	5.0
Abrasion resistance					
Taber, times MMA polymer	30	3	1	9	--
Heat distortion					
distortion temp.,°C, 10mils	69	63	67	60	120
distortion at 130°C, mils	40	192	500	140	24
Water absorption, %					
24 hr at 25°C	0.2	0.2	0.2	0.2	0.2
Acetone absorption, %					
7 days at 25°C	0.4	6.1	crumbled	2.7	0.6

Copolymers of CR-39 using 25% diallyl o-phthalate (DAP) had softening points only slightly higher than CR-39 homopolymer (11). Copolymer of 25% triethylene glycol bisallyl carbonate with 75% DAP had a higher softening range. Copolymers with diallyl isophthalate made using isopropyl percarbonate were studied in Germany (12). Copolymers of CR-39 with 40% triallyl cyanurate (TAC) underwent little heat distortion up to 210°C under load but were rather brittle. The use of only 10% TAC with CR-39 and a heating cycle of 42 hr at 60°C, 24 hr at 90°C, and 1 hr at 110°C gave heat-resistant, less brittle copolymers. Commercial cast copolymers of CR-39-TAC show characteristic infrared bands at 6.4, 8.8, and 12.2 microns.

Additions of comonomers to CR-39 in order to accelerate polymerization rates and improve toughness and scratch resistance continue to be of interest. Strain discovered that 20% of N-ethylmaleimide accelerated CR-39 polymerization and gave copolymers of promising properties (13). In the structure of this copolymer presented here, the broken bonds represent continuation of the crosslinked three-dimensional network.

$$--CH_2CHCH_2O\overset{O}{\overset{\|}{C}}OCH_2CH_2OCH_2CH_2O\overset{O}{\overset{\|}{C}}OCH_2CHCH_2CHCH--$$

Using 2% benzoyl peroxide at 70°C, the following times were observed until gelation of mixtures of 80% CR-39 and 20% comonomer: CR-39 (no comonomer), 3 hr; maleimide, 6 min; N-phenylmaleimide, 15 min; N-allylmaleimide, 8 min; N-butylmaleimide, 6 min. Carlson found that adding 26 mole % maleic anhydride reduced the gel time of peroxide-catalyzed CR-39 heated at 75°C from 3 hr to 15 min (14). Such proportions of maleic anhydride with CR-39 give copolymers which are sensitive to water and to alkali.

Some references to other copolymerizations using CR-39 and related monomers are listed below.

Comonomers	References
Triallyl cyanurate	U.S. 3,171,869 (Bausch & Lomb
MMA + TAC	U.S. 2,910,456 (Peterlite
MMA prepolymer	Ger. 1,016,446 (Peterlite); CA 54, 13727 (1958)
MMA, also TAC	Brit. 796,867 (Peterlite)
MMA delayed addition	U.S. 2,774,697 (Bjorksten)
MMA + vinyl siloxane	U.S. 2,787,568 (Bjorksten)
Acrylic monomers	U.S. 3,031,347 (Aerojet)

Comonomers	References
Unsaturated polyesters	U.S. 3,150,018 (Aerojet)
Allyl carbamate	U.S. 2,483,194 (Wingfoot)
Drying oils	Ger. 1,099,174 (PPG); CA 55, 26475 (1959)
Diallyl o-phthalate	U.S. 2,964,501 (Titmus Optical)
Vinyl chloride	U.S. 3,012,009 (Monsanto)

Hard, mar-resistant surfaces on MMA polymer castings were obtained by coating the inside surfaces of a mold with CR-39 monomer and initiator (15). This layer is largely polymerized in the absence of oxygen. The mold is then filled with MMA and/or other acrylic monomers with catalyst and heated to complete polymerization.

CR-39 IN OPTICAL APPLICATIONS

Not only is Columbia Resin or CR-39 polymer the most widely used optical plastic for lenses, safety glasses and guards, watch crystals, and instrument windows, but also compositions based largely on the catalyzed monomer are widely employed for cementing glass optical devices. The bonds are unaffected by severe conditions of tempera-ture or temperature change, or by exposure to weather, solvents, and many chemicals. Filtered monomer containing 3% benzoyl peroxide may be used without modification, except that slight thickening by warming for an hour up to 70°C improves the application properties and rates of cure (16). Two hours at 70°C was suggested for setting the cement with lenses. The cemented lens components are then gradually heated to 80 or 85°C and kept at this temperature for at least 8 hr. Advantages in optical applications include mild odor and freedom from toxicity, high optical clarity, abrasion resistance, heat resis-tance, and resistance of the polymers to crazing on contact with solvents.

Coles and co-workers reviewed organic optical cements (17). Unmodified CR-39 monomer does not "set up" as rapidly as might be desired, and the polymerization is retarded by air. Additions of diallyl phthalate, e.g., 12%, improved flow properties and gave less blistering and eccentricity. Addition of 30% n-butyl methacrylate improves performance for some types of work (18). Another product, CR-149, a diallyl urethane carbonate, gave good adhesion to glass and desirable high refractive index of 1.5193 but had a yellow color. CR-149 polymers, discussed on page 316, are harder than those of CR-39; they have less water absorption and less acetone absorption. In the use of diallyl esters as cements and coatings, it must be remembered that polymerization is retarded not only by dissolved oxygen but also by phenols, many amines, sulfur, copper, lead, and many of their compounds. The monomer forms peroxides on storing in contact with air, but these are not efficient as polymerization catalysts.

Special procedures for casting optical lenses from CR-39 and other ethylenic monomers have been suggested (19, 20). Spectacle lenses of CR-39 and modified CR-39 have become of major importance, especially for safe glasses for school children and for thick eyeglass lenses of high curvature where low density is an advantage.

Armorlite and Plastolite lenses are examples of commercial optical materials based on CR-39. These provide overall protection of the eyes against small missiles. The CR-39 polymer lenses are more resistant to misting than glass, and they are lighter in weight. Resistance to heat, solvents, and scratching are superior to that of lenses based on methyl methacrylate cast polymers (21). Fragmentation gives more obtuse, less sharp pieces than inorganic glasses.

Copolymers of CR-39 with maleic anhydride or glycidyl methacrylate have been evaluated for hydrophilic contact lenses (22). In one example, a solution of 90% CR-39 monomer and 10% maleic anhydride was heated for 17 hr from 44 to 105°C with gradual rise of temperature. The surface of the casting was hydrolyzed in 10% aqueous sodium hydroxide for one minute at room temperature. In another example, a copolymer cast by a similar heating schedule from equal amounts of CR-39 and glycidyl methacrylate was hydrolyzed by 15-min contact of the surface with NaOH solution. Hydrophilic copolymers with N-vinyl pyrrolidone have been studied (23).

CR-39 gave promise as a lens coating more resistant to marring than other organic lens coatings (24). A limitation was the requirement of an inert atmosphere for curing. Special compositions less sensitive to inhibition have been supplied, but some of these have yellow color. Addition of paraffin wax to exclude air and surface oxidation has many disadvantages. One faster-curing composition contained the following: CR-39, 80%; diethylene glycol maleate polyester, 20%; cobalt resinate, 0.1%; hydroquinone, 0.1%. Small additions of diallyl phthalate and dibutyl maleate to CR-39 gave greater toughness in lens coatings (25). In Germany small amounts of MMA have been added to CR-39 to accelerate polymerization in optical applications. Aromatic hydrocarbons or chlorinated aromatic hydrocarbons may be added as plasticizers and to raise the refractive index in optical applications. Coloring by salts of transition metals with organic acids has been suggested, but some of these salts retard poly-

merization (26). Blends of diallyl compounds with unsaturated polyesters and chlorinated biphenyl, together
with minor proportions of vinyl monomers, have been proposed as optical cements (27). CR-39 has been blended
with such other monomers as triallyl cyanurate and
styrenes for forming spectacle lenses (28).

PROPERTIES AND APPLICATIONS OF
DIALLYL DIGLYCOL CARBONATE POLYMERS

Thermosetting composition based on CR-39 diallyl diethylene glycol carbonate were introduced commercially by
Pittsburgh Plate Glass Company in 1942. The polymer properties were discussed by Strain (29). The favorable
properties of heat resistance, color and clarity, and mar
and scratch resistance have promoted optical and other
applications that tolerate relatively high costs. Unfavorable properties include tendencies to yellow on long
outdoor exposure and to become brownish when heated above
130°C, a degree of plasticity or creep above 100°C, and
limited impact resistance. Clear cast sheets, rods,
tubes, and cementing compositions have remained the most
important outlets. Cast clear transparent sheets ranging
from 1/32 to 2 in. in thickness are supplied by Cast
Optics Corporation. HT-CR-39 is a copolymer having
improved heat resistance. Copolymers with a little methyl
methacrylate have better forming properties with heat.
CR-39 and copolymers in optical uses have been called
Polycast Resin, Cocor, and other trade names. Methyl
methacrylate, vinyl acetate, and polyester copolymer
sheets have been manufactured.

In resistance to abrasion, to low-temperature impact, to
heat, gamma radiation, and crazing, CR-39 plastic exceeds
Plexiglas methyl methacrylate polymers. Cast sheets can
be heat formed within limits. Since cured CR-39 polymers
are crosslinked they do not dissolve in any solvent. The
polymer is etched by concentrated sulfuric acid and with
discoloration. The polymer is resistant to dilute sulfuric, hydrochloric, nitric, and hydrofluoric acids, to
dilute ammonium hydroxide, as well as to such organic
solvents as ethanol, carbon tetrachloride, gasoline, benzene, and toluene. In acetone about 0.4% increase in
weight may occur and in chloroform about 1.5%. Other
properties of the cured polymers are given in Table 10.2.

Dyeing of cast CR-39 has been discussed with lists of
satisfactory colors (30). The color produced in the
polymer may be different from that applied because of
oxidation by residual peroxide catalyst. Fully cured
CR-39 sheets can be dyed in boiling solutions of Tintex
or Rit textile dyes.

TABLE 10.2

Properties of Clear Fully Cured
Columbia Resin   (CR-39 Polymer)

Property	Value
Specific gravity, 25°C	1.32
Refractive index at 20°C, n_D	1.504
Dispersion factor, $n_D - 1/n_F - n_C$	57.8
Light transmission, 1/4 in. thickness, %	89-92
Hardness, Rockwell	M95-M100
Modulus of Elasticity in flexure, 10^5 psi	
50°C (122°F)	1.6-2.0
25°C (77°F)	2.5-3.3
-10°C (14°F)	4.3-4.6
-57°C (-70°F)	5.5
Tensile strength, psi	5000-6000
Modulus of elasticity in tension, 10^5 psi	3.0
Compressive strength, ultimate, psi	22,500
Modulus of elasticity in compression, 10^5 psi	3.0
Impact strength, 25°C, ft-lb/in.	
Izod, notched	0.2-0.4
Izod, unnotched	2-3
Max. recommended operating temp. under no load	
Continuous service	100°C (212°F)
Intermittent (1 hr duration)	150°C (302°F)
Specific heat, cal/(g)(°C)	0.55
Thermal expansion, linear coefficient/°C	
-40 to +25°C	8.1×10^{-5}
25 tp 75°C	11.4×10^{-5}
75 to 125°C	14.3×10^{-5}
Thermal conductivity	
Btu/(hr(ft^2)(in.)(°F)	1.45
Burning rate, in./min	0.35
Abrasion resistance	
Mar resistance	
Modified Taber, times methyl methacrylate	30-40
Modified Falling Emery:	
Times methyl methacrylate polymer	9-12
Times glass	0.9-1.2
Water absorption, 24 hr/25°C, %	0.2-0.4

DIALLYL URETHANE CARBONATES

In 1944 Strain prepared a diallyl urethane carbonate monomer known as Allymer 149 or CR-149 (31). Allyl chloroformate AOC(O)Cl was reacted with ethanolamine. The allyl urethane formed was then reacted with phosgene to give the CR-149 monomer as follows:

$$2AOCNHCH_2CH_2OH + ClCCl \xrightarrow{\ base\ } (AOCNHCH_2CH_2)_2CO_3 \ \ etc.$$

An alternate synthesis of the allyl urethane intermediate was reaction of ethanolamine with diallyl carbonate. CR-149 monomer with peroxide initiator gave cast sheets of promising hardness, toughness, flexural strength, and resistance to abrasion. However, much research failed to overcome the yellowish color and high costs of these polymers which are no longer supplied. Some properties of CR-149 monomer and the cured polymers are recorded in Table 10.3.

TABLE 10.3

Allymer CR-149 [a]

Monomer $(AOCNHCH_2CH_2O)_2CO$	Unfilled cast polymer of bis(2-carballyloxyaminoethyl) carbonate
Melting point, 45°C	Max. use temperature, 100°C
Boiling point at 760 mm, 250°C (with decomposition)	Intermittent use, 150°C
n_D^{20}, 1.478	Hardness, Rockwell M, 117
Specific gravity 25/4, 1.22	Density, 1.34 g/ml
Storage life without catalyst, several months	Water absorption at 25°C, 0.2%
Solubility in water, hydrocarbons, none	Tensile strength, 11,000-13,000 psi
Solubility in alcohol, acetone, benzene, CCl_4, soluble	Flexural strength, 25,000-31,000 psi
Odor, almost none	Impact, notched Izod, 0.3-0.6 ft-lb/in.
Effect on skin, non-irritating	Linear coefficient of expansion, 7-8 10^{-5}
Shrinkage in cure, 9%	Solvent and chemical absorption at 25°C, <1%
Polymerization in 1/8 to 1/4 in. thickness:	Resin content with glass cloth, 40%
Clear, unmodified, 2-4 hr, 80-115°C	Tensile strength with glass cloth, 56,000 psi
Fiber-glass laminates, 1-2 hr, 90-115°C	

[a] From data sheets and research reports of PPG Industries.

The melted crystalline monomer could be used for casting crosslinked sheets of high toughness and abrasion resistance by techniques similar to those described for CR-39 monomer. From 3 to 5% organic peroxide was needed, and air was excluded to avoid inhibition by oxygen. This monomer was considered promising for glass cloth laminates before the development of the less expensive unsaturated polyester resins. Abrasion resistance of the cured polymers far exceeded that of Plexiglas MMA polymers but was not as good as that of glass. The polymers were attacked by concentrated nitric acid, hydrochloric acid, and ammonium hydroxide. Use of CR-149 in dental plastics was suggested (32).

Strain also prepared the diallyl urethane carbonate of ethylene glycol by reacting the N-(carballyloxy)aminoethanol with the dichloroformate of ethylene glycol with cooling below 5°C (33). The viscous liquid monomer, $n_D^{20} = 1.4778$, was heated with 5% BP at 75°C for 36 hr to form a casting that was clear, hard, and tough. Similarly, bis N-(carballyloxy)aminoethyl-o-phthalate, $n_D^{20} = 1.227$, was prepared and polymerized to a slightly yellow sheet. Other diallyl dicarbonate derivatives of carbamates (or urethanes) were also polymerized to yellowish, thermoset castings (33).

OTHER ALLYL CARBONATE MONOMERS AND RELATED

The methallyl analog of CR-39 is unstable and difficult to make. Methallyl chloroformate has an extremely irritating, more-than-pungent odor (34).

Bis(carballyloxymethyl) carbonate was prepared by reacting phosgene with 2 moles of allyl glycolate in the presence of pyridine with cooling (35). Clear tough polymers were prepared. The corresponding dicrotyl compound, on heating for 2 hr at 60°C, gave a soft gel. Diallyl dilactate carbonate (bp 180°C at 4 mm) readily formed fusible prepolymers that could be cured during molding (36). The lower valence cations of Cu, Mg, Co, and Ni acted as inhibitors of polymerization. Moldable prepolymers were made also from the corresponding dimethallyl ester.

Preparation of diallyl carbonate AOC(O)OA has been described by Schulthess (37). A solution of 1 mole of phosgene in 200 ml of benzene was added dropwise to 3 moles of dry allyl alcohol. After a short time, heat was evolved and HCl liberated. When the temperature had fallen, the mixture was heated on an oil bath for 12 hr with slow refluxing. Most of the benzene was distilled off under vacuum, and the liquid was washed and dried. Diallyl carbonate could be distilled at atmospheric pressure at

166°C. The refractive index was n_D^{20} = 1.4320. Diallyl
carbonate has been prepared also by reacting allyl alcohol
with diethyl carbonate in benzene using $NaOC_2H_5$ as catalyst at reflux temperature (38). In bulk with 1% BP at
70°C, diallyl carbonate polymerized somewhat slower than
diallyl o-phthalate (37). Gelation occurred in about 14
hr after the soluble polymer content had reached about
20%. The monomer was epoxidized, and copolymerization
with other epoxides was suggested (39).

Diallyl carbonate prepared by alcoholysis was polymerized on long heating with BP at 65°C to give a soft insoluble solid (40). Dimethallyl carbonate also prepared
by alcoholysis of diethyl carbonate was observed to
polymerize. Kinetics of polymerization of diallyl carbonate in solution at 60°C with BP catalyst was studied
by Matsumoto and co-workers (41). Evidence of cyclopolymerization was observed. As expected residual polymer
unsaturation and DP decreased with decreasing initial
monomer concentration. Rates of unimolecular cyclization
and of bimolecular propagation were determined. Prepolymers were prepared from diallyl carbonate by heating the
monomer with BP in dioxane 2 hr at 80°C (42). After 30%
conversion the prepolymer was precipitated by addition of
methanol. Molding formulations containing diallyl carbonate prepolymer, alpha-cellulose filler and t-butyl
perbenzoate were cured in a mold at 300°F for 8 min. The
thermoset moldings were said to have high stain resistance
and impact strength suitable for use as dinnerware. Diallyl carbonate monomer has been supplied by Polysciences,
Incorporated.

Allyl alcohol polymers of low DP were reacted with ethyl
chloroformate in anhydrous solvents to form insoluble
polymer with a high content of carbonate groups, apparently with cyclic structure (43). The polymers could be
used to bind water-soluble enzymes. Allyl allyloxy cyclic
carbonates were prepared by phosgenation of glycols (44).

Glycol bis(allylglycolyl carbonate) polymerized at 65°C
with 5% BP to transparent castings free from cracks (45).
Glyceryl tris(allylcarbonate) n_D^{20} = 1.4558 gave solid
polymers by heating with 5% BP for 18 hr at 75°C. Preparation of tetraallyl carbonate of pentaerythritol also
was attempted. The carbonate diester of allyl salicylate
melted at 53°C (46). The same reference gave refractive
indices and densities of diallyl carbonate monomers
obtained by reacting allyl chloroformate with triethylene
glycol, tetraethylene glycol, as well as 1,2-propanediol.
The monomer from the reaction of methallyl chloroformate
(bp 130°C, n_D^{20} = 1.427) with diethylene glycol was not
distilled (n_D^{20} = 1.453).

Monomers bearing an allyl carbonate ester group and
also an allyl ether group have been studied (47). Allyl
2-allyloxyethyl carbonate boiled at 133°C (29 mm) and had
n_D^{20} = 1.4382. The diallyl dicarbonate ester of diethanol-
amine when heated with 1% BP at 75°C required more than
3 hr for gelation (48). When this monomer was copolymerized with 25 mole % of maleic anhydride under similar
conditions, gelation occurred in less than 15 min.

Allyl chloroformate AOC(O)Cl (bp 114°C) is toxic and
unstable. Useful polymers have not been developed from
it. The compound was polymerized by long heating with 3%
isopropyl percarbonate at 45°C, forming yellow, viscous
liquid polymers of low molecular weight (49). The latter
were reacted with allyl alcohol to give polyfunctional
allyl derivatives that, when heated with peroxide, formed
crosslinked polymers. Allyl chloroformate was reacted
with 2-allyloxy-1-ethanol (50). Minor proportions of
allyl chloroformate were copolymerized with thiocarbonyl
fluoride at -78°C using boron triethyl and oxygen for
initiation, along with rigorous exclusion of moisture (51).
A copolymer containing 2 mole % allyl chloroformate units
gave tough elastomeric films and had a molecular weight of
about 400,000. The copolymers were crosslinked by heating
with zinc oxide or other polyvalent metallic salt. Copolymers of allyl chloroformate, 1-hexene, and SO_2 were
prepared in benzene solution using t-butyl peroxypivalate
as initiator (52). Films were water resistant.

Allyl mercaptan has been reacted with $COCl_2$ over activated charcoal to form allyl chlorothiolformate (bp 61°C,
n_D^{30} = 1.4976). The compound was suggested for control of
nematodes and fungi (53).

Organosiloxane polymers bearing terminal allyl carbonate
groups could be cured by blending with polysiloxanes containing active mercaptan groups (54).

Table 10.4 provides additional data and references; the
allyl group is represented by A and the methallyl group
by MA.

TABLE 10.4

Allyl and Methallyl Compounds

Compound	bp, °C (mm)	Remarks	References
AOCOA $\overset{\|\|}{O}$	97 (61)	FMC supplied; n_D^{20} = 1.4288	U.S. 2,514,354 (Shell)
MAOCOMA $\overset{\|\|}{O}$	101 (28)	FMC supplied	U.S. 2,410,305 (Du Pont); U.S. 2,514,354 (Shell)
AOCCH$_2$OCOA $\overset{\|\|}{O}$ $\overset{\|\|}{O}$	--	Crosslinked polymers	U.S. 2,370,573 (PPG)

TABLE 10.4

Allyl and Methallyl Compounds (continued)

Compound	bp, °C(mm)	Remarks	References
AOOCCH$_2$O$\overset{\text{O}}{\underset{}{\text{C}}}OCH_2$COOA	150(1-2)	Cured castings	Brit. 588,660 (ICI)
AO$\overset{\text{O}}{\underset{}{\text{C}}}CH_2$OH	50(20)	--	Brit. 599,837
(AO$\overset{\text{O}}{\underset{}{\text{C}}}CH_2$OOC(CH$_2$)$_2$COOCH$_2$)$_2$	--	Tough, clear polymers	Brit. 599,837
(AOOCCH$_2$OCH$_2$)$_2$	133(1)	d$_{25}^{25}$ = 1.1227	Sax
(AO$\overset{\text{O}}{\underset{}{\text{C}}}OCH_2$CH=CH)$_2$	--	--	U.S. 2,563,771 (Shell)
AO$\overset{\text{O}}{\underset{}{\text{C}}}OCH_2$$\overset{\text{R}}{\underset{}{\text{C}}}$(CH$_3$)CH$_2O\overset{\text{O}}{\underset{}{\text{C}}}$OA	--	--	U.S. 3,497,478 (Atlantic-Richfield)

DIALLYL CARBONATES FROM BIS-PHENOLS

Goodrich investigated thermosetting monomers derived from bis-phenols, and one cast polymer was offered commercially under the name Kriston (55). Allyl chloroformate reacted with bis-phenol A in acetone (56):

$$CH_2=CHCH_2O\overset{O}{\underset{}{C}}Cl + HO\langle\bigcirc\rangle\overset{CH_3}{\underset{CH_3}{C}}\langle\bigcirc\rangle OH \xrightarrow{NaOH}$$

$$CH_2=CHCH_2O\overset{O}{\underset{}{C}}O\langle\bigcirc\rangle\overset{CH_3}{\underset{CH_3}{C}}\langle\bigcirc\rangle O\overset{O}{\underset{}{C}}OCH_2CH=CH_2 + NaCl + H_2O$$

The refractive index of the resulting aromatic diallyl carbonate monomer was n_D^{20} = 1.5422. Heating with 2% BP at 75°C for 24 hr gave a hard, clear, thermoset polymer of n_D^{20} = 1.5662, Barcol hardness of 25, and promising flexural strength. Several related aromatic diallyl carbonates also were made. The ring-chlorinated diallyl carbonates of bis-phenol A gave harder polymers and were found to be faster in polymerization rates (57). A bis-(3,5-dichlorophenyl) propane derivative containing 1% BP polymerized to a gel at 70°C in 145 min (58). Adding 20% maleic anhydride to this diallyl carbonate monomer gave a

gel in about 45 min and increased the hardness. Mixtures
of chlorine-containing monomer and fine polymer could be
molded to strain-free thermoset shapes (59). Kriston A2
monomer was 2,2-bis(3-chlorophenyl-4-allyl carbonate)
propane. The 3,3'-dichloro-bis-phenol A could be prepared
by reaction of o-chlorophenol with acetone and sulfuric
acid. The diallyl carbonate monomer therefrom is a low
melting solid or viscous, almost odorless liquid.

Diallyl carbonates of bis-phenol A and of chlorinated
bis-phenol A were copolymerized with vinyl benzoate and
with vinyl acetate by free radical methods to give clear,
infusible castings of good color (60). Diallyl carbonates
from resorcinol, hydroquinone, and pyrogallol were suggested as crosslinking comonomers (61).

Kriston monomer, based upon the diallyl carbonate of
bis-phenol A, gave polymer of specific gravity 1.30,
while the refractive index increased in polymerization
from n_D^{20} = 1.55 to 1.57 (62). The purified monomer melted
at 44°C. The shrinkage in polymerization was less than
4%, and the polymer coefficient of thermal expansion about
9×10^{-5} cm/(cm)(°C). Air inhibited the monomer and contributed to discoloration. Surfaces covered by cellophane
polymerized to highly glossy hard surfaces. Cast sheets
of Barcol hardness as high as 60 could be obtained by
complete polymerization. For larger castings, 3% acetyl
benzoyl peroxide could be used with gradual rise of polymerization temperature from 50 to 100°C over 24 hr. The
monomer showed little tendency to polymerize on storage
without added inhibitor.

Kriston polymer was resistant to most acids but was
charred by sulfuric acid. The cured polymers resisted
most solvents but disintegrated to a slush of flakes in
acetone or vinyl acetate. Long exposure to UV light
caused yellowing and further hardening of the polymer,
but no crazing or cracking occurred (62). Generally lower
temperatures for initial polymerization with low-temperature catalysts, such as o-chlorobenzoyl peroxide or acetyl
peroxide, resulted in harder polymers after final curing
at higher temperatures (63). Plots of time for gelation
versus peroxide concentration were nearly linear in the
range of 65-85°C.

The diallyl dicarbonate of monochlorobisphenol A gave
gelation at about 20% conversion at 70 to 80°C (Fig. 10.5).
The acetone-soluble polymer isolated just before gelation
had an average DP of 9 to 10 and melting point of 105 to
110°C (64). Using 1% BP at 75°C, the rate of polymerization reached a maximum near the gel point. Around 35%
conversion the rate had fallen off sharply, and in 24 hr
it reached a final level of 83% conversion. The gel times
such as 100 min at 75°C allowed ample opportunity to study

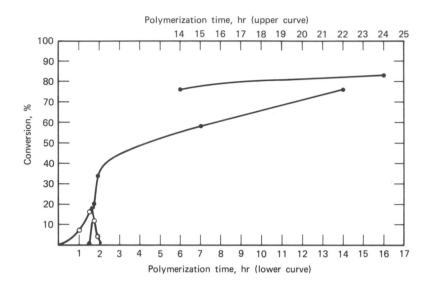

Fig. 10.5. Bulk or cast polymerization of a diallyl carbonate
derivative from a chlorinated bisphenol A using 1% initial benzoyl
peroxide at 75°C. Open circles show conversion to soluble polymer;
solid circles show conversion to insoluble, crosslinked polymer.
Note that higher concentrations of peroxide catalysts and higher
temperatures are normally used in commercial polymerizations of diallyl ester monomers. E. E. Gruber, Goodrich Research Report.

the formation of soluble prepolymer, which could be precipitated by methanol. Saponification and other tests
indicated more linear prepolymers than in the soluble
prepolymers of diallyl o-phthalate.

With 1% BP at 75°C no soluble polymer solids were
present soon after gelation (except for oily oligomers
and peroxide residues). The soluble prepolymer was about
39% as unsaturated as the monomer. By the same method
the gelled polymer formed after 7 hr at 75°C was 31% as
unsaturated as the monomer. The remaining liquid with
the gel was only 63% as unsaturated as the initial monomer, and some amber liquid dimer was observed.

The high refractive index, high boiling point, and low
concentration of diallyl carbonates of bis-phenols might
find application in optical and adhesive applications if
the polymers could be made free of discoloration on aging
in light. In photoelastic studies, stresses in Kriston

moldings were retained up to 135°C (65).

Recently the p,p'-diallyl carbonate of biphenyl and related monomers have been investigated (66). Among other diallyl aromatic monomers is the allyl ester-allyl urethane prepared by reaction of allyl chloroformate with allyl anthranilate in the presence of pyridine at 10°C (67). This allyl N-carballoxy anthranilate boiled at 167°C at 2 mm and had n_D^{28} = 1.5410. Heating with 5% BP at 65°C for 43 hr gave a hard, clear polymer.

Monoallyl alkyl carbonate esters AOCOR do not give homo-
  $\overset{\shortparallel}{O}$
polymers of high molecular weight and they have been little studied. However, allyl butyl carbonate has been copolymerized with SO_2 at -40°C under high-energy radiation (68). The product could be molded at 140°C but it decomposed at 215°C.

References

1. M. A. Pollack, I. E. Muskat, and F. Strain, U.S. 2,273,891, appl. 1939 (Columbia Chemical Division of PPG).
2. H. S. Rothrock, U.S. 2,221,663 (Du Pont).
3. I. E. Muskat and F. Strain, U.S. 2,592,058 (Columbia Southern); cf. U.S. 2,370,565-8 (PPG) and 2,385,932 (PPG).
4. I. E. Muskat and F. Strain, U.S. 2,592,058 (PPG).
5. W. R. Dial et al, Ind. Eng. Chem. 47, 2447 (1955).
6. W. R. Dial and C. Gould, U.S. 2,379,218 (PPG).
7. R. L. McCombie, Encyclopedia of Polymer Technology, N. M. Bikales, Ed., Vol. 1, p. 800, Wiley.
8. J. O. Beattie, Mod. Plastics 33, 109-117 (July 1956).
9. C. E. Schildknecht, Vinyl and Related Polymers, Wiley, 1952.
10. I. E. Muskat and F. Strain, U.S. 2,370,573 (PPG).
11. H. W. Starkweather et al, Ind. Eng. Chem. 47, 302 (1955).
12. Ursula Herrmann, German application (Zeiss, Jena).
13. F. Strain, U.S. 2,650,215 (PPG).
14. E. J. Carlson, U.S. 2,587,442 (Goodrich); cf. U.S. 3,256,113 (PPG).
15. K. Asami and H. Hariu, U.S. 3,465,076 (Mitsubishi).
16. H. W. Coles et al, J. Opt. Soc. Am. 34, 623 (1944).
17. H. W. Coles et al, Mod. Plastics 24, 141-148 (June 1947).
18. W. F. Parsons and J. R. Dann, U.S. 2,445,535-6 (Eastman).
19. A. O. Hungerford and P. J. Mullane, U.S. 3,038,210 (Parmelee); M. Greshes, Plast. Technol. 11/7, 38-39 (1965); G. H. Bowser, U.S. 3,422,168 and 3,423,488 (PPG).
20. J. O. Beattie, U.S. 2,542,386; A. A. Naujokas, U.S. 3,248,460 (Bausch & Lomb); R. Grandperret, U.S. 3,222,432 and 3,278,654 (Lentilles); H. Sauer et al, Ger. 1,155,911 (Ruetgerswerke).
21. M. Lasch et al, J. Appl. Polym. Sci. 11, 369 (1967).
22. H. D. Crandon, U.S. 3,221,083 (American Optical).
23. J. W. Breitenbach and A. Schmidt, Monatsh. 85, 52 (1954); CA 49, 6184 (1955).

24. H. W. Coles et al, Mod. Plastics 25, 123 (July 1948).
25. G. M. J. Sarofeen, U.S. 2,964,501 (Titmus Optical).
26. C. A. Sheld, U.S. 3,216,958 (Bausch & Lomb).
27. E. Carnall and J. J. Lugert, U.S. 2,856,379 (Eastman).
28. H. Sauer and H. Binder, Ger. 1,155,911 and U.S. 3,228,915
     (Ruetgerswerke).
29. F. Strain, Mod. Plastics 21, 97 (August 1944).
30. Bulletin of PPG Industries.
31. F. Strain, U.S. 2,397,631 (PPG).
32. J. A. Saffir, U.S. 2,477,268 (Dentists' Supply).
33. F. Strain, U.S. 2,379,261 and 2,441,298 (PPG); cf. A. G.
     Chenicek, U.S. 2,401,549 (PPG).
34. I. E. Muskat, U.S. 2,370,570 (PPG).
35. I. E. Muskat and F. Strain, U.S. 2,370,569 (PPG); cf.
     U.S. 2,370,574 (PPG).
36. I. E. Muskat and F. Strain, U.S. 2,370,566 and Brit. 586,520
     (PPG).
37. A. Schulthess, Promotionsarbeit 3004 Zurich 1960; cf.
     H. E. Fierz-David and W. Mueller, J. Chem. Soc. 125, 26 (1924);
     U.S. 2,648,697 (Ohio-Apex).
38. D. E. Adelson and H. Dannenberg, U.S. 2,514,354 (Shell).
39. A. C. Mueller and E. C. Shokal, U.S. 2,795,572 (Shell).
40. H. J. Richter and H. S. Rothrock, U.S. 2,410,305 (Du Pont).
41. A. Matsumoto et al, Bull. Chem. Soc. Japan 42, 1959 (1969);
     CA 71, 70965 (1969).
42. D. A. Berry and G. M. Gynn, U.S. 3,644,242 (Dart).
43. S. A. Barker et al, J. Chem. Soc. C 2726 (1971).
44. F. Hostettler and E. F. Cox, U.S. 3,532,715 (Carbide).
45. F. J. H. Mackereth, Brit. 596,467 (ICI).
46. I. E. Muskat and F. Strain, U.S. 2,592,058 (PPG); CA 46, 8417
     (1952).
47. M. A. Pollack, U.S. 2,407,446 (PPG).
48. E. J. Carlson, U.S. 2,587,442 (Goodrich).
49. A. Pechukas, U.S. 2,464,056 (PPG).
50. C. D. Hurd et al, J. Am. Chem. Soc. 74, 5128 (1952).
51. J. N. Coker, U.S. 3,275,607 (Du Pont).
52. W. S. Pickle and N. B. Lorette, U.S. 3,563,961 (Dow).
53. Brit. 948,831 (Stauffer); CA 60, 11902 (1964).
54. R. V. Viventi, U.S. 3,600,288; H. Vaugh, U.S. 3,419,634 (G.E.).
55. R. B. Rice et al, Mod. Plastics 24, 156 (May 1947).
56. J. A. Bralley and F. B. Pope, U.S. 2,587,437 (Goodrich).
57. J. A. Bralley, U.S. 2,548,141 and 2,455,652 (Goodrich).
58. E. J. Carlson, U.S. 2,587,442 (Goodrich).
59. F. B. Pope, U.S. 2,568,658 (Goodrich).
60. E. J. Carlson, U.S. 2,529,866 (Goodrich).
61. Brit. 585,775 (Wingfoot).
62. J. A. Bralley and F. B. Pope, Goodrich report 1945.
63. E. J. Carlson, Goodrich report, 1945.
64. E. E. Gruber, Goodrich report, 1949.

65. M. Leven and R. C. Sampson, Mod. Plastics $\underline{34}$, 151 (May 1957).
66. S. Schmucker, U.S. 3,328,455 (G.E.).
67. C. R. Milone, U.S. 2,522,393 (Goodrich).
68. B. G. Harper et al, U.S. 3,365,430 (Dow).

11. DIALLYL PHTHALATES

Of all the diallyl monomers, diallyl orthophthalate and isophthalate find the greatest use in polymerizations. The principal application of the polymers of diallyl phthalates is in thermoset-reinforced plastics having outstanding electrical properties, heat resistance, and dimensional stability. Monomer-polymer solutions are used to impregnate glass, graphite, and organic fibers as well as textiles, papers, and nonwoven webs. These premium materials are used especially with fibrous reinforcement in high-performance electronic, aircraft, and space hardware applications. Already in 1963 Air Force helicopters made by Sikorsky Aircraft Company included nine parts based on diallyl phthalate polymers. Smaller amounts of diallyl phthalate are used for copolymerization with unsaturated polyesters, in sealants and adhesives.

The principal supplier of the polymers and monomers in America has been Food Machinery Corporation or FMC. In recent years diallyl o-phthalate has been produced by Sumitomo Chemicals and Daiso Kasei (Osaka Soda) in Japan, by the Division Ftalital of the Swiss Aluminum Company in Italy (Società Alluminia Veneto per Azioni), and by Hardwicke Chemical Company in South Carolina. Diallyl iso or metaphthalate has been developed recently in America for improvement in polymer heat resistance, but diallyl terephthalate polymers have had negligible applications as yet.

Studies of diallyl phthalate isomers by Simpson and Haward in England provided the first examples of cyclic allyl polymerizations. Such cyclopolymerizations were later found with many other monomers, especially by Butler and co-workers. The diallyl isophthalate (DAIP) and diallyl terephthalate (DATP) show less cyclization in their polymerizations than diallyl orthophthalate.

The applications of diallyl phthalate monomers have contributed to advanced polymer technology unique methods of preparing and formulating prepolymer compositions which can cure by crosslinking during molding. These methods permit rapid production of shapes of outstanding dimen-

326

sional stability and of good electrical properties, maintained even under prolonged high humidity and elevated
temperatures. Specifications for military, computer, and
space requirements have stimulated development of a
variety of fiber-reinforced and flame-resistant proprietary formulations. However, much basic research, for
example, in copolymerization and chemical stabilization
remains to be carried out.

DIALLYL PHTHALATE MONOMERS

   The ready availability of phthalic anhydride together
with development of allyl alcohol by Shell Chemical Company led to commercial production of diallyl orthophthalate (DAP) beginning in the 1950s. Diallyl isophthalate
or diallyl metaphthalate became available much later. The
monomers have outstanding advantages in mild odor and
relatively low toxicity. Tests with rabbits, guinea pigs,
and mice showed that absorption of DAP through the skin
produced the only harm. There was no harm from ingestion
of moderate amounts. In contact with the skin, DAP is
comparable to tri-o-tolyl phosphate and less harmful than
formalin and acrylonitrile. Although DAP and DAIP are not
primary irritants to the human skin, particular individuals
do show reactions. Moreover, the liquid monomers are
irritants of the eyes and sensitive skin areas such as the
lips. DAP monomer can be prepared by esterification of
phthalic anhydride by an excess of allyl alcohol:

$$2CH_2=CHCH_2OH \; + \; \text{(phthalic anhydride)} \; \xrightarrow{H^+} \; \text{(diallyl phthalate)} \quad + \; H_2O$$

In a patent example, 2.5 moles of allyl alcohol and 1.0
mole of phthalic anhydride with 0.5% sulfuric acid (based
on anhydride) were heated under reflux at 120°C (1). From
a distillation column connected with the reactor, an azeotrope of water, allyl alcohol, and diallyl ether was removed at the top and DAP was withdrawn below. After
removal of water, the allyl alcohol-diallyl ether could
be returned continuously to the reactor. In continuous
operation during 100 hr, the conversions to DAP, monoallyl
phthalate, and diallyl ether were 85.6, 14, and 9.7%,
respectively. In the preparation of diallyl phthalates
and dimethallyl phthalates by esterification with mineral
acid catalyst at 80 to 100°C, it was reported that passing
some SO_2 through the reactor continuously minimized discoloration (2). The reducing action of SO_2 seems to oppose
the darkening observed when DAP is heated with certain

peroxides or other oxidizing agents. Diallyl phthalate
monomers also can be prepared by alcoholysis of dimethyl
phthalates by allyl alcohol.

DAP monomer can be prepared from reaction of phthalic
anhydride, allyl chloride, and sodium hydroxide or sodium
carbonate (3,4). A small amount of tertiary amine or
allyltrimethylammonium chloride catalyst may be added and
the mixture heated at 130°C for 4 hr. In order to prevent polymerization during monomer synthesis, about 0.2%
of a t-butylphenol inhibitor was added. Copper or cuprous
compounds may be present also as inhibitors of polymerization (5). Small amounts of amides such as dimethylformamide act as promoters, apparently through increasing the
miscibility of reactants (6). Some o-dimethyl phthalate
may be added to the initial charge of anhydride, alkenyl
chloride, and water (7). Heating for 2 hr in the range
of 100-130°C was suitable without added solvents. Ciba
has patented a two-stage process using both allyl alcohol
and allyl chloride with phthalic anhydride (8).

DAP monomer very slowly undergoes autoxidation in air.
The peroxides formed are relatively stable and do not lead
to oxidative polymerization as in the "drying" of unsaturated oils and certain polyfunctional allyl ethers. The
peroxides have been observed to contribute to initiation
of polymerization at temperatures above 150°C (9).

DAP is relatively stable against hydrolysis and surprisingly stable against polymerization on storage at normal
temperatures. At higher temperatures, free-radical-cata-
lyzed polymerizations of DAP seem to be less inhibited by
atmospheric oxygen than are typical free-radical polymerizations of acrylic and vinyl ester monomers. Peroxide
initiators, especially the higher temperature types, are
almost always added to DAP for polymerization. Dicumyl
peroxide is a suitable initiator (10), as well as t-butyl
perbenzoate. High-temperature azo initiators are available from the Lucidol Division of Pennwalt Corporation.

To the polymerization technologist, outstanding DAP
monomer properties include mild odor, low volatility (2.4
mm at 150°C), low shrinkage of 12% on polymerization,
stability on storage, and ability to form relatively
stable thermoplastic prepolymers. Table 11.1 lists some
specific properties.

The two monomers are soluble in most common organic
solvents but have limited solubility in hydrocarbons,
glycols, and amines. They are only slightly soluble in
water. About 0.6% water dissolves in DAP. At 25°C the
vapor pressure of DAP is less than 0.001 mm. DAIP exceeds
DAP in solubility in hydrocarbons such as gasoline and
mineral oil. The monomers can be tested for polymer
content by adding 1 cc to 10 cc of methanol. Traces of

TABLE 11.1

Properties of Diallyl Phthalate Monomers

Properties	(DAP) $\overset{COOCH_2CH=CH_2}{\underset{COOCH_2CH=CH_2}{\bigcirc}}$	(DAIP) $\overset{COOCH_2CH=CH_2}{\bigcirc}$ $COOCH_2CH=CH_2$
Boiling points, °C at 4 mm	161	181
Freezing point, °C	below -70	-3
Density, g/ml	1.117(25°C)	1.124(20°C)
Refractive index, n_D^{25}	1.518	1.521
Viscosity, cp at 25°C	12	17
Shrinkage in polymerization, %	11.8	--
Specific heat, cal/(g)(°C)	0.50	--
Thermal expansion at 10-40°C in./(in.)(°C)	0.00076	--
Surface tension at 20°C, dynes/cm	39	35.4
Solubility in gasoline at 25°C, %	24	miscible

polymer produce white turbidity, since the polymers are not soluble in methanol (Fig. 11.1).

As supplied by FMC, diallyl o-phthalate monomer catalyzed with 1.0% benzoyl peroxide and heated at 100°C under standard conditions undergoes gelation in about 41 min, whereas DAIP may show a gel time of 36 min. DAP monomer containing 2 to 4% peroxide catalyst may be stored for several weeks at room temperature before gelation. Molecular oxygen and phenolic compounds may be applied to retard polymerization. Since high concentrations of active peroxide catalysts in some cases may cause spontaneous polymerization, the catalysts always should be added to the monomer rather than monomer to catalyst.

Early research on cast polymerizations of DAP was recorded in a laboratory report of Shell Development Company in 1945 (11). With 2% BP at 65°C, excessive long times, e.g., 144 hr, were needed for polymerization. The greater effectiveness of t-butyl perbenzoate in accelerating reaction above 100°C to form cast polymers of good color and clarity was demonstrated. In thick castings, uncontrolled exothermic polymerizations often liberated some allyl alcohol. Catalyzed solutions of prepolymer (syrups) in casting processes were shown to be less sensitive to inhibition by oxygen of air than was catalyzed monomer. Glass cloth-DAP cured laminates gave surprisingly high tensile strengths, e.g., 90,000 psi.

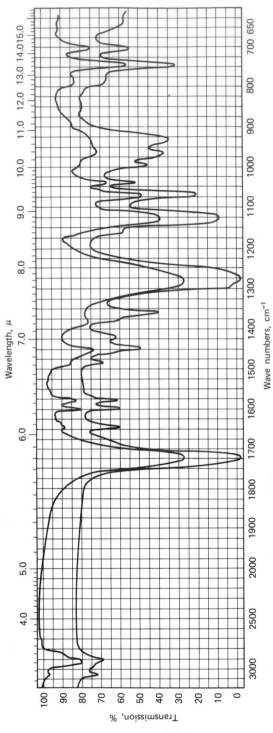

Fig. 11.1. Infrared spectra of diallyl orthophthalate monomer, below, and largely polymerized monomer, above. The absorption at 1645 cm^{-1} is associated with the allyl double bond of the monomer. R. A. Spurr, Betty M. Hankin, and J. W. Rowen, Preprints, ACS Division of Paint, Plastics, and Printing Ink Chem. $\underline{17}$, 446 (September 1957).

330

DAP POLYMERIZATIONS

Bradley in 1937 reported polymerization of diallyl o-phthalate by heating with benzoyl peroxide at 115 to 125°C (12). Only the insoluble crosslinked DAP polymers were claimed in this patent. In one early example, DAP was heated with 1% BP and 5 to 10% of copper (13). After removal of undissolved copper, more benzoyl peroxide was added and more heat was applied until a viscous syrup was formed. Copper was said to act as an inhibitor at lower temperatures but as a catalyst on heating to higher temperatures. Vaughn and Rust heated 100 parts of DAP containing 2 parts of di-t-butyl peroxide until a refractive index of 1.5313 was shown by the syrup (14). At that point there was added gradually with stirring an excess of methanol in order to precipitate the prepolymer and remove unreacted monomer. After drying under vacuum, the prepolymer was dissolved in toluene-xylene and used to coat steel. After baking an hour at 150°C, a hard film was obtained. In the polymerization step, peroxides such as benzoyl peroxide caused polymer discoloration. Pollack, Muskat, and Strain heated the soluble prepolymer in a mold at 150°C under 1000 psi for 4 hr, producing hard, clear insoluble sheets (15).

Wagers and Shokal studied the preparation and separation of DAP prepolymer (16). The monomer was heated at 200 to 250°C without added catalyst to a conversion just short of the gel point as indicated by refractive index. Solvents were added to arrest polymerization and to remove remaining monomer. The substantially linear prepolymer could be dissolved in solvents such as acetone, toluene, or chloroform; then 0.1 to 5% peroxide catalyst was added and films were spread. After the solvents had evaporated, the prepolymer could be heated to cure it to thermoset coatings. The fusible prepolymers were suggested also for compression and injection molding, for extrusion, for laminates, and for castings. During distillation at atmospheric pressure, DAP polymerized to 20 to 40% conversion. Remaining monomer was extracted out by acetone-water mixture (17).

Simpson in England made careful studies of polymerization of DAP in bulk at 80°C with benzoyl peroxide (18). To conversions of around 20% to the soluble or beta polymer, the rates were nearly linear with time and with concentration of initiator (Fig. 11.2). When only 1% BP was added, rates fell off markedly as the initiator was consumed. As shown in Fig. 11.3, this occurred at lower conversion at 100°C than at 80°C. As a result, polymerization at 80°C for 24 hr with 1% added BP gave a harder product than was obtained when 100°C was used. Simpson

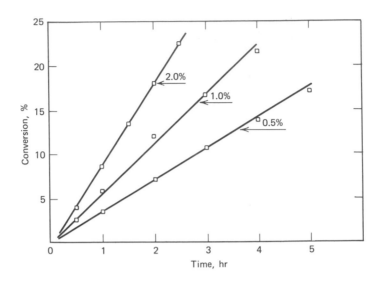

Fig. 11.2. Bulk polymerization of diallyl orthophthalate at 80°C using different concentrations of benzoyl peroxide. At these low conversions to soluble polymer, rates of polymerization are nearly linear with time and initiator concentration. W. Simpson, J. Soc. Chem. Ind. 65, 107 (1946).

proved the need for high initiator concentrations in allyl polymerizations and demonstrated that gelation occurs near 25% conversion with DAP at 80°C (Fig. 11.4). The acetone-soluble prepolymer melted near 90°C when pure.

DAP was polymerized using t-butyl alkenyl peroxides (19). DAP also was polymerized in several stages initiated by two or more peroxides having their maximum effectiveness at different temperatures, for example, t-butyl hydroperoxide and di-t-butyl peroxide (20).

Shokal and Bent found 25 to 40% soluble prepolymer in DAP syrups (21). They found metallic copper to be a retarder of polymerization at 205°C or below but reported that it acted as an accelerator at 225°C. The patent claims comprise polymerization of DAP or diallyl glycolate under nonoxidizing conditions. An increase of 0.0005 in refractive index at 25°C was observed for each percent polymerization of DAP to linear prepolymer. Up to 22% conversion to soluble polymer the rates of DAP polymerization increased nearly linearly with concentration of

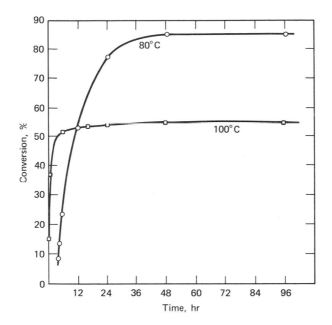

Fig. 11.3. The rates of bulk polymerization of DAP fall off sharply after the initiator has been largely consumed. The 1% benzoyl peroxide added here was used up in polymerization more rapidly at 100°C than at 80°C. In industrial polymerizations of diallyl monomers, higher concentrations of free radical initiators are used. W. Simpson, J. Soc. Chem. Ind. 65, 107 (1946).

peroxide initiator, in contrast to typical vinyl-type polymerizations (22).

For a higher conversion to DAP prepolymer it was suggested that monomer-polymer-acetone solution containing t-butyl perbenzoate catalyst might be sprayed into a dry gas space at 100°C or above (23). Small additions of hydrogen halides or alkyl halides were observed to reduce discoloration during polymerization (24). DAP prepolymers prepared at lower temperatures were said to be more branched and less unsaturated judging from IR absorption at 1644 cm^{-1} (25).

Although homogeneous polymerizations of DAP initiated by hydrogen peroxide in the presence of 1 to 2% water were difficult to control, they produced desirable lower molecular weight polymers at higher conversions (26). In one

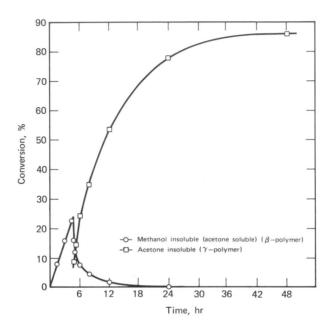

Fig. 11.4. Conversion to soluble or beta-diallyl orthophthalate
polymer reaches a maximum near the gel point at which the concentration of crosslinked, three-dimensional polymer is increasing rapidly.
Benzoyl peroxide was the initiator at 80°C. W. Simpson, J. Soc.
Chem. Ind. 65, 107 (1946).

patent example, 8860 lb of DAP monomer, 622 lb of isopropanol, and 75 lb of hydrogen peroxide (50%) were heated
at 108°C under reflux. After 10 hr the viscosity was 27
cp at 106°C. The resulting solution contained 27% polymer
and when cooled to 25°C had a viscosity of 425 cp. The
prepolymer was then precipitated by adding excess isopropanol at 0°C. The precipitated solid prepolymer was
filtered and dried. Polymerization may be carried to a
high viscosity but not to a gelled state, after which the
prepolymer may be precipitated in a shearing zone. This
may be carried out in the presence of unreactive non-
solvents for the polymer which are, however, miscible
with the monomer (27).
    Willard gave examples of homopolymerization and a pro-
cedure for mixing the viscous polymer solutions with
precipitant in a colloid mill in such a way that unreacted

monomer was removed completely, thus giving a nonsticky powder of prepolymer after drying (27). In a stainless steel reactor was charged 8860 lb DAP monomer, 622 lb isopropanol (91% by vol.), and 75 lb of 50.4% hydrogen peroxide. The mixture was agitated and heated to a temperature of 104 to 108°C. After 10 hr the viscosity had reached 27 cps at 106°C, corresponding to about 27% conversion to prepolymer. In precipitating the prepolymer, the solution from the reaction vessel was cooled and pumped with about 5 times its volume of isopropanol into a colloid mill; the mill was set to operate at 3600 rpm with 0.001 in. clearance between rotor and stator and a flow of 200 gal/hr. The procedure used a total of 48,000 lb of isopropanol which had been cooled so that precipitation took place at 0°C. The resulting suspension of prepolymer in isopropanol-monomer was agitated 45 min at about 15°C in order to coagulate the smaller particles and to aid filtration. The filter cake was reslurried in cold isopropanol and refiltered. A third washing and filtration was carried out from cold water. After drying at 80°C there was obtained a free-flowing white powder of prepolymer which was soluble in ketones and aromatic solvents but insoluble in alcohols and in petroleum ether.

A later patent disclosed polymerization at higher temperatures under conditions to yield lower polymer molecular weights, higher conversions, and less residual catalyst in the prepolymer (28). In one example, 11.6 kg of DAP (n_D^{25} = 1.5181) containing 13 ppm t-butyl hydroperoxide was heated at 200°C for 780 min to reach n_D^{25} = 1.5339 and a viscosity of 450 centistokes (25°C). The cooled syrup was transferred gradually to a wiped film-type vacuum still to remove monomer and to recover the molten prepolymer. Completion of polymerization to crosslinked reinforced moldings was suggested with the following example:

DAP prepolymer	190 parts
DAP monomer	10
Asbestos	200
t-Butyl perbenzoate	6
Calcium stearate	4

In the polymerization to stable prepolymer, the free-radical-producing catalyst may be added portionwise at temperatures below 225°C, thus minimizing catalyst residues in the prepolymer product (29). The same prepolymer-monomer ratio was used with a filler mixture ($CaCO_3$, TiO_2, and $CaSO_4$) along with 2% lauric acid as mold release (30).

Other methods have been developed abroad to obtain high conversions of DAP to prepolymer of controlled moderate

molecular weight. Porret of Ciba polymerized DAP in
acetone with H_2O_2 catalyst and small additions of an acid
such as p-toluene sulfonic acid (31). Swiss Aluminum
obtained more than 25% conversion to prepolymer with a
peroxide catalyst and up to 15% of added alcohol-chlori-
nated hydrocarbon mixture (32). In one example, the
following mixture was heated for 3.5 hr at 110°C, after
which the prepolymer was precipitated by adding excess
methanol:

DAP	100	parts
Methanol	2	
Carbon tetrachloride	2	
Benzoyl peroxide	0.4	

There was 27% conversion to prepolymer of iodine #58,
soluble in acetone. Residual monomer may be extracted by
a nonsolvent for the polymer at temperatures where both
monomer and polymer are fluids (33). A Japanese process
claimed 67% conversion to prepolymer of mp 89-92°C and
iodine value 52.8 (34). To 100 p DAP was added 30 p
acetic acid and 1.2 p t-butyl hydroperoxide. Polymerization required 6 hr at 120°C under nitrogen. Carbon tetrachloride was useful as a regulator. Telomerization was
observed by Porret and Leumann when chloroalkanes or
ketones were used as regulators to limit polymer molecular
weight (35).

DAP polymerization at 70°C was reported to accelerate
under pressures of 5000 kg/cm^2, but the tendency toward
cyclization also increased at high pressures (36). The
course of isothermal bulk polymerization of DAP using 2%
BP was followed by electrical resistivity (37). Plots of
resistivity with time were nearly linear, and gelation
did not affect the slope. Prolonged post curing at 120°C
increased resistivity to 5×10^{14} at 25°C.

In general, DAP monomer compositions cannot be trigger
polymerized or "set up" easily from ordinary temperatures
as can the methacrylate esters. With added sensitizers,
however, polymerization in UV or high-energy radiation
can be carried out readily at moderate temperatures. FMC
has developed printing plates based on photosensitized
DAP polymers. Both letterpress and offset plates can be
made. Tertiary amines such as dibutyl phenylethylamine
with chlorobenzoyl peroxide were suggested as promoters
giving little discoloration to DAP polymers (38). For
high-temperature final cure of DAP, around 200°C, free
radicals from thermal fission of symmetrical polyaryl-
alkanes have been suggested (39).

The molecular weights of DAP prepolymers are normally
so low, 10,000 or below, that 25% solutions of polymer in

monomer may be used for viscosity tests. The following
viscosities of prepolymers were obtained using different
concentrations of hydrogen peroxide and different conversions in solution with alcohol (40):

	I	II	III	IV
Hydrogen peroxide, %	0.11	0.27	0.51	0.54
Conversion, %	24.5	25.8	28.0	29.3
Viscosity of 25% solution in DAP at 25°C, cps	550	390	220	179

Viscosities of 25% solutions of common commercial prepolymers in monomer at 25°C are said to range from 200-700 cp.

Little has been disclosed about polymerization of diallyl phthalates in aqueous medium. However, one patent
example described DAP polymerization by heating with
benzoyl peroxide for 21 hr at 80°C in aqueous suspension,
using 0.33% glue as suspending agent (41). A dimethyl-
benzyl alcohol was present as regulator to avoid crosslinking in the prepolymer. After the prepolymer had been
washed with methanol, it gave clear solutions in acetone.
After storage for some time t-butyl perbenzoate could be
added and thermoset moldings obtained by heating under
pressure above 100°C.

Aqueous dispersions of DAP prepolymer with nonionic
emulsifiers have been applied for binding together random
glass fiber mats (42). The binder on the glass was heat-
cured before impregnation with unsaturated polyester-
monomer solutions. Polymerization of DAP along with a
little CCl_4 as regulator has been suggested (43). The
prepolymer precipitated by methanol, after drying, had a
refractive index of n_D^{25} = 1.535, and analysis showed 5.2%
chlorine. The partial polymer as an aqueous dispersion
was applied to glass fiber webs and the water was evaporated. DAP polymerization in aqueous suspension in the
presence of mercapto-organic acids as regulators has been
proposed (44). Prepolymers have been prepared by heating
at 165°C in the presence of an aldehyde (45). DAP has
been polymerized in aqueous dispersion using $MgCO_3$ powder
as suspending agent (46). It polymerized faster under
nitrogen than under air. Under these conditions of peroxide initiation, DAP polymerized faster than triallyl
citrate. Aqueous emulsions of DAP prepolymer and monomer
containing t-butyl perbenzoate were applied to paper and
textiles (47).

There has been little success in ionic polymerizations
of diallyl phthalates. Boron alkyl catalysts in the
presence of oxygen probably initiate DAP by free radical
reactions (48).

Little has been published on chemical reactions of DAP prepolymer other than curing. However, the prepolymer has been partially epoxidized (0.6-3.6% oxirane or epoxy groups)(49). DAP polymers have been sulfonated (50). Little has been reported on nonvinyl polymerizations involving diallyl phthalates. However, they have been observed to react with SCl_2 and with S_2Cl_2 to form resinous products of uncertain structure (51).

CYCLIZATION IN DIALLYL PHTHALATE POLYMERIZATIONS

In the early 1940s, certain polyfunctional ethylenic monomers were observed to undergo gelation at conversions higher than predicted by theoretical considerations. Partial cyclization had been considered as a possible wastage of double bonds and of crosslinking, which might explain such delays in gelation. Therefore, the question arose as to the extent to which such cyclization reactions might be expected in polymerizing systems.

Haward (52) considered the possibilities of intramolecular reactions in ethylenic polymerizations, where he showed that the concentrations of radicals present in a normal free radical polymerization were so low that self-termination of biradicals would be expected to take place. Although the conclusion reached was that biradicals were probably absent, the calculations emphasized the possible importance of cyclization; indeed, Simpson and Haward showed later that intramolecular crosslinking probably occurred during the early stages of the copolymerization of styrene and divinyl benzene (53).

However, the first conclusive evidence for intramolecular cyclization was obtained from a study of diallyl o-phthalate polymerization by Simpson and co-workers in England (54). On heating with a free radical catalyst, DAP polymerizes to form an initially soluble polymer (beta polymer). At about 25% conversion, the system gels and the beta polymer and monomer are thereafter converted to insoluble, infusible crosslinked polymers (Fig. 11.4). The structure of the soluble polymer was studied by examining the degree of polymerization and residual unsaturation, followed by saponification and examination of degree of polymerization of the resulting polyallyl alcohol, which represented the linear portions of the branched polymer (54). There was no evidence that the saponification treatment affected the degree of polymerization of the polymer. The beta polymer from DAP has a residual unsaturation of about 26%, very nearly independent of conversion up to the gel point. Saponification and acetylation to polyallyl acetate showed that the DP of the individual chains was about 20 allyl groups, again

independent of conversion. The results were consistent
with the idea that a considerable fraction, about 40%, of
the DAP monomer units contribute both of their allyl
groups to each polymer chain. This intramolecular cyclization was called "incestuous" polymerization by Gordon
(55).

These results raised the question of the behavior of
other diallyl esters during polymerization. A systematic
study of cyclization in diallyl ester polymerization was
therefore carried out by Simpson and Holt (56), who found
that the ratio of cyclization to propagation reaction coefficients could be readily measured and provided a
quantitative measurement of the tendency of the monomer
to undergo cyclization. Of the three isomeric phthalate
esters, only the ortho-ester showed a marked tendency to
cyclize. In the case of diallyl terephthalate, the evidence suggested that the initial polymer was not cyclized;
further polymerization resulted in polymers containing
more than one crosslink per chain. Other workers confirmed that the isophthalate and terephthalate monomers
showed less tendency to cyclize than DAP (57). The homologous series of alpha, omega esters decreased in tendency
to cyclize as the distance between allyl groups increased
from diallyl oxalate to diallyl sebacate, except in the
case of diallyl succinate, which showed an abnormally high
tendency to cyclize. This was in accordance with measurements of other physical properties of the homologous ester
series which indicated abnormal behavior of the succinate.

Following Simpson's work, Haward considered the competition between the intra- and intermolecular polymerization
of diallyl phthalate (58, 59). Using a geometric model
which assumed flexible linkages connecting the double
bonds to the benzene nucleus and restricting the calcula-
tion to the contribution of the two nearest double bonds,
he showed that about 31% of intramolecular polymerization
would be predicted compared with 41% found by Simpson;
that is, the extent of intramolecular reaction corre-
sponded with what might reasonably have been expected
from a rational model.

The effect of cyclization on conversion at gelation has
been considered theoretically by Gordon (55), but surprisingly, the experimental evidence is that although diallyl
o-phthalate polymer contains considerably more cyclized
units than other diallyl ester polymers, all of the monomers appear to gel at roughly the same conversion (60).

Butler and co-workers observed alternating cyclization
and propagation in the polymerization of diallylquater-
nary ammonium salts (61). See Chapter 19. Here and in
other cases, a portion of the unsaturation lost may

possibly result from mechanisms other than cyclization, depending on the experimental conditions.

Holt (62) showed that the high-temperature mechanical properties of copolymers of DAP with unsaturated polyesters are inferior to those of comparable copolymers prepared from the diallyl isophthalate and terephthalate esters.

The following equation represents a possible cyclization step in polymerization of diallyl o-phthalate at low conversion. Free allyl groups in the chain molecules are sources of branching and crosslinking at higher conversions.

Simpson and Holt reported that about 40% of reacted DAP monomer double bonds were used up in cyclic reactions. They used bromide-bromate for estimating residual double bonds. Using 1% benzoyl peroxide as initiator at 80°C, the conversions at first gelation were near 25% for all three monomers, DAP, DAIP, and DATP. The prepolymers of DAIP and DATP had greater proportions of active unsaturation. The chain lengths of the prepolymers were nearly independent of the conversions at which they were obtained. It is surprising that, despite more cyclization, DAP did not gel later than the other phthalate monomers. Other authors have given higher estimates of cyclized structure in DAP polymers (63). The fraction of internally cyclized unsaturation in DAP prepolymer was found by Gordon to be about one-half, and of the same order of magnitude as estimated for ethylene dimethacrylate-methyl methacrylate copolymers (64). High-temperature peroxides gave DAP polymers having higher second-order transitions, and curing with electron irradiation further raised the glass transition temperature (65). The precise structures of DAP prepolymers remain to be elucidated.

Gelation and unsaturation data for other diallyl and some dimethallyl monomers were reported by Simpson and Holt, but few details of purification, physical properties, and reaction conditions were recorded (56). DAP was believed to give degradative chain transfer to monomer, although a few of the resulting radicals may initiate new chain growth. Most of the monomers were estimated to use

up about 20 reacted allyl groups per macromolecule, little
affected by conversion. Exceptions were diallyl diphenate
and several dimethallyl esters. All monomers except DATP
showed some loss of unsaturation by cyclization.

Later both DAIP and DATP at 80 and at 100°C were reported to give initial soluble polymers containing about
50% of residual double bonds (66). D'Alelio copolymerized
DAP with divinyl benzenes, along with CCl₄ as a regulator,
to obtain larger proportions of soluble partial copolymer
(67). In the future, advantages may be expected from the
use of DAIP and DATP as curing agents because of less
tendency for cyclization. This can be attributed in part
to the greater distance between allyl groups in these
monomers.

POLYMERS FROM DIALLYL PHTHALATE

The DAP polymers introduced commercially in the United
States in 1951 have been most successful in filled electrical moldings, but the number of other applications is
growing. New features of polymer technology which have
been possible with DAP include unsaturated prepolymer
molding compositions, capable of curing during molding to
give solvent-resistant thermoset shapes; catalyzed fiber
premixes; and "prepreg" impregnated fabrics of good
storage stability. Whereas phenolic moldings cure by
condensation polymerization with loss of water, diallyl
phthalate prepolymers cure by ethylenic addition polymerization, essentially without formation of small-molecule
byproducts. In many of the first commercial applications,
DAP prepolymer compositions replaced phenolics which cured
at comparable rates and temperatures. Retention of good
electrical properties even at high humidities and elevated
temperatures encouraged development of special DAP resin
compositions for use in guided missiles, radomes, and
other sophisticated military hardware. Arc resistance
unaffected by water and chemicals has been most important
of all.

DAP molding resins have made possible water- and solvent-
resistant molded shapes for small electrical devices re-
quiring high dimensional stability, such as multipronged
electrical connectors. In electrical moldings advantages
include low shrinkage in molding with exceptional reproduction of details and compatibility with metal inserts.
High volume and surface resistance, low dissipation loss,
and low dielectric constant are possible even after long
conditioning at high humidity and elevated temperatures.

The molded commercial polymers usually are not quite
colorless. However, DAP polymer moldings have very much
better color and color stability than competitive phenol-
formaldehyde moldings. Brilliant, permanent colored

effects are possible. Chemical resistance to water,
acids, and dilute alkali is superior to that of phenolics
and polyester-styrene plastics. Heat resistance is good
(about 180°C) and somewhat better in more recently developed polymers based on diallyl isophthalate (DAIP).
Cured moldings may lose 16% in weight by heating for
1000 hr at 400°F.

For curing prepolymers, 2 to 4% of peroxides and temperatures above 300°F are required. The good storage
stability of catalyzed prepolymer-monomer compositions is
an advantage in many techniques of fabrication. The DAP
prepolymers, free-flowing white powders or granules, are
soluble in ketones, in aromatic hydrocarbons, and in
acetic acid. They are little soluble in lower alcohols,
ethylene glycol, diethyl ether, carbon tetrachloride, and
in aliphatic hydrocarbons. The tightly crosslinked cured
polymers, however, are not attacked by solvents and little
swollen even in chloroform.

Solutions of about 70% partial polymer to 30% DAP monomer are sufficiently fluid for application to glass fiber
and coatings. Solutions of prepolymer in acetone or methyl
ethyl ketone have been used for Dapon prepregs and coatings. From 5 to 30% monomer is often added to Dapon prepolymers to increase flow in molding. For fully curing DAP
prepolymers, relatively high-temperature peroxides are
used (e.g., 2-3% t-butyl perbenzoate at 120°C or 275°F or
above). At 400°F substantially full cure can be obtained
in 15 to 20 sec. Although the process is seldom used, it
has been reported that cure without added catalyst can be
obtained by heating for 15 min at 246°C.

When fully cured, diallyl phthalate polymers are resistant to most organic solvents, but they may increase up to
2% in weight in acetone. The polymers can be clear and
nearly colorless but have been little used without pigments
or fiber-reinforcing agents needed for reducing brittleness. Flame-resistant moldings can be obtained by incorporation of antimony oxide and by copolymerization with
diallyl chlorendate.

Table 11.2 presents some properties of DAP prepolymer,
cured polymer, and a cured proprietary polymer formulation
of high specific gravity made from Dapon D of FMC. Viscosities of solutions of a Dapon prepolymer in different
solvents are shown in Table 11.3.

There have been comparatively few fundamental studies of
pure unfilled DAP homopolymers. Polymers prepared by heating DAP with 2% benzoyl peroxide for 20 hr at 80°C have
been examined by the electron microscope (69). A micellar
structure of globular texture was detected which suggested
that polymerization did not take place uniformly through
the mass but more rapidly from specific scattered points.

TABLE 11.2

Properties of Diallyl o-Phthalate Polymers [a]

	Partial polymers	Cured polymers	Cured Dapon D (low viscosity composition)
Specific gravity at 25°C	1.259	1.270	1.745
Refractive index at 25°C	--	1.571	--
Iodine number	55	--	--
Softening range, °C	75-90	thermoset	thermoset
Shrinkage in cure, in./in.	ca. 0.05	--	0.06
Dielectric strength, volts/mil	--	450	--
Dielectric constant, 60 to 10^6	--	3.5-3.4	4.3-4.1
Dissipation factor, 60 to 10^6		0.010-0.011	0.009-0.007
Water absorption in 24 hr at 25°C, %	--	to 0.2	0.17 (50°C)
Rockwell hardness (M)	--	114-116	113
Izod impact, ft-lb/in.	--	0.3	0.49
Heat distortion at 264 psi, °C	--	155	242
Compression strength, psi	--	$22-24 \times 10^3$	29×10^3
Thermal conductivity, cal/(sec)(cm^2)(°C)(cm)	--	--	16×10^{-4}
Linear coefficient of expansion $\times 10^{-6}$ in./(in.)(°C), at 25 to 160°C	--	--	28.9

[a] Descriptions of tests in Ref. 68

TABLE 11.3

Viscosities of Solutions of a Dapon Prepolymer, Poises at 25°C [a]

Dapon, %	In DAP monomer	In benzene	In methyl ethyl ketone
20	2.2	0.03	0.02
30	20.0	0.27	0.09
35	60.0	0.75	0.21
40	--	2.1	0.48
50	--	17.0	2.50

[a] From data bulletin, FMC Corporation.

Molecular weight distributions of DAP prepolymers have
been determined by turbidimetric titration of solutions
in acetone by a mixture of methanol-water (2 to 3) as
precipitant (70). Curing of DAP containing dicumyl peroxide was followed by differential calorimetry and found
to be a first order reaction incomplete even after 20 min
at 160°C (71). An infrared band at 1598 cm⁻ was used to
estimate residual unsaturation.

DIALLYL PHTHALATE MOLDING MATERIALS

Because of the premium properties of thermoset, fully
cured DAP polymer moldings, much effort has been devoted
to developing formulations suitable for injection, compression, and especially for transfer molding (72). The
rheology or flow characteristics in molding depend upon
such factors as molecular weight distribution of the prepolymer, reactivity and percentage of added monomer, proportions of peroxide initiators of different temperature
characteristics, and small amounts of retarding agents.
Fillers influence flow, but they cannot be relied upon as
major controls. The degree of "advance" of polymerization
brought about by heat in the blending equipment also
greatly affects molding rheology. Small amounts of
selected phenolic inhibitors or metal compounds may be
added to control viscosity during molding (73). Mixtures
of both low and high temperature peroxide or azo initiators may be used--for example, BP along with t-butyl perbenzoate or dicumyl peroxide.

Temperatures in the range of 300-375°F or 150-190°C may
be employed for compression and transfer molding with
cycles of 60 to 90 sec. For compression molding, ram
pressures of 2500 to 7500 psi are suitable, but transfer
molding may require pressures of 5000 to 15,000 psi. Although granular compositions are preferred, long glass
fiber formulations also have been molded satisfactorily.
Chromium plated or stainless steel molds are employed.

Some experience of the U.S. Polymeric Corporation in
injection molding of short glass fiber compositions was
published in 1969 (74). A heat deflection test was devised to judge rates of cure. Concentrations of catalyst
and retarding agents were designed for successful injection in a particular mold. One composition required a
cycle of 35 sec at 340°F in order that the molded shape
passed a cure test of 3 hr in boiling chloroform without
attack. Without addition of retarder, cure required only
about 2 sec at 380°F and 8 sec at 340°F. However, a
15-sec cycle was necessary to prevent formation of blisters (attributed to inclusion of air). Minimum volatiles
and occluded gases must be maintained for success in in-

jection molding of DAP resins. It is recommended that the
plasticizing barrel be maintained at 225°F, permitting
volatiles to escape from the melt while the rate of cure
is relatively slow. Other aspects of molding allylic
resins were discussed by Raech (75).

A typical prepolymer for molding is Dapon 35 of FMC Corporation. Compositions including clays, fibers, or other
fillers together with specific peroxides give commercial
moldings having outstanding electrical properties and
dimensional stability. Careful control of flow characteristics and the low contraction during completion of
polymerization permit molding about metal inserts and
accurate reproduction of even small mold details of complicated shapes. Most metals are not attacked by the DAP
molding compositions. From molding data the activation
energy of the second stage polymerization or cure was
estimated as 13 kcal/mole (76).

Like most thermosetting plastics, DAP molding formulations need fillers for improving impact strength, for
lower cost, and for reducing shrinkage. Diallyl prepolymer, diallyl ester monomer, peroxide catalyst, pgiment,
filler, and other modifiers are blended. Relatively long,
chopped glass fibers provide best impact but give molding
difficulties. For best impact strength, glass fibers are
treated with finishes such as a vinyl silane (A-172 or
Garan). Acrylic and polyester fibers have been used to
advantage. Asbestos fillers give moderate strength at
low cost but inferior electrical properties.

Pigments used in DAP molding compositions include clays,
calcium carbonate, calcium silicates, barytes, and silicas. Small proportions of high surface silicas such as
Cab-o-sil and Hi-sil impart desirable thixotropic qualities for controlling flow. Small amounts of metal
stearates or long-chain acids may be added as internal
mold lubricants. DAP prepolymers with kaolin, t-butyl
perbenzoate, and zinc stearate or lauric acid as mold
releases have been proposed for electrical molding compositions (77). Coating the kaolin filler particles with
melamine-formaldehyde resin may improve resistance to
hydrolysis of the ester polymer and give better retention
of electrical properties (78). Besides copolymerization
with diallyl chlorendate, additions of antimony oxide and
hydrated alumina improve flame resistance (79).

An example of diallyl-phthalate-reinforced thermosetting
resin is the following flame-retardant, high-flow, encapsulating composition (parts by weight) (80):

Dapon EF partial polymer	45 parts
Wollastonite P-1	20
Antimony trioxide	10

Lauric acid	1	parts
Diallyl phthalate monomer	2	
Vinyl silane A-172	1	
t-Butyl perbenzoate	1	
1/4 in. glass fibers	20	
Silica (Cab-o-sil)	0.3	
Hydroquinone	0.002	

The first four solids may be blended in a ball mill, a
Hobart mixer, or a ribbon-type blender. Other items are
added into the mixing equipment, and finally the glass is
added and mixing continued for 10 to 30 min. Blending
may be completed on a two-roll rubber mill where fric-
tional heat is sufficient to soften and densify the mass
into a sheet. After cooling the sheets may be granulated
using a Wiley mill.

For higher impact glass-reinforced DAP moldings, mechanical mixing is minimized and oven drying is employed
instead of milling; this removes any acetone or other
solvent employed in blending. Dispersing the glass
fibers without breaking them is not easy. Using more
than 8% DAP monomer based on DAP prepolymer may cause
cold flow of granules during storage. For filler loadings
higher than 50%, the t-butyl perbenzoate concentration
should be raised to 3% based on the resin content. In
short glass fiber prepolymer molding compositions, economy
can be obtained without sacrifice of electrical and
strength properties by replacement of a part of the glass
by Wollastonite (naturally occurring calcium silicate
having acicular crystal structure). Properties of some
filled and unfilled DAP polymer moldings are shown in
Table 11.4.

A clay and asbestos filled DAP molding formulation
recommended for steam and flame resistant moldings has
been suggested by FMC (81):

Dapon 35 partial polymer	90	parts
Diallyl chlorendate	10	
t-Butyl perbenzoate	3	
Titanium dioxide	20	
Antimony oxide	10	
Silica (HiSil 233)	10	
Clay (ASP-403)	80	
Chrome yellow	3	
Lauric acid	2	
1/2 in. white asbestos (or		
1/4 in. glass fibers)	40	

Research is underway to obtain more complete curing and
more heat-resistant molding plastics. One approach is to
add other polyfunctional ethylenic monomers of high reac-

TABLE 11.4

Properties of Unmodified and Reinforced
Diallyl Phthalate Resin Moldings

Property	Unreinforced DAP polymer	Asbestos-filled DAP resin [a]	Orlon-filled DAP resin [a]	Short glass reinforced DAIP polymer (flameproof)
Tensile strength, psi	4000	6700	6000	10,000
Flexural strength, psi	9000	11,400	11,000	16,000
Compressive strength, psi	24,000	22,000	27,000	25,000
Rockwell hardness	115	--	--	--
Impact, Izod, ft-lb/in. notch	0.2-0.3	0.4	0.85	0.6-0.9
Heat distortion temp. at 264 psi, °C	155	163	129	>200
Specific gravity at 25°C	1.270	1.68	1.34	1.85
Refractive index (25°C)	1.571	--	--	--
Water absorption in 24 hr, 25°C, %	0.2	0.49	0.20	0.25(50°C)
Dielectric constant, 25°C, 10^3 cp	3.4	4.8	3.8	4.4
10^6 cp	3.4	4.4	3.5	4.4
Dissipation factor, 10^3 cp	0.011	0.030	0.020	0.007
10^6 cp	0.011	0.035	0.017	0.012
Dielectric strength, 25°C, V/mil	450	395+	398+	390
Arc resistance, sec	118	140	115	125
Surface resistivity, ohms	10^{15}	10^7+	10^7+	10^{13}

[a] Products of Parr Corp. (U.S. Polymerics Inc.).

tivity and low volatility. Softening of moldings by
chloroform has been used as a measure of degree of cure.
Another indication of undercure is the degree of staining
when heated with a 1% solution of a Rhodamine B fluorescent
dye (82).

Among recent developments are Dapon D and MD prepolymers
having much lower molecular weights and melt viscosities
than earlier molding grade Dapon M. Dapon D and MD, along
with suitable concentrations of reactive monomer and radical initiators, provide the "soft flow" necessary for
encapsulation of fragile, pressure-sensitive electronic
devices such as integrated electronic circuits (83).
Pressures substantially below the range of ordinary transfer molding of conventional thermosetting materials are
adequate for encapsulation of such sensitive electronic

devices. For example, pressures on the molding ram of
1000 to 100 psi were used.

The tight crosslinking of completely cured DAP polymers
almost completely excludes water. In tests of encapsulating cured polymer compositions at 100% relative humidity
and 70°C for 1000 hr, the following initial and final
volume resistances were observed. A silicone resin was
more resistant to moisture.

Cured polymer	Initial resistance, ohms	Final resistance, ohms
Silicone type	10^{14}	10^{13}
Dapon MD type	10^{14}	10^{12}
Epoxy type	10^{14}	10^{7}

OTHER APPLICATIONS OF DIALLYL PHTHALATE POLYMERS

DAP prepolymers have found application for impregnating
glass cloth or other textiles, which then can be laminated
or molded into rigid structures of high strength. In an
early patent, cotton duck cloth was impregnated with a
solution of prepolymer containing t-butyl hydroperoxide
and benzoyl peroxide (84). Eight layers of the treated
cloth were cured under pressure between platens at 90°C
for 30 min, followed by curing at 90 to 115°C for 30 min
and at 115°C for an hour. Muskat heated DAP containing 5%
BP at 85°C until a syrup was formed (85). Glass cloth or
muslin was impregnated and curing was advanced by heating
above 100°C.

Woven or nonwoven glass cloth may be preimpregnated with
mixtures of DAP partial polymer, DAP monomer, and peroxide
catalyst by drawing the cloth through a ketone solution.
After removal of solvent in an oven or drying tower, the
flexible somewhat tacky sheet with a film interlayer can
be rolled up for storage. Such glass cloth prepregs are
more stable than epoxy and unsaturated polyester types can
be stored for years before curing. They can be formed by
vacuum bag, pressure bag, or matched die molding. Curing
may be at 200 to 300°F for 30 min. Matched die molding
may use higher temperatures such as 350°F for a few
minutes at pressures up to 100 psi. DAP prepolymer-
monomer catalyzed solutions may be applied as a finish
coat to such articles (86).

For a Dapon-glass cloth laminate there was applied to
the glass 37.7% DAP resin (of which 20% was added as
monomer). The resin contained 2% t-butyl perbenzoate.
The glass cloth had been pretreated with Volan A finish.
After removal of solvent, the laminate was molded at 280°F

for 30 min at 100 psi. An FMC data sheet showed the performance of 1/8 in. moldings at elevated temperatures:

	Flexural strength, psi	Modulus, psi $\times 10^6$
As received at room temp.	76,030	2.88
at 200°F	61,200	2.56
at 300°F	45,200	2.21
at 400°F	19,500	1.40

By replacing DAP by diallyl isophthalate in the formulation just cited, the heat resistance of the glass cloth
laminate could be raised to flex strengths of 22,300 psi
at 450°F and 15,400 at 500°F. Strengths can be improved
in some cases by postcuring for hours to days at 300 to
350°F. Applications of molded prepreg glass cloth-DAP
resin have included rigid tubing, ducts, radomes, junction
boxes, aircraft and missile parts, nuclear devices, and
chemical equipment. Reinforcement with synthetic fibers
such as polyamides, polyethylene terephthalates, and
acrylic fibers has been developed (87). Acrylic fibers
give best electrical properties under high humidity and
have been used for insulating undersea cable amplifiers.
Orlon cloth laminates have been employed for chemical-
resistant equipment. Fiber-reinforced molding compositions have been supplied by Allied Chemical (Mesa), U.S.
Polymeric, and Rogers Companies.

Impregnated papers have been developed for decorative,
stain-resistant overlays for wall panels, tables, and
furniture, using in some cases pressures as low as 100 to
200 psi for lamination (88). In one example, the acetone
solution used for impregnation contained 96 parts of DAP
prepolymer, 6 parts of DAP monomer, 1 part of t-butyl
perbenzoate, and 1.5 parts of long-chain alcohol (parting
agent) (89). Decorative surface finishes have been obtained from DAP-impregnated synthetic fiber textiles (90).
DAP polymers have been used along with melamine-formaldehyde and other aminoplasts in laminates and coatings (91).
In decorative laminates over wood veneer, 1 to 3% of added
glycerol or diallylmelamine may improve mar resistance
(92). Fire-retardant laminates have been developed (93).
Polyvinyl alcohol papers have been used with DAP polymers
for decorative laminates in Japan (94).

Solutions of DAP polymer and monomer are used in protective coatings, in sealing agents, and for encapsulation. Most applications so far have been in electrical
and metal fields. Cured coatings have been used on

capacitors, resistors, transformers, coils, motor wind-
ings, and transistors. DAP polymer-monomer solutions have
been applied to render concrete surfaces impervious to
water and resistant to staining (95). Powdered compositions have been applied for coatings; for example, 100
parts DAP prepolymer, 5 to 50 parts unsaturated solid
polyester, up to 10 parts DAP monomer, and peroxide catalyst (96).

One simple DAP sealant for imperfect metal castings contained 25% prepolymer dissolved in 75% monomer catalyzed
by 4% t-butyl perbenzoate. The solution was stable for a
year at normal temperatures. Voids in aluminum or other
metal castings were given a vacuum treatment before applying the DAP solution and heating with 100 psi pressure
(97). The polymer filled casting was heated at 300°F for
an hour or longer for curing. The main purpose of this
filling of metal castings is to exclude gases and corrosive liquids from the castings. Solutions of DAP polymer
in monomer have been used as sealants for wood, e.g., for
textile bobbins resistant to textile finishes. Approximate viscosities of solutions of one DAP prepolymer in
monomer are shown in Fig. 11.5. DAP prepolymer compositions photosensitized with aromatic azides have been
evaluated in photographic films (98).

Diallyl phthalate polymers have limited compatibility
with other polymer systems. Mixtures of bisphenol A-
epichlorohydrin-epoxy resins with phthalic anhydride
curing agent and peroxide-catalyzed DAP have been cured
at 130°C for 4 hr (99). DAP-modified epoxidized polybutadiene low polymer blends have been cured by heating at
temperatures of up to 200°C (100). In Europe polyvinyl
chloride plastisols have been formulated with 30 parts of
DAP to 30 parts of phthalate ester for use in coatings and
adhesives cured at 205°C for 3 min. As catalyst, 0.2 part
of benzoyl peroxide and 0.4 part of dicumyl peroxide may
be used. The following composition was suggested as a
coating for steel to be cured for 30 min at 160°C (101):

Vinyl chloride polymer	100 parts
Plasticizer	40
DAP	40
TiO_2	20
Stabilizer	3
t-Butyl perbenzoate	0.08
Xylene-toluene (2:1)	90

Coatings have been made from blends of vinylidene fluoride
polymers with DAP prepolymer (102). DAP resin compositions
cured at 130°C were used as adhesives for bonding metals
such as dichromate-treated copper (103). Solutions of DAP

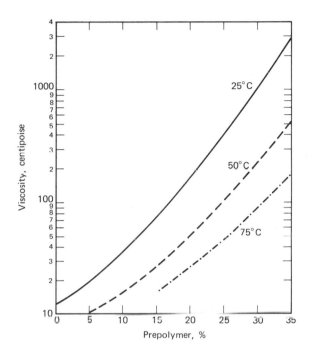

Fig. 11.5. Approximate viscosities of solutions of a commercial DAP polymer in DAP monomer at three temperatures. Data from FMC Corp.

prepolymer in tetraethylene glycol dimethacrylate containing a little acrylic acid were employed in bonding metals (104). Promoters permitted cure at room temperature.

UNSATURATED POLYESTER-DIALLYL PHTHALATE COPOLYMERS

In comparison to styrenes and styrene-methyl methacrylate mixtures the use of diallyl phthalates for copolymerization with unsaturated polyester resins offers advantages of much longer stability of the catalyzed syrups, less odor and volatility, lower shrinkage during cure, and higher heat resistance in the thermoset articles. Disadvantages include slower cure rates, somewhat higher syrup viscosities, and higher costs. The DAP-polyester or DAP-alkyd resin systems require higher catalyst concentrations, longer curing cycles, and in some cases higher curing temperatures. The usual range is 20 to 40% DAP in polyester-DAP syrups.

As in the case of the polyester-styrene resins, the
polyester-DAP resins generally are used with fiber reinforcement, especially glass fibers or glass cloth. In
addition, DAP monomer has found wide use as a copolymerizable, yet storage-stable, solvent for peroxide catalysts
for all types of polyester syrups and for other ethylenic
polymerizations (105). In spray molding of thermosetting
polyesters DAP is particularly useful as a non-volatile
catalyst carrier.

In early work on copolymerization of DAP with unsaturated
ester polymers, application of DAP-drying oil modified
alkyds was suggested in coatings (106). DAP-polyglycol
maleates were proposed by Kropa as aqueous dispersions in
printing and finishing of textiles (107). Kropa disclosed
early cast polymerizations of a number of high-boiling
diallyl esters including diallyl fumarates, maleates,
phthalates, and succinates with viscous liquid unsaturated
polyester resins (108). Organic peroxides, cobalt salts,
as well as Lewis acids, heat, and light were suggested as
promoters of copolymerizations. In one example, 40 parts
of DAP and 60 parts of viscous liquid ethylene glycol
maleate polymer were copolymerized in the presence of 0.2%
benzoyl peroxide using temperatures up to 150°C. A solid
clear copolymer was obtained in 5 min (109). This was an
early example of high temperatures useful for fast curing.
Kropa stabilized the stored solutions of polyester-DAP by
0.01 to 0.1% hydroquinone.

In another patent, an unsaturated polyester was prepared
by heating the following mixture for 8 hr at 180°C in the
absence of gaseous oxygen (110): 3.6 moles ethylene
glycol, 3.0 moles diethylene glycol, 4.0 moles fumaric
acid, and 2.0 moles phthalic anhydride. To 4 parts of the
viscous liquid product was added 1 part of DAP, and the
resulting homogeneous solution, with added peroxide, was
heated to form a thermoset clear copolymer. Heat resistance was favored by reactive fumarate unsaturation in the
unsaturated polyesters (111).

Preparation of a series of unsaturated polyesters for
experimental copolymerization with DAP was described (112).
The ester polymers were formulated using 10% excess diol
with maleic anhydride and other diacids to produce molecular weights of 800 to 900. At the conclusion of the
polyesterification at 210 to 225°C, the viscous liquid
was cooled to 135°C before diluting to 25% DAP monomer.
The solution contained 0.01 to 0.02% hydroquinone inhibitor based on polyester. Portions of the warm solution
were further diluted by DAP to give 30, 35, and 50% DAP
for evaluation in clear castings and in glass cloth laminates. In all cases 2% t-butyl perbenzoate was added as
catalyst. Slow curing cycles at 180 to 300°F were followed

by 2 hr postcure at 300°F. The advantages of DAP over styrene with ordinary polyesters in strength properties were not so outstanding, but DAP gave better electrical properties at higher temperatures, especially when relatively high proportions of maleic anhydride and of DAP were used.

Inhibitors or stabilizers against polymerization of polyester-DAP solutions during storage have been chosen for minimum retardation of polymerization during cure. Quaternary ammonium salts such as quaternary ammonium acetates and oxalates at concentrations of the order of 0.1% have been suggested (113). Some t-arylamines can accelerate polymerization at optimum low concentrations. See Fig. 11.6. Rates and exotherms of several polyesters with DAP and a series of peroxides have been investigated (114). A polyester from phthalic anhydride, maleic anhydride, and propylene glycol with an equal weight of DAP

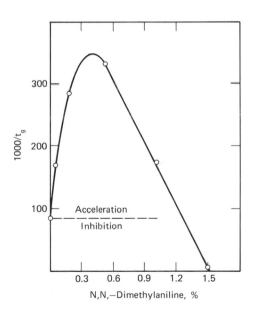

Fig. 11.6. Small additions of N,N-dimethylaniline accelerated copolymerization of a 1:1 mixture of DAP and diethylene glycol maleate at 77°C; higher concentrations produced inhibition. The t-amine triggered the initiation by benzoyl peroxide. Extent of polymerization is plotted as 1000/time for gelation. W. E. Cass and R. E. Burnett, Ind. Eng. Chem. <u>46</u>, 1623 (1954).

cured twice as fast as similarly catalyzed DAP at 280°F.
At 50% DAP the high temperature catalysts such as t-butyl
perbenzoate gave maximum exotherm and minimum time to
peak exotherm. Benzoyl peroxide gave faster gelation but
incomplete cure.

Beacham and Litwin found that 50 ppm of 2,6-di-t-butyl-
p-cresol in DAIP gave good storage stability up to 50°C
without impairing the exotherm in curing. This is a lower
concentration of inhibitor than is usually effective with
polyester-styrene. Difunctional peroxides gave special
promise in faster cure of unsaturated polyester-DAP solutions at 280 to 300°F.

Diallyl phthalate monomers copolymerized with polyesters
containing symmetrical 1,4-aromatic segments derived from
bisphenol A give outstanding heat resistance. Such a
polyester is the fumarate polyester from 2,2-bis(hydroxy-
propoxyphenyl) propane (115). Holland suggested triallyl
cyanurate, triallyl isocyanurate, diallyl maleate, diallyl
chloromethyl phosphonate, diallyl benzene phosphonate, and
triallyl trimesate as high-boiling comonomers along with
diallyl phthalates in copolymerizations with unsaturated
polyesters (109).

Laminac Polyester 4204 of American Cyanamid is an
example of a commercial unsaturated polyester-diallyl
phthalate syrup. It has a viscosity at room temperature
above 1000 poises and a specific gravity of 1.22. It has
been used for production of fiber-reinforced structures
by hand lay-up, open assembly methods where the odor of
styrene had proved objectionable to operators. The useful
life after catalyzing with 1% benzoyl peroxide is 7 days
or longer at 77°F. Another advantage is the low volatility
of DAP, making possible the fabrication of large sections
by hand lay-up followed by oven cure without appreciable
weight losses. The high viscosities reduce drainage during
curing. Although not capable of complete cure at room
temperatures, the syrups can be gelled with methyl ethyl
ketone peroxide activated by cobalt naphthenate. The
minimum temperature for complete cure is 200°F. The low
volatility of the monomer makes possible the use of hot
air for curing large structures. Aircraft radomes are
typical military applications. Laminac 4202 showed a gel
time of about 14 min at 90°C in a test using 1% benzoyl
peroxide. For curing small glass fiber-reinforced moldings of 1/8 in. thickness with 1% benzoyl peroxide, 7 to
8 min at 220°F or 1 to 2 min at 280°F were the conditions
recommended. Advantages and properties of cured DAP-
polyester molding formulations, prepregs, wet lay-ups, and
matched metal die moldings have been reviewed (117).

Fiber-reinforced DAP syrups without polyester, such as
disclosed in an early patent by Swedlow, have had limited

application in casting processes (118).

Formulations and technology of DAP-polyester premixes
have been discussed (119). Syrups of 1000 to 3000 cps are
generally used, but higher viscosities up to 50,000 cps
find special applications. Glass fibers are most often
used as reinforcement. Sisal fibers lend economy and
surface smoothness but less durability in the molded part.
Acrylic, nylon, and other synthetic fibers, as well as
asbestos, impart medium strength and surface smoothness.
Special mixers have been designed for gentle final mixing
in of glass fibers without disastrous fracture of the
glass. China clay, coarse talcs, and calcium salts are
most employed as pigment fillers. DAP-polyesters provide
better pigment compatibility than do styrene-polyesters.
The premix must be "dry" enough to leave the mixer clean
and to be handled without excessive tack, but it must be
sufficiently fluid to flow into the mold without separation of the components. The composition ought not to be
so abrasive as to cause excessive wear of mold surfaces.
Interesting studies have been made of the shapes and behavior of different inorganic filler particles. China
clay and talc have plate-shaped silicate particles, Wollastonite is needlelike, and calcium carbonate is nearly
spherical (120).

Somewhat different types of formulations have been
emphasized in Europe where DAP resins under the name Poly-
dap have been supplied. For example, approximately equal
weights of glass fiber mat and DAP resin (75% prepolymer
and 25% monomer) have been used as prepregs for resistant
laminates replacing stainless steel in chemical industry
and for heat-resistant electrical insulation. Durapreg
formulations are preimpregnated glass fiber mats containing special unsaturated polyesters without styrene. One
formulation designed for engineering applications was the
following:

Material	parts
Glass fiber mat	42
Mineral filler	8
DAP monomer	15
DAP prepolymer	50
Unsaturated polyester	30

Aqueous emulsions of polyester-DAP have had only limited
use, e.g., as binder for glass fiber preforms. A mixture
of 70 parts unsaturated polyester, 30 parts DAP, and 4
parts benzoyl peroxide monomer was emulsified by shaking
with water and sprayed on glass fiber mat (121). After
curing at 300°F, the treated mat was used for impregnation
with polyester-styrene syrup.

Minor proportions of DAP have been copolymerized in
thermosetting unsaturated polyester alkyds for nonflammable electrical plastics. In one patent example, one
part of DAP was used with 9 parts of solid unsaturated
alkyd. The latter was prepared by heating at 175 to 235°C
a mixture of maleic anhydride, tetrachlorophthalic anhydride, ethylene glycol, and propylene glycol under carbon
dioxide atmosphere (122). Small proportions of methyl
methacrylate have been included with unsaturated polyester-DAP in filled electrical molding plastics (123). In
one example, 125 parts of polyester, 48 parts of DAP, 8
parts of MMA, and 3.5 parts of benzoyl peroxide with pigment and filler were milled to a sheet and then cut to
nontacky granules. Addition of $Zn(OH)_2$ was said to improve water resistance of electrical thermosetting moldings
based on unsaturated polyesters with DAP and styrene (124).
One patent example includes 2% benzoyl peroxide, 6% zinc
hydroxide, and 2% zinc stearate, with kaolin and asbestos
as fillers. DAP and vinyl acetate were copolymerized with
a mixture of unsaturated polyesters (125). Clay and
asbestos-filled unsaturated alkyd-DAP molding granules
may be favorably stabilized against premature cure during
storage by a small amount of an azo-naphthol derivative
(126).
Unsaturated polyesters blended with DAP have been cured
by high-energy electrons applied at rates below 10^6 roent-
gens per second (127). Numerous patents disclose copolymerization of DAP with chlorine-containing polyesters to
give fire-resistant polymers (128). Kaolin-filled polyester-DAP solutions were proposed for dip coating (129).
A thermosetting adhesive for Mylar type polyethylene
terephthalate films has been formulated from 100 parts of
polyvinyl butyral, 100 parts of DAP, 150 parts of diethylene glycol maleate polyester, 6 parts of benzoyl peroxide,
and 1000 parts of ethylene dichloride solvent (130).
Laminates of outstanding strength and tensile properties
were molded from alternating layers of adhesive-coated
Mylar film and glass cloth that had been "wetted" with a
solution of peroxide-catalyzed diallyl phthalate. The
lamination process required 45 min under pressure at 200°F.
Other applications of DAP-unsaturated polyester formulations have been discussed by Nowlin and Burnett (131).
Partially epoxidized alkyd resin was blended with DAP or
with diallyl cyanamide and the mixture was cured with both
a peroxide and a curing agent for epoxy resins (132).
Additional U.S. patents on copolymerizations of DAP monomers with unsaturated polyesters show the interest in this
field:

         2,532,498 (L-O-F)   2,726,177 (Atlas)
         2,543,635 (G.E.)   2,819,243 (Allied)

2,549,732 (L-O-F)   2,840,535 (Crandon)
2,583,150 (Westinghouse)   3,216,877 (Manville)
2,584,315-6 (Texas)   3,464,942 (Ferro)
2,652,382 (Am.Cyan.)

OTHER DIALLYL PHTHALATE COPOLYMERIZATIONS

Allyl esters tend to copolymerize rather reluctantly
under many conditions, described in the patent literature.
DAP was found to copolymerize only slowly with monoallyl
esters of long-chain acids, e.g., 1-30% allyl stearate
(133). Low yields of soft gels were obtained after long
heating. DAP copolymers with allyl propionate were found
to be compatible with cellulose nitrate within limits
(134). DAP was copolymerized with vinyl acetate on heating with benzoyl peroxide (135). DAP copolymerized with
diallyl sebacate at 250°C in the presence of ceric
linoleate (136).

Just as mixtures of vinyl acetate and styrene resist
copolymerization and may remain mutually inhibited, diallyl phthalates do not copolymerize readily with styrene
under mild free radical conditions used for styrene homopolymerization. At higher temperatures, copolymerizations
of DAP with styrene were slightly more favorable. With
t-butyl hydroperoxide as catalyst, reactivity ratios for
DAP + styrene at 60°C were $r_1 = 0.057$ and $r_2 = 32.8$; at
120°C, $r_1 = 0.105$ and $r_2 = 16.5$ (137,138). Less cyclic
polymerization occurred as the proportion of styrene was
increased. In some cases monomers which resist copolymerization can be added gradually to the other monomer
heated with initiator. Thus small successive proportions
of vinylidene chloride have been added to polymerizing
DAP (139).

Of special interest are copolymerizations producing
precopolymers, which cure substantially faster than DAP
fusible homopolymers. In one patent example, a mixture
of 185 parts of DAP, 187 parts of triallyl cyanurate, 45
parts of isopropanol (91% by volume), and 3.7 parts of 50%
hydrogen peroxide was agitated and heated at 110°C for
4.3 hr (140). The viscous solution was precipitated by
10 times its volume of methanol using a colloid mill
operating at 3000 rpm. After washing and filtering with
methanol a second time, and by water a third time, the
white filter cake of copolymer was dried at 80°C, giving
a 31.5% yield from the initial monomers. The melting
range of the soluble copolymer was 100-120°C. The copolymer powder mixed with 2% t-butyl perbenzoate could be
molded at 175°C and 6000 psi for 15 min to produce a clear
infusible disk. Combinations of hydrogen peroxide with
organic peroxides may be chosen for copolymerization of

DAP with triallyl cyanurate (141).

Solid diallyl phthalate prepolymers may be copolymerized with polyfunctional, high-boiling monomers for easier molding and better heat resistance by tighter crosslinking. In one example, 80 parts of DAP prepolymer, 20 parts of dimethallyl maleate, and 3 parts of t-butyl perbenzoate were mixed and molded for 15 min at 150°C at 8000 psi (142). A postcure of 12 hr at 115°C was applied for encapsulating electrical components. DAP has been observed to copolymerize rapidly with minor proportions of "oxygenated" tetraethylene glycol dimethacrylate (143) and with minor proportions of allyl vinyl phthalate (144). Additions of diallyl maleate or related polyfunctional monomers can accelerate completion of curing (145). Minor proportions of diallyl melamine were added to DAP prepolymer to give molding compositions of low melt viscosity (146).

DAP copolymerizes also with ethylenic monomers bearing electron-attracting or negative groups, such as maleic anhydride (147), maleate and fumarate esters (148), acrylates (149), methacrylates (150), acrylonitrile (151), and acrylamide. Copolymerization of 35 parts of DAP with 65 parts of diallyl diglycolate within layers of glass cloth gave a nearly transparent laminate by matching refractive indices of the glass and the copolymer (n_D^{20} = 1.553) (152). It was necessary to add 5% benzoyl peroxide and to heat for 5 hr at 75°C followed by 3 hr at 115°C. Diallyl isophthalate was copolymerized with diallyl diglycolate under similar conditions. DAP copolymerized with CR-39 when relatively high BP concentrations were employed (153).

Other references disclosing copolymerizations of diallyl phthalate monomers are given below:

Styrene, acrylonitrile, maleic anhydride: U.S. 2,542,827 (Westinghouse.

Styrene: U.S. 2,341,175 (Dow) and 2,578,770 (Nash-Kelvinator).

Ethylene: U.S. 2,495,286 (Du Pont); cf. CA 78, 17045 (1973).

Drying oils: Brit. 916,090 (Resines et Vernis); Belg. 668,854 (Plastugil).

Vinyl chloride: Matsumoto, CA 76, 154181 (1972).

Vinyl chloride and vinyl acetate: U.S. 2,609,355 (Shell).

Monoallyl esters: Swern and Jordan, J. Am. Chem. Soc. 70, 2334 (1948).

Allyl isobutyl ether: Japan 18,541 (Mitsubishi).

N-Vinyl pyrrolidone: U.S. 2,762,735 and 2,831,836 (GAF).

Maleic anhydride: U.S. 2,479,522 (PPG).

Maleic or fumaric acid with prepolymer: Belg. 551,285 (Solvay).

Phenol-formaldehyde resin: U.S. 2,985,614 (Air Logistics).

Melamine- or urea-formaldehyde: U.S. 2,813,844 (GAF).

In order to improve the toughness of DAP polymer films, it is possible to copolymerize with minor proportions of selected acrylic monomers. Precopolymers of DAP with methylolacrylamide have been studied, for example (154). Gel times of mixtures of DAP with 20% of lower alkyl methacrylates at 65°C using 2% benzoyl peroxide were shorter than for homopolymerization of DAP (155). The copolymers would be capable of less crosslinking and should be less brittle than unfilled homopolymers. Copolymerization of DAP with methacrylonitrile gave relatively fast formation of hard, infusible products resistant to organic solvents (156). Copolymerization of DAP with minor proportions of diallyl esters of linseed dimer acids was tried for overcoming the brittleness of DAP homopolymers in films and coatings (157). A solution of t-butyl hydroperoxide was added in portions during the copolymerization.

Diallyl phthalate monomers have had interest as crosslinking agents in optical cements. In one example, a mixture was suggested of 40 to 60% DAP with diallyl benzene phosphonate, unsaturated polyester, n-butyl methacrylate, and a divinyl monomer (158).

Copolymerization of diallyl phthalates with selected chain-transferring or regulating monomers will give soluble polymers at higher conversions than are possible in homopolymerization of DAP. Curable soluble copolymers were obtained from heating catalyzed mixtures of DAP and methallyl chloride (159). DAP and terpenes have been heated above 100°C in the presence of air and a heavy metal naphthenate (160). The soluble polymeric products were copolymerized with acrylate esters or diallyl fumarate to give insoluble films.

There have been numerous studies of copolymerization of DAP with minor proportions of halogen-containing monomers for developing practical flame-resistant molding materials. In one example, 95 parts of DAP prepolymer was mixed with 5 parts of diallyl chlorendate (Dapon FR) (161), 3 parts of t-butyl perbenzoate, as well as pigments, calcium carbonate filler, and lauric acid lubricant (162). During the last stage of mixing on the two-roll mill, there was added 20 parts of asbestos fiber. Test bars, transfer molded at 150°C under 8000 psi for 2 min, were found to be slow burning, but the weathering properties and costs of such flame-resistant copolymers have been limitations. Dapon FR flame-resistant copolymers of DAP and diallyl chlorendate were discussed by Thomas of FMC (163). Use of antimony oxide gave additional fire resistance. Diallyl chlorendate may be gelled upon the surface of antimony oxide powder before incorporation into DAP prepolymer resins (164). Small amounts of crosslinking monomers such as DAP may increase polymer molecular weights or give desirable chain branching in vinyl copolymers. Addition of 0.02% of a diallyl phthalate monomer to vinyl chloride before polymerization was said to be beneficial in raising molecular weight, tensile strength, and modulus (165). Up to 10% DAP with N-vinyl pyrrolidone gave soluble copolymers which probably were branched (166). Higher proportions of DAP with vinyl pyrrolidone gave copolymers which were insoluble in water and lower alcohols. Azobisisobutyronitrile was used as initiator.

There has been recent interest in adding minor proportions of diallyl phthalates with selected high-temperature peroxides to preformed polymers for curing by crosslinking. In most cases it has not been proved to what extent actual graft copolymerization occurs. Such graft copolymerizations with polyethylenes and with acrylic polymers were studied (167). In one example, a dispersion of finely divided vinyl chloroacetate copolymer, DAP, glycerol cyclohexylmaleate, di-2-ethylhexyl tetrachlorophthalate, and t-butyl peroxide was used to coat wires (168). Heating produced sufficient fusion and curing without volatile solvents. Several di- and triallyl esters were added to enhance radiation crosslinking of vinyl chloride polymers (169).

In order to improve the strength and rigidity of polyethylenes and other polyolefins in the temperature range above the crystal melting points, diallyl phthalate as well as triallyl citrate and triallyl cyanurate have been evaluated. Polyethylene strips were allowed to absorb several percent DAP and then were irradiated for 3 sec using a high voltage electron accelerator (167). The polymer strips were then treated with methyl acrylate for 2 hr at 70°C. The weight increase suggested that a crosslinked graft copolymer of 69% polyethylene, 29% methyl acrylate, and 2% DAP units had been formed. Such DAP treatment before irradiation increased the percentage of grafting and the elastic modulus of the copolymer.

DAP is an interesting crosslinking monomer in polyvinyl chloride (PVC) plastisol compositions (170). One formulation included 100 parts PVC, 35 parts DAP, 15 parts epoxidized linseed oil, 0.2 part BP and 0.5 part di-t-butyl peroxide. PVC-DAP molding plastics have been studied (171). In one patent example, the following were mixed thoroughly and heated at 65°C:

PVC	100 parts
DAP	60
Dibasic lead phosphite	10
$CaCO_3$	10
Dicumyl peroxide	1.2

The powdered composition could be molded above 200°C for 15 min.

DAP has been used in forming sprayed-in-place rigid polyurethane foam shelters (172). DAP was added as a crosslinker of polyamides (173). Heat was applied until products of higher softening temperature and improved solvent resistance were obtained.

DIALLYL ISOPHTHALATE (DAIP)

This interesting monomer is also known as diallyl meta-phthalate; like its isomer DAP, it is a liquid of mild odor, low volatility, and relatively slow polymerization. It is significant that the pure iso monomer polymerizes on heating with peroxides more rapidly than the ortho monomer, that it undergoes less cyclization and yields cured polymers of somewhat higher heat resistance. DAIP also yields solid, nonsticky prepolymers or partial polymers such as Dapon M of FMC. The development of DAIP molding materials came after that of DAP products. Research and commercial utilization have been stimulated in recent years by the need for better heat resistant materials for electronic and space applications. The cured polymers from DAIP seem to show less degradation with loss of strength on heat aging than is encountered with the DAP polymers. While the DAP polymer formulations are suitable in some cases for continuous use at 175°C, the DAIP polymers may be used in some cases up to 200°C (174).

DAIP monomer can be prepared by methods similar to those used for DAP. A laboratory synthesis has been based on reaction of isophthalyl chloride with allyl alcohol for an hour at reflux (175). The DAIP after purification had refractive index $n_D^{25} = 1.5183$, compared with $n_D^{25} = 1.5240$ for diallyl terephthalate prepared under similar conditions. The monomer has been prepared by reaction of sodium isophthalate with allyl chloride in dimethyl formamide solvent (176). Zinc-promoted copper catalyst was used to prepare DAIP having $n_D^{25} = 1.522$ (177). Physical properties of DAIP monomer were given on page 329.

When test samples of the two diallyl phthalate monomers were heated with 1% benzoyl peroxide at 100°C, gelation occurred after 18 to 21 min in DAIP, compared with 32 to 40 min with samples of DAP (175). DAIP containing 5% BP polymerized through a syrup to a hard, clear solid after exposure for 24 hr to a mercury arc lamp at 1 ft (178).

The more rapid polymerization of diallyl isophthalate compared to diallyl orthophthalate was confirmed in polymerizations using DAP monomer samples from two American manufacturers (178). The following yields of dried prepolymers, insoluble in methanol, were obtained after

irradiation in bulk with 2% initial benzoyl peroxide at
30-40°C. The radiation was from a mercury arc lamp with
3660 Å and longer wavelengths as transmitted through the
Pyrex test tubes and cooling water:

| | Time of light exposure | |
	6.5 hr	14.5 hr
DAP (Mfgr. #1)	8.0%	15.7%
DAP (Mfgr. #2)	10.0	17.0
DAIP (Mfgr. #1)	19.0	30.0
Diallyl adipate	17.0	17.3

Exhaustion of the BP photosensitizer and initiator apparently began to reduce rates of conversion toward the end
of irradiation. The syrups formed were smooth flowing
and free from crosslinking and from particles of gel.
However, the DAIP viscous syrup obtained after 14.5 hr
showed a somewhat "jellyfish-like" type of flow (Gardiner
viscosity tube with flow time 4.5 sec) indicating approach
of gelation. The above prepolymers after precipitation by
methanol and drying overnight at about 80°C in contact
with glass were largely crosslinked but were much swollen
in toluene. Films of DAP polymer were less attacked by
toluene than those from DAIP. The DAP and DAIP polymers
were quite resistant to saponification by 5% aqueous triethylene tetramine at 60°C overnight. Under the same
treatment the diallyl adipate polymer dissolved to form a
viscous aqueous solution of copolymer. The DAP polymer
films were nearly clear and were fairly brittle after contact with the aqueous base. The DAIP polymer was more
white opaque and remained more tenaciously adherent to the
glass beaker than the DAP polymer.
When the monomers above were irradiated under similar
conditions without BP, conversions were low, e.g., below
1.0% (178). It was surprising that the DAP prepolymer
thus formed without initiator from DAP #1 monomer could
not be precipitated by addition of excess methanol. The
stable milky-white dispersions resulting suggested the
behavior of certain block and graft copolymers. Perhaps
the length and distribution of cyclic, linear, and branched
units in these macromolecules may be involved. The three
isomeric diallyl phthalate monomers offer interesting
possibilities for future structure-property studies.
Prepolymers of both DAIP and DAP are white, nonadhering
powders, soluble in ketones, aromatic hydrocarbons, lower
esters, dioxane, chloroform, and ethylene dichloride. The
isophthalate prepolymers show higher iodine numbers and
higher unsaturation than the orthophthalate prepolymers,
which seems to result from less cyclication. For this

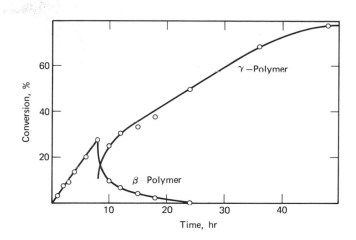

Fig. 11.7. Polymerization of diallyl isophthalate: prepolymer syrup formation to about 27% conversion was followed by gelation in the bulk polymerization with 1% BP at 70°C. Under these conditions DAIP and DATP polymerized faster than DAP. After gelation, soluble beta-polymer decreases as insoluble gamma-polymer increases. Alex Schulthess, Promotionsarbeit, Zurich ETH, 1960.

reason the DAIP polymers may be more highly crosslinked after curing. DAIP undergoes first gelation at slightly higher conversions than DAP. See Fig. 11.7. DAIP prepolymer is said to be less stable on storage than DAP prepolymer. Dapon M molding resins may contain several percent of DAIP monomer based on the polymer, along with peroxides for initiating crosslinking.

Electrical properties and resistance to chemicals and water of the cure polymers are outstanding, but low impact strength limits uses without reinforcement by fillers. Properties of typical cured, mineral-filled resin are presented in Table 11.5. Besides higher heat distortion temperatures, the diallyl isophthalate cured polymers show less weight loss on prolonged heating than DAP polymers. Molding conditions for both diallyl phthalate prepolymer compositions are similar, e.g., 270-320°F and pressures of 500 and 8000 psi for transfer and compression molding, respectively. Cured moldings from the diallyl isophthalate prepolymers have greater hot strength and can be ejected from the mold with less warping and deformation.

Neither prepolymers nor crosslinked polymers of high

TABLE 11.5

Properties of a Mineral-Filled, Cured
Diallyl Isophthalate Resin [a]
(Compression and Transfer Molding)

Property	Value
Specific gravity	1.82
Mold shrinkage, in./in.	0.003-0.006
Molding temperatures, °F	280-350
Tensile strength, psi	6500
Flexural strength, psi	10,600
Compressive strength, psi	27,200
Hardness (Barcol)	68
Weight loss in 100 hr at 400°F, %	0.75
Deflection temperature under load, °C	301
Water absorption in 48 hr at 50°C, %	0.26
Arc resistance, sec	197
Dielectric strength, volts/mil	
dry	480
wet	424
Dielectric constant	
1 kilocycle, dry or wet	4.0
1 megacycle, dry or wet	3.9
Dissipation factor	
1 kilocycle, wet	0.010
1 megacycle, wet	0.017
Surface resistance, megohms	3.0×10^3

[a] Poly-Dap 6260 (U.S. Polymeric Inc.) used for
switches, terminals, and insulators exposed to high
temperatures with high humidities.

molecular weight have been reported from any of the three
isomeric diallyl phthalates by ionic polymerizations with
Lewis acid or Lewis basic catalysts. Only brownish
viscous liquids were formed observed from treatment with
boron fluoride-etherate catalyst under various conditions
(179). Alkaline agents may cause saponification instead
of polymerization.

APPLICATIONS OF DIALLYL ISOPHTHALATE POLYMERS

Fiber-reinforced moldings of outstanding electrical properties have been developed from the filled prepolymers.
For example, orlon acrylonitrile copolymer fibers give
excellent electrical properties even under prolonged extreme conditions of high temperature and humidity. Even
after long exposure at 100% relative humidity at 80°F

orlon-filled moldings have shown insulation resistance of
5×10^6 megohms. Dacron, polyethylene terephthalate,
fiber reinforcement also has provided good impact strength
and heat resistance. Properties of a mineral-filled resin
after cure are shown in Table 11.3. The filled electrical
moldings based on DAIP have lower dielectric loss and
higher arc resistance than those based on DAP polymers
(180). Copolymers with minor proportions of diallyl
maleate, with diallyl chlorendate, as well as with diallyl
diglycolate or diallyl adipate, have been evaluated in
thermosetting moldings for electrical applications. Allied-
Mesa Plastics, Acme, and U.S. Polymeric Companies have
supplied mineral and fiber-reinforced electrical molding
materials for specific applications. Diall 1610 is an
electrically conducting DAIP plastic.

DAIP monomer-polymer mixtures have found use for preimpregnated glass cloth and for wet lay-up of glass cloth
reinforced plastics. FMC has supplied a blend of equal
parts monomer and polymer (Dapon M-50). These viscous
solutions can be heated or diluted for incorporating t-
butyl perbenzoate catalyst. The storage stability is not
as good as that of DAP formulations. The resin solutions
containing 2% t-butyl perbenzoate have been usable after
six weeks at 70°C, although viscosity had increased considerably. As much as 30% DAIP monomer based on total
resin may be used in preimpregnated glass cloth laminates.

High-temperature-resistant radomes have been made from
glass cloth impregnated with 37 to 63% diallyl isophthalate
monomer-polymer. A pressure of 800 psi was applied to 175
plies of cloth stacked in the cold press. The molding
press was heated to 220°F and held for 1 hr. The pressure
was increased to 1700 psi and the temperature to 240°F for
an additional hour. Finally 280°F and 1700 psi were
applied for an hour. Airborne radar for jet aircraft has
employed DAIP polymer-glass cloth where change of dielectric constant between room temperature and 500°F may be
only 4.3% and maximum loss factor only 0.020 at 500°F.

Remarkable strength properties can be obtained from com-
posited diallyl isophthalate prepolymer-glass cloth. The
following data were reported by FMC for Garan-finished
glass cloth with the 37% type Dapon M laminated at 100 psi
at 300°F:

                Flexural strength
                  Ambient   68,000 psi
                  Wet, 30 days   63,000
                  3-Month weathering   61,000
                Tensile strength
                  Ambient   50,000 psi
                  Wet, 30 days   49,000
                Specific gravity, 1.8

Asbestos-Dapon M has given promising ablative performance in rocket motor linings compared to reinforced
phenolics (180). As little as 25% DAIP with 75% DAP gave
improved heat-distortion temperature in asbestos-filled
molding formulations (181). There was used as catalyst
2% t-butyl perbenzoate based upon the prepolymer. Nonwoven glass fabric was saturated with a mixture of 60
parts of DAIP prepolymer, 40 parts of diallyl maleate,
3 parts of dicumyl peroxide, 80 parts of acetone, 75 ppm
hydroquinone (182). Layers were laminated by molding in
a bag under pressure at 82°C rising to 149°C.

Molding materials of glass fiber, as well as asbestos-
reinforced diallyl phthalate, are supplied in many formulations. Molding materials have been available with long
glass fibers as well as with glass flakes or spheroids.
Short glass molding compositions have been supplied as
granules, kernels, or beads. These molding materials
have been used in military and nonmilitary applications
such as connectors, switch bases, containers, insulators,
and other electronic components. Recently DAIP polymers
have been applied as binders for graphite and special
carbon-filled systems where low moisture absorption, high
wear resistance, and outstanding chemical resistance are
required. Applications include potentiometers, resistors,
pump bearings, and seals. For wet-lay-up fiber-reinforced
systems, approximately equal weights of monomer and prepolymer have advantages. The monomer imparts fluidity to
the compositions during forming. However, excessive
monomer retards curing and may contribute to shrinkage
and deformation of structures during molding and cure.

Polymers of diallyl phthalates and of other diallyl
esters have been suggested for lithography based upon
heat or radiation patterns produced by degradation of
polymer in the presence of peroxides (183).

Diallyl isophthalate solutions have given promise as
coatings resistant to heat and solvents. They have found
application as varnishes for dip encapsulation of small
high-performance electronic parts. A typical solution
contained: 100 parts DAIP prepolymer, 114 parts methyl
isobutyl ketone, 6 parts n-butyl acetate, 5 parts t-butyl
perbenzoate. One dip in this solution followed by drying
at 32°C for 10 min and curing for 15 min at 160°C formed
a 2.5-mil coating suitable for continuous use at 180°C.
For temperature-sensitive parts, curing could be accomplished at 100°C for 16 hr. If greater film flexibility
or tackiness was desired in the films before cure, some
monomer was added, e.g., 10% based on prepolymer. In dip
coating of carbon resistors, for example, solutions of
225 cp viscosity were suitable. The coatings could be
handled before cure, and after cure the films withstood
soldering of adjacent metal connectors.

DAIP prepolymers show promise in the preparation of
photosensitive coatings (184). Soluble prepolymer was
isolated at about 25% conversion by polymerization with
BP above 82°C. The prepolymers of DP 40 to 100 melted
below 95°C. Shrinkage of the prepolymer coatings in cure
was less than 1%. Mixtures of benzophenone or benzil with
Michler's ketone were effective sensitizers to 3200 to
4000 Å UV light. DAIP prepolymer and technical information have been supplied by Polysciences, Inc.

In copolymers with unsaturated polyesters the advantages
of DAIP over DAP are not outstanding. However, polyesters
from propylene glycol and maleic anhydride containing
fumarate double bonds copolymerize readily with DAIP (185).
An hour or longer at 160°C was required with peroxide
catalyst for full cure.

The applications of DAIP will be expected to grow substantially as the price of monomers is reduced and as more
research is done toward further improvement of heat resistance by copolymerization and by addition of stabilizing
agents.

Reactivity ratios have been reported from radical copolymerizations with styrene at 60°C (186):

	r_1	r_2
Diallyl isophthalate	0.09	27
Diallyl terephthalate	0.07	26
o-Dimethallyl phthalate	0.06	25
m-Dimethallyl phthalate	0.07	23

DIALLYL TEREPHTHALATE (DATP)

Although exploratory studies of DATP and its polymers
have been promising (187, 188), the practical advantages
of this monomer over diallyl isophthalate and the unique
properties of the polymers and copolymers remain to be
fully demonstrated. The monomer can be made by methods
similar to those employed for DAP and DAIP. A novel
method is reaction of dimethyl terephthalate with allyl
acetate in the presence of titanium n-propoxide at 58 to
75°C (189). The DATP monomer boiled near 160°C at 1 mm.
Diallyl terephthalate monomer was prepared by dropping a
solution of 152 g of distilled terephthalyl dichloride in
300 ml of benzene into 205 ml of dry allyl alcohol at
65°C (190). After addition of 50 ml of the acid chloride
at about 2 drops per second, the mixture began to boil.
When the addition had been completed, the mixture was
cooled, neutralized with aqueous sodium carbonate, and
the organic layer was distilled under vacuum. The diallyl
terephthalate boiled at 121°C at 0.02 mm (n_D^{20} = 1.5280).

Heating DATP without added catalyst under air for 5 days

at 80°C gave a viscous polymer syrup, and heating for 5 more days gave a soft, translucent gel (188). The thermal polymerization of this sample was faster than that of DAP. With 5% benzoyl peroxide and heating for 7 hr at 70°C, DATP became much more viscous than did DAP. Around part of the benzoyl peroxide which had not completely dissolved, polymerization was especially rapid. Some methyl methacrylate was added to accelerate polymerization by the Trommsdorff effect. With t-butyl perbenzoate DATP gave hard, clear, colorless polymers.

Swelling tests on diallyl phthalate crosslinked polymers gave surprising results (188). The DATP polymers absorbed much more o-xylene than DAP polymers. However, DATP polymers absorbed less toluene than DAP polymers. Such specificity in swelling might be useful for identification and characterization of polymer networks and in the design of semipermeable membranes and gels.

Both DAP and DATP monomers, when heated with 50% sulfuric acid, formed dark red solutions that gave strong green fluorescence when added to dilute aqueous alkali (188). DATP polymer was slowly hydrolyzed by sulfuric acid, and positive fluorescein tests after heating with resorcinol were obtained.

Few references deal with DATP polymerizations specifically. Matsumoto and co-workers studied copolymerization of DATP with allyl benzoate in bulk at 60°C with BP catalyst (191). Copolymerization ratios $r_1 = 0.77$ and $r_2 = 0.89$ were observed compared to $r_1 = 0.69$ and $r_2 = 1.02$ for DAIP and allyl benzoate. Higher proportions of initial allyl benzoate gave copolymers of higher residual unsaturation. DATP was copolymerized with a fumarate polyester in the presence of methyl ethyl ketone peroxide and cobalt naphthenate (192). Gelation occurred at 20°C and complete polymerization at 100°C. Dimethallyl terephthalate and an unsaturated polyester containing asbestos powder were copolymerized in a mold at 150°C (193).

DATP has been suggested to accelerate curing of fluoropolymers such as terpolymers of tetrafluoroethylene, vinylidene fluoride, and perfluoropropylene (194). The elastomer may be mixed on a cold rubber mill with 2 parts of dicumyl peroxide and 2 parts of DATP followed by curing for 18 hr at 200°C. The N-allyl monomer tetraallyl terephthalamide also gave vulcanized fluororubbers having promising physical properties. Use of allyl phthalates and cyanurates might be explored in dental plastics; indeed, DAP has been so disclosed in a Japanese patent (195). Diallyl monomers where the allyl groups are separated by at least three carbon atoms have been applied to crosslinking preformed polypropylene plastics in the presence of free radical initiators (196).

Courtaulds patented novel addition copolymerizations of DATP with silicon and tin compounds containing two reactive hydrogen atoms (197). Substantially equimolar proportions of DATP were reacted in dioxane solution with $(C_2H_5)_2SnH_2$ by heating for 5 hr at 60°C followed by 11 hr at 100°C. In another patent example, 0.25 mole DATP, 1 mole diphenylsilane, and 0.75 mole diphenyldivinyl sulfone were heated under nitrogen at 100°C for 48 hr using platinum-on-charcoal catalyst (198). A polymer of softening point near 80°C was obtained. The low softening temperatures may have resulted from the low molecular weight so far attained by this type of hydrogen transfer polymerization.

Among other monomers related to the diallyl phthalates are glycidyl allyl phthalate (199) and diallyl benzophenone carboxylates (200).

References

1. T. F. Bradley, U.S. 2,311,327 and Brit. 645,218 (Shell).
2. Dutch 71,996 (1953) (Shell).
3. A. DeBenedictis, U.S. 2,939,879 (Shell).
4. H. Stange et al, U.S. 3,086,985 and 3,250,801 (FMC).
5. G. Borsini et al, Fr. 1,372,967 (Sicedison); CA 62, 2742 (1965).
6. K. S. Tsou et al, U.S. 3,069,459 (Borden).
7. J. H. Lux et al, U.S. 3,306,928 (Hercules).
8. E. Leumann and D. Porret, U.S. 3,465,030 (Ciba).
9. H. H. Beacham, FMC Corp., private communication.
10. C. L. Wright and H. H. Beacham, U.S. 3,527,665 (FMC).
11. Diallyl Thermosetting Plastics, Shell Dev. Co. Report S-9837.
12. T. F. Bradley, U.S. 2,311,327 (Am. Cyan.), appl. 1937.
13. Brit. 571,496 (Thomson-Houston).
14. W. E. Vaughn and F. F. Rust, U.S. 2,426,476 (Shell); cf. C. A. Heiberger, U.S. 3,096,310 (FMC).
15. M. A. Pollack, I. E. Muskat, and F. Strain, U.S. 2,273,891 (PPG), appl. 1939.
16. J. K. Wagers and E. C. Shokal, U.S. 2,446,314, appl. August 1948, CA 42, 8527 (1948).
17. J. Anderson et al, U.S. 2,613,201 (Shell).
18. W. Simpson, J. Soc. Chem. Ind. 65, 107 (1946).
19. F. F. Rust and F. H. Dickey, U.S. 2,516,649 (Shell).
20. E. C. Shokal, Brit. 604,544 (Shell).
21. E. C. Shokal and F. A. Bent, U.S. 2,475,296 (Shell); cf. U.S. 2,475,297 (Shell).
22. W. E. Cass and R. E. Burnett, Ind. Eng. Chem. 46, 1619 (1954).
23. F. C. Hopper and Q. T. Wiles, U.S. 2,595,852 (Shell).
24. E. C. Shokal, U.S. 2,445,189; Shokal and K. E. Marple, U.S. 2,433,616 (Shell).
25. F. Lalau-Keraly, Compt. Rend. 249, 1213 (1959).

26.   C. A. Heiberger, U.S. 3,096,310 (FMC).
27.   P. E. Willard, U.S. 3,030,341 and Brit. 894,239 (FMC).
28.   S. A. Mednick et al, U.S. 3,385,836 (FMC).
29.   S. A. Mednick, U.S. 3,398,125 (FMC).
30.   C. W. Johnston and D. Warren, U.S. 2,990,388 (FMC).
31.   D. Porret, U.S. 3,390,116 (Ciba).
32.   A. Neri and L. Capitano, U.S. 3,474,127 (Swiss Al.);
      Belg. 666,639 (Swiss Al.); CA 65, 9053 (1966).
33.   T. Lanaka, U.S. 3,424,729 (Osaka Soda).
34.   T. Horio et al, CA 68, 50350 (1968).
35.   D. Porret and L. Leumann, U.S. 3,366,667; French 1,412,673;
      Belg. 608,489 (Ciba).
36.   T. Imoto and T. Nakajima, J. Chem. Soc. Japan, Ind. Chem.
      Sect. 69/3, 520-524 (1966).
37.   R. W. Warfield and M. C. Petree, J. Polym. Sci. 37, 305 (1959).
38.   O. Schweitzer and W. Zuerfurth, Ger. 928,734 (Degussa); cf.
      W. C. Heraeus, Ger. 963,995 (Degussa); CA 53, 20914 (1959).
39.   R. R. Smith et al, Brit. 864,675 (BX Plastics).
40.   C. A. Heiberger, U.S. 3,096,310 (FMC).
41.   C. A. Heiberger and J. L. Thomas, U.S. 2,832,758 (FMC).
42.   R. R. Hough and R. Barone, U.S. 3,002,869 (U.S. Rubber).
43.   W. Festag, U.S. 3,131,088 and Ger. 1,123,111 (Goldschmidt).
44.   German patent application 1962 (Henkel); cf. CA 78, 30789 (1973).
45.   A. Polis, Ger. 1,053,786 (Solvay).
46.   K.-S. Chia and F.-Y. Chao, Chemistry (Taiwan) 31 (1956);
      CA 51, 1645 (1957).
47.   W. Festag, Ger. 1,163,774 (Goldschmidt); CA 60, 13452 (1964).
48.   Belg. 562,701 (Solvay); CA 53, 10845 (1959).
49.   M. B. Mueller et al, U.S. 3,155,638 (Allied).
50.   I. P. Losev, USSR Patent, CA 53, 17373 (1959).
51.   J. L. Lang, U.S. 3,055,872 (Dow).
52.   R. N. Haward, Trans. Faraday Soc. 46, 204 (1950).
53.   W. Simpson and R. N. Haward, J. Polym. Sci. 16, 440 (1955).
54.   W. Simpson, T. Holt and R. J. Zetie, J. Polym. Sci. 10, 489
      (1953).
55.   M. Gordon, J. Phys. Chem. 22, 610 (1954).
56.   T. Holt and W. Simpson, Proc. Roy. Soc. (London) A, 238, 154
      (1956).
57.   M. Oiwa and Y. Ogata, Nippon Kagaku Zasshi 79, 1506 (1958);
      CA 54, 4488 (1960).
58.   R. N. Haward, Lecture at Polytechnic Institute of Brooklyn,
      October 1953.
59.   R. N. Haward, J. Polym. Sci. 14, 535 (1954).
60.   W. Simpson and T. Holt, J. Polym. Sci. 18, 335 (1955).
61.   G. B. Butler and R. J. Angelo, J. Am. Chem. Soc. 79, 3128 (1957):
      cf. G. B. Butler, J. Macromol. Sci.-Chem. A5, 219-227 (1971)
      (Ref. 5 cites A.E.C. Report of September 1953).
62.   T. Holt, unpublished.
63.   M. Gordon and R.-J. Roe, J. Polym. Sci. 21, 27 (1956).
64.   S. Loshaek, J. Polym. Sci. 15, 391 (1955).

65. W. Simpson and T. Holt, J. Polym. Sci. 28, 445 (1958).
66. F. Lalau-Keraly, Compt. Rend. 253, 2975 (1961).
67. G. F. D'Alelio, U.S. 2,378,197.
68. G. Nowlin and L. S. Burnett, SPE J. 17, 1093 (October 1961).
69. E. H. Erath and R. A. Spurr, J. Polym. Sci. 35, 391 (1959).
70. Y. P. Vyraskii et al, CA 76, 141450 (1972).
71. P. E. Willard, Polym. Eng. and Sci. 12, 125 (1972).
72. J. N. Jessup and H. H. Beacham, Mod. Plastics 45, 129 (December 1968).
73. J. L. Thomas, U.S. 3,579,484 (FMC).
74. K. R. Hoffman and R. E. Nanfeldt, SPE J. 25, 31 (1969).
75. H. Raech, Allylic Resins and Monomers, Reinhold 1965.
76. S. Y. Choi, SPE J. 26, 51 (1970).
77. T. F. Anderson, U.S. 2,680,722 and Brit. 737,507 (Allied); cf. U.S. 2,811,500 (Allied); C. W. Johnson and D. Warren, U.S. 2,990,388 (FMC).
78. D. E. Cordier, U.S. 2,811,500 and Brit. 752,805 (Allied).
79. A. J. Dontje and R. C. Berry, U.S. 3,362,928 (Rogers Corp.).
80. Technical literature, FMC Corp.
81. Cf. J. L. Thomas, SPE J. 23, 30-5 (October 1967).
82. A. L. Ashton and J. B. Pratt, Brit. Plastics 37, 310 (June 1964).
83. N. R. Segro and H. H. Beacham, Proc. Electrical Insulation Conf. September, 1969, p. 98; cf. R. F. Zecher, SPE J. 26, 45 (May 1970).
84. J. K. Wagers and E. C. Shokal, Can. 435,549 (Shell).
85. I. E. Muskat, U.S. 2,517,698 and 2,596,162 (Marco Chemicals).
86. B. S. Taylor, U.S. 2,965,532 (FMC).
87. Brit. 720,270 and 722,267 (Bakelite Ltd.).
88. P. E. Willard, U.S. 3,049,458 (FMC); B. S. Taylor, U.S. 3,108,030 (FMC); A. V. Dupuis, U.S. 3,154,454 (FMC); T. Wakayoshi et al, U.S. 3,484,334 (Kurashiki).
89. Belg. 614,143 (Goldschmidt); CA 57, 14004 (1962).
90. W. A. Kelley and A. V. Dupuis, U.S. 3,208,901 (FMC).
91. H. V. Boenig and J. A. Waters, U.S. 3,464,884 (Brookpark); F. P. Greenspan and A. V. Dupuis, U.S. 3,468,754 (FMC).
92. H. H. Beacham, U.S. 3,509,019 (FMC); F. P. Greenspan, U.S. 3,547,770 (FMC).
93. A. J. Heeb and E. L. Chalmers, U.S. 3,511,748 (Formica).
94. T. Wakayoshi et al, U.S. 3,484,334 (Kurashiki); T. Tsumaki, Japan Plastics 2, #2, 27-38 (April 1968).
95. C. L. Rohn, U.S. 3,197,331 (Johns-Manville).
96. J. L. Thomas and H. H. Beacham, U.S. 3,331,891 (FMC).
97. H. Raech, U.S. 3,345,205 (FMC).
98. F. J. Rauner, U.S. 3,475,176 (Eastman).
99. F. Meyer and K. Demmler, U.S. 3,009,898 (Badische).
100. G. Nowlin et al, U.S. 3,084,137 (FMC).
101. Neth. appl., Distillers, CA 63, 15107 (1965); cf. P. R. Graham, Fr. 1,349,169 (Monsanto); CA 61, 8492 (1964).
102. A. A. Dukert and A. Christofas, U.S. 3,607,827 (Pennwalt).
103. S. Nara and K. Matsuyama, J. Appl. Polym. Sci. 15, 477 (1971).

104. L. W. Kalinowski, Belg. 657,691, CA 64, 19923 (1966); CA 65, 15618 (1968).
105. Cf. L. Mageli et al, U.S. 3,326,809 (W and T).
106. E. L. Kropa, U.S. 2,443,740 (Am. Cyan.); E. H. Dafter, U.S. 2,562,140 (Am. Cyan.); J. B. Rust and W. B. Canfield, U.S. 2,530,315 (Montclair Res. and Foster-Ellis).
107. E. L. Kropa, U.S. 2,473,801 (Am. Cyan.).
108. E. L. Kropa, U.S. 2,409,633 (Am. Cyan.).
109. E. L. Kropa, U.S. 2,443,740 (Am. Cyan.).
110. R. R. Harris, U.S. 2,529,214 (Am. Cyan.).
111. G. Nischk and K. H. Andres, Ger. 1,052,115 (Bayer).
112. J. Litwin, H. H. Beacham, and C. W. Johnston, Plastics Technol. 9, 44 (May 1963).
113. E. E. Parker, U.S. 2,593,787 (PPG).
114. H. H. Beacham and J. Litwin, Mod. Plastics 42, 133 (August 1965).
115. H. M. Richardson, U.S. 2,809,911; cf. U.S. 2,748,028 and Ger. 924,587 (Atlas).
116. W. D. Holland, U.S. 3,036,031 (Am. Cyan.).
117. Booklet, Diallyl Phthalate Monomer in Polyester Resin Systems, FMC Corp., 1965.
118. D. A. Swedlow, U.S. 2,456,093 (Shellmar Products).
119. Booklet, Premix Polyester Resins, Interchem, 1965; cf. J. B. Crenshaw, Mod. Plastics 34, 133 (August 1957).
120. R. L. Tracey, Brit. Plastics 39/1, 40-44 (1966).
121. J. R. Guenther, U.S. 2,855,373 (PPG).
122. J. W. Hyland, U.S. 2,871,215 (Allied Chemical); cf. D. E. Cordier, U.S. 2,623,030 (L-O-F).
123. T. F. Anderson, U.S. 2,757,160 (Allied).
124. A. M. Howald, U.S. 2,665,263 (Allied); cf. W. L. Weaver, Brit. 727,566 (Allied).
125. D. A. Rogers, U.S. 3,054,770 (Westinghouse).
126. T. F. Anderson, U.S. 2,607,756 (L-O-F).
127. Brit. 762,953 (G.E.).
128. For example, G. Nischk and E. Mueller, U.S. 2,912,409 (Mobay and Bayer).
129. B. W. Nordlander and J. A. Loritsch, U.S. 2,641,586 (G.E.).
130. F. B. Shaw and J. B. Merriam, U.S. 3,000,775 (Continental Can).
131. G. Nowlin and L. S. Burnett, SPE J. 17, 1093 (October 1961).
132. M. Yusem, U.S. 2,895,929 (Bradley and Vrooman).
133. D. Swern and E. F. Jordan, U.S. 2,631,141 (USA).
134. R. R. Whetstone and T. W. Evans, U.S. 2,585,359 (Shell).
135. G. Takahashi, CA 52, 1670 (1958).
136. A. G. Chenicek, U.S. 2,522,254 (Interchemical).
137. A. Matsumoto and M. Oiwa, CA 68, 13435 (1968).
138. Cf. E. C. Britton et al, U.S. 2,341,175 (Dow); L. L. Yeager, U.S. 2,578,770.
139. Brit. 793,481 (Solvay); cf. E. C. Britton et al, U.S. 2,160,940-1, U.S. 2,160,943 and U.S. 2,160,946 (Dow).
140. P. E. Willard, U.S. 3,030,341 (FMC).
141. C. A. Heiberger, U.S. 3,096,310 (FMC).

142. C. A. Heiberger and J. L. Thomas, U.S. 3,087,915 (FMC).
143. P. T. Etchason and H. F. Jones, U.S. 2,628,210 (G.E.).
144. T. W. Evans et al, U.S. 2,500,607 (Shell).
145. C. A. Heiberger and J. L. Thomas, U.S. 3,113,123 (FMC).
146. J. L. Thomas, U.S. 3,455,888 (FMC).
147. F. Strain, U.S. 2,479,522 (PPG).
148. E. L. Kropa, U.S. 2,437,962 (Am. Cyan.).
149. B. Phillips and W. M. Quattlebaum, U.S. 2,543,335 (Carbide); A. Appelbaum, U.S. 3,460,982 (Du Pont).
150. E. P. Irany et al, U.S. 2,557,189 (Celanese); H. Mark, J. Polym. Sci. 1, 135 (1946); cf. CA 78, 17787 (1973).
151. E. C. Shokal and C. W. Schroeder, U.S. 2,596,945 (Shell).
152. K. E. Marple and E. C. Shokal, U.S. 2,567,675 (Shell).
153. H. W. Starkweather et al, Ind. Eng. Chem. 47, 302 (1955).
154. German appl. 1963 (Goldschmidt).
155. E. C. Shokal and C. W. Schroeder, Brit. 618,295 (Shell); cf. K. Kawota, J. Polym. Sci. 32, 27 (1958).
156. E. C. Shokal and C. W. Schroeder, U.S. 2,596,945 (Shell).
157. H. Dannenberg and T. F. Bradley, U.S. 2,564,395 (Shell).
158. E. Carnall and J. J. Lugert, U.S. 2,856,379 (Eastman).
159. P. O. Tawney, U.S. 2,568,872 (U.S. Rubber).
160. P. O. Tawney, Can. 498,072 (Dominion Rubber).
161. Diallyl hexachlorobicyclo-5-heptene-2,3-dicarboxylate.
162. B. S. Taylor et al, U.S. 3,093,619 (FMC).
163. J. L. Thomas, SPE J. 23, 30 (October 1967).
164. C. L. Wright and J. L. Thomas, U.S. 3,557,047 (FMC).
165. T. Morikawa and K. Takiguchi, U.S. 3,012,013 (Monsanto).
166. A. L. Forchielli, U.S. 2,831,836 (GAF).
167. N. S. Marans and W. D. Addy, U.S. 2,137,674 (Grace).
168. J. A. Loritsch and P. M. DiCerbo, U.S. 2,567,719 (G.E.).
169. S. H. Pinner, Nature 183, 1108 (1959).
170. D. A. Lima and J. P. Hamilton, U.S. 3,329,642 and 3,481,894 (FMC).
171. J. P. Hamilton and D. A. Lima, U.S. 3,496,253 (FMC).
172. PB 161,790, U.S. Dept. of Commerce, Office of Tech. Service.
173. C. E. Frank and S. P. Rowland, U.S. 3,061,582 (Nat. Distillers).
174. FMC Corp., private communication.
175. W. Simpson and T. Holt, J. Polym. Sci. 18, 335 (1955).
176. K. C. Tsou et al, U.S. 3,069,459 (Borden).
177. G. B. Linden and W. Brooks, U.S. 3,574,705 (Allied).
178. C. E. Schildknecht, unpublished.
179. E. Hagmann and C. E. Schildknecht, unpublished.
180. H. Raech and J. L. Thomas, Mod. Plastics 38, 144 (June 1961).
181. L. W. Hartzel et al, Mod. Plastics 44, 137 (July 1967).
182. H. H. Beacham et al, U.S. 3,441,535 (FMC).
183. L. S. Burnett, U.S. 3,574,657; 3,586,507; CA 78, 36322 (1973).
184. M. N. Gilano and M. A. Lipson, SPE Mid-Hudson Regional Conf., Oct. 15, 1970.
185. G. A. Cypher and M. Cohen, U.S. 2,959,564 (G.E.).
186. A. Matsumoto and M. Oiwa, CA 72, 121965 (1970).

187. W. Simpson and T. Holt, J. Polym. Sci. 18, 335 (1955).
188. Nancy Maddock and C. E. Schildknecht, unpublished.
189. Netherlands appl. (ICI); CA 66, 85609 (1967).
190. Alex Schulthess, Promotionsarbeit No. 3004 Polytechnic, Zurich (1960).
191. A. Matsumoto et al, CA 76, 154179 (1972).
192. H. Willersinn et al, Ger. 1,194,141 (Badische).
193. Ger. 1,010,272 (Hoechst); CA 53, 23073 (1959).
194. J. F. Smith, U.S. 3,011,995 (Du Pont).
195. T. Nishimura (Fuji Kaser Co.), CA 52, 1670 (1958).
196. A. E. Robinson, U.S. 3,294,869 (Hercules).
197. H. R. Niebergall, Brit. 916,260 (Courtaulds).
198. L. J. Reihs, Brit. 935,740 (Courtaulds).
199. E. C. Shokal, U.S. 2,476,922 (Shell).
200. J. W. Hirzy, U.S. 3,389,168 (Monsanto).

12. OTHER POLYFUNCTIONAL ALLYL ESTERS

None of the other allyl esters has had major commercial development in polymers comparable to CR-39 and the diallyl phthalates. Of the monomers discussed in this chapter, however, a number have found industrial applications. The diallyl esters of chlorine-containing cyclic acids have been used for flame-resistant copolymer plastics. Diallyl esters of aliphatic dicarboxylic acids such as diallyl adipate and diallyl sebacate have found use as crosslinking and branching comonomers. The possibilities of preparing useful prepolymers and precopolymers bearing reactive allyl groups give the polyfunctional allyl esters continuing interest in technology.

The experiments of Rothrock reported in a patent application of 1937 were early examples of formation of prepolymers (using dimethallyl and diallyl esters) that could be cured to useful insoluble forms in a second-stage polymerization (1). Highly polyfunctional allyl esters can be prepared by esterifying polymers bearing carboxyl groups with allyl alcohol. These then can be cured during casting or molding with heat and radical initiators.

Polyfunctional allyl esters have been applied as minor comonomers to impart heat resistance, photosensitivity, solvent resistance, weather resistance, flexibility, dyeability, and memory properties to polymer systems. Triallyl cyanurate, an allyl ester of the cyclic compound cyanuric acid, is discussed in Chapter 24. Some polyfunctional monomers give interesting adhesives rapidly curable by high-energy radiation and copolymer plastics of exceptional heat resistance.

The diallyl esters of alpha-omega dicarboxylic acids can be estimated by saponification with excess KOH in mixtures of diethylene glycol and water heated an hour or longer at 100°C. Many of the polyfunctional allyl esters copolymerize with styrene and related monomers more readily than diallyl acetals (2), allyl or vinyl ethers. Although there has been little study of cyclization in polymerizations of diallyl esters other than the diallyl phthalates, as expected, cyclopolymerization seems to decrease with increasing distance between allyl groups in the monomers (3).

OTHER ALLYLIC PHTHALATE DERIVATIVES

Remarkably little has been published about the isomeric dimethallyl phthalates. They may be made by reacting alkali metal phthalates with 2 molar equivalents of methallyl chloride in the presence of allyltriethylammonium chloride as a catalyst (4).

The preparation of a free-flowing prepolymer powder from dimethallyl isophthalate has been described (5). A mixture of 100 parts of dimethallyl isophthalate, 12 parts of methanol, and 0.38 part of t-butyl perbenzoate was agitated at reflux temperature of 110°C for 3.7 hr at a pressure of 50 psi. After the viscosity had reached 300 cps (25°C), the mixture was cooled to -10°C and pumped, along with 4 times its volume of methanol, into a colloid mill. The resulting polymer suspension was agitated for 30 min at 15°C, cooled to 0°C, and filtered. A second washing in methanol and filtering was followed by a third washing with water. The filter cake was dried at 80°C to give a nonadhesive white powder of dimethallyl isophthalate prepolymer. The powder mixed with 2% t-butyl perbenzoate could be molded at 175°C and 6000 psi for 15 min to give a clear, infusible, insoluble disk having a Rockwell hardness of 117 (M scale). The prepolymers melt at 80 to 115°C, depending on reaction conditions (6). Films of dimethallyl phthalate were cured by heating in air with 0.7% cerium naphthenate as catalyst (7). Copolymerizations of dimethallyl orthophthalate with vinyl acetate at 60°C initiated by BP occurred with partial cyclization (8). The proportion of residual copolymer unsaturation was little affected by comonomer feed ratio. The reactivity ratio of the cyclized radical was estimated as 0.73 compared to 1.08 for the uncyclized radical and 0.99 for the vinyl acetate radical.

Diallyl tetrachlorophthalate (mp 74°C) was copolymerized with unsaturated polyesters such as diethylene glycol maleate (9). To prepare the monomer, tetrachlorophthalic anhydride was heated at 90°C with excess allyl alcohol and about 5% camphor sulfonic acid for 16 hr. The mixture was washed with water and sodium carbonate until neutral. The monomer was extracted by ether and distilled at 190°C at 1.6 mm. Dimethallyl 3-chlorophthalate was suggested as an herbicide (10). Dimethallyl epoxyhexahydrophthalate was tried as a curing agent for epoxidized viscous liquid polybutadiene (11). Diallyl esters of epoxidized acids were polymerized to thermoset polymers (12). Diallyl ester of 3,5-dimethyl phthalic acid has been prepared (13). The physical properties were: boiling point at 1 mm = 144°C, n_D^{20} = 1.5080, and d_4^{20} = 1.0854. Heating with benzoyl peroxide gave liquid partial polymer solutions which could be

baked to give hard clear films. Polymers from diallyl
o-tetrachlorophthalate were studied by Shell (14) and by
General Electric (15). Dimethallyl 3-chlorophthalates
were suggested as herbicides (16). Some properties and
references to other diallyl esters of cyclic acids are
given in Table 12.1.

Isomerization and polymerization of allyl esters of
cyclohexadiene dicarboxylic acids were examined (17).
Diels-Alder products from cyclopentadiene and maleic anhydride were esterified by allyl or methallyl alcohol (18).
Polymers prepared by heating with peroxide catalysts were
suggested for coatings and electrical insulating resins.
Allyl esters of oxabicycloheptane dicarboxylic acids were
studied (19).

The following dimethallyl esters of 1,2-dicarboxylic
acids were prepared (20):

Ester	bp, °C	n_D^{25}
Phthalate	175(1)	1.5130
Tetrahydrophthalate	122(0.5)	1.4770
Endomethylene		
tetrahydrophthalate	260(0.1)	--
Hexahydrophthalate	119(0.3)	1.4930
Maleate	101(0.7)	1.4668

DIALLYL CHLORENDATE (DAC) AND ITS COPOLYMERS

This monomer which suffers the name 1,4,5,6,7,7-hexa-
chlorobicyclo[2.2.1]hept-5-ene-2,3-dicarboxylate has been
developed by Hooker Chemical Corporation and FMC as a comonomer especially for flame-resistant Dapon and polyester
copolymers. Diallyl chlorendate can be prepared by esterification of the chlorinated acid anhydride (Het anhydride),
which is obtained by Diels-Alder reaction of hexachloro-
cyclopentadiene with maleic anhydride (21):

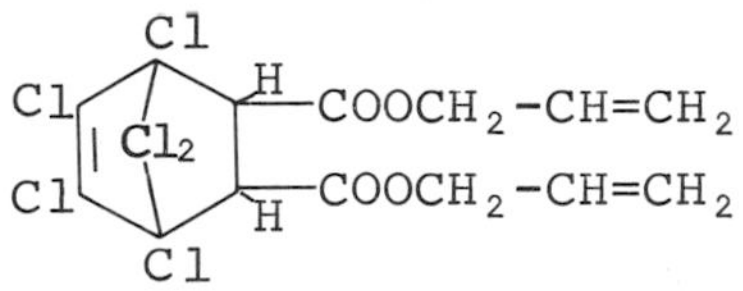

In one example, 5577 g of allyl alcohol, 50 g of p-toluene sulfonic acid monohydrate, 1084 g of anhydride, and
1000 ml of benzene were refluxed until the free acid content of the mixture was less than 2%. After washing with
water, with 5% aqueous alkali, and again with water, the
organic layer was distilled, giving crude diallyl ester
boiling to 175°C at 0.24 mm (n_D^{20} = 1.5360). Although the

TABLE 12.1

Miscellaneous Diallyl Esters of Cyclic Acids

Diallyl Ester	Properties	References
Methoxyterephthalate	mp 108°C, n_D^{20} = 1.5237; heated to hard polymer	Burnett, U.S. 2,821,520 (G.E.); J. Org. Chem. 21, 1226 (1956)
3,5-Dimethyl-o-phthalate	n_D^{20} = 1.5080; polymerized	Morris, U.S. 2,501,610 (Shell)
Dimethyl tetrahydrophthalates (also chloroallyl esters)	--	Morris et al, U.S. 2,489,103 (Shell)
Methyl trimellitate	bp 169°C(0.6); n_D = 1.522; (clear yellow polymer)	Hodes et al, U.S. 3,046,258 (Std. Oil Ind.)
Bicycloheptene dicarboxylate	--	Willard, U.S. 3,030,341 (FMC)
Endomethylene tetrahydrophthalate	Copolymers with TAC	Cummings et al, Ger. 1,007,059 (U.S. Rubber)
Oxabicycloheptane dicarboxylate (also methallyl esters)	Hard polymers by peroxide	Brit. 662,913 (Poulenc)
Phenylene-di-p-toluates	--	Brit. 730,890 (Shell)
Dihydronaphthoate	--	Rothrock, U.S. 2,221,663 (Du Pont)
Chloronaphthoates	--	U.S. 3,152,103 (Dow)
Dehydroabietates	mp 46°C; peroxide and UV polymerization	L.-V. Thoi et al, CA 49, 14697 (1955)

monomer contains a ring double bond, this has little reactivity in polymerizations. Highly purified DAC melts
at 27°C.

The diallyl ester was homopolymerized by heating for 4
hr at 80°C with 0.5% BP (21). There was formed a hard,
transparent, infusible, and insoluble polymer with outstanding flame resistance. After being "heat aged" for
3 days at 200°C, the resin had a crushing strength of
1300 lb; after one week at 200°C, the strength had risen
to 23,000 lb. This showed the slowness of completion of
polymerization. Monomer of d_{25} = 1.486 g/ml gave polymer
of d_{25} = 1.601 g/ml. DAC is resistant to hydrolysis.

A sample of commercial DAC heated at 90°C with 4%
t-butyl perbenzoate slowly passed through a viscous syrup
to form a gel of crosslinked polymer-monomer (22). However, with 5% azobis a bubbled mass of polymer-monomer
formed more rapidly at 90°C. Copolymerizations with MMA
and with maleic anhydride were faster than homopolymerizations. The purified chlorendic acid may be used to
prepare diallyl ester or for the preparation of flame-
resistant unsaturated polyesters (23). The latter copolymerized with divinyl or diallyl compounds (24).

Typical properties of commercial diallyl chlorendate
were given in a data sheet of FMC:

> Specific gravity 20/20 = 1.47
> n_D^{25} = 1.533
> Gel time   (1% BP at 100°C) = 15 min
> Viscosity at 25°C, 3-5 poise

Added DAC may contribute to good electrical properties,
improving hardness, heat resistance, and flame resistance
of diallyl phthalate (Dapon) and polyester resins as
shown below:

	Dapon (unfilled)	20% DAC in Dapon (unfilled)	Polyester styrene	35% DAC in polyester
Hardness (Rockwell M)	114–116	118	115	117
Heat distortion, °F	310	324	280	345
Impact, ft-lb/in. notch	0.25	0.27	0.23	0.27
Dissipation factor, 10^3 cycles	0.9	1.0	1.06	--
Volume resistivity × 10^5	1.8	8.9	2.5	--
Burning rate (ASTM)	0.5–0.6	Self-ext.	1.1 to self-ext.	Self-ext.

The DAC monomer may be added to granular DAP prepolymer
molding materials where low shrinkage of 6.5% in cure is
an advantage. Unfilled Dapon moldings containing up to
20% DAC and 3% t-butyl perbenzoate have been cured at
320°F for 15 min. Disadvantages of diallyl chlorendate

products include discoloration on exposure to light and
high cost. DAC polymers containing Sb_2O_3 (called Dapon FR
by FMC) are flame resistant and can impart low flammability to other resins such as polyesters and epoxies (25).
Military thermosetting formulations have employed 45%
glass fibers, 40 to 45% DAP and 5 to 10% DAC.

Dow has patented a chlorine-containing diallyl ester of
structure somewhat similar to that of diallyl chlorendate
(26). The Diels-Alder addition compound of hexachloro-
cyclopentadiene with cis-4-cyclohexene-1,2-carboxylic
anhydride was esterified with excess allyl alcohol. The
diester is a white crystalline solid melting at 76°C.
Heating the melted monomer at 95°C with 0.5% BP gave a
yellow transparent polymer. Copolymerization occurred
readily with methyl methacrylate, but the products were
less promising in properties other than flame resistance.
Diallyl tetrabromophthalate has been supplied by Monomer-
Polymer Laboratories of Borden Chemical Company.

ALLYL ESTERS OF ALIPHATIC DICARBOXYLIC ACIDS

Diallyl esters can be prepared conveniently by direct
esterification, normally under mild conditions of low concentration of mineral acid catalyst and moderate temperatures, in order to avoid losses by rearrangement of allyl
alcohol to propionaldehyde. Dimethallyl esters can be
made by alcoholysis of dialkyl esters by methallyl alcohol
in the presence of basic catalysts:

$$CH_3OOC(CH_2)_nCOOCH_3 + CH_2{=}\underset{CH_3}{C}CH_2OH \longrightarrow (CH_2{=}\underset{CH_3}{C}CH_2OOC)_2(CH_2)_n + 2CH_3OH$$

Diallyl and dimethallyl esters are hydrolyzed by aqueous
mineral acids and saponified by alkalies on heating.
Especially in the case of dimethallyl esters, the alcohol
may rearrange to aldehyde in the presence of acids.

Diallyl oxalate was prepared in 1856 by reaction of
silver oxalate with allyl iodide (27). It has the odor
of lower oxalate esters alone with a suggestion of mustard
odor. It was slowly hydrolyzed by water. Pyrolysis of
diallyl oxalate at 130 to 190°C formed initially carbon
dioxide and allyl radicals, which led to propylene and
1,5-hexadiene (28).

Dimethallyl oxalate was prepared by heating oxalic acid
with methallyl alcohol (29). Isobutyraldehyde was formed
as a byproduct. Preparations of dichlorallyl oxalate,
succinate, and adipate by esterification with mineral
acid catalyst were reported (30).

Diallyl malonate was prepared by esterification (31).
Kinetics of addition of bromine in acetic acid were
studied (32). In very dilute solution the additions were

higher than second order and were accelerated by small
amounts of water. Diallyl malonate is one of the diallyl
esters reported to have insecticidal activity. Diallyl
sebacate was prepared by esterification in the presence of
finely divided copper as an inhibitor of polymerization
(33). These diallyl esters are soluble in common organic
solvents but little soluble in water and in petroleum
ethers. Diallyl adipate and sebacate have strong IR absorptions at 1730 and 1170 cm^{-1}. DAA has a doublet of
equal intensity near 1440 cm^{-1}, whereas DAS has a doublet
with the stronger component at 1450 cm^{-1}. Table 12.2
gives properties and other references to syntheses of such
diallyl esters.

A number of polyfunctional allyl esters of aliphatic dicarboxylic acids were prepared and studied by Schulthess,
working with H. Hopff at Zurich (34). Most of the monomers were prepared by direct esterification using p-toluene
sulfonic acid catalyst (less darkening than with sulfuric
acid). For the esterifications there was used 100% molar
excess of allyl alcohol in the presence of a water-
immiscible solvent such as benzene or carbon tetrachloride.
Water could be distilled off azeotropically during esterification, along with excess allyl alcohol. The diallyl
esters of oxalic acid, malonic acid, succinic acid, and
maleic acid could be steam distilled. For example, a
mixture of 1 mole of oxalic acid, 5.1 moles of allyl
alcohol, 400 ml of benzene, and 2 ml of concentrated sulfuric acid was heated at reflux until separated water
showed the end of reaction. The mixture was then steam
distilled. The crude ester was extracted by ether and
dried and distilled under vacuum.

Diallyl adipate boiling 133 to 140°C at 4 mm, having
specific gravity 1.025, refractive index $n_D^{25} = 1.4506$, and
viscosity 3.8 cp at 25°C has been supplied by FMC. It has
been used as a crosslinking agent especially in reinforced
thermosetting plastics. The cured resins showed relatively
good flexibility.

POLYMERIZATIONS OF DIALLYL ESTERS
OF ALIPHATIC DICARBOXYLIC ACIDS

In addition to their pioneer work with the diallyl
phthalates, Simpson and Holt showed that intramolecular
cyclization and delayed gelation occurred in the polymerization of a number of nonaromatic diallyl esters (35).
These included diallyl carbonate, oxalate, malonate,
succinate, glutarate, adipate, azelate, and sebacate, as
well as dimethallyl carbonate. In general the diallyl
esters of the alpha-omega-dicarboxylic acids showed decreasing tendency to cyclize as the distance between allyl

TABLE 12.2

Properties of Diallyl Esters of Dibasic Acids,
AOOC(CH$_2$)$_n$COOA

	bp, °C at mm	n$_D^{20}$	References Preparation	Polymerization
Diallyl Esters				
Oxalate	217	1.4481		U.S. 2,541,957
	86(3)	1.4472 [a]		U.S. 2,528,773
				U.S. 3,035,085
Malonate	112(9)[b]	1.4478		U.S. 3,012,011
		1.4488 [a]		U.S. 2,437,962
Succinate	94(2)	1.4554	U.S. 2,217,673	U.S. 2,311,327
	106(3)[b]		U.S. 2,249,768	U.S. 2,712,025
			U.S. 2,275,467	U.S. 2,428,787
Dichlorallyl succinate	145(4)	1.4820	U.S. 2,159,008	U.S. 2,217,673
Glutarate	98(0.1)	1.4524 [a]		CA 52, 1105
Adipate	140(4)	1.4544 [a]	U.S. 2,217,673 (insecticide)	U.S. 2,712,025
				U.S. 2,160,940
				U.S. 2,202,846
				U.S. 2,610,161
Dichlorallyl adipate	175(4)	1.4769	U.S. 2,159,008 U.S. 2,217,673 (insecticide)	
Pimelate	--	1.4550 [a]		U.S. 3,068,210
Suberate	143(3)[b]	1.4548	CA 4559 (1946)	
Azelate	140(2)	1.4539(22)	CA 44,13675 (1950)	
Sebacate	163(3)[b]	1.4551	CA 40, 4559 (1946)	
		1.4558 [a]		
Dimethallyl Esters				
Oxalate	113(8)	--	Rothrock, U.S. 2,218,439 (DuPont)	
Malonate	130(9)	--	Rothrock	
Succinate	140(8)	--	Rothrock	
Adipate	163(9)	--	Rothrock	

[a] A. Schulthess, Reference 34.
[b] Borden Chemical Company data.

groups increased. However, the sample of diallyl succinate studied showed an exceptional tendency to cyclize.
The ratios of rate constants for cyclization to propagation were measured and they explained, at least qualita-
tively, why gelation occurred only at conversions higher
than predicted from the simple network theory of gelation.
Simpson and Holt suggested that an important factor in
determining cyclization may be steric interference in the
addition of a free radical in the transition state to a
monomer unit. In some cases this might mask the influence
of the number of atoms in the ring being formed.

In the formation of useful prepolymer syrups in the
presence of allyl monomers, small amounts of chain-transferring impurities or additives can change the DP and
conversion at which gelation begins. Along with cyclization, both degradative and effective chain transfers
normally can occur with diallyl monomers. There are
formed from monoallyl esters ordinarily linear chains of
DP about 20 or below. Theoretical discussions of cyclization of diallyl esters have been published by Butler
(36), but thorough quantitative studies of these complex
polymerizations remain to be undertaken.

In general the rates of homopolymerization at 70°C with
BP decrease with increasing molecular weights of diallyl
ester monomers. Polar groups such as halogen may cause
acceleration in rates. Diallyl esters of branched dicarboxylic acids bearing electron-donating methyl groups
generally polymerize more slowly than the saturated
straight-chain esters. The more rapidly polymerizing
diallyl esters of unsaturated dicarboxylic acids such as
diallyl maleate and diallyl fumarate are discussed in
Chapter 26. As expected, the presence of three or four
allyl groups in esters leads to very rapid polymerization
and difficulties in purifying and storing the monomers.

In bulk polymerization at 70°C with 1% BP, Schulthess
observed the following gelation times: DA carbonate,
14 hr; DA oxalate, 9 hr; and DA sebacate, 17 hr. Introducing polar groups into diallyl succinate accelerated
homopolymerization and gave gel times as follows: succinate, 12.0 hr; malate, 11.6 hr; tartrate, 3.75 hr; α,α-
dichlorosuccinate, 3.75 hr. These compared to gel times
for diallyl maleate and diallyl fumarate of 85 and 22 min,
respectively. Figure 12.1 shows conversion of diallyl
sebacate to soluble and crosslinked polymer, which
resembles that for diallyl o-phthalate.

Diallyl oxalate polymers were prepared and saponified
to give soluble products (37). A series of diallyl esters
including the oxalate, succinate, and adipate was prepared
by Kardashev and co-workers in Russia (38). The crosslinked, somewhat brittle polymers made by heating with

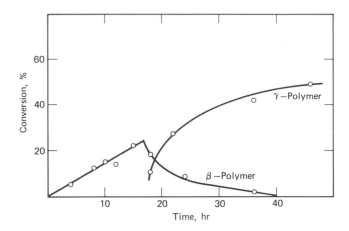

Fig. 12.1. Conversions of diallyl sebacate to soluble or beta-
polymer and crosslinked gamma-polymer on heating at 70°C with 1%
BP catalyst. Other monomers that behave similarly in giving syrup
of about 25% polymer before gelation include diallyl carbonate,
oxalate, succinate, glutarate, and adipate. Alex Schulthess, Ueber
die Polymerisation von Allylestern, Promotionsarbeit, Zurich ETH,
1960.

peroxide were saponified rather easily by aqueous alkali.
D'Alelio copolymerized diallyl succinate with different
proportions of styrene at 90°C with 0.1% BP (39). Compared to the homopolymerization of diallyl succinate with
gelation at 36% conversion, the copolymerization of 20%
diallyl succinate with styrene gave gelation at 71% conversion.

Copolymerization of a number of diallyl esters with
vinylidene chloride was studied by Dow (40). Diallyl
succinate has been copolymerized with maleate and fumarate
polyesters (41). Diallyl esters copolymerize with vinyl
esters and with acrylic monomers (42). Allyl esters are
reluctant to copolymerize with styrenes or other conjugated unsaturated hydrocarbons. Diallyl malonate, diallyl
oxalate and other diallyl esters of dicarboxylic acids
were copolymerized with styrene in Japan (43). The
addition of the diallyl esters retarded the rates of
styrene polymerization.

Diallyl succinate (365 g) and diethyl fumarate (635 g)
with benzoyl peroxide catalyst gelled in 25 min at 90°C

and gave a hard solid in 3 hr (44). A similar catalyzed
mixture of diallyl sebacate and diethyl fumarate gelled
after 4 hr. Copolymers of 82% vinyl chloride with 13%
diallyl succinate units could be cured by molding at 150°C
for 10 min in the presence of t-butyl perbenzoate and
transition metal compound accelerators (45). In vinyl
chloride suspension, polymerization with lauroyl peroxide
at 51°C adding as little as 0.1% diallyl succinate increased the specific viscosity of the polymer product
beneficially (46). The polymers also showed desirable
nongelling properties. Adding 1% or less of diallyl succinate or diallyl itaconate to styrene in peroxide suspension polymerization gave lightly crosslinked expandable
polystyrene beads (47). The foamed moldings showed
improved heat stability in a 2-hr test at 150°C. Diallyl
succinate copolymerizations with crotonate esters (48) and
with linseed oil and styrene (49) were examined briefly.

One of the most significant developments from polymerization studies of diallyl esters has been prepolymerization to fusible soluble polymers which then can be cured
in useful shapes. Rothrock made a diallyl ester of a
1,4-dihydronaphthalene dicarboxylic acid as well as other
polyfunctional monomers; some of these gave viscous liquid
products curable in coatings (50). Kropa heated a mixture
of diallyl succinate and diethyl fumarate at 135°C for
2 hr to form a viscous liquid (51). Benzoyl peroxide was
added to this and polymerization was completed by heating.
Addition of hydrocarbon to the viscous syrup gave a precipitate of prepolymer which could be molded. Other
studies of prepolymer formation were made using diallyl
maleate by Strain and co-workers and using DAP by Simpson
(Chapters 11 and 26).

Nowlin and co-workers studied polymerizations of a
series of 2-substituted diallyl succinates, RCHCOOA ,
                                             |
                                             CH$_2$COOA
obtained by addition of alkenes and alkyl aromatic compounds to maleic anhydride (52). Their rates of polymerization to fusible prepolymers were inferior to those of
DAP. The degree of cyclization in radical polymerizations
decreased in the following order of substituents: aralkyl
and isobutyl > unhindered alkenyl > alkyl > hindered
alkenyl. In general, thermoset moldings from the substituted diallyl succinates were inferior to those based
on DAP. One of the monomers studied was diallyl benzyl
succinate (bp 178°C at 0.5 mm). With 1% BP it gave a gel
time of 123 min at 100°C. At 130°C gelation occurred at
about 33% conversion.

A commercial diallyl adipate (DAA) had a freezing point
of -33°C, surface tension 32 dynes/cm, and viscosity of
4.1 cp; it was miscible with gasoline and was only 0.2%

soluble in water. Diallyl adipate can be made readily by
esterification of adipic acid with excess allyl alcohol in
the presence of small concentrations of toluene sulfonic
acid and hydroquinone inhibitor. Heating the colorless
commercial monomer with 2% t-butyl perbenzoate in test
tubes for 21 hr at 100°C gave had, clear, relatively tough
homopolymers (53). Uncatalyzed monomer under the same
conditions increased very little in viscosity. Higher
concentrations of peroxide gave more yellowish cast
polymers. Small cast rods had surprisingly few bubbles
or other defects. Under similar conditions with t-butyl
perbenzoate, adding 5% methyl methacrylate accelerated
early polymerization, but the final polymer was soft.
Experimental castings from DAA were free from soft layers
when polymerized in contact with air. A gel time of 60
min at 100°C with 1.0% BP was found for DAA by FMC. Diallyl adipate monomer has been reported to have value as
an insecticide (lice, mites, house flies) and has served
as a stabilizer against discoloration of rubber hydrochloride packaging films (54). Diallyl adipate fusible
prepolymers have been patented as thermosetting adhesives
in electrical applications (55). Prepolymers have been
prepared in a solvent with stannous chloride as accelerator and a mercaptan as regulator (56).

Diallyl adipate homopolymers have desirable flexibility
in films. More flexible films can be obtained by copolymerization of DAA with minor proportions of vinyl n-butyl
ether (57). However, solvent resistance may be sacrificed
thereby. Monomers containing 4% BP without solvent at
35 to 40°C were irradiated for 14 hr at 1 ft from a mercury arc lamp. The radiation passing through Pyrex glass
and cooling water was 3660 Å and longer wavelengths.
Slight gelation was observed first near the air-monomer
interface of the first three polymerizations:

| Monomer feed | | Extent of gelation | Conversion (polymer in- | Polymer, swelling |
DAA	VNBE	observed	soluble in CH_3OH)	in toluene
20 g	0 g	slight	27 %	low
30	0.1	slight	28	moderate
19	1.0	traces	25	high
15	5	none	21	high
10	10	none	10	low*

*Films were attacked and partially dissolved by toluene.

In the above experiments the polymers were precipitated
from the viscous syrups by addition of 100 ml methanol
with stirring. The precipitates were washed with a second

portion of methanol and dried overnight at 85°C. The
resulting films were clear, colorless, flexible, and
largely crosslinked; they were exposed to boiling toluene
for 5 min before the observations in the last column above.

In Japan small additions of diallyl adipate were found
to increase the rates of polymerization of vinyl chloride,
vinylidene chloride, vinyl acetate, or acrylic monomers
(58). Diallyl adipate has been copolymerized with various
unsaturated polyesters (59). Diallyl adipate is of
interest for curing preformed polymers. In one example, a
vinyl chloride-vinyl acetate copolymer composition containing DAA and a peroxide was heated below 100°C under
pressure (60).

Diallyl sebacate has been emulsion polymerized using
activated hydrogen peroxide added during 2 hr at 95°C (61).
Nonionic and hexadecylpyridinium emulsifiers gave positively charged latices that were stable for 3 months. Diallyl sebacate and diallyl azelate were homopolymerized
as well as copolymerized with diallyl orthophthalate (62).
The prepolymers were cured by heating with cerium naphthenate. Methyl methacrylate crosslinked in polymerization
with 1% or less diallyl sebacate has been powdered and
then swollen with vinyl or acrylic monomers before molding
to strain-free shapes (63). This resembled the monomer-
polymer slurry technique used for dental plastics.

Diallyl sebacate (DAS) was one of the polyfunctional
monomers found useful by Pinner and co-workers in the
curing with peroxides of preformed polymers (so-called
polymer-extended networks)(64). DAS is partially compatible with polyethylenes. For example, low-density
polyethylene could be blended with up to 30 parts of DAS
per 100 on the two-roll mill at 125°C during 15 min. There
was added 0.1 to 3 parts of dicumyl peroxide, and the mixture was immediately molded at 160°C for 45 min. The
homogeneous, rubbery sheets obtained had improved tensile
strength and elongation at 100°C, compared with the un-
treated polyethylene. High-density polyethylene, polypropylene, and natural rubber were also cured by diallyl
sebacate and peroxide to give extended network polymers
having modified properties. Methyl methacrylate and vinyl
chloride polymers were crosslinked using diallyl sebacate
and high-energy radiation (2 MeV electrons from a Van de
Graaf accelerator)(65). The diallyl monomers have advantages of stability to incorporation by hot milling, yet
the response in forming crosslinks under radiation is good.

Rothrock made dimethallyl sebacate by heating diethyl
sebacate, methallyl alcohol, and benzene in the presence of
a little sodium ethoxide (66). Ethanol and ethanol-benzene
binary were distilled off. The monomer, boiling at 170°C
(1 mm), was heated at 260°C under air with 1% BP for 30 min

to form a viscous syrup. A solution of the latter in
butyl acetate with a little cobalt drier was baked for an
hour at 100°C, producing water-resistant solid films.
Rothrock prepared soluble prepolymers by interrupting the
polymerization of a number of other dimethallyl esters.
He also made a tetraallyl ester of $\alpha,\alpha,\beta,\delta$-propane tetracarboxylic acid which distilled at 230°C (8 mm). This
monomer gelled quickly on heating at 100°C with 1% BP.

   Table 12.3 lists some diallyl esters of substituted dicarboxylic acids.

   Diallyl diglycolate $AOOCCH_2OCH_2COOA$ was prepared by
Cupery and Rothrock by distilling off water-benzene azeotrope from a mixture of 144 parts diglycolic acid, 174
parts allyl alcohol, 2.7 parts sulfuric acid, and 180 parts
benzene (67). After washing with sodium carbonate solution
and water, the product was dried over magnesium sulfate and
distilled at 111°C at 1 mm. Diallyl diglycolate is a mobile
liquid (n_D^{24} = 1.4548, d_{20}^{20} = 1.113, and mp -31°C); it is
insoluble in petroleum ether and cyclohexane but miscible
with common alcohols, ketones, esters, aromatic hydrocarbons, and ethers. About 1% dissolves in water. The surface tension of commercial monomer was 34.4 dynes/cm, the
viscosity 7.8 cp, and the "gel time" with 1% BP at 100°C
was 49 min.

   When this monomer containing 1% BP was heated 16 hr at
60°C, it gave a viscous prepolymer solution that was
diluted with xylene and polymerized in films on an iron
panel by heating for 20 min at 165°C (67). The films were
tough, scratch resistant, and unaffected by soaking in

TABLE 12.3

Some Diallyl Esters of
Substituted Dicarboxylic Acids

Diallyl Ester	Properties	References
Acetyl malonate	Polymerized	Brit. 592,421 (Goodyear)
Diethyl malonate	--	Levina, CA 49, 11667 (1955); Matsoyan, CA 58, 14107 (1963)
Allyl malonate	bp 93°C at 0.5 mm; n_D^{20} = 1.4545	Croxall, U.S. 2,541,957 (R & H)
Acetyl methyl pimelate	Polymerized	Brit. 593,475 (Goodyear)
2,5-Diphenyl adipate	n_D^{25} = 1.5344; tough polymers	Rowland, U.S. 3,062,769 (Natl. Distillers)
Diacetate tartrate	bp 127°C at 0.4 mm; n_D^{20} = 1.4529	Whitehill, U.S. 2,545,184 (Shell)
Saccharate, etc.	--	Whitehill

water. In contrast to films from DAP and diallyl maleate,
the diallyl diglycolate films were not cracked by bending
on metals. Another patent example suggested "drying" type
polymerization. A slow stream of air was bubbled through
diallyl diglycolate heated at 170 to 180°C. The viscosity
had increased after 40 min and a little hydroquinone was
added to delay gelation while the syrup cooled. Applications in impregnation, laminating, and surface coatings
were suggested with final curing at higher temperatures.

Colorless, clear castings were made from diallyl diglycolate in glass cells using 2% BP and heating for 144 hr
at 65°C. They showed Rockwell hardness of H-94 together
with scratch resistance (even to steel wool). The monomer
also was polymerized in t-butanol with 2% BP at 80°C for
5.5 hr producing high yields of soluble partial polymer.
Cupery and Rothrock made a viscous precopolymer solution
from 80 parts of diallyl diglycolate, 20 parts of styrene,
and 1 part of BP by heating for 48 hr at 80°C. This could
be baked on metal surfaces to give clear, tough insoluble
coatings. Promising films were also obtained from mixtures
of diallyl diglycolate and methyl methacrylate. Diallyl
diglycolate has been supplied by Ohio-Apex and FMC. The
properties of monomer were as follows:

> Boiling point, to 138°C at 4 mm
> Freezing point, -31°C
> Solubility in water, 1.1%
> Solubility in gasoline, 4.8%
> Specific gravity = 1.113
> n_D^{25} = 1.4540
> Surface tension, 34.4 dynes/cm
> Viscosity, 7.8 cp
> Soluble in glycols, and most organic
>   liquids except hydrocarbons

The diallyl ester, $AOOCCH_2OCH_2COOCH_2COOA$ (boiling at
165°C at 0.5 mm), gave crosslinked cast polymers having
favorable flexural and impact strength compared with MMA
polymers (68), but it was not developed commercially.

TRIALLYL AND TETRAALLYL ESTERS

The tri and tetraallyl esters give difficulties in purification because of their great tendency to polymerize.
For example, the tetraallyl ester of 1,2,3,4-butane tetracarboxylic acid could not be vacuum distilled. Surprisingly, additions of copper powder at the oil bath temperature of 230°C were reported to accelerate polymerization. The raw ester could be steam distilled. The
monomers listed in Table 12.4 were prepared by Schulthess
(34) unless otherwise indicated.

TABLE 12.4

Triallyl Esters

Triallyl esters	n_D^{20}	bp, °C	References
Tricarballylate	1.4660	182(0.01)[a]	cf. U.S. 2,249,768 (Am. Cyan.)
Citrate	1.4715	150(0.2)[a]	cf. U.S. 2,385,931 (PPG); U.S. 2,936,297 (Pfizer); Brit. 948,762 (BX Plastics)
Trimellitate	1.5229 1.5240[a]	160(0.1)[a]	Shih, CA <u>63</u>, 2920 (1965)
Orthoformate	1.4380	205	Beilstein, Ber. <u>18</u>, 482 (1965) (supplied by Aldrich)
Acetyl citrate	1.4665(25)	143(0.2)	Sumner Div., Miles Labs; also Borden Chemical
Aconitate	n = 1.4807	108(0.1)	Roberts Rubber World <u>130</u>, 801 (1954); Tawney, CA <u>50</u>, 1372 & 2207 (styrene copolymers)
Butanetricarboxy- lates (and epoxidized esters)	--	--	Lynn et al, J. Org. Chem. <u>26</u>, 3757 (1961)

[a]A. Schulthess, Ref. 34.

Tawney prepared soluble prepolymers from triallyl aconitate (TAA) AOOCCH=C-COOA by using allyl alcohol as regu-
$\qquad\qquad\qquad\quad$ CH$_2$COOA
lator and comonomer (69). The monomer homopolymerized slowly but gave crosslinked polymers even at low conversion. When equal parts of TAA and allyl alcohol were heated at 60°C for 336 hr with daily additions of BP, there was obtained 23 parts of polymer (precipitated by n-hexane-diethyl ether 1:1). Some allyl alcohol had copolymerized (3.5% OH). The fusible, soluble copolymer could be cured by heating above 120°C. A similar polymerization with equal parts of TAA and methallyl alcohol gave 36.6 parts of solid fusible copolymer after 144 hr. The OH content was 2.1%, indicating 9.6% methallyl alcohol groups. Heating the product at 100°C with 1% BP formed a solvent-resistant and heat-resistant crosslinked polymer.

Trimethallyl orthoformate HC(OMA)$_3$ (bp 120°C at 13 mm) was copolymerized with MMA in the presence of peroxide catalyst (70). The same patent discloses polymerization of dimethallyl carbonate to an insoluble gel by heating for 60 hr with 1% BP at 65°C. Triallyl orthoformate containing 5% BP slowly formed a viscous prepolymer solution under light from a hot mercury arc lamp (57). Cured films

are hard and brittle. Surprisingly, bulk polymerizations
at 70°C with 1% BP gave syrups of about 20% soluble prepolymer before gelation in cases of triallyl esters of
tricarballylic and trimellitic acids, and also tetraallyl
esters of butanetetracarboxylic acid and pyromellitic
acid (34).
    Methyl diallyl trimellitate has been polymerized by
peroxide catalysts and heating to give thermoset polymers
(71). With styrene the monomer copolymerized to a partially crosslinked product. Allyl-terminated polyesters
from glycols and trimellitic anhydride were suggested in
coatings (72). Triallyl trimellitate, a yellowish oil,
was made by refluxing the anhydride, allyl alcohol, and
benzene along with a little sulfuric acid and hydroquinone (73). The IR and UV spectra were reported. Triallyl
trimellitate and tetraallyl pyromellitate were made by
reacting sodium or potassium salts of the acids with allyl
chloride in an autoclave (74). The monomers were evaluated
as curing agents for unsaturated polyesters in order to
obtain improved heat resistance. Cast copolymers of triallyl trimesate with unsaturated polyesters were reported
to compare favorably with those from triallyl cyanurate-
polyesters in heat resistance (75).
    Triallyl tricarballylate was copolymerized with diethy-
lene glycol maleate ester (76). A small casting made with
0.4% BP was heated for 24 hr at 40 to 60°C, followed by
several hours at 100°C. Triallyl citrate has been available from Pfizer for copolymerization with unsaturated
polyesters. The monomer is miscible with acetone, benzene,
chloroform, dioxane, ethanol, and glacial acetic acid, but
not with water. Soluble prepolymers of acetyl triallyl
citrate (77) and terpolymers with maleic anhydride and
styrene have been studied at Miles Laboratories (78).
Methallyl orthoformate was made by alcoholysis of ethyl
orthoformate and was used to crosslink methyl methacrylate
(79). Triallyl citrate was used as a "reactive plasticizer" to cure cellulose acetate. See Fig. 12.2.
    Properties of some tetraallyl esters are given below.
These reactive monomers are very difficult to purify by
distillation.

Tetraallyl esters	n_D^{20}	bp, °C	References
1,2,3,4-Butane- tetracarboxylate	1.4736	187(0.3)	Reference 34
Pyromellitate	1.5218	195(0.02)	CA 65, 20365
Allyl carbonate of triallyl citrate	--	--	Strain, U.S. 2,385,931 (PPG)

MISCELLANEOUS POLYFUNCTIONAL ALLYL ESTERS

Early work of Kronstein in Germany on polymerization of allyl cinnamate $CH_2=CHCH_2OOCCH=CHC_6H_5$ is disclosed in a patent of 1907 (80). Going against prevailing opinion, Kronstein doubted that the thickening of drying oils and resins was necessarily an oxidation. His process consisted in heating the monomer in closed vessels with little contact with air. Heating allyl cinnamate for 36 hr at 230°C transformed it through a viscous liquid into a clear solid of color and appearance resembling amber. Kronstein also heated diallyl malonate at 170°C for 24 hr in a closed vessel, gradually thickening it to a balsam-like mass. He generalized that besides the double bonds, negative groups such as COOH or COOR groups could promote thickening or resinification; he also thought that two negative groups in unsymmetrical arrangement were more favorable than in symmetrical arrangement.

At about the same time, the thermal and photopolymerizations of allyl cinnamate were also studied by Liebermann and Kardos (81). The allyl double bond was recognized to add bromine more rapidly than the cinnamate double bond. The polymer content of the syrup was estimated by precipitation with methanol. Heating the monomer for 6 hr in a sealed tube at 210°C gave a viscous polymer solution; 15 hr yielded an insoluble glassy mass. Molecular weights of 1500 to 2300 found for the soluble polymers were not considered significant because of the prevailing belief that such polymers were dimers or trimers. On heating up to 300°C the polymer did not melt but sintered. In Russia, changes from the pleasant peach odor of allyl cinnamate with changes in chemical structure were studied (82). The monomer boiled at 163°C at 17 mm and had refractive index $n_D^{20} = 1.530$.

Methallyl cinnamate was prepared by Britton but its homopolymerization was not described (83). Allyl crotonate, cinnamate, and some diallyl esters were slowly copolymerized with vinylidene chloride using peroxide catalyst at 30 to 40°C (84). Allyl crotonate, as well as methallyl crotonate, can form solid polymers by peroxide initiation (85). Ethylene bis(allyl oxalate) (bp 186°C, mp 43.5°C) was polymerized by peroxide initiation (86).

Polyesters prepared to contain an excess of free carboxyl groups may be esterified with allyl alcohol. Thus a polyester from citric acid and ethylene glycol gave a polyfunctional allyl ester polymer that was cured by heating with peroxide catalyst (87). A polyester from excess phthalic anhydride and ethylene glycol was reacted with allyl alcohol in xylene in the presence of p-toluene sulfonic acid catalyst (88). The yellow syrup obtained required heating for 4 hr at 100°C with 1% BP in order to form an infusible, crosslinked polymer.

Allyl esters of polymers bearing carboxyl groups may be regarded as polymeric monomers for further polymerization. Examples of such disclosures in patents include allyl esters of copolymers of vinyl acetate-maleic anhydride (89), styrene-maleic anhydride (90), vinyl alkyl ether-maleic anhydride (91), polyacrylic acid (92), and allyl esters of polycarbonate resins based on derivatives of bisphenol A (93).

Some miscellaneous polyfunctional allyl esters are listed in Table 12.5 with references.

TABLE 12.5

Miscellaneous Polyfunctional Allyl Esters

	bp, °C and refractive index	References, etc.
Diallyl esters of:		
sulfone diacids	--	Whitehill, U.S. 2,427,640 (Shell) (Hard polymers)
diacid from bisphenol A		Rust, U.S. 2,672,478 (Ellis)
8,12 eicosadiene diacid	236(0.3) $n_D^{20}=1.4726$	Payne, U.S. 2,957,907 (Shell) (Polymers, also epoxides)
dimerized linoleic acid	--	Payne, U.S. 2,783,250 (Shell); U.S. 2,870,170 (Shell)
Allyl esters of drying oil acids	--	Bradley, U.S. 2,378,827 (Am. Cyan.)
Allyl-2-hydroxybutenoate	--	Stansbury, U.S. 2,786,073 (Carbide)
Allyl allyloxyhydroxy-stearate	200(0.5) $n_D^{30}=1.4589$	Swern, U.S. 2,516,928 and U.S. 2,542,062 (USA)
Allyl p-methallyl-benzoate	101(0.02) $n_D^{20}=1.5523$	Morris, U.S. 2,677,678 (Shell)
Allyl allyloxyacetate	--	D'Alelio, U.S. 2,437,508 (G.E.)
Allyl allylacetate	--	Croxall, U.S. 2,541,957 (R & H)
Allylsuccinyl allylglycolate	--	Neher, U.S. 2,521,203 (R & H) (Copolymers)

References

1. H. S. Rothrock, U.S. 2,218,439 (Du Pont).
2. A. Matsumoto and M. Oiwa, J. Polym. Sci. A1, 10, 103 (1972).
3. E. N. Rostovskii and I. A. Arbuzova, CA 76, 113669 (1972).
4. J. A. Garman and W. B. Tuemmler, U.S. 3,035,084 (FMC); cf.
    U.S. 2,939,879 (Shell) and CA 58, 12474 (1963).
5. P. E. Willard, U.S. 3,030,341 (FMC).
6. Brit. 863,573 (FMC); cf. Brit. 572,858.
7. A. G. Chenicek, U.S. 2,522,254 (Interchemical); cf. H. S. Roth-
    rock, U.S. 2,218,439 (Du Pont).
8. A. Matsumoto and M. Oiwa, CA 76, 113644 (1972).
9. B. W. Nordlander, U.S. 2,450,682 (G.E.).
10. French 1,387,635 (FMC); CA 62, 13091 (1965).
11. G. Nowlin et al, Brit. 922,356 (FMC).
12. M. L. A. Fluchaire and G. Collardeau, Fr. 977,285-6 (Rhône-Poulenc)
13. R. C. Morris and A. V. Snider, U.S. 2,501,610 (Shell).
14. L. N. Whitehill et al, Brit. 595,758 (Shell).
15. B. W. Nordlander, U.S. 2,450,682 (G.E.).
16. J. R. Willard and K. P. Dorschner, U.S. 3,232,737; cf. C. E.
    Rehberg et al, U.S. 2,712,025 (USA).
17. W. E. Elwell, U.S. 2,502,645.
18. J. A. Cottrell and D. H. Hewitt, Brit. 612,311-2 (Berger, Lewis).
19. M. L. A. Fluchaire and G. Collardeau, U.S. 2,570,029 (Rhône-Poulenc)
20. H. Stange and W. B. Tuemmler, U.S. 3,250,801 (FMC).
21. C. F. Baranauckas, U.S. 2,810,712 and 2,903,463 (Hooker);
    cf. U.S. 2,650,942 and CA 57, 698 (1962).
22. Richard Mao and C. E. Schildknecht, unpublished.
23. R. H. Kimball and G. W. Darling, Ger. 1,045,394 (Hooker);
    CA 55, 418; cf. U.S. 2,931,746 and 2,863,794-5.
24. P. Robitschek and C. T. Bean, U.S. 2,783,215 (Hooker).
25. J. L. Thomas, SPE J. 23, (10), 30 (1967).
26. D. H. Haigh et al, U.S. 3,152,103 (Dow).
27. A. Cahours and A. W. Hofmann, Ann. 102, 294 (1857).
28. D. G. L. James and S. M. Kambanis, Trans. Faraday Soc. 65,
    1350 (1969).
29. H. P. A. Groll and G. Hearne, U.S. 2,164,188 (Shell).
30. G. H. Coleman and B. C. Hadler, U.S. 2,159,008 (Dow).
31. M. S. Newman et al, J. Am. Chem. Soc. 68, 2113 (1946).
32. I. K. Walker and P. W. Robertson, J. Chem. Soc. 1517 (1939).
33. E. L. Kropa, U.S. 2,249,768 (Am. Cyan.).
34. Alex Schulthess, Promotionsarbeit No. 3004, Zurich 1960.
35. T. Holt and W. Simpson, Proc. Roy. Soc. (London) A 238, 154 (1956)
    cf. J. Polym. Sci. 18, 335 (1955).
36. G. B. Butler, see references in Chapters 1, 11, and 19.
37. I. E. Muskat and M. A. Pollack, U.S. 2,332,460.
38. D. A. Kardashev et al, CA 40, 4559 (1946); cf. U.S. 2,467,105
    (Shell).
39. G. F. D'Alelio, Fundamental Principles of Polymerization, Wiley,
    1952, p. 95.

40. E. C. Britton et al, U.S. 2,160,940 (Dow).
41. E. L. Kropa, U.S. 2,409,633 and 2,437,962 (Am. Cyan.).
42. H. W. Coover and J. B. Dickey, U.S. 2,759,912 (Eastman).
43. A. Matsumoto and M. Oiwa, J. Chem. Soc. Japan (Ind. Chem. Sect.) 71, 2063-6 (1968).
44. E. L. Kropa, U.S. 2,437,962 (Am. Cyan.).
45. S. D. Douglas, U.S. 3,074,905 (Carbide); cf. U.S. 3,068,210.
46. R. H. Martin, U.S. 3,012,011 (Monsanto).
47. H. A. Wright, U.S. 3,259,594-5 (Koppers).
48. T. F. Bradley, U.S. 2,575,440 (Shell).
49. O. L. Polly, U.S. 2,610,161 (Union Oil).
50. H. S. Rothrock, U.S. 2,221,663 (Du Pont).
51. E. L. Kropa, U.S. 2,437,962 (Am. Cyan.); cf. U.S. 2,443,740.
52. G. Nowlin et al, J. Appl. Polym. Sci. 13, 463 (1969).
53. J. A. Seckar and C. E. Schildknecht, unpublished.
54. J. E. Snyder and G. W. Ferner, U.S. 2,900,261 (Goodyear).
55. Fr. 1,142,677 (Usines Dielectriques), CA 53, 20632 (1959).
56. P. P. W. Varlet, U.S. 3,133,826 (Chausson).
57. C. E. Schildknecht, unpublished, September 1972.
58. S. Matsuoka and T. Kaku, Japan 297 (January 1958); CA 53, 8715 (1959).
59. P. P. William, Fr. 1,258,405 (Chausson); E. Behnke and H. Wulff, Ger. 1,138,217.
60. H. S. Busby and W. L. Ward, U.S. 2,587,591 (USA).
61. W. S. Reid, Brit. 821,093 (Geigy).
62. A. G. Chenicek, U.S. 2,522,254 (Interchemical).
63. J. J. P. Staudinger, U.S. 2,539,376 (Nat. Distillers).
64. S. H. Pinner et al, Brit. 948,302 and U.S. 3,125,546 (BX Plastics), appl. April 1954.
65. S. H. Pinner and V. Wycherley, J. Appl. Polym. Sci. 3, 338 (1960).
66. H. S. Rothrock, U.S. 2,218,439 (Du Pont), appl. 1937.
67. M. E. Cupery and H. S. Rothrock, U.S. 2,451,536 (Du Pont); cf. Brit. 572,858 (Shell).
68. H. T. Neher et al, U.S. 2,474,686 (R & H).
69. P. O. Tawney, U.S. 2,599,027 (U.S. Rubber).
70. H. J. Richter and H. S. Rothrock, U.S. 2,410,305 (Du Pont).
71. W. Hodes and R. E. VanStrien, U.S. 3,046,258 (Std. Oil Indiana).
72. J. R. Stephens et al, U.S. 3,040,000 (Std. Oil Indiana).
73. J-W. Shih and E. Ogata, CA 63, 2920 (1965).
74. L. A. Naumets et al, CA 77, 6200 (1972); cf. CA 78, 4972 (1973).
75. E. M. Beavers et al, U.S. 2,806,014 and Brit. 754,537 (R & H).
76. E. L. Kropa, U.S. 2,443,740 (Am. Cyan.), example 26.
77. P. J. Borchert, U.S. 3,095,444 and 3,242,143 (Miles Labs.).
78. J. H. Jaspers, U.S. 3,167,465 (Miles Labs.); cf. CA 57, 6137 (1962).
79. H. J. Richter and H. S. Rothrock, U.S. 2,410,305 (Du Pont).
80. A. Kronstein, U.S. 843,401; Ber. 46, 1812-1814 (1913).
81. C. Liebermann and M. Kardos, Ber. 46, 1055-1066 (1913).
82. N. A. Derbentseva, CA 46, 10229 (1952).

83. E. C. Britton and C. L. Moyes, U.S. 2,155,856 (Dow).
84. E. C. Britton et al, U.S. 2,160,940-1 (Dow); cf. Brit. 513,221.
85. G. B. Butler and M. D. Barnett, CA 55, 24083 (1961).
86. J. C. W. Crawford, Brit. 581,251 (ICI).
87. C. J. Knuth and A. Barley, Plastics Technol. 3, 555 (1957);
     CA 51, 14315 (1957).
88. Fr. 1,232,633 (Usines Dielectriques); CA 55, 16021 (1961).
89. R. A. Jacobson, U.S. 2,519,764 (Du Pont).
90. J. A. Verdol and B. G. Gower, U.S. 3,429,946 (Sinclair).
91. G. Klement et al, Ger. 1,207,532 (Henkel).
92. F. Engelhardt et al, U.S. 3,622,546 (Cassella).
93. R. Butterworth and J. A. Parker, U.S. 3,164,564 (Armstrong).

13. MONOALLYL ALKYL ETHERS AND RELATED

Of the many monoallyl ether compounds prepared, there has been most interest and use of monoallyl ethers of polyols, such as glycerol and trimethylolpropane, and of allyl epoxyalkyl ethers, particularly allyl glycidyl ether. In spite of company efforts, however, these allyl compounds have had only limited commercial use. Rates of oxidative polymerization in films are much slower than those of drying oils and the polyfunctional allyl ethers discussed in the next chapter. Practical methods for preparing useful vinyl-type homopolymers and copolymers of high molecular weight have been achieved in only a few cases. Using allyl glycidyl ether as an epoxy comonomer, copolyether rubbers capable of curing were developed to a semicommercial stage. Minor proportions of monoallyl ethers of polyols have been used in manufacturing air-drying alkyds for coatings. The greater reactivity in polymerizations of polyfunctional allyl ethers has made them of greater interest for technological evaluation than the monoallyl ethers.

ALLYL ALKYL ETHERS

The compounds of the type $CH_2=CHCH_2OR$ can be made by Williamson synthesis from allyl halides reacting with alkali metal alcoholates or with alkali-alcohol mixtures (1):

$$CH_2=CHCH_2Cl + NaOR \longrightarrow CH_2=CHCH_2OR + NaCl$$

In the case of allyl chloride, the proportion of byproduct diallyl ether formed increases with higher concentrations of reactants. Ethyl allyl ether has been prepared from allyl chloride and aqueous alcoholic sodium hydroxide (2). Allyl n-butyl ether was made by heating allyl chloride with KOH and butanol for 2 hr at 75°C and distilling off the azeotrope with butanol (3). Cetyl allyl ether, which melts at 25°C, was prepared by reacting allyl alcohol with sodium and cetyl chloride (4). Methyl allyl ether has been made from sodium allylate, dimethyl sulfate, and excess allyl alcohol (5). Other

references to the preparation of allyl alkyl ethers were
given in the review of De Wolfe and Young (6). Allyl
ethers of lower alcohols have anesthetic properties, but
tests with animals have shown them to be unpromising because of irritation and toxicity.

Allyl alcohol is more stable toward rearrangement in
the presence of acids than is methallyl alcohol. Under
certain conditions allyl ethers can be prepared in good
yields using acid systems. Allyl cyclohexyl ether and
other allyl ethers have been made by allylation of alcohols with diallyl ether or allyl alcohol in the presence
of mercuric acetate and boron fluoride-etherate in benzene
at reflux for several hours (7). The mercuric salt had to
be added portionwise because of reduction to metallic
mercury during reaction. In one patent example, 99 parts
cyclohexanol, 70 parts allyl alcohol, 57 parts benzene,
2 parts of a 45% solution of BF_3 in dibutyl ether, and
4.6 parts mercuric acetate were heated under reflux with
continuous separation of water. After an hour, second
portions of BF_3 etherate and mercuric acetate were added.
After 2 hr more, the mixture was flash distilled at low
pressure, giving allyl cyclohexyl ether. Cuprous salts
together with aryl sulfonic acid or acid ion-exchange
resins were used as catalysts in preparation of a number
of allyl alkyl ethers and monoallyl ethers of polyols (8).
Allyl esters were prepared from reacting an allyl ester
with an alcohol in the presence of palladium compounds
(9). Allyl bornyl ether and other terpene ethers were
made (10). Other early references have been given to the
preparation of monoallyl ethers (11).

Allyl n-butyl ether on heating with oxygen near 100°C
formed hydroperoxide more than twice as fast as benzyl
ethers (12). About 17% conversion to hydroperoxide,
believed to be $CH_2=CHCHOC_4H_9$, was reached after 2.2 hr.
$\overset{|}{O}OH$
Some polymer and acrolein also were formed. On heating
with aqueous ferrous sulfate, the hydroperoxide gave low
yields of butyl acrylate.

Hypochlorous acid was added to methyl allyl ether to
give the 2-chloro-3-hydroxy derivative (13). Ethyl allyl
ether, on heating with 2% aqueous sulfuric acid, formed
largely ethanol and allyl alcohol (14). Allyl alkyl
ethers with 0.66 molar potassium t-butoxide rearranged 10^3
faster in dimethyl sulfoxide than in 1,2-dimethoxyethane
to form propenyl ethers (15). Kinetics of the hydrolysis
of ethyl allyl ether in dilute acid solution have been
studied (16). Lower alkyl allyl ethers have had interest
as anesthesia (17). Butyl allyl ether, on heating with
oxygen and a cobalt salt, has given acrylic acid, n-butanol, and n-butyl acrylate (18). Alkyl allyl ethers on

treatment with N_2O_3 in ether at -10°C give pseudo-nitro-
sites (19). Pyrolyses near 550°C of allyl ethyl, allyl
benzyl, and allyl isopropyl ethers have given propylene
and, respectively, acetaldehyde, benzaldehyde, and
acetone (20). Cleavage of allyl alkyl ethers by boron
halides depends on the degree of electron donation of the
alkyl groups (21).

With benzoyl peroxide and heating, ethyl and butyl allyl
ethers gave even lower DP liquid polymers than did allyl
acetate because of excessive degradative chain transfer
(22). Allyl alkyl ethers have not given solid high polymers by conventional conditions of free radical and ionic
polymerizations. For example, allyl methyl ether did not
respond on heating at 100°C nor did it respond in light
or to treatment with stannic chloride (23). Lewis-acid
catalysts at low temperatures and transition metal catalysts of the Ziegler type may be expected to yield interesting polymers from allyl alkyl ethers. Natta's group
reported some rather inconclusive studies of homopolymerization of isopropyl, isobutyl, and n-butyl allyl ethers
(24).

Ethyl allyl ether has been reported to copolymerize with
vinylidene chloride (25). The ether formed on standing
with m-cresol and concentrated sulfuric acid a polymer of
2-isopropenyl-5-methyl phenol (26). Allyl tetradecyl
ether reacted with maleic anhydride and an alcohol to give
copolymers suggested as oil modifiers (27). Methyl phen-
allyl ether failed to homopolymerize by free radical
initiation and it copolymerized slowly with MMA and with
styrene (28). A compound related to allyl ethers, 2,5-
dihydrofuran, formed 1:1 copolymers with SO_2 melting at
195°C (29). Ammonium nitrate in methanol acted as a
catalyst. Recently fluoralkyl allyl ether polymers have
been evaluated in making textiles repellent to oils and
water (30). Allyl heptafluoroisopropyl ether, boiling at
65°C, is available from Pierce Chemical Company. Only
homopolymers of low molecular weight were obtained by
free radical initiation (31).

Recently heptafluoroisopropyl methallyl ether (bp 82°C)
was copolymerized with maleic anhydride in presence of a
little lauroyl peroxide to give 1:1 molar copolymers of
high molecular weight which were soluble in acetone and
in dimethyl formamide (32). The corresponding allyl ether
gave lower conversions and lower molecular weight copolymers. Allyl heptafluoroisopropyl ether was also copolymerized with ethylene and fluoroethylenes in presence of
fluoroacetylperoxide (33).

The literature contains examples of allyl ether esters;
for example, trimethylolpropane allyl ether esters of
dimerized linoleic acid (34). Rearrangement on distilla-

tion of the allyl ether of acetoacetic ester to the keto
isomer was the first Claisen rearrangement (35):

$$CH_3C=CHCOOC_2H_5 \longrightarrow CH_3C-CHCOOC_2H_5$$
$$\quad\ |\qquad\qquad\qquad\qquad\ \|\ \ |$$
$$\quad OA \qquad\qquad\qquad\qquad O\ \ A$$

Some additional references to monoallyl ethers are given
in Table 13.1.

TABLE 13.1

Allyl Ethers

Ether	bp, °C	n_D^{20}	References
Allyl methyl ether	43 or 46	1.3764(25)	Henry, Ber. 5, 455 (1872); Bailey, J. Org. Chem. 21, 648 (1956)
Allyl ethyl ether	66	1.3877 $d_4^{20}=0.7651$	Berthelot, Ann. 100, 359 (1856); Adkins, JACS 71, 3053 (1949)
Allyl n-propyl ether	92	1.3919	Kurssanow, CA 43, 2159 (1949)
Allyl isopropyl ether	84	1.3946	Skrabal, Z. Phys-Chem. A 185, 92 (1940)
Allyl n-butyl ether	119	1.4057	Talley, JACS 73, 3528 (1951)
Allyl isobutyl ether	107	1.4008	Talley
Allyl sec-butyl ether	108	1.4023	Talley
Allyl t-butyl ether	100	1.4011	Talley
Allyl isoamyl ether	120	--	Talley
Allyl n-octyl ether	206	1.4267	Devaney, JACS 75, 4836 (1953)
Allyl dodecyl ether	--	--	Watanabe, U.S. 2,847,477-8 (R & H)
Allyl octadecyl ether	152(0.3)	1.444(32) (mp, 29°C)	Kornblum, JACS 64, 3045 (1942); Watanabe, J. Org. Chem. 23, 1666 (1958)
Allyl cyclohexyl ether	89(37)	--	Watanabe
Allyl o-methyl-benzyl ether	98(9)	1.4091	Newman, JACS 68, 2113 (1946)
Allyl bornyl ether	107(17)	$d_4^{21}=0.9221$	Haller, Compt. Rend. 138, 1665 (1904)
Allyl menthyl ether	104(13)	$d_4^0=0.8830$	Haller
Allyl linalyl ether	105(15)	$d_4^0=0.8722$	Haller

METHALLYL ALKYL ETHERS

The first methallyl alkyl ether $CH_2=C(CH_3)CH_2OC_2H_5$ was prepared by Sheshukov in Russia by reacting methallyl chloride with sodium ethylate or alcoholic KOH (36). So reactive is methallyl chloride that alkali in aqueous alcohol can be used for preparation of methyl, ethyl, and isopropyl methallyl ethers (37). Seven hours of reflux- ing has given satisfactory yields of ethyl methallyl ether. With sec-butanol, however, rates were unsatisfac- tory in the presence of water; use of sodium sec-butylate was more satisfactory. Dimethallyl ether forms as a by- product at higher temperatures. Methallyl t-butyl ether was best prepared by portionwise addition of methallyl chloride during 2 hr to hot sodium t-butylate solution (38). It formed an explosive peroxide on warming in air. However, peroxide formation after short storage was not observed as in the case of benzyl ethers studied at the same time. Hexadecyl methallyl ether and related com- pounds were prepared from long alkyl bromides and sodium methylate (39). Methallyl alkyl ethers cannot be made in acidic medium because of rapid rearrangement of methallyl alcohol to isobutyraldehyde.

Propyl and butyl methallyl ethers, along with byproducts, have been prepared by heating 1,3-dichloro-2-methyl pro- pane with alcoholic sodium hydroxide near 125°C (40). One of the products of the reaction of butadiene with n-butanol in the presence of cuprous chloride was butyl methallyl ether (41). Properties of some methallyl ethers are shown in Table 13.2.

TABLE 13.2

Methallyl Alkyl Ethers

Ether	bp, °C	n_D^{20}	References
Methyl	66 or 70	1.3943 (mp, -113°C)	U.S. 2,886,660; 2,148,437; 2,153,513 (Dow)
Ethyl	86	1.4067	Tamele, Ind. Eng. Chem. 33, 116 (1941)
Propyl	114	--	Dow patents
Isopropyl	103	1.4014	Dow patents
Butyl	137	1.4045(25)	Dow patents; Brit. 943,160
sec-Butyl	131	1.4136	U.S. 2,042,219 (Shell)
t-Butyl	--	--	Olson, JACS 69, 2451 (1947)
Hexyl	120(60)	--	Dow patents; U.S. 2,241,421
Hexadecyl	159(2)	--	U.S. 2,241,421
Octadecyl	--	1.444(32) (mp, 29°C)	Kornblum, JACS 64, 3045 (1942)

Attempts to prepare homopolymers of known structure and
high molecular weight from methallyl alkyl ethers by conventional polymerization methods have not been fruitful.
This is not surprising, since many other isopropenyl compounds (without activating electron-attracting groups)
have given high polymers only under special conditions
with complex catalysts such as Ziegler-type activated
transition metal compounds. Treatment of methallyl ethers
with strong Lewis-acid catalysts such as concentrated
sulfuric acid may give isobutyraldehyde and polymers therefrom. Copolymerizations of methallyl alkyl ethers by
radical initiation with vinylidene chloride were investigated by Dow (42); copolymerizations with diallyl fumarate
and styrene were studied by U.S. Rubber (43). Methallyl
ethyl ether was copolymerized with maleic anhydride by
heating with benzoyl peroxide in a sealed tube at 60°C for
16 hr (44). Methallyl n-butyl ether, on treatment with
KNH_2 in liquid ammonia, has given 1-butoxy-2-methyl-1-
propene (45).

MONOALLYL ETHERS OF POLYOLS

There has been application of the alpha-allyl ether of
glycerol in the modification of unsaturated polyesters
(drying alkyds) for coatings. Technical bulletins issued
by Shell in 1953 and 1958 described preparation of
phthalate and succinate polyesters using allyl ether of
glycerol (AEG) as a polyol for use in "air-drying" coatings. The commercial monomer (bp 84.5°C at 1 mm;
n_D^{20} = 1.4627) is miscible with water, acetone, or toluene
and 0.6% soluble in octane. The product has low toxicity
and freezes near -100°C. For preparation of polyesters
bearing allyl groups, it was suggested that equal weights
of AEG and phthalic anhydride be heated up to 240°C during
4 hr under a stream of inert gas until an acid number of
30 was reached (46). Succinate polyesters could be prepared at lower temperatures. Alcoholysis of dimethyl
phthalate by AEG at 230°C with 0.2% CaO catalyst also was
used to prepare viscous liquid unsaturated polyesters.

Early allyl-modified alkyds required aromatic hydrocarbon solvents, since they did not dissolve well in aliphatic hydrocarbons. They also suffered from limited compatibility with other polymers used in coatings and showed
slow rates of oxidative polymerization, even in the
presence of cobalt driers and peroxide catalysts. For
example, 6 hr under air was required for films to become
nontacky, and a week was needed for curing to a water-
resistant state. Even after baking for 45 min at 150°C,
the films were attacked by toluene and by 5% aqueous KOH.
However, such films showed good color stability after

heating at 200°C for 4 hr. Allyl ether-modified polyesters based on AEG or trimethylolpropane allyl ethers were
compared with drying-oil-modified polyesters with more
optimism by some authors than by others (47). Practical
formulations for preparing the unsaturated polyesters
used both glycerol and its monoallyl ether, along with
maleic anhydride and phthalic anhydride. Various combinations of allyl ethers of alkanols with drying oils such
as linseed and soya oils were tried with polyols and
maleic and phthalic anhydrides for making modified unsaturated polyesters (48). The next chapter discusses
polyfunctional allyl ethers evaluated in polyester
coatings.

Recently adipate polyesters from ethylene glycol and
AEG were reacted with toluene diisocyanates to give
allyl-containing polyurethanes (49). Copolymers of glycol
monoallyl ether with vinyl chloride were crosslinked by
tolylene diisocyanate (50). Homopolymers of high molecular weight have not been reported from AEG. Esters of
viscous liquid homopolymers were suggested as drying oils
by Shell.

Allyl ethers of polyols have been added to avoid air
inhibition of the polymerization of certain unsaturated
polyester-styrene finishes used for furniture in Germany.
The allyl ether may be incorporated into the maleate-
fumarate polyester during condensation polymerization, or
it may be added later, along with about 25% styrene (51).
These exceedingly glossy furniture finishes have a number
of limitations, including alkali sensitivity, the color
of the cobalt drier, and the odor of styrene monomer
during curing.

The alpha-allyl ether of glycerol alone was heated with
air near 190°C for 7 hr; during this time, the refractive
index increased from 1.462 to 1.492 (52). This was interpreted as 56% conversion to polymer of molecular weight
about 1000. Such products of oxidation and polymerization
must not be regarded as pure homopolymers of allyl ethers.
The oxygen absorptions of mono- and diallyl ethers of
glycerol were studied by Mleziva and co-workers. Little
was absorbed at 20°C without catalyst; but with cobalt
octanoate, considerable oxygen absorption occurred, but
less than by linoleic acid. See Fig. 13.1.

AEG has been added to "moderate" the polymerization of
acrolein (53). Soluble copolymers were obtained, apparently by radical initiation, and these were suggested as
tanning agents, crosslinking agents for polyvinyl and
alkyd resins, and as modifiers for paper. The monoallyl
ether of pentaerythritol has been copolymerized with
methacrylate and styrene in hydrocarbon-ester solution
using peroxide initiators (54).

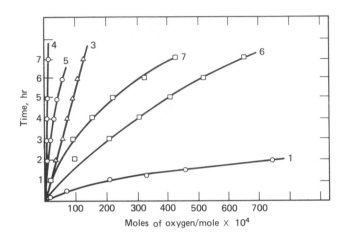

Fig. 13.1. Of a group of allyl compounds the rate of pure oxygen
absorption by autoxidation at 20°C in the presence of 0.03 mole %
cobalt octanoate was highest for linoleic acid followed by the
monoallyl ether of glycerol. Curves: 1, linoleic acid; 3, allyl
n-butyl ether; 4, allyl n-butyrate; 5, diallyl formal; 6, glycerol
monoallyl ether; 7, glycerol diallyl ether. J. Mleziva, J. Jarusek,
J. Thoms, and M. Bleha, Dtsch. Farben-Z. 21, #3 127 (1967).

Possibilities of the monoallyl ether of trimethylolpropane $CH_3CH_2C(OH)_2OA$ have been discussed in bulletins of
the Celanese Corporation and Proctor Chemical Company.
This monomer has low reactivity in polymerization compared
to the diallyl ether of trimethylolpropane discussed in
the next chapter. The monoallyl compound polymerizes to a
syrup on long heating at 150°C. The monomer has been
tried as a diol for preparation of unsaturated polyesters
for "air-drying" films (55), and it has been copolymerized
with maleic anhydride and 2-ethylhexyl acrylate at 100°C
(56).
A number of monoallyl ethers of other polyols have been
made (57), but high polymers have not been prepared from
these or their simple esters by vinyl-type homopolymerization. Monoallyl ethers of the type $CH_2=CHCH_2OCH_2CH_2OR$
have been synthesized (58). Monoallyl ethers of sugars
have been studied (59), but these have had less industrial
interest than polyfunctional allyl ethers. Polymers of
low DP were reported from a monoallyl ether of α-D-glucose
(60).

Swern and co-workers reacted allyl alcohol with the un-symmetrical epoxy compounds propylene oxide, glycidol, 3,4-epoxy-1-butene, and epichlorohydrin to form allyl hydroxyalkyl ethers (61). With propylene oxide and basic catalyst at reflux temperatures, the secondary alcohol was obtained:

$$CH_2CHCH_2 + AOH \xrightarrow{\text{base}} CH_2CHCH_2OA$$
$$\underset{O}{} \qquad\qquad \underset{OH}{}$$

With acid catalyst, however, addition occurred both ways to give a mixture of the two isomeric hydroxypropyl allyl ethers.

Allyl cellosolve has been employed to etherify melamine-formaldehyde resins in benzene at 80°C (62). The pH was adjusted to 4 by addition of formic acid. Mercuration of 2-allyloxyethanol has been studied (63), as well as its copolymerization with acrylonitrile by bisulfite-activated persulfate (64). Some properties and additional referen-ces to monoallyl ethers of polyols are given in Table 13.3.

TABLE 13.3

Monoallyl Ethers of Polyols

Polyol	bp, °C	Miscellaneous properties	References
Ethylene glycol	170	$n_D^{17} = 1.4367$	Harding, J. Chem. Soc. 1528 (1954); Gaylord, U.S. 2,853,462
1,2-Propanediol	85	low toxicity	Hine, CA 45, 1683 (1951)
Glycerol (alpha-ether)	85(1)	$n_D^{25} = 1.4589$ sp.gr.= 1.0679	Morris, U.S. 2,634,296 (Shell) (Viscosity 41 cp at 20°C)
Mononitroglycerol	80(1)	$n_D^{20} = 1.4586$	Ingham, JACS 75, 4255 (1953)
Trimethylol-propane	265	$n_D^{25} = 1.4650$ sp.gr.=1.020	Celanese literature
Trimethylol-ethane	140	$n_D^{20} = 1.4628$	Pattison, U.S. 2,808,391 (Du Pont)
Glucose	--	--	Freudenberg, Ber. 58B 666 (1925)
Mannitol and derivatives	--	--	Wrigley, JACS 70, 2194 (1948); Wolfrom, JACS 81, 5701 (1959)
Xylitols, etc.	--	--	Werner, U.S. 2,790,816 (Ciba)
Starch	--	--	Nichols, Ind. Eng. Chem. 37, 201 (1945)

Table 13.4 lists some additional monoallylic ethers;
few of these have been studied in polymerization. In
addition to the monoallyl ethers of this chapter, some
allyloxy acids are included in Chapter 6 and allyloxy
ketones in Chapter 7. Allyl aryl ethers are discussed
in Chapter 16 and allyl vinyl ethers in Chapter 27.

ALLYL GLYCIDYL ETHER

Although $CH_2=CHCH_2OCH_2\underset{\underset{O}{\smile}}{C}HCH_2$ has been available from

Shell since 1956, it has had only limited use in polymer
products. Properties include bp 154°C, n_D^{20} = 1.4348,
d_4^{20} = 0.9678, solubility in water 14% at 20°C, and miscible with acetone and toluene. The monomer must be
handled with special precautions because of toxicity and
irritation to eyes and skin. The two different functional
groups of such allyl epoxy compounds permit two-stage
polymerizations with different types of catalysts. Allyl
glycidyl ether (AGE) can be prepared by reaction of allyl
alcohol with epichlorohydrin or by action of alkali upon
the monoallyl ether of glycerol monochlorohydrin (65).
AGE is supplied by Shell, Dow, Borden, and others.
   The epoxy group reacts with alcohols in the presence of
acid catalysts to give terminal OR groups; it reacts with
amines to give terminal amino groups and with organic
acids to form one or two ester groups. Allyl glycidyl
ether can be reacted with o-phthalic acid to form polyesters bearing side allyl ether groups, which were suggested as heat and oxygen-curable alkyd coatings (66).
With maleic anhydride allyl glycidyl ether gives soluble
prepolymers followed by crosslinked polymers (67). AGE
has been added to promote curing of unsaturated polyesters in coatings (68), but toxicity is a limitation.
Diels-Alder adducts of dienes and allyl glycidyl ether
were reacted with unsaturated polyesters before copolymerization with styrene (69). AGE was reacted with
hydroxyethyl ethylene urea (70). The product was copolymerized in emulsion with acrylate esters for use in
baking enamels. AGE has been used as a modifier in the
curing of epoxy resins with precautions against its
toxicity.
   Partial vinyl-type homopolymerization of allyl glycidyl
ether was described by air blowing at 220°C or by heating
with peroxide in an autoclave (71). Conversion to syrup
of 20 poise containing 40% polymer of molecular weight
500 occurred at least partially by addition at double
bonds. Low polymers also could be made by portionwise
addition of BF_3 etherate to AGE in isopentane at 30°C.
   Preparation of homopolymers of high molecular weight

TABLE 13.4

Other Monoallyl Ether Compounds

Compound	bp, °C	Refractive index	References
Allyl-2-butoxy ethyl ether	85	--	Liston, JACS 60, 1264 (1938); Brit. 916,561 (Monsanto)
Methallyl-1-methoxy-2-propyl ether	160	$n_D^{25}= 1.4160$	Horsley, U.S. 2,886,600 (Dow)
Allyl ethyl ether of diethylene glycol	--	--	Moyle, U.S. 2,244,308 (Dow)
Allyl-1,4-pentadienyl ether	55(25)	$n_D^{20}= 1.4504$	Montagna, U.S. 2,962,534 (Carbide)
Allyloxy tetrahydrofurans	--	--	Kratochvil, CA 56, 15452 (1962)
Allyloxy hexyne derivatives	--	--	Colaianni, Ger. 1,080,098 (LaRoche) (good odors)
Allyl cyclopentenyl ether	--	--	Young, U.S. 2,898,378 (Carbide)
Allyl bicyclic ethers	--	--	Bruson, U.S. 2,421,597 (R & H)
Allyl chloromethyl ether	108	$n_D^{23}= 1.431$	Hurd, U.S. 2,847,456 (Columbia Southern)
Allyl 2-bromoethyl ether	69	$n_D^{20}= 1.4668$	Hurd, JACS 60, 1908 (1938)
Allyl 2-chloroethyl ether	135	--	Liston; Ham, U.S. 2,643,986 (Chemstrand) (acrylonitrile copolymers)
Allyl chloroalkyl ethers	--	--	Ballard, U.S. 2,608,586 (Shell)
Allyloxy hydroxybutenes and esters	--	--	Swern, U.S. 2,715,132 (USA)
2-Allyloxyethyl acetate	--	--	Jenkins, J. Oil Colour Chem. Assoc. 44, 50 (1961) (copolymers)
Allyloxy esters (various)	--	--	Hurd, JACS 74, 5128 (1952); D'Alelio, U.S. 2,406,590 (GE)
β-Allyloxycrotonic acid	--	--	LeNoble, J. Org. Chem. 27, 3875 (1962) (mp, 81°C)
Allyl ethers of aminoalcohols	--	--	Bruson, U.S. 2,712,015 (Ind. Rayon)
Allyl ether of adiponitrile	--	--	Hager, U.S. 2,500,942 (Du Pont)
2-Allyloxytetrahydropyran	--	--	Brannen, U.S. 3,311,556 (std. (Std. Oil Ind.)

from allyl glycidyl ether through reaction of ethylenic
double bonds has proved to be difficult. Other examples
have been given of formation of viscous liquid polymers
of low molecular weight by free radical reactions. For
example, allyl glycidyl ether was polymerized in benzene
solution using 3% di-t-butyl peroxide at 155°C (72). The
polymer of about 500 average mol. wt. recovered by distilling off the volatiles was reacted with hydrogen sulfide to form harder polymer which could be melt spun to
fibers. The low polymer also could be reacted with 1,5-
pentanedithiol to form rubberlike products. Liquid low
polymers bearing epoxy groups have been suggested as
stabilizers for vinyl halide polymers (73).

EPOXIDE POLYMERIZATION

   Polymers of low molecular weight were obtained by
epoxy-type polymerization of allyl glycidyl ether in the
presence of Lewis-acid catalysts (74). Vandenberg first
prepared polymers of high molecular weight from allyl
glycidyl ether using as catalyst reaction products of
aluminum alkyls with small amounts of water (75). In one
example, 17.2 parts of diethyl ether and 10 parts of allyl
glycidyl ether were treated at room temperature under
nitrogen. There was added 5.6 parts of a solution prepared from diluting a 1.53 molar solution of triethyl
aluminum in n-heptane to 0.5 molar by adding diethyl ether
and then injecting 0.5 mole of water per mole of aluminum
alkyl. The catalyst solution had been aged with agitation for 22 hr at 35°C. After this catalyst had been
added, the polymerization reaction proceeded for 19 hr at
30°C, after which the catalyst was quenched by adding
4 parts of ethanol and 40 parts of ether. The ether-
insoluble polymer was washed with fresh ether containing
0.2% 4,4'-thiobis(6-t-butyl-m-cresol) stabilizer and then
was dried 16 hr at 50°C under vacuum. The ether polymer
recovered was a tacky rubbery solid having a reduced
viscosity of 0.86 measured at 0.1% solution in cyclohexanone at 50°C. Infrared absorption indicated 33% allyl
groups, compared with 36% if no allyl groups had reacted.
This showed that the polymerization had occurred predominantly through reaction of epoxy groups. The polymer
was vulcanized by sulfur along with carbon black and
mercaptobenzothiazole by molding at 300°F for 45 min. It
gave tensile strength of 500 psi, tensile modulus of 400
psi at 100% elongation, Shore A hardness of 50, and a
break set of 5%.
   In a second example of AGE polymerization, 0.6 mole of
water was added per mole of aluminum triethyl, and the
polymer was purified by precipitation from methanol by

adding water. The tacky rubber obtained had a reduced
viscosity of 1.2. Vandenberg made copolymers of allyl
glycidyl ether with epichlorohydrin at 20°C by means of
catalysts prepared by adding aluminum triethyl to diethyl
ether and heptane (76). Copolymers of AGE with alkylene
oxides were crosslinked to form membranes evaluated for
desalination of water (77). Copolymers with propylene
oxide were also studied by Dow as vulcanizable elastomers
(78).

Copolymers of propylene oxide with minor proportions of
allyl glycidyl ether as prepared by Vandenberg gave promise
as strong, snappy synthetic rubbers (79). In one example,
90 parts of propylene oxide and 10 parts of allyl glycidyl
ether were reacted for 19 hr at 30°C with a catalyst
obtained by hydrolyzing diethyl zinc by ether containing
0.2% water. A fraction of about 20% had stereoregularity
and gave a crystalline x-ray pattern. Epoxy copolymerization of allyl glycidyl ether with ethylene oxide and
with propylene oxide was studied at 90°C, using as catalysts calcium organic compounds, zinc carbonate, and zinc
butyl (80). Copolymers with propylene oxide had tear
strength and aging properties equal to those of natural
rubber (81). Parel-50 of Hercules is cured using sulfur.

In 1964 rubbers based on copolymerization of propylene
oxide with about 5% allyl glycidyl ether were available
from General Tire and Rubber Company. Such elastomers,
made by use of $FeCl_3$ as catalyst contained both amorphous
and crystalline fractions. They could be vulcanized by
sulfur. Tetrapolyether rubbers based on copolymerization
of epichlorohydrin, ethylene oxide, propylene oxide, and
allyl glycidyl ether have been prepared from solution
polymerization using aluminum alkyl catalyst (82). They
were reinforced by carbon black and could be vulcanized
using peroxides or sulfur. These terpolymer elastomers
had better properties at low temperatures than copolymers
of epichlorohydrin and ethylene oxide. For copolymerization of alkylene oxides with allyl glycidyl ether,
aluminum chloride-amine complex catalysts were employed
in Italy (83). Copolymerization of tetrahydrofuran with
AGE with an antimony chloride derivative as catalyst was
believed to occur by an oxonium ion mechanism (84). Alkylene oxide polymers have been specially terminated by
allyl epoxy monomers (85). Water-dispersible waxes were
thus prepared.

RADICAL-INITIATED COPOLYMERIZATIONS
OF ALLYL GLYCIDYL ETHER

Copolymers of allyl glycidyl ether with vinyl type
monomers made by free radical initiation have been evalua-

ted especially for curable coatings. Cupery of Du Pont
heated equal weights of styrene and allyl glycidyl ether
with 0.5% benzoyl peroxide at 90°C for 1.5 hr (86). After
the volatiles had been distilled off, a copolymer containing only 0.9% epoxy or oxirane oxygen was recovered.
The polymer was then heated with phosphoric acid in dioxane solution. Films deposited from organic solvents
slowly became insoluble. In Shell laboratories, low copolymers of styrene with allyl glycidyl ether were made
by heating with cumene hydroperoxide at high temperatures
(87). Curing was tried with polyamines.

ICI copolymerized AGE with styrene by portionwise addition of the latter to the refluxing ether over a period
of 10 hr (88). A resinous copolymer containing 86.7%
styrene units resulted. With 600 parts of copolymer was
reacted 395 parts of linseed oil at 225°C to give a product of interest for coatings. Styrene-AGE copolymers
were reacted with phosphoric acid to give phosphated
copolymer coatings (89). Copolymerization of allyl glycidyl ether by continuous addition of n-butyl methacrylate
along with t-butyl peroxide at 130°C gave polymers of
molecular weight 1900 and 4% epoxy oxygen (90).

Copolymers of low molecular weight were obtained by
heating AGE with acrylate esters and azobis (91). They
were suggested as polymeric epoxy stabilizers for vinyl
chloride polymer coatings. AGE, maleic anhydride, and
cyclohexyl acrylate were copolymerized in benzene solution (92). Copolymerization was largely vinyl-type with
a little by epoxy groups. Elastomeric copolymers from
ethyl acrylate with minor proportions of AGE and allyl
methacrylate were crosslinked by heating with succinic
anhydride (93).

Copolymers of vinyl acetate with minor proportions of
allyl glycidyl ether were prepared in isopropanol at 83°C
with 3% benzoyl peroxide (94). Over 20% of combined allyl
glycidyl ether units were present in the copolymers of
average molecular weight 1400. Vinyl acetate-ethylene-AGE
interpolymer latices were prepared in emulsion and applied
as binders for nonwoven fabrics (95). One copolymerization began with the following composition:

Ingredient	parts
Water	33,000
Vinyl acetate	36,000
Igepal 887	1,020
Igepal 630	510
Ferric ammonium sulfate	0.05
Sodium lauryl sulfate	5.0
Potassium persulfate	105

The reactor and charge were purged by nitrogen, pressurized with ethylene at 40 atm, and stirred at 50°C. There
was added 20 ml of a 0.5% aqueous solution of the activator sodium formaldehyde sulfoxylate (Formopon). During
the course of polymerization (4.5 hr) there was added
slowly 139 g maleic acid, 1800 g allyl glycidyl ether, and
136 g persulfate. The latex was finally cooled and neutralized to pH 5.2 with ammonium hydroxide. Zinc fluoroborate could be added to the latex as a curing agent for
binding the fibers of nonwoven fabrics.

Copolymerization with vinyl chloride was studied in free
radical systems (96). Terpolymers were prepared by
radical polymerization incorporating units from 60 to 80%
vinylidene chloride, 12 to 37% vinyl chloride, and 3 to
8% AGE (97). Phosphated with 0.5 to 1.0 mole of phosphoric acid per mole of combined allyl glycidyl ether,
these terpolymers were evaluated in coatings.

Vulcanizable interpolymers of tetrafluoroethylene, ethyl
vinyl ether, and AGE were prepared in heptane-tetrahydro-
furan solvent using benzoyl peroxide as initiator (98). A
little calcium carbonate was present in the copolymerization as acid acceptor. The terpolymers showed good
adhesion to a range of surfaces, including tetrafluoroethylene polymer. Copolymers of 2 to 20% allyl, chlorallyl, or methallyl glycidyl ether with acrylonitrile were
reacted with amines or thioureas to form dyeable fibers
(99). Reaction products of alkanolamines with allyl
glycidyl ether were copolymerized with acrylonitrile (100).
AGE copolymerized slowly with maleic anhydride at 60°C in
acetone with azobis catalyst (101).

There has been interest in allyl glycidyl ether for the
preparation of thermosetting acrylic coatings. In one
example, the following solution was heated at reflux for
copolymerization (102):

Ingredient	parts
n-Butyl methacrylate	426
Vinyltoluene	708
Methacrylic acid	72
AGE	29
Benzoyl peroxide	12.4
Cumene hydroperoxide	12.4
Xylene solvent	618

Sprayed coatings on steel were cured for 20 min at 350°F
to produce crosslinking through reaction of epoxy with
carboxy groups. Part of the styrene in unsaturated polyester casting syrups may be replaced by AGE (103). However, in most free radical copolymerizations with vinyl

monomers, the retardation of reaction rates by AGE, low
content of epoxy groups in the products, and toxicity of
the monomer have been unfavorable factors.

Allyl derivatives of rings containing oxygen have been
studied. Sakurada and co-workers carried out a two-stage
polymerization of 3-chloromethyl-3-allyloxy methyloxa-
cyclobutane (104). The allyl groups were first reacted
by initiation with azobis and then by added boron fluoride-etherate the ring opening was catalyzed. In a patent
of 1952 Thomas showed that 4-allyloxymethyl-1,3-dioxolane
could be polymerized either through its ethylenic bonds
by heating with acyl peroxides or through ring opening by
means of acids or amines (105). Allyl epithiopropyl ether
$CH_2=CHCH_2OCH_2CHCH_2$ was prepared from allyl glycidyl ether
                    $\overset{\diagdown S \diagup}{}$
by reacting with sodium thiocyanate under nitrogen in the
presence of diethyl zinc (106). Elastomeric vulcanizable
copolymers were made. A few additional allyl epoxy compounds are listed in Table 13.5. The products of epoxidation of dienes also have had interest in vinyl copolymerizations. Examples are $CH_2=CHCHCH_2$ (107) and
                                    $\overset{\diagdown O \diagup}{}$
$CH_2=C(CH_3)CHCH_2$ (108).
          $\overset{\diagdown O \diagup}{}$

TABLE 13.5

Other Allyl Epoxy Compounds

Compound	bp, °C	Other properties, etc.	References
3,4-Epoxy-1-butene	70	$n_D^{20} = 1.416$	--
4-Vinylcyclohexene oxide	169	sp. gr. 0.9598; copolymers	Union Carbide data sheet; U.S. 2,687,404 (Du Pont)
2-Allylphenyl glycidyl ether	103(0.6)	$n_D^{20} = 1.5310$; low polymers by radical initiation	Brit. 770,080 (Shell)
Allyl 2,3-epoxy-cyclopentyl ether	--	Cured by tin organic salts	U.S. 3,117,099 (Carbide)
Allyl ether of o-glycidylphenol	--	--	Brit. 841,589 (General Mills), CA 55, 6496
Allylaryl glycidyl ethers	--	Styrene copolymers	U.S. 2,983,703 (Koppers)
Allyl ether of oxetane	--	--	U.S. 2,924,607 (Du Pont)
Other allyloxy derivatives of epoxy compounds	--	--	U.S. 3,226,401 (Carbide)

1. L. Henry, Ber. 5, 449 (1872); CZ 468 (1872); H. P. A. Groll and C. J. Ott, U.S. 2,042,219 (Shell).
2. R. Skrabal, Z. Physik. Chem. A 185, 92 (1939).
3. D. B. Sharp and T. M. Patrick, J. Org. Chem. 26, 1389 (1961).
4. J. S. H. Davies et al, J. Chem. Soc. 2545 (1930).
5. A. Fairbourne et al, J. Chem. Soc. 452 (1931).
6. R. H. DeWolfe and W. G. Young, Chem. Rev. 56, 828 (1956).
7. W. H. Watanabe et al, J. Org. Chem. 23, 1666 (1958); U.S. 2,847,477-8 (R & H).
8. R. J. Stephenson, Brit. 913,919 and 916,566 (Monsanto).
9. D. Clark and P. Hayden, U.S. 3,591,640 (ICI).
10. A. Haller and F. March, Compt. Rend. 138, 1665 (1904).
11. Allyl Alcohol, booklet, Shell Chemical Co. 1946.
12. D. B. Sharp and T. M. Patrick, J. Org. Chem. 26, 1389 (1961).
13. L. Henry, CZ II, 302 (1904); Compt. Rend. 123, 351 (1896).
14. M. Sheshukov and Eltekov, Ber. 10, 1903 (1877); (correspondence from St. Petersburg).
15. C. C. Price and W. H. Snyder, J. Am. Chem. Soc. 83, 1773 (1961).
16. R. Skrabal, Z. Physik. Chem. A 185, 92 (1939).
17. C. D. Leake and M.-Y. Chen-Mai, Proc. Soc. Exp. Biol. Med. 28, 152 (1930, 1931).
18. H. Grimm and W. Flemming, Ger. 728,834 (IG); CA 38, 379 (1944).
19. N. Y. Maslov, CA 35, 3962 (1941); CA 40, 1778 (1946).
20. A. Malzahn, CA 59, 491 (1963); cf. R. C. Cookson and S. R. Wallis, J. Chem. Soc. B 1245 (1966).
21. W. Gerrard et al, J. Chem. Soc. 4007 (1956).
22. I. Sakurada and G. Takahashi, Chem. High Polym. (Japan) 11, 348 (1954); CA 50, 602 (1956).
23. H. Staudinger and T. Fleitmann, Ann. 480, 93 (1930).
24. G. Natta et al, CA 59, 15393 (1963).
25. E. C. Britton and C. W. Davis, U.S. 2,160,943 (Dow).
26. J. B. Niederl et al, J. Am. Chem. Soc. 53, 3393 (1931).
27. J. J. Giammaria, U.S. 2,704,277 (Socony).
28. M. G. Baldwin and S. F. Reed, J. Polym. Sci. A 6, 2627 (1968).
29. M. Slovinsky, U.S. 3,297,329 (Celanese).
30. A. G. Pittman and W. L. Wasley, U.S. 3,522,084 and 3,541,159 (USA).
31. A. G. Pittman et al, U.S. 3,382,222 and 3,437,692 (USA); J. Polym. Sci. A 1, 1741 (1968).
32. W. L. Wasley and A. G. Pittman, Polymer Lett. 10, 279 (1972); Polymer Preprints 12, 445 (1971); CA 78, 30257 (1973).
33. D. P. Carlson, Ger. Offen. 2,012,069 (Du Pont); CA 76, 113862 (1972).
34. W. E. Parker and R. E. Koos, U.S. 3,536,738 (USA).
35. L. Claisen, Ber. 45B, 3157 (1912).
36. M. Sheshukov, J. Russ. Phys.-Chem. Soc. 16, 478 (1884).
37. M. Tamele et al, Ind. Eng. Chem. 33, 119 (1941); H. P. A. Groll and C. J. Ott, U.S. 2,042,219 (Shell).
38. W. T. Olson et al, J. Am. Chem. Soc. 69, 2453 (1947).
39. D. Price and B. A. Dombrow, U.S. 2,241,421 (Nat. Oil Products).

40.   G. H. Coleman and G. V. Moore, U.S. 2,148,437; cf.
      U.S. 2,153,513 (Dow).
41.   R. J. Stephenson, U.S. 3,271,461 (Monsanto).
42.   E. C. Britton and C. W. Davis, U.S. 2,160,943 (Dow).
43.   P. O. Tawney, U.S. 2,561,153-4 (U.S. Rubber).
44.   R. T. Armstrong, U.S. 2,415,400 (U.S. Rubber).
45.   A. J. Birch, J. Chem. Soc., 1647 (1947).
46.   cf. T. W. Evans and D. E. Adelson, U.S. 2,488,258 and 2,399,214
      (Shell); also E. L. Kropa and T. F. Bradley, U.S. 2,280,242
      (Am. Cyan.).
47.   H. Dannenburg et al, Ind. Eng. Chem. $\underline{41}$, 1709 (1949); H. W.
      Chatfield, Paint Technol. $\underline{26}$, (4), 13-21 (1962); Celanese
      bulletins on allyl ethers of trimethylolpropane.
48.   T. L. Phillips, Brit. Plastics $\underline{34}$, 69 (February 1961);
      G. R. Svoboda, Offic. Dig. 1104 (1962); D. E. Cordier, U.S.
      2,623,030 (L-O-F); T. F. Anderson, U.S. 2,632,753 and 2,635,089
      (L-O-F).
49.   G. A. Kuhar, U.S. 3,577,389 (Goodyear).
50.   F. Wingler and H. Barth, Ger. appl. (Bayer); CA $\underline{75}$, 6642 (1971).
51.   V. F. Jenkins et al, J. Oil Colour Chemists' Assoc. $\underline{44}$ (1),
      42-60 (1961); cf. Brit. 810,222 and 869,298; Belg. 545,201
      (Bayer).
52.   H. Dannenberg and D. E. Adelson, U.S. 2,545,689 (Shell).
53.   H. C. Miller and H. S. Rothrock, U.S. 2,657,192 (Du Pont).
54.   H. J. Wright and R. D. Kincheloe, U.S. 3,288,736 (Cook Paint).
55.   K. Raichle and W. Biederman, Ger. 1,024,654 (Bayer).
56.   J. R. Costanza et al, U.S. 3,268,492 (Celanese).
57.   C. D. Hurd and M. O. Pollack, J. Am. Chem. Soc. $\underline{60}$, 1905 (1938);
      R. Skrabal, Z. Physik. Chem. A $\underline{185}$, 81 (1939); CA $\underline{34}$, 1235 (1940);
      V. N. Kotrelev and I. K. Rubtsova, CA $\underline{50}$, 6384 (1956).
58.   L. Liston and W. M. Dehn, J. Am. Chem. Soc. $\underline{60}$, 1264 (1938);
      C. L. Moyle and G. H. Coleman, U.S. 2,244,308 (Dow).
59.   K. Freudenberg et al, Ber. $\underline{58B}$, 666 (1925); A. N. Wrigley and
      E. Yanovsky, J. Am. Chem. Soc. $\underline{70}$, 2194 (1948); also references
      in Chapter 14.
60.   R. C. Schweiger, U.S. 3,340,239 (Kelco).
61.   D. Swern, V. Billen, and H. B. Knight, J. Am. Chem. Soc. $\underline{71}$,
      1152 (1949).
62.   E. Ishizuka et al, Japan 17,746 (1961)(Reichhold); CA $\underline{57}$, 1069
      (1962).
63.   L. H. Werner and C. R. Scholz, J. Am. Chem. Soc. $\underline{76}$, 2701 (1954).
64.   H. S. Rothrock, U.S. 2,605,258 (Du Pont).
65.   T. W. Evans et al, U.S. 2,314,039 (Shell).
66.   T. W. Evans and D. E. Adelson, U.S. 2,399,214 (Shell).
67.   R. F. Fischer, J. Appl. Polym. Sci. $\underline{7}$, 1451 (1963).
68.   H. Delius and W. Becker, U.S. 3,006,876 (Reichhold).
69.   H. Wulff and E. D. Behnke, U.S. 3,236,916 (Witten).
70.   K. Sekmakas, U.S. 3,509,085 (DeSoto).
71.   Shell Data Booklet 1956; cf. U.S. 2,752,269.
72.   Brit. 730,670 (Shell).

73. D. E. Winkler et al, U.S. 2,585,506 and 2,609,355 (Shell).
74. T. W. Evans and E. C. Shokal, U.S. 2,450,234 (Shell); cf. U.S. 2,464,753 (Shell).
75. E. J. Vandenberg, U.S. 3,065,213 (Hercules); cf. U.S. 3,219,591 (Hercules).
76. E. J. Vandenberg, U.S. 3,285,893 (Hercules).
77. C. A. Luklach, U.S. 3,567,630 (Hercules).
78. A. F. Gurgiolo and R. W. McAda, U.S. 3,591,570; 3,616,462 (Dow).
79. Brit. 927,817 (Hercules) appl. 1960; CA 59, 12999 (1963); E. J. Vandenberg, J. Polym. Sci. 47, 486 (1960).
80. F. E. Bailey and H. G. France, J. Polym. Sci. 45, 243 (1960).
81. Brit. 614,808 (Dunlop); CA 57, 12685 (1962); J. G. Hendrickson et al, Ind. Eng. Chem. Prod. Res. Develop. 2, 199 (1963).
82. Z. T. Ossefort et al, Ind. Eng. Chem. Prod. Res. Develop. 7, 17 (1968).
83. W. Marconi et al, U.S. 3,394,088 (SNAM, Milan).
84. I. Kuntz (Esso), CA 76, 25591 (1972); cf. C. E. Snyder and J. A. Lovell, U.S. 3,133,905 (Goodyear).
85. R. L. McKellar, U.S. 3,447,271 (Dow Corning).
86. M. E. Cupery, U.S. 2,692,876 (Du Pont).
87. E. C. Shokal et al, U.S. 2,839,514 (Shell).
88. Brit. 730,397 (ICI); cf. U.S. 2,839,514 and 3,040,010 (Shell).
89. Brit. 757,043 (Du Pont).
90. M. E. Cupery, U.S. 2,723,971 (Du Pont); cf. S. Kordzinski and M. B. Horn, U.S. 3,579,490 (Ashland Oil).
91. H. S. Rothrock and W. K. Wilkinson, U.S. 2,687,405 (Du Pont).
92. W. S. Pickle, U.S. 3,527,738 (Dow).
93. R. E. Lauer et al, Belg. 664,688 (Thiokol); CA 65, 2453 (1966).
94. H. S. Rothrock and W. K. Wilkinson, U.S. 2,788,339 (Du Pont); cf. U.S. 2,781,335 (Du Pont).
95. M. K. Lindemann and R. P. Volpe, U.S. 3,526,538 and 3,526,540 (Airco).
96. E. K. Ellingboe, U.S. 2,562,897; 2,607,754 and 2,589,237 (Du Pont).
97. J. A. Robertson, U.S. 3,291,781 (Du Pont); cf. U.S. 3,018,197 (Du Pont).
98. Brit. 948,998 (Du Pont).
99. G. E. Ham, U.S. 2,650,151 (Chemstrand).
100. H. A. Bruson, U.S. 2,712,015; cf. U.S. 2,631,995 (Ind. Rayon).
101. K. Noma and M. Niwa, CA 68, 115001 (1968).
102. H. C. Woodruff, U.S. 3,052,659.
103. R. E. Davies and L. G. Rosen, U.S. 3,011,994 (Celanese).
104. I. Sakurada et al, Chem. High Polym. (Tokyo) 23, 172 (1966).
105. W. M. Thomas, U.S. 2,601,572 (Am. Cyan.).
106. J. Lal, J. Polym. Sci. B 3, 969 (1965).
107. F. E. Bailey, U.S. 3,417,064 (Carbide).
108. M. N. Sheng and J. G. Zajacek, U.S. 3,538,124 (Atlantic Richfield).

Diallyl ethers and polyfunctional allyl ethers derived
from allylation of polyols and carbohydrates with allyl
chloride under alkaline conditions have been much investigated; only limited commercial use has been found for
these, however. Of greatest interest have been allylated
pentaerythritol, trimethylolpropane, starches, and sucrose, which undergo autoxidative polymerization in contact
with air to form coatings. Like linseed oil, these allyl
derivatives can be "bodied" to soluble prepolymers by
heating in the presence of oxygen, and they cure to solid
nontacky films in the presence of cobalt compounds or
other driers. However, the slow rates of curing, the
color, uniformity, and strength properties of the films
generally have not proved adequate. A number of the
monomers continue to have possibilities in specific applications, particularly as branching and crosslinking agents
in copolymerizations with vinyl monomers. Relatively
small amounts of allylated starches and sugars have been
used from time to time. Many of the monomers and their
partial polymers decompose when overheated, giving the
characteristic piercing odor of acrolein. Allyl ethers
of polyols such as pentaerythritol and trimethylol propane
have been used in manufacture of film-forming polyester-
styrene resins which cure uninhibited by air.
In this chapter a name such as allyl sucrose is meant to
include partially allylated sucroses. Many of the polyfunctional allyl ether monomers of this chapter have not
been studied as pure, fully allylated compounds.

DIALLYL ETHERS

Diallyl ethers may be prepared by reacting allyl chloride with alkali. Diallyl ether and allyl alkyl ethers
also are readily prepared from allyl alcohol (1). Allyl
alcohol was heated above 60°C in the presence of a cuprous
salt and sulfuric or phosphoric acid (2). The alcohol
vapors may be passed over contact catalysts such as $CuCl_2$
in the presence of a promoter such as HCl or allyl chloride (3). Commercial diallyl ether (DAE) is available as

a byproduct from the Shell process for manufacture of allyl alcohol. DAE can form explosive peroxides on storage under air, especially in light. The monomer has a mild ethereal odor producing very little sting in the nose.

Diallyl ether adds chlorine rapidly even below 0°C (4). Diallyl ether reacts with alcohols in the presence of mercuric salts and sulfuric acid to give alkyl allyl ethers and allyl alcohol (5). Allyl ethers with basic catalysts can isomerize to propenyl ethers. Diallyl ether reacts with sodium to give allyl sodium and sodium allyloxide (6). With boron trichloride even at low temperatures, diallyl ether and dimethallyl ether were cleaved to form the organic halides and boron-oxygen compounds (7). There was evidence that the methallyl group is more electron releasing than the allyl group. Allyl methallyl ether has been reacted with boron trichloride to give triallyl borate (8). Diallyl ether can be estimated by reaction with mercuric acetate and methanol followed by titration of liberated acetic acid (9).

Diallyl ether forms acrolein on pyrolysis (10). It forms a diepoxide by way of dichlorohydrin (11). Diallyl ether is an irritant of the skin and eyes. Its high toxicity may be related to the formation of acrolein. In the presence of iron pentacarbonyl, irradiation of diallyl ether (as well as allyl ethyl ether, allyl phenyl ether, and allyl benzene) gave the corresponding propenyl compounds (12).

Davison and Bates studied infrared spectra of diallyl ether and a number of other allyl compounds (13). They found consistent displacements of the band associated with the C=C group. As shown below, this absorption is high when electron-releasing, nonconjugated groups are attached to the ethylenic nucleus but low when the stretching force constant is lowered by conjugation.

CH_2=CHR	1642 cm^{-1}
Diallyl ether	1648
1,1-Diallyloxybutane	1644
1,1,3-Triallyloxypropane	1644
Diallyl adipate	1650
Diallyl sebacate	1654
Vinyl acetate	1647
n-Butyl methacrylate	1640
Methyl acrylate	1637
Allyl acrylate	1635
Allyl methacrylate	1635
Methacrylic acid	1632
Methacrylonitrile	1620
Vinyl methyl ketone	1618

In contrast to vinyl alkyl ether monomers diallyl ether does not polymerize readily when solutions in hydrocarbon are treated with boron fluoride etherate (14). On prolonged contact with air, the liquid DAE becomes viscous, and nontacky films are formed finally. DAE exposed to a hot mercury lamp for 300 hr formed a barely flowing viscous polymer syrup. Polymerization in UV light and on heating at 60°C was accelerated by adding 5% benzoyl peroxide. Crosslinking does not occur at low conversion. Cured films were not tacky at 80°C. Diallyl ether reacted violently with m-cresol in the presence of sulfuric acid to form a polymer believed to be derived from 5-methyl-2-isopropenylphenol (15).

Recently diallyl ether has been found to have favorable reactivity in copolymerizations with unsaturated polyesters. It may be used along with or replacing diallyl phthalates. In Europe DAE has been used in coatings replacing drying oils and modifying unsaturated polyesters. Diallyl ether, maleic anhydride, and ethylene were copolymerized under high pressure with peroxide catalyst (16). A 1% aqueous dispersion of the ammonium salt of the terpolymer had a gel-like consistency. Whether true interpolymers resulted with appreciable ether content in some of the references cited in Table 14.1 may be open to question.

TABLE 14.1

Diallyl Ether Copolymerization [a]

Ether	Comonomers	References
Diallyl	Acrylonitrile, also telomers with CCl$_4$	Aso, J. Chem. Soc. Japan (Ind. Chem.), 68, 1970 (1965)
Diallyl	Vinyl chloride	Martin, U.S. 3,025,280 (Monsanto)
Diallyl	Acrylate esters	Fisher, U.S. 2,492,169 (USA)
Diallyl	MMA and diethyl fumarate	Tawney, U.S. 2,426,325 (U.S. Rubber)
Dimethallyl	Acrylates	Snyder, U.S. 2,539,706 (U.S. Rubber)
Dimethallyl	Acrylate esters, etc.	Sparks, U.S. 2,411,599 (Std. Oil of N.J.)

[a] Radical catalysts were used.

Alkali metal or ammonium salts of ethylene-diallyl ether-maleic anhydride copolymers have been added to oil drilling muds (17).

The diallyl ether of 1,4-butynediol is said to polymerize by addition of free radical or Lewis acid catalysts (18).

Diallyl ethers can participate in polymer formation by certain methods other than addition polymerization. Reaction with alkali cellulose was suggested for preparation of curable allyl cellulose partial ethers (19).

As in the case of nonconjugated dienes, active hydrogen compounds can add to allyl ethers. These additions, catalyzed by light and free radical catalysts, generally occur in a reverse Markovnikov way; that is, in accord with Posner's rule (20). Viscous liquid low polymers obtained from heating diallyl ether with H_2S under pressure with peroxides at 65 or 100°C were suggested as lubricant and rubber additives (21). By reacting equimolar proportions of diallyl ether and H_2S in liquid phase at 5°C, irradiated by a quartz mercury light, Vaughn obtained a distillable mixture of allyl ether terminated ether-sulfide low polymers (22). Free radical emulsion polymerization of dithiols with diallyl ether can give polymers of higher molecular weight (23). In another example of hydrogen migration copolymerization, 50 parts diallyl ether, 55 parts phenyl phosphine, and 0.5 part BP were heated under nitrogen at 70°C for 90 hr (24). A tough, flexible copolymer, recovered by vacuum distillation, contained 14% phosphorus and was soluble in methanol.

Di-2-chloroallyl ether (bp 173°C, n_D^{20} = 1.4781) can be made by heating 1,2,3-trichloropropane with aqueous NaOH at 100°C for 10 hr (25). On heating, the compound polymerized to black, rubberlike to brittle solids, insoluble in common solvents (25). The 1,2-divinyl derivative of ethylene oxide (110°C) is related to diallyl ether. It formed ether polymers by treatment with phosphorous trifluoride-tetrahydrofuran (26). The free allyl groups of the polymer added N_2F_4 readily.

Dimethallyl ether $(CH_2=CHCH_2)_2O$ cannot be prepared from
                                                  $\overset{|}{C}H_3$
methallyl alcohol in strong acidic medium because of the rapid rearrangement of the alcohol to isobutyraldehyde. Dimethallyl ether may be obtained from neutral hydrolysis of methallyl chloride and, in higher yields, by use of aqueous alkali. By refluxing 2 moles of methallyl chloride with 3 moles of methallyl alcohol and 2.8 moles of KOH for an hour, dimethallyl ether was obtained in 91% yield (27). Even when dilute aqueous alkali is used, some dimethallyl ether is formed along with methallyl alcohol. Dimethallyl ether forms azeotropes with methallyl alcohol and with water. Preparation and properties of a number of polyfunctional allyl ethers will be found in the references of Table 14.2.

Methallyl ethers can slowly isomerize to isobutenyl ethers when heated with alkali above 10-°C (28). Dimethallyl ether does not homopolymerize readily. Copolymeri-

TABLE 14.2

Some Polyfunctional Allyl Ethers

Ether	bp, °C	n_D^{20}	References
Diallyl	94.8 50(152)	1.4165 $n_D^{25} =$ 1.4138	Williams et al, CA 15, 1657; 3801 (1941); Blitz et al, CZ I, 1666 (1928); Shell
Allyl methallyl	115	1.4236	Tamele et al, Ind. Eng. Chem. 33, 116 (1941)
Dimethallyl	134	1.4276	Tamele et al.
Allyl vinyl	150	1.4315	
Diallyl of ethylene glycol	37(1)	1.4340	Yanovsky, JACS 67, 46 (1945)
Dimethallyl of ethylene glycol	50(0.4)	1.4383	Yanovsky, JACS 68, 2020 (1946)
Diallyl of 1,3-butanediol	50(1)	1.4330	Yanovsky, JACS 67, 46 (1945)
Diallyl of propylene glycol	77(1)	1.4380	Yanovsky, JACS 67, 46 (1945)
Diallyl of poly- ethylene glycol	--	--	Alpern, ACS Div. Paint, Varnish Plastics Preprint p. 39 (spring 1947)
Dichloroallyl of 1,6-hexanediol	--	--	Pechukas, U.S. 2,684,380 (PPG)
1,4-Diallyloxy- 2-butene	79(4)	1.4565	G. F. Drebel, U.S. 2,426,863 (Monsanto)
Diallyloxytetra- hydropyran	80(3.3)	1.4575	C. W. Smith et al, JACS 74, 2018 (1952); U.S. 2,619,491 (Shell)
Diallyl ether of 1,4-cyclohexane dimethanol	95(0.8)	1.4662	R. L. McConnell, U.S. 3,236,900 (Eastman)
Polyallyloxy- octadiene, etc.	--	--	L. J. Colaianni, U.S. 2,841,620 (LaRoche)(flowery perfume)
1,1,2,2-Tetra- allyloxyethane	--	--	F. Jonas Corp. (for cross- linking); cf. A. Ribba, CA 78, 16997 (1973).
1,1,3,3-Tetra- allyloxypropane	104(0.7)	1.4530	Kay-Fries Chem.

zation with isobutene in liquid ethylene at -103°C using
AlCl$_3$ catalyst was reported (29). The product was insoluble in solvents and not reactive with sulfur.

ALLYL ETHERS OF GLYCEROL AND RELATED COMPOUNDS

Symmetrical diallyl glycerol (CH$_2$=CHCH$_2$OCH$_2$)$_2$CHOH can be
prepared from allyl alcohol and glyceryl dichlorohydrin by

heating in the presence of 50% NaOH (30). Hydroxylation
of the diallyl compound by potassium permanganate was
studied. Esters were prepared from the partial allyl
ethers of glycerol and linseed acids. The mono- and diallyl ethers of glycerol resist homopolymerization. Long
heating with high-temperature peroxides caused some increase in viscosity. The diallyl ether of glycerol did
not homopolymerize even at 200°C in the absence of air.

Following the discovery by Yanovsky in 1944 that the
polyfunctional allyl ethers of carbohydrates may undergo
"drying" or oxidative polymerization in films in contact
with air, there has been considerable interest in developing this feature in commercial coatings. Glycerol monoallyl ether has been supplied by Shell and has been
evaluated as the polyol in preparation of phthalate,
succinate, and medium oil alkyd finishes (31). Sensitivity
of the resins to water and to alkali were limitations.

It had been observed in Europe that unsaturated poly-
ether-esters based in part on tri- or tetraethylene
glycols were not inhibited by oxygen against polymerization at temperatures above 100°C. Apparently these ether
groups contributed to the removal of inhibiting oxygen by
peroxide formation. Allyl ethers of glycerol were used
in preparing unsaturated polyester-styrene solvent-free
finishes in order to overcome inhibition by air and to
accelerate "drying" (32). By adding 5 to 8% of allyl
ether, the use of paraffin wax to exclude oxygen could be
avoided and coatings on wood became tack free more quickly.
About the same time, the accelerating effect of allyl
ether groups on air-curing polyester finishes was observed
by Maker in America (33). In one example, half of the
usual propylene glycol was replaced by monoallyl glycerol
in preparing a polyester. In another process as much as
30% triallyl ether of glycerol was added to preformed
polyesters (34). Derivatives such as glycerol diallyl
ether adipate also were used as so-called skin-forming
agents to accelerate polymerization in unsaturated polyester-styrene coatings (35, 37). The latter finishes as
developed in Germany particularly for furniture were
sanded and polished to spectacular luster. The disadvantages of adding about 0.1% of a wax of limited compatibility so that its surface exudation protected the curing
film from air inhibition were discussed (36). Under most
favorable conditions these glossy polyester-styrene
finishes are resistant to cigarette burns, marring,
scratching, nail polish, and foods. Thicknesses of 0.006
to 0.01 in. were attained by a single coat (with 0.2%
cobalt naphthenate and 4% hydroperoxide added).

Jenkins in England evaluated the incorporation of allyl
groups into unsaturated polyesters (37). Polyesters from

monoallyl glycerol and from diallyl glycerol with phthalic anhydride had "air-drying" properties, and as with drying oils, the film-forming properties could be improved by heating with heavy metal soaps. The presence of the allyl ethers in the polyesterification unfortunately reduced the polymer molecular weights which could be obtained before gelation. The high monoallyl ether-polyesters showed faster "drying" times than the polyesters only terminated by diallyl ether of glycerol. In the coatings a large proportion of the styrene was lost by evaporation since styrene did not copolymerize very readily with the allyl ether groups. For mar-resistant coatings, molecular weights needed to be at least 800 before curing. The diallyl ether of glycerol was chosen for evaluation because of low odor and low volatility.

Besides built-in allyl ether groups, Jenkins added to conventional unsaturated polyesters up to 40% of glycerol diallyl adipate to accelerate cure along with cobalt naphthenate and hydroperoxide. An oxime such as acetaldoxime was added also as a temporary volatile inhibitor lost in spraying. Accelerated weathering tests showed the allyl ether-polyesters to be as good as conventional styrene-polyesters. A polyester was prepared from a hexallyl ether of tetraglycerol, propylene glycol, maleic anhydride, and phthalic anhydride (38). It was polymerized in films along with 55% styrene and cobalt drier.

Free radical copolymerizations of allyl ethers of glycerol occur reluctantly but have been suggested in minor proportions with acrylonitrile (39), with N-vinyl monomers (40), and with vinyl chloride (41). Polyallyl ethers of polyols were copolymerized with acrylamide and its derivatives (42).

Bis(glycerin monoallyl ether) fumarate $AOCH_2CHOHCH_2OOCCH=CHCOOCH_2CHOHCH_2OA$, prepared by reaction of fumaric acid with excess allyl glycidyl ether, was evaluated in blends with unsaturated polyesters and styrene for "air-drying" films (43). With hydroperoxide initiation the allyl and fumarate double bonds were found to polymerize with each other. Figure 14.1 shows the much faster polymerization of fumarate double bonds (absorption at 773 cm^{-1}) than of allyl double bonds (930 cm^{-1}) in the absence of air. In the presence of air, however, the fumarate double bonds were used up by polymerization more slowly than the allyl double bonds.

ALLYL ETHERS OF TRIMETHYLOLPROPANE

Trimethylolpropane is a trifunctional primary polyol derived from formaldehyde and n-butyraldehyde. Both the monoallyl and diallyl ethers were made available by

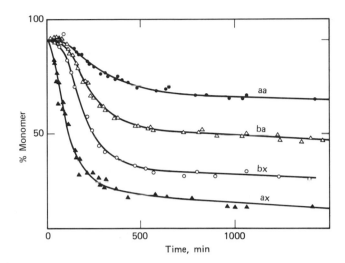

Fig. 14.1. Kinetics of polymerization of bis(glycerol monoallylether) fumarate followed by reduction of double bond concentrations by IR absorptions. Without air, aa = allyl and ax = fumarate; with air, ba = allyl, bx = fumarate. J. K. J. Vlcek and J. Mleziva, J. Polym. Sci. C No. 16, 252 (1967).

Celanese and later by Proctor Chemical Company. The formula of DATMP is

$$CH_2=CHCH_2OCH_2-\overset{\displaystyle CH_2CH_3}{\underset{\displaystyle CH_2OH}{C}}-CH_2OCH_2CH=CH_2$$

The two monomers can be distilled at atmospheric pressure at about 250 and 258°C. The triallyl ether is not well known. DATMP is a liquid of mild odor, soluble in acetone, methanol, n-butanol, and benzene. It is substantially insoluble in water and only moderately toxic. DATMP readily adds $2Br_2$ in CCl_4 at 10°C.

Diallyl trimethylolpropane alone does not air "dry" in films and it homopolymerizes reluctantly. Heating for two days at 100°C with 5% t-butyl perbenzoate gave a viscous syrup, and after a week there was formed a nearly color-less, nontacky, crosslinked, clear solid that crumbled easily to small fragments (44). With 11% t-butyl perben-zoate the result was similar, except the polymer casting was yellow and it cracked spontaneously on cooling. Benzoyl peroxide under similar conditions gave only a viscous prepolymer solution. The diallyl monomer did not polymerize on addition of a variety of Lewis acid and basic catalysts. On adding sulfuric acid, a viscous brown solution was formed. Waters and co-workers reported that the monoallyl ether heated with peroxide at 150°C gave a yellow viscous polymer solution at 74% conversion. The polymer had a cryoscopic molecular weight of about 1500.

Bayer laboratories (45) gave an example of preparing a polyester from 870 parts DATMP, 392 parts maleic anhydride, 888 parts phthalic anhydride, and 410 parts ethylene glycol under nitrogen, with temperature rising to 190°C. When the acid number was 25 and the viscous solution had been cooled to 130°C, there was stirred in 0.25% hydroquinone. Styrene was added to give 30% of the total, as well as about 0.1% cobalt naphthenate and 2% dicyclohexyl hydroperoxide to formulate an "air-drying" lacquer. In another patent example, 45% vinyl toluene was used and coatings sprayed on wood were said to be polishable after 7 hr at 23°C (46).

A recipe for a DATMP-polyester was suggested from Celanese work (47):

DATMP	18.1%
1,3-Butylene glycol	36.2
Phthalic anhydride	12.5
Maleic anhydride	33.2

Under nitrogen at reflux, the acid number was brought to 90 before the diallyl ether monomer was added. Then heating was continued to an acid number of 40, followed by cooling to 180°C and addition of 0.3% hydroquinone. Finally the viscous solution at 100°C was diluted with styrene. Air "dried" films had resistance to acids but were only fair to ketones and alcohols in prolonged exposures. Polyester paper coatings made with DATMP were said to have gloss and chemical resistance. Air-drying alkyds based on DATMP had outstanding color retention after light exposure but were inferior to drying oil alkyds in flexibility after heat aging (48).

Celanese studied vinyl-type copolymerization of allyl ethers of trimethylolpropane (TMP) in solution and in emulsion systems. Rates and molecular weights of products generally were low. Waters and co-workers suggested preparation of terpolymers from equimolar proportions of monoallyl ether of TMP, maleic anhydride, and 2-ethylhexyl acrylate by heating in dioxane with 1% BP. The catalyzed monomer mixture was added over an hour to the refluxing solvent followed by heating for 2 hr at reflux. Ethyl acetate or benzene also could be used as solvents. Films were tested. Copolymerization of DATMP with vinyl and

acrylic monomers has been suggested as a way to introduce
primary OH groups for use in curing films.

The following composition was suggested for preparing a
vehicle for industrial metal finishes (49):

Monoallyl ether TMP	4.9%	Di-t-butyl peroxide	1.0%
Methyl methacrylate	21.0	Cumene hydroperoxide	0.5
2-Ethylhexyl acrylate	20.8	Xylene	37.3
Acrylic acid	2.1	Cellosolve acetate	12.4

The xylene and allyl ether were refluxed and then the pre-
mixed other monomers containing di-t-butyl peroxide were
added dropwise over a 2-hr period. Then refluxing was
continued for 3 hr. At this point the cellosolve acetate-
cumene hydroperoxide mixture was added during 15 min.
Refluxing was continued for 11 hr to desired conversion.
The solution had a viscosity of 230 cp at 25°C and was
clear and colorless. Other copolymerizations in solution
were suggested using 10 to 13% allyl ether of TMP with
acrylate, methacrylate, and styrene monomers. The copolymer solutions were blended with formaldehyde resins such
as Cymel 300 (hexamethoxymethylol melamine) for use as
thermosetting baking enamels for metals (49). DATMP and
other polyfunctional allyl ethers have been used in
England with long oil alkyds to give promising "drying"
properties in films (50). However, faster drying and
better film properties are needed.

An unsaturated polyester was prepared by heating a mixture of trimellitic anhydride, trimethylolpropane, and
DATMP under nitrogen for 3 hr at 165°C, then for 2 hr at
185°C (51). Water was removed azeotropically by distilling off with xylene-methyl isobutyl ketone solvent
mixture. The ammonium salt of the allylic polyester at
45°C in t-butanol-isopropanol had a viscosity of 70 poises.
Nonyellowing films were obtained after curing by baking.

Allyl ethers of trimethylolpropane have been suggested
for other copolymerizations; for example, with itaconic
anhydride (52) and methacrylate esters (53). Allyl ethers
of TMP have been proposed to improve dyeability of polypropylene fibers (54). DATMP copolymerized with SO_2 to
give an insoluble product which decomposed without melting
about 200°C (55). Copolymerization of the monoallyl ether
with vinyl acetate and with MMA was studied by electron-
spin resonance (56). Some polyfunctional allyl ethers
from other polyols are listed in Table 14.3.

ALLYL ETHERS OF PENTAERYTHRITOL (PE)

Allylated compounds from polyol of formula $C(CH_2OH)_4$
became available commercially in the 1960s and have been

TABLE 14.3

Some Polyallyl Ethers of Polyols

Polyol	Approx. no. of allyl groups	bp, °C at mm	n_D^{20}	References
Glycerol	1	124(12)	1.4605	Fairbourne, Chem. & Ind. 49, 1021 (1930)
Glycerol	2	106(7)	1.4580	Alpern, ACS Paint, Varnish, Plastics Preprint, p. 44 (April 1947)
Glycerol	3	100(5)	1.4501	Alpern; Yanovsky, JACS 67, 46 (1945)
Diglycerol	4	--	--	Wittcoff, JACS 69, 2655 (1947)
Trimethylol propane	2	258	1.4560	Celanese data; (Sp. gr. = 0.95)
Erythritol	4	104(0.01)	1.4590	Yanovsky, JACS 70, 2194 (1948)
Pentaerythritol	2	174(10)	1.4695	Celanese data
Pentaerythritol	3	161(10)	1.4627	--
Pentaerythritol	4	125(1)	1.4595	Yanovsky, JACS 67, 46 (1945)
D-Mannitol	4.6	160(0.5)	--	Yanovsky
	6	172(1)	1.4710	(14.5 cps viscosity)
Inositol	6	170(1)	1.4788	Yanovsky
D-Arabitol	5	131(0.01)	1.4662	Yanovsky, JACS 70, 2194 (1948)
D-Galactose	4.3	128(0.1)	1.4760	Talley, JACS 67, 2037 (1945)
D-Glucose	4.9	159(0.7)	--	Talley; (yellow liquid of 19 cps at 25°)
Trimethallyl glycerol	--	167(0.3)	1.4532	Yanovsky, JACS 68, 2020 (1946)

used for unsaturated polyester-styrene coatings of high gloss in Europe and America. Yanovsky and co-workers had described tri- and tetraallyl ethers of PE boiling at 129°C at 3 mm and 140°C at 0.8 mm, respectively (57). Copolymerizations with acrylic and vinyl monomers have been described (58). Reaction products of PE and glycerol dichlorohydrin at 90°C in the presence of 50% aqueous sodium hydroxide have been allylated with allyl bromide during 6 hr at 70°C (59).

Diallyl ethers of PE were used in the preparation of water-soluble polyesters evaluated as curable coatings (60). In one patent example, 4.4 parts ethylene glycol, 21.5 parts diallyl ether of pentaerythritol, 44.7 parts pentaerythritol, 22.3 parts maleic anhydride, 23 parts boric acid, and 18.6 parts water were heated under CO_2 in a pressure vessel at 190°C for 3 hr. Films deposited from aqueous solutions containing potassium persulfate could be

crosslinked on heating. Allyl pentaerythritols were
added in polyesterifications with maleic anhydride and
glycol to give unsaturated polyesters for copolymerization
with styrene or vinyl toluene (61). These allyl ether-
containing polyesters gave fair curing rates in coatings
in the presence of air.

Hercules developed the following procedure for preparation of "air-drying" polyesters containing units from
diallyl ether of pentaerythritol:

        Isophthalic acid (6.0 moles)   996 g
        Maleic anhydride (6.0 moles)   588
        Xylene   90
        Propylene glycol (10.1 moles)   766
        DAPE (2.52 moles)   625
        t-Butylcatechol   0.28

The isophthalic acid, propylene glycol, and xylene were
heated at reflux, and water from esterification was removed by azeotropic distillation. Any glycol distilled
was returned to the reaction vessel. When no more azeotrope was coming over, the temperature was lowered to
350°F, the mixture was blanketed with nitrogen, and the
maleic anhydride, t-butyl catechol and diallyl ether were
added in that order. While the nitrogen blanket was
maintained, the temperature was raised to 320°F in 1 hr,
to 392°F in the next 3-hr period, and finally to 410°F in
a 30-min interval. After holding at 410°F for 1 hr, 95%
of the water of esterification had been removed by azeotropic distillation and the product was allowed to cool
under nitrogen. The solid resin of acid number 30 was
dissolved in styrene to give a 50% solution for spray
coating. Rates of oxidation polymerization of DAPE-
modified polyesters to form hard nontacky films were
slower than desired. Other methods of curing, e.g., by
acid catalysts, were explored (62), but products of importance have not developed.

Two viscous liquid products have been supplied by
Hercules (63):

Property	G39 (DAPE)	G40 (TAPE)
Number of allyl groups	2.0	2.6
Color of liquid	pale yellow	pale yellow
Boiling points at 1 mm, °C	120	128
Hydroxyl, %	12.5 - 14.5	6.5 - 7.5
Bromine number	145 - 155	165 - 175

The diallyl and triallyl ethers are soluble at 25°C in
benzene, carbon tetrachloride, common alcohols, ethers,
and ketones. They are only slightly soluble in water.

Like many other allyl compounds, when heated in the presence of oxygen or acids, allylated pentaerythritol can
evolve formaldehyde and acrolein. Besides their air-drying
function, such monomers are of interest as minor comonomers
to produce branching in polymers.

The author heated the Hercules liquid products for a week
at 100°C in half-inch depth in vials open to air. DAPE
became a viscous syrup containing some gel particles; TAPE
was transformed to a nontacky, rubbery, but brittle, clear
solid. Irradiation by UV light of TAPE containing 5% BP
gave increase in viscosity during 50 hr without discoloration.

Acrylonitrile has been copolymerized in aqueous persulfate with small proportions of allyl derivatives of PE to
give thickening agents for nonaqueous solvents (paint
rqmovers, etc.)(64). Vinyl pyrrolidone was copolymerized
with diallyl ethers of PE and other polyols to give agents
for improving dye receptivity of acrylonitrile polymer
fibers (65). After epoxidation of tetraallyl PE in chloroform, peracetic acid, and acetic acid, polymerizations may
be carried out through the ether groups (66). Copolymers
of TAPE with 2-hydroxyethyl methacrylate were crosslinked
by baking in the presence of an azidoformate (67). Aromatic polyazides also were used to crosslink these polymers
(68).

Prosser made interesting studies of the rearrangement of
polyfunctional allyl ethers to cis-propenyl ethers (69).
This reaction with strong alkali catalysts is only appreciable above 125°C. When tetraallyl pentaerythritol was
heated under nitrogen with 5% sodium methacrylate at 175°C,
conversion was 10.7% after 1 hr, and only 50% after 50 hr.
TAPE was 34.6% converted to cis-propenyl structure in 1 hr
and 57% after 52 hr. Other allyl ethers were partially
rearranged. The reduction in allyl groups was followed by
weakening of the IR band at 6.07 microns and at 10.8
microns, along with the increase in 5.98 indicative of
propenyl C=C stretching, and a broad band about 13.8
attributed to the C-H out-of-plane bending of the cis-
propenyl ether. The propenyl group could be estimated by
acid hydrolysis giving propionaldehyde. Methallyl ethers
were similarly rearranged and could be hydrolyzed to form
isobutyraldehyde. The partial rearrangement reactions of
a number of polyfunctional allyl ethers exhibited a remarkable color cycle. Yellow colors were produced on
adding alkali. During the initial heating this deepened
to orange and orange-red followed slowly by return to a
straw yellow color. The cis-propenyl ethers of pentaerythritol could be polymerized by aluminum alkoxide-sulfuric
acid complex catalysts.

Polyesters from pentaerythritol and phthalic anhydride
were terminated by TAPE (70). These with styrene and
cobalt naphthenate gave tack-free films after 4 hr and
fully hardened films after 20 hr at room temperature.

The paint research laboratories at Teddington, England,
evaluated tetraallyl ethers of sorbitol and of dipentaerythritol, as well as tetramethallyl dipentaerythritol
in so-called air-drying coatings (71). Their behavior was
similar to that of nonconjugated natural drying oils except there was less tendency to yellow, and the films
became more brittle on aging. These synthetics absorb
less oxygen and form a smaller amount of volatile scission
products during curing than do drying oils. The maximum
absorption of oxygen was 12% for the tetraallyl ether of
sorbitol (before becoming dry to touch). With linseed
oil, the oxygen content was still rising after 30% absorption. The methallyl ether films were softer and were
resistant to yellowing (even in the presence of ammonia).
Crosslinking was less complete in the synthetic films than
in drying oil films, judging from the fractions extractable by solvents. Soluble cobalt compounds catalyzed
polymerization, but lead and manganese compounds retarded
cure. Acrolein and some acrolein dimer were scission
products of tetraallyl sorbitol when heated in air.

Tetraallyl ethers of polyols in varnish films produced
relatively poor pigment dispersion; poor "wetting" of substrates, brittleness, and detachment on weathering were
observed. Combinations with oil-modified alkyds seemed
most practical. Rates of absorption of oxygen by allyl
and methallyl ethers of four different polyols were
studied by Nichols and co-workers. Cobalt naphthenate
accelerated oxygen absorption and addition of NaOH retarded it as shown in Fig. 14.2.

PARTIAL ALLYL ETHERS OF STARCHES

Partial allyl ethers of starches were supplied commercially in the 1950s. They were used as air curing and
baked coatings, as textile finishes, in printing inks, and
in other applications. As manufactured at that time, they
gave water-sensitive films that discolored in light, and
the products were not very uniform from lot to lot. Allyl
starches were available as solids at about $0.50/lb and as
concentrated solutions in toluene-butanol at $0.25/lb.
Allyl starch preparation and properties were studied by
Yanovsky and co-workers of the U.S. Department of Agriculture, investigators at General Mills (72), and others.
This work revealed polyfunctional allyl ethers as a new
class of "air-drying" compounds. The best films were
glossy, fairly hard, and resistant to many organic sol-

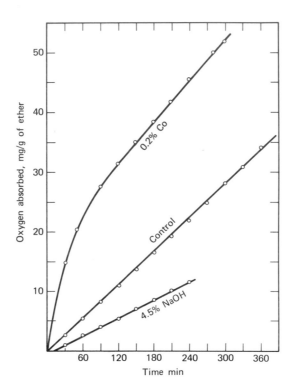

Fig. 14.2.   Influence of cobalt naphthenate and sodium hydroxide on oxygen absorption of tetramethallyl pentaerythritol at 80°C.   Tetra-allyl pentaerythritol absorbed oxygen less rapidly.   P. L. Nichols, A. N. Wrigley, and E. Yanovsky, J. Am. Chem. Soc. **68**, 2020 (1946).

vents, but softened by water, aqueous cleaning agents and dilute alkali.

Allyl starches are made by Williamson-type reaction by heating allyl chloride with starch in the presence of concentrated aqueous alkali. Reaction conditions resembled those developed for the manufacture of ethyl cellulose (73) except that allyl halides are faster reacting than ethyl halides. The preferred range was 1.5 to 1.7 allyl groups per glucose unit (74). Early patent references on etherification of carbohydrates were given by Tomecko and Adams, who used allyl bromide to prepare some partial allyl ethers of alpha-methyl glucoside, sucrose, and starches (75).

Yanovsky and co-workers gave European references to allylation of polyols and made significant studies of preparation and polymerization of numerous polyallyl ethers of carbohydrates (76). High concentrations of NaOH, e.g., 50% aqueous, were necessary for allylation; otherwise, losses of allyl bromide were excessive. Allyl bromide with starch and alkali at 80°C tended to give insoluble polymers. Allylation then occurred rapidly before the degradation of molecular weight required for solubility. The following method for preparing soluble allyl starch illustrates the high consumption of alkali.

The following mixture was heated in a stirred autoclave at 86°C under about 2 atm pressure:

| Dry starch | 500 g | Acetone | 2500 cc |
| 50% NaOH | 2000 g | Allyl chloride | 3000 cc |

Ordinary starch required 10 hr reaction, amylopectin about 20 hr; but dextrin (degraded starch) needed only 4 hr to give degrees of substitution or DS of 2.0 to 2.4 allyl groups per glucose unit. After the desired polymer DS and satisfactory solubility properties had been reached, the mixture was steamed to remove acetone, allyl chloride, and byproduct diallyl ether. The gummy allylated polymer was washed with water to remove alkali and dried. The faster reaction of dextrin to give "allyl starch" soluble in organic solvents led to its use in the General Mills process.

Allylation of starches using allyl chloride was discussed in a review by Yanovsky (77). The allyl starch formed in earlier stages of allylation had very high molecular weight which contributed to insolubility in organic solvents. Further reaction more completely etherified the polymer but also partially degrades it to the molecular weight range of 15,000-20,000 required for good solubility. The favored process produced polymer containing 1.5 to 1.8 allyl groups per glucose unit. Theories for the mechanism of polymerizations of allyl starch also were discussed.

Yanovsky found that the allylated polyols, including diallyl ethers of glycols, containing at least 2 allyl ether groups per molecule, increased in viscosity and hardness on standing in air and eventually became brittle, crosslinked polymers. For example, a solution of allyl methyl glucoside of DS about 3 changed from viscosity of 2.7 to 200 centistokes on heating for 200 min at 97°C while air was bubbled through it. Benzoyl peroxide was not effective in promoting the curing polymerizations. However, the changes in air to form a gel polymer were accompanied by peroxide formation. It was necessary to supply oxygen throughout the curing process. Even heating

an allylated pentaerythritol at 120°C for 24 hr in the
absence of air would not convert the soft gel state into
hard polymer. An odor of acrolein, and enough acrolein
for identification by 2,4-dinitrophenyl hydrazone, accompanied the polymerization of the polyallyl polyols in air.
It was suggested that allylic hydroperoxides might epoxidize neighboring allyl groups leaving hemiacetal groups
which might liberate acrolein

$$\underset{\text{ROCHCH=CH}_2}{\overset{\text{OH}}{|}} \longrightarrow \text{ROH} + \underset{\text{O=CCH=CH}_2}{\overset{\text{H}}{|}}$$

Difficulties were encountered with technical allylated
carbohydrates. Traces of mineral acid impurities caused
discoloration on heating; traces of alkali retarded the
polymerization--perhaps by decomposing hydroperoxides
prematurely.

Wrigley and Yanovsky suggested an "acrylate theory" of
autoxidative polymerization of allyl ethers (78). There
was evidence that hydroperoxides of allyl ethers may
decompose to water and the corresponding acrylate ester.
With initiation by free radicals, the acrylate groups were
believed to participate in polymerization of the allyl
ethers, e.g., in films hardening under air. Later alpha-
butoxyallyl hydroperoxide was reacted with ferrous sulfate
to form a 27% yield of butyl acrylate monomer (79). Evidence for ester formation in allyl starch was observed by
infrared (80).

PROPERTIES OF TECHNICAL ALLYL STARCH

The commercial allyl starch supplied by General Mills
from allylation of dextrin was a yellowish solid that
polymerized slowly in contact with air and therefore
generally was transported and stored in solution or under
water in tight containers. Solutions in toluene-isobuta-
nol (95-5) of concentrations 40 to 50% were stable against
polymerization. They were light amber colored (Gardner 10
to 11). The allyl starch was soluble in alcohols, ketones,
esters, halogenated hydrocarbons, nitroparaffins, ethers,
and in aromatic hydrocarbons when these contained some
hydrogen-bonding solvent such as isobutanol. The product
was insoluble in aliphatic hydrocarbons and substantially
insoluble, but swollen to a soft gummy state by water. A
commercial solution at 40% concentration in xylene-iso-
butanol (95-5) had 360 centipoise viscosity at 25°C, but
in a lacquer solvent the viscosity was only 83 centipoise.
Solutions supplied by General Mills were substantially
neutral and free of driers. They could be stored with
little or no gelation. Technical allyl starch was compatible with numerous liquid plasticizers such as phthalate

sebacate and phosphate esters, as well as with some of
the Paraplex and alkyd ester resins. In order to avoid
excessive brittleness in cured films, 20% or more of
plasticizer was found necessary. The product was incompatible with drying oils and with most cellulose derivatives.

The curing of allyl starch films in air was a slow
process; it took place over several weeks and, unfortunately, increasing brittleness continued thereafter.
After partial evaporation of solvent, heating for 2 hr at
95°C or 1 hr at 130 to 150°C was recommended for curing
films (preferably without added catalyst). Such films
were resistant to many organic solvents. Films that had
been baked for an hour or longer at 150°C in the absence
of added accelerators had somewhat improved water resistance. Discoloration and brittleness of films on aging
were not overcome.

The principal use of allyl starch has been in printing
inks. It was well received as a printing ink vehicle,
especially for use on cellophane and glassine paper, and
also for overprint varnishes. Recently allyl starches
have been manufactured in Europe.

MODIFICATION OF ALLYL STARCHES

Allyl potato starches with a degree of substitution of
2 gave films that became hard and brittle with 0.2% cobalt
naphthenate (81). The alkaline allylation reaction was
facilitated by adding a little alkali iodide as a catalyst
(82). Better polymer solubility in alcohol-aromatic hydrocarbon solvents could be obtained if an acidic after-
treatment was applied following allylation. For example,
the batch was brought to pH 3 by addition of a large
amount of hydrochloric acid and heated at 90°C for 45 min.
From 1.6 to 2.0 allyl groups was the range of substitution
obtained. Curing by sulfur, accelerators, and zinc oxide
was proposed (83). Aqueous polymer emulsions were prepared (84). Some efforts to improve on partial allyl
starches and dextrins by graft copolymerization with
acrylamide or acrylonitrile were reported (85).

As the limitations of allyl starches became evident,
research was directed toward other related products. Allyl
dextrins of 1.6 to 1.85 allyl groups per glucose unit were
prepared by reacting alkali and allyl chloride for 3 to
5 hr at 85 to 100°C (86). Mixed allyl-higher alkyl ethers
of starch were made in efforts to improve resistance to
aqueous agents (87). Higher alkyl groups retarded polymerization. Allyl starch propionates and other allyl ether
esters of starch were prepared (88). Films of styrenated
allyl starch cured at 150°C had increased resistance to

alcohol and soapy water (89). Glycidyl allyl ether was
used to partially etherify starch in the presence of a
quaternary ammonium hydroxide (90). Allyl alcohol and
dialdehyde starch were reacted in dioxane containing HCl
to give allyl acetal groups (91). Films cured at 150°C
were said to resist boiling water.

PARTIAL ALLYL ETHERS OF SUGARS

Publications of Yanovsky already cited included references and observations on allylated sugars. There have
been continuing efforts to develop uses for sugars other
than food. Preparation of allylated sucrose on a small
pilot-plant scale using allyl chloride and alkali was
described (92). Controlling the degree of substitution
and the state of cure proved difficult. Film properties
on wood and rates of polymerization were not promising
(93). An octaallyloxycarbonyl sucrose was described as a
viscous syrup of $n_D^{25} = 1.4778$ (94). With 1% benzoyl
peroxide at 100°C, the gel time was as long as 18 hr. Copolymerizations were studied.

Polyallyl ethers of sucrose and of sugar alcohols and
their solutions increase in viscosity by polymerization
when heated in the presence of oxygen. The "bodied" prepolymer solutions free of gel when coated upon test panels
can be cured either slowly at room temperature (cobalt
drier) or faster by baking. Polymerization rates are
about the same as those observed with allyl starches (95).
Figure 14.3 compares rates of polymerization of a number
of allylated sugar derivatives and other polyfunctional
allyl ethers.

Various modifications of allyl sucroses have been suggested for applications, but many of these were not
realistic. Copolymerizations of allyl sugar derivatives
with drying oils (96), acrylonitrile (97), and ether
monomers (98) have been reported. Partial allyl ethers
of sucrose and of mannitol were esterified by methacrylic
acid in order to give cured films of better water resistance (99). Polymerized allyl sucrose was reacted with
ethylene oxide (100). Allyl sucroses were incompatible
with most drying oils at room temperature but could be
blown with air at 100°C to improve compatibility (101).
Styrenation of allyl sucrose was described (102).

Of practical importance has been an invention of Brown
of Goodrich: the use of partial allyl ethers of sucrose
as branching and crosslinking comonomers with acrylic
acid to give commercially successful synthetic thickening
agents and gums. Since many natural gums are branched
macromolecules, it is not so surprising that a comonomer
providing many points of branching, e.g., five or more

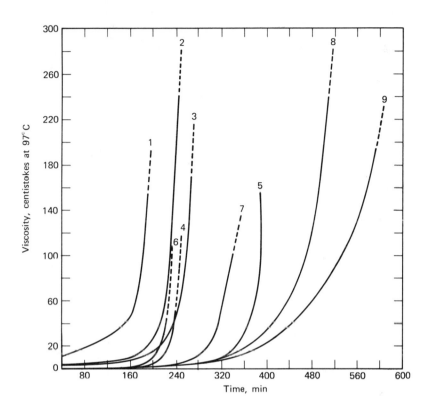

Fig. 14.3.   Polymerization of polyallyl ethers under oxygen at 97°C:

| | | Gelation |
Compound	Curve	time, min
Triallyl glycerol	1	207
Hexaallyl mannitol	2	220
Hexaallyl sorbitol	3	240
Heptaallyl sucrose	4	207
Hexaallyl inositol	5	350
Tetraallyl pentaerythritol	6	188
Diallyl ethylene glycol	7	315
Diallyl 1,3-butylene glycol	8	550
Diallyl dipropylene glycol	9	765

P. L. Nichols and E. Yanovsky, J. Am. Chem. Soc., <u>67</u>, 46 (1945).

allyl groups, would be effective. In one patent example,
15 parts of allyl sucrose and 87 parts of acrylic acid
were copolymerized in 1000 parts of heptane with 0.5 part
of caproyl peroxide catalyst at 75°C (103). The copolymer, which formed as a separate solid phase, could be
neutralized by NaOH by shaking for 4 hr. The hydrogen or
free acid form of the copolymer could be regenerated by
steeping in aqueous HCl or acetic acid. These hydrophilic
polymers of very high molecular weight and high swelling
capacity are useful at low concentrations in aqueous solutions or "mucilaginous aqueous dispersions" for denti-
frices, surgical jellies, creams, ointments, bulk laxatives, ion-exchange resins, thickeners for printing pastes,
and many applications in which natural gums have been
employed. Terpolymers such as obtained from acrylic acid,
a vinyl ether, and allyl sucrose (more than three allyl
groups) have been studied as synthetic gums (104). When
acrylic anhydride is copolymerized in such systems, the
initial product may contain some cyclic glutaric anhydride
units (105). An acrylic acid copolymer suggested along
with polyvinyl alcohol in hair wave-set solutions was said
to contain about 1% of polyallyl sucrose units (106). A
"Carbopol" acrylic acid copolymer containing 1 to 2% of
octaallyl sucrose units was particularly effective as a
thickening agent in formulation of pigment printing pastes
for textiles (107).
    Salts of copolymers of acrylic acid with small amounts
of diallyl sucrose have found applications as aqueous dispersing and thickening agents in shaving creams, clay
dispersions, and cosmetics (108). From 0.1 to 1.0% of
these branched mucilaginous copolymers in water have been
used to deposit fluorescent pigments within lighting
tubes (109). Acrylic acid copolymers partially crosslinked by polyallyl sucrose have been used in coatings
for medicinal pills and tablets (110). Copolymers of
acrylic acid with 0.75 to 1.5% polyallyl sucrose were
used to gel hydrogen peroxide solutions (111).
    There has been some work on partial allyl ethers of
monosaccharides and their derivatives. Reaction of alkali
diacetoneglucose with allyl chloride gave 3-allylglucose,
a colorless crystalline solid melting at 132°C (112).
Methyl mono- and diallyl glucosides were prepared and
copolymerized (113). Diallyl and dimethallyl derivatives
of sorbitol and mannitol were characterized by boiling
points at 10 to 20 mm and by refractive index (114). They
polymerized reluctantly. A tetraallyl ether of mannitol
was copolymerized with methacrylate esters (115). Polyfunctional partial allyl ethers of erythritol, arabitol,
and related sugar erivatives were prepared and found to
polymerize reluctantly (116). Recently 3,5,6-triallyl-D-

glucose and related derivatives were prepared (117).
Methallyl ethers of mannitol, sorbitol, glycerol, ethylene
glycol, pentaerythritol, and sucrose were reported (118).

PARTIAL ALLYL ETHERS OF CELLULOSE

In contrast to the commercially important water-soluble
cellulose derivatives such as carboxymethyl cellulose and
hydroxyethyl cellulose made by Williamson-type etherification, the allyl cellulose ethers have not achieved commerical importance. Although soluble in organic solvents,
they are only swollen by water and have some of the same
limitations in coatings and plastics as allyl starches.
Very strong alkali at temperatures of 110 to 140°C
generally has been needed for the allylation reaction,
causing variable amounts of degradation of the cellulose
macromolecules.

Research emphasized water-dispersible mixed type polymers which might be cured in films through allyl groups.
Water-soluble allyl ethyl cellulose (119), allyl cellulose
of only 0.4 degree of substitution (120), allyl carboxymethyl cellulose (121), and allyl hydroxyethyl cellulose
(122) have been patented. Graft copolymerization of
vinyl-type monomers on allyl cellulose has been disclosed
(123). Cellulose fibers were allylated superficially and
styrene was polymerized thereon (124). Reactivity of
vinyl-type monomers in forming crosslinked copolymers with
allyl celluloses increases with polarity (e values) of the
monomers (125).

METHALLYL ETHERS OF CARBOHYDRATES

Yanovsky and co-workers compared the oxidation and
polymerization of the partial methallyl ethers with those
of corresponding partial allyl ethers (126). Methallyl
starch, for example, was prepared by stirring for 10 hr
at 85°C a mixture of 34 g of dried potato starch, 3 times
the stoichiometric equivalent of methallyl bromide, and
50% sodium hydroxide dispersed in 150 ml of methyl ethyl
ketone. The reaction mixture was then steam distilled
and the gummy derivative was kneaded with water until it
was free of alkali. The methallyl starch was crumbly
when dry. The methallyl ethers of carbohydrates absorbed
oxygen, as did the allyl ethers. The oxidation was
accelerated by cobalt naphthenate and retarded by alkali.
Oxidative attack occurs at least in part at the alpha-
methylenic carbon atom of the methallyl group, since
methacrolein was identified. Peroxide content as well as
epoxide content were estimated after passing oxygen for
7 hr at 80°C through methallyl mannitol and allyl pentaerythritol. Boiling points and refractive indices of some
methallyl ethers of polyols are given in Table 14.4.

TABLE 14.4

Polyfunctional Methallyl Ethers of Polyols [a]

Ether	bp, °C	n_D^{20}
Dimethallyl ethylene glycol	50(0.4)	1.4383
Trimethallyl glycerol	96(0.5)	1.4532
Hexamethallyl-D-mannitol	174(0.8)	1.4713
Hexamethallyl-D-sorbitol	167(0.3)	1.4703
Hexamethallyl sucrose	--	1.4835
Tetramethallyl pentaerythritol	129(0.3)	1.4621
Hexamethallyl dipentaerythritol	185(0.3)	1.4682
Dimethallyl starch (2.25 methallyl)	--	--

[a] P. L. Nichols et al, J. Am. Chem. Soc. 68, 2020 (1946)

ALLYL ETHERS OF FORMALDEHYDE CONDENSATE RESINS

The free methylol groups in resins from condensation of formaldehyde with melamine, urea derivatives, or phenol have been etherified with allyl alcohol with the hope of promoting oxidative polymerization or "drying" in coatings.

$$-CH_2OH + HOCH_2CH=CH_2 \longrightarrow -CH_2OCH_2CH=CH_2 + H_2O$$

Such polyallyl ether products have not been very successful because further polymerization occurred only slowly, even with catalysts.

Attempts were made to prepare useful resins for coatings directly from allyl alcohol, urea or melamine, and formalin (127). Dimethylolurea was allylated in the presence of acid catalyst (128). Allylation of water-soluble methylolmelamine (2.5 methylol groups) is difficult without immediate crosslinking. A tetramethylolmelamine was allylated by reacting allyl alcohol with phosphoric acid catalyst at 44°C for 15 min (129). Side reactions occurred and the products were not homogeneous. The polyallyl ether of tetramethylolmelamine could be precipitated by dilution with water. Films cast from solutions of the dried polymer in chloroform with added cobalt naphthenate only became 70% insoluble in ethanol after 6 weeks at room temperature in diffused light. Rates of oxidative polymerization were not improved by introduction of more than four methylol groups per melamine molecule. Reaction products of urea-formaldehyde prepolymers with allyl alcohol were described as hornlike masses (130).

Allylated melamine-formaldehyde resins such as pentaallyl ethers were studied by Ciba (131), but rates of oxidative polymerization were slower than those of drying oils. Allylated methylol-melamines were copolymerized with maleate polyesters before oxidative polymerization in films (132). An allyl-melamine-ester resin known as Ciba-dur A was available for finishing furniture and for water-proofing (133). A fully allylated hexamethylolmelamine has been added to unsaturated polyesters to overcome inhibition by molecular oxygen. Allylated melamine-formaldehyde resin has been reacted with soya bean unsaturated acids for application as pigment binding polymers (134).

Badische investigated allylated resins from glyoxal diureine and formaldehyde (135). The following patent example was said to form a tetraallyl ether of tetrakis-(hydroxymethyl) glyoxal diureine. Acetylene diurea was reacted with formalin at 80°C at pH 6.5 to 7.5 (trace of sodium carbonate). The diureine obtained was reacted with allyl alcohol acidified with aqueous HCl. After distilling off most of the water, reaction was continued at 140°C for 2 hr. A mixture of 30.5 parts of the product with 37 parts of unsaturated polyester and 33 parts of styrene, catalyzed by cyclohexanone peroxide and cobalt naphthenate, was heated in air to produce hard films. Allylated condensation products of formaldehyde with imidazolidinone were evaluated as crease-resistant finishes for cotton (136).

Allylated phenol-formaldehyde resins have not been commercially successful; the allyl ether groups have not enhanced curing or other properties. In a patent example, equal parts of allyl ether modified-phenolic resin and conventional phenol-formaldehyde in a glass laminate were molded at 165°C and postcured for 24 hr at 130°C (137). A polyallyl ether of a cresol-formaldehyde condensate was reported to have "air-drying" properties (138) but the products have not been manufactured. Diallyl ethers of bisphenol A-formaldehyde condensates have been evaluated (139). Formaldehyde reacts with certain aromatic hydrocarbons and halogenated aromatic hydrocarbons to give resin intermediates that can be allylated. Thus bis-(allyloxymethyl) tetrachlorobenzenes have been studied in coatings (140). Bis(allyloxymethyl)xylene has been epoxidized (141). Naphthalene was reacted with formalin and HCl forming about equal amounts of 1,4 and 1,5-bis-(chloromethyl)naphthalenes (142). This mixture was reacted with NaOH and allyl alcohol, giving allyloxy-methylnaphthalenes boiling at 170 to 177°C at 6 mm. The latter and styrene at -50°C were treated with BF_3 etherate as copolymerization catalyst yielding a solvent-resistant polymer.

References

1. Allyl Alcohol, publication of Shell Chemical Co. 1946.
2. R. J. Stephenson, U.S. 3,250,813-4 (Monsanto).
3. K. Nozaki, U.S. 3,577,466 (Shell); cf. J. Milgrom and W. H. Urry, U.S. 3,173,958 (Std. Oil Ind.).
4. P. W. Williams and T. W. Evans, U.S. 2,464,758 (Shell).
5. W. H. Watanabe et al, U.S. 2,847,477-8 (R & H).
6. R. L. Letsinger and J. G. Traynham, J. Am. Chem. Soc. 70, 3342 (1948).
7. W. Gerrard et al, J. Chem. Soc. 3285 (1956).
8. M. W. Tamele et al, Ind. Eng. Chem. 33, 115 (1941).
9. R. W. Martin, Anal. Chem. 21, 922 (1949).
10. W. B. Converse, U.S. 2,309,576 (Shell).
11. G. H. Segall, U.S. 2,543,992 (Can. Inds.).
12. P. W. Jolly and F. G. A. Stone, J. Chem. Soc. 6416 (1965).
13. W. H. T. Davison and G. R. Bates, J. Chem. Soc. 2607 (1953).
14. C. E. Schildknecht, David Walborn, and L. D. Rinehart, unpublished.
15. J. B. Niederl et al, J. Am. Chem. Soc. 53, 3393 (1931).
16. R. H. Reinhard, U.S. 3,073,805 (Monsanto).
17. W. R. Johnson and D. T. Oakes, U.S. 3,070,544 (Monsanto).
18. E. L. Karlan and D. D. Perry, U.S. 3,149,168 (Thiokol).
19. J. B. Rust, U.S. 2,415,041 (Montclair Res.).
20. T. Posner, Ber. 38, 646 (1905).
21. D. Harman and W. E. Vaughan, U.S. 2,522,512 (Shell); Brit. 896,250; F. M. McMillan, U.S. 2,514,661 (Shell).
22. W. E. Vaughan and F. F. Rust, U.S. 2,563,383 (Shell); J. Org. Chem. 7, 472 (1942).
23. C. S. Marvel and H. E. Baumgarten, J. Polym. Sci. 6, 127 (1951).
24. A. Y. Garner, U.S. 3,010,946 (Monsanto).
25. G. H. Coleman and R. W. Sapp, U.S. 2,285,329 (Dow); cf. U.S. 2,161,737 (Dow) and 2,339,476 (Shell).
26. A. J. Passannate and E. L. Stogryn, U.S. 3,433,776 (Esso).
27. M. Tamele, C. J. Ott, K. E. Marple, and G. Hearne, Ind. Eng. Chem. 33, 115 (1941); U.S. 2,042,219 (Shell).
28. T. J. Prosser, U.S. 3,168,575 (Hercules).
29. W. J. Sparks and R. M. Thomas, U.S. 2,412,921 (Esso).
30. J. R. Roach and H. Wittcoff, J. Am. Chem. Soc. 71, 3944 (1949); U.S. 2,527,853, 2,559,171 and 2,585,035 (Gen. Mills); cf. Zunino, J. Chem. Soc. 410 (1899).
31. H. Dannenberg et al, Ind. Eng. Chem. 41, 1709 (1949); cf. U.S. 2,288,315; Brit. 810,222; Fr. 1,334,978 (Sorepe).
32. W. Gumlich et al, Ger. 1,019,421; CA 54, 8110 (Huels); cf. Brit. 810,222 and Ger. 1,024,654 (Bayer).
33. W. J. Maker, U.S. 2,852,487 (Glidden).
34. V. F. Jenkins et al, Brit. 928,801 (Howards) appl. 1958.
35. T. L. Phillips, Brit. Plastics 34, 69 (1961).
36. Brit. 713,332, 744,468·, and 744,807.

37. V. F. Jenkins, J. Oil & Colour Chem. Assoc. __44__, 42 (January 1961).
38. R. Behar et al, U.S. 3,226,451 (Francaise Duco).
39. Brit. 747,798 (Chemstrand).
40. F. A. Ehlers, U.S. 3,017,390 (Dow).
41. R. H. Martin, U.S. 3,025,280 (Monsanto).
42. W. F. Rutherford, U.S. 3,298,978 (Freeman Chemical).
43. J. K. Vlcek and J. Mleziva, J. Polym. Sci. C No. 16, 247 (1967).
44. C. E. Schildknecht and E. Hagmann, unpublished.
45. Brit. 810,222 (Bayer).
46. K. Raichle and C. Niehaus, Ger. 1,129,688 (Bayer); U.S. 3,576,772
     (Bayer); cf. H. W. Chatfield, Paint Technol. __26__/4, 17 (1962);
     U.S. 3,316,193 (Vantorex).
47. E. E. Waters et al, Paint Varnish Prod. __53__, 97 (November 1963);
     cf. W. A. Ledger, U.S. 3,491,065 (Coates).
48. J. Jarusek and J. Mleziva, Dtsch. Farben-Z. __22__/4, 156 (1968).
49. Celanese Product Development Dept., 1963.
50. L. A. O'Neill and R. A. Brett (Teddington), J. Oil & Colour Chem.
     Assoc. __48__, 1025 (November 1965).
51. A. G. Ghosh, U.S. 3,463,750 (Perstorp, Sweden).
52. J. R. Constanza, U.S. 3,268,484 (Celanese).
53. H. Dalibor, Ger. 1,162,568 and 1,163,550 (Reichhold).
54. V. Cappuccio, U.S. 3,195,975 (Monte).
55. Celanese Chemical Co. data from Bulletin of Proctor Chemical Co.
56. Y. Doi and B. Rånby, J. Polym. Sci. C No. 3, 231 (1970).
57. E. Yanovsky et al, Paint Ind. Mag. __60__, 84 (1945); R. Evans and
     J. A. Gallaghan, J. Am. Chem. Soc. __75__, 1248 (1953).
58. L. T. Smith, U.S. 2,462,817 (USA).
59. J. R. Roach and H. Wittcoff, U.S. 2,585,035 (Gen. Mills).
60. R. P. Silver, U.S. 2,884,394 (Hercules).
61. Ger. 962,009 (Bayer) and W. J. Maker, U.S. 2,852,487 (Glidden);
     cf. V. F. Jenkins et al, Brit. 939,939 (Howards); C. J. Campbell,
     U.S. 2,885,375 (Hercules).
62. R. Zimmerman, U.S. 3,296,337 (Albert).
63. Data sheet and Hercules Magazine, May 1963.
64. J. A. Holloway, U.S. 2,978,421 (Goodrich).
65. Brit. 837,982 (Dow).
66. K. Ito et al, Japan 19,104 (1963), CA __60__, 4112.
67. N. C. MacArthur, U.S. 3,369,009 and 3,369,030 (Hercules).
68. D. S. Breslow and F. E. Piech, U.S. 3,328,324 (Hercules).
69. T. J. Prosser, U.S. 3,168,575 (Hercules).
70. F. Engelhardt, U.S. 3,296,336 (Bergbau & Chemie).
71. L. A. O'Neill and R. A. Brett, J. Oil & Colour Chem. Assoc. __48__,
     1025 (November 1965).
72. Allyl Starch, Data sheet, General Mills, Minneapolis, 1952.
73. L. Lilienfeld, U.S. 1,350,820; cf. Irvine and McDonald, J. Chem.
     Soc. 2681 (1928).
74. E. Yanovsky et al, U.S. 2,413,463 (USA).
75. C. G. Tomecko and Roger Adams, J. Am. Chem. Soc. __45__, 2698 (1923).
76. E. Yanovsky et al, J. Am. Chem. Soc. __66__, 1625 (1944); __67__, 46
     (1945); __70__, 2194 (1948); Ind. Eng. Chem. __37__, 201 (1945).

77. E. Yanovsky, Allyl Starch Review, AIC-362, September 1953.
78. A. N. Wrigley and E. Yanovsky, J. Am. Chem. Soc. 70, 2194 (1948).
79. D. B. Sharp and T. R. Patrick, J. Org. Chem. 26, 1386 (1961).
80. C. A. Wilham et al, J. Appl. Polym. Sci. 7, 1403 (1963).
81. G. Champetier, Bull. Soc. Chim. France 185 (1948).
82. R. M. Hamilton and E. Yanovsky, U.S. 2,524,792 (USA).
83. P. L. Nichols and R. M. Hamilton, U.S. 2,449,816 (USA).
84. A. N. Wrigley and J. H. Schwartz, U.S. 2,740,724 (USA).
85. C. A. Wilham et al, J. Appl. Polym. Sci. 7, 1403 (1963)
    (Northern Regional Lab, Dept. of Agriculture).
86. J. R. Roach et al, U.S. 2,676,172 (Gen. Mills).
87. R. M. Hamilton and E. Yanovsky, U.S. 2,463,869 (USA); Ind. Eng.
    Chem. 38, 864 (1946).
88. J. H. Schwartz et al, U.S. 2,602,789 (USA); Dept. of Agricul-
    ture, Bureau of Agricultural and Industrial Chemistry Report
    AIC 339 (1952); CA 47, 10486 (1953).
89. A. N. Wrigley et al, U.S. Dept. of Agriculture Report AIC 266,7
    (1950); CA 45, 5422 (1951).
90. R. W. Kerr and W. A. Faucette, U.S. 2,733,238 (Corn Products).
91. L. A. Gugliemelli et al, U.S. 3,032,550 and 3,063,855 (USA);
    Preprints ACS Div. Org. Coatings and Plastics Chem. 26, 266
    (March 1966).
92. E. L. Griffin et al, Ind. Eng. Chem. 43, 2629 (1951); cf.
    U.S. 2,719,970 (USA).
93. M. Zief and E. Yanovsky, Ind. Eng. Chem. 41, 1697 (1949).
94. M. Zief, J. Am. Chem. Soc. 72, 1137 (1950).
95. A. N. Wrigley, private communication.
96. J. R. Roach, U.S. 2,562,537 and 2,555,834 (Gen. Mills);
    M. Zief et al, U.S. 2,594,303 (USA).
97. W. A. P. Black et al, Makromol. Chem. 71, 189 (1964).
98. M. Zief and E. Yanovsky, Ind. Eng. Chem. 41, 1697 (1949).
99. M. Zief and E. Yanovsky, U.S. 2,541,142 (USA).
100. M. DeGroote, U.S. 2,574,544 (Petrolite).
101. M. Zief, Official Dig. 297, 711 (1949); 303, 302 (1950).
102. A. N. Wrigley and M. Zief, Official Dig., Fed. Paint & Varnish
     Prod. Clubs, 302 (April 1950).
103. H. P. Brown, U.S. 2,798,053 (Goodrich).
104. J. F. Jones, U.S. 2,985,625 and Brit. 800,011 (Goodrich).
105. J. F. Jones, U.S. 3,137,660 (Goodrich).
106. E. L. Richardson and G. C. Hoss, U.S. 3,133,865 (Am. Home
     Products).
107. J. R. Abrams et al, U.S. 3,138,567 (Interchemical).
108. U.S. 3,312,640 (Mineral & Chemicals); U.S. 3,314,857 (Clairol);
     U.S. 3,330,731 (Colgate).
109. A. K. Reed and R. W. Repsher, U.S. 3,303,042 (Westinghouse).
110. N. E. Brindamour, U.S. 3,461,089 (Merck).
111. W. H. Kibbel and J. A. Shepherd, U.S. 3,499,844 (FMC).
112. W. L. Glen et al, Can. 521,245 (Ayerst).
113. R. G. Schweiger, J. Polym. Sci. A 2, 2471 (1964).
114. H. Gregory and L. F. Wiggins, J. Chem. Soc. 1405 (1947).

115. M. Zief and E. Yanovsky, U.S. 2,606,881 (USA).
116. A. N. Wrigley and E. Yanovsky, J. Am. Chem. Soc. 70, 2194 (1948).
117. V. L. Lapenko et al, CA 75, 130159 (1971).
118. E. Yanovsky et al, J. Am. Chem. Soc. 68, 2020 (1946).
119. R. B. Wagner, U.S. 2,891,056 (Hercules); cf. U.S. 3,251,825
      (Kalle).
120. I. Haidasch and J. Voss, Ger. 1,065,828 (Kalle).
121. J. Voss et al, U.S. 3,143,516 and 3,071,572 (Kalle).
122. H. H. Grün, Ger. 1,065,720 (Henkel).
123. Fr. 1,211,629 and 1,222,453; CA 55, 20450 and 18107 (1961).
124. M. Lagache et al, Compt. Rend. 251, 2353 (1960).
125. S. Yoshimura, CA 64, 8471 (1966).
126. P. L. Nichols, A. N. Wrigley, and E. Yanovsky, J. Am. Chem.
      Soc. 68, 2020 (1946).
127. R. W. Auden and W. L. Evers, U.S. 2,442,747 (R & H); A. Brookes
      et al, Brit. 830,751 (Brit. Ind. Plastics).
128. P. Talet, U.S. 2,726,223 (Nobel); T. Oshima, Ger. 1,123,334
      (Sumitomo); CA 57, 3460.
129. E. R. Atkinson and A. H. Bump, Ind. Eng. Chem. 44, 333 (1952);
      cf. U.S. 2,082,797 and 2,764,574.
130. E. E. Novotny et al, U.S. 2,512,672 (Borden).
131. Brit. 746,394 (Ciba); CA 50, 12494 (1956).
132. P. Zuppinger and G. Widmer, U.S. 2,986,541 and Brit. 861,852
      (Ciba).
133. Plastverarb. 9/6, 234 (1958).
134. G. Sulzer, U.S. 2,898,239.
135. H. Willersinn et al, Brit. 848,400; Brit. 838,007;
      Ger. 1,049,572 and 1,082,735 (Badische).
136. G. Gruber and H. A. Wannow, Ger. 1,134,655 (Cassella).
137. G. Alexander, U.S. 2,936,260 (G.E.); cf. L. E. Cockerham,
      U.S. 3,517,082 (Commercial Solvents).
138. Ger. 436,445 (Riebeck).
139. P. Bruin and J. Selman, CA 52, 1680 (1958)(Shell); Brit.
      777,385 (Shell).
140. S. D. Ross and M. Markarian, U.S. 2,564,214 and 2,799,694
      (Sprague Elec.); J. P. Stallings, ACS Org. Coatings and
      Plastics Preprint 29, 357 (1969).
141. A. I. Medalia and H. H. Freedman, U.S. 2,874,151 (PPG).
142. J. O. Brochta, U.S. 3,290,385 (Koppers).

15. ALLYLIC ACETALS AND KETALS

The allylic acetals and ketals and their derivatives
contain pairs of oxygen atoms joined by a single carbon
atom, e.g., -OCH(R)O- or -OC(R)O-, making them much less
stable than typical ethers in presence of acids. Although
many of the allylic acetals and ketals can be made from
readily available intermediates, the industrial development of their polymers has had little success. Many of
the monomers and polymers are unstable with evolution of
acrolein, and polymer properties have not been attractive
and reproducible. Most attention in industrial research
has been given to diallylic cyclic acetals such as reaction products of acrolein with polyols. Some of these
polymerize under the action of Lewis-acid catalysts to
give viscous liquid to solid products. A few derivatives
polymerize slowly by oxidative polymerization in air but
do not respond well to conventional radical initiators.
It is convenient to name a number of these compounds as
vinyl derivatives; but this is misleading since they contain reactive allylic hydrogen atoms and have low activity
in polymerization. The methallyl and chlorallyl compounds
have received little study. We begin with the acetals of
acrolein which may be regarded as allylidene ethers.

ACETALS OF ACROLEIN

Although acrolein, having the electron-attracting aldehyde group, behaves as an acrylic compound and polymerizes
very readily, the acetals of acrolein, methacrolein, and
chloroacrolein behave as allyl compounds and polymerize
reluctantly. The liquid acetals may be prepared by warming alcohol and acrolein with acid catalyst:

$$2ROH + CH_2=CHCHO \xrightarrow{H^+} CH_2=CHCH(OR)_2$$

Hemiacetals RCH(OH)OR are unstable intermediates seldom
isolated. The preparation of acetals from lower alcohols
and acrolein in good yield is not easy (1); a fume hood
is necessary to avoid exposure to the dreadful vapors of
acrolein. Trace amounts of acid catalyst give best
yields. The following properties are typical:

444

Acetals	bp, °C	n_D^{20}	Properties and references
Acrolein dimethyl	90 or 40 (120)	1.3962	5.1% soluble in water at 20°C; ref. (1)
Acrolein diethyl	125 or 63 (9)	1.4014	Ref. (1)
Acrolein di-n-butyl	86(10)	1.4205	Ref. (1)
Acrolein diallyl	75(28)	1.4380	Ref. (1)
Acrolein dibenzyl	120(0.005)	1.5469	Ref. (1)
Chloroacrolein dimethyl	28(12)	1.4305	Hearne, U.S. 2,508,257 (Shell)
Methacrolein diethyl	52(30)	1.4081	Ref. (2)

The dimethyl acetal of acrolein is miscible with alcohols, ketones, and hydrocarbons. The compound is toxic by oral, skin, and vapor inhalation. This volatile compound does not provide adequate warning by odor at low concentrations.

Acrolein dimethyl acetal is stable in neutral and alkaline media but with aqueous mineral acid at pH 4 it can hydrolyze to acrolein and methanol in about 15 min. The tertiary hydrogen atom was rapidly peroxidized in air to give a polymerization catalyst (3). As a stabilizer 0.1% hydroquinone may be added. Acrolein dimethyl acetal reacted with aqueous permanganate to give glyceraldehyde diethyl acetal. The acrolein dimethyl and diethyl acetals can be reacted with hypochlorous acid followed by treatment with alkali to give glycidaldehyde acetals (4). Direct epoxidation can be carried out with hydrogen peroxide in acetonitrile (5). Peroxidation in acid solution is difficult because of hydrolysis of acetal groups. Both acetaldehyde and butyraldehyde have been added to the double bond of acrolein dimethyl acetal at 80°C using azobis as catalyst (6). Low molecular weight addition products with CCl_4 have been prepared (7).

Acrolein diethyl acetal was made by reaction of potassium t-butylate with 3-chloropropionaldehyde diethyl acetal (8) and by reaction of acrolein with ethyl orthoformate (9). The acrolein acetals find use as sources of acrolein and methacrolein in situ by acidification, thus avoiding the physiological hazards of handling the acrylic aldehydes.

Direct homopolymerizations of acrolein dimethyl acetal and related allylidene compounds have not given promising products of high molecular weight. Acid catalysts, of course, can generate acrolein which may polymerize readily. Suitable catalysts and conditions for preparing homogeneous homopolymers of high molecular weight from these acetal monomers remain to be discovered.

Minor proportions of acetals of acrolein have been copolymerized with vinyl acetate and with methyl methacrylate

in bulk with free radical initiators (10). On heating
with traces of mineral acids, alcohol solutions of the
vinyl acetate copolymers gave insoluble crosslinked copolymers through reaction of intermediate aldehyde groups.
Applications were suggested in textile, paper, and tanning
technologies. Acrolein acetals were copolymerized with
m-vinylphenol using peroxide catalysts (11). These
soluble copolymers could be insolubilized by heating or
by acid catalyst, apparently by reaction of aldehyde
groups at active ortho and para sites of the phenol ring.
Copolymerization with isobutene was disclosed (12). Addition of acrolein diethyl acetal greatly retarded polymerizations of vinyl chloride and of acrylonitrile (13).

DIALLYL ACETALS

Diallyl acetal compounds $RCH(OCH_2CH=CH_2)_2$ are obtained
by reaction of aldehydes with allyl alcohol in the presence of acid catalysts; they are highly reactive and
unstable. Although they are hardly promising for industrial polymer products, their complexities offer challenge
for research. Since allyl alcohol and substituted allyl
alcohols can rearrange to aldehydes in the presence of
Lewis acids, the acetals of lower aldehydes can be obtained
only by use of very weak acidic catalysts such as calcium
chloride acting at moderate temperatures (14). The beta-
substituted allyl alcohols such as methallyl alcohol are
even more prone to rearrangement, and syntheses of the
dimethallyl acetals are difficult. In this area it must
be remembered that the polymers, the monomers, and some of
the intermediates are all unstable to heating with acids.
Moreover, possible cyclization and air oxidation at active
hydrogen atoms also may contribute complexity. Pyrolysis
of diallyl acetals can yield allyl esters (14). Properties
of some diallyl and dimethallyl acetals are given in
Table 15.1.

Adelson and Gray showed that diallyl acetals could polymerize slowly through air oxidation (15). The diallyl
acetal of formaldehyde was prepared from heating a mixture
of 120 g paraformaldehyde, 464 g allyl alcohol, 12 g
anhydrous ferric chloride, and 50 ml benzene. After 104 g
of water had been removed in the azeotrope, the diallyl
acetal was vacuum distilled and then redistilled dry at
atmospheric pressure. The monomer reacted quantitatively
with bromine solution. It polymerized slowly to a gel on
heating for 20 hr at 130°C with air bubbling through. A
gel fraction formed in a few days on standing at room
temperature in contact with air. Finally a hard infusible
polymer resulted. The polymer contained more than 5%
excess oxygen. Diallyl acetals reacted with SO_2 to give

TABLE 15.1

Properties of Some Diallyl and Dimethallyl Acetals

Aldehyde used	bp, °C (at mm)	n_D^{20}	References
Diallyl acetals			
HCHO	140	1.4232	Adelson, U.S. 2,469,288 (Shell)
CH_3CHO	152	1.4235	Adelson
$(CH_3)_2CHCHO$	52(6)	1.4263	Panradl, CA 50, 13796
$CH_3CH_2CH_2CHO$	129(150)	1.4300	Brannock, JACS 81, 3379 (1959)
CH_2ClCHO	--	1.4525	Fr. 1,011,278 (St. Gobain); CA 52, 6394
Dimethallyl acetals			
HCHO	176	1.4339	Adelson
CH_3CHO	184	1.4324	Adelson
$CH_3CH=CHCHO$	164(720)	--	Herrmann, Br. 757,907 (Wacker); Ger. 940,293
$CH_2=CHCH_2OCH_2CH_2CHO$, etc.	--	--	Whetstone, U.S. 2,561,564 (Shell)
$CH_3CH_2CH_2CHO$	97(16)	1.4371	Brannock

insoluble polymers (16). Dimethallyl acetal of formaldehyde was slower to polymerize than the diallyl compound; 39 hr at 130°C with air gave only a viscous solution and no gel. Analysis of the polymeric product showed only 2.3% additional combined oxygen.

The allyl acetal of acetaldehyde was made from reaction of allyl acetate with allyl alcohol in the presence of HgO and BF_3 (17). The tetrafunctional allyl acetal of glyoxal, which boiled at 160°C and 25 mm, gave only polymers of low molecular weight when heated with peroxides (18). Hard films formed on baking in air. With vinyl monomers crosslinked copolymers formed.

From viscous liquid low polymers of allyl alcohol and several aldehydes, Adelson and Gray prepared low molecular weight polyallyl acetals (15). A partial polymer fraction that was alcohol soluble was molded at 147°C in the presence of 1% hexamethylenetetramine as catalyst and stabilizer to give an infusible crosslinked product. Possibilities of cyclic polymer structures and the nature of the oxidations were not discussed. Evidence of partial cyclization in free radical polymerizations of diallyl acetals was reported in Russia (19). Vinyl allyl acetal polymerized more readily than diallyl acetal on heating at 130°C with di-t-butyl peroxides.

For complexity in polymerization, consider the diallyl acetal of acrolein prepared by Fischer and Smith in 75%

yield (20). This trifunctional monomer boiled at 75°C at
28 mm and had n_D^{20} = 1.4380. Allyl-terminated acetal low
polymers from formaldehyde were prepared, and properties
were given of members up to AO(CH$_2$O)$_6$A (mp 22.5°C.;.
n_D^{25} = 1.4411) (21). Diallyl acetals have been epoxidized
(22). Diallyl acetals of the type (AOCH$_2$OÇH$_2$)$_2$ were undistillable oils suggested for resins, pesticides, and
herbicides (23). Copolymerizations of allyl acetals and
of methallyl acetals have been suggested with methacrylate
esters (24), vinyl acetate, and other monomers to partially crosslinked products (25). Recently Matsumoto and
Oiwa found chain transfer constants of diallyl acetals to
be greater than those of allyl esters in copolymerization
with styrene at 60°C using azobis as initiator (26).

The polyfunctional acetal monomer sym. tetraallyloxy-
ethane or allyl acetal of glyoxal, supplied by Nobel-Bozel
(Paris) and by Jonas Corp. (New York), was said to be a
"stable product for controlled crosslinking" (27). A
sample did not polymerize on storage 6 years but gave a
strong sting in the nose. Triallyl orthoformate
CH(OCH$_2$CH=CH$_2$)$_3$ has been available as a crosslinking
monomer from Kay-Fries Chemicals (New York). In contact
with air it polymerized readily in UV light to nontacky,
hard, clear films which, on further polymerization, became brittle and yellowish (28). The monomer gives a
delayed sting in the nose but much less than allyl alcohol.

DIALLYL KETALS

Diallyl ketals were made long ago; for example, the
diallyl ketal of cyclohexane (29). Lorette and Howard
prepared a series of ketals from different ketones including diallyl and allyl alkyl mixed ketals (30). It was
convenient to use interchange or transketalization reactions; that is, reaction of a dimethyl acetal of acetone
or butanone with the allyl alcohol or an alcohol mixture.
Benzene was used as solvent and p-toluene sulfonic acid
served as catalyst:

$$(CH_3)_2C(OCH_3)_2 + 2CH_2=CHCH_2OH \longrightarrow (CH_3)_2C(OCH_2CH=CH_2)_2 + 2CH_3OH$$

During reaction, methanol-benzene azeotrope was distilled
off. Then the system was made alkaline by adding sodium
methylate before distilling off benzene and excess allyl
alcohol. Finally the diallyl ketal was distilled at low
pressure. Allyl acetate may be used instead of allyl
alcohol (31). A procedure for preparing cyclohexanone
diallyl ketal used p-toluene sulfonic acid monohydrate as
weak catalyst (32). Thus polymerization catalyzed by dry
Lewis acids is avoided. Diallyl ketals on heating in the
presence of acids undergo rearrangement to alpha-allyl
substituted ketones (31):

$$CH_3CH_2\overset{\displaystyle CH_2CH=CH_2}{\underset{\displaystyle CH_3}{CO(CH_2CH=CH_2)_2}} \xrightarrow[\text{toluene}]{H^+} CH_3\overset{|}{\underset{\displaystyle O}{CHCCH_3}} + CH_2=CHCH_2OH$$

This is another example of the familiar end-for-end migration of allyl groups. Properties of allyl ketals are
shown in Table 15.2. Little has been reported about the
polymerizations of diallyl ketals. The instability of the
monomers and polymers even to dilute aqueous acids dis-
courages technical interest.

TABLE 15.2

Allyl Ketals [a]

Ketone	Type ketal	bp, °C at mm	n_D at °C
Acetone	Methyl allyl	49(50)	1.4040(24)
Acetone	Diallyl	59(20)	1.4228(25)
Butanone	Methyl allyl	68(50)	1.4146(24)
Butanone	Diallyl	94(50)	1.4300(24)
3-Pentanone	Diallyl	87(20)	1.4343(25)
Cyclopentanone	Diallyl	98(20)	1.4535(24)
Cyclohexanone	Diallyl	98(10)	1.4603(24)

[a] Reference 30.

DIALLYLIDENE PENTAERYTHRITOL
AND OTHER SPIRO COMPOUNDS

Some companies which have had abundant formaldehyde and
polyols have tried to develop commercially acceptable
synthetic materials based upon diallylidene pentaerythritol (DAPE), the acrolein acetal of pentaerythritol, also
known as 3,9-divinyl spirobi(m-dioxane) (33). Note that
acrolein is not an allyl compound and the cyclic acetals
of acrolein are not vinyl compounds.

$$2CH_2=CHCHO + \overset{\displaystyle HOH_2C \quad CH_2OH}{\underset{\displaystyle HOCH_2 \quad CH_2OH}{>\!\!C\!\!<}} \xrightarrow[\text{catalyst}]{\text{acid}} CH_2=CHCH\overset{\displaystyle OCH_2 \quad CH_2O}{\underset{\displaystyle OCH_2 \quad CH_2O}{>\!\!C\!\!<}}CHCH=CH_2 + H_2O$$

DAPE

Generally the product contains some monoallylidene PE as
well as allylidene derivatives of some dipentaerythritol
present in commercial PE. Unfortunately, the chemists who
have suffered the vapors of acrolein and the naming of
such diallylidene acetal compounds have not been rewarded
with successful plastics. The polymerization reactions
and products in general are very complex and variable.

The colors and heat distortion temperatures of the polymers are inferior and they differ often from lot to lot.
It is common experience that acetal structures are relatively unstable, especially in the presence of acid.
Nevertheless some of the patents state that these products
are easy to make and have outstandingly useful properties.
DAPE does not homopolymerize readily by ordinary ethylenic
addition polymerization (34), but it reacts by hydrogen
migration or H-addition polymerization with compounds
bearing two or more active hydrogen atoms, e.g., polyols
or polythiols, as represented:

$$CH_2=CHCH\underset{OCH_2}{\overset{OCH_2}{\diagup}}\underset{CH_2O}{C}\overset{CH_2O}{\diagdown}CHCH=CH_2 + HOROH \longrightarrow$$

$$--OROCH_2-CH_2CH\underset{OCH_2}{\overset{OCH_2}{\diagup}}\underset{CH_2O}{C}\overset{CH_2O}{\diagdown}CHCH_2CH_2OR--$$

This type of polyaddition, which generally has been written in a reverse Markovnikov way, occurs under the influence of Lewis-acid catalysts which, at the same time, may
produce some vinyl addition cationic polymerization.
Generally after the early stages crosslinked products
result. The presence of up to about 7% dipentaerythritol
in commercial pentaerythritol apparently contributes to
crosslinking. Diallylidene pentaerythritol has been
available from Union Carbide. The commercial material
melts at 42°C, boils at 120°C at 2 mm, has specific gravity
of 1.250, and is 1.4% soluble in water at 20°C. Such monomers have been called spiroacetals.
    Before 1946 Rothrock made coatings by reacting polyols
with acrolein in the presence of strong acid catalysts (35).
In Germany diallylidene pentaerythritol was isolated by
using mild acid catalysts and could be distilled at 115°C
at 2 mm (36). The process is difficult to repeat without
encountering undistillable viscous syrups (37). Degussa
workers suggested addition of triethanolamine to check the
polymerization at desired viscosity before curing as shaped
plastics. Most other bases caused yellow discoloration.
Orth carried out the polymerization of DAPE as a separate
step with two molar proportions of a polyol and 0.3 to
0.6% acid catalyst, followed by curing in molds at 70 to
80°C (38). Peroxide catalysts and heating would not promote homopolymerization, and copolymerizations were slow.
    Celanese chemists improved the preparation of DAPE and
evaluated a variety of addition products (39). They
heated a mixture of 139 g pentaerythritol, an excess of
456 g acrolein, and 5.9 g phosphoric acid (85%) at reflux

with stirring for 3 hr.   Then the azeotrope of water and
acrolein was distilled off while more acrolein was being
added.   At the end of the reaction, 10.3 g of NaHCO$_3$ was
added to neutralize the catalyst.   Acrolein was stripped
off at 20 mm pressure and then 700 ml of cold water was
added.   The DAPE began to crystallize at once.   After
drying under vacuum over KOH a yield of 194 g melted at
43°C.   The monoacetal byproduct remained in water solution.
DAPE could be recrystallized from water-methanol.   The
preparation of dimethallylidene PE (mp 116°C) was said to
be less troublesome because the compound did not tend to
homopolymerize as readily under the influence of acids.
    The best mild Lewis acid for catalyzing H-addition of
DAPE with polyols was reported to be boron fluoride-
etherate (39).   With 1,2-propanediol and heating for 6 hr
at 80°C an amber-colored, crosslinked rubber was obtained
that decomposed at about 270°C.   Addition of DAPE to
maleic anhydride gave an insoluble rubbery mass.   Best in
color, heat resistance, and hardness from Celanese work
were the addition products of DAPE with sorbitol.   They
were catalyzed by p-toluene sulfonic acid and were cured
in molds for 20 hr at 80°C (with evolution of vapors of
acrolein).   Surprisingly, DAPE could react with water on
standing at room temperature to give polymers almost as
favorable in properties as those from sorbitol.   Among
the hydroxyl-containing compounds with which DAPE and its
prepolymers have been reacted are drying oil partial
esters of polyols (40) and butylated melamine-formaldehyde
resin (41).
    A number of variations of H-addition polymerizations of
DAPE have been proposed.   Acrolein and PE were heated with
Lewis-acid catalyst until a viscous solution was obtained,
and then more acrolein and more catalyst were added before
curing in molds by heating (42).   Monofunctional as well
as difunctional phenols have been reacted with DAPE (43).
In one patent example, 945 g of DAPE without purification
was heated with 280 g of phenol along with diethyl sulfate
catalyst (44).   To 133 g of this liquid precondensate was
added 7 g of a partial phosphite ester of PE.   After heating for 40 min at 115°C, the viscous liquid was cured in
a mold for 16 hr at 150°C to a relatively hard material.
DAPE was reacted with phosphoric acid polymers (45).   DAPE
also has been reacted by H-addition with hydrogen sulfide
and with dimercaptans.   Heating with H$_2$S at 200°C without
catalyst only gave polymers of molecular weight 1600 and
lower (46).   DAPE has been reacted with hydroxyl-terminated maleate polyesters in the presence of boron fluoride
etherate (47), with polyfunctional organic acids (48), and
with cyclic formals (49).   Aqueous dispersions of the
thermosetting polyspiroacetals were tried with melamine

ether resins for baked coatings (50). DAPE has been reacted with aqueous solutions of polysulfides of alkali
metals to give rubbery polysulfide polymers (51), and DAPE
with sulfur chlorides formed various low molecular weight
products (52). DAPE was reacted with acrylic acids (53)
and with partial esters from polyols and dry oil fatty
acids (54).

Diallylidene pentaerythritol homopolymerizes extremely
slowly on heating with free radical catalysts (55). However, soluble solid low polymers were reported by heating
at 90°C for 3 hr under nitrogen (56). In contrast to
such ethylenic compounds as drying oils, certain allyl
ethers of carbohydrates, allylidene dioxanes, and some
olefinic hydrocarbons, DAPE does not show polymerization
by way of air autoxidation or "drying" in films. Attempts
to copolymerize DAPE with styrene and other vinyl-type
monomers have been frustrating (57). A review of attempts
to develop synthetic resins based on cyclic acetals of
acrolein was written (58). DAPE greatly retarded the
radical polymerizations of styrene, acrylic and vinyl
monomers (59). Both the Q and e values of DAPE were found
to be near zero. Copolymers with maleic anhydride of 1:1
molar ratio were reported (60). Diallylidene spiro compounds more complex than DAPE also have given only low
polymers (61).

OTHER CYCLIC ACETALS FROM UNSATURATED ALDEHYDES

Hibbert and Whelen in Canada made early studies of
allylidene derivatives of glycerol (62). They showed that,
on heating acrolein and glycerol without added catalyst,
the main product is the five-membered allylidene acetal:

$$CH_2OH \atop CHOH \atop CH_2OH \quad + \quad OCHCH=CH_2 \quad \longrightarrow \quad {CH_2 \atop CHOH \atop CH_2}\!\!>\!\!{O \atop O}\!\!>\!\!CHCH=CH_2 \quad + \quad H_2O$$

Attempts to distill the product invariably resulted in
polymerization to a viscous, involatile polymer mass. In
contrast, the 1,2-allylidene acetal of glycerol, a derivative of 2-vinyl-1,3-dioxolane, showed less tendency to
polymerize on standing but gave a dark brown residue on
distillation. This unstable compound had a strong odor
and hydrolyzed readily in dilute aqueous acids. Surprisingly, the methyl ether $CH_3OCH_2CH\!\!-\!\!CH_2$ showed more

$$O \quad O \atop CH\!\!-\!\!CH=CH_2$$

tendency to polymerize, becoming a viscous mass on standing at room temperature. This behavior may have involved

oxidative polymerization or partial hydrolysis to acrolein.
An allyloxymethyl dioxolane prepared from glycerol alpha-
allylether was copolymerized with styrene and acrylic
monomers (63). Cyclic acetals from acrolein were copolymerized with acrylic acid derivatives in free radical
suspension systems (64).

The allyl compound 2-vinyl-1,3-dioxolane $CH_2=CHCH \begin{array}{c} O-CH_2 \\ | \\ O-CH_2 \end{array}$

was copolymerized with acrylic monomers (65). It could
be made by dropping acrolein into a mixture of HCl and
ethylene glycol cooled to 0°C (66). It reacted readily
with alkaline permanganate to give the glycol. A number
of derivatives of 2-vinyl-1,3-dioxolane were prepared by
Wacker chemists and physical constants were reported (67).
Cationic polymerization of 2-vinyl-1,3-dioxolane at low
temperatures gave oily to half-solid polymers showing
evidence of complex structures (68). Polymerization by
gamma radiation gave white powders with isomerization (69).
    Most of the polymers from other cyclic acetals of acro-
lein seem to have limitations similar to those of DAPE
polymers with respect to development of useful synthetic
materials. The partial acetals containing residual
hydroxyl groups would be expected to show even less tendency to react by free radical or ionic vinyl-type polymerizations. Few analogous compounds from methacrolein
have been reported.
    Polymerization of 2-vinyl-4,5-dimethyl-1,3-dioxolane in
presence of BF_3 etherate or $SnCl_4$ catalyst gave polymers
of low molecular weight (70). In the presence of epichlorohydrin, crosslinked polymers were formed containing
chlorine. Triallylidene sorbitol was distilled at 151°C
at 0.9 mm and had n_D^{20} of 1.4865 (66). Both monoallylidene
and polyfunctional allylidene derivatives of polyols were
discussed with physical properties, but statements about
the polymer applications were too optimistic (71). A
number of the compounds have been epoxidized in Shell
laboratories (72). References to some di- and triallylidene acetals follow:

Polyol	References, etc.
Aromatic polyols	Brit. 717,418 (Dynamit)
Sorbitol, etc.	Fischer and Smith, J. Org. Chem. 25, 319 (1960); plus triamines to polymers, U.S. 2,895,962 (Shell); cured by $C_2H_5SO_3H$, U.S. 2,974,128 (Celanese); reluctant copolymerization, Ouchi, CA 75, 110641 (1971)
Sucrose, polyvinyl alcohol	Polymers, U.S. 3,208,993 (Shell)
Polypentaerythritol	Polymerized, U.S. 2,895,945 (Shell)

A methacryloxybutyl derivative of so-called 2-vinyl-1,3-dioxolane was polymerized with BF_3 etherate catalyst to a yellow syrup (73). Addition of cobalt naphthenate and methyl ethyl ketone peroxide permitted oxidative polymerization to nontacky films after 6 hr and films resistant to methyl ethyl ketone after 24 hr. Polymerizations of the allylic compounds 2-vinyl-1,3-dioxolane and 2-vinyl-4-methyl-1,3-dioxolane were made with BF_3 etherate at room temperature and below. Dimers were formed along with complex polymers of molecular weight 2000 to 5000 having at least three types of chain segments. Ring opening and rearrangements contributed to the heterogeneity of these polymers.

Among other acetals of acrolein which have not proved promising for useful polymers is so-called 2-vinyl-m-dioxane. Some yellow solid homopolymers of low molecular weight and uncertain structure were made from this compound by use of fluoroborate catalysts (74). Copolymers with formaldehyde or trioxane were prepared using electrophilic catalysts (75). In contrast vinyl-p-dioxane is an allyl ether derivative (76).

DIALLYLIDENE CYCLIC ACETAL ESTERS

Esters of cyclic acetals of acrolein, which have been named vinyl dioxolane esters in patents, were evaluated as synthetic drying oils by Du Pont. These can be prepared by heating acrolein with triols in the presence of p-toluene sulfonic acid (77, 78), followed by transesterification of the cyclic alcohol with methyl esters of dicarboxylic acids; for example

The tertiary hydrogen of the allylidene group of such compounds undergoes peroxidation in air when catalyzed by cobalt compounds, and this leads to crosslinked polymers. Ikeda isolated the hydroperoxide formed under oxygen at 85°C from 2-vinyl-4-methyl dioxolane (77). On reaction

with water, the hydroperoxide gave 2-hydroxypropyl acrylate by breaking of the dioxolane ring. Oxidation of this
allylidene acetal in air at room temperature for 6 hr, in
the presence of a little cobalt butyl phthalate, yielded
polymeric products believed to be derived from 2-hydroxy-
propyl acrylate. Oxidation in the presence of toluene
sulfonic acid was reported to give a different spirocyclic
peroxide. Oxidation of methallylidene cyclic acetals in
presence of cobalt compounds was believed to form analogous methacrylate polymers. Oxidations of the allylidene
acetals gave some acidic products also. Ikeda and co-
workers reported boiling points and refractive indices of
a number of the vinyl dioxolanes and m- or 1,3-dioxanes.

The "drying" properties and formulation of coatings from
allylidene acetal esters of phthalic, sebacic, fumaric,
itaconate, and other dicarboxylic acids were discussed
(79). For example, 73.2 parts of bis(2-vinyl-1,3-dioxo-
lane-4-butyl) isophthalate was mixed with 1.46 parts of
coconut oil alkyd resin and 7.0 parts solvent, as well as
cobalt naphthenate and t-butyl hydroperoxide. Films cast
on glass required nearly 28 hr at 75°F to become tack-
free. The same diallylidene acetal ester in films containing 0.025% cobalt as the butyl phthalate and 0.5%
t-butyl hydroperoxide became tack-free after 6 hr or after
12 hr without peroxide. Infrared absorption bands showed
complex reactions in the "drying" process, including the
formation of hydroxyl groups and acrylate ester polymer
structures.

The diallylidene acetal esters did not polymerize
readily on heating with radical initiators. Addition of
transition metal catalysts to hydroperoxides of the monomers caused evolution of heat and resin formation. For
example, bis(2-vinyl-1,3-dioxolane-4-butyl) carbonate
shaken with oxygen for an hour at 86°C absorbed enough
oxygen to be half-transformed into a hydroperoxide at the
tertiary hydrogen atom. Adding 2% of copper butyl phthalate caused rapid polymerization with evolution of heat.
Different methods of accelerating the "drying" of these
so-called vinyl cyclic acetals in films were proposed,
such as adding t-amines, preoxidation, and copolymerization (80).

The slow air-oxidative polymerizations of diallylidene
cyclic acetal esters were discussed by Hochberg (81).
The compounds most evaluated in paint films were made from
acrolein and 1,2,6-hexanetriol, followed by preparation
of the bifunctional esters. Since the cyclic acetal rings
are unstable to acids, it was necessary to prepare the
esters by alkali-catalyzed alcoholysis (ester interchange)
using methyl esters such as dimethyl-o-phthalate, dimethyl
adipate, and dimethyl sebacate. Esters of unsaturated

acids such as methacrylic and itaconate also were tested. Some of these "dried" in films under air as fast as natural drying oils, apparently by way of initiation by oxidation at the tertiary allylic hydrogen atoms to form hydroperoxides. Unfortunately, the synthetic films became more brittle than those from linseed oil.

Oxygen-heterocyclic rings larger than the five-membered dioxolanes gave slower "drying" diallylidene ester compounds (81). The o-phthalate and sebacate esters of analogous dioxane derivatives were slower than the dioxolanes. The particular acid group of the diallylidene ester considerably affected the "drying" behavior in films. The best formulations were based on blending two esters. The o-phthalate ester formed soft, flexible films. The itaconate ester formed relatively brittle films. The phthalate ester films tended to wrinkle from faster polymerization at the air interface.

Many limitations in the properties of the diallylidene cyclic acetal ester films have discouraged commercial development. Both the ester and acetal linkages were slowly broken by hydrolytic agents. In the presence of dilute aqueous acids acrolein was evolved. Gloss and pigment wetting properties were not good, but pigment wetting was improved by partial oxidation. Discoloration and brittleness on aging the films were unfavorable. Suggestions have been made to use the diallylidene cyclic acetal esters along with vinyl-type polymers (82).

Isopropenyl derivatives of 2-vinyl-1,3-dioxolane have been studied in copolymerization (83). Polyfunctional monomers having allylidene groups attached directly to heterocyclic rings have been considered for coatings (84). A tetrafunctional allylic dioxolane monomer was prepared by heating divinyl ethylene glycol with 1,1,3-trimethoxy-3-ethoxypropane at 90°C using H_3PO_4 on silica-alumina as catalyst (85). With cobalt butyl phthalate drier, several days were required to form tack-free films from the monomer.

METHYLENE DIOXOLANES (1,2-ALLYLIDENE COMPOUNDS)

Fischer and co-workers heated acetone glycerol chlorohydrin with KOH to obtain $CH_2=CH-CH_2$ which boiled at 106°C

$$\begin{array}{c} CH_2=CH-CH_2 \\ \underset{O}{|} \quad \underset{O}{|} \\ \diagdown C \diagup \\ H_3C \quad CH_3 \end{array}$$

and had n_D^{20} = 1.4221. The compound decolorized bromine solution instantly and it hydrolyzed rapidly in 0.1% N sulfuric acid at room temperature (86). The compounds of such structure were studied briefly by Eastman Kodak, by Firestone, and later by Nobel. Soft, discolored polymers

were obtained from 4-methylene dioxolanes by use of Lewis
acid catalysts (87). Copolymerizations of 4-methylene-
1,3-dioxolane using free radical initiation were described
with acrylic, vinyl ketone, and vinyl ester monomers (88).
Heating 2-phenyl-4-methylene-1,3-dioxolane for 48 hr at
60°C with 1% BP gave a hard, clear polymer which was
soluble in acetone (89). A homopolymerization of 2,2-
dimethyl-4-methylene-1,3-dioxolane may have involved
cationic polymerization. Maleic anhydride copolymerized
vigorously with these monomers.

Orth reviewed early work on unsaturated acetals and
ketals of 1,2-alkendiols (90). He found that these compounds polymerize to colored products by action of weak
Lewis acid catalysts, such as 25% zinc chloride in ethanol.
This behavior is not surprising considering the similarity
of these monomers to vinyl ethers (91). The following
synthesis was carried out from glycerol monochlorohydrin
and cyclohexanone:

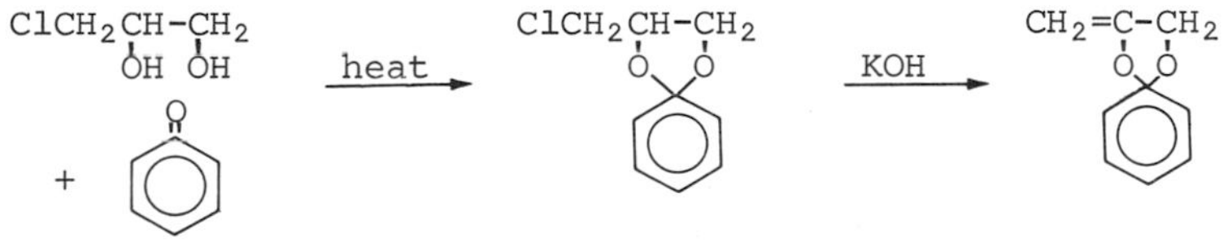

A number of such substituted 4-methylene dioxolanes were
prepared from different aldehydes and ketones; their
polymerizations were tested briefly with dilute zinc
chloride catalyst, giving in most cases soft yellow resins.
Low-temperature cationic polymerizations were not reported.
Because of electron withdrawal from the vinyl carbon,
these monomers are not typical allyl compounds; they seem
to be closer relatives of vinyl ethers. Properties of the
high polymers would have interest because of the unusual
relation of the rings to the main chain.

Orth prepared a number of bifunctional methylene dioxolane monomers but did not publish many details of purification or polymerizations. The following monomers gave
polymers with yellow coloration and low heat resistance.
For example, glyoxal and glycerol chlorohydrin were
reacted:

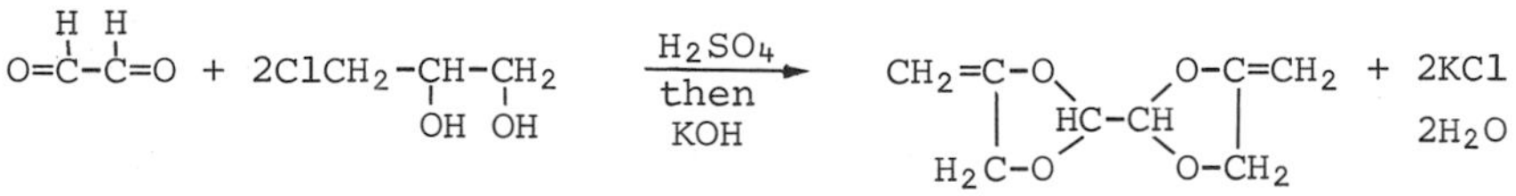

bp 100°C(12 mm)

For polymerization, the 25% zinc chloride in alcohol was added to stirred monomer at room temperature. Polymerization along with yellow coloration began in a few minutes.

  Recently the polymerizations of 4-methylene-1,3-dioxolane and related derivatives in solvents using BF_3 or $AlCl_3$ catalyst have been reported to occur, with partial ring opening, giving ether oxygen and carbonyl groups in the main chain (92). Polymers from 4-methylene dioxolanes containing chlorine have been prepared starting with chloroacetaldehydes (93). Mono-, di-, and trichloro-methyl-4-methylene-1,3-dioxolanes were polymerized by cationic catalysts through the methylene group, apparently without opening the dioxolane ring. Solid high-melting polymers were reported. Little has been published on the polymerization of methylene or allylidene derivatives of larger heterocyclic rings.

References

1. C. W. Smith, Acrolein, Wiley, 1962, p. 121.
2. S. M. McElvain, J. Am. Chem. Soc. 64, 1968 (1942).
3. R. H. Wiley, U.S. 2,432,601 (Du Pont).
4. D. I. Weisblat et al, J. Am. Chem. Soc. 75, 5893 (1953).
5. G. B. Payne and P. H. Deming, U.S. 3,053,856 (Shell); G. B. Payne et al, J. Org. Chem. 26, 659 (1961).
6. A. Mondon, Angew. Chem. 64, 224 (1952); Brit. 635,934 (U.S. Rubber).
7. R. H. Hall and D. I. H. Jacobs, J. Chem. Soc. 2034 (1954).
8. E. Rothstein, J. Chem. Soc. 1559 (1940).
9. J. A. VanAllan, Org. Syn. 32, 5 (1952).
10. E. F. Izard, U.S. 2,467,430 (1949); cf. K. Ohyanagi, CA 54, 1934 (1960).
11. E. M. Evans et al, Brit. 679,374 (British Resin Products).
12. C. G. Brannen and J. A. Wuellner, U.S. 3,269,952 (Std. Oil Ind.).
13. T. Oota et al, CA 69, 67781 (1968).
14. H. Chafetz, U.S. 3,079,429 (Texaco).
15. D. A. Adelson and H. F. Gray, U.S. 2,469,288 (Shell).
16. S. N. Ushakov et al, CA 44, 1746 (1950).
17. F. J. Glavis, J. Am. Chem. Soc. 70, 2805 (1948); W. Croxall and H. T. Neher, U.S. 2,446,171 (R & H).
18. Belg. 609,343 (Nobel); CA 57, 10037 (1962).
19. I. A. Arbuzova et al, CA 61, 1946 (1963); 60, 5648 (1964).
20. R. F. Fischer and C. W. Smith, J. Org. Chem. 25, 319 (1960).
21. R. F. Webb et al, J. Chem. Soc. 4307 (1962).
22. B. Phillips and P. S. Starcher, U.S. 3,018,294 (Carbide).
23. L. Orthner and H. Koch, Ger. 831,992 (Hoechst); CA 49, 15962 (1955).
24. W. O. Kenyon and T. F. Murray, U.S. 2,487,879 (Eastman).
25. I. Szanto et al (Hungary), CA 54, 20297 and 55, 8283.

26. A. Matsumoto and M. Oiwa, J. Polym. Sci. Al, <u>10</u>, 103 (1972).
27. P. Talet, U.S. 3,197,447 (Nobel-Bozel).
28. C. E. Schildknecht, unpublished.
29. B. Helfrich and J. Hausen, Ber. <u>57B</u>, 795 (1924).
30. N. B. Lorette and W. L. Howard, J. Org. Chem. <u>25</u>, 521 (1960); U.S. 3,127,450 (Dow); cf. F. J. Glavis, J. Am. Chem. Soc. <u>70</u>, 2805 (1948).
31. N. B. Lorette and W. L. Howard, J. Org. Chem. <u>26</u>, 3112 (1961).
32. W. L. Howard and N. B. Lorette, Org. Syn. <u>42</u>, 34.
33. Also called 3,9-divinyl-2,4,8,10-tetraoxaspiro[5.5]undecane.
34. Cf. T. Ouchi (Osaka Univ.), CA <u>74</u>, 76,708 (1971).
35. H. S. Rothrock, U.S. 2,401,776 (Du Pont).
36. H. Schultz and H. Wagner, Angew. Chem. <u>62</u>, 105-132 (1950); Ger. 870,032 (Degussa); cf. Ger. 858,406 and 885,006.
37. J. E. Paustian and C. E. Schildknecht, unpublished; cf. R. B. Mesrobian, Polymer Laboratory Manual (Polytech. Inst. Brooklyn).
38. H. Orth, U.S. 2,687,407 and Ger. 852,301 (Dynamit); Kunststoffe <u>41</u>, 454; cf. U.S. 2,895,945 (Shell).
39. F. Brown, D. E. Hudgin and R. J. Kray, J. Chem. Eng. Data <u>4</u>, 182-187 (1959); cf. U.S. 2,915,530, 2,955,092, 2,974,127 (Cela-nese); cf. U.S. 2,870,121 and 3,109,830 (Heyden-Newport).
40. N. C. MacArthur, U.S. 3,311,580 (Hercules).
41. N. C. MacArthur, U.S. 3,311,674 (Hercules).
42. H. R. Guest et al, U.S. 2,909,506 (Carbide .
43. H. R. Guest et al, U.S. 2,915,501, 3,010,942, and 3,022,273; J. E. Wilson and R. K. Walton, U.S. 2,915,492, 2,915,499 and 2,915,500 (all to Carbide).
44. H. R. Guest and B. W. Kiff, U.S. 3,010,941 (Carbide).
45. H. R. Guest et al, U.S. 3,110,703; cf. U.S. 2,951,826 and 2,957,856 (all to Carbide).
46. H. R. Guest et al, U.S. 2,998,427 and 2,996,516 (Carbide).
47. H. Orth, Ger. 855,165 (Dynamit); J. A. Parker et al, U.S. 2,974,116 (Armstrong Cork); A. Englisch et al, U.S. 3,209,054 (Albert).
48. R. J. Kray and F. Brown, U.S. 2,917,484 (Celanese).
49. H. I. Berman, U.S. 3,132,126 (Armstrong Cork).
50. E. M. Cohen et al, Offic. Dig. <u>37</u>, 1215 (1965).
51. H. R. Guest et al, U.S. 2,960,495 (Carbide).
52. H. R. Guest and H. A. Stansbury, U.S. 2,992,233 (Carbide).
53. F. Fekete, U.S. 2,975,156 (PPG).
54. N. C. McArthur, U.S. 3,311,580 and 3,311,674 (Hercules).
55. Brit. 733,996 (Degussa).
56. W. Kern and G. Dall'Asta, Ger. 923,393 (Degussa); cf. U.S. 2,902,476 (Hercules); Brit. 757,573 (Degussa).
57. Ger. 895,529 (Degussa); U.S. 3,247,282, Ber. 1,099,163 and 1,159,639 (all to Albert).
58. R. Zimmerman and F. Reiners, Kunst. <u>56</u>, 395-402 (1966).
59. T. Ouchi et al, CA <u>69</u>, 97225 (1968).
60. T. Ouchi and M. Oiwa, CA <u>73</u>, 120944 (1970).
61. Belg. 634,626 (Hoechst); CA <u>61</u>, 3120 (1964).

62.   H. Hibbert and M. S. Whelen, J. Am. Chem. Soc. 51, 3115 (1929);
      cf. J. U. Nef, Ann. 335, 224 (1904).
63.   E. L. Kropa and W. M. Thomas, U.S. 2,522,680 and 2,578,861 (Am.
      Cyan.); cf. U.S. 2,500,155 (R & H); U.S. 2,418,297 (Shell).
64.   R. Zimmerman and H. Hotze, U.S. 3,471,430 (Albert).
65.   R. H. Wiley, U.S. 2,432,601 (Du Pont).
66.   R. F. Fischer and C. W. Smith, U.S. 2,888,492 (Shell); J. Org.
      Chem. 25, 319 (1960); cf. Brit. 717,418 (Dynamit); U. Faas and
      H. Hilgert, Chem. Ber. 87, 1343 (1954).
67.   Brit. 948,084 (Wacker); CA 60, 13253 (1964).
68.   K. Tada et al, Makromol. Chem. 95, 168 (1966); J. Jedlinski et
      al, J. Polym. Sci. A1, 6, 1382 (1968).
69.   T. Kagiya et al, J. Polym. Sci. A1, 5, 2351 (1967).
70.   J. Jedlinski et al, Makromol. Chem. 114, 226 (1968); cf.
      N. Yamashita et al, CA 76, 60127 (1972).
71.   Acrolein, C. W. Smith, Ed., Wiley, 1962.
72.   R. F. Fischer, U.S. 2,895,962 (Shell).
73.   J. D. Nordstrom, J. Polym. Sci. A1, 7, 1349 (1969).
74.   H. Sumitomo et al, J. Polym. Sci. A1, 9, 3115 (1971).
75.   W. Wilson and H. May, Brit. 1,022,562 and 1,034,282 (Brit. Ind.
      Plastics); cf. CA 65, 13885 (1966).
76.   N. I. Shuikin et al, CA 62, 3998 (1965).
77.   C. K. Ikeda et al, J. Org. Chem. 29, 286 (1964).
78.   R. F. Fischer and C. W. Smith, J. Org. Chem. 25, 319 (1960);
      U.S. 2,987,524 (Shell); cf. U.S. 3,250,788 (Carbide).
79.   C. K. Ikeda, U.S. 3,010,918, 3,010,923-4, 3,190,878, 3,197,484,
      3,245,927 and 3,301,693 (Du Pont).
80.   C. J. A. Peters, U.S. 3,230,188 and 3,232,894; cf. C. K. Ikeda,
      U.S. 3,010,945, 3,190,878 and 3,373,160; R. A. Braun, U.S.
      3,258,439 (all to Du Pont).
81.   S. Hochberg, J. Oil and Colour Chem. Assoc. 48, 1043-1068
      (November 1965).
82.   C. K. Ikeda, U.S. 3,058,933; J. G. McNally, U.S. 3,242,144 and
      Brit. 1,005,205 (Du Pont).
83.   H. A. Stansbury et al, U.S. 2,862,007 (Carbide); cf. S. W.
      Tinsley, Belg. 632,164 (Carbide).
84.   R. A. Braun, U.S. 3,133,087; cf. U.S. 3,090,790 and 3,157,525
      (Du Pont).
85.   H. F. Reinhardt, U.S. 3,441,423 (Du Pont).
86.   H. O. L. Fischer et al, Ber. 63B, 1732 (1930); C. C. Price et
      al, J. Am. Chem. Soc. 72, 5335 (1950).
87.   M. R. Radcliffe and W. G. Mayes, U.S. 2,445,733 (Firestone);
      cf. A. Omori et al (Kyoto), CA 77, 5821 (1972).
88.   W. O. Kenyon and T. F. Murray, U.S. 2,382,640 (Eastman).
89.   W. O. Kenyon and T. F. Murray, U.S. 2,415,638 (Eastman).
90.   H. Orth, Angew. Chem. 64, 544 (1952); Ger. 926,937 (Nobel).
91.   C. E. Schildknecht, Vinyl and Related Polymers, Wiley, 1952.
92.   M. Goodman and A. Abe, J. Polym. Sci. A2, 8, 3471 (1964).
93.   H. J. Dietrich, J. Polym. Sci. A1, 6, 2255 (1968).

16. ALLYL ARYL ETHERS

The allyl aryl ethers have received much attention because of the famous Claisen rearrangement to allylphenols. Detailed studies of polymerizations have not been made, but some scattered observations of the formation of polymers, usually of low molecular weight, have been recorded. References on allyl aryl ethers and allylphenols until 1948 have been reviewed (1). Allyl aryl ethers can be made by reaction of allyl bromide, phenol, and alkali (1). Most of these compounds are pleasant-smelling liquids which can be distilled at atmospheric pressure without polymerization.

In relatively polar solvents such as acetone, allyl halides react with sodium phenoxide or derivatives to give allyl phenyl ethers, whereas in less polar media such as benzene, o-allylphenols may form, even at temperatures below those normally giving Claisen rearrangement. Allyl bromide reacts more readily with phenols and alkali than does allyl chloride. Rates of etherification of phenols by allyl chloride can be promoted by adding some sodium iodide (2). Claisen refluxed phenols in acetone with allyl bromide and anhydrous potassium carbonate for several hours to obtain allyl aryl ethers. Aqueous acetone with sodium hydroxide and allyl bromide also may be used (3).

Allyl phenyl ether and 2-allyl phenol both have pungent odors, although much milder than allyl alcohol. Properties of some of these compounds are given in Tables 16.1 and 16.2. An infrared spectrum of allyl phenyl ether is shown in Fig. 17.1a.

Claisen and co-workers discovered that the allyl groups of many allyl aryl ethers migrate, on heating in the range of 200°C, to the ortho positions, or if these are occupied, to the para position (4). Migration to the ortho group has long been believed to involve cyclic inversion of the allyl double bond (5). Arylamines have often been used as solvents for such Claisen rearrangements. A cyclic intermediate has been suggested as the first step in the mechanism following.

461

Migrations to the para position seem not to involve inversion, perhaps because of two successive cyclic stages. By successive etherifications followed by Claisen rearrangements, useful polyfunctional allylphenols and their derivatives can be synthesized. Vinyl aryl ethers do not undergo such rearrangement. The α,α-disubstituted allyl ethers also fail to rearrange (6).

Tarbell found the intramolecular rearrangements in diphenyl ether solution to be first order and little catalyzed by acids or bases. Electron-donating groups such as CH_3O on the ring promote rearrangements as in the transformation from the allyl ether of eugenol to diallyl guaiacol (7). However, ortho or para nitro or COOR groups (electron attracting) retarded rearrangement so that instead such allyl aryl ether derivatives have often polymerized or resinified on heating at 140°C or higher. Neither $C_6H_5OCH_2CH_2CH=CH_2$ nor $C_6H_5OCH=CH_2$ rearranged readily in the Claisen manner (8). Claisen found that heating in dimethyl- or diethylaniline under CO_2 facilitated rearrangements. Allyl ethers of naphthalene rearranged exceptionally fast (25 min at 174°C) (9), and allyl ethers of phenanthrols rearranged even at 100°C (10).

Reaction of guaiacol with allyl bromide and potassium carbonate in acetone yielded allyl o-methoxyphenyl ether (11).

Heating this ether in dimethylaniline led to rearrangement to 6-methoxy-2-allylphenol. Warming the allyl ether with boron fluoride-diacetic acid gave the same phenol, along with some 2-methoxy-4-allyl phenol. Lithium phenyl at 38°C split allyl o-methoxyphenyl ether to give guaiacol and catechol.

There has been much research on the effects of different substituents, solvents, and reaction conditions on Claisen rearrangements. White and co-workers were able to correlate rates of rearrangement of different p- and m-substituted phenyl ethers in Carbitol near 180°C to Hammett's sigma values (12). See Fig. 16.1. In general, electron-

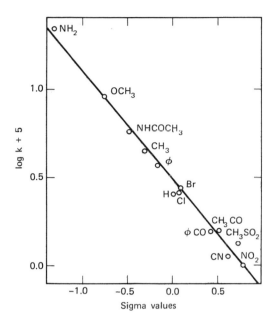

Fig. 16.1. Hammett sigma values are plotted against rate constants for Claisen rearrangement at 181°C of p-substituted allyl phenyl ethers in Carbitol (ethyl ether of diethylene glycol). W. N. White, D. Gwynn, R. Schlitt, C. Girard and W. Fife, J. Am. Chem. Soc. 80, 3275 (1958).

attracting groups retarded rearrangement. The series of p-substituted phenyl ethers was prepared by refluxing for 8 hr a mixture of 0.2 mole of substituted phenol, 0.23 mole allyl bromide, 0.20 mole dry potassium carbonate, and 50 ml of acetone. Water was added and then the solution was extracted with ether. The ether solution was washed twice by 10% aqueous NaOH and distilled. Methallyl phenyl ethers were prepared by a similar process and were found to have similar rates of rearrangement (13). White and co-workers studied the rearrangement of a deuterated allyl p-tolyl ether in Carbitol by a spectrophotometric method and concluded that the breaking of the ether bond was the slower rate-determining step (14). The following rate constants were observed for rearrangement of allyl m-substituted phenyl ethers to 2-allyl-5-substituted phenols in Carbitol at 181°C, based on changes in UV absorption (15):

m-Substituent	Rate constant	m-Substituent	Rate constant
CH_3O	82	F	33
CH_3	50	H	25
Br	45	CN	21
Cl	39	CF_3	16

Rearrangement of allyl-3-methyl phenyl ether gave the following mixture, the more activated ortho position being favored (16):

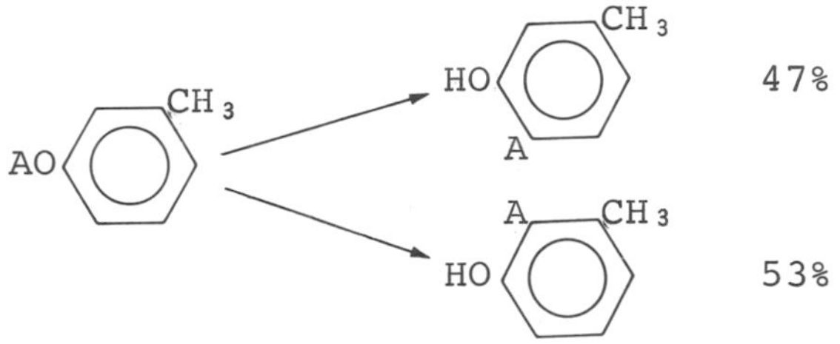

Lewis acids catalyzed rearrangement of allyl 2,6-dichlorophenyl ether with simultaneous rearrangement of halogen (17). The thermal uncatalyzed rearrangements occurred better in polar solvents. Water, phenol, fluorinated alcohols, and use of high pressures favored rearrangement to C-allyl aryl compounds even at moderate temperatures (18). With strong Lewis acids such as BF_3 allyl phenyl ether can be rearranged even at -78°C (19).

The rearrangement of allyl-2,6-dimethylphenyl ether at 168°C was found to be reversible by ^{14}C tagging (20). Even though both ortho positions in allyl aryl ethers are free, steric factors may force migration to a para-position (21). Methallyl phenyl ether derivatives were heated for 16 hr in N,N-diethylaniline in these studies. Novel Claisen rearrangements were carried out at 20°C or below in chlorobenzene with BCl_3 and borate catalyst mixtures (22). An out-of-ring migration of an allyl group of allyl aryl ether derivative to a p-propenyl side chain was reported (23). Allyl ethers of some phenyl benzyl glyoxals were rearranged on heating to form C-allyl glyoxals (24). Electron-donating methoxy groups in ortho and in para positions of the benzyl nucleus promoted rearrangement, but electron-withdrawing nitro groups retarded rearrangement. Allyl 2,4,6-trimethyl phenyl ether could be cleaved to give allyl bromide by heating with HBr in acetic acid at 90°C (25).

Allyl phenyl ethers, on warming with potassium t-butoxide, or with alkali hydroxides at higher temperatures, can rearrange to propenyl phenyl ethers (26). Heating allyl phenyl ether with magnesium bromide resulted in phenol and allyl bromide (27). Allyl phenyl ether was isomerized

to propiophenone with n-butyllithium in 1,2-dimethoxy-
ethane (4 hr at room temperature)(28). However, in
pentane-diethyl ether the main product was vinyl phenyl
carbinol.
  It is doubtful that simple allyl aryl ethers have been
polymerized without rearrangement to give polymers of
high molecular weight. Hui and Yip reported that polymerization at 50°C with BF$_3$ etherate involved rearrangement giving polymers of o-allylphenol (29). With C$_6$H$_5$MgBr
or n-BuMgI in ether at room temperature, allyl phenyl
ether gave small yields of gummy to brittle polymers (30).
Allyl ethers of bisphenol A and its derivatives have been
suggested for copolymerization with styrene by heating
with peroxide catalysts (31). Coating compositions have
been based on mixtures of 1-allyloxy-2,4,6-tris(hydroxymethyl) benzene, polyethers, and a little phosphoric acid
catalyst dissolved in toluene-cellosolve acetate (32).
Long exposure of 2,6-dimethylphenyl allyl ether to UV
light gave a polymeric material along with a phenol (33).
  Methallyl phenyl ether was prepared in 70% yield by refluxing for 22 hr a mixture of 10 moles of phenol, 11
moles of potassium carbonate, 11 moles of methallyl chloride, and 1250 ml of acetone (34). The rearrangement to
methallylphenol was studied and the latter was hydrogenated to o-isobutylphenol, a compound not easily made
by alkylation (35). Methallyl ethers of substituted
phenols have been prepared (36). The Claisen rearrangement of methallyl phenyl ether was studied in different
solvents at 200 to 225°C (37). N,N-Diethylaniline gave
best yields of o-methallylphenol not only by preventing
polymerization, but also by minimizing further reaction
of the phenol to isobutenyl isomer and to a dihydrobenzo-
furan. The latter predominated when an acidic solvent
2,6-xylenol was used. Alpha, omega-dimethylallyl phenyl
ether largely gave ortho-Claisen rearrangement on heating
(38). Compounds of the type C$_6$H$_5$OCH(CH$_3$)CH=CH$_2$ have been
studied (39). References to other allyl aryl ethers and
their rearrangements are given in Table 16.1.
  Allyl dinitrophenyl ether has been suggested as a
plasticizer for polyvinyl chloride in rocket propellant
compositions (40). Picryl allyl ether, which melts at
86°C, has been evaluated as an explosive (41). Alkyl
allyloxybenzene sulfonates have been copolymerized with
acrylonitrile for fibers of good receptivity to basic
dyes (42).
  Among allyl ethers of other aryl derivatives which have
been subjects of rearrangement studies have been those of
hydroxybenzoates (43), of pyridines (44), and pyrimidines
(45). A number of allyl ethers of tri-, tetra-, and
pentachlorobenzenes were prepared and their rearrangements

TABLE 16.1

Allyl Aryl Ethers

	bp, °C	Refractive index, etc.	References
Allyl ethers			
Phenyl	94(29)	$n_D^{25}=1.5210$	White, JACS <u>80</u>, 3272 (1958)
	195	$D_{15}^{15}=0.9856$	Pfaltz & Bauer Co.
o-Tolyl	82(12)	$n_D^{20}=1.5179$	Aldrich Chemical Co;
	192	mp = 30°C	Pfaltz & Bauer Co.
m-Tolyl	65(4)	--	White, J. Org. Chem. <u>26</u>, 3631 (1961)
p-Tolyl	92(10)	$n_D^{25}=1.5168$	Aldrich Chemical; Pfaltz
	202	mp = 34°C	& Bauer
p-Methylbenzyl	108(14)	$n_D^{25}=1.5067$	Okawara, CA <u>57</u>, 4853
o-Methoxyphenyl	90(7)	--	Werner, Fr. 1,344,997; CA <u>60</u>, 14475
m-Methoxyphenyl	73(1)	--	White, J. Org. Chem.
p-Methoxyphenyl	117(11)	$n_D^{25}=1.5265$	White, JACS
o-Glycidylphenyl	96(0.74)	$n_D^{30}=1.5282$	Aelony, J. Org. Chem. <u>27</u>, 3311 (1962)
o-Aminophenyl	85(0.6)	$n_D^{25}=1.5628$	Tiffany, JACS <u>70</u>, 592 (1948)
p-Aminophenyl	112(2)	--	White, JACS
m-Benzyl	164(1)	--	White, J. Org. Chem.
p-Phenylphenyl	--	mp = 85°C	White, JACS
m-Chlorophenyl	80(2.5)	--	White, J. Org. Chem.
p-Chlorophenyl	110(15)	--	White, JACS
2,4-Dichlorophenyl	99(2)	$n_D^{25}=1.5522$	Tarbell, JACS <u>64</u>, 602 (1942)
p-Bromophenyl	111(7)	--	White, JACS
m-Bromophenyl	105(5)	--	White, J. Org. Chem.
2,4-Dibromophenyl	134(0.5)	$n_D^{20}=1.5988$	Hurd, JACS <u>58</u>, 941 (1936)
Pentabromophenyl	--	--	Sims, U.S. 3,507,816 (Monsanto)
p-Cyanophenyl	--	mp = 48°C	White, JACS
m-Cyanophenyl	129(6)	--	White, J. Org. Chem.
2,4,6-Tris(hydroxy-methyl)phenyl in resin compositions	--	--	Phillips, U.S. 2,890,195 (Carbide); Somerville, U.S. 2,894,931 (Shell); Masters, U.S. 2,917,481 (Am. Cyan.)
p-Nitrophenyl	126(2)	--	White, JACS
2,4-Dinitrophenyl	--	mp = 44°C	Verkade, Rec. Trav. Chim. <u>65</u>, 346 (1946)
Pentachlorophenyl	--	mp = 108°C	Rocklin, U.S. 2,949,488 (Dow)

TABLE 16.1 (continued)

2-Allyloxy-naphthalene	$60(10^{-2})$	$n_D^{21} = 1.6078$	Marcinkiewicz, Tetrahedron, 14, 208 (1961)
1-Allyloxy-naphthalene	$60(10^{-2})$	$n_D^{23} = 1.6055$	Marcinkiewicz
9-Allyloxy-phenanthrene	--	mp = 139°C	Marcinkiewicz
6-Allylkojic acid and allyl kojate	--	mp = 125°C	McLamore, JACS 78, 2816
	--	mp = 68°C	(1956)
Methallyl phenyl ether	54(2)	$n_D^{25} = 1.5148$	White, JACS 83, 3265 (1961); Schales, Ber. 70, 119 (1937)
Methallyl p-methoxy-phenyl ether	90(2)	$n_D^{25} = 1.5210$	White, JACS 83, 3265 (1961)
Methallyl 4-nitro-phenyl ether	137(1)	mp = 40°C	U.S. 2,233,080 (R & H)
3-Methallyloxy-phenol	128(3)	$n_D^{20} = 1.5451$	Bartz, JACS 57, 371 (1935)
2-Allyloxyphenol	104(8)	$n_D^{21} = 1.5408$	Hurd, J. Org. Chem. 2, 381 (1937)
3-Allyloxy-6-nitrophenol	158(10)	--	Baker, J. Chem. Soc. 274 (1936)
4-Allyloxy-2-nitrophenol	--	mp = 48°C	Baker
4-Allyloxy-2,3,6-trimethyl phenol	--	mp = 84°C	Werder, Z. Physiol. Chem. 257, 129 (1939)

were studied, e.g., in tetralin at 210°C (46). There was some loss of halogen in rearrangement. For example, 2,4,6-tribromophenyl allyl ether gave largely 4,6-dibromo-2-allyl phenol. Bruson copolymerized allyl ethers of chlorinated phenols with maleic anhydride for use as resinous fungicides (47). Claisen rearrangements of allyl ethers of heterocyclic compounds with few exceptions seem to occur in the normal fashion (48). Thus 2-allyloxy-pyridine gave expected products in about equal proportions (49). Impurities promoted polymerization.

POLYFUNCTIONAL ALLYL ARYL ETHERS

The literature on Claisen rearrangements contains some observations of dark tarry masses and "resinification" of polyfunctional aryl ethers. Details of polymerization behavior and polymer structure were not studied. For example, o- and p-nitrophenyl allyl ethers polymerized on heating to 180°C, as did 3-allylsalicylic acid (49). The allyl ether of o-allylcresol did not rearrange well for Claisen but gave high boiling products. When an allyl o,o-diallylphenyl ether was heated under CO_2 at 250°C,

rapid spontaneous boiling occurred with temperature rise
to 290°C. About half of the liquid distilled off and the
remainder was a bubbled polymer mass. Claisen carried
out this rearrangement successfully at 248°C for half an
hour in diethylaniline, a known inhibitor for radical
polymerization. Note that partial rearrangement on heating allyl aryl ethers gives phenols which are inhibitors
of radical polymerization.

The triallyl ester compounds 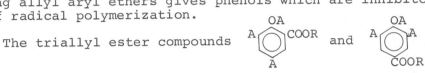 and

did not rearrange when heated by Claisen but gave polymer
masses (resinified). However, if saponified, the free
acids rearranged to 2,4,6-triallyl phenol with decarboxylation. When the monoallyl ether of resorcinol was heated
at 235°C, exothermic polymerization occurred to form an
amber resin (50). The polymer melted near 100°C and was
soluble in acetone and in 10% aqueous NaOH solution.

Some exploratory polymerizations of allyl ethers of
allyl-substituted phenols were made by Butler and Ingley
(51). Such allyl and methallyl phenyl ethers gave evidence of polymerization under the action of boron
fluoride-diethyl ether complex. Heating with benzoyl or
other peroxide initiators surprisingly gave no polymerization nor copolymerization with styrene or diallyl
maleate, even when two or three allyl groups were present
in the ether. Adding 10% of BF_3 etherate dropwise at
room temperature gave dark, brittle, soluble low polymers
from methallyl 2-allylphenyl ether and methallyl 2,6-
diallylphenyl ether. The polymers melted below 80°C.
Only viscous liquid polymers were obtained from chloroallyl 2-allylphenyl ether and from methallyl 2,6-diallyl-
phenyl ether. Butler was interested in crosslinking the
polymers through reaction of the allyl groups attached to
the ring, but he found them very reluctant to become
thermosetting on heating with peroxides. Partial Claisen
rearrangement to polymerization-inhibiting phenol may
have occurred (52).

Polyfunctional allyl chloroaryl ethers such as bis-
(allyloxy)octachlorodiphenyl were suggested for making
fire-resistant plastics (53). A diallyl ether of tetra-
chloroxylene was heated with 2% t-butyl perbenzoate for
an hour at 138°C, forming a soluble prepolymer (54).
Copolymers were also prepared by free radical initiation.

There has been considerable interest in preparing
phenolic and other polymers containing many reactive allyl
ether groups. In Japan the Claisen rearrangement of polyfunctional allyl ethers of novolac phenol-formaldehyde
resin was studied by infrared (55). The reaction was

TABLE 16.2

Some Polyfunctional Allyl Aromatic Ethers

Ether	References, etc.
2-Allylphenyl allyl ether	Weizmann, J. Chem. Soc. Ind. 67, 203 (1948); bp 114°C at 19 mm
2,6-Diallylphenyl allyl ether	Claisen, Ann. 418, 69, 120
2-Allyl-6-methoxyphenyl allyl ether	Claisen, Ann. 401, 26, 119 (1913)
2-Allyloxyphenyl allyl ether	Hurd, JACS 52, 1700 (1930)
2-Allyloxy-3-allylphenyl allyl ether	Hurd
1,3-Dimethallyloxybenzene	Bartz, JACS 59, 371 (1935); bp 147°C at 6 mm
Diallyl ethers from chlorinated xylenes	Stallings, Preprints ACS Div. Org. Coatings and Plastics Chem. 29, 356 (September 1969)

first order and had an energy of activation of about
25 kcal/mole. The rearrangement of the allyl novolac
ether occurred faster than that of allyl phenyl ether.
Under the experimental conditions rearrangement to propenyl groups did not occur. References to other polyfunctional allyl aryl ethers are given in Table 16.2.

Many of the allylphenyl ether compounds that have been
studied are not allyl ethers. For example, $A\langle\bigcirc\rangle OCH_2CHCH_2$
with an epoxide O bridging the CHCH$_2$,
was said to form linear soluble polymers containing epoxy
groups by heating for 18 hr at 60°C with BP catalyst under
nitrogen (56).

The diallyl ether of bisphenol A, $AO\langle\bigcirc\rangle\overset{CH_3}{\underset{CH_3}{C}}\langle\bigcirc\rangle OA$, can be
made by heating ACl, concentrated NaOH, the phenol, and
$Na_2S_2O_4$ at 100°C for an hour (57). This diallyl ether has
had interest as an insecticide (58) and as an intermediate
for polymers (59). Diallyl ethers of hydroxy-methylated
bisphenol A have been evaluated as resin intermediates
(60). Bifunctional siloxanes can add to the double bonds
of the diallyl ether of bisphenol A to form symmetrical
linear copolymers (61).

References

1. Allyl Chloride book, Shell Chemical Corp., New York and San
   Francisco, 1949; D. S. Tarbell, Chem. Rev. 27, 495 (1940);
   Organic Reactions, Vol. II, Wiley, 1944; cf. Tarbell and S. J.
   Rhoads in Molecular Rearrangements, Vol. I, E. P. DeMayo, Ed.,
   Wiley, 1963.

2. L. I. Smith et al, J. Am. Chem. Soc. 62, 1863 (1940).
3. C. D. Hurd and W. A. Yarnall, J. Am. Chem. Soc. 59, 1686 (1937).
4. L. Claisen, Ber. 45, 3157 (1912), CA 7, 1016 (1913); Angew.
    Chem. 36, 478 (1923); CA 18, 234 (1924); Ann. 449, 81 (1926);
    CA 21, 71 (1927).
5. L. Claisen and E. Tietze, Ber. 58, 275 (1925); cf. C. D. Hurd
    and M. A. Pollack, J. Org. Chem. 3, 550 (1939); cf. D. L.
    Dalrymple et al, in The Chemistry of the Ether Linkage, S. Patai,
    Ed., Interscience, 1967.
6. L. Claisen et al, J. Prakt. Chem. 105, 67 (1922); C. D. Hurd and
    F. L. Cohen, J. Am. Chem. Soc. 53, 1919 (1931).
7. L. Claisen et al, Ann. 401, 21, 46 (1913).
8. S. G. Powell and R. Adams, J. Am. Chem. Soc. 42, 646 (1920).
9. J. F. Kincaid, in D. S. Tarbell, Ref. 1.
10. L. F. Fieser and M. W. Young, J. Am. Chem. Soc. 53, 4120 (1931).
11. C. F. H. Allen and J. W. Gates, Org. Syn. Coll. Vol. III, 418
    (1955).
12. W. N. White et al, J. Am. Chem. Soc. 80, 3271 (1958); J. Org.
    Chem. 26, 3631 (1961); cf. H. Schmid, Helv. Chim. Acta 20, 13
    (1957); H. L. Goering and R. R. Jacobson, J. Am. Chem. Soc. 80,
    3277 (1958).
13. W. N. White and B. E. Norcross, J. Am. Chem. Soc. 83, 3265 (1961).
14. W. N. White and E. F. Wolfarth, J. Org. Chem. 26, 3509 (1961).
15. W. N. White and C. D. Slater, J. Org. Chem. 27, 2908 (1962).
16. D. S. Tarbell and S. S. Stradling, J. Org. Chem. 27, 2724 (1962).
17. E. Piers and R. K. Brown, Can. J. Chem. 41, 2917 (1963).
18. W. J. Noble, J. Am. Chem. Soc. 85, 1470 (1963).
19. W. Gerrard et al, Proc. Chem. Soc. 19 (1957); cf. P. Fahrni et
    al., Helv. Chim. Acta. 43, 448 (1960).
20. H. Schmid et al, Helv. Chim. Acta 39, 555 (1956); CA 50, 13898
    (1956).
21. J. Borgulya et al, Helv. Chim. Acta 46, 2444 (1963); CA 60,
    2808 (1964).
22. H. Schmid, Gazz. Chim. Ital. 92, 968 (1962); CA 58, 8882 (1963).
23. A. Nickon and B. R. Aaronoff, J. Org. Chem. 27, 3379 (1962).
24. R. D. Barnes and F. E. Chigbo, J. Org. Chem. 28, 1644 (1963).
25. J. Segers et al, Helv. Chim. Acta 41, 1198 (1958); CA 56, 4706
    (1962).
26. W. H. Snyder, Diss. Abstr. 22, 1020 (1961).
27. A. Schoenberg and R. Moubasher, J. Chem. Soc. 462 (1944).
28. D. R. Dimmel and S. B. Gharpure, J. Am. Chem. Soc. 93, 3991 (1971).
29. K. M. Hui and L. C. Yip, CA 75, 141212 (1971).
30. J. W. Hafey and C. E. Schildknecht, unpublished.
31. E. C. Shokal and R. W. H. Tess, U.S. 2,864,804 (Shell).
32. H. W. Howard et al, U.S. 2,774,748 (Shell); cf. U.S. 2,843,503
    (Eastman).
33. K. and H. Schmid, Helv. Chim. Acta 36, 687 (1953).
34. W. T. Olson et al, J. Am. Chem. Soc. 69, 2452 (1947).
35. Q. R. Bartz et al, J. Am. Chem. Soc. 57, 371 (1935).
36. W. Bradley et al, J. Chem. Soc. 2877 (1951).

37. A. T. Shulgin and A. W. Baker, J. Org. Chem. *28*, 2468 (1963).
38. H. L. Goering and W. I. Kimoto, J. Am. Chem. Soc. *87*, 1748 (1965).
39. H. L. Goering and R. R. Jacobson, J. Am. Chem. Soc. *80*, 3277 (1958).
40. J. R. Eiszner, U.S. 2,973,255 (Std. Oil Ind.).
41. A. H. Platt and A. W. Rytina, J. Am. Chem. Soc. *72*, 3274 (1950).
42. J. C. Masson, U.S. 3,426,104 (Monsanto).
43. W. M. Lauer and R. M. Leekley, J. Am. Chem. Soc. *61*, 3043 (1939).
44. F. J. Dinan and H. Tieckelman, J. Org. Chem. *29*, 892 (1964).
45. F. J. Dinan et al, J. Org. Chem. *28*, 1015 (1963).
46. L. C. Felton and C. B. McLaughlin, J. Org. Chem. *17*, 298 (1947).
47. H. A. Bruson, U.S. 2,497,927 (R & H).
48. F. J. Dinan and H. Tieckelman, J. Org. Chem. *29*, 892 (1964).
49. L. Claisen and O. Eisleb, Ann. *401*, 21 (1913); *418*, 74 and 96 (1919).
50. C. D. Hurd et al, J. Am. Chem. Soc. *52*, 1700 (1930).
51. G. B. Butler and F. L. Ingley, J. Am. Chem. Soc. *73*, 1512 (1951).
52. G. B. Butler, private communication.
53. J. Vuillemenot et al, U.S. 3,282,882 (Soc. Electro-Chimie).
54. J. P. Stallings, J. Polym. Sci. A 1, *8*, 1557 (1970).
55. M. Morinaga et al, Chem. High Polym. (Tokyo) *22*, 615-18 (1965).
56. G. F. D'Alelio, U.S. 2,983,703 (Koppers); cf. U.S. 2,851,440.
57. R. M. Christenson and W. C. Bean, U.S. 2,910,455 (PPG); cf. G. E. Ham, U.S. 3,060,243 (Dow).
58. A. G. Jelinek, U.S. 2,560,350 (Du Pont).
59. J. B. Rust and W. B. Canfield, U.S. 2,708,662 (Ellis-Foster); P. Bruin and J. Selman, U.S. 3,024,285 (Shell).
60. P. Bruin, CA *52*, 1680 (1958).
61. H. A. Clark, U.S. 3,176,034 (Dow).

17. ALLYL PHENOLS AND RELATED

Allylic phenols are well known because of their formation by Claisen rearrangement of allyl aryl ethers and
because a number of naturally-occurring allylphenols and
their derivatives, such as eugenol, are used widely in
flavors, perfumes, and medicinal products. Some of their
odors have fine, spicy, pungence in contrast to the unpleasant odor of unsubstituted phenol.

In view of the inhibiting effects of phenols upon free-
radical polymerizations, it is not surprising that allylic
phenols do not polymerize or copolymerize readily on
heating with peroxide or azo catalysts. A few copolymers
of interest have been reported. Monoallylic phenols also
have not given homopolymers of high molecular weight by
ionic mechanisms. Unstable, colored resins have been
obtained by reaction of allyl-substituted phenols with
formaldehyde, but these have not attained commercial
importance.

MONOALLYL PHENOLS

The preparation of allyl phenols by Claisen rearrangements has been discussed in Chapter 16 (See also Fig. 17.1.)
Phenols may be allylated directly by allyl chloride by use
of aqueous cuprous chloride catalyst at 35 to 75°C (1).
Vapor-phase allylation with allyl alcohol over zinc
chloride-alumina gave allyl phenols along with corresponding coumarans (2). Treating phenols or cresols with NaOH
and allyl chloride at 110°C in toluene solution for 12 to
30 hr introduced two or more allyl groups into the ring,
along with formation of some allyl aryl ether (3). Use of
polar solvents favored formation of allyl aryl ethers
instead of polyallyl phenols. One-step polyallylation and
polymethallylation processes of Aelony were disclosed (4).
Allyl ethers of glycidyl phenols also were prepared (5).
Meta-allylphenol has been made by heating allyl m-bromo-
phenyl ether with magnesium in ether for 8 hr and then
treating with acetic acid (6). The infrared spectrum of
o-allylphenol is shown in Fig. 17.1 b.

Low polymers of allyl phenols (below DP 10) have been suggested as antioxidants for diene copolymer rubbers (7). Boiling 2-allyl phenyl ether for 5 hr has given as one product a solid trimer of 2-allylphenol (8). Ortho-allyl phenol in hexane at 0 to 27°C gave violet colors immediately on treatment with boron fluoride etherate and slowly formed sticky brown solid polymers (9). Ortho-allyl phenol as well as 2-allylanisole were reacted with SO_2 in the presence of ascaridole catalyst to give soluble polysulfones melting above 100°C (10). Small amounts of o-allylphenol were copolymerized with ethylene and propylene, using a catalyst from reaction of diethyl aluminum chloride, vanadium tetrachloride, and perchloropropylene (11). The elastomers obtained were expected to be internally stabilized against oxidation.

When heated with strong alkali, allyl phenols may rearrange to propenyl phenols (12). The o-propenyl phenols, like o-allyl phenols, give coumarans under the influence of acid catalysts (13). Heating o-methallylphenol with KOH at 140°C gave methylpropenylphenol (14).

Isopropenylphenol has been reported from reaction of allyl chloride with phenol in the presence of sulfuric acid (15). Isopropenyl phenols, which are derivatives of alpha-methylstyrene, are not discussed in this book. However, like monoallyl compounds, they do not polymerize readily and they inhibit free radical polymerizations.

Allyl phenols and allyl ethers of phenols have been suggested as additives to fuel oils (16). Attempts to prepare unsaturated phenolic resins from o-allylphenol and formaldehyde with NaOH as catalyst have given only viscous oils (17). Emulsions of the product were proposed as adhesives for bonding rayon tire cord to rubber. Ortho-allylphenol also has been reacted with NaOH and HCHO for 72 hr at 42°C (18). Novolacs prepared from reaction of o-allylphenol and formaldehyde with HCl catalyst turned red in color in a few hours (19).

The 2,6-dimethallylphenol can be prepared by thermal rearrangement of 2-methallylphenyl methallyl ether in the presence of anhydrous sodium carbonate (20). The methyl group in o-methallylphenol increased the strength of hydrogen bonding compared with o-allylphenol, as indicated by infrared hydroxyl bands. The bonded OH band (longer wavelength component) was stronger than that of the unbonded.

Ortho-allylphenol in the presence of acid catalysts can undergo ring closure to form 2-methylcoumaran; such compounds are often observed as byproducts in Claisen rearrangements (21). Allyl chlorophenols have been suggested as bactericides and fungicides (22). Other allyl phenols are listed in Table 17.1.

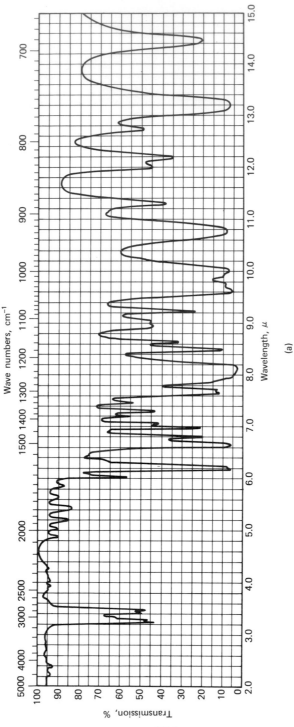

Wave numbers, cm⁻¹

Transmission, %

Wavelength, μ

(a)

Fig. 17.1(a). Infrared spectrum of a sample of allyl phenyl ether using cell thickness of 0.01 mm. Courtesy of Sadtler Research Laboratories. CH bands occur at 3.3 to 3.5 microns and ethylenic double bond absorptions at 6.0 to 6.3 microns. Bands at 12 to 13.3 microns are characteristic of the substituted benzene ring.

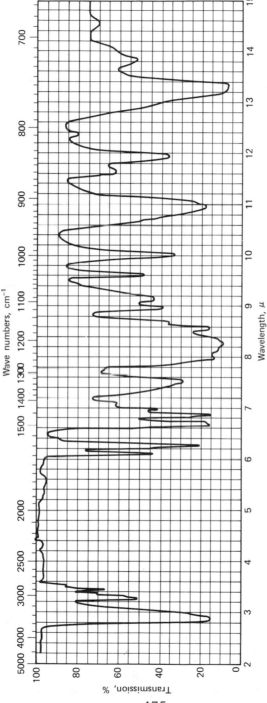

Fig. 17.1(b). Infrared spectrum of o-allylphenol (cell thickness 0.03 mm) obtained by Claisen rearrangement of allyl phenyl ether. Courtesy of Sadtler Research Laboratories. Among the characteristic bands are those of the OH group at 2.9 microns and the separation of the bands in the region of 7 to 10 microns.

Table 17.1

Some Allyl Phenols

Phenol	bp, °C	Refractive index, etc.	References
o-Allyl	103(14) or 220	n_D^{24}=1.5443	Tarbell, JACS 64, 607 (1942)
m-Allyl	115(13)	--	Luettringhaus, Ann. 597, 46 (1947)
2-Methyl-6-allyl	122(26)	--	Hurd, J. Org. Chem. 5, 212 (1940)
2-Allyl-4-methyl	--	n_D^{20}=1.5385	Aldrich Chemical
2-Allyl-4-t-butyl	132(15)	n_D^{20}=1.5245	Belg. 615,700 (Shell)
2-Allyl-4-methyl	118(14)	n_D^{24}=1.535	Hurd, JACS 59, 1686 (1937)
2-Allyl-6-methyl	--	n_D^{20}=1.5381	Aldrich Chemical
o-Amino-p-allyl	--	mp = 123°C	Tiffany, JACS 70, 592 (1948)
2-Allyl-6-methoxy-phenol (3-allylguaiacol)	115(9) 250	n_D^{20}=1.5393	Pfaltz & Bauer Co.
6-Chloro-2-allyl	220	n_D^{25}=1.5447	Tarbell, JACS 64, 1066 (1942)
2,6-Dichloro-4-allyl	108(3)	mp = 35°C	Tarbell
6-bromo-2-allyl	88(2)	n_D^{20}=1.5706	Hurd, JACS 58, 941 (1936)
o-Methallyl	103(11)	n_D^{20}=1.5534	Bartz, JACS 57, 371 (1935); Shulgin, J. Org. Chem. 28, 2468 (1963)
o-Methallyl-p-methoxy	105(1.5)	n_D^{25}=1.5535	White, JACS 83, 3265 (1961)
o-t-Butallyl-p-methoxy	90(1)	n_D^{25}=1.5423	White
o-Chlorallyl	134(12)	n_D^{20}=1.5778	Hurd, JACS 58, 2190 (1936)
1-Allyl-2-naphthol	--	mp = 55°C	Hurd, JACS 59, 107 (1937)
2-Allyl-1-naphthol	144(4)	--	Fieser, JACS 61, 2206 (1939)

Esters have been prepared from a number of allyl phenols, for example, by using ketene in the presence of H_2SO_4 (23). By reacting with liquid SO_2 in the presence of ascaridole as catalyst, 2-allylanisole gave a solid polysulfone of high molecular weight (24).

Most phenols deactivate catalysts of the Ziegler type. However, hindered phenols such as 4-but-3-enyl-2,6-di-t-butylphenol were copolymerized in small proportions with 1-olefins such as propylene (25). From 0.007 to 0.5% of

the sterically hindered phenolic groups entered into copolymerization.

Phenylene-oxide-type polymers (PPO) for molding have
been developed by General Electric from 2,6-dimethylphenol
and related compounds. Analogous polymers bearing allyl
groups have been prepared by oxidative coupling polymerization (26):

bp 61°C (1 mm)

The oxidative catalyst in one example consisted of $CuCl_2$,
pyridine, nitrobenzene, and oxygen. An hour was sufficient for polymerization. Curing the polymer through the
allyl groups was suggested by heating with peroxides in
order to obtain crosslinked moldings of superior heat
resistance. The presence of the allyl group adds to the
difficulties in preparing polymers of high enough molecular weight for plastics by the oxidative coupling process
(27).

NATURAL ALLYL PHENOLS AND THEIR ETHERS

Chavicol or p-allylphenol is a natural product from
betel leaves. It has been made by heating 4-allylanisole
with methyl magnesium iodide at 170°C (28), and also by
treating safrole with sodium in liquid ammonia and ethanol (29). Methylchavicol or estragole is p-allylanisole,
a compound that has received considerable study. It can
be made by Grignard coupling of 4-methoxyphenyl magnesium
bromide with allyl bromide (30), or by methylation of
4-allylphenol using dimethyl sulfate and alkali or methyl
iodide in methanolic KOH (31). The isomer of chavicol,
m-allylphenol, was made by heating allyl 3-bromophenyl
ether with magnesium in ether for 8 hr to form a Grignard
reagent, followed by treatment with acetic acid (32). A
number of methyl-substituted allylanisoles have been prepared (33).

Naturally occurring allylphenols and ethers of allylphenols are listed in Table 17.2. A number of these have
been synthesized by way of Claisen rearrangements (34).
Allyl phenols on treatment with alkali may rearrange to
propenyl phenols, for example

TABLE 17.2

Some Allyl Phenols and Their Ethers

Compound	bp, °C at mm	Refractive index, etc.	Occurrence	References, etc.
Chavicol	235	$n_D^{20}=1.5448$ mp = 16°C	Pimento-tarragon, betel, bay oil, scent glands of beavers, plum branches	Quelet, Bull. Soc. Chim. [4], 45, 265 (1929); Palkin, JACS 55, 1556 (1933)
Estragole or esdragol	114(25) or 215	$n_D^{18}=1.523$	Aniselike odor; in sweet goldenrod, basil, fennel, pine gums	Nelsov, Am. Perfumer 31, 69 (1935); Hasselstrom, JACS 60, 3086 (1938)
Eugenol	253	$n_D^{20}=1.541$ mp = -10°C	Sweet basil, oil of cloves, burley tobacco, hyacinth, rose	Meyer, CA 71, 128774 (1969); Gilson, J. Dent. Res. 48, 366 (1969) (dental)
4-Allyl-anisole	62(2)	$n_D^{20}=1.5232$	Pinewood, plant oils	Sanvordeker, CA 71, 128576 (1969)
2-Allyl-phenetole	71(2)	$n_D^{20}=1.5110$	Synthetic	Shamshurin, CA 44, 1443 (1950)
3-Allyl-anisole	216	$n_D^{20}=1.5245$	Synthetic	Martin, JACS 86, 233 (1964)

Name	Structure	b.p. (mm)	Physical constants	Source	Reference
4-Allyl-veratrole or "methyl eugenol"	A — OCH_3, OCH_3	123(14) or 248	$n_D^{20} = 1.533$	Anemopsis eremophilia; rose oil (spicy odor)	Mauthner, J. Prakt. Chem. [2] 148, 95 (1937); cf. U.S. 2,457,074 (Monsanto); (fruit fly attractant)
Safrole	A — O—CH_2, O	64(1) or 233	$n_D^{20} = 1.535$ mp = 10°C	Sassafras; camphor wood; parsley; nutmeg; cinnamon	Hagan, CA 62, 13704 (1965); Taylor 62, 5789 (1965); (toxicity)
Apiole	A — OCH_3, O—CH_2, O; CH_3O	292	$n_D^{20} = 1.537$ mp = 29°C	Parsley seed (with tetramethoxy-allylbenzene)	Stahl, CA 61, 2901 (1964); Rajkowski, CA 60, 13091 (1964)
Myristicin	A — O—CH_2, O; CH_3O	150(15)	Sp. gr. — 1.142	Nutmeg, parsley seed	Psychopharmacology; toxic
Elemicin	A — OCH_3, OCH_3, OCH_3; CH_3O	— —	Sp. gr. — 1.063	Backhousia myrtifolia; ziera rue	Nagasawa, CA 55, 13353 (1961)

479

Sodium polyanetholsulfonate finds medical use as an antico-
agulant and diagnostic reagent. Ortho-allylphenol is isomerized to o-propenylphenol by KOH in methanol at 110°C
during 6 hr (35). The o-allylphenol, when heated with
acids, can give 2-methyl-dihydrobenzofuran (36). Such
compounds also can be formed as byproducts in Claisen
rearrangements. Commercial isomerization of eugenol to
isoeugenol at 220°C is said to require two equivalents of
KOH; NaOH is less effective. Crotyl phenols rearranged
less readily than allyl phenols on heating with KOH (37).
With sulfuric acid or boron fluoride-etherate, anethole
gives orange to brown viscous liquid polymers with blue
fluorescence.

Methylcoumarans are encountered as byproducts in the
synthesis of allyl phenols, and they are obtained by
reaction of 2-allyl phenols with acids.

$$\text{o-allylphenol} \xrightarrow{\text{HBr}} \text{2-methyl-dihydrobenzofuran}$$

Allyl phenols and their acetates heated in the presence
of peroxides can form chromans (38).

Attempts to epoxidize o-allylphenol with peracetic acid
gave 2-hydroxymethyl coumaran (39).

Of the three isomeric allylanisoles or methyl ethers of
allylphenols, the para isomer known as estragole or methyl
chavicol occurs most widely in aromatic oils of plants.
The ortho isomer, as well as the corresponding ethoxy
compound, gave viscous polymeric oils on heating at 250°C
(40). The ortho isomer has formed polysulfones by reaction
with SO_2 and a peroxide (41). The meta-allylanisole has
been made by warming 3-methoxy-phenyl magnesium bromide
with allyl iodide or bromide (42).

Eugenol (o-methoxy-p-allylphenol)(43), comprising 85%
of natural oil of cloves, has important applications in
dentistry and medicine as an antiseptic and anesthetic;
it also is used in adhesives, in foods as a flavor, and
as an intermediate for synthesis of vanillin. It is much
less toxic than phenol. Eugenol darkens and becomes more
viscous on exposure to air and light. It does not homopolymerize readily by cationic or radical methods. The
oxidative polymerization of eugenol has been studied to a
limited extent (44). Bromine water near 0°C has polymerized eugenol to tars (45). Eugenol is miscible with
alcohol, ether, and chloroform but is only slightly

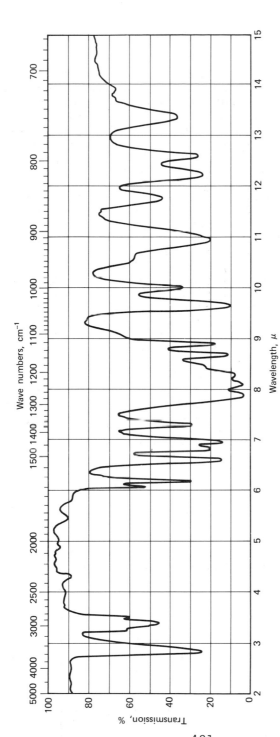

Fig. 17.2. Infrared spectrum of eugenol, a naturally occurring allyl phenol of remarkably diverse applications. Courtesy of Sadtler Research Laboratories. The following bands are similar but not identical to those of o-allylphenol in Fig. 17.1(b):   OH near 2.85 microns, CH at 3.3 to 3.5, ethylenic bands near 6.1 and 6.2, and multiple bands near 8 and 10 microns. The band associated with the monosubstituted benzene ring about 13.2 has moved nearly to 13.4 in this trisubstituted benzene compound.

soluble in water. Toxic effects of eugenol with rats were
low (46). With nitric acid in aqueous solution eugenol
gives orange to red colored products. Dental cements containing about 1 part of eugenol to 6 parts of zinc oxide
harden in half an hour or longer. A more recent formulation was made by adding 0.2 ml of a mixture of eugenol and
o-ethoxybenzoic acid (37.5 to 62.5 by weight) to 1.7 g of
prepared powder containing 30% alumina, 6% hydrogenated
rosin, and 64% zinc oxide (47). Eugenol acetate is used
in perfumes, and related derivatives have been evaluated
(48). An improved attractant for Japanese beetles containing eugenol and phenylethyl propionate was announced
by the U.S. Department of Agriculture in 1972. The
infrared spectrum of eugenol is shown in Fig. 17.2.
    Methyl chavicol has been observed to dimerize on treat-
ment with hot sulfuric acid (49). p-Allylanisole and
p-allylphenetole gave dimers and trimers on heating above
150°C (50). Safrole was used as a flavoring agent until
the discovery that it produced liver damage and other
harm in animals (51). This fine-smelling compound 1-allyl-
3,4-methylenedioxbenzene is toxic as a local irritant and
also by ingestion. Nevertheless, as the chief component
of oil of sassafras it was present in the famous sassafras
tea of the Pennsylvania Germans, an alleged cleanser of
the blood. Myristicin or 5-allyl-1-methoxy-2,3-(methylene-
dioxy) benzene has been observed to produce mild cerebral
stimulation in man and animals, but it was less active
than the nutmeg from which it was obtained (52).
    The allyl ether of thymol is a liquid (bp 122 at 17 mm;
n_D^{22} = 1.5131) (53). It can be rearranged to 2-allylthymol,
a liquid with a fruity odor (bp 139°C at 17 mm; n_D^{25} =
1.5240). On heating with pyridine hydrochloride, both
compounds gave 2,3-dihydrobenzofuran derivatives. An
allyl derivative of benzoyl resorcinol was added to polyvinyl chloride as a stabilizer against light (54). Some
allyl hydroxyaryl ketones and their copolymers were cited
in Chapter 7. Allyl-substituted phenols have been used
to prepare allyl-substituted phenylene oxide polymers
which crosslinked slowly in air (55).

POLYFUNCTIONAL ALLYL PHENOLS

    Since the polyfunctional allyl phenols contain phenolic
hydroxyl groups that retard free radical polymerizations,
it is not surprising that they do not polymerize readily
on heating or under conditions of formation by Claisen
rearrangement of allyl aryl ethers. The compounds formed
by rearrangement of ethers are largely ortho isomers; if
these positions are already occupied, however, para isomers
form; for example (56):

(in diethylaniline)   2,4,6-triallylphenol

This triallylphenol also has been made by heating 2-allyl-
oxy-3,5-diallyl benzoic acid (57). Strong alkali may
rearrange allyl phenols to propenyl phenols. Both compounds may form coumarans by treatment with acid catalysts.
Isopropenyl phenols were formed by reaction of allyl
chloride with phenol or cresol in the presence of sulfuric
acid (58). References to polyfunctional allyl phenols are
given in Table 17.3.

TABLE 17.3

Some Polyfunctional Allyl Phenols

Compound	bp, °C	Refractive index	References
2,6-Diallylphenol	121(11)	$n_D^{20}=1.5400$	Claisen, Ann. 401, 102 (1914); Belg. 615,700 (Shell)
2,6-Diallyl-p-cresol	145(14)	--	Claisen, Ann. 401, 45 (1914)
4,6-Diallyl-resorcinol	--	--	Monroe, U.S. 2,459,835 (Koppers)
2,4,6-triallyl-phenol	295	$n_D^{13}=1.5445$	Auwers, Ann. 422, 174 (1921)
3,4,6-Triallyl-catechol	--	--	Hurd, JACS 52, 1700 (1930)
Tetraallylcatechol	--	--	Hurd
3-Allyloxy-2,4,6-triallylphenol	--	--	Hurd
o,o'-Diallyl-bis-phenol	--	--	Brit. 512,485 (IG); Kaiser, JACS 68, 636 (1946)
2,3-Diallyl-hydroquinone	--	--	Fieser, JACS 2206 (1939)

Polyesters were prepared by heating isophthalyl chloride,
phenolphthalein, and diallyldian, a ring-diallyl-substi-
tuted bisphenol A (59). Films cured by heating for 3 hr
at 200°C were said to be strong even at 300°C.
Hexamethylene dithiol has been copolymerized by hydrogen
migration with 2,6-diallylphenol in aqueous bisulfite-
persulfate (60). Polyesters bearing allyl groups were
prepared from adipic acid and m,m'-diallyl-substituted bis-
hydroxyethyl derivative of bisphenol A (61).

References

1. I. S. Nikiforova et al, CA 66, 65219 and 94755 (1967).
2. E. A. Viktorova et al, CA 66, 94753 and 104,472 (1967).
3. D. Aelony, J. Appl. Polym. Sci. 2, 1509 (1963); cf. W. H. Starnes and T. L. Patton, U.S. 3,526,668 (Esso).
4. D. Aelony, U.S. 2,968,679; Brit. 850,252 (Gen. Mills); cf. P. J. Berrigan, U.S. 3,198,842 (Shell); W. J. LeNoble et al, J. Org. Chem. 36, 193 (1971).
5. D. Aelony, Brit. 841,589 (1960) (Gen. Mills).
6. A. Luettringhaus, Ann. 557, 46 (1947).
7. R. B. Spacht, U.S. 2,801,980-1 (Goodyear).
8. Hurd and Schmerling, J. Am. Chem. Soc. 59, 107 (1937).
9. C. E. Schildknecht and D. Walborn, unpublished.
10. L. L. Ryden et al, J. Am. Chem. Soc. 59, 1014 (1937).
11. J. W. Collette et al, German appl., CA 74, 64683 (1971).
12. L. Claisen and E. Tietze, Ann. 449, 81 (1926); CA 21, 71 (1927); C. S. Marvel and N. A. Higgins, J. Am. Chem. Soc. 70, 2218 (1948).
13. L. Claisen and O. Eisleb, Ann. 401, 21 (1913); CA 8, 64 (1914).
14. Q. R. Bartz et al, J. Am. Chem. Soc. 57, 371 (1935).
15. R. A. Smith and J. B. Niederl, J. Am. Chem. Soc. 55, 4151 (1933).
16. D. Stevens and E. L. Fareri, U.S. 2,699,385 (Gulf Res.).
17. E. F. Tollis, U.S. 2,716,083 (Courtaulds).
18. R. W. Martin, U.S. 2,707,715 (G.E.); cf. M. T. Harvey, U.S. 2,460,255 (Harvel); CA 78, 4849 (1973).
19. K. M. Hui and L. C. Yip, J. Chem. Soc. D7, 402 (1970).
20. A. W. Baker and A. T. Shulgin, J. Am. Chem. Soc. 81, 4524 (1959).
21. L. Claisen et al, Ann. 418, 69 (1919); CA 13, 2340 (1919); C. D. Hurd and C. N. Webb, J. Am. Chem. Soc. 58, 2190 (1936).
22. D. H. Spalding, U.S. 2,922,736 (Dow).
23. C. D. Hurd and W. A. Hoffmann, J. Org. Chem. 5, 212 (1940).
24. L. L. Ryden et al, J. Am. Chem. Soc. 59, 1014 (1937).
25. T. L. Patton and J. T. Horeczy, U.S. 3,477,991 (Esso).
26. C. L. Slegal and D. Philip, U.S. 3,422,062 (N. A. Rockwell).
27. A. S. Hay, J. Polym. Sci. 58, 593 (1961).
28. G. Zemplen and A. Gerecs, Ber. 70, 1098 (1937).
29. A. J. Birch, J. Chem. Soc. 102 (1947).
30. J. M. Van der Zanden, Rec. Trav. Chim. Pays-Bas 57, 233 (1938); M. Tiffeneau, Compt. Rend. 139, 482 (1904).
31. J. F. Eykman, Ber. 22, 2743 (1889).
32. A. Luettringhaus, Ann. 557, 46 (1947).
33. A. Higginbottom, J. Chem. Soc. 263 (1937); P. Hill et al, J. Chem. Soc. 510 (1937).
34. See D. S. Tarbell, Chem. Rev. 27, 534 (1940); A. Wagner, Mfg. Chem. 22, 271 (1951).
35. H. Pines and J. A. Vesely, U.S. 2,553,470 and 2,578,206.
36. L. Claisen and E. Tietze, Ber. 59B, 2344 (1926).
37. A. R. Bader, J. Am. Chem. Soc. 78, 1709 (1956).
38. C. D. Hurd and W. A. Hoffmann, J. Org. Chem. 5, 212 (1940).

39. S. W. Tinsley, J. Org. Chem. 24, 1197 (1959).
40. A. A. Shamshurin, CA 44, 1443 (1950).
41. L. L. Ryden et al, J. Am. Chem. Soc. 59, 1014 (1937).
42. B. M. Dubinin, CA 44, 1060 (1950); M. M. Martin and G. J. Gleicher, J. Am. Chem. Soc. 86, 233 (1964).
43. Also known as 4-allyl-2-methoxyphenol or caryophyllic acid.
44. S. M. Siegel, J. Am. Chem. Soc. 78, 1753 (1956).
45. J. Read and W. G. Reid, J. Chem. Soc. 1491 (1928).
46. H. A. Sober et al, Proc. Soc. Exptl. Biol. Med. 73, 148 (1950).
47. G. M. Brauer et al, J. Dental Res. 47, 622 (1968); cf. J. Res. Natl. Bur. Std. 68A, 619 (1964).
48. C. Morel, Soap, Perfum. Cos. 24, 1113 (1951).
49. J. M. Van der Zanden and G. deVries, Rec. Trav. Chim. Pays-Bas 71, 879 (1952).
50. J. M. Van der Zanden, Rec. Trav. Chim. Pays-Bas 58, 181 (1939).
51. F. Homburger et al, Med. Exptl. 4, 1 (1961); M. B. Jacobs, Am. Perfumer Aromat. 71, 57 (1958).
52. E. B. Truitt et al, J. Neuropsych. 2, 205 (1961).
53. R. Royer and E. Bisagni, Bull. Soc. Chim. France 521 (1959); CA 54, 8883 (1960).
54. J. A. Clark, U.S. 2,947,723 (Dow).
55. C. J. Kurian and C. C. Price, J. Polym. Sci. 49, 267 (1961).
56. L. Claisen et al, Ann. 418, 69 (1919).
57. L. Claisen and O. Eisleb, Ann. 401, 79 (1913).
58. R. A. Smith and J. B. Niederl, J. Am. Chem. Soc. 55, 4151 (1933).
59. V. V. Korshak et al, Siling. Khim. Volokna (1) 16 (1965); Plastics (London) 741 (June 1966).
60. C. S. Marvel and H. E. Baumgarten, J. Polym. Sci. 6, 127 (1951).
61. H. Schnell et al, Ger. 1,031,965 (Bayer); CA 54, 16919 (1960).

18. ALLYL SULFUR COMPOUNDS

The allyl sulfur compounds have exceptional interest because of their varied chemistry, wide occurrence in plants, novel biological activities, and numerous uses in food flavors, repellents, perfumes, medicine, photography, and copolymer fibers. They are generally used at low concentrations where their odor and toxicity are not objectionable. Although a number of the naturally occurring allyl sulfur compounds, such as those in mustard and onions, are toxic at higher concentrations, people ordinarily do not fear them as they do new synthetic chemicals.

The repellent, bacteriostatic and antioxidant properties of some of the allyl sulfur compounds suggest future possibilities for controlling harmful bacteria, fungi, and insects by agents which are completely harmless to higher forms of life. By new techniques of analysis, such as gas chromatography, knowledge of allyl sulfur compounds has expanded recently. However, amazing pioneering research was achieved already in the 19th century by Europeans such as Dumas, Will, Claisen, Billeter, Wertheim, Semmler, and their co-workers.

ALLYL ISOTHIOCYANATE

Reactions of inorganic thiocyanates with allyl halides under mild conditions can give liquid allyl thiocyanate, but this rearranges readily to give allyl isothiocyanate:

$$CH_2=CHCH_2SCN \longrightarrow CH_2=CHCH_2NCS \quad \text{(mustard oil)}$$

A solution of ammonium thiocyanate in 85% ethanol was mixed with allyl bromide and stored at 0°C (1). Adding water precipitated allyl thiocyanate (ASCN). Billeter and others observed that allyl thiocyanate, with onion-like odor, was stable only near 0°C and out of light and air. At room temperature it slowly isomerized to much stronger smelling mustard oil, allyl isothiocyanate (ANCS). The isomerization was accelerated by heating. ASCN was hydrogenated to ASH and HCN by treating with zinc and hydrochloric acid. Allyl bromide and lithium thiocyanate

486

were reacted in acetone solution at 25°C for 30 hr (2).
There was recovered ASCN (bp 36°C at 3 mm, n_D^{20} = 1.5153)
containing less than 2% ANCS. Repeated distillation of
ASCN at atmospheric pressure resulted in ANCS. Purified
allyl isothiocyanate has been reported to boil at 151°C
or 38°C at 10 mm, to melt at -80°C, and to have refractive
index n_D^{20} = 1.5306. Similarly methallyl thiocyanate
(bp 30°C at 0.15 mm, n_D^{25} = 1.4892) was prepared and isomerized to methallyl isothiocyanate (bp 38°C at 2.5 mm,
n_D^{25} = 1.5216). In polar solvents allyl thiocyanate
appears to be more stable, and as much as 10% may remain
in equilibrium with the isothiocyanate even after heating.
ASCN appears to be more stable than AOCN while ANCS is
more stable and better known than ANCO. Allyl selenocyanate was prepared by heating allyl bromide with potassium selenocyanate in methanol (3). It is an oil that
smells like garlic. It quickly becomes red-brown on
standing in air. Bromallyl selenocyanate (bp 88°C at
22 mm) also was made.

Allyl isothiocyanate, the major constituent of the
volatile oil of black mustard, may be obtained syntheti-
cally by heating under pressure a mixture of fine sodium
thiocyanate and allyl chloride (4). Mustard oil may be
prepared also by the following reactions (5):

$$ACl + NaCN + S \xrightarrow[CH_3OH]{\Delta} ASCN + NaCl \xrightarrow{\Delta} ANCS$$

ANCS also can be made by addition of sulfur to allyl isocyanide and by reaction of allylamine with CS_2. Recently
Emerson investigated the preparation and rearrangement of
ASCN (6). A mixture of allyl bromide and sodium thiosul-
fate pentahydrate in 95% ethanol was refluxed while
additional water was added during 45 min. To the cold
solution of $ASSO_3Na$ that was obtained, a cold solution of
NaCN was added. After a minute this mixture was extracted
with diethyl ether. Allyl thiocyanate having a garlic-
like odor was recovered by evaporation of the ether
extract. The experiment must not be delayed since ASCN
rearranges to ANCS at room temperature. During rearrangement, the thiocyanate stretch band at 2160 cm^{-1} is replaced
by the isothiocyanate stretch absorption at 2081 cm^{-1} in
the IR spectra. A dramatic increase in odor develops as
ANCS provides the pungent odor and "bite" of freshly
macerated horseradish. ANCS was distilled at 150 to 152°C
and was reacted with aniline to form the solid N-allyl-
N'-phenylthiourea derivative.

Steam distillation of seeds of cultivated Brassica
nigra (black mustard, a Crucifera) remains the source of
much mustard flavoring in Europe. Allyl isothiocyanate

results from hydrolysis of glucosides such as sinigrin in
turnip, cabbage, radish and horseradish seeds, both
leaves and roots. The hydrolysis of the glycoside sinigrin is promoted by heat or by an enzyme liberated when
plant cells of Cruciferae are crushed. The formation of
ANCS is believed to involve rearrangement of an intermediate thiohydroxamic acid as in the Lossen rearrangement:

$$ACS\text{-}glucosyl \xrightarrow{\text{H}_2\text{O}} glucose + ACSH \longrightarrow ANCS + KHSO_4$$
$$\overset{\shortparallel}{N}OSO_3K \qquad\qquad\qquad \overset{\shortparallel}{N}OSO_3K$$

In very low concentrations, allyl isothiocyanate is
widely used as mustard food flavor; its extreme pungence
and the irritation it produces in larger amounts led to
consideration as a war gas. It is slightly soluble in
water and miscible with alcohol and common organic solvents. The commercial synthetic mustard oil is colorless
to light yellow but on aging turns reddish brown (7). It
is supplied by Greef, Millmaster, Morton Chemical, Ritter,
and others in America.

In high concentrations allyl isothiocyanate is an acute
and chronic irritant. Care must be exercised even in
smelling the liquid to avoid irritation of the lungs. It
may produce burns, dermatitis, and blisters on the skin.
It should be tasted only when very highly diluted. When
hog feed contains synthetic or natural mustard oil, it
should be apportioned so that not more than 400 mg are
consumed per day. Mustard oil may cause allergenic reactions with liberation of histamine. This may be inhibited by injections of paraldehyde or calcium salts (8).
Excessive inhalation of ANCS may cause asthma, watery
eyes, and sneezing. Toxic cyanide fumes may be produced
by heating ANCS or by its contact with acids.

Allyl isothiocyanate has been reported to inactivate
enzymes such as papain and urease. Concentrations of 0.15
to 0.23% ANCS have killed staphylococci and streptococci,
also proteus, typhoid, and colon bacilli. Allyl mustard
oil has been proposed as a bacteriocide in disinfectants
and as a preservative of foods. Laboratory experiments
have shown mustard oil to be effective against May beetle
grubs, wireworms, potato root eelworms, and certain fish,
tadpoles, and snails (9). ANCS has been suggested as a
mothproofing agent for wool. It has been used as an
insecticide, fumigant and as a repellent for cats and dogs
(10). Mustard oil at 2% or less alcohol-water solution
has long found use as a counterirritant against pain in
human muscles and joints.

The odor of allyl isothiocyanate from plants of the
Cruciferae family attract cabbage white butterflies

(Pieris rapae), and the larvae of the butterflies contain
this mustard oil, which may act as a repellent for some
predators (11). However, some birds of the flycatcher
family relish these butterflies. Bromallyl isothiocyanate
(bp 79°C at 10 mm) is perhaps a stronger repellent (12).
Allyl isothiocyanate reacts with ammonia in alcohol-
water to give N-allylthiourea, one of the oldest synthetic
urea derivatives (13):

$$CH_2=CHCH_2NCS + NH_3 \longrightarrow CH_2=CHCH_2NCNH_2$$
$$S$$

On standing with water in light, allyl isothiocyanate was
reported to form allylamine, diallyl ether, thiocyanic
acid, HCN, H_2S, and CO_2. Distillation of oil of mustard
left a brown, resinous residue (14). Warming mustard oil
with sulfuric acid followed by dilution with water gives
1-amino-2-propanol. Allyl isothiocyanate reacted with
sodium urethane in ether to form N-allyl-N-carbethoxy-
thiourea (15). Heating the compound with guanidine in
dioxane gave N-allyl-N'-guanylthiourea (16). Heating
allyl isothiocyanate with salicylic acid in sunlight gave
N-allyl thiocarbamoyl cystein (18). ANCS reacted with
phenyl magnesium bromide to give N-allylthiobenzamide
(bp 215°C at 17 mm), an unstable oil of garlic odor (19);
with thioglycolic acid it reacted to form N-allylrhoda-
nine (20).
Allyl isothiocyanate has been estimated iodometrically
or bromometrically (21). One drop of solution containing
0.1 ppm mustard oil is said to give a visible deposit of
crystalline derivative with phenylhydrazine in alcohol
(22). A blue color results when to 0.5 ml of 1% solution
of ANCS is added 2 drops of 50% hexylresorcinol in alcohol
and 1.0 ml of concentrated sulfuric acid (23). Allyl
isothiocyanate in plant material can be identified by
reacting with benzylamine for 5 min in boiling ethanol
(24). When water is added, crystals of allyl benzylthio-
urea (mp 94.5°C) form. ACNS has been titrated potentio-
metrically with KIO_4 solution (25).
Plants yielding mustard oils have found use in European
folk medicine since early times. In the middle ages such
plants were used to combat urinary infections and for
disinfecting wounds. Wagner, Hoerhammer, and Nufer of
Munich stated that not less than 65% of these plants have
been proved to possess antibiotic action (26). Mustard
oils and the glucosides from which they are derived are
not limited to the Cruciferae as often supposed; they
occur in other plant families, many of them totally un-
related to Cruciferae according to modern concepts of
taxonomy. Plants yielding much ANCS, such as black

mustard, irritate the mucous membranes and were not favored for medical use. Mustard oils from garden cress (Lepidium sativum), nasturtium (Tropaeolum mojus), and white mustard (Sinapis alba) showed more favorable antibiotic action (27). Injections of mustard oils in animals showed changes in capillary resistance and capillary permeability, as well as rise in blood leucocyte number (28).

Wagner and co-workers identified allyl, alkyl, aryl, and benzyl isothiocyanates from black mustard, white mustard, horseradish, and garden cress by means of layer chromatography (26). The isothiocyanates were first converted to substituted ureas and spots were run on plates coated with silica gel using ethyl acetate-chloroform-water. The spots were developed by a $K_3Fe(CN)_6$-$FeCl_3$ aqueous solution.

Allyl mustard oil has been used as a counterirritant in treatment of rheumatism and pleurisy. Mustard plasters of the past have been largely succeeded by synthetic mustard ointments which may contain 1 to 4% ANCS. Mustard condiments in Britain may contain at least 0.35% ANCS; in the United States the amount may be 0.6% (29). Both allyl isothiocyanate and its glucoside sinigrin can act as antioxidants even at low concentrations (30). In Turkey ANCS has been added as a preservative to grape juice. It has been used also to prevent mold in fodder.

OTHER ALLYLIC ISOTHIOCYANATES

Bruson and Eastes reacted sodium thiocyanate with methallyl chloride in boiling methanol for 3 hr to form methallyl isothiocyanate $CH_2=C(CH_3)CH_2NCS$ (bp 170°C) having an onion odor (31). On standing with 27% ammonium hydroxide solution for 24 hr, this compound gave white crystals of N-methallylthiourea melting at 94°C. Heating methallyl-thiourea with concentrated HCl for 8 hr at 140°C gave 5,5-dimethyl-2-aminothiazoline hydrochloride. Reaction of MANCS with H_2SO_4 gave a mercaptothiazoline derivative.

The compound $CH_2=CHCH(CH_3)NCS$ (bp 72°C at 34 mm) was reacted with concentrated ammonium hydroxide to form a thiourea derivative melting at 110°C (32). Cleavage of glucosides in rape seed (Brassica napus) by steam distillation yields ACH_2NCS and ACH_2CH_2NCS (33). From turnip (Brassica rapa) $CH_2=CHCH(OH)NCS$ was isolated (34).

Methallyl isothiocyanate (bp 69°C at 20 mm, $n_D^{20} = 1.5220$) was prepared by Tamele and co-workers by slowly adding methallyl chloride to a boiling, stirred solution of NH_4SCN in ethanol (35). The product was washed with aqueous sodium chloride solution to remove ethanol. Steam distillation gave a water-methallyl isothiocyanate azeo-

trope boiling at 97°C. Allyl carbinyl isothiocyanate, ACH_2NCS (bp 77.5°C at 28 mm; $n_D^{26.5} = 1.5182$) was prepared from CS_2 and allyl carbinylamine (36). The mustard oil obtained by steam distillation of macerated rape seed (Brassica napus) was found to contain allyl carbinyl isothiocyanate and the thiooxazolidone $CH_2=CHCHCH_2NH$. No
$$O\text{——}C=S$$
ANCS was found from hydrolysis of the glucosides of rape seed. The ACH_2NCS was identified by reacting with ammonium hydroxide to give allyl carbinyl thiourea needles (mp 67°C) and by reaction with aniline at 100°C to give N-allylcarbinyl-N'-phenylthiourea (mp 57°C).

ALLYL THIOUREAS

Monoallylthiourea or thiosinamine was made by reaction of allyl isothiocyanate (mustard oil) with ammonia. White crystals of the compound have a faint garlic odor and 3% will dissolve in water. Allylthiourea occurs in fresh white cabbage and certain plant seeds. It has biological activity, retarding growth of beans, but its toxicity seems uncertain. It has been used for sensitization of photographic films, and in medicine it has served to minimize scar tissue and to combat some types of dermatitis. Crystallization from water gives the stable monoclinic crystal form (mp 78°C); rapid cooling from organic solvent or from the melt gives a second crystal form of lower melting point 71°C. Because of its use with gelatins for activating photographic films, its photochemical reactions (37) and reaction with silver halides (38) are of interest. Thiosinamine was estimated by oxidation with H_2O_2 and KOH to form K_2SO_4 and monoallylurea (39). Color and precipitation reactions with inorganic ions have been studied (40). The picrate melts at 162°C. Many derivatives of allylthiourea have been reported recently in Chemical Abstracts and have been evaluated in fields of photography, medicine, agriculture, and control of corrosion.

The photographic sensitivity of films from particular gelatins derived from cattle depend upon the presence of traces of allyl isothiocyanate and allylthiourea from pasture plants, as investigated by Eastman laboratories. The sensitization process was believed to involve reactions with silver halide, forming Ag_2S and complexes (41). Allyldiethylthiourea and acetylallylthiourea also have been added as sensitizers along with special dyes in photographic gelatins. Studies were made of the reactions of allylthiourea on the surface of silver bromide in fine-grain photographic emulsions (42). Allyl heptadecyl thiourea was more effective than allylthiourea in promoting

sensitivity by the "ripening" of silver chloride grains
during 32 min at pH 3 (43). A number of allylthioureas
are supplied by the Industrial Dyestuff and other companies. Allylthiourea was added in recent photographic
polymerization recipes (44).

Copolymers of monoallylthiourea with SO_2 were made in
ethanol at -40°C under activation by ^{60}Co radiation (45).
A total dose of 0.1 Mrad was used during 30 min. The 1:1
molar copolymers melted near 180°C. When swollen by water
they showed chelation with cupric ions.

Addition of allylthiourea to polyethylene and to methacrylate polymers was found to give protection against
oxidative degradation under high-energy radiation (46).
Allylthiourea was observed to inhibit growth of trans-
planted tumors in mice (47). The compound acted as a
water-soluble antioxidant, e.g., retarding degradation of
sodium polymethacrylate in solution (48). Small amounts
of thiosinamine were toxic to growing beans (49).

Dry allylthiourea was observed in 1840 to react violently with an excess of mercuric oxide (50). A similar
reaction occurred with lead oxide or hydroxide (51).
Sulfur was removed from nitrogen and a basic substance
was formed along with some dark viscous polymer, and much
heat was evolved. Other references to allylthiourea and
derivatives are given in Table 18.1.

Symmetrical or N,N'-diallylthiourea was prepared by
Hecht by reaction of allyl isothiocyanate with allylamine
(52). Symmetrical diallylthiourea has been used as a
rubber accelerator and corrosion inhibitor. It has been
available from Roberts Chemicals of Nitro, West Virginia.

Allyl halides can react with the sulfur of thiourea to
form S-allyl isothiourea hydrohalides (53):

$$AX + (H_2N)_2CS \longrightarrow \left[\begin{array}{c} NH_2\underset{\|}{C}SA \\ H_2N+ \end{array} \right] \bar{X}$$

The N-phenyl- and N,N'-diphenylthioureas react similarly.
Free S-allyl isothiourea apparently has not been isolated.
Addition of alkali to the halides gives cyanamide and
allylmercaptan. Phenyl-substituted S-allylthiourea derivatives were prepared by Werner. The picrate of S-allyl
thiourea, melting at 155°C, was used in identification.
Analogous to the equation just given is the behavior of
allyl bromide or allyl iodide, which react with the sodium
salt of 4-methyl-2-thiouracil to give S-allyl derivatives
(54). An N,S-diallyl isothiourea was prepared (55).
S-Allylcysteine is offered by Sigma Chemical Company.

Allylic thiocarbamates and thiocarbonates have received
little attention as monomers. Fluoroallylic thiocarbamates
and thiocarbonates have been evaluated as herbicides (56).

TABLE 18.1

Allyl Thioureas and Related Compounds

Compound	mp, °C	References
Allylthiourea, ANHCSNH$_2$	74	Gadamer, J. Chem. Soc. $\underline{70}$, 414 (1896); McCrone, Anal. Chem. $\underline{21}$, 421 (1949)
Methallylthiourea, CH$_2$=C(CH$_3$)CH$_2$NHCSNH$_2$	94	Bruson, JACS $\underline{59}$, 2012 (1937)
1-Methylallylthiourea, CH$_2$=CHCH(CH$_3$)NHCSNH$_2$	108	Krueger, JACS $\underline{63}$, 2512 (1941)
N-butenylthiourea, ACH$_2$NHCSNH$_2$	113	Ettlinger, JACS $\underline{77}$, 1832 (1955)
N-allyl-N'(β-hydroxylethyl)-thiourea	83	Available from Aldrich
N,N-dibutyl-N'-allylthiourea, ANHCSN(CH$_2$CH$_2$CH$_2$CH$_3$)$_2$	83	Dehn, JACS $\underline{62}$, 3190 (1940)
N-allyl-N'-dodecylthiourea	57	Erickson, J. Org. Chem. $\underline{21}$, 483 (1956)
N-allyl-N-octodecylthiourea	80	Erickson
4-Allyl thiosemicarbazide	96	Canadian J. Chem. $\underline{35}$, 832 (1957); available from Aldrich
N,N'-diallylthiourea, ANHCSNHA	49	Hecht, Ber. $\underline{23}$, 287 (1890); Kitamura, CZ 4608 (1939)
N-allyl-N'-phenylthiourea	98	Weith, Ber. 8, 1529 (1875); Dains, JACS 21, 164 (1899)
N-allyl-N'-o-tolylthiourea	98	Dixon, J. Chem. Soc. $\underline{55}$, 622; Prager, Ber. $\underline{22}$, 2998 (1889)
Diallyl thiuram disulfide ANHC(S)SSC(S)NHA	--	U.S. 1,762,531; CZ II 2585 (1930)
Bromallylalkylthioureas	--	Schmidt, Ann. $\underline{560}$, 229 (1948)
N-allyl-N'-arylthioureas, etc.	--	Dieke, J. Pharm. Expt. Therap. 90, 260 (1947); CA $\underline{41}$, 6662 (1947)
N,N'-diallyl dithiooxamide (ANHC=S)$_2$	-- --	Peacock, Brit. 1,065,004 (ICI); CA $\underline{67}$, 10698 (1967)

ALLYL ALKYL SULFIDES

Few of these compounds have been studied in polymerization, and high polymers have not been prepared. Methyl allyl sulfide has been made from allyl bromide and Pb(SCH$_3$)$_2$ (57). Such volatile sulfur compounds have extremely repulsive odors which discourage research and preclude many uses except in very low concentrations. Sodium methyl thiol was reacted with allyl bromide under reflux for 30 min (58). On pouring the solution into water, CH$_3$SCH$_2$CH=CH$_2$ separated. After drying it was distilled at 93°C.

Rearrangement of allyl hexyl sulfide and allyl t-butyl sulfide to cis- and trans-propenyl sulfides occurs in the presence of concentrated KOH, NaOCH$_3$, or KOCH$_3$ (59). Allyl n-dodecyl sulfide isomerized largely to the trans-propenyl compound on treatment with sodium ethylate (60). Mercaptans and allyl or methallyl halides may be heated with catalysts or UV radiation to give allyl and propenyl alkyl sulfides (61). Allyl alkyl and aryl sulfides can add a molecule of mercaptan readily in the reverse way: RSCH$_2$CH=CH$_2$ + HSR $\longrightarrow$ RSCH$_2$CHCH$_2$SR.

Methyl allyl disulfide and isopropyl allyl disulfide are formed on maceration of garlics, onions, leeks, and chives. Allyl n-propyl disulfide from onions was estimated by oxidation by bromine to form sulfate (62). Allyl n-propyl sulfide, along with diallyl sulfide and allyl alcohol, resulted from radiolysis of S-allyl-L-cysteine sulfoxide in oxygen-free aqueous solution (63).

Allyl phenyl sulfide (prepared from thiophenol and allene) was reacted with thiophenol in benzene at 80°C for 6 hr to give a 4:1 mixture of "normal" and reverse addition products (64). When acetic acid was used as solvent and when light and radical initiators were excluded, reverse addition predominated and an ionic mechanism was indicated. Reaction of allene with methyl mercaptan gave allyl methyl sulfide and dithioether products directly (65). Allyl benzyl sulfide (bp 130°C at 14 mm; n_D^{20} = 1.4725) was prepared from allyl chloride and sodium thiobenzylate (66). The sulfide was oxidized to the sulfone, which was isomerized to CH$_3$CH=CHSO$_2$CH$_2$C$_6$H$_5$ by heating for 6 hr with triethylamine at 120°C.

Allyl butyl sulfide and methallyl butyl sulfide were prepared from the allyl chloride or bromide and sodium thioalcoholates by Bateman and Cunneen (67). Oxidation in light to sulfoxides and to hydroperoxides was observed without appreciable polymerization. Treatment of allyl thioethers with NaOC$_2$H$_5$ or potassium butoxide causes isomerization to propenyl thioether. References to other monoallyl sulfides are given in Table 18.2.

In the presence of palladium-charcoal catalyst, simple allyl sulfides and vinyl sulfide are hydrogenated less rapidly than related olefins (68). Bateman and Shipley found that methyl substituents in the allyl group further reduce rates of hydrogenation. Thus allyl n-butyl sulfide required an hour for hydrogenation, whereas methallyl n-butyl sulfide required about 23 hr under similar conditions. Diallyl sulfide was only half hydrogenated in 18 hr under such conditions, and some hydrogenolysis to thiol also occurred. Diallyl sulfone was completely hydrogenated in 15 min. The electron-donating effects of SR groups are weaker than those of OR groups, and rates

TABLE 18.2

Monoallyl Sulfur Compounds

Compound	bp, °C	Other properties	References
Allyl methyl sulfide	93	Comp'd with $HgCl_2$ mp = 115°C; $n_D^{20} = 1.4714$	Challenger, Biochem. J. 44, 89 (1949); Pfaltz and Bauer (supplier)
Allyl ethyl sulfide	115	$d_4^{20} = 0.8676$	Dawson, JACS 55, 2073 (1933)
Allyl propyl sulfide	140	--	Farmer, J. Chem. Soc. 1529 (1947)
Allyl propyl disulfide (onion oil)	69(16)	From garlic, etc.; $n_D^{25} = 1.4730$	Semmler, Arch. Phys. 230, 438; Merck Index
Allyl butyl sulfide	160	--	Luettringhaus, Ann. 557, 62 (1947)
Allyl t-butyl sulfide	143	$n_D^{20} = 1.4633$	Price, J. Org. Chem. 27, 4639 (1962)
Methallyl methyl sulfide	113	$n_D^{20} = 1.4712$	Barnard et al, J. Chem. Soc. 2443 (1949)
Methallyl butyl sulfide	71(15)	Addition reactions	Olsen et al, Ind. Eng. Chem. 38, 1273 (1946); Bost et al, Org. Syn. 15, 72 (1935)
Allyl thioacetic acid	138(15)	$n_D^{25} = 1.5054$	Larsson, Acta Chem. Scand. 14, 768 (1960)
Allyl alkyl xanthates	--	--	Brit. 700,334 (Am. Cyan.)
4-Allylthiopentanol	--	--	Morris, U.S. 2,645,659 (Shell)
1-Allylthio-3-chloro-2-propanol (as anthelmintic)	--	--	Levine, U.S. 3,579,592 (Dow)
Allyl thienyl sulfide	54(0.4)	--	Brooks, U.S. 2,577,566 (Socony)
Allyl phenyl sulfide	86(4.9)	$n_D^{25} = 1.5732$	Cope, JACS 72, 59 (1950); Kwart et al, J. Org. Chem. 31, 413 (1966)
Methallyl phenyl sulfide	89(3.4)	$n_D^{23.5} = 1.5605$	Cope; Kwart
o-Allylthiophenol	--	mp, 49°C	Cope; Kwart

of electrophilic addition to allyl and vinyl alkyl sulfides are slower.

ALLYL ARYL SULFIDES

Allyl aryl sulfides or thioethers can be made by reacting allyl chloride or bromide with alcoholic sodium ethoxide and thiophenols (69). For example, to 0.1 mole sodium ethoxide in 30 ml ethanol was added 0.1 mole thiophenol followed by 0.11 mole allyl bromide. There was appreciable heat of reaction and an immediate precipitate of sodium bromide formed. The reaction was completed by allowing the mixture to stand overnight, after which the mixture was no longer alkaline to litmus. Residual volatiles were distilled off, water was added, and the oily layer of allyl phenyl sulfide was recovered by ether extraction. After drying with $CaCl_2$ the yield recovered by vacuum distillation was nearly quantitative (bp 106°C at 25 mm). A similar procedure was used except that the reaction was begun at 0°C (70). The purified ASC_6H_5 distilled at 56°C (4.9 mm) and showed $n_D^{25} = 1.5732$.

The isomerizations of allyl aryl sulfides under normal conditions used for Claisen rearrangement of allyl aryl ethers do not occur very readily, and they tend to give cyclic products rather than allyl thiophenols.

Hurd and Greengard refluxed phenyl allyl sulfide for 6 hr; during this time the temperature rose to 240°C and the liquid darkened (69). The mixture acquired a mercaptan odor, in contrast to the odor of the original thioether. Apparently only about one-third of the sulfide was isomerized to o-allylthiophenol, which has a decidedly unpleasant odor (bp 190°C at 17 mm; $n_D^{21} = 1.6098$). The identity of the product was supported by the analysis of its lead salt and by oxidation to o-sulfobenzoic acid.

The alleged o-allylthiophenol did not dissolve in alkali. There was some evidence that a small amount of thiacoumaran was formed along with Claisen rearrangement. Partial isomerization on refluxing allyl p-tolyl sulfide apparently formed some 2-allyl-4-methylthiophenol.

Karaulova and co-workers were able to obtain only traces of o-allylthiophenol by heating allyl phenyl sulfide, the main product being propenyl phenyl ether (71). This was supported by work of Meyers and co-workers, who observed that the presence of quinoline or other secondary amine catalyzed isomerization of allyl phenyl sulfide to 2-methylthiacoumaran and thiachroman (72):

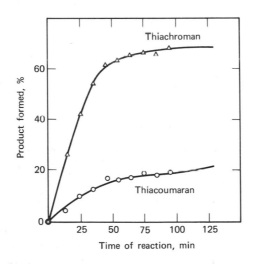

Kwart and Evans rearranged allyl phenyl sulfide by heating 30 g with 26 g of quinoline under nitrogen at 229 to 241°C for 7 hr (73). By column chromatography they isolated 11.7 g 2-methyl-1-thiacoumaran, melting at 116°C, 3.5 g thiachroman (n_D^{27} = 1.6106), 0.9 g black polymer, and 1.1 g thiophenol. Preparation and purification of the very reactive o-allyl thiophenol proved difficult. The compound was best made by way of O,S-bis(o-allylphenyl)thiol carbonate.

Kwart and Cohen rearranged methallyl phenyl sulfide by heating at 300°C with quinoline to form thiachroman and thiacoumaran derivatives along with phenyl isobutyl sulfide, diphenyl sulfide, polymer, and other products (Fig. 18.1) (74). A mechanism for pyrolysis via initial isomerization to isobutenyl phenyl sulfide was suggested.

Fig. 18.1. Heating methallyl phenyl sulfide in quinoline at 237°C gave largely cyclization to thiachroman and thiacoumaran derivatives, and only a little methallylthiophenol was formed by Claisen rearrangement. H. Kwart and M. H. Cohen, J. Org. Chem. 32, 3137 (1967).

Reactions of allyl phenyl sulfide and of methallyl phenyl sulfide can be catalyzed by octanoic acid at 300°C to give thiocoumarans and thiachromans (75). Chloroallyl phenyl sulfide gave the thiocoumaran exclusively, with either quinoline or octanoic acid as catalyst.

Hydrogenation of unsaturated sulfur compounds over heavy metal catalysts may involve partial hydrogenolysis. However, allyl phenyl sulfide has been reduced without appreciable C-S bond fission using rhenium sulfide catalyst above 150°C and hydrogen at 2000 psi (76). Methallyl phenyl sulfide has been reduced using palladium on charcoal at moderate temperatures (77). References to a few other allyl aromatic sulfides are given in Table 18.3.

TABLE 18.3

Other Allyl Aromatic Sulfides, etc.

Compound	Properties	References
Allyl-2,4-dinitro-phenyl sulfide	mp, 71°C	Bost, JACS 73, 1968 (1950)
Allyl-2,4-diamino-phenyl sulfide	Epoxy curing agent	Nischk et al, Ger. 1,129,281 (Bayer); CA 57, 7466 (1962)
Allyl phenoxy-phenyl sulfide	bp 140°C at 0.8 mm; $n_D^{25} = 1.6067$; anthelmintic agent	Reifschneider, U.S. 3,155,733 (Dow)
Allyl anthraquinone sulfide derivatives	--	Gattermann, Ann. 393, 119 (1913)
Allyl phenyl selenide	bp 83°C at 3 mm; $n_D^{25} = 1.5948$	Kataev, CA 66, 75790 (1967)

ALLYL AND METHALLYL MERCAPTANS

Allyl mercaptan $CH_2=CHCH_2SH$ can be prepared from allyl halides and NaSH, but diallyl sulfide is also formed, with elimination of H_2S (78). Purer allyl mercaptan can be prepared by heating the dithiourethane $H_2NCS_2CH_2-CH=CH_2$ with excess NaOH (78). Allyl mercaptan has been made by reduction of diallyl sulfide (79). The compound boils lower than allyl alcohol and has an extremely offensive odor somewhat resembling decaying onions even at low concentrations. Allene and H_2S in the presence of radical initiators form allyl mercaptan and diallyl sulfide (80). Allyl mercaptan (bp 68°C) of Aldrich Chemical has $n_D^{20} = 1.4832$. It is supplied also by Fairfield Chemical, Ritter & Company, and Medical Chemicals Corporation and has been used in very low concentrations in foods and perfumes.

Allyl mercaptan reacts slowly with sodium to form sodium allyl mercaptide. It forms insoluble silver allyl mercaptide (mp 115°C) and mercury dimercaptide $Hg(SCH_2CH=CH_2)_2$ (mp 74°C). The tin compound $Sn(SC_3H_5)_4$ is an oil that forms solid polymers on standing (81). Reaction of allyl mercaptan with formaldehyde gave a diallyl sulfide derivative (82).

Little research has been done on allyl sulfonium salts. Allyl iodide and dipropyl sulfide form allyl dipropyl sulfonium iodide on standing at room temperature (83). Allyl chloride and dimethyl sulfide agitated in water for 8 days gave 89% conversion to $CH_2=CHCH_2S(CH_3)_2{}^+$ $\bar{C}l$ (84).

Several monoallyl sulfur compounds have been observed to rise gradually in boiling point during distillation and to form oily residues of polymer of relatively low molecular weight. With allyl mercaptans, where SH groups would be expected to prevent free radical polymerization, hydrogen-migration or H-polymerizations may occur, especially in the presence of alkaline catalysts or impurities; for example:

$$CH_2=CHCH_2SH \longrightarrow --CH_2CH_2CH_2S--$$

While distilling allyl mercaptan prepared from allyl iodide and KSH (85), Braun and Hahn obtained some diallyl sulfide with evolution of H_2S and an oily polymer residue (86). Long refluxing of allyl mercaptan gave almost complete polymerization to an oil with only slight odor which boiled under vacuum over the range 71-200°C. Cinnamyl mercaptan also was observed to polymerize on heating at 100°C.

Braun and Plate compared polymerization tendencies of allylic mercaptans on heating (87):

Polymerized readily	Did not polymerize
$CH_2=CHCH_2SH$	$CH_2=CHCH_2CH_2CH_2SH$
$CH_3CH=CHCH_2SH$	$CH_2{\Large\diagdown}\begin{array}{c}CH_2CSH\\ \parallel\\ CH_2CH\end{array}$
$CH_2=CHCH_2CH_2SH$	

The compounds in the right-hand column were heated for several hours at 100°C without much increase in density or formation of high boiling polymer residues. Formation of polymers from allyl mercaptan also was observed by Backer (88). The homopolymerization of ASH has been promoted by UV light (89). The formation of dimer was rapid, but further polymerization was slow. More uniform, semisolid polymers were obtained by treating the distilled dimer. Recently p-vinylthiocyclohexanol has been reported to polymerize on storage (90).

Methallyl mercaptan (bp 92°C, n_D^{20} = 1.4872) was prepared by heating the chloride with excess aqueous or
alcoholic NaSH at 70°C for several hours (91). It is
available from Aldrich as a colorless liquid (n_D^{20} = 1.4803).
Methallyl mercaptan gives liquid polymers by self-reverse
addition, but they are rather unstable.

DIALLYL SULFIDES

Diallyl sulfide (ASA abbreviated DAS) can be prepared
by reaction of allyl halides with potassium or sodium
sulfide (85, 92). It is slightly soluble in water and is
miscible with alcohol and with ether. With alcoholic
silver nitrate it gives white needles. Diallyl sulfide
also can be prepared by heating bis(2-hydroxy-ethyl) sulfide with KOH (93). Some polymerization was observed
during distillation. Diallyl sulfides react readily with
bromine and with iodine chloride. Diallyl sulfide was
prepared from allyl alcohol and hydrogen sulfide (94);
also by treating ASH with NOCl (95). DAS has an odor
suggesting a mixture of garlic, onion, and cooked cabbage,
less pungent and disagreeable than the odor of ASSA or
diallyl disulfide.

Diallyl sulfide is present in garlic oil obtained by
steam distillation of Allium sativum. However, most
members of the onion family have been found to yield
larger proportions of diallyl disulfide and propyl allyl
disulfide on steam distillation. Synthetic diallyl sulfide, which has a less pungent odor than onions and
garlic, is used in low concentrations in flavoring food-
stuffs. Diallyl sulfide may occur as an impurity in
petroleum naphthas. The toxicity, taste, and odor or
diallyl sulfides are less severe than those of allyl
isothiocyanate from mustard oils (96). Commercial allyl
mercaptan may contain 20% or more diallyl sulfide.

In 1844 Wertheim suggested the name allyl for the
radical attached to sulfur in garlic oil, and he seems to
have observed polymerization also (97). In the laboratory of Redtenbacher he steam distilled crushed bulbs of
the European garlic (Allium sativum). When the oily distillate was heated to 150°C, sudden "decomposition"
occurred accompanied by evolution of heat and formation
of a brown adhesive mass. However, some carefully purified fractions of the oil could be redistilled without
decomposition. There was evidence of several different
liquids of boiling points and densities different from
those of the purified garlic oil. The crude oil with
sulfuric acid gave a purple-red product; and with dry
hydrogen chloride an indigo blue coloration was obtained.
Wertheim believed that diallyl sulfide was the principal

volatile sulfur compound obtained by steam distillation
of garlic, but later work showed that larger amounts of
allyl disulfides are present. Small amounts of diallyl
monosulfide were confirmed, along with allyl alcohol in
garlic oil, by gas chromatography (98). Diallyl disulfide
was confirmed as the main component, however.

DAS may be characterized by reaction with chloramine T
to give crystalline sulfilimine (99). On standing, this
formed an oil believed to be $CH_3C_6H_4SO_2NSA$. Hydrolysis
                                                    A
of the latter in hot aqueous NaOH followed by acidification gave H_2S, ASSA, $CH_3C_6H_4SO_2NHA$, and some polymer.
Diallyl sulfide can be estimated by the double replacement reaction with silver nitrate solution and weighing
the Ag_2S formed. Diallyl sulfide was reacted with hydrogen peroxide in acetic acid to form diallyl sulfoxide
(100). Rates of oxidation of DAS and of allyl n-butyl
sulfide in air were slower than those of methyl- and
phenyl-substituted allylic sulfides (Fig. 18.2).

Diallyl disulfide was isolated by Semmler from oil of
garlic, and its distillation with zinc dust gave diallyl
monosulfide (101). In the steam-distilled oil, Semmler
found 60% ASSA, 6% $ASSCH_2CH_2CH_3$, and 34% of a mixture of
AS_3A and AS_4A. Diallyl disulfide was prepared by heating
allyl chloride with alcoholic potassium sulfide at 50°C
(102). Diallyl trisulfide was isolated free of the disulfide. ASSA has a very strong, pungent, and garliclike
odor. To some persons it is irritating even in minute
concentrations and can produce headache.

The characteristically pungent odors of the Allium
genus often are not very strong from living intact plant
tissues; they are produced enzymatically very rapidly
when plant injury occurs. Typical substrates for producing the volatile sulfides and disulfides (called alliins)
are derivatives of the amino acid cysteine. Among the
reactions occurring when cloves of garlic or onion are
chopped in air are the following (103):

$$2ASCH_2CHCOOH + H_2O \xrightarrow{\text{alliinase}} ASSA + 2CH_3CCOOH + 2NH_3$$
$$\;\;\;\;\;\; O \;\; NH_2 \qquad\qquad\qquad\qquad O \qquad\qquad O$$

$$\qquad\qquad\qquad\qquad\qquad\qquad\qquad \text{Allicin} \qquad \text{pyruvic acid}$$

$$\qquad\qquad\qquad\qquad O$$
$$ASSA \longrightarrow ASSA + ASSA + SO_2 + \text{polymers}$$
$$\;\; O \qquad\qquad\qquad\qquad\quad O$$

Some diallyl monosulfide may also form from diallyl disulfide. Allicin (see p. 505) does not form readily when
onion extracts are stabilized by alcohol.

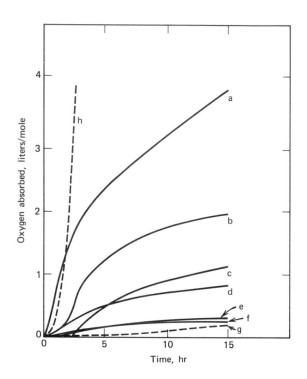

Fig. 18.2. Comparative rates of autoxidation of a series of allylic sulfides and two unsaturated esters at 75°C:

a, $C_6H_5CH=CHCH_2SBu$ e, $(CH_2=CHCH_2)_2S$
b, $CH(CH_3)=CHCH(CH_3)SBu$ f, $CH_2=CHCH_2SBu$
c, $CH_2=CHCH(CH_3)SBu$ g, Methyl oleate
d, $CH(CH_3)=CHCH_2SBu$ h, Ethyl linoleate

The unsaturated long chain esters were the least and most reactive, respectively, with oxygen. Conjugation in the aromatic sulfide apparently promoted oxidation. Allylic sulfur compounds with electron-releasing methyl groups adjacent to the carbon-sulfur bonds oxidized more rapidly than unsubstituted diallyl sulfide and allyl butyl sulfide. L. Bateman and J. I. Cunneen, J. Chem. Soc., 1598 (1955).

Allyl n-propyl sulfide was identified by IR absorption bands from great-headed garlic and methyl allyl sulfide from this garlic, as well as from common garlic (A. sativum) (104). A number of allium species of the western United States gave similar proportions of ASA, CH_3SSA, $CH_3CH_2CH_2SSA$, and ASSA (105). Neither the habitat, the stage of growth, nor the plant part affected the proportions of sulfur compounds very much. However, fully developed leaves produced most of these volatiles.

Bernhard and co-workers used gas chromatography to estimate the allyl sulfur compounds taken from chopped bulb tissue held at 40°C for 30 min (106). In addition to the five allyl compounds tabulated below, dimethyl disulfide and methyl propyl disulfide also were detected from these plants.

	A_2S	A_2S_2	CH_3S_2A	$n-C_3H_7S_2A$	$(n-C_3H_7)_2S_2$
Great-headed garlic	<1	55	31	5	<1
A. sativum (California early garlic)	3	74	22	<1	<1
A. cepa (3 types of onions)	0	<1	<1 to 1	4-6	80-93
A. fistulosum (Japanese bunch onions)	5	<1	2	4	65
A. ampeloprasum (leek)	?	<1	2	3	38
A. schoenoprasum (chive)	0	<1	4	4	<1
A. tuberosum (Chinese chive)	<1	<1	16	<1	<1

Garlics gave principally diallyl disulfide and methyl allyl disulfide. Milder smelling chopped onions yielded largely dipropyl sulfide. Leeks and chives also are low in unsaturated sulfides, giving principally methyl propyl disulfide and dipropyl sulfide. However, the relative contributions of different sulfur compounds to odor, as well as to the lachrymatory action of the different alliums, remain to be more fully explored. In addition to the pungent allyl sulfides, it is believed that methyl and n-propyl-1-propenyl disulfides may be formed. Bernhard and co-workers also found about 3% of allyl alcohol in volatiles from A. sativum and A. ampeloprasum. Diallyl sulfide has been reported from horseradish (Armoracia rusticana) (107). Recent research suggests that the diallyl monosulfide from garlic and onions is formed by secondary reactions, and the proportion depends very much on processing conditions.

The aroma and taste of the onion genus have been highly prized since antiquity, especially in southern and eastern sections of the Old World. Leeks and onions of Egypt delighted the Israelites (Numbers 11:5) as they did the builders of the Pyramids according to Herodotus. Pliny tells us that the ancient Egyptians even placed garlic among their deities.

Allyl sulfides apparently do not occur as widely in plant species as allyl isothiocyanate and related mustard oils, but there is a large literature on alleged medical applications of onions, garlic, and related species. Two reviews are available at the National Library of Medicine, Bethesda, Maryland (108). Since bacteriostatic effects of allicin have been confirmed, perhaps some claims in the folk medicine for onions deserve further study. Onions are believed to yield substances having cardiac activity and ability to widen capillaries. Eating onions has been recommended to lower blood pressure and to relieve arteriosclerosis. Onions have been said to alleviate diarrhea and dysentery, to promote gall secretion, to reduce blood sugar, and to act as a vermifuge. Conflicting results have been published on interaction of onion oils with viruses. Crushed onions and garlic have been applied to promote healing of wounds and as insecticides. A publication on Russian folk medicine states that fresh onion juice loses its curative effectiveness after 10 to 15 min in air. Onion juice has been used as a hair tonic and against dandruff. Garlic has been added to vodka to combat kidney and bladder stones, arthritis, and rheumatism. Further research is encouraged by the well-known low toxicity of onions and possibilities of improving the social acceptability of garlic and onions. Aqueous garlic and onion juices supplied by McCormick and Company are stabilized by 0.1% sodium benzoate and 0.1% potassium sorbate.

Among medical preparations available are tinctures of garlic and onion (Galenic preparations) prepared by maceration with ethyl alcohol, which is said to prevent fermentation induced by water. Garlic oil capsules as well as dry garlic tablets are supplied, e.g., by Richards Laboratories, Babylon, New York.

Recently it has been found that from Allium cepa di-n-propyl disulfide and diallyl disulfide (but not ASA) can inhibit thyroid function in rats (109). Reduced intake of iodine may lead to goiter if iodine in the diet is low. ASA was less toxic than ASSA to salamander tadpoles and was nongoitrogenic (110).

Small amounts of DAS were added to suspension polymerizations of vinyl chloride in order to raise the specific viscosity of the polymer products (111). Mixtures of

triphenyl phosphine and DAS with nitrogen under UV light
gave viscous liquid polymers after 38 hr by H-migration
polymerization (112). Diallyl sulfides also react with
dithiols by H-polymerization to give polymers of moderate
molecular weight (113). The reactions were promoted by
UV light at 16°C and were believed to occur by a slow
stepwise mechanism. Diallyl sulfide reacts with SO_2 to
give equimolar sulfone copolymers (114). In one experiment, 11.4 g DAS and 6.4 g liquid SO_2 in 95% ethanol
saturated with silver nitrate were heated 65 hr at 50°C.
The polymer product melted at 155°C and had inherent
viscosity of only 0.05. References to other work on
diallyl sulfur compounds are given in Table 18.4.

Many allyl compounds react with sulfur dioxide to give
polysulfones having limited heat stability (low ceiling
temperatures)(115). In one example, 11.4 p diallyl sulfide and 0.75 p by volume of saturated solution of silver
nitrate in 95% ethanol were reacted with 6.4 p liquid SO_2
in a sealed tube at 50°C for 50 hr (116). The tan-colored
polymer which was isolated melted near 150°C and showed
an inherent viscosity in dimethyl formamide of 0.05 dl/g.

Reaction of allyl iodide with dipropyl sulfide during
several weeks at room temperature formed allyl dipropyl
sulfonium iodide $A(C_3H_7)_2SI$ (117). Diallyl alkyl and
triallyl sulfonium halides are quite unstable in light,
but it is feasible to prepare their mercuric halide complexes (118). Methyl diallyl sulfonium methosulfate has
been submitted to copolymerization with acrylamide in
aqueous persulfate-bisulfite (119). Allyl chloride was
reacted with dimethyl sulfide at room temperature during
8 days to prepare allyl dimethyl sulfonium chloride in
89% conversion (120). Such allyl sulfonium halides of
the type $CH_2=CRCH_2\overset{+}{S}RR(\overline{X})$ were reacted with aldehydes to
form vinylidene epoxides.

ALLICIN

The thiosulfinate AS̈SA, obtainable by partial oxidation
of diallyl disulfide, has most interesting biological
activity. Its roles as a bacteriostat and in the flavor
of the onion genus in relation to the sulfides and disulfides remain to be completely clarified. As shown on
page 501, allicin seems to be formed by enzymatic hydrolysis of S-allyl-L-cysteine sulfoxide $AS(O)CH_2CH(NH_2)COOH$,
when onion tissues are macerated. Allicin was discovered
to be a bacteriostatic agent by Cavallito and Bailey (121).
About 0.3 to 0.5% allicin was obtained from garlic cloves.
Even after the cloves had been stored for a year, allicin
could be obtained by extraction or steam distillation. It
was suggested that much of the odor of macerated garlic

TABLE 18.4

S,S-Diallyl Sulfur Compounds

Compound	bp, °C (at mm)	Other properties	References
Diallyl sulfide	140 40(10)	$n_D^{20}=1.4889$ $d_{27}=0.888$ g/cc mp, 83°C	Aldrich; cf. Challenger and Greenwood, J. Chem. Soc. 26 (1950)
Diallyl disulfide (garlic oil)	83(13)	$d_{15}=1.0237$	Milligan, J. Chem. Soc. 683 (1962); Twiss, J. Chem. Soc. 105, 39 (1914)
Diallyl trisulfide	122(16)	$n_D^{20}=1.5896$ $d_{15}=1.085$	Semmler, Arch. Phar. 230, 443 (1892)
Diallyl tetra-sulfide	Decomposes	--	--
Diallyl penta-sulfide	Decomposes	Soluble alcohol	Thomas, J. Chem. Soc. 125, 2215
Di-(3-chloroallyl) sulfide, etc.		Evaluated as vesicants in warfare	Childs, J. Chem. Soc. 2181 (1948)
Allicin	--	$n_D^{25}=1.5660$	Cavallito, et al.
Diallyl sulfoxide	115(12)	$n_D^{20}=1.5115$ mp 23°C	Lewin, J. Prakt. Chem. [2], 127, 87 (1930)
Diallyl sulfone	128(10)	$n_D^{20}=1.4893$ slightly soluble in water	Backer et al, Rec. Trav. 67, 456 (1948)
Diallyl sulphili-mine	--	From diallyl sulfide + chloramine T (mp 71°C)	Challenger, J. Chem. Soc. 26 (1950)
Diallyl methyl sulfonium hydroxide	--	--	U.S. 2,316,152 (Battelle)
Diallyl sulfate	--	--	Braun, Ber. 50, 293 (1917)
Diallyl sulfite	--	--	Carre, Bull. Soc. Chim. (France) [5] 1, 1249 (1934)
Triallylsulfonium hydroxide	--	Yellow needles (mp 84°C)	Steinkopf, J. Prakt. Chem. [2] 109, 251 (1925)
Dimethallyl sulfide	60(10) 173	Addition of Br_2 was fast $n_D^{20}=1.4862$	Olsen, Ind. Eng. Chem. 38, 1273 (1946)

may result from allicin. ASSA from leek oil was reported to inhibit mitosis even in low concentrations, whereas ASA had much less effect (122). Allicin may have contributed to this result. Low concentrations of allicin extracted by chloroform from ground garlic root displayed effective bacteriostatic action in low concentrations against a number of Gram-positive and Gram-negative bacteria (123). Research on the sulfur compounds from onions and garlic was reviewed and the antibacterial sulfoxide compounds were included (124). Stoll and Seebeck made allicin by oxidation of ASSA by perbenzoic acid or hydrogen peroxide. Allicin is said to have a pleasant garlic-like odor (125). It is unstable and apparently forms ASSA as a product. When two or more alliin homologs are present together in plants, different allicins may be formed.

Little basic research has been reported on the role of allicin and its control in the flavor of foods. However, Bockman and co-workers have been assigned an interesting patent (126). Flavors of allium condiments are modified by treating plants such as garlic, onions, and leeks to control formation of allicin. The allicin is formed from alliins by the enzymatic action of alliinase controlled by macerating the plants in the presence of citrus juice.

Allicin is soluble in benzene, alcohol, and ether, and about 2.5% dissolves in water. Permanganate and bromine solutions in water were rapidly decolorized by allicin. The compound is fairly stable to dilute acids, but when it is attacked by alkali it gives alkali sulfite and some polymeric material.

ALLYL SULFOXIDES AND SULFONES

Diallyl sulfide has been oxidized by benzoyl peroxide to give diallyl sulfoxide and diallyl sulfone (127):

$$(CH_2=CHCH_2)_2S \longrightarrow (CH_2=CHCH_2)_2SO \longrightarrow (CH_2=CHCH_2)_2SO_2$$

The sulfoxide also was obtained by oxidation with hydrogen peroxide at low temperatures (128). Oxidation in acetic acid by air gave largely the sulfone (129). Diallyl sulfoxide is a hygroscopic liquid melting near room temperature. Stoll and co-workers observed that on standing it became yellow and less soluble in water. On treatment with P_2O_5 even at 0°C violent reaction occurred. Whether these changes involved polymerization was not clarified. Reaction of phenyl allyl sulfoxide with butyl lithium and methyl iodide gave trans-crotyl alcohol by 1,3-transposition (130).

Diallyl sulfone is less soluble in water than diallyl sulfoxide. On heating with aqueous alkali the cyclic

ether sulfone is formed by way of normal addition of
water (131). Diallyl sulfone resists allylic halogenation
by N-bromosuccinimide, which is an indication of electron
withdrawal by the sulfone group. Diallyl sulfone showed
reverse addition of HBr in the presence of peroxides in
CCl_4 solution (132). Allyl sulfones do not add bromine
readily. Allyl alkyl sulfones ASO_2R can be pyrolyzed to
AR and SO_2 (133).

Allyl aryl sulfones have been prepared from reaction of
allyl halides with sodium aryl sulfonates (134). Similarly
1,3-bis(allylsulfonyl)benzene was prepared (135). Cope
and co-workers studied allyl phenyl sulfoxide and sulfone
as well as allyl vinyl sulfone; of these compounds only
the last was observed to polymerize on heating at 150°C
(136). Heating allyl p-tolyl sulfone with KOH in ethanol
was observed to give a brown resin (137).

The hydrogen atoms in the cyclic compound trimethylene
sulfone are quite acidic, and this compound reacts with
allyl chloride when heated for 45 min at reflux in the
presence of aqueous alkali, forming a diallyl trisulfone.

$$O_2 S-CH_2-SO_2$$
$$ACHSO_2HCA$$

Solutions of this monomer in diethylene glycol or dimethyl
formamide have given polymer on heating (138). Pentaallyl
trimethylene sulfone (mp 102°C) was copolymerized with
unsaturated polyesters and with styrene (139). Hexaallyl
trimethylene sulfone has been suggested as a crosslinking
agent for acrylic acid in the preparation of high-viscosity
mucilages (140).

Ethyl allyl sulfone and a series of aryl allyl sulfones
were copolymerized with acrylonitrile using persulfate
emulsion (141). Slow rates of copolymerization were encountered, since temperatures as high as 80°C for 20 hr
were employed (beyond the range normally applied with
persulfates). References to allyl sulfoxides and sulfones
are given in Table 18.5 and to some miscellaneous allyl
sulfur compounds in Table 18.6.

The existence of the thiosulfoxides has been doubted
since disulfides are more stable. Recent evidence suggests that thiosulfoxides may occur as intermediates in
rearrangements of allylic disulfides (142). Allyl methane-
thiosulfinate is said to occur in certain varieties of
leeks (143). Among other allyl compounds reported have
been diallyl sulfine derivatives (144), allyl chloro-
sulfinates (145), and methallyl sulfinite salts (146).

In addition to possibilities as comonomers, some of the
compounds in Table 18.6 have been suggested as rubber
auxiliaries, oil modifiers, herbicides, insecticides, and
reagents in analytical chemistry.

TABLE 18.5

Allyl Sulfoxides and Sulfones
and Related Compounds

Compound	bp, °C (or mp)	n_D^{20}	References
Diallyl sulfoxide	mp 23	1.5115	Barnard et al, J. Chem. Soc. 2444 (1949)
Methallyl methyl-sulfoxide	99(13)	1.4996[a]	Barnard et al.
Diallyl sulfone	128(10)	1.4893	Ford-Moore, J. Chem. Soc. 2436 (1949)
Allyl methyl sulfone	130(15)	--	Rothstein, J. Chem. Soc. 686 (1934)
Allyl ethyl sulfone	129(11)	--	Rothstein, J. Chem. Soc. 312 (1937)
Diallyl methyl sul-fonium hydroxide	--	--	U.S. 2,316,152 (Battelle)
Allyl p-chlorophenyl sulfone	mp 42	--	Ger. 878,718 (Cassella)
Allyl tolyl sulfone	mp 53	--	Otto, Ber. __24__, 1510 (1891)
Allyl benzyl sulfone	mp 65		

[a] Hydroscopic

ALLYL SULFONATES

Soluble salts of allyl and methallyl sulfonic acid have interest as copolymerizing surface active monomers. When 10 parts of vinyl acetate and 1 part of sodium methallyl sulfonate were polymerized in ethanol at reflux for 6 hr with azobis as catalyst, water-dispersible copolymers were obtained containing 6% sodium methallyl sulfonate units (147). Sodium allyl sulfonate in small proportions can act as an emulsifying monomer, as does sodium vinyl sulfonate in aqueous copolymerizations with vinyl acetate. In one example, vinyl acetate was added dropwise to agitated aqueous persulfate and ASO_3Na at 80°C, forming high-solids latices of small particle size (148). Vinyl acetate with less than 1% sodium methallyl sulfonate in emulsion copolymerization gave latices of good stability, Copolymerizations of vinyl acetate, sodium allyl sulfonate, and triethylene glycol dimethacrylate in light have been used in preparing photographic images (149).

Copolymerization of sodium allyl sulfonate with methyl acrylate was used to prepare high polymers bearing sulfonic acid groups (150). Copolymers of sodium allyl sulfonate with minor proportions of acrylamide have been

TABLE 18.6

Miscellaneous Allyl Sulfur Compounds

Compound	References
Monoallyl Derivatives	
Thiourethanes	Allewelt, U.S. 2,748,109 (Am. Viscose)
Thiocarbamates	Batty et al, Brit. 599,178 (ICI); Tilles et al, U.S. 2,916,369 (Stauffer); Harman et al, U.S. 3,382,144 (Monsanto)
Thionocarbamates	U.S. 3,288,782 (Monsanto)
Thiosemicarbazide	Stolle, J. Prakt. Chem. [2] 132, 220 (1932)
Homocysteine	Frankel et al, J. Chem. Soc. 463 (1961)
Sulfolanes	Morris, U.S. 2,439,345 (Shell)
Esters of Sulfoaliphatic acids	Caldwell et al, U.S. 3,260,707 (Eastman)
Dithiocarbanilate	Braun, Ber. 35, 3368 (1902); 36, 2259 (1903)
Rhodanine (mp 46°C, bp 186°C at 12 mm)	Beilstein 27, 243
Diallyl Derivatives	
Dithiocarbamates	Compin, CA 14, 3025 (1920); Weiss, U.S. 3,225,078 (Monsanto); D'Amico, U.S. 3,051,735 (Monsanto)
Thiocarbamate esters	Weiss et al, U.S. 3,166,401 (Monsanto)
Thionocarbamates	Weiss et al, U.S. 3,205,248 (Monsanto)
Dithiodiglycolate	Alfrey et al, J. Polym. Sci. 11, 61 (1953); Evans Research literature; Brit. 659,751
Thiolactams (polymerized)	Sidel'kovskaya, CA 63, 2895 (1965)
Aryl sulfonamides	Ash et al, J. Chem. Soc. 1877 (1951)
Xanthate	Oddo, CA 3, 1004 (1909)

suggested in antistatic finishes for textiles (151). Small amounts of alkali or ammonium allyl sulfonates were copolymerized with acrylonitrile to give acrylic fibers with better dye receptivity (152). In copolymerizations of acrylonitrile with sodium allyl sulfonate, different reactivity ratios were reported in different solvents (153). Using azobis initiator at 30 to 60°C, dimethyl sulfoxide gave $r_1 = 1.00$ and $r_2 = 0.38$, and a mixture of DMSO and water gave $r_1 = 1.25$ and $r_2 = 0.28$. It was suggested that different electron distributions in the monomer resulted from degrees of solvation. Copolymerizations occurred more readily in homogeneous solutions than in heterogeneous dispersions.

Bruson prepared sodium allyl sulfonate by boiling a mixture of allyl chloride, sodium sulfite, water, and ethanol under reflux for 12 hr. The distillation residue was crystallized from ethanol. Bruson copolymerized up to 15% allyl sodium sulfonate or methallyl sodium sulfonate with acrylonitrile in aqueous persulfate (154). A number of allyl sulfate and sulfonate monomers were described. Allyl and methallyl sulfonate salts have been copolymerized with acrylonitrile, along with vinyl acetate or methyl acrylate, for preparation of terpolymer acrylic fibers (155). Dyestuff receptivity and antistatic qualities have been evaluated in copolymers of acrylonitrile with minor proportions of sodium allyl sulfonate (156). Copolymerizations of minor proportions of allyl sulfonate salts with acrylonitrile have received attention also in Germany. The ammonium and substituted ammonium salts of allyl sulfonic acid were compared in rates of radical copolymerization and in molecular weight of the copolymer products (157). Copolymerization parameters for glycidyl allyl sulfonate with acrylonitrile were unfavorable for preparation of fibers (158). Acrylonitrile, methyl methacrylate, and sodium allyl sulfonate were copolymerized in dimethyl formamide using free radical initiators (159). Fibers were spun from solutions. From 0.8 to 5.7% allyl sulfonate groups have been identified in acrylic fibers by IR bands at 1043 cm^{-1} (SO) and at 2240 cm^{-1} (CN) (160).

Allyl sulfonic acid was heated with preformed acrylonitrile polymers in order to obtain acrylic fibers of improved dye receptivity (161). Allyl sulfonates such as allyl p-toluene sulfonate were suggested for copolymerization in minor proportions with acrylonitrile by persulfate-sulfite aqueous emulsion (162). N-Allyl and N-methallyl derivatives of methylaminoethane sulfonates (taurines) were also copolymerized with acrylonitrile (163).

Up to 0.3% sodium methallyl sulfonate was copolymerized with acrylonitrile (164). One copolymer from 76% conversion in aqueous thiocyanate salt solution contained 1.2% methallyl sulfonate units and had an intrinsic viscosity near 1.5. Methallyl sulfonate salts were prepared by reacting alkali metal sulfites with methallyl chloride in aqueous medium (165). By control of water content and molar ratio of the reactants, high yields were obtained at 8 to 15°C. The monomers were of interest for preparing copolymer textile finishes. Copolymerizations with acrylonitrile may be carried out in dimethyl sulfoxide and mixtures of dimethyl sulfoxide with water (166). Sodium allyl sulfonate has been used to improve the electrodeposition of bright nickel (167).

Allyl aryl sulfonates such as p-$CH_3C_6H_4SO_3A$ can be made from allyl halides and silver aryl sulfonates (168). They may be copolymerized with acrylonitrile for preparing dyeable fibers (169). Salts of N-allyl taurine $ANHCH_2CH_2SO_3H$ (mp 195°C) have been tried as copolymerizing emulsifiers (170).

Allyl vinyl sulfonate (bp 62°C at 0.2 mm) was prepared by reacting 2-chloroethane-1-sulfonyl chloride and allyl alcohol in methylene chloride and pyridine at 0°C (171). Heating the monomer with azo catalyst without solvent gave crosslinked heat-resistant polymers.

Allyl vinyl sulfonate (AVS) gave soluble polymers by free radical methods in solution (172). From 67 to 86% cyclic chain segments were indicated from polymerization in 0.24 mole/liter concentration. The following copolymerization reactivity ratios were observed where r_1 was AVS:

$$r_1 = 0.23 \qquad r_2 \text{ (styrene)} = 1.6$$
$$r_1 = 0.07 \qquad r_2 \text{ (methyl acrylate)} = 10.7$$
$$r_1 = 3.6 \qquad r_2 \text{ (vinyl acetate)} = 0.38$$

With vinyl acetate AVS showed more favorable reactive ratios than with butyl vinyl sulfonate. Allyl al yl sulfonate ASO_2OA was polymerized with azobis in benzene at 60°C, forming soluble polymers of mixed cyclic and open chain segments (173). The proportion of cyclic groups ranged from 66 to 47% as the concentration in benzene was raised from 0.28 mole/liter to bulk polymerization. The proportions of sulfonate groups incorporated into copolymers with styrene, with methyl acrylate, and with vinyl acetate were greater than when propyl allyl sulfonate was copolymerized with these three monomers. (See Fig. 18.3.) Allyl ethane sulfonate and allyl propyl sulfonate would not homopolymerize readily by radical initiation. Telomerizations of allyl sulfonates with n-butyl mercaptan and with $CBrCl_3$ were studied.

N-Allyl ethylene sulfonamide (bp 96°C at 0.5 mm) was prepared by Goethals and co-workers as follows (174):

$$ClCH_2CH_2SO_2Cl + CH_2=CHCH_2NH_2 \xrightarrow[0°C]{2N(C_2H_5)_3}$$

$$CH_2=CHSO_2NHCH_2CH=CH_2, \text{ etc.}$$

The IR spectrum of the monomer showed an allyl absorption at 1650 cm^{-1} and a weak vinyl absorption at 1620 cm^{-1}. Polymerizations in benzene with azobis at 45 and 25°C gave soluble polymers of reduced viscosities 0.15 and 0.5 g/dl in dimethyl formamide. Polymerization probably occurs partly by intramolecular cyclopolymerization of vinyl

and allyl groups to give cyclic sultam units $--\overset{}{C}H_2CH_2\overset{}{C}H--$.
$\overset{}{S}O_2-\overset{}{N}-\overset{}{C}H_2$
$\overset{}{H}$

The white polymers, softening at about 235°C, were soluble in aqueous alkali. The radical polymerizations of $CH_2=CHSO_2NHA$ were considerably faster than those of various N-alkyl vinyl sulfonamides.

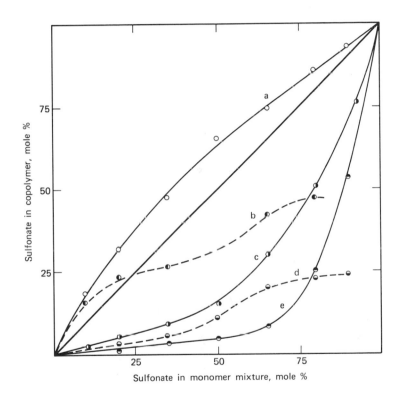

Fig. 18.3. Copolymerizations of the allyl ester of allyl sulfonic acid ASO_2OA in benzene at 60°C with azobis initiator occurred more readily with vinyl acetate (curve a) than with methyl acrylate (curve c) or styrene (curve e). Dotted curves b and c give mole % unsaturation in the vinyl acetate and the methyl acrylate copolymers, respectively. E. J. Goethals and E. DeWitte, J. Macromol. Sci. Chem. A5, 68 (1971).

References

1. G. Gerlich, Ann. <u>178</u>, 85 (1875); cf. O. Billeter, Ber. <u>8</u>, 464 (1875); L. Claisen et al, Ann. <u>401</u>, 21 (1913); for earlier work see H. Will, Ann. <u>52</u>, 1 (1844).
2. A. Iliceto et al, Gazz. Chim. Ital. <u>90</u>, 919 (1960); CA <u>55</u>, 21046 (1961).
3. H. Brintzinger et al, Z. Anorg. Chem. <u>256</u>, 86 (1948).
4. An explosion is recorded from this reaction at 80 psi: Chem. Eng. News <u>19</u>, 1408 (1941).
5. N. E. Searle, U.S. 2,462,433 (Du Pont).
6. D. W. Emerson, J. Chem. Educ. <u>48</u>, 81 (1971).
7. Data sheet, Morton Chemical Co.
8. H. T. A. Haas, Arch. Exptl. Path. Pharmacol. <u>199</u>, 637-641 (1942).
9. Communication from J. T. Venerable, Morton Chemical Co., Woodstock, Ill.
10. D. E. H. Frear, Pesticide Handbook, 17th ed., College Science Publications, State College, Pa., 1965.
11. V. G. Dethier, Insect Attractants and Repellents, Blakiston, 1947.
12. E. Schmidt, Ann. <u>560</u>, 228 (1948).
13. J. B. A. Dumas and J. Pelouze, Ann. <u>10</u>, 326 (1834).
14. H. Will, Ann. <u>52</u>, 1-51 (1844).
15. T. N. Gosh and P. C. Guha, J. Indian Chem. Soc. <u>7</u>, 267 (1930).
16. J. D. Riedel and E. deHaën, Ger. 504,996; Slotta et al, Ber. <u>63</u>, 214 (1930).
17. M. Asano and Y. Kameda, J. Pharm. Soc. Japan <u>55</u>, 8 (1935).
18. A. Todrick and E. Walker, Biochem. J. <u>31</u>, 297 (1937).
19. F. Sachs and H. Loevy, Ber. <u>37</u>, 874 (1904).
20. R. Andreasch and A. Zipser, Monatsh. <u>24</u>, 504 (1903).
21. G. Cavicchi, CA 4184 (1949); V. Madis, Pharmacia <u>18</u>, 252 (1938); E. Andre et al, CA 3888 (1949).
22. A. Pietschmann, CA <u>18</u>, 2486 (1924).
23. L. Rosenthaler, Pharm.-Z. <u>100</u>, 20 (1955); CA <u>49</u>, 8044 (1955).
24. L. E. Weller et al, J. Am. Chem. Soc. <u>74</u>, 1104 (1952).
25. A. Berka and J. Zijka, CA <u>53</u>, 1634 (1959).
26. H. Wagner, L. Hoerhammer, and H. Nufer, Arzneimittelforsch. <u>15</u>, 453 (1965).
27. A. G. Winter and L. Welleke, Naturwiss. <u>38</u>, 354 and 437 (1952).
28. G. Reis and F. Sjoestrand, Skand. Arch. Physiol. <u>79</u>, 139 (1938); H. T. A. Haas et al, Arch. Exp. Path. Pharmak. <u>209</u>, 138 (1959).
29. Methods of analysis in H. E. Cox and D. Pearson, The Chemical Analysis of Foods, Chemical Publishing Company, 1962.
30. Erna Bach, CA <u>38</u>, 3697 (1944).
31. H. A. Bruson and J. W. Eastes, J. Am. Chem. Soc. <u>59</u>, 2011 (1937).
32. A. Kjaer et al, Acta. Chem. Scand. <u>7</u>, 518 and 1271 (1953); CA <u>48</u>, 2598 (1954).
33. A. Kjaer et al, Acta Chem. Scand. <u>10</u>, 432 and 1365 (1956).
34. M. A. Greer, J. Am. Chem. Soc. <u>78</u>, 1260 (1956).
35. M. W. Tamele et al, Ind. Eng. Chem. <u>33</u>, 116 (1941); cf. A. Kjaer et al, CA <u>48</u>, 2598 (1954).

36. M. G. Ettingler and J. E. Hodgkins, J. Am. Chem. Soc. 77, 1831 (1955).
37. F. Hurd and R. Livingston, J. Phys. Chem. 44, 870 (1940).
38. B. H. Carroll and D. Hubbard, J. Res. Natl. Bur. Std. (U.S.) 12, 334 (1934).
39. R. Kitamura, J. Pharm. Soc. Japan 54, 11 and 110 (1934); CZ I, 3345 (1934).
40. J. H. Yoe and L. G. Overholser, Ind. Eng. Chem. (Anal.) 14, 435 (1942); J. E. Currah et al, Ind. Eng. Chem. (Anal.) 18, 120 (1946).
41. S. E. Sheppard and H. Hudson, J. Am. Chem. Soc. 49, 1814 (1927); Photogr. J. 65, 380 (1925).
42. E. A. Suthern and E. E. Loening, J. Photogr. Sci. 4, 154 (1956); CA 51, 3331 (1957); cf. CA 48, 1865 (1954).
43. A. Narath and E. Borcke, Sci. Ind. Photogr. 28, 104 (1957); CA 51, 6407 (1957).
44. L. D. Taylor, U.S. 3,578,458 (Polaroid).
45. J. B. Gardner and G. B. Harper, U.S. 3,386,972 (Dow).
46. P. Alexander and D. Toms, J. Polym. Sci. 22, 343 (1956); CA 48, 8080 (1954).
47. H. Koenigsfeld and C. Prausnitz, CA 8, 742 (1914).
48. P. Alexander and M. Fox, J. Polym. Sci. 12, 533 (1954).
49. E. G. Nicolas, Compt. Rend. 180, 1288 (1925).
50. A. Bussy and P. J. Robiquet, Compt. Rend. 10, 4 (1840).
51. H. Will, Ann. 52, 1 (1844).
52. O. Hecht, Ber. 23, 287 (1890); CA 78, 8946 (1973).
53. E. A. Werner, J. Chem. Soc. 283 (1890); E. L. Brown and N. J. Campbell, J. Chem. Soc. 1699 (1937); H. J. Backer and J. Kramer, Rec. Trav. Chim. Pays-Bas 53, 1101 (1934); CZ I, 1530 (1935); W. J. Levy and N. J. Campbell, J. Chem. Soc. 1442 (1939).
54. T. B. Johnson and H. W. Haggard, J. Am. Chem. Soc. 37, 177 (1915).
55. O. Hecht, Ber. 23, 1663 (1890); cf. H. L. Wheeler and G. S. Jamieson, J. Am. Chem. Soc. 25, 719 (1903).
56. M. E. Brokke et al, U.S. 3,510,503 (Stauffer).
57. J. Obermeyer, Ber. 20, 2925 (1887).
58. F. Challenger and D. Greenwood, Biochem. J. 44, 87 (1949).
59. C. C. Price and W. H. Snyder, J. Org. Chem. 27, 4639 (1962); cf. D. S. Tarbell and W. E. Lovett, J. Am. Chem. Soc. 78, 2259 (1956).
60. W. E. Parkham et al, J. Org. Chem. 27, 2415 (1962).
61. R. P. Louthan, U.S. 3,031,391 and 3,084,116 (Phillips).
62. E. F. Kohman, Food Technol. 6, 288 (1952); CA 46, 11496 (1952).
63. H. Nishimura et al, Tetrahedron 27 (2) 307 (1971).
64. H. J. Van Der Ploeg et al, Rec. Trav. Chim. Pays-Bas 81, 775 (1962).
65. K. Griesbaum et al, J. Org. Chem. 28, 1952 (1963); 30, 3829 (1965).
66. C. S. Marvel and E. D. Weil, J. Am. Chem. Soc. 76, 61 (1954).
67. L. Bateman and J. I. Cuneen, J. Chem. Soc. 1596 (1955); cf. R. W. Bost and M. W. Conn, Org. Syn. 15, 72 (1935).
68. L. Bateman and F. W. Shipley, J. Chem. Soc. 2888 (1958).
69. C. D. Hurd and H. Greengard, J. Am. Chem. Soc. 52, 3356 (1930).

70. H. Kwart and E. R. Evans, J. Org. Chem. _31_, 413 (1966).
71. E. N. Karaulova et al, Zh. Obshch. Khim. _27_, 3034 (1957).
72. C. Y. Meyers et al, J. Org. Chem. _28_, 2440 (1963).
73. H. Kwart and E. R. Evans, J. Org. Chem. _31_, 410 and 413 (1966).
74. H. Kwart and M. H. Cohen, J. Org. Chem. _32_, 3135 (1967).
75. H. Kwart and M. H. Cohen, Chem. Commun. No. 6, 319 (1968).
76. H. S. Broadbent et al, J. Am. Chem. Soc. _76_, 1519 (1954);
    A. C. Cope et al, J. Am. Chem. Soc. _72_, 66 (1950).
77. D. S. Tarbell et al, J. Am. Chem. Soc. _74_, 48 (1952).
78. J. Braun and R. Murjahn, Ber. _59B_, 1202 (1926); cf. J. F. Carson
    and F. F. Wong, J. Org. Chem. _24_, 175 (1959).
79. R. C. Krieg and S. Tocker, J. Org. Chem. _20_, 1 (1955).
80. A. A. Oswald et al, U.S. 3,488,270 (Esso); J. Polym. Sci. C
    No. 24, 113 (1968).
81. H. J. Backer and J. Kramer, Rec. Trav. Chim. Pays-Bas _53_,
    1106 (1934).
82. J. C. Patrick, U.S. 2,527,377 (Thiokol).
83. W. Steinkopf and R. Bessaritsch, J. Prakt. Chem. [2], _109_,
    230 (1925); CZ I, 1871 (1925).
84. M. J. Hatch, U.S. 3,426,046 (Dow).
85. A. W. Hofmann and A. Cahours, Ann. _102_, 292 (1857).
86. J. Braun and R. Murjahn, Ber. _59B_, 1202 (1926).
87. J. Braun and T. Plate, Ber. _67B_, 281 (1934).
88. H. J. Backer and P. L. Stedehouder, Rec. Trav. Chim. Pays-Bas
    _52_, 453 (1933).
89. A. A. Oswald et al, J. Polym. Sci. C No. 24, 113 (1968).
90. V. R. Ratzlow and C. A. Ray, U.S. 3,156,731 (Phillips).
91. M. Tamele et al, Ind. Eng. Chem. _33_, 115 (1941).
92. H. J. Backer et al, Rec. Trav. Chim. Pays-Bas _67_, 451 (1948).
93. T. F. Doumani, U.S. 2,532,612 (Union Oil).
94. P. Sabatier and A. Mailke, Compt. Rend. _150_, 1217 (1910);
    CZ II, 288 (1910).
95. H. Rheinboldt and O. Diepenbruck, Ber. _59_, 1311 (1926).
96. W. H. Peterson, J. Biol. Chem. _34_, 583 (1918).
97. Theodor Wertheim, Ann. _51_, 289 (1844).
98. R. A. Bernhard, Arch. Biochem. Biophysics _107_, 137 (1964);
    O. E. Schultz and H. L. Mohrman, CA _63_, 9744 and 8116 (1965).
99. F. Challenger et al, J. Chem. Soc. 26 (1950) and 1877 (1951).
100. A. Stoll and E. Seebeck, Helv. Chim. Acta _31_, 199 (1948);
     D. Barnhard et al, J. Chem. Soc. 2443 (1949); cf. Y. O. Gabel,
     CA _47_, 2683 (1953).
101. F. W. Semmler, Arch. Pharm. _230_, 434 (1892); cf. C. G. Moore
     and B. T. Trego, CA _57_, 2062 and 14924 (1962).
102. H. J. Backer et al, Rec. Trav. Chim. Pays-Bas _67_, 456 (1948);
     cf. F. Challenger and G. Greenwood, Biochem. J. _44_, 90 (1949).
103. R. A. Bernhard (University of California, Davis), Qual. Plant
     Mater. Veg. XVIII, _1-3_, 72-84 (1969).
104. J. V. Jacobsen et al, Arch. Biochem. Biophys. _104_, 473 (1964);
     cf. C. F. Carson and F. Wong, J. Agr. Food Chem. _9_, 140 (1961).
105. A. R. Sagir et al, Plant Physiol. _40_, 681 (1965).

106. R. A. Bernhard et al, Am. Soc. for Hort. Sci. 84, 386 (1964);
     Arch. Biochem. Biophys. 107, 137-140 (1964); cf. O. E. Schultz
     and H. L. Mohrman, CA 63, 8116 and 9744 (1965).
107. A. Guillaume and A. Shajik, Tunisie Med. 39, 951 (1951).
108. Edith Schnauffer, Die Kuechenzweibel in der Medezin, Thesis,
     Munich, 1954; A. B. D. Enalak, Contribution to the Study of
     Garlic and Onions (including "preparations galenique"),
     Strasbourg, 1950.
109. A. R. Saghir, J. W. Cowan, and J. Salji, Nature 211, 87 (1966);
     Eur. J. Pharmacol. 2, 399 (1968).
110. Helen H. Darrah and Gregory T. Heyl, Gettysburg College, un-
     published.
111. R. H. Martin, U.S. 2,996,484 (Monsanto).
112. H. Niebergall, Ger. 1,116,411; CA 56, 7523 (1962); CA 55,
     20503 (1961).
113. A. A. Oswald et al, J. Polym. Sci. C No. 24, 113 (1968);
     cf. references in Coffman, U.S. 2,347,182 (Du Pont).
114. C. D. Wright and W. S. Friedlander, U.S. 3,072,616 (3M).
115. F. S. Dainton et al, J. Polym. Sci. 26, 351 (1957).
116. C. D. Wright and W. S. Friedlander, U.S. 3,072,616 (3M).
117. W. Steinkopf and R. Bessaritsch, J. Prakt. Chem. [2] 109, 230
     (1925); CZ I, 1871 (1925).
118. G. B. Butler and B. M. Benjamin, J. Am. Chem. Soc. 74, 1846 (1952).
119. J. A. Price and W. H. Schuller, U.S. 2,923,700 (Am. Cyan.).
120. M. J. Hatch, U.S. 3,462,462 (Dow); cf. V. Franzen and H.-E.
     Driesen, Ber. 96, 1881 (1963).
121. C. J. Cavallito et al, J. Am. Chem. Soc. 66, 1950 (1944);
     69, 1710 (1947).
122. O. Hoffman-Ostenhoff and K. Keck, Monatsh. 82, 562 (1951).
123. Anna Szitagyi, CA 49, 14275 (1955).
124. A. Stoll and E. Seebeck, Helv. Chim. Acta 31, 189 (1948);
     F. Challenger and D. Greenwood, Biochem. J. 44, 87 (1949).
125. A. R. Sagir et al, Am. Soc. Hort. Sci. 84, 386 (1964).
126. C. Bockman, R. S. Nelson, and W. A. Klein, U.S. 3,424,593.
127. L. N. Levin, J. Prakt. Chem. 127, 79 (1930); CA 24, 4257 (1930).
128. A. Stoll and E. Seebeck, Helv. Chim. Acta 31, 209 (1948).
129. H. J. Backer et al, Rec. Trav. Chim. Pays-Bas 67, 451 (1948).
130. D. A. Evans et al, J. Am. Chem. Soc. 93, 495 (1971).
131. A. H. Ford-Moore, J. Chem. Soc. 2436 (1949).
132. D. Edwards and J. B. Stenlake, CA 50, 10094 (1956).
133. E. M. LaCombe and B. Stewart, J. Am. Chem. Soc. 83, 3457 (1961).
134. R. Otto, Ann. 283, 181 (1894); J. Troeger and A. Hinze, J.
     Prakt. Chem. [2] 55, 202 (1897).
135. J. Troeger and W. Meine, J. Prakt. Chem. [2] 68, 313 (1903).
136. A. C. Cope et al, J. Am. Chem. Soc. 72, 59 (1950).
137. H. J. Backer et al, Rec. Trav. Chim. Pays-Bas 70, 365 (1951);
     CA 45, 8470 (1951).
138. E. M. Evans and H. T. Hookway, Brit. 614,538 and 2,535,533-4
     (Brit. Resin Products).
139. A. E. Smith, U.S. 2,591,020 (USA).

140. J. F. Jones, U.S. 2,958,679 (Goodrich).

141. W. Zerweck and W. Kunze, Ger. 878,718 (Cassella).

142. G. Hoefle and J. E. Baldwin, J. Am. Chem. Soc. 93, 6307 (1971).

143. M. Yashimura, CA 55, 12568 (1961).

144. A. S. F. Ash et al, J. Chem. Soc. 2792 (1952).

145. S. H. Sharman et al, J. Am. Chem. Soc. 80, 5965 (1958).

146. E. E. Wellisch et al, J. Polym. Sci. B 2, 35 (1964).

147. P. R. Austin, U.S. 2,834,759 (Du Pont).

148. N. Turnbull, U.S. 2,859,191 (Du Pont).

149. Brit. 835,849 (Du Pont).

150. H. H. Hoffman and W. K. Wilkinson, U.S. 2,849,330 (Du Pont).

151. R. L. Baechtold, U.S. 3,134,686 (Am. Cyan.).

152. Fr. 1,349,974 (Asahi Chem. Ind.); CA 61, 1999 (1964).

153. Z. Izumi et al, J. Polym. Sci. A 3, 2965 (1965); Chem. High Polym. (Japan) 21, 79-82 (1964).

154. H. A. Bruson, U.S. 2,601,256 (Ind. Rayon).

155. V. Groebe and H. Reichert, Faserforsch. Textiltech. 16, 339 (1965); K. Morida et al, CA 78, 5335 (1973).

156. R. N. Blomberg, U.S. 3,123,434 (Du Pont).

157. V. Groebe et al, Faserforsch. Textiltech. 18, 573 (1967); CA 68, 40124 (1968); CA 68, 59916 (1968).

158. Y. Iwakura et al, Makromol. Chem. 104, 37 (1967).

159. D. Corradi and A. Pasin, U.S. 3,463,616 (Snia Viscosa).

160. W. Groebe et al, CA 77, 89057 (1972).

161. T. G. Traylor and A. Armen, U.S. 3,043,811 (Dow).

162. A. G. Lowther and F. Reeder, Ger. 975,540 (Courtaulds); CA 57, 2458 (1962).

163. V. Groebe and H. Reichert, CA 70, 97227 (1969).

164. R. A. Davison and F. Reeder, Brit. 923,377; CA 59, 1779 (1963).

165. M. O. Robeson, U.S. 3,453,320 (Procter Chemical).

166. Z. Izumi and H. Kitagawa, Chem. High Polym. (Japan) 26/285, 153 (1969); cf. CA 78, 31315 (1973).

167. D. G. Foulke, U.S. 3,366,577; CA 78, 10905 (1973).

168. W. Stoll, Z. Physiol. Chem. 246, 1 (1937); CZ I 4646 (1937).

169. A. G. Lowther and F. Reeder, U.S. 2,796,414 (Courtaulds).

170. J. W. James, J. Prakt. Chem. [2] 31, 415 (1885).

171. E. J. Goethals, Polym. Lett. 4, 691 (1966).

172. E. DeWitte and E. J. Goethals, Makromol. Chem. 115, 234 (1968); CA 69, 36491 (1968).

173. E. J. Goethals and E. DeWitte, J. Macromol. Sci.-Chem. A 5, 63, 73, and 15 (1971).

174. E. J. Goethals, J. Bombeke, and E. DeWitte, Makromol. Chem. 108, 312 (1967).

19. ALLYL AMINES AND THEIR SALTS

Allyl amines are useful organic intermediates for
synthesis of pharmaceuticals, textile and agricultural
products. Many of the free amines are quite toxic and
irritating and they do not polymerize readily. However,
the allyl ammonium halides and other salts formed by
neutralization of allylic amines are more readily handled
and some homopolymerize and copolymerize fairly readily
in suitable free radical systems. A number of diallyl
alkyl ammonium halides can be polymerized to form soluble
high polymers which contain cyclic units. Polymers from
diallyl dimethyl ammonium chloride (DDAC) and related
monomers recently have found applications in conductive
papers, water purification, sewage disposal, cosmetics,
textile and ion-exchange products. Free amine and amphoteric polymers can be obtained from the copolymerized
salts. Nitrogen-allyl drugs which have found medical use
include narcotic antagonists and diallyl toxifcrinc (1),
a curare-type muscle relaxant.

ALLYLAMINE

Allylamine or 3-aminopropylene $CH_2=CHCH_2NH_2$ is a colorless or yellowish liquid having a very strong ammonialike
odor. It is quite toxic. In animal tests inhalation
produced irritation of nose and mouth, irregular respi-
ration, cyanosis, and convulsions. Allylamine has been
used in the manufacture of diuretics, in the preparation
of anionic copolymers, and as an intermediate in many
organic syntheses. It is a weaker base than ethylamine
and ammonia but stronger than aniline. Allyl amines have
been reported among decomposition products from blood,
urine, feces, and pus, as well as synthetic polyamides.
Allylamine has been identified as a degradation product
of allyl mustard oils. In animals injections of allylamine have been observed to produce lesions in the brain,
vascular system and heart similar in some cases to those
occurring in certain human diseases (2).

Allylamine can be prepared from hydrolysis by hydrochloric acid of allyl isothiocyanate (mustard oil)(3) or

by reaction of allyl halides with excess ammonia under
pressure (4). Earlier procedures with high conversion
gave considerable diallylamine byproduct (5). Reaction
of ammonia with allyl chloride at about 1000°C for less
than 0.1 sec gives allylamine with little secondary or
tertiary amine (6). A Shell pilot plant for manufacture
of allyl and methallylamine has been described (7). In a
recently patented process, ammonia and allyl chloride are
reacted at a hot platinum filament at 600 to 1500°C during
0.001 to 0.1 sec, with 10% conversion and 90% yield (8).

Allylamine was hydrogenated in the presence of rhodium
catalyst more rapidly than other less Lewis-basic olefinic
compounds studied (10). Values of the rate constants
$K \times 10^5$ were reported as follows:

Compound	Value
Allylamine	3.12
Acrylic acid	2.63
Acrylonitrile	2.12
Allyl alcohol	2.08
Allyl acetate	1.94
Allyl ethyl ether	0.97
Acrolein	0.28

Allylamines form pi-complexes with platinum(II) compounds.
From studies of pi-complexes of different substituted
allyl amines, it was concluded that both steric and electronic effects are important in determining the strengths
of Pt^{II}-olefin bonds (11).

The alkenyl amines, besides reactivity of the allyl
double bond, show expected formation of salts, amides,
imines, and imides. For example, allylamine reacts with
esters to give N-allyl amides (Figs. 19.1 and 19.2). In
addition reactions of active hydrogen compounds to allyl
amines, Markovnikov normal additions seem to predominate,
but reverse addition has been observed with H_2S (12).
Reaction of $CH_2=CHCH_2NH_2HCl$ with HCl gave both normal and
reverse addition (13). Allylamine added to benzene in
the presence of aluminum chloride (14).

$$NH_2CH_2CH=CH_2 \; + \; \bigcirc \; \longrightarrow \; \begin{matrix} H_2NCH_2 \\ HC- \\ CH_3 \end{matrix} \bigcirc$$

Heating allylamine with dilute hydrochloric or sulfuric
acid at 200°C for 10 hr resulted in normal addition of
water to give a 1-amino-2-propanol salt. Addition to
phosphines was studied (15). Addition of lithium alkyls
or other alkali metal organic compounds to allyl amines

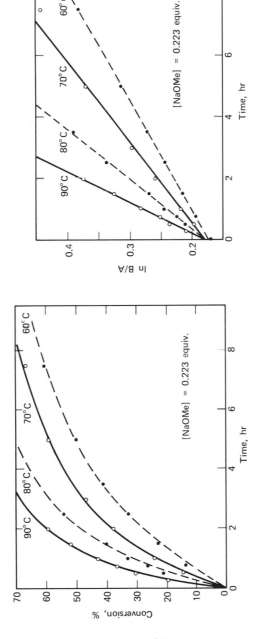

Fig. 19.2. Reactions of allylamine with glyceryl stearate as in Fig. 19.1. Second-order kinetics of the reactions were indicated by straight line plots of the log of concentration of amine over concentration of ester against time. E. F. Jordan et al.

Fig. 19.1. Reactions of allylamine with tallow (glyceryl stearate) in presence of sodium methylate to form N-allylstearamide at four temperatures. E. F. Jordan, B. Artymyshyn, C. R. Eddy, and A. N. Wrigley, J. Am. Oil Chem. Soc. 43, 76 (February 1966).

gives orange to red colors. A crystalline picrate of allylamine obtained from chloroform-alcohol solution melted at 141°C.

Two molecules of ethylene oxide have been added to allylamine to form di(2-hydroxyethyl)allylamine (16). Reaction of ANH_2 with formaldehyde in ether gives 1,3,5-triallyl hexahydro-s-triazine (17). This, like the similar reaction of allylamine with chloromethyl methyl ether, probably occurs by trimerization of $CH_2=CHCH_2N=CH_2$ (18). Allylamine has been reacted with carbon monoxide in the presence of copper cyanide or azobis catalyst to form N-allyl formamide (19).

Allylamine has reacted with chloracetic acid in the presence of HCl to give a hydrochloride of allylamino-acetic acid (20). Alkyl allyl amines can be prepared readily by alkaline alkylation (21):

$$H_2NCH_2CH=CH_2 + CH_3I \xrightarrow{\text{KOH}} CH_3NHCH_2CH=CH_2 + KI + H_2O$$

Allylamine has added to ethylene sulfide by refluxing for 4 days in ethanol to give 2,2'-(allylimino)diethanethiol (22). Oxidation of allylamine over silver catalyst at 500°C was evaluated as a synthesis of acrylonitrile (23).

Allylamine and tetrahydrofuran over Al_2O_3 at 400°C formed 1-allylpyrrolidone (24). Allylamine reacted with gamma-butyrolactone at 200°C to give 4-hydroxy-butyrallyl-amide and at 270°C to give largely 1-allyl-2-pyrrolidone (25). Rates of reaction of allylamine with phthalic anhydride in acetic acid solution to form N-allyl phthali-mide were studied (26). Allyl and methallyl amines have been added to benzene, naphthalene, and other aromatic hydrocarbons, yielding largely 1,4- or 4,4'-isomers (27). Thus methallyl amine and benzene with $AlCl_3$ catalyst in cyclohexane solution formed bis(amino-t-butyl)benzene. Allylamine hydrochloride (or allyl ammonium chloride) reacts with bentonite in water to give an organophilic filler of interest for rubber compounding (28).

Relatively little work has been reported with methallyl amine. It has been made by saponification of N-methallyl acetamide (29). Among the methallyl amine derivatives prepared have been epoxy compounds (30).

The following properties of commercial allyl amines were given in a bulletin of Shell Chemical Company:

Amine	Monoallyl	Diallyl	Triallyl
Boiling point, °C	52.9	110.4	149.5
Melting point, °C	-88.2	-88.4	<-70
Specific gravity, 20/4°C	0.7627	0.7874	0.800
Solubility in water, 20°C, %	100	8.6	0.25
pH of 0.1M aqueous solution	11.2	11.5	--

The three allyl amines are miscible in all proportions
with toluene, with acetone, and with n-octane.

Allyl amines are much used as intermediates for organic
synthesis. For example, diallylamine was reacted with
propylene sulfide to form 4-allyl-1,4-thiazine derivative
that was found to be promising as an antioxidant and
insecticide (31). Allylamine was reacted at 130°C with
N-vinyl pyrrolidone polymer in the presence of di-t-butyl
peroxide to form aminoalkylated polymers (32). The treatment facilitated crosslinking of the polymers. Mono- and
diallyl amines can be condensed to form pyridine derivatives (33).

POLYMERIZATIONS OF ALLYLAMINE

Homopolymerization of liquid allylamine does not occur
with conventional free radical or ionic initiators to
give high polymers. Allylamine inhibits or retards
radical polymerizations of vinyl monomers such as acrylonitrile (34). Surprisingly the gaseous catalyst tetrafluorohydrazine (N_2F_4) gave hard, brown polymers from
allylamine (35). In a patent example of micro-polymeri-
zation, 0.007 mole of pure allylamine was charged into a
pressure reactor at -76°C. After evacuation, 0.006 mole
of N_2F_4 was charged into the reactor which was then
heated at 100°C for 70 min. Conversion was near 100% to
brown resinous polyallylamine, and N_2F_4 was recovered
practically unchanged. Numerous other allyl compounds
were said to polymerize by this process, but it was not
successful with allyl halides. Allylamine polymer of
molecular weight 1000, prepared by use of hydrogen peroxide, was reacted with diglycidyl ether giving polymer
gels (36). A polyallylamine was said to result from
direct hydrogenation of polyacrylonitrile latex, but this
seems doubtful (37).

Devlin of Shell gave examples of the reluctance of
allylamine to copolymerize by free radical conditions (38).
When 80 parts allylamine, 20 parts styrene, and 2 parts
di-t-butyl peroxide were heated for 5 hr at 140°C, there
was obtained only a low yield of amber solid copolymer of
molecular weight near 1200 and containing about 12% allylamine groups. The polyamine product was used to cure an
epoxy resin with heating for several hours at 100°C. By
heating 150 parts allylamine with 20 parts butadiene and
4 parts di-t-butyl peroxide for 4 hr at 140°C and then
distilling off the residual monomers under vacuum, an
amber, rubbery solid was formed. It was soluble in
chloroform and contained about 17% allylamine groups. By
heating 80 parts allylamine with 20 parts methyl methacrylate and 2 parts of di-t-butyl peroxide for 5 hr at
140°C, a copolymeric amine was recovered having about 32%

allylamine groups. The patent claims were restricted to copolymers of low (500 to 1000) molecular weight. Soluble prepolymers from diallylamine copolymerization to low conversions were disclosed, as well as crosslinked amine copolymer ion-exchange resins, but no examples of preparation of the latter were described. A copolymer latex of low molecular weight was prepared using 15 parts allylamine, 15 parts styrene, and 70 parts butadiene as monomers (39). The product was mixed with resorcinol-formaldehyde resin to promote tire cord adhesion.

There has been considerable interest in copolymerizations of allyl amines and their salts with unsaturated acidic monomers to form useful amphoteric copolymers. Allylamine hydrochloride ($ANH_2 \cdot HCl$ or ANH_3Cl) was copolymerized with acrylic acid using bisulfite-activated, aqueous persulfate under nitrogen at 0 to 35°C for several days (40). A polyampholyte was reported in Russia by copolymerization of allylamine and potassium methacrylate in aqueous solution using ^{60}Co initiation, followed by treatment of the product with mineral acid (41). Copper complexes of approximately equimolar copolymers of allylamine with methacrylic acid and having molecular weights of 10,000 or above were suggested as oral deodorants (42). Neutralized allylamine was copolymerized with methacrylic acid in aqueous solution, forming high molecular weight products containing 6.9% nitrogen (43). Solid copolymers of allylamine and acrylic acid were formed by use of gamma radiation (44). Polymerizations of preformed allyl ammonium salts are discussed later.

Copolymers of styrene with methallylamine were acetylated by heating for an hour near 100°C with acetic anhydride and pyridine (45). The products remained soluble in acetone. Copolymers of allylamine with long alkyl fumarates and vinyl acetate, prepared in toluene using ^{60}Co radiation, were proposed as lubricating oil additives (46). Graft copolymers with polybutene also were disclosed. Copolymers of allylamine with vinyl acetate gave allylamine-vinyl alcohol copolymers having possibilities as additives to synthetic fibers for improving dye receptivity (47). Allylamine derivatives have been reacted with chlorosulfonated polyethylene (48). Allylamine was copolymerized with ethylene in persulfate emulsion for 18 hr at 75°C above 200 atm (49). Copolymers containing combined allylamine could be coagulated by addition of acid to the latex. Unsaturated polyamides have been reacted with allylamine in the presence of peroxide catalyst and cobalt naphthenate to form basic ion-exchange resins (50). In many of the examples above it was not clear whether the free amine or the amine salt (acidic medium) was copolymerizing.

ALLYL ALKYL AMINES

The sec- and t-amines $CH_2=CHCH_2NHR$ and $CH_2=CHCH_2NRR'$ can be prepared by reaction of allyl halides with alkyl amines, followed by treatment of the allyl ammonium salts with alkali, for example:

$$ACl + HN(CH_3)_2 \longrightarrow [H\overset{A}{\underset{CH_3}{N}}CH_3]^+Cl^- \xrightarrow{\ NaOH\ } AN(CH_3)_2 + NaCl + H_2O$$

Allylethylamine has been made by reaction of allyl chloride with 33% ethylamine solution (51). Allylmethylamine was made by reacting the N-methyl-N-allyl amide of p-toluene sulfonic acid with sodium and butanol. Preparations of allyldimethylamine and allyldibenzylamine have been described (52). Allyldimethylamine has been made by reaction of allyl chloride and dimethylamine in the presence of aqueous NaOH (53). Allyldimethylamine formed a picrate from ethanol solution which melted at 116°C. Cope and Towle reacted allyldimethylamine with hydrogen peroxide to form allyl dimethyl aminoxide which, on heating under nitrogen above 105°C, gave an N,N-dimethyl-O-allyl derivative.

In a commercial process methylamine was heated with allyl chloride at 180°C, 80 atm, for 30 min (54). After cooling, excess aqueous KOH was added and an 86% yield of methylallylamine was distilled off followed by some dimethylallylamine. Allylation of dimethylamine by use of an allyl ether was carried out in the presence of a palladium complex (55).

In 1906 Menschutkin of the Polytechnic of St. Petersburg reported pioneer quantitative studies of rates of reaction of allyl bromide with cyclohexylamine (56). Cycloalkyl amines reacted much faster than isomeric open-chain primary amines. The hydrogen bromide liberated was titrated with alcoholic sodium hydroxide.

Diethyl allyl ammonium chloride adds HCl only very slowly (57). Heating for 48 hr with concentrated HCl at 120°C formed about equal amounts of diethyl-2-chloropropyl and diethyl-3-chloropropyl ammonium chlorides. However, heating for only 12 hr at 155°C caused the second addition product to predominate. Kharasch and Fuchs found that adding HBr in the presence of peroxide catalyst (ascaridole) formed 2-bromo to 3-bromo addition compounds in ratio of 2 to 3 (57). Diethyl allyl ammonium bromide has been observed to melt about 190°C, and a picrate of diethylallylamine melted at 91°C. Platinum halide complexes are formed of the type $CH_2=CHCH_2NHR \cdot PtCl_3$, and these appear to be dimeric (58). Isopropylallylamine as well as alkyl chloroallyl amines have been reacted with CS_2 to

form allyl dithiocarbamate acids and esters (59). Base-
catalyzed isomerization of tertiary allylamines
$CH_2=CHCH_2NRR'$ to give cis-propenylamines has been compared
with similar isomerizations of allyl ethers (60).

A series of allyl secondary amines $RN(CH_3)A$ was pyrolyzed to form azomethenes or methylene imines; some of
them polymerized immediately (mol. wt. 1000-6000) (61).

Allyl alkyl amines and allyl dialkyl amines give unsubstituted ammonium chloride salts or amine hydrochlorides
at once when hydrochloric acid is added. Trimethyl allyl
ammonium chloride has been prepared by action of sodium
ethylate solution on trimethyl 3-chloropropyl ammonium
chloride (62). The free amine forms a picrate melting at
220°C and the trimethyl allyl ammonium iodide melted near
108°C (63). Allyl trimethyl ammonium hydroxide is homo-
neurine. Dibutyl dodecyl allyl ammonium bromide was made
by reaction of allyl bromide with dibutyldodecylamine
(64). Tris(hydroxymethyl) allyl ammonium bromide was
prepared as a viscous red oil (65). On treatment with
strong bases, triethyl allyl ammonium bromide rearranged
to give diethyl gamma-ethyl allyl amine (66). Properties
of a number of allyl and methallyl amines and more
references are listed in Table 19.1. Monoallyl amines
are useful intermediates in the preparation of sedatives,
diuretics, antiseptics, and other pharmaceuticals.

Most free allyl amines like aryl amines inhibit or
retard free radical polymerizations. With peroxides and
other oxidizing agents they form red to brown colored
products rather than forming high polymers. When polymerizations and copolymerizations have been most successful, the substituted ammonium salts formed from reaction
of allylic amines with acids probably were undergoing
polymerization. Minor proportions of amine groups can be
incorporated by suitable copolymerizations to give interesting basic or amphoteric polymers. Of particular
interest have been copolymers of acrylonitrile $CH_2=CHCN$
with minor proportions of allyl amines for use as dye-
receptive acrylic fibers. Chaney of Chemstrand described
copolymerizations of acrylonitrile with minor proportions
of allyl alkyl amines or methallyl alkyl amines in acidic
aqueous persulfate (67). The acrylonitrile and vinyl
acetate could be added portionwise in order to control
the rate at the relatively high temperatures, e.g., 71°C,
and in order to promote copolymerization of the sluggishly
reacting allylamine. Since the copolymerization systems
were acidic, the allylamines must have reacted as allyl
ammonium salts. Price and co-workers of American Cyanamid
suggested copolymerization of 1.5% or more of a monoallyl
amine with acrylonitrile (68). Hydrochlorides of allyl
amines were copolymerized in Britain with acrylonitrile

TABLE 19.1

Some Mono-N-allyl Amines

Compound	bp, °C	Refractive index	References
Allylamine	55	$n_D^{20} = 1.4205$	Vogel, J. Chem. Soc. 1830 (1948)
Allylmethylamine	65	$n_D^{20} = 1.4065$	Weston, JACS 65, 676 (1943)
Allylethylamine	84	$n_D^{20} = 1.4145$	Weston
Allyl-n-butylamine	134	$n_D^{20} = 1.4260$	Burnett, JACS 59, 2249 (1937)
Allyldimethylamine	64	$n_D^{20} = 1.3998$	Burnett
Allyldiethylamine	110	$n_D^{20} - 1.4209$	Kharasch and Fuchs, J. Org. Chem. 10, 163 (1945); Shryne, U.S. 3,493,617
Allyldipropylamine	153	$n_D^{25} = 1.4239$	Cope, JACS 71, 3424 (1949)
Allyldiisopropyl-amine	147	$n_D^{25} = 1.4258$	Cope; also Kuffner, CA 57, 4526 (1963)
Allyldi-n-butylamine	187	--	Burnett
Allyl-n-butyl-dodecylamine	143	--	U.S. 2,367,878 (LaRoche); CA 39, 3634 (1945)
Allylcyclohexyl-amine	66(12)	$n_D^{20} = 1.4664$	Aldrich supplies
Allylpyrrolidine	--	--	Braun, Ber. 50, 292 (1917); chloroplatinate mp 205°C
Allylpiperidine	152	--	Menschutkin, CZ I, 1066 (1899); Hoerlein, Ber. 39, 1432 (1906); Borden supplies
Allylpyrrole	105(48)	--	Ciamician, Ber. 15, 2581 (1882); darkened and resinified
Allylhydrazine	--	--	Wegner, Angew. Chem. 72, 129 (1960); Heubusch, U.S. 3,099,629 (Bell Aerospace) and Brit. 941,067 (U.S. Rubber)
Methallylamine	78	$n_D^{20} = 1.431$	Tamele, Ind. Eng. Chem. 33, 116 (1941)
Methallyl-n-butyl-amine	155	--	Brit. 642,836 (Nopco)
Methallylphenyl-amine	122(17)	--	Brit. 642,836 (Nopco)

in aqueous persulfate to give dye-receptive copolymer
fibers (69). Allyl amines such as l-allylpiperidine, and
also trimethyl allyl ammonium chloride in small proportions, were copolymerized with acrylonitrile as well as
with vinyl acetate and methyl methacrylate (67). Copolymers of acrylonitrile with allyl or methallylmorpholine
were added to polyacrylonitrile to give fibers receptive
to acid wool dyes (70). In copolymerizations of allyldimethylamine in polar solvents such as dimethylformamide
the reactivity ratios were 2 to 4 times greater than in
non-polar solvents (71).

Allyl dimethylamine oxide (mp 70°C) was prepared by
vigorous reaction of the free amine with 10% H_2O_2 (72).
The amine oxide (13 parts) and acrylonitrile (87 parts)
dispersed in water were heated for 3 hr at 73°C using
azobisisobutyronitrile catalyst. The resulting copolymer
contained nearly 11% allyl dimethylamine oxide units.
This suggests that the amine oxide copolymerizes more
readily than monoallylamines. Polymers containing allyl-
diethylamine units were partially N-oxidized (73). The
products showed heparin antagonism and anticoagulation
actions similar to those of protamine sulfate. Allyldimethylamine and methallyldimethylamine were copolymerized with vinyl acetate by heating in t-butanol with
azobis catalyst (74). The quaternary salts of the copolymers were suggested as antistatic agents for fibers
and films and as dyeing auxiliaries. Dimethyl allylamine
was copolymerized with ethyl acrylate and N-butylacryl-
amide (75).

Amphoteric copolymers were prepared from allyldiethyl-
amine and acrylic acid (77). For example, 28 g of amine
and 18 g of acrylic acid were treated with aqueous
potassium persulfate at 50°C for 3 days in the absence of
oxygen. The neutralized copolymer recovered by precipitation by acetone showed a ratio of 2.8 sodium acrylate
units to one amine unit. Titration with 1.0M HCl indicated an isoelectric point at pH 4 near also a minimum
value of viscosity (η_{sp}/C). The amino groups did not
interfere with the cation-binding capacity of the carboxyl
groups in the pH range where the COOH groups were un-
charged.

Portionwise addition of a number of liquid allyl, methallyl, and polyfunctional allyl amines to solutions of
maleic anhydride in toluene (1 to 2) in presence of 1% BP
gave in most cases only liquid low polymers with red-brown
color and weak fluorescence in UV light (78). However,
methallylamine showed active copolymerization with
temperature rise from 35 to 74°C. After evaporation of
the toluene clear, light yellow copolymers were obtained
which were tacky, balsamlike to form-stable. These maleic

copolymers did not dissolve readily in water but in acetone solution showed strong blue-white fluorescence under 3660 Å radiation. A similar procedure gave rapid copolymerization of N-allylimidazole with maleic anhydride accompanied by temperature rise from 30°C to 60°C during 5 minutes. The brown copolymers could be drawn out to fibers while hot, but were brittle solids at room temperature. These copolymers were completely soluble in water to clear, brown solutions which fluoresced strongly yellow-white. In these copolymerizations the maleic anhydride was in excess and the maleate salts of the amines apparently were reacting.

From 11-allylaminoundecanoic acid, Lavalou of Pechiney prepared an interesting polyamide that could be cured by crosslinking when heated with peroxides or treated with high-energy radiation (79). Amino acids can react on heating with allyl bromide or allyl chloride to form N-allyl amino acids. Allyl amino acids have been used in research, but they have not become important commercially. N-Allylglycine was prepared as a powder (mp 159°C) by crystallization from methanol (80). The ethyl ester boiled at 78°C (15 mm). Allylglycine and S-allylcysteine have been available from Nutritional Biochemicals Corporation. In the development of dye-receptive acrylic fibers, acrylonitrile was copolymerized with minor proportions of N-allylglycine (mp 170°C) and N-allyl-β-alanine (mp 140°C) in aqueous zinc chloride solution (81). Copolymerization rates and parameters were reported.

Among other N-allyl amino acids prepared were allyl aminobenzoic acid and allyl anthranilic acid (mp 115°C) (82). The ethyl ester of N,N-diallyl DL-alanine was reported (83). A polypeptide containing C-allyl glycine units hydrolyzed to give aminovalerolactone hydrochloride (84):

$$CH_2=CHCH_2CHNH-- \quad \xrightarrow[HCl]{H_2O} \quad CH_3CHCH_2 \overset{+}{C}HNH_3 \overset{-}{C}l$$
$$--C=O \qquad\qquad\qquad OC=O$$

2-HALOALLYL AMINES

The 2-haloallyl amines $CH_2=CClCH_2NRR'$ are extremely reactive and troublesome to handle because of severe attack of the eyes and nose by vapors and of the skin by contact with the liquids. Rubber gloves, goggles, and a good hood exhaust are recommended. An example is 2-bromoallyl-amine, which was made by reacting hexamethylene tetraamine in chloroform at reflux with 2,3-dibromopropylene added dropwise during an hour (85). The quaternary chloride melted at 186°C. The salt in aqueous solution was treated with NaOH to give a red-brown organic phase which was

removed with ether. The 2-bromoallylamine distilled at
67°C (100 mm) and had $n_D^{25} = 1.5081$. A p-bromobenzene
sulfonamide derivative melted at 92°C. The purified
compound discolored slowly even when stored at 0°C in a
dark bottle. Long ago bromoallylamine was prepared by
reacting 2,3-dibromopropylamine hydrochloride with alcoholic KOH and by treating 1,2,3-tribromopropane with
alcoholic ammonia at 100°C (86). Dimethyl bromoallylamine,
boiling at 134°C, forms a picrate melting at 95°C (87).

A number of 2-bromoallyl alkyl amines were prepared by
slow addition of 2,3-dibromopropylene to alkyl amines with
cooling (88). After addition of NaOH the bromoallylalkyl-
amine was distilled off. Pollard and Porcell believed
that they obtained N-allylidene alkyl amines $RN=CHCH=CH_2$
by slow dehydrohalogenation of 2-haloallyl alkyl amines
by sodamide in liquid ammonia. However, Bottini and
Roberts concluded that N-alkyl allenimines such as
$CH_2=C-CH_2$ (bp 78°C; $n_D^{25} = 1.4283$) are formed (89). The
$\quad$ $N—C_2H_5$
latter compounds are lachrymators; they are extremely
reactive and can be rapidly hydrolyzed by dilute acids to
give alkylamine and aldehyde. Alleged ethyl and also
n-butyl allylidene amines were stored with $NaOCH_3$ to prevent polymerization (90).

Polymerization occurred when distillation of 2-bromoallyl
n-propylamine was attempted at 173°C and atmospheric
pressure (91). Unfortunately no properties of the polymer
were recorded. The monomer distilled at 53°C(20 mm) and
had $n_D^{25} = 1.4734$. Precautions against polymerization were
given also for 2-bromoallylethylamine [bp 55°C(27 mm);
$n_D^{25} = 1.4770$]. Treatment of N-2-chloroallyl-N-alkyl
ethanolamine with $NaNH_2$ in ether gave 3-alkyl-2-vinyl
oxazolidines (92). However, the analogous 2-bromoallyl
compounds with $NaNH_2$ formed only a propargyl derivative
by loss of HCl. A number of bromoallyl dialkyl amines
$RR'NCH_2CBr=CH_2$ were prepared by Ficini (93). Surprisingly,
some of these amines formed Grignard reagents.

Since the 2-chlorallyl amines are related to vinyl
chloride they polymerize more readily than allyl amines.
Bruson found tertiary 2-chloroallyl amines reactive in
free radical copolymerization with acrylonitrile to give
fibers of good affinity for acid dyes (94). For example,
N-chloroallylmorpholine (bp 99°C at 17 mm) was reacted
with acrylonitrile in bisulfite-activated aqueous persulfate for 4.5 hr at 60°C. There was formed a copolymer of
molecular weight about 30,000 containing 3.1% of the
tertiary chloroallylamine units. Among the 2-chloroallyl
amines prepared were the following, with boiling points:
dimethyl (110°C), diethyl 62°C(34 mm), methyl aniline

132°C(18 mm), bis(2-hydroxyethyl) 130°C(0.4 mm). Bruson
also prepared diethyl-3-chloroallylamine (62°C at 20 mm)
and its copolymers with acrylonitrile.

N-(2-chloroallyl) derivatives of morpholine and piperidine have been available from Borden. Allyl bis(2-chloro-
ethyl)amine was observed to dimerize and turn brown in
the presence of $SOCl_2$ and HCl (95). Several bromoallyl
carbodiimide derivatives were converted to yellow to brown
solid polymers on storage or, more rapidly, in the
presence of sodium (96).

Allyl haloamines have had little use. Recently ANF_2
(bp 41°C) was evaluated as an oxidizing agent in rocket
propellants (97). It was made by reacting propylene with
tetrafluorohydrazine at 250°C.

POLYFUNCTIONAL ALLYL AMINES AND THEIR DERIVATIVES

Diallyl and triallyl amines are strong smelling, irritating liquids; they do not polymerize readily and they
have not given homopolymers of high molecular weight. The
compounds with active hydrogen atoms might be expected to
give low polymers by H-polymerization. The N-allylamines
are less basic than the corresponding N-propylamines.
Copolymerizations with free polyfunctional allyl amines
have been little studied but appear to be difficult because of inhibition of free radical polymerizations by
even small concentrations. These amines appear to be
quite toxic, but there has been limited laboratory and
technical experience. Polymers from the polyfunctional
allyl ammonium halides can be reacted with strong alkali
to give basic allylamine high polymers which cannot readily be prepared directly.

Ladenburg prepared diallylamine from allylamine and
allyl bromide (98). Diallylamine and triallylamine were
made by heating allylamine and allyl chloride for 10 hr
near 100°C (99). The free amines are liberated by heating the corresponding allyl ammonium halides with excess
alkali. At atmospheric pressure the two allyl amines
distilled over wide ranges of 130-170 and 170-200°C,
suggesting that some short-chain polymerization might
have occurred. With H_2PtCl_6 they give characteristic
crystalline double salts. Diallylamine reacts with
lithium alkyls and some other organometallic compounds
to form orange to red products. Vapors from diallylamine
and sulfuric acid gave tests for pyrrole by the pine
chip-HCl method (99). Diallylamine free of triallylamine
can be prepared by refluxing diallyl cyanamide with dilute
aqueous sulfuric acid followed by addition of NaOH (100):

$$(CH_2=CHCH_2)_2NCN \xrightarrow[H_2O]{H_2SO_4} (CH_2=CHCH_2)NH + CO_2 + NH_4HSO_4$$

Ammonia and allyl iodide at ordinary temperatures form
principally tetraallyl ammonium iodide (101).

Tetraallylammonium hydroxide can be prepared by bubbling
ammonia into an alcoholic solution of allyl bromide and
purifying the resultant mass by recrystallization from
absolute ethanol (102). The quaternary bromide salt may
be treated with silver oxide to obtain the free base.
Tetraallyl ammonium hydroxide is a strongly alkaline
liquid which, on heating with sodium hydroxide gives
triallylamine.

Diallylamine and N-allyl-N'-β-hydroxyethylthiourea were
about equally active as bacteriocides (103).

Diallyl and triallyl amines react with water to form
substituted ammonium hydroxide bases and with mineral
acids to form the allylammonium salts. Hydrogenation of
triallylamine at 150°C over Raney nickel was slow (104).
Diallylamine adds H_2S in the reverse way to give the ring
compound 2,6-dimethyl-1,4-thiazane (105). Physical properties and references to a number of polyfunctional
allyl amines are given in Tables 19.2 and 19.3.

Triallylamine hydrochloride was polymerized in an unusual "reverse suspension" system in South Africa (106).
In one example, 40 parts of the amine hydrochloride and
6 parts of ammonium persulfate were dissolved in 20 parts
of water at low pH. This aqueous phase was dispersed as
droplets by stirring into 400 parts of a mixture of xylene
and ethylene dibromide (82-18 by volume) containing 4
parts of dissolved ethyl cellulose. After heating 4 hr
at 80°C polymer beads were separated. Polymerizations
were disclosed also of tetraallyl ammonium chloride and
tetraallyl ethylenediamine dihydrochloride. Such polymers
were evaluated as anion perm-selective membranes and as
modifiers for paper.

Polymers of triallylamine hydrohalides have been evaluated in Australia for desalination of water (Sirotherm
process)(107). For example, the hydrochloride was formed
at 0°C or lower in order to minimize addition of HCl to
the allyl double bonds. In aqueous solution homopolymer
gels were obtained by use of gamma radiation. From mixtures of dioxane and water soluble polymers resulted. For
polymerization in water-acetone 5 to 15 megarads were
adequate at 10 to 40°C (108). The weakly basic anionic
resins prepared were used in mixed beds for water treatment.

There have been few examples of products of high molecular weight prepared directly from N,N-diallyl amines or
from triallyl amines. However, brown resinous solids
have been reported by treating diallylamine or triallylamine with tetrafluorohydrazine for 90 min at 100°C under

TABLE 19.2

Properties of Diallyl and Triallyl Amines

N-Allyl Compound	bp, °C	Refractive index, etc.	References
Diallylamine	112	$n_D^{20} = 1.4405$ picrate, mp 31°C	Ladenburg, Ber. 14, 1879; Tsuda, CA 45, 9467 (1951)
Triallylamine	156	$d_{14} = 0.08094$ picrate, mp 94°C	Grosheintz, Bull. Soc. Chim. France [2], 31, 391
Diallylmethylamine	112	--	Broich, Ber. 30, 618 (1897); Borden supplies
Diallylethylamine	130	$n_D^{20} = 1.4360$	Weston, JACS 65, 676 (1943)
Diallylbutylamine	170	$n_D^{25} = 1.4389$	Butler, JACS 71, 3120 (1949)
Diallyldodecylamine	--	--	Ger. 681,850
Diallylaminoacetone	81(22)	$n_D^{20} = 1.4586$	Magee et al, JACS 60, 2149 (1938)
Diallylpiperazine	213 64(3)	$n_D^{25} = 1.4761$	Butler; Borden supplies
Dimethallyl-piperazine	235	$n_D^{25} = 1.4710$	Butler
Diallyl-n-butoxy-methylamine	89(11)	--	Stewart et al, JACS 54, 4177 (1932)
Diallylethylamine	167	--	Oberreit, Ber. 29, 2006 (1886)
Diallylaminoethanol	197	--	Ladenburg; Borden supplies
Triallylamine	150	picrate, mp 94°C	Tsuda, CA 9467 (1951); Pinner, Ber. 12, 2054
Dimethallylamine	149	$n_D^{20} = 1.446$	Tamele et al, Ind. Eng. Chem. 33, 119 (1941)
Trimethallylamine	72(10)	$n_D^{20} = 1.457$	Tamele; U.S. 2,172,822
Dimethallyl-methylamine	145	$n_D^{20} = 1.4372$	Weston et al, JACS 65, 676 (1943)

pressure (109). The remaining gas was reported to be pure tetrafluorohydrazine. At Gettysburg we have observed only brown viscous liquid polymers by heating diallylamine with catalysts at 50°C. In Russia attempts to prepare homopolymers of high molecular weight from numerous diallyl amines and diallyl amides were unsuccessful (110). Diallyl- and triallylamine act as retarding and chain transfer agents when added to free radical polymerizations. Allyl amines normally do not homopolymerize but react with electrophilic catalysts such as Ziegler-Natta catalysts. Certain C-allyl amines are exceptions noted later.

TABLE 19.3

Properties of Other N-Allyl Compounds

Compound	Properties and uses	References
Diallylglycidyl-amine	bp 56°C(4); $n_D^{29}=1.4579$	F. W. Michelotti, J. Polym. Sci. 59/167, S1 (1962)
Diallyltetra-hydrodipyridyl	Inhibited polymeriza-tion	Kramer and Joo, U.S. 3,168,503 (Pure Oil)
Tetraallyl-1-propene-1,3-diamine	bp 99°C(0.08); rocket fuel additive	Smith, U.S. 2,565,529 (Shell); Brit. 808,590 (Phillips)
Hexaallyl-1,2,3-propanetriamine	Rocket fuel	Mahan, U.S. 2,987,547 (Phillips)
Allylated diethy-lenetriamine	--	Agnew, J. Chem. Soc. C, 203 (1966)
Dichlorallylamine	bp 64°C(2)	Becker, Ger. 801,330 (Badische)
Diallylamine oxides	Acrylamide copolymers	Price, U.S. 2,871,229 (Am. Cyan.)

The basic monomers may react violently with acidic mono-mers such as maleic anhydride to give neutralization and in some cases resinous products. An N-allyl aminoallyloxy propanol has been reported to copolymerize with maleic anhydride in benzene solution to give crosslinked infusi-ble copolymers (111). Copolymerizations were also suggested with styrene and with acrylic monomers. Soluble copolymers of diallyl amines with acrylamide and with acrylonitrile have been studied (112). The aqueous per-sulfate-bisulfite system was brought to pH 2 by sulfuric acid so that it actually contained diallyl ammonium sulfate as reactant. Acrylonitrile was copolymerized with a minor proportion of diallyl ammonium nitrate using aqueous chlorite-bisulfite initiation at 40°C. The co-polymer fibers were readily dyed.

Copolymerizations of minor proportions of triallylamine with acrylic acid in benzene with azobis must be consi-dered actually to be copolymerizations of triallyl ammon-ium acrylate (113). 1-Diallylamino-2,3-epoxypropane has been copolymerized with styrene (114). Acrylamide has been copolymerized with a diallylmethylamine salt (115).

Diallyl amines have interest as weakly basic, anti-oxidant stabilizers in vapor and liquid phase, for example, in polychlorinated hydrocarbon liquids used as dry clean-ing agents (116, 117). In spite of odor and toxicity, diallylamine has been proposed for incorporation as a

curing agent into vinylidene chloride-hexafluoropropylene
copolymer elastomers on the rubber mill (118). Diallylmethylamine and epichlorohydrin have been reacted with
granular starch to form a gelatinizable quaternary ammonium starch (119).

Low-molecular-weight polymers or "adducts" from reaction
of hydrogen sulfide and diallylamine have been suggested
as lubricant additives (120). More favorable conditions
should give high molecular amine-sulfide polymers by H-
polymerization. Diallylmethylamine oxide has been polymerized in acidic bisulfite-activated aqueous persulfate
solution (3.5 hr at 40°C)(121). Propylene has been
reported to be copolymerized with small proportions of
diallylamine for preparing dyeable fibers (122). One
product obtained by Ziegler-Natta catalyst contained 1.86%
nitrogen. Addition of diallyl or triallylamine was found
to increase the efficiency of crosslinking of vinyl toluene sulfonate polymers under high-energy radiation (123).
Polyfunctional allyl amines have been used as chain
branching agents in acrylic acid polymerization (124).
Diallylamine copolymerized 1:1 molar with SO_2 in methanol
at 20°C using t-butyl hydroperoxide initiator (125). The
soluble copolymers believed to contain some cyclic chain
segments gave an inherent viscosity of 0.45 (0.5% in 0.1N
aqueous NaCl at 30°C).

QUATERNARY ALLYL AMMONIUM HALIDES

Polyfunctional allyl ammonium hydroxides and halides
have been known for a long time. It is remarkable that
few, if any, observations of homopolymerization of the
halides occur in the literature before the work of Butler
and co-workers. Cahours and Hofmann reacted allyl iodide
and ammonium hydroxide at room temperature to obtain A_4NI
(126). Grosheintz passed ammonia into an alcohol solution
of allyl bromide and purified the quaternary allyl ammonium bromide by recrystallization from alcohol-ether (127).
The free base was obtained by treating the salt with Ag_2O.
Cahours and Hofmann pyrolyzed the free base, obtaining
triallylamine (126). Tetraallyl ammonium iodide in needlelike crystals from alcohol-ether melted at 128°C (128).
The quaternary iodide formed with iodoform red crystals
that melted near 161°C.

Allyl trimethyl ammonium chloride can be prepared by
heating allyl chloride with trimethylamine for 6 hr at
100°C (129). The base and salts are less toxic than
analogous vinyl compounds (neurine and neurine hydrochloride). Butler and Bunch prepared polyfunctional allyl
quaternary bromides by reacting a series of unsaturated
amines with allyl bromide (130). After 2 hours refluxing

in acetophenone the halides crystallized out on cooling. Most of the salts were hygroscopic and soluble in water, in lower alcohols, and in formamide. They were moderately soluble in hot ketones. Attempts to polymerize the poly- functional allyl quaternary halides in aqueous solution with hydrogen peroxide and with persulfate salts as ini- tiators were not very successful at first.

The best method found for polymerization was to add only enough high boiling solvent to cause the monomer to liquefy at the polymerization temperature. Thus a mixture of 10 g monomer, 0.94 g formamide, and 0.22 g t-butyl hy- droperoxide (60%) was heated at 75°C. Soon after fusion, a dark brown solid, insoluble polymer formed rapidly. Treat- ing the quaternary bromide polymers with KOH solution gave evidence by release of bromide ions of the formation of crosslinked quaternary ammonium hydroxide polymers which were shown to have ion-exchange properties. Copolymeriza- tion of melted mixtures of triallyl and tetraallyl ammo- nium bromides heated in an oven at 95°C occurred rapidly, with the temperature rising inside to 140°C. The products were red-brown, rubbery masses that were insoluble in solvents.

Butler and co-workers, using t-BHP in concentrated aqueous monomer solutions, found that three allyl groups per molecule generally were necessary in order for cross- linked polymers to form. After studying these low- efficiency polymerizations, Butler advanced the theory of intramolecular cyclization earlier applied by Simpson and Haward to explain the loss of double bonds in diallyl phthalate polymerizations.

Water-insoluble polymers were obtained from triallyl 2-vinyloxyethyl ammonium bromide and from allyl dimethallyl 2-vinyloxyethyl ammonium bromide by heating with t-BHP for 24 hr at 60°C (131). The vinyl ether groups apparently were not reactive under these conditions. The free poly- mer bases were prepared by reaction with alkali and were demonstrated to possess ion-exchange activity. The quaternary allyl ammonium halides did not respond to polymerization by boron fluoride catalyst. Diallyl butyl 2-vinyloxyethyl ammonium bromide gave water-soluble poly- mers by heating with t-BHP at 60°C. No data on polymer molecular weights or viscosities were reported. (Table 19.4 lists properties of allyl quaternary ammonium halides in- cluding those polymerized by Butler and Goette.)

Bis(allyl ammonium) alkane dihalides were prepared (132). For example, diallyl amine and ethylene dibromide in water were refluxed with sodium bicarbonate for 9 hr to give

$$(CH_2=CHCH_2)_2 \overset{+}{N}CH_2CH_2 \overset{+}{N}(CH_2CH=CH_2)_2 \cdot$$
$$\overset{-}{Br} \qquad \overset{-}{Br}$$

TABLE 19.4

Allyl Quaternary Ammonium Halides

Compound	mp, °C	Solu-bility[a]	Other references
Allyl trimethyl ammonium chloride	--	WS	Rothstein, J. Chem. Soc. 1559 (1940)
Allyl trimethyl ammonium bromide	170	WS	Eastman supplies
Allyl trimethyl ammonium iodide	108	WS	Lukes, Collect. Trav. Chim. Tchecosl. 10, 73 (1938)
Allyl triethyl ammonium bromide	238(dec)	WS	Aldrich supplies
Allyl triethyl ammonium iodide	202(dec)	WS	Eastman supplies
Diallyl dimethyl ammonium bromide	--	WS	b (footnote)
Triallyl methyl ammonium bromide	91	WI	b
Triallyl 2-hydroxyethyl ammonium bromide	--	WI	b
Tetraallyldiamino-methane·2HBr	--	WI	b; L. Henry, Ber. 26, 934 (1893)
1,4-Bis-(triallyl-ammonium)-2-butene· 2HBr	157	WI	b
1,4-Bis-(allyldiethyl-ammonium)-2-butene· 2HBr	173	WS	b
Triallyl butanonyl ammonium bromide	--	WI	b; A. H. Green, U.S. 2,801,222 (Permutit)

[a]WI = water insoluble; WS = water soluble. Most polymers were made by peroxide polymerization for 24 hr at 65°C.

[b]Polymerized by Butler and Goette, J. Am. Chem. Soc. 76, 2418 (1954).

Another series of polymerizations employed t-BHP at 60°C with polyallyl ammonium bromides; here three or more allyl groups per molecule were again necessary for forming crosslinked polymers (133). The double bond of a 2-butene-substituted quaternary bromide monomer did not participate in polymerization. The ion-exchange behaviors of the free polymer bases were studied; anion capacities were from 50 to 75% of theoretical, depending on the tightness of crosslinking. The triple bonds of propargyl diallyl ammonium bromide did not participate in free radical poly-merizations (134). For example, 1 g of triallyl propar-gyl ammonium bromide was polymerized by heating for 24 hr

at 65°C using 0.012 g of t-BHP (60%) and one drop of water
to liquefy the mixture. In order to demonstrate the ion-
exchange properties of the resulting crosslinked polymer,
it was treated with an excess of 4% NaOH solution.

A series of tetraallyl diammonium bromides of the type
$BrA_2R\overset{+}{N}(CH_2)_n\overset{+}{N}RA_2\overset{-}{Br}$ was prepared where n ranged from 3 to
10 (135). The bromides were polymerized in 67% aqueous
solution with t-BHP suspended in ethyl benzene-mineral oil
with stirring and heating for 48 hr at 50°C. The anion-
exchange rates and fraction of theoretical ion capacity of
the products were highest for n = 5. In 1957 Butler and
Angelo suggested that the following type of intramolecular
cyclization occurred along with normal propagation, thus
explaining that three allyl groups in ammonium halide
monomers were necessary for crosslinking (136):

$$--CH_2\overset{.}{C}H \overset{CH_2}{\underset{CH_2}{\overset{||}{C}}}\overset{}{\underset{CH_2}{C}H} \underset{\overset{|}{N}}{\underset{R \quad R \quad \overset{-}{X}}{}} \longrightarrow --CH_2CH \overset{CH_2}{\underset{CH_2}{\overset{}{C}}}\overset{}{\underset{CH_2}{\overset{.}{C}}H} \underset{\overset{|}{N}}{\underset{R \quad R \quad \overset{-}{X}}{}}$$

They proposed such six-membered piperidine rings analogous
to the larger rings suggested by Simpson in diallyl phthalate polymerizations. A number of diallyl dialkyl ammonium bromides polymerized in water with t-BHP to form
soluble polymers free of crosslinking (linear or with
branched chains). The only diallyl ammonium halide giving
crosslinked homopolymers was diallyl ammonium bromide. Its
solvent-swollen polymer showed elastic properties. Diallyl dimethyl ammonium bromide gave substantially saturated, soluble polymers (137). Monoallyl ammonium chloride showed little reactivity in copolymerization with
diallyl ammonium chloride. Theoretical aspects of cyclopolymerization of diallyl quaternary ammonium compounds
have been discussed (138). These polymer structures and
mechanisms remain to be elucidated in detail.

TECHNICAL DEVELOPMENT OF
DIALLYL AMMONIUM HALIDE POLYMERS

Butler and Angelo prepared diallyl diethyl ammonium
bromide by dissolving 60.5 g of allyl bromide and 50.0 g
of diethyl allyl amine in 100 ml of acetone. Crystals of
the quaternary salt began to form at once. The monomer
(8.0 g), with 4.0 ml of water and 0.24 g of t-BHP (60%),
was polymerized by heating at 60°C for 48 hr under air.
There was formed 8 g of white polymer powder soluble in
water and in ethanol. Similar soluble polymer was obtained
from diallyl dimethyl ammonium bromide. Di-t-BP and BP

were ineffectual as polymerization initiators, but azobis
could be used at high concentrations. The soluble polymers still contained some double bonds (IR at 6.10μ).
Hydrogenation showed that some of the polymers contained
one residual double bond for every five monomer units.
Butler patented diallyl ammonium halides symmetrical with
respect to the double bonds in aqueous polymerization at
concentrations below 80% (139). Highest concentrations
tended to give some crosslinking. The free amine polymers
could be made from the polymeric halides. Diallyl dimethyl ammonium chloride in concentrated aqueous solution
was polymerized with t-BHP under nitrogen at 50°C (24 hr)
followed by 75°C (24 hr)(140). A polymer precipitated by
acetone had intrinsic viscosity of 1.35 in 0.1N aqueous
potassium chloride.

Trifan and Hoglen of Princeton polymerized triallyl
ethyl ammonium bromide and tetraallyl ammonium bromide in
water at different concentrations using t-BHP at 60°C
(141). The polymers had low molecular weights, e.g.,
$\eta_{int} = 0.14$, and they were soluble in water and also in
methanol. From the lowest monomer concentration of 0.17%
there was only 0.011 to 0.018 double bond per monomer unit
in the polymer. However, using 50% monomer the triallyl
compound gave 0.30 double bond and tetraallyl compound
gave 0.69 residual double bond as estimated by hydrogenation. A partial bicyclic ring structure was proposed for
the triallyl ethyl ammonium bromide polymer. On polymerization with azobis, triallyl orthoformate also gave
evidence of some ring formation.

Diallyl ammonium halides were polymerized using ammonium
persulfate catalyst in dimethyl sulfoxide (DMSO) solution
(142). The chlorides polymerized somewhat better than
the bromides. Conversion was faster in water, but molecular weights were lower than in DMSO. Diallyl diethyl
ammonium chloride gave nearly 100% conversion after 20 hr
at 50°C. The polymers were brittle, white, hygroscopic
solids soluble in water.

Harada studied copolymerization of diallyl dimethyl
ammonium chloride (DDAC) with sulfur dioxide in DMSO using
persulfate catalyst at 30°C for 20 hr or longer (143).
The copolymers were soluble in water over the whole pH
range but were degraded by alkali. The copolymer, believed to have a cyclic structure, decomposed at about
180°C without melting. Ultraviolet light in polymerization gave higher intrinsic viscosities (above 1.0 in 0.1N
NaCl aqueous) than did radical catalysts. Methanol,
acetone, or DMSO-water also could be used as solvents for
polymerization but gave polymers of lower molecular weight.
Later SO_2 copolymers of inherent viscosity 1.98 were obtained by 2% ammonium persulfate and 50% diallyl monomer

in methanol at 30°C for 93 hr (144). Four different
sulfone copolymers were found to have similar capacity
for flocculating kaolin aqueous suspensions. At higher
copolymer concentrations, however, the clay redispersed,
apparently by changing its charge from negative to positive. The copolymer could be fractionated by additions
of 35% hydrochloric acid to water-methanol solutions.
Boothe prepared DDAC of high purity by reacting excess
dimethylamine with allyl chloride and an inorganic chloride at pH 12 to 14 (145). Monomer free of residual allyl
chloride gave polymers of higher molecular weight.

 Boothe, Flock, and Hoover of Calgon Corporation studied
polymerizations of dimethyl diallyl ammonium chloride as
well as the coagulation-flocculation characteristics of
dilute aqueous solutions of the homopolymers and of copolymers (146). Persulfates gave homopolymers of higher
molecular weight than did diazo and hydroperoxide initiators. Addition of a little ethylene diamine tetraacetate
salt raised the viscosity and effectiveness of the polymers formed. As an example, 65% recrystallized monomer
in 35% water with 200 ppm EDTA sodium salt was brought to
pH 6.5 and purged with nitrogen at 80°C for an hour.
Ammonium persulfate solution was added gradually during
100 min. Within 5 min after the first addition of catalyst, the inside temperature rose from exothermic polymerization and it remained above 95°C throughout 100 min.
Then the reaction mass was heated an additional half-hour
at 90 to 100°C, diluted to 40% solids, and cooled to room
temperature. Polymer solutions made by this method were
stable on storage.

 With increasing persulfate concentration, the viscosity
of 5% polymer solutions passed through a maximum at a
persulfate to monomer molar ratio of 5×10^{-5} and then
declined. At the same time the amount of residual monomer
decreased from 14% to a minimum of 4%. The Brookfield
viscosities of the 40% polymer solutions passed through a
maximum as the initial addition of EDTA salt was increased
at the same time residual monomer declined to a plateau at
4%. The molar ratio of persulfate to monomer was held at
5×10^{-5}. Above 68% initial monomer concentration the
homopolymer of DDAC precipitated as a waxy solid during
reaction. This precipitate largely dissolved when more
water was added, but some remained as a hydrated gel,
apparently crosslinked. Residual $(CH_3)_2NA$ and allyl
alcohol from the monomer synthesis had to be removed carefully before polymerization in order to obtain high polymer viscosities. Allyl alcohol was found to be 10 times
as active as $(CH_3)_2NA$ in reducing viscosity by chain
transfer. Purification of the monomer by treatment with
activated carbon gave polymers of higher mol. wt. (147).

Calgon used a specially prepared kaolinite clay suspension to test flocculation effectiveness. Residual turbidity decreased with increasing polymer addition until a minimum turbidity was reached. Then turbidity began to increase as the excess added polymer exerted an opposite dispersing action. Thus the homopolymer of intrinsic viscosity 1.36 showed an optimum dosage of 60 micrograms per liter and that of viscosity 0.36 an optimum dosage of 125 micrograms/liter. For viscosity determinations, the DDAC polymers were precipitated by addition of methanol-water and dried over phosphorus pentoxide. Solutions in 1.0N aqueous sodium chloride were used for capillary viscometer measurements.

Cat-Floc cyclic polymers of DDAC of highest molecular weight (above 300,000) were most effective in flocculating kaolin suspensions of particle size below 0.5 micron (148). Concentrations of the cationic polyelectrolyte above 200 micrograms per liter gave best flocculation apparently by adsorption upon particles of kaolin or other insoluble contaminants of water. The DDAC homopolymers were fractionated by precipitation by acetone or by dia-filtration through Amicon TC-1 membranes.

American Cyanamid workers found DDAC copolymers with acrylamide improved the settling rate of raw sewage and the filtration rate of settled sludge (149). Hoover and co-workers prepared such copolymers from 30% monomer concentration in water under nitrogen at 50°C, using $(NH_4)_2S_2O_8$ initiator. Copolymers of a range of known compositions were prepared by terminating reactions below 10% conversion by addition of methanol. The precipitated copolymers were tested in a standard way in filtration of treated sewage sludge (dewatering). Figure 19.3 shows the effectiveness of acrylamide-DDAC copolymers in flocculating and dewatering sewage sludge. The activity of the copolymers was approximately proportional to the content of diallylammonium polyelectrolyte units. Hoover and co-workers found that the monomer reactivity ratios for copolymerization of DDAC (M_1) with acrylamide (M_2) were $r_1 = 0.30$ and $r_2 = 1.95$. Copolymers of DDAC with diacetone acrylamide were prepared under nitrogen at 40°C by aqueous bisulfite-activated persulfate (150). Copolymers from 3 parts of acrylamide to 1 part of diallyl ammonium chloride were made also in persulfate-bisulfite aqueous solution starting from room temperature at pH 4.5 (151). The viscous copolymer solutions, diluted to 5% solids, were effective in breaking hydrocarbon-water emulsions in a standard test for water clarification.

Polymers based upon diallyl dimethyl ammonium chloride polymers and believed to contain cyclic polypiperidium cations have found use as polymeric electroconductive and

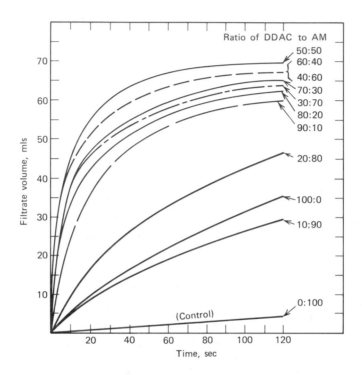

Fig. 19.3.   Sewage sludge dewatering curves showing maximum effectiveness of a copolymer from equal weights of diallyl dimethyl
ammonium chloride (DDAC) and acrylamide (AM) in flocculation.   Unmodified polyacrylamide had little action.   J. E. Boothe, H. G.
Flock, and M. F. Hoover, paper presented at Div. Polymer Chemistry
ACS Meeting, New York, September 1969.

humectant agents in electrostatic printing and copying
processes (152).   Electrophotographic reproduction processes depend on rapid dissipation of electrical charge
on exposure of the photosensitive coatings to light (153).
The allyl polyelectrolyte promotes this by an ionic
mechanism and by maintaining optimum water content of the
paper or other substrate.   See Fig. 19.4.   The polyelectrolyte (e.g., Calgon's Conductive Polymer 261) is added
along with other coating polymers to facilitate printing
processes under adverse conditions of low humidity.   The
diallyl polyelectrolyte was found to hold 7.4% water even
at 10% relative humidity.   Polymer 261 may also be used

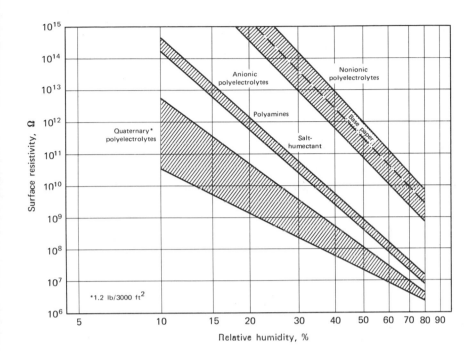

Fig. 19.4. Quaternary ammonium halide polymers offer outstanding electrical conductivity (low resistivity) on paper even at low humidities. M. F. Hoover and H. E. Carr, Tappi 51, 555 (December 1968).

as an antistatic additive for conventional printing papers and in coatings. Polymers evaluated for conductive papers include copolymers of DDAC with n-dodecylammonium chloride, with acrylamide, and with 5% diacetone acrylamide (154).

Polymer 261 is supplied in 40% aqueous solution having a viscosity of about 100 poises and also as a low-viscosity grade LVF. This cationic polymer was said to be stable up to 280°C and resistant to acid and alkali degradation. It is insoluble in most organic solvents but gives somewhat hazy dispersions in methanol. Such polymers of diallyl dimethyl ammonium chloride have been used to provide fibers with electroconductive coatings (155). Conductivity facilitates the orientation of fibers in producing plush fabrics.

Calgon research showed that surface resistivity, film formation, tack, and solvent solubility were little

changed by copolymerization of DDAC with up to 15% of
comonomers such as acrylamide, N-isopropylacrylamide,
N-methylolacrylamide, acrylonitrile, ethyl acrylate, and
methyl methacrylate. For example, 15% diacetone acrylamide available from Lubrizol Company resulted in little
sacrifice in surface resistivity and other properties.

In formulating paper coatings, 10% or more of the
electroconductive polymer may be added to starches, natural gums, polyvinyl alcohols, methyl cellulose, hydroxyethyl cellulose, vinyl acetate, or styrene copolymer
latices along with pigments. The DDAC polymers are incompatible and give precipitation with such anionic polyelectrolytes as carboxymethyl cellulose, sodium alginate,
Tamol, styrene-maleate copolymer salts, and polyphosphates.
A series of data sheets on application properties of
Polymer 261 and Cat-Floc solutions has been supplied by
Calgon. The cationic polyelectrolytes can be estimated in
alkaline aqueous medium by reduction in color of indigo
carmine dye. As little as 0.5 mg/liter can be detected.

Calgon's PCL-7001 is similar to Polymer 261 except that
it can be cured or crosslinked to make it insoluble in
water. For example, the water solution of PCL-7001 was
adjusted to pH 8.6 by addition of ammonium hydroxide, and
from 0.5 to 2.5% hydrazine hydrate based on the polyelectrolyte was added. This formulation is stable for at
least 24 hr. In starch-polyvinyl latex paper coatings,
crosslinking of the DDAC-based polymer occurred on drying.

Copolymers of allyl ammonium chlorides are useful as
coagulants with alum or ferric salts in the purification
of drinking water. Calgon's Cat-Floc is a soluble, nontoxic polymer of diallyl dimethyl ammonium chloride (156).
As a primary flocculant in water clarification, 0.2 to
7.0 ppm can replace 50 to 60 ppm alum. Such polymers
generally are sold as aqueous solutions. Other applications suggested for water-soluble cationic polymers
include antistatic materials, spinning aids, bacteriostatic and fungistatic agents (157), rubber accelerators,
curing agents for epoxy resins, repellent coatings (158),
sizes for paper, and surface-active agents. Hoover reviewed literature on cationic quaternary polyelectrolytes
(158).

Copolymers of diallyl dimethyl ammonium chloride with
small amounts of triallylamine have been suggested for
electrically conductive papers for electrostatic printing
and electrophotography (159). The trifunctional crosslinking agent minimizes both penetration into the paper
and sensitivity to changes in humidity. Diallyl ammonium
halide copolymers have been used with gelatin and silver
halides in photographic emulsions (160). DDAC and vinyl
acetate were copolymerized by portionwise addition to

refluxing isopropanol containing ammonium persulfate (161).
If the copolymer solution is saponified by heating with
NaOH, a cationic derivative of polyvinyl alcohol is obtained. This could be crosslinked by glyoxal for improving the wet strength of paper.

Allyl stearoylamine propyl piperidinum bromide and other
long alkyl-allyl-substituted ammonium halides were copolymerized in minor proportions with acrylonitrile, methyl
methacrylate, styrene, and other monomers (162). Graft
copolymers of diallyl dialkyl ammonium salts were prepared
with styrene and acrylonitrile (163). Tri- and tetraallyl
ammonium chlorides were polymerized in aqueous persulfate
at 50°C during 20 hr to give crosslinked ion-exchange
resins (164). Resins of high anion-exchange capacity
were compared in rates of exchange by measurements of
electrical conductivity (165). Diallylamine was copolymerized with acrylamide in aqueous persulfate-bisulfite
system (166). Copolymers of acrolein with allyl ammonium
halides were reacted with sulfur dioxide to form water-
soluble polysulfonic acids (167).

Drucker and Morawetz prepared copolymers of relatively
high intrinsic viscosity and 4 to 5% nitrogen content from
reaction of allyl ammonium chloride and methacrylic acid
in methanol-water solution with azobis catalyst at 60°C
(168). The copolymers from low conversion were precipitated by pouring the solutions into dioxane. Intrinsic
viscosities were measured in 2N aqueous NaOH at 30°C. The
composition of the copolymer appeared to be sensitive to
the degree of ionization of the methacrylic acid, the
amine content passing through a minimum when the acid
monomer was about half-ionized. Copolymers obtained from
equimolar allylamine salt and methacrylic acid in neutral
solution contained about 15% of lactam groups, judging
from differences in analysis between total nitrogen and
amine nitrogen.

D'Alelio reported that copolymers of 90 parts of allyl
trimethyl ammonium chloride and 10 parts of divinylbenzene
show promising efficiency as ion-exchange resins (169).
Allyl ammonium chloride copolymers were suggested for use
in solid propellants for space craft (170). Allyl triethyl ammonium bromide was employed as electrolyte in the
electropolymerization of styrene in dimethyl formamide
solution (171).

Crosslinked copolymers of allyl-substituted ammonium
salts are expected to find applications as membranes,
gels, and beads in water purification, medicine and
analysis. Gels were prepared from diallyl dimethyl
ammonium chloride with minor proportions of methylenebisacrylamide in an aqueous persulfate system (172).

Commercial uses of copolymers from diallyl ammonium salts have stimulated research. Triallyl ammonium compounds have been prepared from isomers of amphetamine and menthylamine (173). Optically active polymers and copolymers with methacrylic acid were prepared. Dimethallyl and trimethallyl ammonium chlorides were polymerized in water, dimethyl formamide, or ethanol at 80°C (174). N,N'-Diallyl diammonium dihalides of the type $Br\bar{A}\overset{+}{N}(CH_3)_2(CH_2)_n(CH_3)_2\overset{+}{N}\bar{A}Br$ were suggested as dicationic crosslinking agents in radical initiated copolymerizations (175). Tri- and tetraallyl ammonium pentaborates were studied (176). Ammonium halides from N,N-diallyl amino acids were polymerized in water using t-butyl hydroperoxide at 65 to 80°C (177). The soluble low polymers obtained were believed to contain cyclic structures. Research has been done with N-allyl pyridinium bromide available from Borden and N-allyl pyridinium-3-sulfonate (mp 180°C) supplied by Aldrich. Both of these compounds are highly hygroscopic.

N-ALLYL AROMATIC AMINES

Most of the N-allyl aromatic amines are oils that become orange to red-brown on heating or standing in air and light. They normally resist polymerization and inhibit free radical polymerizations of other ethylenic compounds. Schiff prepared N-allylaniline from aniline and allyl iodide more than a century ago (178). The compound reacts with benzyl chloride at 100°C to give a mixture of allylbenzylaniline and dibenzylamine. N-Methyl-N-allylaniline and many other N-allyl amines and their ammonium derivatives were prepared by Wedekind (179).

Allyl aryl amines readily form substituted ammonium halides. Thus allylbenzyl-o-anisidine (bp 206°C at 50 mm) reacts with methyl iodide:

$$C_6H_5CH_2N\overset{A}{\underset{OCH_3}{\bigcirc}} + CH_3I \longrightarrow \left[C_6H_5CH_2\overset{CH_3}{\underset{A}{N}}\overset{}{\underset{OCH_3}{\bigcirc}}\right]^+ I^-$$

The ammonium halides can be reacted with Ag_2O to give the quaternary ammonium hydroxides which are stable, strong bases. Much research was done near the turn of the century on substituted ammonium compounds bearing four different groups attached to nitrogen (180). These can be resolved into stereoisomers showing rotation of polarized light.

N-Allylaniline and derivatives do not readily give rearrangements analogous to Claisen rearrangements (181).

N,N-Diallylaniline gave some monoallylamine on pyrolysis but no o-allylaniline (182). However, Hurd and Jenkins heated N-allylaniline in the presence of zinc chloride to form o-allylaniline, p-propylaniline, propylene, hydrogen, and tarry polymer (183). More recently N-allyl-1-naphthylamine (bp 120°C at 0.5 mm) heated at 260°C gave 2-allyl-1-naphthylamine (184). Pyrolysis of N-allylaniline at 275°C gave largely propylene and aniline. Diallylaniline (bp 245°C) was prepared by adding 2 moles of allyl chloride dropwise with stirring to 1 mole of aniline diluted with benzene and kept at reflux temperature (185). The compound quickly becomes deep red in air.

In 1899 Wedekind observed stereoisomerism in quaternary ammonium salts such as methyl allyl phenyl benzyl ammonium iodide (186). The free substituted amines do not show stereoisomerism; they are very weak bases because both the resonating allyl and phenyl groups attract electron density from the nitrogen atom. The quaternary hydroxides are relatively strong bases but weaker than quaternary alkyl ammonium hydroxides. N-Allyl-N-p-tolylhydrazine (bp 170°C at 90 mm) gave a hydrochloride melting about 129°C with decomposition (187). The free amine was oxidized by $FeCl_3$ to form a diallyl di-p-tolyl tetrazene.

Acrolein was reacted with p-toluidine, apparently forming the Schiff base $H_3C\langle\bigcirc\rangle N=CHCH=CH_2$ or allylidene p-toluidine (188). This was described as a low-melting mass of weakly basic properties, having an odor of turpentine.

Allyl alkyl aryl amine oxides such as $A(CH_3)C_6H_5NO$ can isomerize to o-allyl-substituted hydroxyl amines, e.g., $CH_3(C_6H_5)NOA$ (189).

Menschutkin found that allyl iodide reacts very much faster than methyl iodide with aryl amines (190). The three isomeric N,N-diallyltoluidines were prepared by refluxing 1 mole of toluidine with 2 moles of allyl iodide and aqueous sodium hydroxide. The ortho, meta, and para diallyl toluidines boiled at 232, 249, and 257°C, respectively; and Menschutkin found the respective densities at 19°C to be 0.9392, 0.9430, and 0.9442. Properties and references to N-allyl aromatic amines are listed in Table 19.5.

While most free aryl amines inhibit free radical polymerization, some of the salts or zwitterions can be copolymerized. Thus p-diallylaminobenzoic acid (or its esters) were copolymerized with 2-hydroxypropyl acrylate and methyl acrylate by heating with azobis in isopropanol solution (191). A suntan lotion was formulated from this UV-absorbing terpolymer by addition of ethanol, glycerin, isopropyl myristate, and perfume.

TABLE 19.5

Some N-Allyl Aromatic Amines

	bp, °C	Other properties	References
Monoallyl derivatives:			
A-Aniline	219 or 95(7)	$n_D^{20}=1.5636$	Hurd, J. Org. Chem. 22, 1418 (1957); Wedekind, Ber. 36, 3791 (1903)
A-o-Toluidine	236	--	Wedekind, Ber. 37, 3896 (1904); Aldrich supplies
A-N-Methyl aniline	215 100(19)	picrate, m 92°C; $n_D^{20}=1.5564$	Wedekind, Ber. 32, 524 (1899); Aldrich supplies
A-N-Ethyl aniline	227	--	Wedekind, Ann. 318, 97 (1901); Komatsu, CZ I, 799 (1913)
A-Isopropyl aniline	227	--	Braun, Ber. 33, 2734 (1900)
A-N-alkylaryl amines	--	--	Tweedie, J. Org. Chem. 26, 3676 (1961)
A-Benzyl amine	--	--	Paal, Ber. 32, 80 (1899)
A-p-tolyl amine	230	--	Wedekind, Ber. 37, 3896 (1904)
A-Phenyl benzyl amine	225(42)	--	Wedekind, Ber. 32, 521 (1899); Hantzsch, Ber. 35, 885 (1902)
A-o-Tolyl benzylamine	183(27)	picrate, m 50°C	Wedekind, Ber. 37, 3896 (1904)
A-p-Tolyl benzyl amine	215(31)	picrate, m 138°C	Wedekind, Ber. 37, 2721 (1904)
A-Dibenzyl amine	170(10)	Chloroplatinate, m 169°C	Braun, Ber. 35, 1284 (1902)
A-Carbazole	[mp 56°C]	red picrate, m 86°C	Levy, Monatsh. 33, 182 (1912)
Diallyl Amines:			
A_2-Phenyl	245	--	Zander, Ann. 214, 149 (1882)
A_2-o-Tolyl	232	--	Menschutkin, CZ II, 28 (1903)
A_2-Benzyl	89(4)	$n_D^{25}=1.5122$	Weston, JACS 65, 676 (1943)

Amines such as ACH_2NH_2 are difficult to prepare because alkali catalyzes rearrangement by movement of the double bond to form crotylamine or analogs. Heating the hydrochloride of gamma-chlorobutylamine with alcoholic KOH at 180°C gave a volatile amine boiling at 90°C; its yellow chloroplatinate $(C_4H_9N)_2PtCl_6$, crystallized from water, melted at 204°C (192). The compound ACH_2NH_2, along with crotylamine, was obtained by reduction of vinylacetonitrile (193). The pure amine boiled at 77°C and had $n_D^{26} = 1.4184$. Heating $CH_3CHOHCH_2CH_2N(CH_3)_2$ with 20% sulfuric acid at 200°C in an autoclave gave the C-allyl t-amine (bp 96°C) having a coninelike odor (194). When heated at 230°C, the hydrochloride gave butadiene and dimethylamine. Reduction of $ACH(NO_2)CH_3$ with zinc and HCl gave the allylic amine, boiling at 85°C (195). Reduction of allylacetoxime with sodium and ethanol gave $ACH_2CH(CH_3)NH_2$. The oil (bp 118°C) has an odor suggesting piperidine (196).

To prepare $A_2CHCH_2NH_2$, diallylacetonitrile was reduced with sodium in ethanol (197). It is an unpleasant smelling liquid boiling at 167°C. The corresponding nitrite decomposed at about 80°C, giving nitrogen and A_2CHCH_2OH. Reducing diallyl acetone phenyl hydrazone by sodium amalgam in ethanol and acetic acid yields the primary amine $A_2CHCH(CH_3)NH_2$ (198). It has a piperidine-like odor and boils at 176°C. The chloroplatinate orange crystals melted at 160°C with decomposition.

Because the amino group is removed from the allyl group, the C-allyl amines may have better possibilities in some cases for synthesis of basic polymers and copolymers. Electron-releasing basic monomers ordinarily decompose the electrophylic Ziegler-Natta catalysts. However, it has been possible to polymerize some secondary and tertiary allylic amines of the type $CH_2=CH(CH_2)_nNRR$, where n = 2 to 9, and where the R groups are branched and appear to shield the π bond from the Ziegler catalyst (199). For example, 4-N,N-diisopropylamino-1-butene gave a polymer reported to melt above 300°C. The polymer from the corresponding 1-pentene derivative gave a somewhat crystalline x-ray pattern, suggesting stereoregular structure. The polymerizations could be carried out using delta-titanium trichloride containing dissolved $AlCl_3$ along with a branched alkyl aluminum monochloride at 50°C for 15 to 137 hr. The tertiary allylic amines polymerized best if first complexed with $Al(C_2H_5)_2Cl$. Intrinsic viscosities of these polymers in benzene at 30°C ranged from 0.49 to 1.95. N-Pentenyl carbazole of structure $NCH_2CH_2CH_2CH=CH_2$ was heated with 1% benzoyl peroxide at 70°C for 24 hr to form a vinyl-type polymer of DP about 10 (200). Reaction of the polymer with electron acceptors such as quinones gave charge-transfer complexes having interesting spectral and electrical properties. The literature contains many other C-allyl nitrogen compounds, but few of these have been investigated for polymerization. Further success in polymerizing basic monomers with Ziegler type catalysts will be expected where the π bonds are sufficiently removed from the donor part of the monomer molecule, protected by branching, and suitably complexed.

Hess made 2-allylpyrrole (bp 83°C at 24 mm) which rapidly became colored in air and finally formed a red polymer (201). By reaction of pyrrole magnesium bromide with allyl bromide followed by Claisen-type rearrangement, some 2,5-diallyl pyrrole was obtained. This melted at 118°C with decomposition. By reaction of alkyl cyanides with dicyandiamide some C-allyl cyclic compounds such as 4-pentenoguanamine (mp 195°C) were prepared (202). These were copolymerized with acrylate esters in radical systems for preparation of ionic textile finishes.

Other allylic amines include allyl nitrosamine (203), allyl ethylene diamine (204), and allyl derivatives of drugs (20) such as the narcotic antagonists N-allylnorcodeine, Levallorphan, Nalorphine and Naloxone made by Merck, LaRoche, and Endo Laboratories and used medically as salts.

References

1. A. R. Hunter, Brit. J. Anaesth. $\underline{36}$, 466 (1964), H. P. Baechtold, et al, Helv. Physiol. Pharmacol. Acta $\underline{22}$, 70 (1964).
2. R. M. Lowman et al, CA $\underline{73}$, 23327 (1970); A. R. Giordano et al, CA $\underline{72}$, 64676 (1970); J. A. Will, CA $\underline{74}$, 85552 (1971).
3. A. Cahours and A. W. Hofmann, Ann. $\underline{102}$, 301 (1857); S. Gabriel and G. Eschenbach, Ber. $\underline{30}$, 1124 (1897); M. T. Leffler, Org. Syn. Coll.Vol. II, 24, Wiley, 1943.
4. L. M. Peters and K. E. Marple, Ind. Eng. Chem. $\underline{40}$, 2046 (1948); cf. U.S. 2,216,548 (Shell).
5. A. Ladenburg, Ber. $\underline{14}$, 1876 (1881); C. Liebermann and A. Hagen, Ber. $\underline{16}$, 1641 (1883).
6. R. E. Koski and P. E. Johnson, U.S. 3,175,009 (Shell); cf. U.S. 2,915,385 and 2,216,548.
7. L. M. Peters et al, Ind. Eng. Chem. $\underline{40}$, 2046 (1948).
8. R. E. Koski et al, U.S. 3,175,009 (Shell).
9. Y. D. Smirnov et al (Leningrad), CA $\underline{73}$, 87351 (1970).
10. L. Hernandez and F. F. Nord, Experimentia $\underline{3}$, 489 (1947); CA $\underline{42}$, 1793 (1948).
11. R. G. Denning et al, J. Chem. Soc. A $\underline{2}$, 324 (1967).
12. W. E. Vaughn and F. F. Rust, J. Org. Chem. $\underline{7}$, 472 (1942); cf. J. Am. Chem. Soc. $\underline{72}$, 631 (1950); U.S. 2,562,145 (Shell).

13. L. Henry, Ber. **8**, 398 (1875); S. Gabriel and J. Weiner, Ber. **21**, 2669 (1888); E. Abderhalden and E. Eichwald, Ber. **51**, 1312 (1918); M. S. Kharasch and C. F. Fuchs, J. Org. Chem. **10**, 159 (1945).

14. A. W. Weston et al, J. Am. Chem. Soc. **65**, 674 (1943); Brit. 610,263 (Du Pont).

15. M. M. Rauhut et al, J. Org. Chem. **26**, 5138 (1961).

16. A. L. Morrison and Rinderknecht, J. Chem. Soc. 1467 (1950).

17. M. Dominikiewicz, CZ II, 3388 (1935).

18. H. Ubuch and J. Rubinfeld, J. Org. Chem. **26**, 1637 (1961).

19. T. Saegusa et al, Tetrahedron Lett. **49**, 6125 (1966); J. Furukawa et al, CA **69**, 97233 (1968).

20. R. S. Long and C. M. Hoffman, U.S. 2,425,283 (Am. Cyan.).

21. A. Rinne, Ann. **168**, 261 (1877).

22. H. V. Boenig, U.S. 2,998,451 (Goodyear).

23. K. E. Marple et al, U.S. 2,375,016 (Shell).

24. J. K. Jurjeiv, CA **44**, 1482 (1950).

25. E. Spaeth and J. Lintner, Ber. **69**, 2730 (1936).

26. G. Wanag, CZ II, 3816 (1939).

27. Brit. 610,263 (Du Pont); CA **43**, 3456 (1949).

28. M. I. Cohn, U.S. 2,697,699 (Elastic Colloid Res.).

29. P. E. Fanta, U.S. 2,766,232.

30. R. R. Whetstone, U.S. 2,566,815 (Shell).

31. L. Levine, U.S. 3,458,509 (Dow).

32. A. Merijan et al, U.S. 3,494,907 and 3,563,968 (GAF).

33. G. Dauphin et al, Tetrahedron **28**, 1055 (1972).

34. K. E. Marples and B. Borders, U.S. 2,375,015 (Shell).

35. J. K. Lovett, U.S. 3,062,798 (Esso).

36. V. V. Zykova et al, CA **66**, 46858 (1967).

37. J. H. Parker, U.S. 2,456,428 (Shell).

38. P. A. Devlin, U.S. 3,057,833 (Shell).

39. E. A. Ney, U.S. 3,396,065; Brit. 1,043,440 (Dunlop).

40. A. C. A. Clavier and J. Pouradier, Fr. 1,104,541 (Kodak-Pathe), CA **53**, 12077 (1959).

41. G. V. Gurash et al, USSR 159,654, appl. 1962; CA **61**, 1972 (1964).

42. R. M. Hainer et al, U.S. 3,104,205 (Warner-Lambert).

43. A. Drucker and H. Morawetz, J. Am. Chem. Soc. **78**, 346 (1956).

44. F. F. Shcherbina et al, CA **64**, 818 (1966).

45. H. L. Cohen and L. M. Minsk, J. Org. Chem. **24**, 1407 (1959).

46. P. J. Aguis and E. B. Evans, Brit. 802,082 (Esso).

47. K. Matsubayashi and O. Fukushima, U.S. 3,137,675 (Kurashiki).

48. G. Thompson, U.S. 2,852,497 (Du Pont).

49. D. Glabisch, U.S. 3,308,081 (Bayer).

50. R. Mihail et al, J. Polym. Sci. **30**, 423 (1958).

51. A. W. Weston et al, J. Am. Chem. Soc. **65**, 676 (1953); cf. A. Partheil, Ber. **30**, 619 (1897).

52. N. H. Cromwell and A. Hassner, J. Am. Chem. Soc. **77**, 1568 (1955).

53. A. C. Cope and P. H. Towle, J. Am. Chem. Soc. **71**, 3425 (1949).

54. J. Falbe and F. Korte, Ger. 1,227,450 (Shell); CA **66**, 2184 (1967).

55. G. Hata et al, J. Chem. Soc. D 1392 (1970).

56. Nicholas A. Menschutkin, J. Chem. Soc. **89**, 1532 (1906).

57. M. S. Kharasch and C. F. Fuchs, J. Org. Chem. 10, 163 (1945).
58. R. G. Denning and L. M. Venanzi, J. Chem. Soc. 3241 (1963).
59. M. W. Harman and J. J. D'Amico, U.S. 2,847,419 (Monsanto).
60. C. C. Price and W. H. Snyder, Tetrahedron Lett. 69 (1962); CA 57, 640 (1962).
61. W. J. Bailey et al, Preprint, ACS Polymer Division 9, 404 (April 1968).
62. R. B. Rothstein and M. Rothstein, J. Chem. Soc. 1559 (1940).
63. D. R. Howton, J. Am. Chem. Soc. 69, 2555 (1947).
64. Ger. 681,850; Fr. 771,746 (IG); CZ I, 2408 (1935).
65. J. S. Pierce and J. Wotiz, J. Am. Chem. Soc. 73, 2594 (1951).
66. H. Hellman and G. M. Scheytt, Ann. 654, 39 (1962); CA 57, 5774 (1962).
67. D. W. Chaney, U.S. 2,662,875 (Chemstrand).
68. J. A. Price et al, U.S. 2,626,946 (Am. Cyan.), cf. Price, U.S. 2,654,729 and 2,883,370.
69. E. J. Kowolik and J. W. Fisher, Brit. 757,203 (Brit. Celanese).
70. H. Gerber and W. Sutter, Ger. 1,004,767 (Bayer); CA 54, 16857 (1960).
71. A. I. Ageev et al (Leningrad), CA 74, 3921 (1971).
72. M. R. Lytton, U.S. 2,586,238 (Chemstrand).
73. M. A. Marchisio et al, CA 74, 11660 (1971).
74. F. D. Marsh, U.S. 2,628,221 (Du Pont).
75. A. Maeder, Ger. 1,102,400 (Ciba).
76. Brit. 925,612 (Pechiney); CA 59, 4120 (1963).
77. F. Ascoli and C. Botre, CA 60, 3113 (1964).
78. C. E. Schildknecht and Courtney S. Palmer, 1972, unpublished.
79. M. Lavalou, Fr. 1,186,236 (Pechiney); CA 54, 20324 (1960); cf. F. Carriere, CA 68, 3439 (1968).
80. W. Gluud, CZ I, 429 (1914); R. Alpern, J. Chem. Soc. 99, 87.
81. H. Reichert et al, Faserforsch. Textiltech. 20/7, 317-321 (1969).
82. J. Houben et al, Ber. 39, 3239 (1906).
83. N. B. Burnett et al, J. Am. Chem. Soc. 59, 2249 (1937).
84. C. D. Hurd and L. Bauer, J. Org. Chem. 18, 1440 (1953).
85. A. T. Bottini and V. Dev J. Org. Chem. 27, 968 (1962); cf. Bottini et al, Org. Syn. 43, 6.
86. C. Paul, Ber. 21, 3190 (1888).
87. J. L. Dunn and T. S. Stevens, J. Chem. Soc. 280 (1934).
88. C. B. Pollard and R. F. Parcell, J. Am. Chem. Soc. 73, 2925 (1951).
89. A. T. Bottini and J. D. Roberts, J. Am. Chem. Soc. 79, 1462 (1957).
90. A. E. Montagna and W. C. Bedot, U.S. 2,721,881 (Carbide).
91. A. T. Bottini and R. E. Olsen, J. Am. Chem. Soc. 84, 195 (1962) and Org. Syn. 44, 7 (1964).
92. A. T. Bottini et al, J. Org. Chem. 29, 373 (1964).
93. J. Ficini et al, Bull. Soc. Chim. France 1219 (1962); CA 57, 14922 (1962).
94. H. A. Bruson, U.S. 2,700,027 (Ind. Rayon).
95. A. H. Ford-Moore, J. Chem. Soc. 819 (1946).
96. E. Schmidt et al, Ann. 560, 222 (1948); CA 43, 1016 (1949).

97. R. P. Rhodes and A. J. Passannante, U.S. 3,347,924 and 3,574,753
     (Esso).
98. A. Ladenburg, Ber. 14, 1879 (1881).
99. C. Liebermann and A. Hagen, Ber. 16, 1641 (1883).
100. E. B. Vliet, J. Am. Chem. Soc. 46, 1307 (1924); Org. Syn. Coll.
     Vol. I, 2nd ed., 201 (1941).
101. A. Cahours and A. W. Hofmann, Ann. 102, 305 (1857).
102. H. Grosheintz, Bull. Soc. Chim. France [2] 31, 391 (1879).
103. G. M. Koons, Gettysburg College, unpublished.
104. D. G. Norton and J. L. VanWinkle, U S. 3,116,331 (Shell).
105. D. Harmon and W. E. Vaughan, J. Am. Chem. Soc. 72, 631 (1950);
     U.S. 2,562,145 (Shell).
106. Brit. 939,518; cf. Brit. 907,079 (S. African Council Sci. and
     Ind. Res.), appl. July, 1960.
107. H. A. J. Battaerd, U.S. 3,619,394 (ICI of Australia); Weis,
     Austral. J. Chem. 21, 2703 (1968).
108. H. A. J. Battaerd, Ger. Offen. 1,964,174 (ICI of Australia).
109. J. R. Lovett, U.S. 3,062,798 (Esso).
110. V. G. Ostroverchov et al, Macromol. Verb. (Moscow) 6 (5),
     925-928 (1964).
111. T. W. Evans et al, U.S. 2,831,837 (Shell).
112. W. H. Schuller and J. A. Price, U.S. 3,032,539 (Am. Cyan.).
113. D. C. Spaulding and S. E. Horne, U.S. 3,032,538 (Goodrich).
114. F. W. Michelotti, U.S. 3,316,225 (Interchemical).
115. T. J. Suen and A. M. Schiller, U.S. 3,171,805 (Am. Cyan.).
116. R. F. Monroe and D. E. Rapp, U.S. 2,997,507 (Dow).
117. Fr. 1,309,472 (Vulcan).
118. W. R. Griffins, U.S. 3,041,316 (U.S. Air Force); cf.
     Brit. 954,835 (3M).
119. E. F. Paschall, U.S. 2,876,217 (Corn Products).
120. D. Harman and H. J. Sommer, U.S. 2,517,564; cf. U.S. 2,554,222
     (Shell).
121. J. A. Price, U.S. 2,871,229 (Am. Cyan.).
122. J. L. Jezyl et al, U.S. 3,293,326 (Sun Oil).
123. E. J. Lawton and J. S. Balwit, U.S. 3,392,096 (G.E.).
124. D. C. Spaulding and S. E. Horne, U.S. 3,032,538 (Goodrich).
125. Belg. 664,427 (Nitto Boseki), CA 65, 2374 (1966).
126. A. Cahours and A. W. Hofmann, Ann. 102, 305 (1857); H. Malbot,
     Ann. Chim. Phys. [6], 13, 487 (1888).
127. H. Grosheintz, Bull. Soc. Chim. France [2] 31, 390 (1879);
     CZ 403 (1879).
128. W. Steinkopf and R. Bessaritsch, J. Prakt. Chim. [2] 109, 242
     (1925); CZ I, 1871 (1925).
129. J. Weiss, Ann. 268, 143 (1892); cf. A. Partheil, Ann. 268, 152
     (1892).
130. G. B. Butler and R. L. Bunch, J. Am. Chem. Soc. 71, 3120 (1949).
131. G. B. Butler and R. L. Goette, J. Am. Chem. Soc. 74, 1939 (1952).
132. G. B. Butler and R. L. Bunch, J. Am. Chem. Soc. 74, 3453 (1952);
     cf. U.S. 2,611,768 and 2,687,382 (State of Florida).

133. G. B. Butler and R. L. Goette, J. Am. Chem. Soc. 76, 2418 (1954).
134. G. B. Butler and R. A. Johnson, J. Am. Chem. Soc. 76, 713 (1954).
135. G. B. Butler and R. J. Angelo, J. Am. Chem. Soc. 78, 4797 (1956).
136. G. B. Butler and R. J. Angelo, J. Am. Chem. Soc. 79, 3128 (1957);
     cf. U.S. 2,926,161 (Peninsular Res.).
137. G. B. Butler et al, PB 145,435; CA 55, 24083; J. Macromol.
     Chem. 1, 231 (1966).
138. Y. Minoura, J. Polym. Sci. A 3, 2162 (1965); G. B. Butler and
     M. A. Raymond, J. Polym. Sci. A 3, 3413 (1965).
139. G. B. Butler et al, U.S. 2,926,161 (Peninsular Res.).
140. G. B. Butler, U.S. 3,288,770 (Peninsular Res.).
141. D. S. Trifan and J. J. Hoglen, J. Am. Chem. Soc. 83, 2021 (1961).
142. S. Harada et al, J. Polym. Sci. A 1, 5, 1951 (1967).
143. S. Harada et al, Makromol. Chem. 90, 177 (1966); 107, 64 and
     78 (1967); U.S. 2,375,233 (Nitto Bosei).
144. T. Ueda and S. Harada, J. Appl. Polym. Sci. 12, 2383 (1968).
145. J. E. Boothe, U.S. 3,461,163 (Calgon).
146. J. E. Boothe, H. G. Flock, and M. F. Hoover, ACS Meeting, New
     York, September 1969; J. Macromol. Sci. Chem. 4, 1419 (1970).
147. J. E. Boothe, U.S. 3,472,740 (Calgon).
148. Monica A. Yorke (Calgon), Preprint Org. Coatings and Plastics
     Chem. 32, #2, 208 (1972); ACS Symposium on Water Soluble
     Polymers, N. M. Bikales, Ed. Plenum (1973).
149. W. H. Schuller and W. M. Thomas, U.S. 2,923,701 (Am. Cyan.);
     cf. Ind. Eng. Chem. 48, 2132 (1956); U.S. 3,562,226 (Calgon).
150. H. J. Zeh, U.S. 3,551,384 (Calgon).
151. J. J. Sackis, U.S. 3,316,181 (Nalco).
152. M. F. Hoover and H. E. Carr, Tappi 51, #12, 552 (December 1968);
     Fr. Demande 2,012,630 (Calgon), CA 74, 8404 (1971).
153. T. J. Kucera, U.S. 2,959,481 (Bruning); cf. CA 62, 11339 (1965):
     CA 64, 7577 (1966).
154. J. E. Boothe and M. F. Hoover, U.S. 3,544,318 (Calgon).
155. M. F. Hoover and R. O. Carothers, U.S. 3,490,938 (Calgon).
156. Chem. Eng. News 46 (Jan. 15, 1968); G. B. Butler, U.S.
     3,288,770 (Peninsular Res.); cf. U.S. 3,412,019 (Calgon).
157. M. F. Hoover, U.S. 3,539,684 (Calgon).
158. J. E. Boothe et al, U.S. 3,678,110 (Calgon); M. F. Hoover,
     J. Macromol. Sci. Chem. 4, 1327 (1970).
159. Netherlands appl. of Calgon, CA 66, 77167 (1967).
160. H. W. Wood, U.S. 3,579,347 and 3,607,286 (Ilford).
161. L. L. Williams and A. T. Coscia, U.S. 3,597,313 (Am. Cyan.).
162. J. A. Price, U.S. 2,883,370; cf. U.S. 2,884,058 (Am. Cyan.).
163. Y. Jen, U.S. 2,958,673 (Am. Cyan.).
164. Brit. 907,079 and 934,518 (S. Africa Res.); cf. U.S.
     2,687,382 and 2,946,757.
165. A. L. Clingman, J. Appl. Chem. (London) 13, 1-6 (1963);
     CA 58, 7398 (1963).
166. G. A. Goldsmith, U.S. 2,234,076 (Nalco).
167. A. C. Nixon et al, U.S. 3,278,474 (Shell).
168. A. Drucker and H. Morawetz, J. Am. Chem. Soc. 78, 346 (1956).

169. G. F. D'Alelio, U.S. 2,697,079 (Koppers).
170. J. Philipson, U.S. 3,092,526 (Aerojet).
171. J. W. Bayer and E. Santiago, U.S. 3,489,663 (Owens-Illinois).
172. M. F. Hoover et al, Fr. 1,494,438 (Calgon), CA 69, 3859 (1968).
173. E. Selegny et al, Compt. Rend. C. Sci. Chim. 264, 184 (1967).
174. S. G. Matsoyan et al, CA 61, 4489 (1964).
175. A. Rembaum et al, Polym. Lett. 7, 395 (1969).
176. S. D. Ross et al, J. Am. Chem. Soc. 81, 3264 (1959).
177. H. K. Reimschuessel and F. Boardman, U.S. 3,515,707 (Allied);
      D. G. Woodward, U.S. 3,574,175 (Grace).
178. H. Schiff, Ann. Spl 3, 364 (1865).
179. E. Wedekind, Ber. 32, 524 (1899); 39, 487 (1906).
180. E. Wedekind and E. Froehlich, Stereochemistry of Pentavalent
      Nitrogen Compounds, Leipzig, 1907, and references in Beil-
      stein 12, 170.
181. Y. Ogata and K. Takagi, Tetrahedron Lett. 27, 1573 (1971).
182. F. L. Carnahan and C. D. Hurd, J. Am. Chem. Soc. 52, 4586 (1930).
183. C. D. Hurd and W. W. Jenkins, J. Org. Chem. 22, 1418 (1957).
184. S. Marcinkiewicz et al, Chem. Ind. (London) 438 (1961);
      CA 56, 425 (1962); Tetrahedron 14, 208 (1961).
185. W. H. Schuller and J. A. Price, U.S. 3,032,539 (Am. Cyan.);
      cf. Zander, Ann. 214, 149 (1882).
186. E. Wedekind, Ber. 32, 517 (1899).
187. A. Michaelis and K. Luxembourg, Ber. 26, 2179 (1893).
188. H. Schiff, Ann. 140, 96 (1866).
189. J. Meisenheimer et al, Ber. 52, 1667 (1919); R. F. Klein-
      schmidt and A. C. Cope, J. Am. Chem. Soc. 66, 1929 (1944).
190. N. Menschutkin and L. Simanowsky, J. Russ. Phys.-Chem. Soc. 35,
      204 (1903).
191. M. Skoultchi and E. A. Meier, U.S. 3,666,732 (Nat. Starch).
192. A. Luchmann, Ber. 29, 1431 (1896).
193. E. Galand, Bull. Soc. Chim. Belg. 39, 536 (1930); J. D. Roberts
      and R. H. Mazur, J. Am. Chem. Soc. 73, 2518 (1951).
194. Ger. 254,529 (Bayer); CZ I, 347 and II 395 (1913); cf.
      von Braun and Lemke, Ber. 55, 3554 (1922); Ann. 472, 130 and
      134 (1929).
195. H. Gal, Compt. Rend. 76, 1355 (1872).
196. J. Braun and F. Stechle, Ber. 33, 1474 (1900); cf. Merling,
      Ann. 264, 324 (1891) and Ger. 261,876.
197. E. Oberreit, Ber. 29, 2006 (1896).
198. W. Jacobi and G. Merling, Ann. 278, 15 (1894).
199. U. Giannini et al, Polym. Lett. 5, 527 (1967); U. Giannini and
      G. Bruckner, U.S. 3,476,726 (Monte); cf. L. Credoli et al,
      Makromol. Chem. 154, 201 (1972).
200. W. E. Hewett and A. H. Sporer, U.S. 3,341,472 (IBM).
201. K. Hess, Ber. 46, 3125 (1913).
202. W. D. Emmons and J. G. Brodnyan, U.S. 3,554,684 (R & H).
203. W. R. Workman, U.S. 3,074,869 (3M).
204. S. Melamed, U.S. 2,824,858 (R & H).
205. Chem. Eng. News, p. 14, July 3, 1972.

The N-allyl amides, most of which are solids, have found limited use in biocides, corrosion inhibitors, in hydrophilic copolymers, and as intermediates for organic synthesis. Formulas of the more important types are shown below, along with formulas of typical N-allyl imides and C-allyl amides:

N-Allyl amides $ARN\overset{\text{C}}{\underset{\text{O}}{}}R$

N,N-Diallyl amides $A_2N\overset{\text{C}}{\underset{\text{O}}{}}R$

N-Allyl imides $AN(\overset{\text{C}}{\underset{\text{O}}{}}R)_2$

N-Allyl ureas $ARN\overset{\text{C}}{\underset{\text{O}}{}}NHR$

N,N'-Diallyl ureas $ARN\overset{\text{C}}{\underset{\text{O}}{}}NHA$

C-Allyl amides $A(CH_2)_nHN\overset{\text{C}}{\underset{\text{O}}{}}R$

Some C-allyl amides have been used as pharmaceuticals. The monoallyl amides have given only homopolymers of low molecular weight. However, there are interesting possibilities for synthesis of useful copolymers from the monoallyl as well as polyfunctional allyl amides and imides.

The simple N-allyl substituted amides are discussed first. N-Allylformamide (bp 109°C at 15 mm) was made by long heating of mustard oil ANCS with formic acid under pressure at 120°C (1). N-Allylacetamide was prepared in a similar way or by reaction of allylamine with excess acetic anhydride (2). Other references to preparations of N-allyl amides are given in Table 20.1.

In the limited work that has been reported, monofunctional N-allyl amides have given only homopolymers of low molecular weight. Swern and co-workers prepared N-substituted amides by adding the acyl chlorides dropwise to allyl-substituted amines in hexane at 0°C (3). Heating with benzoyl peroxide at 90°C produced no evidence of polymerization of the N-allyl compounds, but long heating with 1 to 4% di-t-butyl peroxide at 125°C gave appreciable polymerization of the following (with monomer melting points):

Compound	mp, °C
N-Methallylcaprylamide	43
N-Methallylmyristamide	69
N-Methallylstearamide	82
N,N-Dimethallylmyristamide	15
N,N-Dimethallylstearamide	35
N,N-Diallylmyristamide	23
N,N-Diallyllauramide	12
N,N-Diallylstearamide	43

The polymers of very low molecular weight were solids and all melted below 45°C. The mono-N-allyl amides tested did not show much evidence of polymerization by decrease in unsaturation or increase of refractive index. The polymers from N,N-diallylmyristamide were soluble in organic solvents, perhaps because of partial cyclic polymerization. Several of the N-alkenyl amides were sulfonated to form surface active agents. An aromatic N-methallylamide was polymerized by heating with di-t-butyl peroxide at 90°C, but polymer properties were not given (4).

Jordan and Wrigley obtained polymers of about DP 10 from benzoyl-peroxide-initiated bulk polymerization of N-allylacetamide and N-allylstearamide (5). The monomers were obtained by ester aminolysis (6,7). The investigators found that the ratio of consumption of N-allylamide to peroxide (d[M]/d[P]) had the unusually low value of 2, compared with 14 to 50 for allyl acetate (8,9). This low ratio was due to wastage of benzoyloxy radicals to form low-molecular-weight products (Fig. 20.1). In addition, induced decomposition of the peroxide by amide radicals led to rapid overall reaction rates. Poly-N-allylacet-amide was a viscous semisolid at room temperature, and poly-N-allylstearamide was a soft wax melting at 70 to 80°C. In contrast, medium hard, high-melting waxes (ca. 80°C) of potential utility were obtained from the reaction of 3 to 8 g of benzoyl peroxide with 100 g of N-allyl-stearamide at 90°C (10).

Jordan and Wrigley (11) showed that the Q and e para-meters for N-allylstearamide at 80°C (Q = 0.043, e = -0.59) were similar to those found for many other monoallylic derivatives (12). They also demonstrated that N-allyl-stearamide behaves like both a comonomer and a transfer solvent toward monomer radicals of varying reactivity (13). Transfer constants for the vinyl comonomers increased in the order methyl methacrylate < styrene < diethyl maleate < vinyl acetate. The decrease in molecular weight was proportional to the amount of N-allylamide in the copoly-mer formed. Consequently, with unreactive comonomers,

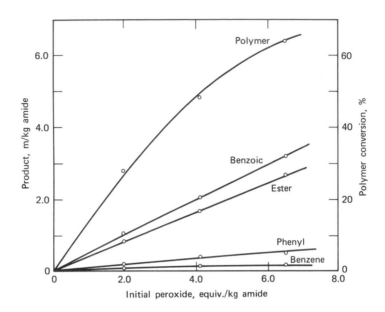

Fig. 20.1.   Low yields of homopolymer of DP about 10 were obtained
from N-allylacetamide at 90°C with benzoyl peroxide as initiator.
Decomposition products of the peroxide, including those from chain
transfer, were estimated.   E. F. Jordan, Jr., and A. N. Wrigley, J.
Polym. Sci. A 2, 3913 (1964).

where the amide entered the copolymer at reasonable rates,
molecular weights were low.   Figure 20.2 shows the compositions of the copolymers formed.
   Weisgerber reacted 110 parts of N-allylacetamide with
172 parts diethyl fumarate in presence of 13.5 parts
benzoyl peroxide by heating 114 hr at 65°C (14).   From
the viscous solution was recovered 150 parts of copolymer
of molar ratio near 1 to 2.   Hydrolysis of the copolymer
by acid or base could be used to prepare amphoteric polymers.   Copolymerizations of N-allylacetamide were also
disclosed with vinyl acetate, styrene, and with methyl
acrylate.
   N-Allyl derivatives of acetamide, chloroacetamide, and
benzamide were homopolymerized to products of DP 4 to 12
in bulk under ^{60}Co radiation (15).   N-Allyl amides such
as N-allylacetamide have been copolymerized with acrylonitrile in activated aqueous persulfate solution for

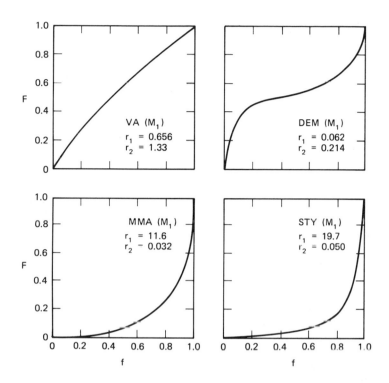

Fig. 20.2. Copolymerizations of vinyl acetate (VA) and of diethyl maleate (DEM) with N-allylstearamide occurred fairly readily to give products with mole fractions of amide units F from monomer mixtures of mole fractions amide f. With DEM there was a tendency to form 1:1 molar copolymers. Additions of N-allylstearamide retarded polymerization of methyl methacrylate (MMA) and of styrene (STY), and the copolymers formed were low in amide groups. E. F. Jordan, Jr., B. Artymyshyn, and A. N. Wrigley, J. Polym. Sci. A 1, 6, 576 (1968).

preparation of dye-receptive fibers (16). Copolymers of N-allylnicotinamide with acrylonitrile were blended with acrylonitrile polymers for improving dye receptivity of fibers (17). N-Allylcyanoacetamide was copolymerized with vinyl acetate in hot methanol solution with benzoyl peroxide (18).

N-Allyllactamide $n_D^{20} = 1.4753$ darkened but showed no evidence of polymerization in contact with cobalt octa-

TABLE 20.1

Some N-Allyl Amides

N-Allyl compound of	bp, °C at mm	n_D, etc.	References
Formamide	216 or 109(15)	--	Gluud, J. Chem. Soc. 103, 941; supplied by Roberts Chemicals
Acetamide	140(10)	$n_D^{20} = 1.4861$	Clayton, Ber. 28, 1666 (1895)
Propionamide	143(11)	$n_D^{17} = 1.4853$	--
t-Butylacetamide	105(7)	$n_D^{20} = 1.4575$	U.S. 2,060,154 (Mallinkrodt)
Ethylisopropyl-acetamide	--	mp 60°C	Volwiler, JACS 58, 1353 (1936)
Caprylamide	128(3)	mp 27°C	Glacet, CA 57, 11147 (1962); Swern, JACS 73, 3642 (1951)
Myristamide	--	mp 71°C	Swern
Stearamide	--	mp 84°C	Swern; also Jordan, J. Am. Oil Chem. Assoc. 38, 600 (1961)
Chloramides and bromamides	--	--	Swenson, JACS 70, 4060 (1948); Behrens, J. Biol. Chem. 175, 780 (1948)
Benzamide	174(14)	--	Kay, Ber. 26, 2848
Benzene sulfonamide	--	--	Gensler, JACS 70, 1843 (1948)
Benzanilide	--	--	Mumm, Ber. 70B, 2214 (1937)
Benzamide derivatives	--	--	U.S. 2,685,594 (Sterling Drug); Goodman, JACS 79, 4788 (1957)
Salicylamide	--	mp 52°C	Diels, Ber. 39, 4127
Acetoacetamides	125(1)	mp 28°C	Jones, U.S. 2,561,205 (GAF) (copolymers)
Oxalamide (mono)	--	mp 98°C	Gluud
Lactamide	116(1)	--	Kenyon, U.S. 2,490,756 (Eastman); cf. U.S. 2,458,423
Pyrrolidone	120(12)	--	Spaeth, Ber. 69B, 2727 (1936)

noate and air during 10 days (19). Heating for 8 days at
100°C with additions of benzoyl peroxide gave a little
evidence of polymerization through increase in viscosity.
N-Methallyllactamide $n_D^{20} = 1.4750$ did not homopolymerize.
N-Allyl amides of halogen containing acids were prepared
(20). N-Allyl-N-butylacetamide was made by heating
allyl-n-butylamine with acetic anhydride (21). N-Allyl-
gluconamide, melting at 27°C, and other N-allyl derivatives of sugar acids were prepared (22).

Polymerization of the N-allyl monoamide of maleic acid
(mp 97°C) under ^{60}Co radiation was studied (23). Irradiation of the crystals in air changed the color from
yellow to blue, but the color was restored to yellow on
storage. Intermittent irradiation increased the radiation
yield of polymer. The solid-state polymerization occurred
by a radical mechanism which continued on stopping irradiation to form crosslinked polymers. The N-allyl monoamide of succinic acid (mp 78°C) also polymerized in the
solid state under high-energy radiation, at the same time
preserving the shape of the monomer crystals. References
to other N-allyl amides are given in Table 20.1.

N,N-DIALLYL AMIDES

Little polymerization work has been reported on N,N-
diallyl amides. $R\underset{O}{C}N(CH_2CH=CH_2)_2$ (24). Diallylurea and
diallylthiourea have been discussed in earlier chapters.
Unless otherwise indicated, the diallyl amides discussed
are solids. Several diallyl amides such as N,N-diallyl-
chloroacetamide have activity as herbicides.

N,N-Diallyl amides of lower acids gave very slow polymerizations with radical catalysts to form low polymer
oils (25). N,N-Diallyl and dimethallyl amides of long-
chain acids have been prepared by reaction of the acyl
chlorides with the dialkenyl amines (26); these included
diallylstearamide and dimethallylstearamide, melting at
43 and 35°C, respectively. Homopolymerizations at 125°C
with 1 to 4% di-t-butyl peroxide were slow. There was
increase in refractive index on prolonged heating and no
crosslinking was recorded. Several of the polymers
isolated melted below 45°C. N,N-Diallylformamide liquid
did not polymerize readily with 5% BP on being irradiated
with UV light (27).

Tetraallyl adipamide has been prepared by portionwise
addition of 1 mole of adipyl chloride to 5 moles of
diallylamine, followed by refluxing for an hour (28). The
brownish monomer that was isolated could be polymerized
on baking in films with polyvinyl butyral. It was copolymerized also with an unsaturated polyester. Hexamethylene

TABLE 20.2

Some Polyfunctional N-Allyl Amides

N,N-Diallyl Compounds of:	mp, °C	Other properties	References
Formamide	--	yellowish liquid; $n_D^{25} = 1.4691$	Aldrich supplies
Phenylacetamide	--	retards seed germination	Pohland, U.S. 3,449,114 (Lilly)
Chloroacetamide	--	bp, 92°C at 2 mm; $n_D^{25} = 1.4932$; irritant	Speziale, JACS 78, 2556 (1956); herbicide (Randox) of Monsanto
Oxamide	154	decomposes, 274°C	Wallach, Ber. 13, 513 (1880)
Succinamide	188	--	Szekeres, Gazz. Chim. It. 79, 58 (1949)
Azeleamide	120	--	Szekeres
Malonamide	149	bitter taste	Pauw, Rec. Trav. 55, 219 (1936)
Malamide	117	--	Franklin, J. Chem. Soc. 89, 1864 (1906)
3,3-Dimethyl glutaramide	--	bp 162°C at 0.5 mm	U.S. 2,447,195 (Geigy)
Long-chain acid derivatives	--	--	Smith, U.S. 2,832,799 (Shell)
Tartaramide	183	--	Franklin, J. Chem. Soc. 89, 1856 (1906)
Gluconamide	174	--	U.S. 2,084,626 (Abbott Labs)

bis(diallyl urea) has been made by the same research worker by reacting diallylamine with hexamethylene diisocyanate (28). Symmetrical diallyl thiourea is available commercially and is used in the dyestuff and pharmaceutical fields. References to polyfunctional N-allyl amides are given in Table 20.2.

N,N'-Diallyl tartardiamide is supplied by Eastman. It can be added in minor proportions to acrylamide for preparing gels that are soluble in 2% periodic acid. These gels are somewhat similar to those made using methylenebisacrylamide as crosslinking agent and which find application in gel chromatography.

N-ALLYL IMIDES

The allyl imides of structure $AN(COR)_2$ have interesting possibilities in copolymerizations. N-allyldiacetamide $AN(COCH_3)_2$ was made by heating equal parts of allyl isothiocyanate and acetic anhydride at 200°C for 8 hr (29). It distilled at 90°C at 14 mm. N-Allylsuccinimide (bp 130°C at 14 mm) was prepared by reaction of ANCS with succinic acid or anhydride (30). N-Allylphthalimide (mp 71°C) was made also using allyl isothiocyanate (30) or by reacting the potassium salt of phthalimide with allyl bromide (31). N-Allyl hexahydrophthalimide was reported to boil at 113°C at 2 mm with $n_D^{20} = 1.5078$ (32). Free radical polymerization of N-allylmaleimide from 10% solution in benzene, dioxane, or anisole gave crosslinked products even at low conversions (33). However, soluble polymer was obtained from 3% solution in tetrahydrofuran. Triallyl isocyanurate, discussed in Chapter 24, is a cyclic N-allyl imide.

A diallyl diimide melting at 110°C was prepared from 1,2,3,4-cyclopentane tetracarboxylic acid and ammonia (34). It was polymerized in a mold with 2% t-butyl perbenzoate as catalyst by heating for 72 hr at 125°C.

ALLYL UREAS

Neither the monoallyl nor diallyl ureas homopolymerize readily, but they can be copolymerized with certain vinyl and acrylic monomers and they have shown interesting biological and chemical properties. The compounds are white solids, and they have shown higher melting points in recent research than in early references.

Mono-N-allylurea, made from allyl isothiocyanate and ammonia, is soluble in water and in alcohol but insoluble in chloroform, ether, and toluene. It has been used as a corrosion inhibitor and is available commercially as white crystals melting at 100°C (35). The compound is one of the urea derivatives reported to inhibit influenza virus. Allylurea is supplied by Chemicals Procurement Laboratories, Pfaltz and Bauer, and others.

Minor proportions of N-allylurea have been copolymerized with acrylonitrile in aqueous persulfate emulsion, and experimental fibers were spun from solutions of the copolymers in dimethyl formamide (36). Another patent discloses copolymerization of 90% acrylonitrile with 10% N-allylurea in aqueous persulfate for the preparation of acrylic fibers receptive to wool dyes (37).

N,N-Diallylurea was prepared by hydrolysis of N,N-diallylcyanamide in the presence of sulfuric acid, and also by reaction of diallyl ammonium chloride with potassium cyanate (38). Crystallization from benzene gave colorless crystals melting at 66°C (39). A methyloldiallyl urea resin, prepared by reacting N,N-diallylurea with formalin, using dilute NaOH catalyst, for 6 hr at 70°C, has been suggested as a shrinkproofing agent for

TABLE 20.3

Allyl Ureas and Related Compounds

Compound	bp or mp, °C	References
Allylurea	mp 100	Kitamura, CZ 3434 (1936); Andreasch, Monatsh 5, 37 (1884); Aldrich supplies
N,N-Diallylurea	mp 66	Yale, U.S. 2,928,855 (Olin); Aeschlimann, CZ II 2943 (1936)
N,N'-Diallylurea	mp 91	Applegath, U.S. 2,857,430 (Monsanto)
Tetraallylurea	bp 123(14)	Aeschlimann, CZ II 2943 (1936)
N-Allyl-N'-phthalyl- urea	--	Drechsel, U.S. 2,516,836 (Am. Cyan.) (copolymers)
N-Allyl ethylene urea AN-C-N-CH₂ H O CH₂	--	Kropa, U.S. 2,655,494 (Am. Cyan.) (Acrylonitrile copolymers)
N,N'-Dimethallyl urea	bp 121	Applegath, Brit. 818,864 (Monsanto); CA 54, 9774 (1960)
N-Allyl-N'-phenylurea	--	Menne, Ber. 33, 661 (1900); Dains, JACS 21, 165 (1899)
N-Allyl-N'-o-tolylurea	mp 152	Menne
m-isomer	mp 115	Menne
p-isomer	mp 139	Menne

textiles. N,N-Diallylurea was prepared also from diallylamine and an isocyanate (40).

N,N'-Diallylurea, long known as sinapolin, can be made by heating allyl isocyanate with water (41) or by heating the mustard oil with aqueous barium hydroxide (42) or lead hydroxide (43). This water-soluble crystalline compound has a sharp odor. The two isomeric diallyl ureas do not homopolymerize readily on heating with free radical catalysts or on exposure to UV light. Their copolymerizations have been studied by Beynon and Hayward with lauryl methacrylate in dioxane solution with benzoyl peroxide at 100°C (39). Infrared investigation revealed that the copolymers of the symmetrical diallylurea contained more residual bonds than copolymers of N,N-diallylurea. The low reactivity of the diallylurea was shown with lauryl methacrylate (for N,N-DAU $r_1 = 27$, $r_2 = 0.014$ and for N,N'-DAU $r_1 = 14$, $r_2 = -0.12$). The data gave evidence for the similarity in reactivity of the allyl groups in the isomeric disubstituted ureas as well as in monoallylurea. Such reactivity ratios provide some evidence for cyclic propagation reactions. References to N-allyl ureas are listed in Table 20.3.

Isoureas are formed by addition of an alcohol to a cyanamide where an allyl group may be attached to nitrogen or to oxygen:

$$R_2NCN + ROH \longrightarrow H_2NC=NH$$
$$OR$$

Price of American Cyanamid copolymerized O-allyl isoureas (44) as well as N-allyl isoureas (45) with acrylonitrile and suggested copolymerization with other $CH_2=C\subset$ monomers.

Among related N-allyl compounds are the N-allyl sulfonamides. For example, N,N-diallyl and N,N-dimethallyl methane sulfonamides were prepared by reacting diallyl or dimethallylamine with methane sulfonyl chloride (46). Radical polymerizations at 80°C gave low conversions to cyclized polymers. A useful fluorescing tagging agent for proteins is N-allyl-DANSA, an N-allyl derivative of 1-dimethylaminonaphthalene-5-sulfonamide (47). Sulfur dioxide was copolymerized with N,N-diallyl perfluorooctane sulfonamide using 5% azobis at 75°C for an hour (48). The products had low intrinsic viscosities in dimethyl formamide.

C-ALLYLIC AMIDES

Some of the C-allylic amides have found use in pharmaceuticals, but their polymerizations have received little attention. Of C-allyl amides of the type formula $CH_2-CHCH_2(CH_2)_nCONH_2$, vinylacetamide or allyl formamide, melting at 72°C, has been prepared from ACN and H_2O_2 in aqueous acetone with sodium carbonate catalyst (49). Rates of ozonolysis of the compound in acetic acid and rates of hydrolysis in acid and alkaline solution were studied (50). Allylacetamide was made by cautious addition of cold concentrated NH_4OH to the acyl chloride (51). It was also prepared by hydrolysis of the nitrile)52). Allylacetamide is relatively unstable. A_3CCONH_2 (mp 66°C) has been taken to induce sleep (53).

The amide of isopropenylacetic acid was made by mixing the nitrile and an acetone solution of 10% hydrogen peroxide and 10% sodium hydroxide with cooling (54). The amide of undecylenic acid was found to have bacteriostatic and fungistatic action (55). The compound $AC(C_2H_5)_2CONH_2$, known as Novonal, was used in the pharmaceuticals Sedormid and Pyramidon (56). Table 20.4 gives a few other C-allylamides chosen from many in the literature.

Thermosetting copolymers have been made from N-methylol ethers of allylacetamide (57). Copolymers of allylacetamide also can be methylolated by formaldehyde. A copolymer of 95% ethyl acrylate, 2% allylacetamide, and 3% N-methylol allylacetamide units was dissolved in butyl acetate-xylene and used as a finish for nylon fabric (58).

TABLE 20.4

C-Allyl Amides

Compound	mp, °C	bp, °C	References
$ACONH_2$	72	--	Brulé, Bull. Soc. Chim. France [4] 5, 1022
ACH_2CONH_2	94 or 106	230	Henry, CZ II, 663 (1898); Paul, Bull. 474 (1949); Cagniant, Compt. Rend. 217, 28 (1943)
$A(CH_2)_7CONH_2$	99	--	Aschan, Ber. 31, 2349 (1898); Kraft, Ber. 33, 3581 (1900); Chuit, Helv. Chim. Acta 9, 1091 (1926)
$ACH_2CONHC_6H_5$	72	--	Wohlgemuth, A. Ch. [9] 2, 331
$ACHCONH_2$ \ $\overset{\|}{C}H(CH_3)_2$	107	--	Riedel, Ger. 461,814; Boedecker, CZ II, 1079 (1927)
$AC(C_2H_5)CONH_2$ \ $\overset{\|}{C}H_3$	51	--	Ger. 412,820 (Hoechst); CZ II, 92 (1925)
$AC(C_2H_5)_2CONH_2$	73 or 80	155(10)	Ger. 412,820; 447,161; 473,329 (Hoechst); CZ I 1679 and 1887 (1928)
$CH_2=C(CH_3)CH_2CONH_2$	120	--	Mooradian, JACS 68, 788 (1946)
$CH_2=CHCH(CH_3)CONH_2$	98	--	Lane, JACS 66, 544 (1944)
$A_2CHCONH_2$	83	--	Ruhkopf, Ber. 73, 940 (1940); Ger. 720,405 (Beiersdorf)
$A_2C(CH_3)CONH_2$	37	--	Ziegler, Ger. 611,374; Ger. 615,399 (Schering) (hypnotic)
$A_2C(C_2H_5)CONH_2$	66	--	U.S. 1,967,388 (Schering)
A_2CCONH_2 \ $\overset{\|}{C}H_2CH_2CH_5$	53	--	Schering
$ACH(CONH_2)_2$	98	289	Henry, Jahresber. 639 (1889)

A small amount of n-butyl acid phosphate catalyzed cure on the fibers when heated for 5 min at 260°F. Such a copolymer was made by portionwise addition during 2 hr of the monomers and azobis initiator to butyl acetate at 80°C.

A diallyl piperidone copolymerized with dithiols by hydrogen transfer using radical-catalyzed emulsion systems at 60°C for 18 hr (59). Molecular weights of the products (20,000-30,000) were high enough for melt spinning to crystalline fibers. The polymers could be hydrolyzed with dilute sulfuric acid to form polyampholytes. The inherent viscosities of the cyclic amide-sulfide polymers were determined in formic acid solutions to be 0.2 to 0.3.

Little research has been carried out with aromatic

C-allyl amides. From 2-allylphenylacetamide of melting
point 54°C, 2-allylbenzylamine was prepared by the Hofmann
degradation (60).

References

1. G. C. Clayton, Ber. $\underline{28}$, 1666 (1895).
2. E. Chiari, Monatsh. Chem. $\underline{19}$, 572 (1898).
3. D. Swern et al, J. Am. Chem. Soc. $\underline{73}$, 3642 (1951).
4. S. A. Ballard et al, U.S. 2,687,403 (Shell).
5. E. F. Jordan, Jr., and A. N. Wrigley, J. Polym. Sci. A $\underline{2}$, 3909
    (1964); U.S. 3,403,172 (USA).
6. E. F. Jordan, Jr., and W. S. Port, J. Am. Oil Chem. Soc. $\underline{38}$,
    600 (1961).
7. E. F. Jordan, Jr. et al, J. Am. Oil Chem. Soc. $\underline{43}$, 75 (1966).
8. P. D. Bartlett and R. Altschul, J. Am. Chem. Soc. $\underline{67}$, 816 (1945).
9. M. Litt and F. R. Eirich, J. Polym. Sci. $\underline{45}$, 379 (1960).
10. E. F. Jordan, Jr. et al, J. Am. Oil Chem. Soc. $\underline{43}$, 452 (1966).
11. E. F. Jordan, Jr., and A. N. Wrigley, J. Appl. Polym. Sci. $\underline{8}$,
    527 (1964).
12. L. J. Young, J. Polym. Sci. $\underline{54}$, 411 (1961).
13. E. F. Jordan, Jr. et al, J. Polym. Sci. A 1 $\underline{6}$, 575 (1968).
14. C. A. Weisgerber, U.S. 2,592,218 (Hercules).
15. F. F. Shcherbina et al, CA $\underline{68}$, 50158 (1968).
16. J. R. Caldwell, U.S. 2,596,650 (Eastman).
17. G. E. Ham and A. B. Craig, U.S. 2,664,412 (Chemstrand).
18. C. C. Unruh et al, U.S. 2,808,331 (Eastman).
19. W. P. Ratchford and C. H. Fisher, J. Org. Chem. $\underline{15}$, 317 (1950).
20. A. D. Swensen and W. E. Weaver, J. Am. Chem. Soc. $\underline{70}$, 4060 (1948);
    O. K. Behrens, J. Biol. Chem. $\underline{175}$, 780 (1948).
21. R. Tiollais, Bull. Soc. Chim. France 962 (1947); cf.
    U.S. 2,060,154.
22. U.S. 2,084,626 (Abbott Labs); CZ II 2210 (1937).
23. F. F. Shcherbina and I. P. Fedorova, Polym. Sci. USSR (in
    English) $\underline{10}$, 56 (1968); CA $\underline{68}$, 69376 (1968).
24. F. H. Adams, U.S. 2,479,890 (Am. Cyan.).
25. G. B. Butler and M. D. Barnett, CA $\underline{55}$, 24083; PB 145,435 (1959).
26. D. Swern et al, J. Am. Chem. Soc. $\underline{73}$, 3642 (1951).
27. C. E. Schildknecht, unpublished.
28. B. R. Franko-Filipasic, U.S. 2,820,020-1 (PPG).
29. P. Kay, Ber. $\underline{26}$, 2851 (1893).
30. F. Moine, Jahresber. 558 (1886); cf. Kay, ref. 29 above.
31. A. Neumann, Ber. $\underline{23}$, 999 (1890); O. Wallach and I. Kamenski,
    Ber. $\underline{14}$, 171 (1881).
32. M. S. Newman et al, J. Am. Chem. Soc. $\underline{68}$, 2113 (1946).
33. T. M. Pyriadi and H. J. Harwood, Polym. Preprints $\underline{11}$, 60
    (February 1970).
34. J. C. Petropoulus and R. R. DiLeone, U.S. 3,444,184 (Am. Cyan.).
35. General Biochemicals, Chagrin Falls, Ohio.

36. J. R. Caldwell, U.S. 2,544,638 (Eastman).

37. D. W. Chaney, U.S. 2,688,010 (Chemstrand).

38. J. P. Milionis and P. Adams, U.S. 2,734,083 (Am. Cyan.).

39. K. I. Beynon and E. J. Hayward, J. Polym. Sci. A $\underline{3}$, 1793 (1965).

40. R. G. Neville, J. Org. Chem. $\underline{23}$, 937 (1958).

41. A. Cahours and A. W. Hofmann, Ann. $\underline{102}$, 300 (1857).

42. O. Hecht, Ber. $\underline{23}$, 287 (1890).

43. J. E. Simon, Ann. Physik. $\underline{50}$, 377 (1840).

44. J. A. Price, U.S. 2,655,493 (Am. Cyan.).

45. J. A. Price, U.S. 2,654,725 (Am. Cyan.).

46. A. Crawshaw and A. G. Jones, J. Macromol. Sci.-Chem. $\underline{A6}$, 65
     (1972).

47. D. P. Borris and J. N. Aronson, Anal. Biochem. $\underline{32}$, 273 (1969).

48. C. D. Wright and W. S. Friedlander, U.S. 3,072,616 (3M).

49. J. V. Murray and J. B. Cloke, J. Am. Chem. Soc. $\underline{56}$, 2751 (1934).

50. P. Bruylants, Bull. Soc. Chim. Belg. $\underline{57}$, 58 (1948); Bull. Acad.
     Belg. [5], $\underline{27}$, 202 (1941); CZ I, 1991 (1944).

51. Buu-Hoi and P. Cogniant, Compt. Rend. $\underline{217}$, 28 (1943).

52. R. Paul et al, Bull. Soc. Chim. France 474 (1949).

53. U.S. 1,958,653; U.S. 1,967,388 (Schering); CZ I, 1197 (1933).

54. A. Mooradian and J. B. Cloke, J. Am. Chem. Soc. $\underline{68}$, 788 (1946).

55. R. Brodersen and A. Kjaer, CA $\underline{42}$, 3803 (1948); cf. O. Wyss,
     Arch. Biochem. $\underline{7}$, 422 (1945).

56. P. Dumont and A. Declerck, J. Pharm. Belg. $\underline{14}$, 179; CZ II, 3742
     (1932).

57. M. Hurwitz, U.S. 3,274,164 (R & H).

58. M. I. Seifer and C.-P. Lo, U.S. 3,307,965 (R & H).

59. C. H. H. Neufeld and C. S. Marvel, J. Polym. Sci. A1 $\underline{5}$, 537
     (1967).

60. R. M. Horowitz and T. A. Geissman, J. Am. Chem. Soc. $\underline{72}$, 1519
     (1950).

21. ALLYL URETHANES

Allyl urethanes, also known as allyl carbamates, comprise N-allyl urethanes such as $CH_2=CHCH_2NHCOOR$ and O-allyl urethanes or allyl carbamate esters such as $RNHCOOCH_2CH=CH_2$. These compounds have not attained commercial importance. However, some of the allyl urethanes have interesting biocidal properties (1). Polymers of allyl urethanes were investigated in laboratories of Pittsburgh Plate Glass (PPG) and Goodyear Rubber Company. Some of the diallyl compounds gave interesting cast polymers. The N-allyl urethanes are discussed first.

The free acids, salts, and esters of N-allyl carbamic acid $ANHCOY$ (where Y is H, M, or R) have been little studied in polymerization. These compounds should not be confused with the allyl esters of carbamic acids. N-Allyl urethane, a high-melting solid, $ANHCOOC_2H_5$, was prepared by reaction of allylamine with ethyl chloroformate (2). It is supplied by Borden. Allyl isocyanate, prepared from allyl chloride and potassium cyanate without isolation, was reacted with alcohols to give alkyl N-allyl carbamates or N-allyl urethanes (3):

$$CH_2=CHCH_2NCO + ROH \longrightarrow CH_2=CHCH_2\overset{H}{\underset{\underset{O}{\|}}{N}COR}$$

Reaction of allyl isocyanate with ethylene glycol yielded ethylene bis-N-allyl carbamate, boiling at 118°C at 0.15 mm and melting at 100°C. The allyl ester of N-allyl carbamate (bp 73°C at 0.5 mm) also was made. Byproducts encountered in this synthesis included diallylurea and triallyl isocyanurate. N-Allyl urethanes have been copolymerized with larger proportions of vinyl acetate using BP catalyst at 50°C (4). The copolymers were saponified by heating at 60°C with sodium hydroxide in methanol, apparently forming vinyl alcohol-allylamine copolymers. Diethylene glycol bis(allyl carbamate) has been polymerized and has been suggested as an oil modifier (5). Some references to N-allyl urethanes and related compounds are given in Table 21.1.

569

TABLE 21.1

Some N-Allyl Urethanes and
Related Compounds, $CH_2=CHCH_2NHC-OR$
                                      $\overset{\|}{O}$

Monomers	bp, °C at mm	References
Esters of N-allyl carbamic acid:		
Methyl	183	Childs, J. Chem. Soc. 2320 (1948)
Ethyl	194 or 92(15)	Hurd, JACS 57, 2656 (1935); Rowland, JACS 72, 3593 (1950); Werner, JACS 76, 2453 (1954)
2-Chlorethyl	130(10)	Pierce, JACS 50, 242 (1928); Haworth, J. Chem. Soc. 179 (1947)
Isopropyl	74	Ger. 652,363 (Ciba)
n-Butyl	86	Ciba patent
Benzyl	150(0.5)	Schleppnik, J. Org. Chem. 25, 1378 (1960)
Related Compounds:		
Ethyl N,N-diallyl carbamate	62(3)	Austrian 154,905 (Ciba); CZ I, 2248 (1939)
Allyl and methallyl thiolcarbamates (e.g., ANRC=OSR)	--	Tilles, JACS 81, 714 (1959)
ANHC=OSA	--	Batty et al, Brit. 589,178
N-Allyl guanidines, $ANHC-NR_2$ $\overset{\|}{\underset{}{N}}H$	--	Mull, U.S. 3,301,753 (Ciba); (Non-toxic oral pharmaceuticals)
A_2N⬡$OCNHCH_3$ $\overset{\|}{O}$	--	Heiss, U.S. 3,210,403 (Bayer)

Allyl dithiocarbamates can be prepared by reaction of
allylamine and CS_2 with alkali (6):

$$CH_2=CHCH_2NH_2 + CS_2 + NaOH \longrightarrow CH_2=CHCH_2NH\overset{\overset{S}{\|}}{C}SNa$$

Diallylamine gave the analogous diallyl dithiocarbamate
salt. Alkali and ammonium salts of N-allyl and N-chlor-
allyl dithiocarbamates have been used as rubber chemicals
(7).

ALLYL CARBAMATE ESTERS

Allyl esters of carbamic acids such as RNHCOOA will be
called hereafter allyl urethanes. These monomers were

studied by Chenicek of PPG (8) before the company selected
the diallyl diglycol carbonate (CR-39) for commercial
development. Allyl urethanes were prepared in the laboratory by dropwise addition of allyl chloroformate to
amino compounds at 0 to 10°C in the presence of pyridine
as an acid acceptor. The reaction mixture was poured into
cold, dilute hydrochloric acid from which the carbamate
was passed into a solution of 368 g of allyl chloroformate
in a liter of benzene at 25 to 35°C until reaction was
complete (9). After the ammonium chloride crystals were
filtered off, the benzene solution was distilled, giving
an 85% yield of allyl carbamate AOC-NH$_2$ (bp 75°C at 2 mm).
                                    ‖
                                    O
Gleim found that allyl carbamate could be polymerized to
solid polymers by long heating with high concentrations
of benzoyl peroxide (10). Because of strong hydrogen
bonding, polymers of low DP can be solids.

Chenicek used acyl chlorides of adipic and other dicarboxylic acids to prepare diallyl dicarbamate derivative
monomers (11). Some of these solid monomers were polymerized by heating the melts with peroxides. A PPG patent
describes the preparation of allyl carbonate carbamate
monomers of the type AOOCORNHCOOR (12). By portionwise
addition of allyl chloroformate to ethylene diamine in
the presence of NaOH at 3 to 18°C, the following diallyl
urethane monomer (mp 84°C) was made (13): (AOOCNHCH$_2$)$_2$.
Heating with benzoyl peroxide gave a crosslinked polymer.

For synthesis of a bifunctional allyl urethane monomer
such as N-carballyl dioxyethyl allyl carbamate, 8 moles
of allyl chloroformate was added at a rate of 2 g per
minute to a solution of 4 moles of monoethanolamine in
9 moles of pyridine at -10°C (14). The mixture was then
poured into cold, dilute HCl. The oily layer was washed
with aqueous sodium bicarbonate and distilled at 151 to
152°C at 3 mm. Bis(N,N'-carballyloxy)urea, a diallyl
urethane, was made by Gleim by passing phosgene into
2 moles of allyl carbamate at 50 to 75°C during 4 hr (15).
The solid product was filtered off and recrystallized
from ethanol. Properties of a number of allyl urethanes
and related compounds prepared by Gleim of Goodyear are
given in Table 21.2.

A diallyl carbamate of formula (CH$_2$=CHCH$_2$OOC)$_2$NH (bp
147°C at 13 mm) gave clear cast copolymers with acrylic,
vinyl ester, and styrene monomers (15). Reaction of
allyl chloroformate with hydroxyurea produced some
(CH$_2$=CHCH$_2$OCONH$_2$)$_2$CO, melting at 63°C (16). It could be
polymerized to clear insoluble polymers resistant to
abrasion. Copolymers of allyl carbamates with lauryl
methacrylate have been suggested as oil modifiers (17).

TABLE 21.2

Some Allyl Urethanes (Allyl Esters of Carbamic Acid)
(where A is Allyl, CH_2=$CHCH_2$) [a]

Compound	bp or mp, °C	n_D	Preparation
$AOOCNH_2$	73(2)	1.4520(28)	A chloroformate + NH_3
AOOCNHCOOA	147(3)	1.4728(23)	A chloroformate + A(N-Na) carbamate
$(AOOCNH)_2$ CO	mp 70	--	A carbamate + $COCl_2$
AOOCNHA	85(3)	1.4566(28)	A chloroformate + monoallyl-amine
$(AOOCNH)_2$	mp 89	--	A chloroformate + hydrazine
$AOOCNHCO_3A$	145(3)	1.4558(28)	A chloroformate + N-hydroxyallyl carbamate
AOOCNHOA	92(2)	1.4569(30)	A chloride + NaONHCOOA
AOOCNHC(O)$NHCO_3A$	mp 63	--	A chloroformate + hydroxyurea
$(AOOCNHCH_2)_2$	mp 90	--	Diallyl carbonate + ethylene diamine
$AOOCNHCH_2CH_2CO_3A$	152(3)	1.4630(28)	A chloroformate + mono-ethanolamine
$AOOCNHC_6H_4CO_3A$	mp 90	--	A chloroformate + p-amino-phenol
$(AOONH)_2CH_2$	mp 109	--	A carbamate + formalin
$(CH_2$=$NCOOA)_x$	185(2) dec.	1.4870(30)	Allyl carbamate, formalin and conc. HCl
$AOOCNHC(CH_3)$=CH_2	60(2)	1.5398(28)	Methacrylamide + A alcohol + KBrO solution
$AOOCN(\overset{\text{O}}{\underset{\text{}}{C}}CH=CH_2)_2$	160(2) dec.	1.4653 (24)	Acrylyl chloride + A(N-Na) carbamate

[a] Data from C. E. Gleim, J. Am. Chem. Soc. 76, 108 (1954) where original references are given.

Allyl chloroformate was reacted with hydrazine hydrochloride in the presence of alkali to give the diallyl urethane AOOCNHNHCOOA (18). With 0.5% t-butyl perbenzoate at 90°C, polymerization required 144 hr.

A triallyl monomer was prepared by reacting allyl chloroformate with hydroxylamine hydrochloride (19). However, in the presence of pyridine as acid acceptor, hydroxylamine reacted with 2 molecules of allyl chloroformate to form AOOCNHOCOOA (20). In the presence of sodium carbonate as acid acceptor only the amino hydrogen reacted, giving AOOCNHOH. Polyfunctional allyl monomers have been prepared by reacting allyl esters of amino acids such as glycine, lysine, and arginine with allyl chloroformate (21). The diallyl monomer from glycine $AOOCCH_2NHCOOA$ boiled at 128°C (1 mm) and had $n_D^{24} = 1.4653$. With 5% BP catalyst and heating for 39 hr at 65° it gave clear, glasslike polymers of Rockwell hardness M = 100 and specific gravity of 1.277. Unfortunately many of the polymers from allyl urethane monomers tend to show yellowish colors. Properties of allyl urethane monomers synthesized by Gleim are given in Table 21.2.

The diallyl monomer AOOCNHCOOA has been prepared by reacting sodium in liquid ammonia with allyl carbamate (22). Ammonia was allowed to evaporate and then there were added successively benzene, dimethylaniline, and allyl chloroformate. The product, diallyl imidodicarboxylate, boiled at 147°C at 3 mm and showed $n_D^{23} = 1.4728$. It was fairly stable at room temperature. Heating with 3% BP for 5 hr at 65°C gave insoluble glasslike polymers of Rockwell M = 92. The polymers were resistant to weak acids and bases. Copolymers formed with allyl methylene carbamate as well as with acrylic and styrene monomers. Diallyl urethanes (0.5%) in natural rubber can accelerate curing under high-energy radiation such as 4×10^7 rads from a ^{60}Co source (23).

Gleim polymerized a series of diallyl carbamates and related diallyl nitrogen compounds by heating with peroxide catalyst to give insoluble crosslinked polymers (24). Rockwell hardness of these products ranged from 85 to 105. Some bifunctional allyl carbamates such as ANHCOOA and AONHCOOA did not polymerize readily as expected. Polymerizations above 60°C tended to cause discoloration and polymer cracking.

From allyl carbamate, formalin, and hydrochloric acid, Chenicek prepared a compound which he believed to be N,N'-methylene bis(allyl carbamate) and which he polymerized (25). An apparently different monomer from the same reactants was obtained by Gleim (24). With 2.5% BP heated for 20 hr at 55°C it gave nearly colorless, insoluble cast polymers of Rockwell hardness 98. Abrasion resistance was between values for CR-39 and those of softer Plexiglas.

Hexamethylene bis(allyl carbamate) $AOOCNH(CH_2)_6NCOOA$ has been prepared by an exothermic reaction between 2 moles of allyl alcohol and 1 mole of hexamethylene diisocyanate (26). The compound is a white solid, melting above 60°C. It was polymerized to a transparent, infusible polymer by heating with t-BHP at 95°C for 24 hr. Ethylene bis(allyl carbamate) was said to polymerize under similar conditions. Hexamethylene bis(methallyl carbamate) was heated with 2% t-BHP at 110° for 24 hr to produce an opaque, thermoset, hard polymer.

A dimethallyl urethane was prepared by adding ethylene

bischloroformate dropwise to a solution of methallyl carbamate in benzene containing metallic sodium (27). The
white solid that was isolated melted at 131°C. On heating
with 5% acetone peroxide at 150°C it polymerized to a
crosslinked solid polymer. Copolymerization of allyl
carbamates with vinyl, acrylic, and styrene monomers by
free radical initiators has been disclosed in patents (28).
Toluylene diallylurethanes have been studied in copolymerization with unsaturated polyesters (29). Copolymers
have been made from 2-phenyl allyl carbamates (30).

Up to 20% of allyl carbanilate $AOOCNHC_6H_5$ has been copolymerized with vinyl acetate in aqueous persulfate
emulsion (31). Homopolymers of low DP were reported by
heating 2-phenyl allyl ester of substituted carbamic
acids with BP under nitrogen at 60°C for 5 hr (32). Allyl
esters of N-acrylcarbamic acids have herbicidal properties
(33). Allyloxyphenyl carbamates were patented as insecticides (34). By reacting 1 mole of allyl alcohol with
hexamethylene diisocyanate, an allyl urethane isocyanate
$AOCOHN(CH_2)_6NCO$ has been made (35). Reaction products
with nitrocellulose gave films that could be graft
copolymerized with acrylonitrile. N-methylol bromallyl
carbamate was reacted with cotton in order to improve
flame resistance (36).

References

1. D. R. Cassaday et al, U.S. 3,531,511 (Eli Lilly).
2. C. Manuelli and E. Comanducci, Gazz. Chim. Ital. 29 II, 146 (1899);
    M. Bergmann, Ber. 54, 2147 (1921); cf. S. Nirdlinger and S. F.
    Acree, Am. Chem. J. 43, 381 (1909).
3. D. W. Kaiser, U.S. 2,697,720; cf. U.S. 2,647,916 (Am. Cyan.);
    cf. U.S. 3,541,031 (G.E.).
4. W. J. Priest, U.S. 2,748,103 (Eastman).
5. S. R. Newman et al, U.S. 2,842,433 (Texas Co.).
6. L. Compin, Bull. Soc. Chim. France [4] 27, 465 (1920).
7. M. W. Harman and J. J. D'Amico, U.S. 2,854,467 (Monsanto); cf.
    U.S. 2,875,260 (Firestone).
8. A. G. Chenicek, U.S. 2,401,549 (PPG).
9. C. E. Gleim, J. Am. Chem. Soc. 76, 107 (1954).
10. C. E. Gleim, U.S. 2,483,194 (Goodyear).
11. A. G. Chenicek, U.S. 2,394,592 (PPG).
12. I. E. Muskat and F. Strain, U.S. 2,390,551 (PPG).
13. I. E. Muskat and F. Strain, U.S. 2,395,750 (PPG).
14. C. E. Gleim, J. Am. Chem. Soc. 76, 107 (1954).
15. C. E. Gleim, U.S. 2,541,646 (Goodyear).
16. C. E. Gleim, U.S. 2,579,427 (Goodyear).
17. P. Kirby and T. Owen, Brit. 961,569 (Shell).
18. L. N. Whitehill and W. M. McLamore, U.S. 2,583,980 (Shell).

19. J. A. Bralley, U.S. 2,568,608 (Goodrich).
20. C. E. Gleim, U.S. 2,579,426 (Goodyear).
21. C. E. Gleim, U.S. 2,508,249 (Goodyear).
22. C. E. Gleim, U.S. 2,541,646 (Goodyear).
23. R. W. Pearson, Brit. 909,061 (Dunlop).
24. C. E. Gleim, J. Am. Chem. Soc. 76, 107 (1954).
25. A. G. Chenicek, U.S. 2,384,074; cf. U.S. 2,385,911 (PPG).
26. J. G. Lichty and N. V. Seeger, U.S. 2,464,519 (Goodyear).
27. A. G. Chenicek, U.S. 2,401,549 (PPG).
28. U.S. 2,598,664 and 2,856,386; Brit. 787,062.
29. H. Delius, U.S. 3,227,778 (Reichhold).
30. J. R. Denchfield, U.S. 3,299,021 (Sinclair).
31. D. T. Mowry and G. E. Ham, U.S. 2,556,437 (Monsanto).
32. J. R. Denchfield and R. P. Zmitrovis, U.S. 3,113,902 (Sinclair).
33. A. L. Abel, Chem. Ind. (London) 1106 (1957).
34. R. W. Addor, U.S. 3,296,068 (Am. Cyan.).
35. A. A. Blagonravova et al, CA 57, 3622 (1962).
36. J. D. Turner, Textile Res. J. 41, (8), 709 (1971).

This chapter includes discussions of allyl isocyanate and allyl cyanamides. Copolymers of allyl isocyanate may be cured through formation of urethane or urea crosslinks. $AN=CH_2$ is an interesting reactive compound that has been polymerized at low temperatures to a syrup. These compounds have had few direct applications.

When urea or cyanuric acid $(HOCN)_3$ is heated and the vapors condensed below 0°C, the volatile liquid obtained is believed to be a tautomeric mixture of cyanic and isocyanic acids. Above 0°C polymerization to the trimer cyanuric acid largely occurs. Inorganic isocyanates do not have separate existence from inorganic cyanates. Until recently organic cyanates were little known because of the much greater stability of the isocyanates. The status can be represented:

$$ROCN \xleftarrow{\hspace{1em}} HOCN \rightleftharpoons HNCO \longrightarrow RNCO$$

However, in 1963 Grigat and Puetter discovered the use of ClCN at below 10°C to make relatively stable aryl cyanates and certain substituted alkyl cyanates, such as 2,2,2-trichloroethyl cyanate. These cyanates can add nucleophilic agents to give iminocarbonates (1):

$$ROCN + HOR' \longrightarrow \underset{\underset{OR'}{|}}{ROC}=NH$$

Organic cyanates rearrange to isocyanates or they can trimerize with acid catalyst to form cyanurates (s-triazine derivatives). More research is needed to confirm conditions under which allyl cyanates may be formed, rearranged, and polymerized.

ALLYL ISOCYANATE

Allyl isocyanate has been prepared from allylamine and phosgene followed by treatment of the resulting carbamyl chloride with lime (2):

$$CH_2=CHCH_2NH_2 + COCl_2 \longrightarrow CH_2=CHCH_2NHCOCl \longrightarrow CH_2=CHCH_2NCO + 2HCl$$

Allyl isocyanate may be made by reaction of an allyl

halide with metallic cyanates followed by rearrangement
(3). In one patent example, allyl chloride and NaOCN were
heated in xylene-N-methylpyrrolidone with Cu_2Cl_2 catalyst
for 5 hr at 130°C (4). Methallyl isocyanate (110°C) was
made under similar conditions. Allyl isocyanate has been
evaluated in polymerization, pesticide, and pharmaceutical applications but comparatively little has been
published yet. Allyl isocyanate supplied by Upjohn and
by Aldrich is an amber liquid boiling at 86°C and having
$n_D^{20} = 1.4190$. This highly reactive monomer is a lachry-
mator which freezes near -80°C. It has been isolated from
horseradish. ANCO may be identified by reaction with
aniline to give allylphenyl urea melting at 108°C. It
reacts with alcohols to form allyl urethanes. The trimerization of allyl isocyanate to triallyl isocyanurate
is discussed in Chapter 24.

Kropa and Nyquist copolymerized allyl isocyanate with
acrylic monomers and suggested the resulting copolymers
containing reactive isocyanate groups as finishes for wool
and other fibers bearing reactive hydrogen atoms (5). In
one example, 45 parts ethyl acrylate, 5 parts allyl isocyanate, and 0.2 part benzoyl peroxide in a quartz flask
under carbon dioxide were irradiated for 162 hr by UV
light. A soft clear mass of copolymer and residual
monomer was obtained. Part of the copolymer was found to
be crosslinked and insoluble in solvents. A portion of
the soluble fraction in toluene crosslinked rapidly to a
gel when treated with ethylene diamine (formation of urea
crosslinks). Cotton treated with soluble copolymer solution was heated at 120°C for 4 min to form polyurethane
crosslinks. Larger proportions of allyl isocyanate (e.g.,
20% with ethyl acrylate and 0.4% BP) gave slow copolymerization to completely soluble, soft copolymer masses.

Copolymerization of allyl isocyanate with N,N-dibutyl-
acrylamide, a monomer free of reactive hydrogen atoms,
gave completely toluene-soluble copolymers in contrast to
N-butyl acrylamide. In examples, 10% allyl isocyanate was
copolymerized using 0.4% BP under carbon dioxide at room
temperature for 4 days or longer.

The addition of ethane dithiol to ANCO under UV irradiation occurred largely in the reverse or Posner way (6).
ANCO and $HOCH_2CH_2SH$ under UV light formed a solid polymer
of molecular weight 7500 containing sulfide and urethane
groups. A mixture of dimethylallyl isocyanates was pre
pared by reaction of KCNO with 2,3-dimethyl butadiene in
presence of p-toluene sulfonic acid at 100°C (7). Copolymers of ethylene with minor proportions of isopropenyl
isocyanate were prepared in laboratories of Badische
Anilin (8).

Shashoua and co-workers were able to make an allyl-
substituted polyamide by polymerization of allyl isocyanate with opening of the N=C linkage (9). Sodium
cyanide was added as the mild catalyst to ANCO in dimethyl
formamide at -40°C. The allyl-substituted 1-nylon obtained softened at 180°C and was soluble in sulfuric acid
and in trifluoroacetic acid.

DIALLYLCYANAMIDE

Diallylcyanamide or N-cyanodiallylamine can be prepared
from crude calcium cyanamide by heating with sodium
hydroxide and allyl bromide in water-ethanol (10).

$$CaCN_2 + NaOH + 2CH_2=CHCH_2Br \longrightarrow (CH_2=CHCH_2)_2NCN + CaBr_2$$

It is a colorless, toxic liquid, boiling at 110°C (18 mm)
or at 222°C with decomposition; it is insoluble in water,
but soluble in ethanol and in benzene. It is readily
hydrolyzed by boiling mineral acids and by alkali. Refluxing with dilute sulfuric acid, followed by adding
excess sodium hydroxide and distilling, gave diallylamine.
Diallylcyanamide has been reacted with starch to give a
cationic starch derivative (11) and has been suggested in
the synthesis of oil modifiers (12). Diallylcyanamide
does not homopolymerize readily but copolymerizes with
vinyl compounds by using radical initiators (13). Recently
soluble polymers of cyclic structure were reported by
radical initiation of diallylcyanamide in solution (14).
Only low polymers (η_{sp}/C, about 0.1) were obtained. Copolymers of diallylcyanamide with vinyl acetate, made in
emulsion or solution, contained 4-12% cyanopiperidine
rings (15). Free radical addition of perfluoroalkyl
iodides to diallyl cyanamide produced pyrrolidine derivatives (16).

Diallylcyanamide, which may be present as an impurity
in diallylmelamine, is a poisonous substance that may be
absorbed through the skin. It is corrosive to animal
body tissue. Monoallylcyanamide $CH_2=CHCH_2NHCN$, melting
at 100°C and known also as sinamine, has been derived
from black mustard seed. Allylcyanamide, prepared by
treating allylthiourea with $Pb(OH)_2$ formed a viscous
liquid trimer, triallylisomelamine (17). Allyl methyl
and allyl isobutyl cyanamide also were reported.

Diallyl carbodiimide AN=C=NA is an interesting reactive
compound (bp 60°C at 10 mm) prepared by action of HgO on
N,N'-diallylthiourea in ether (18). It can be made from
allyl isocyanate (19). It has been observed to stabilize
sodium dioxide-diene copolymer elastomers (20), and it
has possibilities as a crosslinking agent. Schmidt
observed that after standing for several weeks, diallyl

carbodiimide had polymerized to a yellow solid mass.
Bromallyl alkyl carbodiimides also were unstable on
storage (21).

Allyl nitrite $CH_2=CHCH_2ONO$ apparently was formed by
reaction of cold allyl alcohol with glyceryl trinitrite
(22). Boiling point between 43.5 and 44.5 and specific
gravity of 0.9546 (0°C) were reported along with insolubility in water. The vapor exploded on heating to 100°C.
AONO decomposed on shaking with water and it reacted with
ethanol to form ethyl nitrite. Allyl nitrate $CH_2=CHCH_2ONO_2$
was prepared from allyl bromide and silver nitrate by
Henry, who found that the compound boiled at 106°C and
had specific gravity 1.09 (10°C) (23).

Allyl bromide or iodide were found to react vigorously
with silver nitrite to form 3-nitro-1-propene, an oil
with a sharp odor, which could be reduced by zinc and
acetic acid to allylamine (24). Portionwise addition of
silver nitrite to allyl bromide during 24 hr gave 3-nitro-
propylene and a smaller amount of allyl nitrite (25).

Apparently $CH_2=CHCH_2NO_2$ has not given high polymers,
but $CH_3CH=CHNO_2$ (26) and $CH_2=C(NO_2)CH_3$ (27) polymerized
in the presence of aqueous base by anionic mechanisms.
The last two compounds are not allylic in behavior because
of electron withdrawal and resonance of the NO_2 group
attached directly to the ethylenic nucleus. Explosions
have occurred on heating some of these compounds.

The chlorallyl compound $CH_2=\overset{CH_2Cl}{\underset{}{C}}-CH_2NO_2$ (bp 62°C at 4 mm;
$n_D^{20}=1.4742$) was reported from Shell laboratories (28).
Methallyl chloride was nitrated to give $CH_2=C(CH_3)CH_2ONO_2$
by use of silver nitrate or by nitric acid with cooling
(29). Allyl nitroform $CH_2=CHCH_2C(NO_2)_3$ made from silver
nitroform and allyl bromide has been considered for use
as an explosive (30).

N-Allyl methylene imine $AN=CH_2$ was prepared by adding
hexahydro-1,3,5-triallyl-s-triazine portionwide to solid
acidic boron phosphate catalyst at 200°C (31). This
highly reactive compound, melting at -122°C, polymerized
spontaneously to a clear viscous syrup at room temperature.

Allyl azide was made by warming allyl chloride with NaN_3
in ethanol-water (32). The mobile liquid was boiled (at
least once) at 76.5°C at atmospheric pressure. With
sulfuric acid it decomposed violently. Allylic azides
may be made by long reaction of sodium azide with bromo-
alkenes in ethanol-acetone at room temperature. Electron-
withdrawing effects of the azide groups in $CH_2=CHCHN_3CH_2N_3$
retarded allylic rearrangement (33).

Allyl phenyl diimide $AN=NC_6H_5$, a yellow oil (bp 100°C
at 27 mm) was obtained by reaction of yellow mercury oxide
with a solution of allylphenylhydrazine in ether (34). It

is only slightly soluble in water but soluble in alcohol,
ether, or acetic acid. It gives aniline on reduction
with zinc dust and HCl. Allyl p-tolyl diimide was made
in a similar way (35). It melts at 97°C and boils at
110°C at 30 mm.

N-Allyl imides of structure $-\underset{\underset{O}{\|}}{C}N\underset{\underset{O}{\|}}{\overset{A}{C}}-$ have been discussed
on page 562. Triallyl isocyanurate, a triallyl cyclic
imide, is discussed in Chapter 24. References to some
other N-allyl compounds are given in Table 22.1.

TABLE 22.1

Other N-Allyl Compounds

Compound	Selected properties (temp., °C)	References
Allyl isocyanide, ANC	bp, 100; $d_{17}=0.794$; Sl. soluble in water; sol. in ether, alcohol	Lemoult, Compt. Rend. 148, 1602 (1909); cf. Malatesta, Gazz. Chim. Ital. 77, 238 (1947); Millich, Chem. Rev. 72, 101 (1972)
Allyl phenyl carbodimide, $AN=C=NC_6H_5$	bp 165(10)	Dains, JACS 21, 162 (1929)
Allyl Schiff bases, $AN=CHR$	--	Tiollais, Compt. Rend. 224, 1116 (1947); Bull. Soc. Chim. France 714 (1947)
Allyl benztriazole	--	Hopff and Luessi, Helv. Chim. Acta 46, 1052 (1963)
Allyl diethyl amine oxide, $AN=O(C_2H_5)_2$	Picrate mp, 138	Meisenheimer, Ber. 55, 517 (1922)
Allyl methyl nitroamine, $AN(NO_2)CH_3$	bp 96(18)	Umbgrove, Rec. Trav. Chim. 15, 207 (1896)
Allyl amidines, $ANC(=NH)R$	--	Ziegler, Ann. 495, 99 (1932); U.S. 2,049,582 (R & H)
Diallyl diphenyl tetrazone	mp 86°C, dec.	Michaelis, Ber. 26, 2180 (1893)
Diallyl nitrosamine, A_2NNO	bp, 92(20)	Preussmann, CA 57, 7087 (1962)
Triallyl guanidine ·HCl	mp, 176	Connolly, J. Chem. Soc. 828 (1937)
Triallyl hydroxylamine	bp 80(50); $n_D^{30}=1.440$	Dunn, U.S. 3,046,308 (Carbide)

References

1. E. Grigat and R. Puetter, Chem. Ber. $\underline{97}$, 3012 (1964); Angew. Chem. $\underline{79}$, 219 (1967); Ger. 1,195,764; Ger. 1,201,839 (Bayer); R. Kubens et al, Kunst. $\underline{58}$, 827 (1968).
2. V. W. Siefken, Ann. $\underline{562}$, 81 and 111 (1949); cf. Peterson, ibid. $\underline{562}$, 208 and 220 (1949); U.S. 2,640,068 (Am. Cyan.).
3. D. O. DePree, U.S. 2,866,803; cf. U.S. 2,866,801-2 (Ethyl).
4. H. von Brachel and E. Herrmann, U.S. 3,558,684 (Cassella).
5. E. L. Kropa and A. S. Nyquist, U.S. 2,537,064 (Am. Cyan.).
6. A. A. Oswald, CA $\underline{73}$, 120945 (1970) and U.S. 3,597,341 (Esso).
7. F. W. Hoover, Fr. 1,359,098 (Du Pont), CZ #48, 2907 (1967).
8. H. Naarman and E. G. Kastning, Fr. 1,337,619 and Brit. 947,472 (Badische).
9. V. Shashoua et al, J. Am. Chem. Soc. $\underline{82}$, 867 (1960).
10. E. B. Vliet, Org. Syn. $\underline{5}$, 45 (1925) or Coll. Vol. I, 203 (1941); U.S. 1,659,793; cf. A. J. Speziale, U.S. 2,858,338 (Monsanto); H. Staudinger, Ger. 404,174; A. G. Sayadyan et al, CA $\underline{75}$, 7730 (1971).
11. E. F. Paschall, U.S. 2,894,944 (Corn Products).
12. L. V. Mullen and J. M. Boyle, U.S. 2,666,745 (Esso).
13. E. K. Drechsel and J. J. Padbury, U.S. 2,550,652 (Am. Cyan.); M. Mullier and G. Smets, Bull. Soc. Chim. Belg. $\underline{62}$, 491 (1953).
14. K. Uno et al, J. Polym. Sci. A1, $\underline{6}$, 85 (1968); J. P. J. Higgins and K. E. Weale, Polym. Lett. $\underline{7}$, 153 (1969).
15. D. A. Simonyan and A. G. Sayadyan, CA $\underline{76}$, 14987 (1972).
16. N. O. Brace, J. Polym. Sci. A1, $\underline{8}$, 209 (1970).
17. J. Braun and H. Engels, Ann. $\underline{436}$, 315 (1924); U.S. 2,300,597 (Am. Cyan.); cf. H. Will, Ann. $\underline{52}$, 15 (1844).
18. E. Schmidt et al, Ber. $\underline{71}$, 1936 (1938).
19. W. J. Balon, U.S. 2,853,518 (Du Pont).
20. J. M. Goppel et al, U.S. 2,654,680 (Shell).
21. E. Schmidt et al, Ann. $\underline{560}$, 229 (1948).
22. C. Bertoni, Gazz. Chim. Ital. $\underline{15}$, 364 (1885).
23. L. Henry, Ber. $\underline{5}$, 452 (1872).
24. P. Askenasy and V. Meyer, Ber. $\underline{25}$, 1701 (1892); CZ II 154 (1892).
25. R. B. Reynolds and H. Adkins, J. Am. Chem. Soc. $\underline{51}$, 279 (1929).
26. C. W. Scaife and A. E. W. Smith, Brit. 596,303; U.S. 2,460,243 (ICI).
27. A. T. Blomquist et al, J. Am. Chem. Soc. $\underline{67}$, 1522 (1945).
28. E. C. Kooijman and J. Overhoff, U.S. 2,473,341; cf. CA $\underline{45}$, 10478 (1951).
29. M. L. Wolfram et al, J. Org. Chem. $\underline{25}$, 1079 (1960).
30. R. H. Saunders, U.S. 2,993,935 (USA).
31. J. L. Anderson, U.S. 2,729,680 (Du Pont); cf. U.S. 2,729,679 (Du Pont).
32. M. Forster and H. Fierz, J. Chem. Soc. $\underline{93}$, 1177 (1908).
33. C. A. Vanderwerf and V. L. Heasley, J. Or. Chem. $\underline{31}$, 3534 (1966).
34. E. Fischer and O. Knoevenagel, Ann. $\underline{239}$, 205 (1887).
35. A. Michaelis and K. Luxembourg, Ber. $\underline{26}$, 2174 (1893).

Organic nitrogen compounds in which allyl groups are attached to carbon-hydrogen have been utilized for many years by pharmaceutical chemists. The C-allyl barbiturates are among the important sedatives in use. Experimental copolymerizations have been carried out with a number of C-allyl nitrogen compounds, but important industrial polymers have not developed. Allyl cyanide is useful in organic synthesis and its polymerization with basic catalysts is unusual. The allylic cyanides or nitriles are discussed first.

Allyl cyanide or vinyl acetonitrile $CH_2=CHCH_2CN$ can be made by reaction of allyl bromide with aqueous potassium cyanide (1). It can be prepared in better yields by warming dry cuprous cyanide with allyl bromide (2) or with allyl chloride in the presence of a little potassium iodide (3). The liquid has an agreeable odor similar to that of acrylonitrile but slightly suggesting onions and has been found along with allyl isothiocyanate in some mustard oils. Catalytic dehydrogenation of butyronitrile over chromium oxide catalyst has given allyl cyanide along with crotononitrile (4). Allyl cyanide has been reported recently from reaction of allyl chloride and HCN over alumina (5). It can be made also by reaction of allyl acetate with HCN at 80°C, using as catalysts $ZnCl_2$ and $Ni[P(OC_6H_5)_3]_4$ (6). Allyl cyanide and vinyl cyanide were less toxic to guinea pigs than ethyl cyanide (7).

On treatment with alkali, allyl cyanide isomerizes to cis- and trans-crotononitriles, the double bond moving to the conjugated position (8). Allylic monobromination of allyl cyanide does not occur readily when N-bromosuccinimide is used. On reaction with sodium ethylate, normal addition takes place to give 3-ethoxybutyronitrile (9). With molar amounts of aqueous HCl allyl cyanide hydrolyzes to vinylacetic acid (3).

Allyl cyanide adds bromine much faster than do acrylonitrile or crotononitriles (10). The cyanide group retards addition to the double bond, compared with ethylenic compounds bearing electron-donating substituents. Allyl cyanide was saponified by sodium carbonate in aqueous acetone at room temperature, giving vinyl acetamide (11). Allyl cyanide has been reacted with allyl magnesium chloride to produce triallylmethylamine (12). By addition of compounds containing active hydrogen, such as alcohols, to the double bond of allyl cyanide under favorable conditions cyanopropyl groups can be introduced. These can be hydrolyzed to carboxylic acid groups or hydrogenated to amino groups.

Polymerizations of allyl cyanide have special interest. Bruylant and Gevaert reported in 1925 that ethyl magnesium bromide catalyzed formation of dimers and trimers, as had resulted from the use of sodium or sodium ethylate (13). However, 45% of the product resulting from the Grignard reagent was a pitch-like higher polymer. The polymerizations were then explained by initial isomerization to crotononitriles which then polymerized. Dimer and trimer were reported recently by electrolytic reduction (14).

Mazzanti and co-workers confirmed that allyl cyanide polymerizes under action of sodium alkoxides to form polymers of low molecular weight which show no crystallinity in x-ray patterns (15). However, very slow polymerization at low temperatures in presence of free alcohol gave a fraction of stereoregular polymer. For example, a solution of 8.3 g allyl cyanide in 20 ml of anhydrous n-heptane was saturated with nitrogen at -78°C. There was added a millimole of allyl alcohol and a suspension of 2.95 millimoles sodium octylate in 20 ml n-heptane. After 21 hr at -78°C the catalyst was quenched by adding a mixture of methanol and HCl. There was recovered 2.6 g polymer of which 30% was insoluble in boiling acetone and showed crystallinity in its x-ray pattern. Polymer molecular weights above 1000 and stereoregularity were best attained by use of 1 mole alcohol to 3 moles sodium alcoholate.

The author also has observed anionic polymerization of allyl cyanide. Long standing of the monomer with metallic sodium gave viscous to brown, semi-solid polymer masses (16). That allyl cyanide is one of the rare allyl compounds which homopolymerize under the action of Lewis-basic catalysts may be attributed to the exceptionally high electron-attraction of the cyanide group which activates the double bond even from the allylic position. However, homopolymers from allyl cyanide of really high molecular weight have not been clearly disclosed. Allyl cyanide does not respond to cationic polymerization systems, including those with Ziegler-type catalysts.

Sulfur dioxide reacted with allyl cyanide in the presence of alcohol and ascaridole at room temperature to give white, insoluble sulfone copolymers (17). The low reactivity of allyl cyanide in free radical polymerizations is in contrast to the behavior of acrylic monomers such as acrylonitrile and alpha-methylene glutaronitrile (18), where the strong electron-attracting CN group is attached directly to the ethylene nucleus. Allyl cyanide was suggested as an intermediate for synthesis of nitrile-siloxane rubbers. Copolymers of acrylonitrile with minor proportions of allyl or methallyl cyanide have been considered for acrylic fibers (19).

Vinyl diallyl acetonitrile isomerized on heating at 180°C with nitrogen to give 2,4-diallyl crotononitrile (20). Allyl acetonitrile or 4-cyano-1-butene was polymerized in liquid ammonia with sodium metal or amide catalyst at -40°C during 6 hr (21). There was an 84% yield of polymer powder soluble in acetone.

Methallyl cyanide was prepared by heating the chloride with cuprous or sodium cyanide at 120°C in an autoclave (22). On treatment with warm alkali, such as trimethyl benzyl ammonium hydroxide in t-butanol, methallyl cyanide is isomerized to 3-methylcrotononitrile (23). Vinyl methyl acetonitrile (bp 126°C, $n_D^{20} = 1.4063$) was made by heating the chloride or bromide with cuprous cyanide at 60 to 100°C (24). The same authors obtained the compound also by warming the amide with phosphorus pentoxide under vacuum. The compound $A(CH_2)_2CN$ (bp 162°C, $n_D^{25} = 1.4268$) was prepared by heating the bromide with KCN in ethylene glycol at 100°C (25).

Allylacetonitrile was one of the products of reaction of allyl carbinol with HCN over alumina at 455°C (26). It was obtained by addition of $NaNH_2$ into allyl bromide and excess acetonitrile at 60°C (27). This compound, also obtainable from butadiene and HCN, was polymerized at -40°C in liquid ammonia-toluene by portionwise addition of sodium (28). After 6 hr a precipitate of solid yellow polymer was recovered which was soluble in acetone and was believed to have the structure $--CH_2\underset{|}{C}H-- \\ CH_2CH_2CN$.

There has been pharmaceutical interest in $AC(C_2H_5)_2CN$ (bp 72°C at 12 mm). It was made by reaction of allyl bromide or chloride with sodium or mercury derivatives of diethylacetonitrile (29). In one example, a suspension of 1.2 moles of $NaNH_2$ in 1.23 mole of allyl chloride and benzene was added dropwise to diethylacetonitrile and benzene at 70°C; the mixture was then heated (30). On treating diethyl allyl acetamide with $NaNH_2$ in boiling benzene, there was obtained diethyl allyl acetamidine $AC(C_2H_5)_2C(=NH)NH_2$ (31). When vinyl diallyl acetonitrile was heated at 180°C, it formed 2,4-diallylcrotonitrile (32). The nitrile of undecylenic acid was prepared by

heating the amide with $SOCl_2$ (33). A boiling point of
135°C at 15 mm and $n_D^{20} = 1.4442$ were reported.

Triallyl acetonitrile, A_3CCN (bp 95°C at 12 mm), has
been used as a sleep promoting agent. It was made by
addition of a mixture of allyl chloride and CH_3CN to a
suspension of $NaNH_2$ in benzene or ether (34). It was
also formed by reaction of the sodium compound of A_2CHCN
with allyl chloride in benzene (35). Triallyl acetonitrile has a pleasant odor. Additional references to
allylic cyanides are given in Table 23.1.

Allyl alkyl cyanoacetates have been prepared (36). Compounds of the type $A_2C(CN)R$ when heated with benzoyl peroxide gave soluble polymers believed to contain recurring
cyclic units (37).

ALLYL BARBITURIC ACID DERIVATIVES

Diethylbarbituric acid was introduced as Veronal in the
early 1900s to promote sleep. The various allyl barbituric acid derivatives are cyclic compounds, and they are
sedatives and hypnotics of the central nervous system
related to urea. As in cyanuric acid the NH
imide group is acidic and the sodium salts
provide greater water solubility. The allyl
derivatives, such as 5,5-diallyl barbituric
acid (known as Dial or Allobarbital) are most
active and generally have shorter duration than
most alkyl derivatives. The allyl compounds
and their sodium salts are among the barbiturates most
employed in recent years. Unsubstituted barbituric acid
or malonylurea has no sedative or hypnotic action. The
sedative activity and duration are believed to be related
to the greater vulnerability to enzymatic oxidation of
unsaturated groups and also of short-branched alkyl groups
such as isopropyl and sec-butyl. Barbiturates seem to
act upon the cerebral cortex to prevent passage of im-
pulses.

Substituted barbituric acids can be prepared by conden-
sing the substituted malonic esters with urea in the
presence of sodium ethoxide (38). Barbituric acid cannot
be alkylated very readily with alkyl halides, but allyl
bromide reacts faster with barbituric acid in ethanol in
the presence of sodium acetate to give 5,5-diallyl barbituric acid (39). The free acid crystallizes from
alcohol-water as leaflets. It melts at 173°C and is
slightly soluble in water and in benzene. The oral dose
for sedation is about 30 mg; for hypnosis, 100 to 300 mg
is required according to the Merck Index (7th Ed.). The
monoallyl derivative of barbituric acid (mp 69°C) may be
made first and then the second hydrogen on the 5-position

TABLE 23.1

Some Allylic Cyanides

Compound	bp, °C	Refractive indices and other properties	References
Allyl cyanide	119	$n_D^{20} = 1.4060$; $d_{20} = 0.8364$	Merck Index; Jeffrey, J. Chem. Soc. 683 (1948)
Allyl acetonitrile	143	$n_D^{19.5} = 1.4233$	Paul, Bull. Soc. Chim. France 474 (1949)
Allyl malononitrile	218	mp, -12°C	Henry, Jahresber. 640 (1889)
Diallyl aceto-nitrile	85(25)	$n_D^{13.5} = 1.4490$	Cottin, Compt. Rend. 197, 255 (1933); Ger. 473,329 (I.G.)
2,4-Diallyl crotononitrile	--	--	Cope, JACS 65, 2003 (1943)
Triallyl acetonitrile	102(2)	--	Ziegler, Ann. 495, 108 (1932); U.S. 1,958,653, 1,894,301; Ger. 473,329 (I.G.)
Allyl diethyl acetonitrile	72(12)	--	U.S. 1,958,653 (Schering)
Allyl hydroxy nitriles	--	Acrylonitrile copolymers	Price, U.S. 2,812,315 (Am. Cyan.)
Diallyl ethyl acetonitrile	85(13)	--	Ziegler, Ann. 495, 108 (1932); U.S. 1,958,653 (Schering); 1,894,301 (Winthrop)
Diallyl cyano-formamidine	104(16)	$n_D^{20} = 1.4903$	Woodburn, J. Org. Chem. 22, 846 (1957)
Triallyl acetyl-urea	--	Hypnotic and anti-spasmodic	Hildebrandt, U.S. 2,915,553 (Knoll Fabriken)
Chloroallyl cyanide	134 or 40.5(11)	$n_D^{21.5} = 1.450$	Vessiere, CA 51, 2540 (1957); Kurtz, CA 55, 27054 (1961)
Methallyl cyanide	137	$n_D^{20} = 1.4204$	Fuson, JACS 66, 680 (1944); cf. U.S. 2,097,155 (Shell)

replaced by an alkyl group. The soluble sodium salts
generally used for intravenous therapy are easily hydrolyzed to the free barbituric acid derivatives. Detoxi-
fication seems to occur in the liver and kidneys.

In recent years in the United States allyl isobutyl and
allyl l-methybutyl barbituric acid derivatives have been
favored. The January to June 1971 index of Chemical
Abstracts gave 58 references to allyl alkyl derivatives
of barbituric acid and 30 references to 5,5-diallylbarbi-
turic acid derivatives further substituted on the 1 or 3
positions. Other commercial allyl barbiturates include
5-allyl-5-isopropyl (Alurate), 5-allyl-5-sec-butyl
(Lotusate), and 5-allyl-5-phenyl (longer acting). The
bromallyl group $CH_2=CBrCH_2-$ gives barbiturates of even
shorter duration than those containing allyl groups. Such
a product is 5-bromallyl-5-sec-butylbarbituric acid (Per-
noston). Besides their use as sedatives, some of the
allyl barbiturates are applied in treating hypertension
and as anticonvulsants against grand mal seizures.

Diallyl barbituric acid was copolymerized with monomers
such as vinylene carbonate, maleic anhydride, vinyl
acetate, and methyl acrylate in Minnesota Mining laboratories (40). Acetonitrile was a suitable solvent for
copolymerization using azobis or BP catalyst at 80°C.
Soluble copolymer of low inherent viscosity was obtained.
Residual double bonds were virtually absent in IR spectra,
suggesting that cyclic polymerization of the diallyl
compound occurred. Copolymers also were prepared from
diallyl diethyl malonate in the same laboratories (41).
For example, 5.0 g diallyl-substituted diethyl malonate,
2.4 g ethyl acrylate, 5.0 g benzene, and 0.52 g azobis
were heated for 16 hr at 50°C in the absence of air. The
copolymer product was purified by reprecipitation from
methylene chloride solution by addition of heptane. It
was an adhesive material of inherent viscosity 0.4 in
benzene. These diallyl copolymers also showed evidence
of cyclic chain units. Diallyl diethyl malonate copolymerized with SO_2 in ethanol with silver nitrate catalyst
at 50°C (42). A soluble copolymer of inherent viscosity
0.48 in acetone was obtained after 2 hr.

N,N'-Diallylurea heated with diethyl malonyl chloride
at 120°C gave 1,3-diallyl-5,5-diethyl barbituric acid (43).
This compound, boiling at 157°C (9 mm) is insoluble in
aqueous alkali. A series of allyl barbiturates with both
5-positions substituted (with and without N-substitution)
showed the usual sedative and hypnotic properties but
without euphoria or tranquilizing action (44). The
5-allyl-5-bromopropyl barbituric acid had the most favorable narcotic coefficient and the strongest anticonvulsant
action.

Kargin and co-workers have copolymerized N-allyl barbituric acid with acrylic acid and with methacrylic acid
in dioxane solution using benzoyl peroxide at 60°C or
under ^{60}Co radiation (45). The copolymers were white
solids soluble in water, alcohol, dioxane, and dimethyl
formamide. The copolymers decomposed at about 140°C.
From a monomer mixture of 70% AB and 30% acrylic acid, a
copolymer containing 17% AB units was obtained; from 30%
AB there was recovered copolymer containing 11% AB units.
Crosslinked terpolymers were made by including a little
methylene bis(acrylamide). Films of the copolymers complexed Cu^{2+}, Co^{2+}, Ni^{2+}, and Fe^{3+} ions strongly at pH 3
to 7, but did not remove alkali cations. The copolymer
macromolecules formed globular or fibrillar particles,
depending on pH. Copolymers of vinyl acetate with N-allyl
barbituric acid and vinyl alcohol copolymers therefrom
were studied as models of muscle action (46).
Monosulfur analogs of barbituric acids have received
attention. They are stronger acids than the oxygen compounds, have shorter duration as sedatives, and may be
less toxic. Thiamylal is 5-allyl-5-(1-methylbutyl)-2-
thiobarbituric acid or sodium salt. Ultrashort-acting
thiobarbiturates were obtained where one of the 5-substi-
tuents is allyl and the other is $C_2H_5C\equiv C-C(CH_3)H-$ (47).
In Britain 5-allyl-5-(2-cyclohexenyl)-2-thiobarbituric
acid under the name Thialbarbitone has been used intra-
venously as an anesthetic. Aldrich Chemical Company
supplies 1-allyl-6-aminouracil (mp 270°C) which is similar
to allyl barbituric acid except that the 6-position bears
an amino group instead of oxygen.
Interesting allyl derivatives of nitrogen-containing
ring compounds include 1-allyl-1,2,4-triazole, made by
heating triazole with allyl bromide and $NaOC_2H_5$ (48),
1-allylparabanic acid (mp 140°C) (49), and 1-allyltheo-
bromine (50). Note that C-allyl amines were discussed on
page 549 and C-allyl amides on page 565.

References

1. C. Palmer, Am. Chem. J. 11, 89 (1889).
2. P. Bruylants, Bull. Soc. Chim. Belg. 31, 175 (1922); R. A. Letch
    and R. P. Linstead, J. Chem. Soc. 443 (1932).
3. C. W. Smith and H. R. Snyder, Org. Syn. 24, 97 (1944).
4. L. U. Spence and F. O. Haas, U.S. 2,385,552 (R & H).
5. W. L. Fierce and W. J. Sandner, U.S. 3,116,318 (Pure Oil); cf.
    P. Kurtz, Ann. 631, 21 (1960).
6. W. C. Drinkard, U.S. 3,558,688 (Du Pont).
7. L. Ghiringhelli, CA 51, 1461 (1957).
8. L. Falaise and R. Frognier, Bull. Soc. Chim. Belg. 42, 431 (1933);

H. A. Bruson, J. Am. Chem. Soc. <u>65</u>, 22 (1943); U.S. 2,384,630
(Phillips).

9. R. A. Letch and R. P. Linstead, J. Chem. Soc. 455 (1932).

10. G. Heim, Bull. Soc. Chim. Belg. <u>39</u>, 458 (1930); CA <u>25</u>, 2389
    (1931).

11. J. V. Murray and J. B. Cloke, J. Am. Chem. Soc. <u>56</u>, 2751 (1934).

12. H. R. Henze et al, J. Am. Chem. Soc. <u>65</u>, 88 (1943).

13. P. Bruylant and Y. Gevaert, Bull. Soc. Chim. Belg. <u>32</u>, 317 (1925);
    cf. Rathus, Ibid. <u>35</u>, 239 (1926).

14. M. R. Ort and M. M. Baizer, J. Org. Chem. <u>31</u>, 1646 (1966).

15. G. Natta, G. Mazzanti, U. Giannini and G. Brukner, Italian
    718,894 (Monte) appl. April 1964; CA <u>69</u>, 28098 (1968).

16. C. E. Schildknecht and David Walborn, unpublished.

17. L. L. Ryden et al, J. Am. Chem. Soc. <u>59</u>, 1014 (1937).

18. E. G. Pritchett and P. M. Kamath, Polym. Lett. <u>4</u>, 849 (1966).

19. J. R. Caldwell, U.S. 2,529,911 (Eastman).

20. D. E. Whyte and A. C. Cope, J. Am. Chem. Soc. <u>65</u>, 2003 (1943).

21. Brit. 1,132,428 (Bayer).

22. H. P. A. Groll and C. J. Ott, U.S. 2,097,155 (Shell); cf.
    M. W. Tamele et al, Ind. Eng. Chem. <u>33</u>, 116 (1941); W. H. Chel-
    delin and C. A. Schink, J. Am. Chem. Soc. <u>69</u>, 2626 (1947).

23. H. A. Bruson and T. W. Riener, J. Am. Chem. Soc. <u>65</u>, 22 (1943).

24. J. F. Lane et al, J. Am. Chem. Soc. <u>66</u>, 546 (1944).

25. F. B. LaForge et al, J. Am. Chem. Soc. <u>70</u>, 3709 (1948).

26. W. Reppe et al, Ann. <u>596</u>, 91 and 124 (1955); Ger. 743,468-9 (IG).

27. K. Ziegler, Ger. 570,594; U.S. 1,958,653 (Schering).

28. K.-E. Schnalke and N. Schoen, U.S. 3,532,681 (Bayer).

29. K. Ziegler, Ger. 583,561, 616,876, 622,537, 622,875 (IG);
    U.S. 1,894,301; CZ II 3473 (1932) (Winthrop).

30. H. Ohlinger, Ann. <u>495</u>, 101 (1932); cf. Ger. 570,594, U.S.
    1,958,653 (Schering).

31. K. Ziegler, U.S. 2,049,582 (R & H).

32. D. E. Whyte and A. C. Cope, J. Am. Chem. Soc. <u>65</u>, 2003 (1943).

33. J. S. and N. A. Sörensen, Acta. Chem. Scand. <u>2</u>, 174 (1948).

34. K. Ziegler and H. Ohlinger, Ann. <u>495</u>, 108 (1932);
    U.S. 1,958,653 (Schering).

35. W. Bockmuehl and G. Ehrhart, U.S. 1,894,301 (IG, Winthrop);
    Brit. 378,743; CZ II 3473 (1932).

36. T. Cuvigny and H. Norman, Bull. Soc. Chim. France 1872 (1965);
    CA <u>63</u>, 13136 (1965).

37. H-T. Feng et al (Peking), CA <u>60</u>, 3114 (1964).

38. E. Fischer and A. Dilthey, Ann. <u>335</u>, 334 (1904).

39. T. B. Johnson and A. J. Hill, Am. Chem. J. <u>46</u>, 542 (1911);
    E. Preiswerk and E. Grether, U.S. 1,042,265; Ger. 247,952;
    Ger. 268,158 (Ciba) (1912); cf. Ger. 526,854 (LaRoche).

40. C. D. Wright, U.S. 3,057,829 (3M).

41. C. D. Wright, U.S. 3,247,170 (3M).

42. C. D. Wright and W. S. Friedlander, U.S. 3,072,616 (3M).

43. Ger. 258,058 (Merck); CZ I 1374 (1913).

44. W. Prastowski et al, CA 66, 1408 (1967).
45. V. A. Kargin et al, Polym. Sci. (USSR) (in English) 10, 500 (1968).
46. W. Kuhn et al, Fortschr. Hochpolym. Forsch. 1, 540 (1960).
47. W. J. Doran, U.S. 3,172,890 (Eli Lilly).
48. G. Pellizzari, Gazz. Chim. Ital. 35, I 381 (1905); cf. M. Freund, Ber. 29, 2490 (1896).
49. R. L. Maly, Z. Chim. 260 (1869); cf. Rundquist, Arch Pharm. 236, 450.
50. J. Brown, Ber. 50, 292 (1917).

24. TRIALLYL CYANURATE AND RELATED

Triallyl cyanurate (TAC), a relatively new crosslinking monomer of exceptional interest, was developed by American Cyanamid in America. This polyfunctional monomer is applied in copolymerization with unsaturated polyesters for fiber-reinforced structures of high heat resistance. It has found use in small proportions as a crosslinking and hardening agent in other plastics, in coatings, in elastomers, and in tire cord and other adhesives. The less familiar cyclic monomers, triallyl isocyanurate and diallylmelamine, are discussed in this chapter because of their relation to triallyl cyanurate, although they are N-allyl compounds.

The monomer triallyl cyanurate was first prepared by Dudley (1) and was polymerized by Kropa (2). Despite its high cost, triallyl cyanurate has found applications as a polyfunctional crosslinking agent, particularly in free radical copolymerization with unsaturated polyesters, in fiber-reinforced thermoset structures of outstanding heat resistance, and in the curing of elastomers and adhesives. The monomer has found little use in homopolymerization because of cost and difficulties in controlling the polymerization, as well as the brittleness of the homopolymers. The exothermic rearrangement of TAC polymer structures to the more stable isocyanurate structure is an interesting feature. See also page 593.

TAC monomer, also called 2,4,6-tris(allyloxy)-s-triazine, can be prepared by gradual addition of cyanuric chloride to an excess of allyl alcohol in the presence of concentrated aqueous sodium or potassium hydroxide (1,3). Toluene may be present as a diluent with the temperature kept below 10°C.

$$Cl\text{-triazine-}Cl + 3CH_2\text{=}CHCH_2OH \xrightarrow{\text{NaOH}}$$

$$CH_2\text{=}CHCH_2O\text{-triazine-}OCH_2CH\text{=}CH_2 \quad + \text{ NaCl}$$

The literature on preparation of alkoxy-s-triazines has
been reviewed by Dudley and co-workers (4).

Cyanuric chloride is the acid chloride of cyanuric acid.
The latter is acidic because each hydroxyl group is next
to a double bond, e.g., C-OH, just as in phenol. Triallyl cyanurate is therefore an ester of an enol form
stabilized by resonance of the ring. The isomeric imide
compound is triallyl isocyanurate. Ammonia or amines may
be present as catalysts in the allylation reaction (5).
TAC was prepared in good yield by heating the imidocar-
bonate with allyl alcohol in the presence of ferric
chloride (6). TAC was made by reaction of cyanuric
chloride with sodium allyloxide (7) and also by alcoholysis of trialkyl cyanurates by allyl alcohol. Preparation
of trimethallyl cyanurate has been described (6), but it
was not clear whether yields are satisfactory.

Triallyl cyanurate liquid or solid can cause irritation
of the skin and eyes, but it is not a highly toxic substance. It has activity as an insecticide and hypnotic
effects have been observed in mice. TAC hydrolyzes slowly
in water and faster in dilute aqueous acid to give free
allyl alcohol. It is relatively stable to dilute alkali.
Alcoholysis of TAC by glycols in the presence of sodium
methylate can give hydroxyalkyl cyanurate; by further
reaction of these compounds, polymers ranging in properties from viscous oils to rubbery solids were obtained (8).
In one patent example, a TAC-triethylene glycol mixture
was reacted with a little sodium at 80°C for an hour while
distilling off allyl alcohol. To this polyester containing allyl double bonds, Kropa added 1% benzoyl peroxide
and heated 2 hr at 100°C to give a yellow, brittle polymer.
Thus he disclosed a complex substance containing two mixed
polymer structures resulting from condensation and addition polymerization processes.

TAC has been reacted with chlorine until an average of
1.59 double bonds per molecule remained (9). This was
copolymerized with pure TAC and an unsaturated polyester
with the objective of flame-resistant plastics. Triallyl
cyanurate has been added to polyoxymethylenes (formaldehyde high polymers) as a stabilizer against degradation
when heated in air (10). The monomer also has been found
to stabilize chlorinated polyethylenes (11).

Properties of triallyl monomers follow:

	Triallyl cyanurate	Triallyl isocyanurate
Melting point, °C	31	24
Boiling point, °C	140 at 0.5 mm	126 at 0.3 mm
Density, g/cc at 30°C	1.1133	1.1720
Refractive index, n_D^{25}	1.5049	1.5115
Viscosity, cp	12.6 (30°C)	ca. 100 (25°C)

Both TAC and triallyl isocyanurate are soluble in ethanol,
acetone, benzene, chloroform, dioxane, and ethyl acetate.
Both compounds have limited solubility in hexane and are
only very slightly soluble in water (0.6% for TAC). TAC
monomer can be identified by IR absorptions at 6.4, 7.2,
7.6, 8.8, 10.75, and 12.2 microns. From comparison of
heats of combustion, apparently the isomerization of
triallyl cyanurate to triallyl isocyanurate should release
50 kcal/mole.

TAC POLYMERIZATION

Kropa polymerized the monomer containing 2% benzoyl
peroxide in a glass cell by heating for 21.5 hr at 80°C
(12). The resulting sheet was clear light yellow and
hard (Barcol 66). Violent polymerizations were encountered in vacuum distillations of large quantities of
monomer. Triallyl cyanurate monomer was reported to be
relatively stable under nitrogen, even at 125°C for up to
24 hr. At this temperature under air, however, polymerization started within 2 hr. Violent polymerization can
be initiated by heating the monomer in air at 170°C.
Copper and copper salts could promote very rapid polymerization. Nickel, manganese, and mercury also promoted
polymerization of TAC. The liquid monomer becomes more
viscous from increasing soluble polymer after storage for
a year or two.

Heating TAC with peroxide catalyst (e.g., 3% benzoyl
peroxide at 70 to 100°C) gives transparent, nearly colorless, relatively hard and brittle polymers. The changes
in density and the high degree of crosslinking occurring
during polymerization tend to cause cracking and non-
uniformity through polymer cast sheets. Very slow polymerization with little or no added catalyst can give at
first viscous solutions containing soluble prepolymer,
and these can be cured slowly to uniform polymer masses.
Prepolymers, which can be prepared faster by solution
polymerization, are not commercially available as in the
case of diallyl phthalates. As a heat-resistant bonding
agent for electrical condensers, TAC containing 0.5 to 1%
BP or dicumyl peroxide was polymerized during 3 hr at 95°C
followed by 13 hr at 215°C (13).

The heat of complete polymerization of triallyl cyanurate is high. Two distinct exotherms can be observed: an
initial release of about 39 kcal/mole corresponding to
reaction of two allyl groups, and a later surprisingly
great exotherm believed to represent reaction of the third
allyl group together with rearrangement to the more stable
isocyanurate structure (14). The IR spectrum of triallyl
isocyanurate polymer is similar to that of isomerized triallyl cyanurate polymer. Both show a band at 1700 cm^{-1}

of the carbonyl group in contrast to the 1130 cm^{-1} of
C-O-C of TAC polymer. However, the two polymers differ
appreciably in stability on heating and in curves obtained
by differential thermal analysis. Triallyl isocyanurate
polymers were observed to be more stable in air near 400°C
than were triallyl cyanurate polymers at similar temperatures in nitrogen (15). TAIC does not exhibit the violent
exothermic polymerizations with decomposition that can
occur when thick masses of TAC are heated. Certain t-
amine compounds can accelerate polymerization of TAC when
heated with peroxide (16). In the absence of air and
catalyst TAC first isomerized to TAIC on heating and then
polymerized in a single thermal event (17). Thermal
polymerizations of both only gave low conversion compared
to radical-initiated polymerizations.

Radiation polymerization of TAC using ^{60}Co was reported
to avoid rearrangement (18), but conversion was not complete. Temperatures above 200°C produced isomerization
to isocyanurate structure. TAC containing 5% BP was
irradiated for 17 hr at 1 ft from a mercury arc lamp to
form nearly colorless, solid polymer (19).

Homopolymerization or copolymerization of monomer mixtures high in TAC finds application in thin adhesive
layers or filled structures for electronic and other
equipment. Thus with 0.1 to 2% benzoyl or dicumyl peroxide, TAC has been polymerized between metal sheets at
95 to 215°C to form heat-resistant electrical condensers
(20). Violent polymerization may be prevented by good
heat exchange and regulators; cracking (as experienced in
thicker casting) thus may be avoided. Spiral and helical
fractures in the glass-fiber-reinforced polymers have
been studied (21).

Very rapid free radical polymerization of high-boiling
polyfunctional monomers such as TAC in small amounts has
interesting possibilities. TAC may be polymerized on
surfaces of polyester fibers in 20 sec at 200°C. Bonding
of the fibers to rubber was improved (22). Finely divided
TAC polymers were added to polypropylene to promote dyeing
of fibers (23). In electronic applications, a mixture of
TAC with filler, toluene, and high concentrations of
dicumyl peroxide was applied as an adhesive cured at 125°C
for 15 min (24).

Mercury derivatives of TAC have been suggested for
germicidal finishing of textiles (25). Triallyl trithio-
isocyanurate obtained as a yellow oil was not observed to
polymerize (26).

TAC COPOLYMERIZATION WITH UNSATURATED POLYESTERS

Kropa introduced the copolymerization of unsaturated
alkyd-type polyesters with ethylenic monomers in patent

applications made from 1938 to 1944 (27). At first these
copolymers were promoted principally for coatings, but
moldings and fiber-reinforced laminates also received
attention. Diallyl and triallyl monomers, together with
polyesters derived from reaction of ethylene or diethylene
glycols and maleic anhydride, were shown in most patent
examples. Styrene and acrylic monomers were employed
later with unsaturated polyesters as a major industry.
For example, Kropa heated a polydiethylene glycol maleate
viscous liquid resin of acid number 50 with peroxide
catalyst and diallyl maleate, as well as with diallyl-o-
phthalate. Other examples disclosed fumarate polyesters
modified by monofunctional alcohol terminators together
with linseed oil or linseed oil acids. The allyl monomer
such as diallyl phthalate (DAP) was the minor component,
e.g., 40 parts of DAP to 60 parts of unsaturated polyester. Kropa's early patent, although giving no examples
of copolymerization with TAC, opened up the development
that was applied later to polyester-TAC, to polyester-
styrene, and to mixtures of styrene and methyl methacrylate in copolymerization with unsaturated polyesters. TAC
is a useful comonomer in minor proportions with a number
of vinyl-type monomers (Fig. 24.1).

A mixture of 5% TAC and 95% styrene containing 0.5%
benzoyl peroxide was copolymerized by heating for 48 hr
at 100°C (28). The product swelled in toluene but did
not dissolve. Examples of this patent show also copolymerizations of 2-amino-4,6-diallyloxy-1,3,5-triazine. TAC
was copolymerized with diethylene glycol fumarate-sebacate
polyester along with glass-fiber reinforcement. This
composition could be cured during 3 hr at 100°C to give a
hard, stiff laminate. An example was given also of copolymerization of an unsaturated viscous liquid ester
polymer with a mixture of equal parts TAC and methyl
methacrylate. Copolymerizations of unsaturated polyesters
with styrene, methyl methacrylate, vinyl acetate, and
other nonallylic monomers were suggested. Use of inhibitors, separation, and curing of prepolymers and other
features of polyester technology were disclosed (29).

Kropa suggested other ways to employ TAC in polymers
utilizing both condensation and addition polymerizations.
A mixture of TAC and a glycol such as triethylene glycol
could be heated in presence of metallic sodium as catalyst
(e.g., 1 hr at 80°C) to give a low-molecular-weight
polyester by partial alcoholysis of the TAC. This viscous
liquid product of brownish color could be copolymerized
through residual allyl groups with vinyl or allyl monomers (30). Recently this idea has been developed further
(31). Kropa also prepared N-methylol derivatives of
allyl oxytriazines which could be copolymerized with

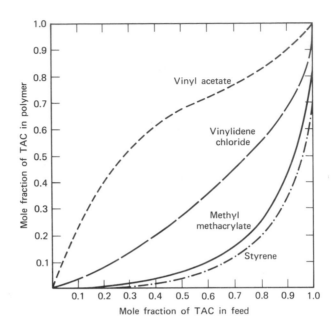

Fig. 24.1. Copolymer composition curves for free-radical-initiated reactions of vinyl monomers with triallyl cyanurate (TAC). Copolymerization occurred most readily with vinyl acetate, with a tendency for alternation of comonomer units. Data sheet of American Cyanamid Company.

formaldehyde condensate resins (32). The methylol derivatives could be polymerized by addition mechanism first. For example, when 2-(N-methylolamino)-4,6-dialloxy-1,3,5-triazine was heated with 5% peroxide, a hard polymer was obtained. Methylol groups in the polymer could react further with formaldehyde condensate resins.

Cummings and Botwick (33) discussed copolymerization of TAC with unsaturated alkyds modified according to the patent of Knapp (34). Styrene was included in most formulations and 2% benzoyl peroxide (from a paste in styrene) served as initiator. One type of fusible alkyd in the patent examples was prepared from ethylene glycol, maleic anhydride, and dicyclopentadiene.

Viscous liquid polyester resins based on triallyl cyanurate and unsaturated alkyd polyesters became available in the early 1950s (35). They have been used with glass fibers where very high heat resistance is required;

radomes and other aerospace devices are among the applications. Vibrin 135 of U.S. Rubber contained about equal
parts of TAC and a modified ethylene glycol maleate polyester of average molecular weight about 750 (34). The
modification process gave rise to isolated double bonds
of the cis configuration. From the heat liberated by
homopolymerization, it was estimated that about 60% conversion of maleate (cis) to fumarate (trans) double bonds
occurred during preparation of a polyester from maleic
anhydride, ethylene glycol, and other reactants (36). The
trans double bonds generally are more reactive both in
homopolymerization and most copolymerizations. Fumarate
double bonds give an IR band at 984 cm^{-1}. The Vibrin
mixtures with about 1% peroxide show two exotherms. The
first was believed to involve most of the polyester double
bonds and two of the double bonds of triallyl cyanurate.
The second may involve isomerization of the crosslinked
copolymer structure to the more stable isocyanurate form.
Curing from 200 up to 300°C was used to give maximum
crosslinking and greatest heat resistance. Vibrin 135
was generally used with 20% styrene addition. Exotherms
in such bulk copolymerizations have been studied by
differential thermal analysis (37). The glass-fiber-
reinforced polymers can retain 19,000 psi flexural
strength after exposure for 192 hr at 500°F (33). With
2% t-butyl perbenzoate curing, cycles as long as 24 hr at
80°C followed by 24 hr at 180°C have been required for
best heat resistance.
    In outdoor applications such as radomes and in struc-
tures exposed to high temperature, Vibrin-type TAC-poly-
ester-glass structures have shown crazing and discoloration. Vibrin 136A gave better translucent, craze-free
structures after long use at 500°F (38). Laminating
methods used with glass fibers have included vacuum bag,
pressure bag, and matched die molding systems. Curing at
180 to 260°F may be followed by postcuring at higher
temperatures depending upon the specific application. For
wire coatings 2% benzoyl peroxide together with ketone
solvents may be added followed by baking at 400 to 500°C
for several hours. Other commercial unsaturated polyester
resin solutions containing TAC have included Laminac 4232
(American Cyanamid) and Selectron 5000 (PPG). Addition
of up to 40% calcium carbonate filler to some of these
resins may reduce crazing without impairing flexural
strength (39).
    Crazing and cracking are encountered in castings and
fiber-reinforced laminates based on TAC-unsaturated polyesters when the copolymerizations are carried out too
rapidly. However, discoloration may occur when curing
is achieved by long heating in the presence of oxygen.

Lundberg sought to overcome these difficulties by adding small proportions of N,N-diallylurea or other amide derivatives and by curing in the absence of air (40). One result of cracking and crazing of fiber-filled laminates is the absorption of several percent of water when the moldings or laminates are exposed to high humidity. Glass-fiber finishes can prevent this impairment of electrical properties. Sometimes reducing the polymerization exotherm peaks by adding an allylic modifier avoids excessive stresses and strains during copolymerization and eliminates cracking.

A suitable acid number range for the polyesters is 35-40. Lundberg (40) described the preparation of a polyester in a stirred reactor under nitrogen from 116 parts of maleic anhydride and 116 parts of ethylene glycol. The mixture was heated to 200°C with the reactor vented through an air-cooled reflux condenser. When the acid number reached 35 (titration of sample with KOH solution) the viscous liquid resin was cooled to 60°C and an equal weight of triallyl cyanurate was mixed in along with 0.01% hydroquinone as storage inhibitor and 1.0% benzoyl peroxide. On heating at 90°C, a 5 g sample of the mixture gave a primary exotherm of 230°C and a secondary exotherm above 250°C. In polymerizations at 90°C, 1.0 or 5.0% N,N-diallylurea was added to 5 g of unsaturated polyester. These gave primary peaks of 178 and 167°C, respectively, and no secondary exotherms were observed. Lamination of 12 plies of glass-fiber cloth using 1% diallylurea as modifier was described. Postcuring of the laminates was carried out for 1 hr at 400°F and for 3 hr at 500°F, using cellophane to minimize contact with air.

Improvement in physical properties of glass-fiber laminates has been suggested by copolymerization of TAC-unsaturated polyester syrups with minor proportions of maleimide (41). It has been proposed to react TAC and the polyester with a small proportion of a phosphorus chloride before adding peroxide and copolymerizing (42). Less flammable are the glass-fiber laminates based on TAC with ethylene glycol tetrachlorophthalate-maleate polyester.

Mar-resistant hard coatings for MMA acrylic polymer plastics have been based on unsaturated polyester from triethylene glycol fumarate with minor proportions of ethyl acrylate, allyl methacrylate, and triallyl cyanurate (43).

TAC IN CURING OF PLASTICS AND ELASTOMERS

Polyfunctional allyl monomers, as well as polyfunctional methacrylates and acrylates of high reactivity and high boiling point, have found application in small concentrations as crosslinking agents for preformed polymers, usually along with peroxides and heating, or with high-energy radiation. Such polyfunctional compatible monomers can promote curing and improvement in physical properties in many plastics and synthetic rubbers. Early work was based on the assumption that the base polymer needed unsaturation for crosslinking. For example, Hewson added TAC to cellulose acetate sorbate and cast films from solution (44). Films were cured for 6 min at 160°C. Pinner discovered that vinyl chloride polymers were particularly responsive to crosslinking by compatible polyfunctional allyl esters, which were later disclosed to be triallyl cyanurate and diallyl sebacate. See Fig. 24.2.

Research workers of Bayer in Germany found that crosslinking and grafting occurred when cellulose acetates containing TAC and peroxides were heated at 120°C for 30 min (45). Fibers of triacetate could be insolubilized by heating for 3 sec at 200°C. From 0.5 to 2% of an allyloxytriazine was used to promote vulcanization of chloroprene polymer synthetic rubbers (46). Ethylene-vinyl acetate, as well as ethylene-propylene copolymer elastomers, were cured by adding about 1% dicumyl peroxide and 3% TAC or triallyl phosphate, followed by heating at 150°C (47). The vulcanizates showed desirable low compression set and rubberlike elasticity. Polysiloxane elastomer moldings also were cured using 0.5 to 5% of TAC or triallyl phosphate (48). Blends of butadiene-acrylonitrile copolymer rubber with TAC have been proposed as binders for extrudable, heat-resistant solid propellants (49). TAC has been added to plasticized PVC for curing with 0.1 to 5% peroxide (50). In one example of the second patent, the following composition was cured by heating in a mold for 30 min at 160°C to give an elastomer of improved elongation properties: 100 parts PVC, 14 parts TAC, 1.0 part dicumyl peroxide, 55 parts polyester plasticizer, and 10 parts lead silicate stabilizer. High-impact blends of vinyl chloride copolymer with butadiene copolymer elastomer cured by heating with peroxide and TAC have shown superior mechanical and thermal properties (51).

Addition of triallyl cyanurate has been studied in the curing of the new ethylene-propylene copolymer rubbers. Efficiency of crosslinking by peroxides could be improved by TAC, DAP, or divinyl adipate, thereby reducing the peroxide requirement to 2.5% or less (52). TAC with dicumyl peroxide has been suggested as a coupling agent in glass cloth laminates (53). An electronic adhesive for bonding to copper was formulated from polyethylene fine powder, TAC, and dicumyl peroxide (54).

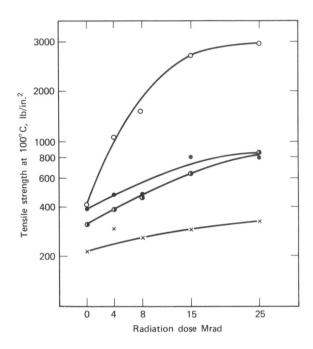

Fig. 24.2. The tensile strength of polyvinyl chloride at 100°C was
increased by adding compatible, polyfunctional allyl monomers and
then crosslinking by high-energy radiation: triallyl cyanurate (open
circles), diallyl sebacate (half-solid circles), unmodified polyvinyl
chloride (solid circles), PVC plasticized with saturated ester
(crosses). S. H. Pinner, Nature, <u>183</u>, 1109 (1959); cf. J. Appl.
Polym. Sci. <u>3</u>, 338 (1960).

   Lyons studied enhancement of high-energy radiation
crosslinking of polymers by adding TAC (55) (Fig. 24.3).
In vinyl alcohol, vinyl formal, vinyl acetate, vinyl
chloride, vinyl pyrrolidone, styrene, vinyl carbazole,
and methyl methacrylate polymers, favorable curing effects
were attributed to one or more of the following mechanisms:
inhibition of chain scission, converting radicals from
scission into crosslinks, and grafting reactions. Acrylate
copolymer elastomers containing chlorine or epoxy groups
were vulcanized by heating with TAC and a triaryl phosphine (56). Estimates of crosslinking efficiency of TAC
in polyethylenes indicated that at concentrations of 0.5
to 5.0%, each monomer molecule crosslinks about 3 polymer
molecules (57). In most of this work, solution cast or

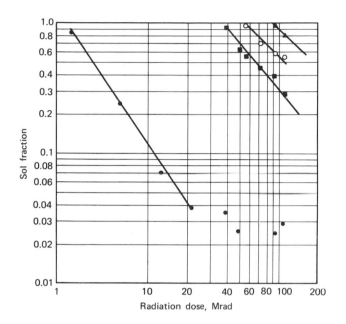

Fig. 24.3. Compared with two other monomers, triallyl cyanurate was more effective in crosslinking polyvinyl acetate: triangles, no additive; open circles, 10% diethyl maleate; squares, 10% dibutyl itaconate; solid circles, 10% TAIC. The amount of remaining soluble fraction was plotted against the dose of high-energy radiation. B. J. Lyons, Nature 185, 605 (1960).

molded films were irradiated with a Van de Graaf accelerator. Radiation curing of polypropylene containing TAC at low doses was promising (58). Ethylene-propylene copolymer rubbers were crosslinked by dicumyl peroxide with TAC (59), as were ethylene-ethyl acrylate copolymers (60). From 2 to 5% TAC was used in polyolefins with high-energy radiation to obtain heat-shrinkable and "self-soldering" insulators (61).

Pinner and co-workers disclosed "extended network polymers" (as distinct from block and graft copolymers) made by polymerizing TAC and other polyfunctional monomers in polyvinyl chloride and in polyethylene (62). By low doses of high-energy radiation, unplasticized polyvinyl chloride containing TAC formed cured products with up to 1000 psi tensile strength at 150°C (63). A mixture of 65 parts vinyl chloride polymer and 22 parts TAC with

stabilizer and pigments has been patented for heat-
resistant wire insulation (64).

A Du Pont patent describes the milling of TAC or diallylmelamine in the range of 1-5% into polyolefins in
order to improve adhesion to polar substrates (65). TAC
has been added to vinyl chloride polymer plastisols in
order to improve adhesion to phosphate or oxychromate
treated metals (66). Films on steel were cured for 15 min
at 175°C. Copolymers of formaldehyde and ethylene oxide
were crosslinked by heating in molds with TAC and dicumyl
peroxide (67). Ring-chlorinated TAC has been added to
tire cord adhesives based upon mixtures of resorcinol-
formaldehyde resin and butadiene-styrene-vinyl pyridine
copolymer latex (68).

OTHER COPOLYMERIZATIONS OF TAC

Although reactivity ratios for copolymerization with
styrene, methacrylate ester and many monomers are rather
low (69), copolymerizations with minor proportions of TAC
are practical. A mixture of TAC and methyl methacrylate
(80:20) gave copolymers to about 40% conversion, after
which TAC homopolymer began to form (69). Methyl methacrylate containing minor proportions of TAC and two free
radical catalysts (effective at different temperatures)
has been polymerized to rigid solids by heating up to
100°C (70). The materials were molded to cured shapes
at 120 to 160°C for 90 min. A heat test at 165°C for an
hour showed little distortion. Cast copolymers of about
5% TAC with MMA have improved scratch resistance (19).
Syrups may be prepared by heating with 0.0025 part of
azobis for 1.5 hr at 85°C (71). For casting there may
be added 0.1 part of benzoyl peroxide and 1.0 part of
t-butyl hydroperoxide for polymerization at 50°C for
10 hr, followed by 0.5 hr at 98°C.

Copolymerizations of maleic anhydride with TAC and with
triallyl isocyanurate (TAIC) in acetone solution with BP
at 60°C gave quite different results (19). The deep red
TAC copolymer solution obtained was heated with a slight
excess of aqueous NaOH over that needed for neutralization. An interesting white opaque microgel dispersion of
remarkable stability resulted which resembled latex and
deposited smooth films on evaporation. In contrast,
similar conditions of copolymerization under air gave
nearly colorless viscous solutions of TAIC-maleic anhydride copolymer. Reaction with aqueous alkali gave
coarse crosslinked masses and only a minor fraction
soluble in water which deposited films.

Other copolymerizations of TAC are listed in Chemical
Abstracts from which some examples follow.

Applications of TAC	References
With CR-39 in clear sheets	Starkweather et al, Ind. Eng. Chem. __47__, 302 (1955)
With acrylic acid for branched copolymers	Jones, U.S. 3,066,118 (Goodrich)
With vinyl chloride (to give nongelling polymers)	Martin, U.S. 3,047,549 (Monsanto) cf. Ital. 611,733 (Edison)
Grafted vinyl chloride polymers	Hardt, Belg. 643,626 (Bayer); Pettit, Brit. 940,139 (BX Plastics)
With butyl acrylate (cellular crosslinked copolymers)	Jefferson, U.S. 3,293,198 (Grace)
With acrylamides in images by photopolymerization	Oster, U.S. 3,097,096; Levinos, Brit. 958,538 and U.S. 3,029,145 (GAF)
With fumarate esters and vinyl esters (oil modifiers)	Brit. 1,038,228 (Esso)
With styrene or acrylic monomers grafted upon polypropylene fibers	Chen, U.S. 3,423,161 (Am. Cyan.)
With vinyl acetate and ethylene in emulsion copolymerization	Lindemann, U.S. 3,404,112 (Airco)
With phosphorus-containing polyesters	Galli et al, U.S. 3,103,987 (Boeing)
With allyl siloxanes for resistant coatings	German 1,139,639 (Witten); CA __58__, 7045 (1963)
With vinyl siloxanes for cured silicones	U.S. 2,899,403 (Westinghouse)
With trivinyl borazoles	U.S. 2,954,361 (Am. Cyan.)
With epoxy resins	Skiff, U.S. 2,707,177 (G.E.); Eirich, U.S. 2,848,433 (Aries)

TRIALLYL ISOCYANURATE

TAIC, a cyclic N-allylimide monomer, has received less study than TAC, but it is of interest because of the mechanical and chemical stability of its polymers at high temperatures. TAIC monomer can be prepared by reaction of allyl chloride with potassium cyanate, a system studied first in efforts to prepare allyl isocyanate (72). The isocyanate trimerizes to triallyl isocyanurate, a cyclic N-allylimide structure (73):

$$CH_2=CHCH_2Cl + 3KOCN \longrightarrow$$

$$CH_2=CHCH_2-N \underset{O}{\overset{CH_2CH=CH_2}{\bigcirc}} N-CH_2CH=CH_2 \quad + 3KCl$$

In one example, acetonitrile was used as solvent for
reaction at 150°C for 3 hr under pressure. A byproduct
was sym-diallylurea (mp 95°C). Triallyl isocyanurate,
along with diallyl isocyanurate, can be prepared by
reaction of allyl chloride with the iso form of cyanuric
acid in the presence of triethylamine, sodium hydroxide,
or other HCl acceptor (74). TAIC has been prepared from
cyanuric acid, allyl chloride, and calcium oxide in dimethyl formamide (75). The same patent used methallyl
chloride to prepare trimethallyl isocyanurate (mp 85°C).
Diallyl and triallyl cyanurate may be obtained by adding
allyl chloride dropwise to an aqueous solution of cyanuric
acid at pH 10 and 50°C in the presence of cuprous ions
(76). Bromides or iodides may be added as catalysts in
the preparation of TAIC monomer (77). In one example,
1.0 kg of cyanuric acid, 1.2 kg of sodium hydroxide and
3 kg of water were stirred for 2 hr. The resulting sodium
cyanurate was heated with a molar excess of allyl chloride,
5 g KI and 200 ml ethylene glycol for 6 hr at 115°C.
After cooling there was extracted by benzene a 51.5% yield
of TAIC.
   The base-catalyzed trimerization of alkyl and allyl
isocyanates to 1,3,5-trisubstituted isocyanurates was
discussed by Saunders and Slocombe (78). Allyl isocyanate
is believed to be a transient intermediate in the preparation of TAIC from cyanates. The reaction of TAIC with
dry HCl gives the exceedingly reactive hydrochloride
adduct of allyl isocyanate (79). Pyrolysis of TAIC under
low pressure has given small amounts of N,N-diallylurea.
   In the preparation of TAIC suitable solvents, inhibitors
and temperatures must be used in order to obtain liquid
monomer of good color without excessive losses by polymerization (80).
   Triallyl isocyanurate is a viscous liquid or white solid
melting at 23.5°C. It has been distilled at 113°C at
0.3 mm. The viscous liquid monomer available from the
Plastics Division of Allied Chemical had the following
properties:

              Boiling point, 152°C at 4 mm
              $d_{25} = 1.1583$
              Color, APHA = 5-35
              Viscosity at 25°C, 100 cp
              $n_D^{25} = 1.5113$
              Solubility in water at 25°C, 0.37%

TAIC is completely miscible with benzene, heptane, ethyl
alcohol, and acetone. The IR spectrum of the liquid
monomer is presented in Fig. 24.4.

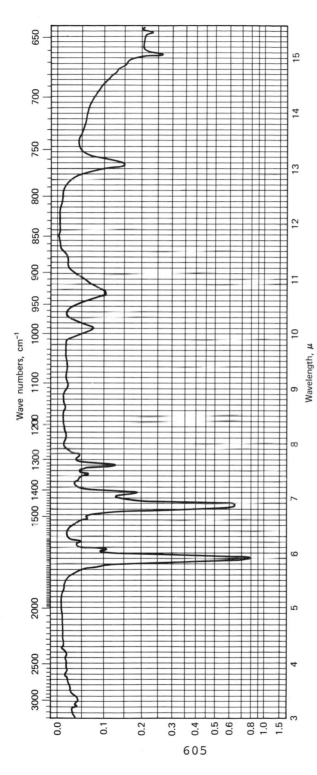

Fig. 24.4. The infrared spectrum of liquid triallyl isocyanurate is characterized by sharp bands near 5.9 and 6.9 microns from carbonyl and double bonds, respectively. Surprisingly weak absorption occurs at 3.4 microns from CH bonds. Product Bulletin, Allied Chemical Corp.

TRIALLYL ISOCYANURATE POLYMERIZATION

Samples of commercial TAC and TAIC monomers behaved differently on storage (19). TAC from American Cyanamid stored as a solid with some remaining liquid became, after 12 years, a soft opaque slurry of crosslinked polymer granules dispersed in residual monomer. Liquid TAIC monomer of Allied Chemical was transformed during 6 years storage into a clear, somewhat rubbery gel which was fairly strong but showed brittle fracture. From viscous syrup of TAIC polymer in monomer formed on standing a year at room temperature, soluble prepolymer was precipitated by adding excess methanol. Heating the prepolymer on aluminum foil on a hot plate rapidly formed clear, colorless films which no longer dissolved in solvents. When prepolymer was dropped directly on a surface at higher temperatures, violent polymerization occurred with decrepitation.

The TAIC polymers formed by isomerization of triallyl cyanurate (TAC) polymers have somewhat different properties than those formed directly from TAC. Understanding of these polymerizations has been advanced by Gillham and co-workers at Princeton (81, 82) but much more remains to be discovered about the structures of the polymers made under different conditions. When TAC is heated in absence of air it may first isomerize to TAIC after which thermal polymerization occurs. The rearrangement of TAC monomer is analogous to the Claisen rearrangement of allyl phenyl ether. TAC is an enol and TAIC is a more stable keto isomer, an N-allyl imide:

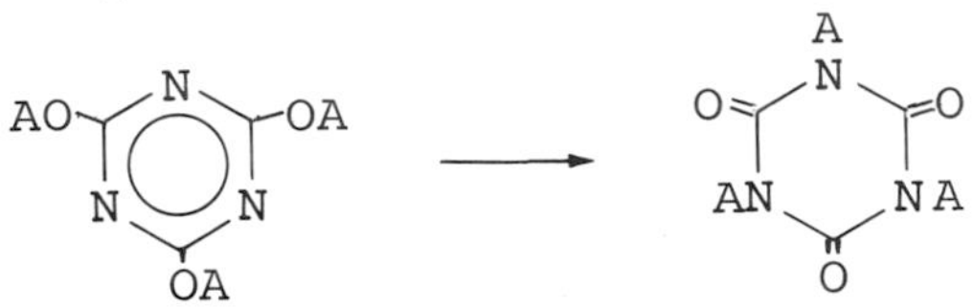

The isomerization can be followed by IR by disappearance of the C-O-C band at 1130 cm^{-1} and appearance of the C=O band at 1700 cm^{-1}. With the same initial temperature and peroxide concentrations in bulk TAC gave higher conversion to polymer than TAIC. When TAC was polymerized a first exotherm of polymerization was followed by a greater exotherm near 300°C from isomerization of TAC polymer to TAIC polymer together with further polymerization. The heat of isomerization is about 35 kcal/mole monomer unit. Conversions of only about 70% resulted, apparently, along with complex cyclization by intramolecular reactions. Complete conversion of all three allyl groups to polymer linkages should liberate 55.5 kcal/mole of TAIC. Even highly crosslinked TAC polymer can rearrange on heating to isocyanurate structure.

Gillham and Mentzer (83) using a differential scanning calorimeter found that the thermal polymerization of TAC (in air without added catalyst) began at lower temperatures (about 130°C) than that of TAIC (about 160°C). The initial small exotherms of polymerizations apparently initiated by small amounts of oxidation products were followed only on reaching much higher temperatures with larger exotherms near 290 and 360°C, respectively. The heat of polymerization of TAIC was about 27 kcal/mole. Most effective in the catalyzed polymerization of TAIC are high temperature peroxides such as 3% or more of 2,5-dimethyl-2,5-bis-(t-butylperoxy)-n-hexane (Lupersol 101). The slow bulk polymerization of TAIC after formation of an initial crosslinked network seems to result from immobilization of the reactive allyl groups. Suitable comonomers, solvents, and plasticizers should give more favorable behavior.

The greater stability of TAIC polymer structures compared to those of TAC polymers has encouraged commercial applications. Triallyl isocyanurate homopolymers are comparatively stable and rigid even near 400°C. The unmodified homopolymers are too brittle for most uses except for some adhesive applications, but copolymers have found applications in automotive and other fields.

TAIC has been polymerized in aqueous dispersion using the following formulation (84):

TAIC	265 g	$K_2S_2O_8$	1.5 g
Water	1000 g	$NaHSO_3$	2.0 g
Triton N 101	30 g	Na_2HPO_4	2.0 g

The above composition was heated 1 hr at 75°C followed by overnight at 50°C. The polymer precipitated overnight to very small particles (< 0.22 μ) which were suggested as an opacifier for polypropylene fibers which were readily dyed. Until recently few allyl monomers have been polymerized in emulsion systems (85).

Although triallyl isocyanurate is of interest for heat-resistant copolymers with unsaturated polyesters and for other crosslinked stable polymers, little has been published as yet. Compatibility with polyester formulations is promising. Higher temperatures are needed for curing than those used with styrene blended polyesters. With ethylene-maleate polyester TAIC was more reactive than MMA but less reactive than TAC and styrene (86).

Reactivity ratios have been estimated for TAC, TAIC, and diallyl melamine (DAML) in bulk copolymerizations with common comonomers to low conversions at 60°C using azobis or peroxide initiators (69). TAIC and DAML were

less reactive than TAC with MMA and with styrene. The
three allyl monomers were more reactive with vinyl acetate than with styrene and MMA. These polyfunctional
allyl monomers have very different e values from typical
monoallyl compounds (87):

	e	Q
Triallyl cyanurate	-1.00	0.020
Triallyl isocyanurate	-0.60	0.011
Diallyl melamine	-0.95	0.017

These monomers resemble vinyl ethyl ether and isobutylene
in e and Q parameters and in ready copolymerization with
maleic anhydride in acetone with peroxide initiator at
60°C (19). TAC gave a red highly viscous solution of
copolymer in an hour while TAIC under similar conditions
with maleic anhydride formed a clear, nearly colorless
gel of copolymer.

TAC and TAIC are not only effective crosslinking agents
with many monomers but find use with high-temperature
peroxides as crosslinking agents for preformed polymers,
including polyurethanes, polyethylenes, vinyl chloride
polymers, and ethylene-propylene terpolymer rubbers. TAIC
may be added to tire cord adhesives and used as a so-
called "polymerizable plasticizer".

Copolymers of vinyl chloride-ethylene-TAIC in foamed
sheets were recommended for packaging fragile glassware
and porcelain (88). TAIC was added as curing agent to
expandable vinyl chloride-propylene copolymer beads (89).
Chlorinated polyethylene was cured by minor amounts of
TAIC, diallyl itaconate, and dicumyl peroxide (90).
Mechanical strength, elastic properties, and solvent
resistance were thereby improved.

The following patents also disclose copolymerizations
of TAIC and related monomers:

Copolymerizations	Patents
Unsaturated polyester copolymers	U.S. 3,108,902 (Boeing)
Trimethallyl IC-vinyl lactams	U.S. 2,848,440 (Dow)
Crosslinking PVC	U.S. 3,392,135;
	U.S. 3,539,488 (Ethyl)
Diallyl hydroxypropyl IC-MMA	Alaminov, CA 73, 77669
	(1970)

A number of allyl triazine compounds related to TAC and
TAIC have been reported (91). FMC workers have prepared
the triallyl ester of tricarboxyhexahydrotriazine (92).
Disubstituted isocyanurates such as diallyl isocyanurate
(mp 148°C) are difficult to make because of competing
formation of trisubstitution products (93). A suspension

of sodium cyanate in dimethyl formamide under nitrogen
was reacted with an organic isocyanate to give mixtures
of di- and trisubstituted isocyanurates (94). In Russia
minor proportions of 1-octyl-3,5-diallyl isocyanurate and
related monomers were copolymerized with methyl methacrylate to give heat-resistant, transparent copolymers (95).
Using 0.2% dicyclohexyl percarbonate cast copolymerizations could be carried out at 20°C for 48 hr followed by
completion at 70 to 150°C.

N,N-DIALLYLMELAMINE

This monomer (DAML) is the N,N-diallyl triamide of
cyanuric acid. As in the case of aniline, the resonating
ring makes this aminotriazine derivative much less basic
than typical amines.

cyanuric acid   N,N-diallylmelamine

Diallylmelamine and other N-allylmelamines can be prepared
from cyanuric chloride by stepwise reaction with allylic
amines and ammonia in the presence of acid acceptors (96).
Because of the unpleasant nature of cyanuric chloride,
preparation of DAML from diallyl cyanamide, dicyandiamide,
and KOH has been proposed (97). DAML is moderately toxic
as a local irritant and by ingestion. It is dangerous
when heated to decomposition or on contact with strong
acids.

DAML is able to polymerize by vinyl-type addition as
well as by formaldehyde condensation reactions. The
monomer has been evaluated by American Cyanamid, who
supply technical literature.

DAML is a white crystalline solid (orthorhombic) melting
at 145°C and having density 1.242 at 30°C. The refractive
index is 1.580 and the optic sign is negative. The
monomer has limited solubility in water, less than 0.01%.
At 2.5×10^{-4} molar in water, the K_b at 30°C is 1.6×10^{-9}
indicating a very weak base, comparable to aromatic
amines. Maximum solubilities (g/100 ml at 30°C) have been
reported by American Cyanamid as follows: dioxane, 21;
methanol, 8.5; acetone, 5.5; ethanol, 4; ethyl acetate, 3;
heptane, 0.03; benzene, 0.15. DAML has limited solubility
in vinyl and allyl monomers. The monomer reacts rapidly
and quantitatively with the bromine from dilute bromide-
bromate solution in the presence of excess HCl.

Diallylmelamine reacts as a monobasic compound with mineral acids. In neutral aqueous solution it combines with only one molecule of formaldehyde in the absence of added catalyst to form methylol DAML. When this compound is heated with removal of water, thermoplastic polymers of molecular weight as high as 4000 are formed. Under alkaline conditions DAML reacts with several molecules of formaldehyde to give hydrophilic low polymers. If the latter are heated at 125°C, hard polymers result; these products are only partially soluble in water-alcohol mixtures or in hot ethanol. Heating at higher temperatures gives completely crosslinked products. At pH 8.0, by addition of NaOH, formalin reacts with DAML in 55 min at 98°C, with introduction of 1.5 to 2.5 CH_2OH groups (98). DAML-formaldehyde resins and their mixtures with silicones or other resins may be cured faster with addition of peroxides or cobalt naphthenate (99). Ether derivatives of methylol DAML can be prepared by reaction with alcohol to give modified solubility (100). Dimethylol DAML as a white powder has been prepared from ethanol-water formaldehyde solutions with traces of NaOH at apparent pH 7.5 by standing for 6 days at room temperature (101). Higher pH values gave syrupy resins of higher methylol content.

In applications with cellulose fibers, DAML-formaldehyde resins can be used to impart water resistance. These diallyl aminotriazine-formaldehyde condensates have been evaluated in filled molding compositions (102). Small amounts of inorganic acidic agents such as phosphoric acid can be present as curing agents at 100°C or higher, but these catalysts may cause corrosion of metals and degradation of cellulose fibers.

In amounts of 5% or more, DAML acts as a curing agent for epoxy resins (e.g., Epon 828 at 100-130°C). Such thermosetting compositions have been used for impregnating paper coatings in electrical applications (103). DAML reacts with alkylene oxides to give cationic surfactant polymers (104). Kaolin particles coated by about 1% DAML (neutralized by acetic acid) have been used as filler in polyester-styrene glass-fiber laminates (105).

ADDITION POLYMERIZATION OF DIALLYLMELAMINE

Polymerizations of the melted monomer above 142°C, with peroxide or azo catalysts, rapidly give crosslinked polymers by three-dimensional polymerization. American Cyanamid found that copolymerizations with acrylates and other vinyl-type monomers in solvents such as butanol could form fusible but curable copolymers. Diallylmelamine can be polymerized in the solid state by high-energy radiation and also in solution with azobis, producing

soluble, fusible polymers with low unsaturation (106).
Partial cyclic intramolecular polymerization was believed
to occur. Prepolymers of DAML can be prepared above 130°C
by using a high-temperature peroxide (107). The monomer
has been polymerized in thin films and from vapor phase
by use of electron beams or UV radiation (108).

In copolymerizations with free radical initiators DAML
behaves somewhat like triallyl cyanurate; this is illustrated in the copolymer compositions of Fig. 24.5. In
bulk copolymerizations of DAML with vinyl acetate, the
fraction of DAML units in the copolymer was nearly as
high as in the monomer mixture. With conjugated monomers
such as methyl methacrylate, however, the proportions of
DAML unit in the copolymer were smaller than in the
monomer feed mixture. The relative reactive ratios in
radical copolymerization (reciprocals of reactivity
ratios) presented in Table 24.1 have been reported for
DAML and other allyl compounds with vinyl monomers.

TABLE 24.1

Growing Radical Chain End [a]

Monomer	Styrene	Vinyl acetate	Allyl chloride	MMA	Acrylo-nitrile
Styrene	1.0	50	30	2.2	20
Methyl methacrylate	1.9	70	--	1.0	5.5
Acrylonitrile	2.4	18	15	0.7	1.0
TAC	0.03	5.4	--	0.06	--
DAML	0.02	1.5	--	0.06	--
Vinyl acetate	0.02	1.0	1.5	0.05	0.20
Allyl chloride	0.03	1.4	1.0	0.02	0.33
Allyl acetate	0.01	1.7	--	0.04	--

[a] DAML and TAC Data Sheets, American Cayanmid

In persulfate emulsion, methyl methacrylate with DAML
(4:1) and a cationic emulsifier gave latices of crosslinked copolymers (109). For preparing soluble copolymers
for casting films, the following solution was heated for
2 hr near 95°C (110): 50 g of DAML, 50 g of MMA, 400 ml
of dioxane, and 0.5 g of BP. The copolymer product by
analysis contained 46.8% DAML units. The copolymers were
thermoplastic at 150°C. Dimethylol DAML was copolymerized
with methyl isobutyl carbinyl acrylate, and films of copolymers containing phosphoric acid catalyst were cured
at 105°C (110).

DAML has been copolymerized with acrylamide in dioxane
using benzoyl peroxide catalyst and heating for 90 min at
reflux (111). The copolymer was then methylolated by

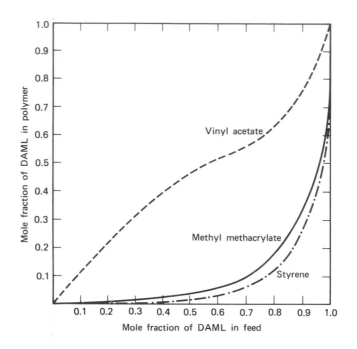

Fig. 24.5. Copolymer compositions from free radical copolymerizations of vinyl monomers with N,N-diallylmelamine. DAML was reluctant
to copolymerize with MMA or with styrene but copolymerized readily
with vinyl acetate. Data bulletin, American Cyanamid Company.

reaction with formaldehyde solution. The resulting
cationic thermosetting, water-soluble resin was suggested
for spot bonding cellulose fibers in high-strength papers.
Acrylamide and acrylonitrile copolymers with DAML were
prepared and applications were suggested in the fields of
paper and dyeable acrylonitrile copolymer fibers (112).

OTHER POLYFUNCTIONAL N-ALLYL TRIAZINES

Hexaallylmelamine has interest in the preparation of
heat-resistant reinforced glass cloth laminates (113).
In nitrogen atmosphere the polymers are stable almost to
400°C. The hexaallylmelamine monomer used was a viscous,
slightly colored liquid; it did not crystallize even at
-78°C but formed a glass. It boils near 160°C at 0.1 mm
(n_D^{25} = 1.5390). The monomer can be prepared by reacting
cyanuric chloride with diallylamine in the presence of

an acid acceptor. Such melamine derivatives may inhibit
the growth of plants.

Hexaallylmelamine was surprisingly reluctant to polymerize (113). It required 2% or more di-t-butyl peroxide
and 24 hr heating at 130°C for polymerization to transparent castings. The homopolymers were brittle and
cracked easily from local strains. High-energy radiation
or heating without peroxide below 200°C caused little
polymerization. A final exotherm observed above 300°C
was attributed to thermal polymerization of residual allyl
groups. The polymers lose little weight on heating below
400°C; they do, however, deteriorate by oxidation, which
is accelerated by cracks.

Gillham and Petropoulos polymerized 2,4-bis(diallyl-
amino)-6-piperidino-s-triazine using di-tBP. This tetra-
allyl-melamine (n_D^{25} - 1.5501) monomer was "bodied" by
preliminary polymerization at 160°C with 0.3% dimethyl
di-t-butylperoxyhexane for an hour and then was applied
to experimental glass-cloth laminates. These materials
after curing have thermal stability with good strength at
300°C or above, and they have good dielectric and arc
resistance when dry.

Allylisomelamine and triallylisomelamine, prepared by
Thomas, are believed to have the allyl groups attached to
the ring nitrogen atoms. The former can be made by heating molar proportions of allylamine hydrochloride,
ammonium chloride, and mono-potassium 1,3 dicyanoguani-
dine for 5 hr at 98°C (114). The monomer melts at 184°C
with decomposition. Copolymers with acrylonitrile were
not as promising for dye receptivity as those from DAML.
Copolymers with acrylamide were prepared by aqueous persulfate.

Polymers were prepared from 2-amino-4,6-diallyloxy-
1,3,5-triazine (mp 61°C and bp 151°C at 1.0 mm) (115).
With 5% BP the monomer gave reaction after heating for a
few minutes at 110°C; on further heating it yielded a
hard polymer mass. Heating the monomer 2.5 hr at 70°C
with 1% BP gave a viscous solution. Heating 19 hr at 70°C
followed by 8 hr at 100°C resulted in the formation of a
clear yellow polymer casting of specific gravity 1.328.
The polymer absorbed 0.17% water after 30 min at 100°C.
It was slow burning and free from crazing on aging. Copolymers were also prepared with styrene and with diethylene glycol fumarate sebacate polyester. Some other
allyl triazine derivatives are listed with references in
Table 24.2.

Related to the triazine monomers is 2,4,6-triallyloxy-
pyrimidine, in which one nitrogen atom of the triazine
ring is replaced by CH. It was prepared by reacting the
trichloride with allyl alcohol in the presence of KOH at

room temperature followed by refluxing for 1 hr (116). The compound ($n_D^{20} = 1.5198$) would not crystallize at $-78°C$ and could not be distilled at 165°C. Differential thermal analysis curves obtained during polymerization showed three peaks. Two were believed to represent reaction of two double bonds; the third peak may represent rearrangement to a barbiturate structure.

TABLE 24.2

Other References to Allyl Triazine
and Isocyanurate Monomers

Compound	bp or mp, °C	References, etc.
Allyl diamino-s-triazine	mp 182	Dudley, JACS 73, 2988 (1951)
Allyl dimethyl cyanurate	bp 145(1)	Cohen and Cypher, Ind. Eng.
Diallyl methyl cyanurate	bp 167(8)	Chem. 50, 154 (1958) (copolymers)
Methallyl dicarboxy-s-triazine	--	Wagner, Ger. 818,581 (Degussa) (low polymers)
Allyl dichloro-s-triazine	76(0.18)	Lundberg, U.S. 3,127,399 (Am. Cyan.) (herbicide)
Allyl isocyanurate	mp 200	Francis, U.S. 3,190,841 (Gulf Oil); Fukui, CA 57, 6128 (1962)
Diallyl isocyanurate	mp 145	Fukui (copolymers); Frazier, J. Org. Chem. 25, 1944 (1960)
Diallyl methyl iso-cyanurate	150(3)	Little, U.S. 3,522,253; $n_D^{25} = 1.5130$ (hard polymer)
Allyl and diallyl hydroxy-allyl isocyanurates	--	Hopkins, U.S. 3,200,119 and 3,215,758 (Gulf Oil); Little, U.S. 3,332,946 (Allied)
Diallyl epoxypropyl isocyanurate	--	Hopkins, U.S. 3,132,142 (Spencer)
Diallyl octadecylamino-s-triazine	--	Holm-Hansen, U.S. 2,513,264
Other triazinyl monomers and copolymers with polyesters	--	Kropa, U.S. 2,496,097; D'Alelio, U.S. 3,047,531-2 and 3,056,760 (Dal Mon)

References

1. J. R. Dudley, U.S. 2,510,564 (Am. Cyan.) appl. 1946; cf. more detailed U.S. 2,537,816.
2. E. L. Kropa, U.S. 2,510,503 (Am. Cyan.) appl. 1946.
3. R. G. Nelb, U.S. 2,631,148 (U.S. Rubber).

4. J. R. Dudley et al, J. Am. Chem. Soc. 73, 2986 (1951).
5. J. R. Dudley, U.S. 2,537,816 (Am. Cyan.).
6. D. W. Kaiser, U.S. 2,682,541 (Am. Cyan.).
7. M. A. Spielman et al, J. Am. Chem. Soc. 73, 1775 (1951).
8. E. L. Kropa, U.S. 2,557,667 (Am. Cyan.).
9. L. A. Lundberg, U.S. 2,930,776 and 2,947,736 (Am. Cyan.).
10. H. Schmidt et al, U.S. 3,240,748 (Hoechst).
11. H. J. Oswald and Edith Turi, U.S. 3,275,592 (Allied).
12. E. L. Kropa, U.S. 2,510,503 (Am. Cyan.).
13. P. S. Dokuchitz and L. H. Segall, Ger. 1,164,571 (Bendix).
14. B. H. Clampitt, D. E. German, and J. R. Galli, J. Polym. Sci. 27, 515 (1958); cf. 38, 433 (1959).
15. J. K. Gillham, Encycl. Polym. Sci. Technol. Vol. 1, 760, Interscience, 1964.
16. Brit. 877,862 (CIL).
17. J. K. Gillham and C. C. Mentzer, CA 77, 00935 (1972); Polymer Preprints 13, #1, 247 (1972).
18. L. Wuckel and H. Wagner, Makromol. Chem. 66, 212 (1963).
19. C. E. Schildknecht, unpublished.
20. P. S. Dokuchitz and L. H. Segall, Ger. 1,164,571 (Bendix).
21. J. K. Gillham et al, Polym. Eng. Sci. 8, 227 (1968).
22. R. G. Aitken, U.S. 3,051,594 (CIL).
23. H. D. Anspon et al, U.S. 3,539,665 (Gulf).
24. A. H. Lybeck, U.S. 3,523,862 (Enka).
25. J. T. Shaw et al, U.S. 3,252,751 (Am. Cyan.).
26. G. Loughran and E. O. Hook, U.S. 2,676,151 (Am. Cyan.).
27. E. L. Kropa, U.S. 2,443,740 (Am. Cyan.).
28. E. L. Kropa, U.S. 2,510,503 (Am. Cyan.).
29. E. L. Kropa, U.S. 2,437,962, 2,443,735-40, 2,473,801 (Am. Cyan.).
30. E. L. Kropa, U.S. 2,557,667 (Am. Cyan.).
31. J. C. Schlegel, U.S. 3,017,381 (Am. Cyan.).
32. E. L. Kropa, U.S. 2,496,097 (Am. Cyan.).
33. W. Cummings and M. Botwick, Ind. Eng. Chem. 47, 1317 (1955).
34. R. L. Knapp, U.S. 2,671,070 (U.S. Rubber).
35. H. M. Day et al, SPE J. 9, 22 (February 1953); P. M. Elliott, Mod. Plastics 29, 113 (July 1952); R. G. Nelb et al, PB 105,570.
36. D. E. German et al, J. Polym. Sci. 38, 434 (1959).
37. C. B. Murphy, Mod. Plastics 37, 125 (August 1960); J. Polym. Sci. 28, 447 (1959).
38. J. P. Walton, SPE J. 15, 567 (July 1959).
39. N. E. Wohl and H. M. Preston, Mod. Plastics 35, 153 (October 1957).
40. L. A. Lundberg, U.S. 3,044,913 (Am. Cyan.).
41. W. Cummings, Ger. 1,083,543 and Brit. 831,246 (U.S. Rubber).
42. J. R. Galli et al, U.S. 3,108,987 (Boeing).
43. A. C. Bristol et al, U.S. 3,294,867 (Am. Cyan.).
44. W. B. Hewson, U.S. 2,749,319 (Hercules).
45. W. Bonin et al, Belg. 616,469 (Bayer).
46. Brit. 775,823 (Bayer); CA 51, 15990 (1957); cf. Brit. 930,761 (Bayer).

47. H. Bartl and J. Peter, Kautschuk Gummi 14, WT23 (1961);
     Brit. 853,640 (Bayer); cf. Brit. 952,336 and 843,974.
48. H. Bartl et al, Ger. 1,136,104; CA 58, 11557 (1963).
49. M. Visnov and J. H. Godsey, U.S. 3,268,377 (USA).
50. M. M. Safford and F. F. Holub, U.S. 3,351,604 and 3,392,135 (G.E.).
51. A. Oth and A. Mathieu, Rev. Gen. Caoutchouc Plast. 45, 971 (1968).
52. A. E. Robinson et al (Hercules), Ind. Eng. Chem. RRD 1, 78 (1962).
53. Brit. 895,971 (Esso).
54. A. H. Lybeck, U.S. 3,523,862 (Am. Enka).
55. B. J. Lyons, Nature 185, 604 (1960); cf. S. H. Pinner, Nature
     183, 1108 (1959).
56. N. P. Ermidis,   U.S. 3,697,620 (Am. Cyan.).
57. P. E. Cross and B. J. Lyons, Trans. Faraday Soc. 59, 2350 (1963).
58. G. G. Odian et al (U.S. AEC, NYU-2481 (1961); CA 55, 25345.
59. L. D. Loan, J. Polym. Sci. A 2, 3053 (1964); W. C. Smith and
     N. F. Newman, Rubber World 153, 79 (1966).
60. S. Bonotto, J. Appl. Polym. Sci. 9, 3819 (1965).
61. P. M. Cook, Belg. 638,274 and Brit. 1,047,053 (Raychem).
62. S. H. Pinner et al, U.S. 3,125,546 and Brit. 948,302 (BX
     Plastics); cf. Brit. 905,711.
63. S. H. Pinner, Proc. Warsaw Conf. (1959), CA 56, 8916 (1962);
     cf. K. Posselt, CA 61, 2010 (1964).
64. Belg. 613,197 (AKU); CA 57, 16903 (1962).
65. W. F. Busse et al, U.S. 3,043,716 (Du Pont).
66. Brit. 923,586 (CIL); CA 59, 1846 (1963).
67. F. F. Holub and M. M. Safford, U.S. 3,494,883 (G.E.).
68. R. Broisman, U.S. 3,644,256 (Am. Cyan.).
69. R. W. Roth and R. F. Church, J. Polym. Sci. 55, 41 (1961).
70. J. A. Caton, U.S. 3,049,517 (ICI) and Brit. 849,048.
71. J. A. Caton, Brit. 849,048; U.S. 3,049,517 (ICI); cf. U.S.
     2,910,456 and other patents to Peterlite Products.
72. D. Kaiser and D. H. H. Church, U.S. 2,536,849 (Am. Cyan.); cf.
     U.S. 2,866,801 (Ethyl).
73. TAIC is named by C.A. triallyl-s-triazine-2,4,6(1H, 3H, 5H)-
     trione.
74. B. E. Lloyd and F. L. Kelly, U.S. 2,894,950; T. C. Frazier and
     B. H. Sherman, U.S. 3,065,231 (Allied); J. Am. Chem. Soc. 25,
     1944 (1960); J. J. Tazuma and R. Miller, U.S. 3,075,979 (FMC).
75. Brit. 912,964 (Spencer).
76. T. C. Frazier et al, J. Org. Chem. 25, 1944 (1960).
77. I. Takeuchi, Japan 70-15,981 (Toho Rayon); CA 73, 87942 (1970).
78. J. H. Saunders and R. J. Slocombe, Chem. Rev. 43, 211 (1948).
79. F. C. Schaefer and E. K. Drechsel, U.S. 2,580,468 (Am. Cyan.).
80. G. R. Muller and W. P. Moore, U.S. 3,330,747, also U.S.
     3,322,761 and 3,065,231 (Allied); J. J. Tazuma et al, U.S.
     3,075,979 and 3,108,101 (FMC); W. C. Francis et al,
     U.S. 3,376,301 (Gulf Oil).
81. J. K. Gillham and C. C. Mentzer, CA 77, 88935 (1972); cf.
     U.S. 3,037,978 (Fukui).
82. B. H. Clampitt et al, J. Polym. Sci. 27, 515 (1958).

83. J. K. Gillham and C. C. Mentzer, Polymer Preprints _13_, #1,
     247 (1972).
84. R. Kochar et al, U.S. 3,576,789 (Gulf).
85. H. Warson, Applications of Synthetic Resin Emulsions, Benn,
     London, 1972.
86. S. I. Omel'chenko et al, CA _77_, 75547 (1972).
87. L. J. Young in Polymer Handbook, J. Brandrup and E. H. Immergut,
     eds., Wiley, 1966.
88. Fr. Demande 2,009,045 (Bayer); CA _73_, 110619 (1970).
89. Fr. Demande 2,009,046 (Bayer); CA _73_, 46199 (1970).
90. H. Higuchi et al, Japan 70-30,220 (Showa Denko), CA _73_, 15742
     (1970).
91. J. R. Dudley et al, J. Am. Chem. Soc. _73_. 2986 (1951);
     Brit. 923,584 (Courtaulds).
92. D. K. George and W. B. Tuemmler, U.S. 3,108,101 (FMC).
93. P. A. Argabright et al, J. Org. Chem. _35_, 2253 (1970); H.
     Priebe, Plaste Kaut _13_ (4), 223 (1966).
94. P. A. Argabright et al, Fr. 1,582,370 (Marathon Oil); CA _73_,
     35412 (1970).
95. K. K. Khomenkova et al, CA _77_, 88885 and 88907 (1972);
     K. Alaminov and N. Andonova, CA _77_, 5824 (1972).
96. The general method of U.S. 2,361,823 is said to be applicable.
97. D. W. Kaiser, U.S. 2,567,847 (Am. Cyan.); cf. B. Bann,
     Brit. 837,167 (Brit. Oxygen).
98. R. Lindenfelser and M. K. Kilthau, Ger. 1,016,928 (Ciba).
99. M. K. Layman, U.S. 2,731,438 (Am. Cyan.).
100. G. Widmer and W. Fisch, U.S. 2,197,357 and 2,448,338 (Ciba).
101. J. R. Dudley and P. Adams, U.S. 2,829,119 (Am. Cyan.).
102. D. W. Kaiser, U.S. 2,567,847 (Am. Cyan.); G. Widmer and
     W. Fisch, U.S. 2,310,004 and 2,328,592-3 (Ciba).
103. R. G. Flowers and P. W. Juneau, U.S. 3,083,119 (G.E.).
104. W. P. Ericks, U.S. 2,414,289 (Am. Cyan.).
105. T. H. Ferrigno, U.S. 3,300,326 (Mineral and Chemical).
106. W. E. Gibbs and R. L. van Deusen, J. Polym. Sci. _54_, No. 159,
     S1 (1961).
107. J. C. Petropoulos and J. K. Gillham, U.S. 3,277,065 (Formica).
108. J. E. Goldmacher and O. E. Dow, U.S. 3,627,599 (RCA).
109. W. M. Thomas, U.S. 2,712,004 (Am. Cyan.).
110. Data Sheet, American Cyanamid.
111. S. T. Moore, U.S. 3,077,432 (Am. Cyan.).
112. W. H. Schuller et al, J. Chem. Eng. Data _4_/3, 273-276 (1959).
113. J. K. Gillham and J. C. Petropoulos, J. Appl. Polym. Sci. _9_,
     2189 (1956); cf. Belg. 616,617 (Am. Cyan.); CA _58_, 6942.
114. W. M. Thomas, U.S. 2,712,004 (Am. Cyan.); cf. U.S. 2,481,758.
115. E. L. Kropa, U.S. 2,510,503 (Am. Cyan.); for monomer prepara-
     tion see Dudley et al, J. Am. Chem. Soc. _73_, 2988 (1951).
116. B. H. Clampitt and A. P. Mueller, J. Polym. Sci. _62_, 15 (1962).

25. ALLYL ACRYLIC MONOMERS

Allyl acrylate and methacrylate have proved to be the most useful of the allyl acrylic monomers. These compounds offer double bonds of different reactivity within a single molecule. Therefore they lend themselves to two-stage copolymerizations, the first forming soluble and formable products by acrylic polymerization and the second producing crosslinking through allyl groups to form thermoset or vulcanized structures. Allyl acrylic monomers are more readily prepared and purified without premature polymerization than are most diacrylic monomers. The dimethacrylates and diacrylates having two double bonds of high reactivity tend to form crosslinked polymers already at low conversions. Formulas of some allyl acrylic monomers follow:

CH_3	$COOA$	CH_2COOA	CH_3
$C-COOA$	CH	$C-COOA$	$C-CONA_2$
CH_2	CH_2	CH_2	CH_2
allyl methacrylate	allyl acrylate	diallyl itaconate	N,N-diallyl-methacrylamide

The brittle insoluble homopolymers of such monomers have had little use, but as minor comonomers there are many applications in dental and optical plastics, synthetic rubber, plastisols, coatings, laminates, and in copolymer latices. Allyl acrylic esters were first employed in modifying methyl methacrylate (MMA) polymers.

Allyl methacrylate and allyl maleate were among monomers used in the early development of prepolymers or partial polymers which retain plasticity and solubility but can be crosslinked by heating with catalysts. In 1939 Pollack, Muskat, and Strain showed that polymerizations of such monomers in solution permitted practical yields of soluble prepolymers (1). Below 30% monomer concentration as much as 75% conversion to soluble polymer was possible in some cases before gelation. In general, polymerizations at higher temperatures (e.g., above 100°C) gave higher yields of prepolymer than slower polymerization at 60°C.

Higher concentrations of peroxide catalyst and other conditions favoring lower molecular weight also favored
yields of soluble prepolymer. In one patent example,
ethyl allyl maleate with 1% benzoyl peroxide was heated
at 130°C for 15 min to form a syrup and then was quickly
cooled to arrest polymerization. The soluble polymer was
precipitated by adding methanol. The dried prepolymer
with 10% dibutyl phthalate plasticizer was heated at 150°C
for 3 hr to give a hard clear molding. Tawney polymerized
allyl methacrylate in the presence of methallyl chloride
to facilitate forming soluble polymers, which could be
cured by heating with peroxide and styrene (2).

ALLYL METHACRYLATE

Allyl methacrylate (AMA) can be prepared best by alcoholysis of MMA by allyl alcohol in the presence of NaOCH$_3$
and a polymerization inhibitor (3). Pyrolysis of the
acetoxy derivative of allyl-alpha-hydroxyisobutyrate also
can be used (4). AMA can be made by passing allyl chloride into aqueous sodium methacrylate at 85°C in the
presence of an arylamine inhibitor and an emulsifying
agent (5). It can also be prepared from methacrylic acid
and allyl alcohol in the presence of benzene, hydroquinone, and p-toluene sulfonic acid held at reflux for 6 hr
(6). Methallyl methacrylate must be prepared in alkaline
systems to prevent the rearrangement of methallyl alcohol
to isobutyraldehyde (7). AMA has moderate toxicity to
rats (LD 0.43 g/kg oral). It is supplied with as much as
60 ppm hydroquinone inhibitor against polymerization on
storage. Shrinkage on homopolymerization is about 23%.
The monomer is supplied by Sartomer Industries, Alcolac
Chemical Corporation, and others, who provide technical
literature. The IR absorption spectra of allyl acrylate
and allyl methacrylate, as well as a number of allyl
ethers, were reported (8). See also Fig. 25.1.

Allyl methacrylate has found use in the preparation of
methacrylate silane derivatives such as trimethoxypropyl
methacrylato silane, a coupling agent for bonding glass
fibers and silicate fillers to unsaturated polyester
laminates and to synthetic rubbers. Water resistance
and strength properties of the reinforced structures are
improved thereby. The silane derivative adds readily to
allyl double bonds in the presence of noble metal halide
catalysts, whereas the methacrylate double bonds bearing
the electron-attracting COOCH$_3$ group does not polymerize
or add to SiH under the conditions chosen (9):

$$(RO)_3 SiH + CH_2=CHCH_2 OOCC=CH_2 \longrightarrow (RO)_3 SiCH_2 CH_2 CH_2 OOCC=CH_2$$
$$\qquad\qquad\qquad CH_3 \qquad\qquad\qquad\qquad\qquad CH_3$$

The methacrylate double bonds can copolymerize with unsaturated polyester or with double bonds of reactive
rubbers; the SiOR linkages may react with the hydroxyl
groups of the glass surfaces during curing thus promoting
adhesion.

ALLYL METHACRYLATE POLYMERIZATION

Blout and co-workers of Polaroid found allyl methacrylate photopolymerization to be unique in giving prepolymer
syrups free of crosslinking (10). When homopolymerized
by heating with benzoyl peroxide, the monomer set to a gel
at conversion of only a few percent. However, irradiation
with an ultraviolet lamp gave 25% conversion to uncrosslinked syrup in 10 hr. Irradiation with 0.3% BP or 0.1%
biacetyl as sensitizers gave syrups in 2 hr. In contrast
to allyl methacrylate the photopolymerization of dimethacrylates produced gelation at low conversions. Other
sensitized photopolymerizations of allyl methacrylate at
25°C gave syrups of 19 to 39% soluble polymer before
gelation began $(11)_0$. In the presence of biacetyl (ab-
sorbing 4000-4600 Å) the conversion at the gel point was
increased as the temperature was lowered and as the light
intensity was increased. The soluble polymer prepared at
75°C had intrinsic viscosities from 3 to 20 times those
of polymer prepared at 25°C. In thermal polymerizations
the reactivity of the allyl group was only about 10^{-2}
that of the vinyl group at 45% conversion, but 4×10^{-2}
at 95% conversion (12).

Copolymerization of 32 parts n-butyl methacrylate with
8 parts allyl methacrylate using 1% azobis in 500 parts
of xylene at 80°C formed 84% copolymer in 2.5 hr without
gelation (13). Similar conditions but with benzoyl peroxide gave 57% conversion before gelation. Photopolymerization of allyl methacrylate using 3650 Å UV radiation
and azobis as sensitizer also largely avoided reaction of
allyl groups (14). As shown in Fig. 25.1, Higgins and
Weale reported evidence of cyclization in the polymerization of allyl methacrylate in solution. Surprisingly
radical polymerizations in diallyl ether and in CCl_4 were
believed to give largely unmodified AMA homopolymer
(Fig. 25.2).

Early research on copolymers of AMA and other bifunctional monomers with larger proportions of MMA and/or
styrene involved efforts to obtain clear, homogeneous
sheets with improved hardness, scratch resistance, and
solvent resistance; numerous difficulties were encountered (15). The onset of gelation often prevented uniform
contraction during polymerization, resulting in bubbles,
voids, and surface defects. The initial mobile monomer
mixtures tended to leak from casting cells more than did

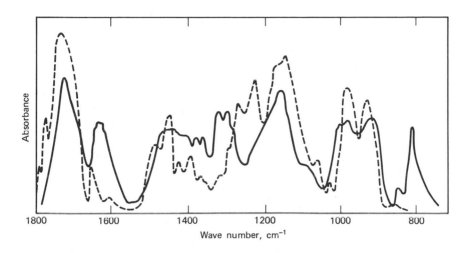

Fig. 25.1. Infrared spectra of allyl methacrylate (solid curve) and soluble poly(allyl methacrylate) (dashed curve) at 800 to 1800 cm^{-1}. In the spectrum of the monomer, the carbonyl stretch frequency absorption exhibits a peak at 1725 cm^{-1}, but in the polymer this peak is broadened and shifted toward 1740^{-1}. This suggested the presence of delta-lactone units (six-membered rings) in the polymer. J. P. J. Higgins and K. E. Weale, J. Polym. Sci., A 1, 6, 3011 (1968).

viscous casting syrups of MMA monomer-polymer. Copolymers with styrene were studied by General Electric and by Norton (16).

There have been numerous attempts to utilize AMA or its prepolymers for forming scratch-resistant surface coatings on Plexiglas or other soft polymer plastics (17). The desired large sheets of optical perfection are very difficult to achieve. Recently it was suggested to over-coat allyl methacrylate polymer coatings by ethylene dimethacrylate-unsaturated polyester before curing (18) and to graft copolymerize to the surface using high-energy radiation (19). From 5 to 15% AMA or dimethacrylate mixed with unsaturated polyesters such as triethylene glycol fumarate have been applied to improve hardness and mar resistance of plastics (20). Similar mixtures have served as binders of ceramic materials (21). AMA pre-polymer with benzoin as photosensitizer has been evaluated in lithographic printing (22). Diallyl succinate and allyl cinnamate polymers also were tested.

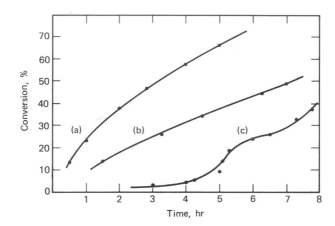

Fig. 25.2. Polymerization of allyl methacrylate in three solvents
at 60°C: curve a, dioxane; curve b, diallylether; curve c, carbon
tetrachloride. BP = 1.1×10^{-2} mole/liter; M = 1 mole/liter. The
soluble polymer formed was believed by the investigators to be substantially unmodified homopolymer. J. P. J. Higgins and K. E.
Weale, J. Polym. Sci. A 1 , 6, 3011 (1968).

   In the development of plastic lenses such as eye
glasses, allyl methacrylate has been copolymerized with
diallyl diglycol carbonate, with MMA, and with styrene.
For ophthalmic lenses 25% AMA with 75% MMA was suggested
for casting (23). Viscous prepolymer syrup containing
25% polymer has been cast to form lenses (24). Homopolymers of AMA have been little studied. However,
Bristow investigated the polymer swelling properties (25).
AMA prepolymers have been cast in cells to form sheets
(26). Allyl methacrylate (5-10%) may be added initially
to MMA monomer or to preformed syrup in preparing thermosetting glass-fiber-reinforced moldings (27).
   Soon after the first development of methyl methacrylate
and ethyl methacrylate polymers and their application as
dental plastics (28), small additions of bifunctional
crosslinking monomers such as allyl methacrylate were
found to impart impart improved dimensional stability and
craze and stain resistance. Most dental plastics, including dentures, teeth, and fillings, have been formed from
slurries or doughs made by mixing monomer with fine polymer. The fine polymer may contain small amounts of allyl
methacrylate units to control swelling in monomer, whereas

the monomer portion may contain some allyl methacrylate
to give a crosslinked polymer network on curing. For
acrylic fillings the monomer portion usually contains in
solution an activator such as N,N-dimethyl-p-toluidine
for "triggering" the peroxide initiator (29). The latter
is present as residual catalyst in the fine polymer. Because of the lower reactivity of the allyl group, the
crosslinking will not occur prematurely but only in the
later stages of the polymerization at elevated temperatures. Adding 5 to 10% AMA or glycol dimethacrylate in
the monomer prolongs the time in the formable dough state
and modifies the optimum ratio of fine polymer to monomer
from the usual 2.5 : 1 ratio (30). The cured dental copolymers may show improved tensile strength, less swelling
in water and alcohol, and better recovery from deformation
than uncrosslinked MMA polymers.

Crosslinking agents such as allyl methacrylate or tetraethylene glycol dimethacrylate may be added to epoxy-
methacrylate dental fillings, based upon addition products
of bisphenol A and glycidyl methacrylate (31). In dental
shapes the proportion and type of crosslinker can be
varied from one part to another of the restoration to
give optimum properties. Molds for cast shapes of MMA
polymer were lined with polyfunctional allyl methacrylate
prepolymer so that the surface was more highly crosslinked
and hardened (32). Minor proportions of AMA have been
copolymerized with ethyl acrylate and MMA using sulfoxy-
late-activated hydroperoxide emulsion with the surfactant
polyoxyethylene lauryl ether (33). Films were cured by
baking at 150 to 200°C. Ethyl acrylate was copolymerized
with AMA within wool fibers (34). Allyl methacrylate-MMA
precopolymers have been used to loosely crosslink methacrylic acid polymer (35). These products,capable of
high swelling in neutral or alkaline aqueous solutions,
are useful as bulk laxatives.

Allyl acrylate and methacrylate have found use as comonomers in the preparation of butadiene-acrylate ester
copolymer rubbers with pendant unsaturation (36). The
allylic double bonds promote cure as well as serving to
graft the rubber in polyblend systems.

A data sheet of Alcolac Chemical Corporation discussed
"case-hardened" plastic castings made by lining a mold
with a prepolymer syrup of AMA containing 25 to 30%
polymer. Polymerization of the syrup on the mold walls
is advanced by heating to a rubbery state. A second
monomer, e.g., MMA, or syrup, is added to fill the mold
and then both polymerizations are completed by heating.
The resulting surfaces are scratch resistant yet the
castings are less brittle than homogeneous cast copolymers. Application is made to plastic lenses.

Water-soluble finishes for textiles can be prepared by partial saponification of terpolymers of acrylonitrile, MMA, and AMA. Vulcanization of acrylic rubbers is improved by adding allyl methacrylate. Copolymers of AMA and diallyl diglycol carbonate together with vinyl poly-siloxanes are useful for adhesives for glass, metals, and acrylic plastics. In the curing of AMA-MMA copolymers with ultraviolet or high energy radiation, allyl radicals formed by attack of methacrylate units may participate (37).

Vinyl chloride was copolymerized with minor proportions of styrene, AMA, and acrylonitrile in ketone solvent for preparing low molecular weight polymers capable of later crosslinking (38). The free radical copolymerization required 75°C and above, higher temperatures than are normally used with vinyl chloride. Copolymers of vinyli-dene chloride with 2% allyl methacrylate were insoluble in o-dichlorobenzene, a solvent for the homopolymer (39). Copolymers of acrylonitrile-MMA-allyl methacrylate (1:69:30) were saponified by aqueous alkali by heating under pressure to give novel water-soluble copolymers (40). In the cast copolymerization of acrylonitrile with minor proportions of methacrylamide and methyl methacrylate, adding several percent of maleic anhydride reduces yellow discoloration; cloudy sheets may result, however. Addi-tion of allyl methacrylate to the monomer mixture was found to reduce cloudiness (41). A copolymer syrup has been made by partial conversion of a mixture of 63 parts acrylonitrile, 32 parts MMA and 5 parts methacrylamide by heating at reflux for 2 hr with 0.05% azobis (42). To the syrup was added 0.5% AMA and a cast sheet was made in a cell mold constructed from two polished glass plates spaced 0.25 in. apart by gaskets. Whether soluble poly-mers from allyl methacrylate and N-allylacrylamide contain cyclic units seems open to question (43). Tacticity of polyallyl methacrylate prepared by anionic polymerization has been studied (44).

After the pioneer work of Pinner and others in cross-linking of preformed polymers by addition of polyfunc-tional allyl compounds, numerous publications appeared. With ^{60}Co radiation, polyethylene and polypropylenes were more efficiently crosslinked when swollen by 5% allyl methacrylate (45). Polyisobutylene is only degraded by high-energy radiation but with 15% allyl acrylate 60% was crosslinked after 1 Mrad. A favorable mixture for cross-linking polyvinyl alcohol by ^{60}Co radiation was methanol-water-AMA (46). Water acted as swelling agent to promote graft copolymerization, and methanol apparently made the mixture more compatible (cf. Fig. 25.3).

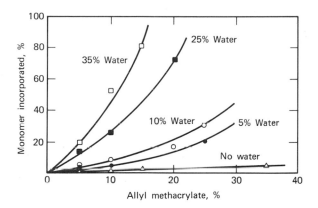

Fig. 25.3.   The grafting of allyl methacrylate upon polyvinyl
alcohol was greatly promoted by water in the presence of radiation
from ^{60}Co (0.7 megarad at 0.04 megarad/hr).   The PVA was Elvanol
72-60 (99% hydrolyzed).   B. S. Bernstein, G. Orban, and G. Odian,
Polym. Preprints 5, 894 (September 1964).

Compared with allyl methacrylate, methallyl methacrylate
has been little used.   It can be prepared by alkali-
catalyzed alcoholysis of MMA by methallyl alcohol in the
presence of a polymerization inhibitor (47).   Methallyl
methacrylate also has been prepared from methacrolein (48).
Methallyl methacrylate was reported to be twice as reactive as allyl methacrylate in crosslinking (49).   Chloroallyl methacrylate was intermediate in reactivity.   From
0.9 to 5.0% methallyl methacrylate was added to MMA
monomer used in monomer-polymer molding of dental plastics (50).
Soluble prepolymers of allyl methacrylate can be prepared by anionic polymerization by use of anhydrous basic
catalysts.   Thus the monomer was added slowly to an
agitated dispersion of finely divided sodium in benzene
under nitrogen at 25°C (51).   After 22 hr, glacial acetic
acid was added to react with remaining metallic sodium.
Evaporation of benzene gave a white polymer of 0.32
intrinsic viscosity.   Allyl acrylate and AMA both showed
reluctance to copolymerize with styrene under anionic
initiation with lithium n-butyl or sodium naphthalene
(52).   However, they did copolymerize with MMA and with
acrylonitrile.

Polymerization of allyl methacrylate with metallic
lithium catalyst in tetrahydrofuran formed soluble polymers bearing free allyl groups (53). Studies were made
of the reactivities of the latter to peroxides and sulfur
for crosslinking as well as in copolymerization. Properties and other references to allylic acrylate and methacrylate compounds are given in Table 25.1.

TABLE 25.1

Some Allylic Acrylic Monomers

Ester Monomer	bp, °C	Refractive index and other properties	References
Allyl methacrylate	43(15)	$n_D^{20}=1.4365$; $d_{20}=0.935$; viscosity, 0.9 cps	Borden, Sartomer, and Alcolac data sheets
Methallyl methacrylate	80(50)	$n_D^{25}=1.4365$	U.S. 2,562,849 (Shell); U.S. 3,002,004 (R & H)
Allyl acrylate	47(40)	$n_D^{20}=1.4320$	Fisher, U.S. 2,456,647 (U.S.); Weiss, Bull. Soc. Chim. (France) (5), 1358 (1965)
Methallyl acrylate	72(50)	$n_D^{20}=1.4372$	Fisher, J. Org. Chem. 12, 227 (1947); U.S. 2,456,647 (U.S.)
Chloroallyl acrylate	74(29)	$n_D^{20}=1.4600$	Fisher references
1-Methallyl acrylate	71(97)	$n_D^{20}=1.4283$	Fisher references
Allyl 2-chloro-acrylate	70(24)	$n_D^{20}=1.4585$	U.S. 2,434,229 (GAF); U.S. 3,446,114 (U.S. Rubber)
2-Allylphenyl acrylate	90(0.9)	$n_D^{20}=1.5225$	Solomon, J. Polym. Sci. A 1, 9, 509 (1971)
2-Allylphenyl methacrylate	102(1.5)	$n_D^{20}=1.5205$	Solomon
Allyl crotonate	89(70)	$n_D^{20}=1.4465$	CZ II, 21994 (1941)
Allyl quaternary ammonium meth-acrylates	--	Polymerized by aqueous persulfate	Batty, Brit. 784,051 (ICI)
Allyl cellosolve methacrylate	--	Polyester copolymers	Sakurada, CA 59, 7756 (1964)
2-Allyloxyethyl methacrylate	--	Polyester copolymers	Collardeau, CA 58, 1548 (1963)
Methacrylate of allyl lactate and glycolate	--	Crosslinked polymers	Filachione, JACS 70, 526 (1948)

ALLYL ACRYLATE POLYMERIZATIONS

The odor of allyl acrylate is bad. Nevertheless the
monomer continues to have applications because of its
uniquely high reactivity. It can be made by alcoholysis
of methyl acrylate. Rehberg and Fisher reacted 3 moles
of methyl acrylate with 2 moles of allyl alcohol in the
presence of 10 g of hydroquinone and 2 g toluene sulfonic
acid in a still pot (54). When the stillhead temperature
at total reflux dropped to 63°C, the azeotrope of
methanol-methyl acrylate was slowly distilled off as
formed. The excess methyl acrylate was then distilled,
followed by the crude allyl acrylate at 120 to 195°C; the
latter was then redistilled at 47°C and 40 mm. When
warmed with peroxide or irradiated with UV light, hard,
brittle, insoluble crosslinked homopolymers were formed.
Methallyl acrylate was prepared in a similar way and was
found to homopolymerize somewhat less readily to crosslinked homopolymers. As little as 0.1% added methallyl
acrylate produced gelation of methyl acrylate. Rehberg
and co-workers also prepared allyl acrylate by pyrolysis
at 550°C of allyl-2-acetoxpropionate (55). Syntheses of
allyl and methallyl acrylates by modified Reppe synthesis
using acetylene, carbon monoxide, nickel carbonyl, and
the alcohols have been patented (56). The odor of methallyl acrylate is surprisingly mild.

The homopolymerization of allyl acrylate is difficult
to prevent by inhibitors during storage. The high reactivity of allyl acrylates makes the preparation of soluble
prepolymers difficult. A PPG patent suggests the possibility of performing an ester interchange between allyl
acrylate and another preformed linear polymer to prepare
fusible allyl acrylate polymers (57). The latter could
be crosslinked on heating at 60 to 125°C with peroxide
catalysts. Soluble prepolymers of styrene, allyl acrylate, or methacrylate and allyl alcohol were prepared by
heating with peroxide initiator (58). The soluble copolymers could be compounded with diallyl adipate or MMA
and molded to thermoset shapes.

Butler and Barnett from solution obtained soluble allyl
acrylate homopolymers below 10% conversion which were
believed to contain about 25% cyclic units (59). Schulz
and co-workers also made soluble prepolymers of allyl
acrylate, and some were believed to contain cyclic
structures (60). Polymerization in dilute benzene solution using 0.2% azobis to 60% conversion produced brittle,
glassy polymers soluble in aromatic hydrocarbons and
recovered by precipitation by methanol. This polymer did
not fuse on heating but charred above 320°C. Assuming no
branching, such soluble polymers might contain 50% cyclic

lactone groups. Polymerizations at higher concentrations gave gelation at low conversions and polymer products containing more free double bonds. In actual fact, acrylate polymers may be considerably branched and yet retain solubility.

Allyl acrylate has been used as minor comonomer in the manufacture of vulcanizable butadiene and acrylic rubbers. Acrylate ester-allyl acrylate copolymer rubbers made using t-amine-modified organoaluminum catalysts could be cured by sulfur (61).

Copolymerizations of allyl acrylate and related monomers have been recorded:

Comonomers	References
With vinyl chloride	Belg. 531,404 (Solvay); CA 53, 7673 (1959)
With alkyl acrylate and itaconic acid	Cruz, U.S. 3,545,972 (Du Pont)
With acrylic acid	D'Alelio, U.S. 3,364,282 (DalMon)
With acrylamide and derivatives	Brit. 826,652-3 (PPG)
With vinyl acetate	Sayadyan, CA 76, 46553 (1972)
With styrene	Roedel, U.S. 2,572,951 (Du Pont)

Allyl acrylate is highly reactive as a crosslinker of preformed polymers. Caldwell and co-workers reacted cellulose in acetone solution with methacrylamide, allyl acrylate, and benzoyl peroxide at 55°C for 18 hr (62). The crosslinked copolymer, in particles of about 3 microns, was used as a pigment in polyethylene terephthalate fibers to promote dye receptivity. Allyl acrylate and methacrylate in polyethylenes were more efficient than diallyl maleate in promoting crosslinking by high-energy radiation (63). As much as 20% monomer could be absorbed by polymer swelling. By using benzophenone as photosensitizer with allyl acrylate, both atactic and isotactic polypropylenes could be crosslinked by UV radiation (64). The IR spectra indicated that both double bonds were reactive in crosslinking. In applications as crosslinking agent, the less objectionable odor of allyl methacrylate is an important consideration.

Anionic homopolymerizations with certain Lewis-basic catalysts have been interesting because some of the linear polymers obtained by Donati and Farina were stereoregular and crystallizable. Allyl acrylate in toluene solution at -50°C with butyl lithium or phenyl magnesium chloride catalyst produced soluble crystalline polymers (65). By IR spectra it was determined that the polymer contained unreacted allyl groups but no acrylic unsaturation. The crystal melting range of the linear homopolymer was 86 to

90°C, and the ternary helix of oriented fibers had an
identity period of 6.5 Å. Heterogeneous saponification
of the polymer gave a water-soluble polymer salt. Allyl
acrylate was polymerized in toluene solution using butyl
lithium and 1,1-diphenyl-n-hexyl lithium as initiators
(66). The products had isotactic linear chains of relatively low molecular weight. On exposure to air the
polymers slowly became crosslinked, presumably through
reaction of allyl groups. Allyl methacrylate polymers
also were studied. Anionic homopolymerization with butyl
lithium or sodium naphthalene gave soluble polymers from
allyl acrylate and allyl methacrylate; when these catalysts were used, the monomers also copolymerized readily
with MMA and with acrylonitrile (67). These allyl
monomers would not copolymerize with styrene under anionic
conditions. Block copolymers would be prepared by reacting the allyl monomers with preformed polystyrene or polymethyl methacrylate anions.

Allyl acrylate was polymerized by an aluminum alkyl
catalyst in the absence of water, oxygen, or other substances reactive with catalyst or growing polymer chains
(68). Allyl acrylate has been copolymerized with
acrylates or vinyl esters in the presence of an aluminum
alkyl and added oxygen (69). It is probable that this
involves radical copolymerization.

Allyl acrylate or methacrylate can be copolymerized with
methyl methacrylate in benzene with sodium or potassium
catalyst or in liquid ammonia with KNH_2 (70). The soluble
copolymer was used to crosslink methacrylic acid by
polymerization in benzene solution with added benzoyl
peroxide. Allyl acrylate copolymers can be used as ion
exchangers in gastrointestinal therapy.

Methallyl acrylate has been prepared by heating methyl
acrylate and methallyl alcohol in the presence of aluminum
isopropylate and phenyl-beta-naphthylamine (71). Other
allylic acrylic esters are listed in Tables 25.1 and 25.2.

OTHER ALLYL ACRYLIC MONOMERS

N-Allyl acrylamide was made from portionwise addition
of allylamine to acrylyl chloride in carbon tetrachloride
at 0°C (72). Polymerization of the purified monomer in
toluene at 60°C for 50 min with azobis as initiator gave
crosslinked polymers, even at 6 to 8% conversion. There
was no evidence of cyclopolymerization. N-Allyl methacrylamide was also prepared, yielding crosslinked polymers
under similar conditions. More evidence of cyclopolymerization at low conversion was reported (73). N-Allyl
acrylamide has been used in small proportions in copolymerizations with vinyl chloride to increase molecular

TABLE 25.2

Other Allylic Acrylate Esters

Monomers	Refractive index and other properties	References
Allyl and methallyl chloroacrylates	--	Brit. 573,596 (Wingfoot); CA $\underline{43}$, 3027 (1949)
Bromoallyl acrylate, etc.	Nonflammable polymers	Waples, U.S. 3,316,329 (Dow)
Allyloxyethyl acrylate	Polymerization partially cyclic	Noma, Chem. High Polym. Japan $\underline{22}$, 1666 (1965)
o-Allylphenyl acrylate	$n_D^{20} = 1.5226;$ bp 78°C, 0.6 mm	Fisher, U.S. 2,477,293 (U.S.) cf. JACS $\underline{72}$, 839 (1950)
m-Allylphenyl acrylate	Prepolymers by Li butyl	Solomon, Polym. Lett. $\underline{6}$, 507 (1968)
1-Methallyl acrylate	$n_D^{20} = 1.4283;$ bp 71°C, 87 mm	Rehberg, U.S. 2,445,925 (U.S.)
Allyl β-furfuryl acrylate	Polymerized on heating	Blicke, Ber. $\underline{47}$, 1352 (1913); JACS $\underline{45}$, 1562 (1923)
Allyl cyanoacrylate (anionic poly-merization)	$n_D^{20} = 1.4586;$ bp 74°C, 4 mm	Leonard, Org. Coatings and Plastics Chem., Preprint $\underline{31}$, 231 (September 1971); (tissue adhesive and hemostatic)

weight (74). The allyl ether of N-methylol acrylamide was prepared by heating 100 parts of acrylamide, 66 parts of paraformaldehyde, 50 parts of allyl alcohol, and 2 parts of hydroquinone (polymerization inhibitor) (75). N-Allyl acrylamide has been made from acetylene, carbon monoxide, $Ni(CO)_4$, and allyl amine (76).

In a British patent it has been disclosed that the allyl ether of methylol methacrylamide copolymerizes with styrene and butyl acrylate in activated aqueous persulfate (77). Films became crosslinked at low pH values. Derivatives of copolymers of styrene with N,N-diallyl acrylamide were suggested as selective anion-exchange resins (78). Some related monomers are listed in Table 25.3. Small amounts of allyl acrylamides were copolymerized with vinyl chloride for obtaining polymers of higher molecular weights and improved strength (79). In one example using a suspension system with lauroyl peroxide as catalyst, incorporation of 0.02% N,N-diallylacrylamide was reported to raise specific viscosity of the vinyl chloride polymer from 0.51 to 0.60.

Viscous liquid low polymers were prepared from 2-vinyl-1,3-dioxolane acrylate, which is an allyl acrylic monomer

(80). Polymerization occurred near 0°C with boron fluoride etherate catalyst. Surprisingly, hydroquinone in small amounts was reported to promote the cationic polymerization. At least three types of structural units were detected in this prepolymer, resulting from 1,2-addition, acetal ring opening, and rearranged ring opening.

Allyl alpha-cyanoacrylate $CH_2=C(CN)COOA$ has special interest because of the ease with which it can polymerize through the acrylic double bond catalyzed by weak bases (cf. polymerization of methyl cyanoacrylate, Adhesive 910, by water). The allyl monomer can be prepared by reaction of formaldehyde with allyl cyanoacetate in benzene to form polymer directly (81). The latter was depolymerized by heating in the presence of phosphorus pentoxide and hydroquinone at 180°C and 6 mm. It could be redistilled at 70°C and 2 mm in a stream of sulfur dioxide as inhibitor. Strong adhesive bonds were prepared by polymerization of the monomer in thin layers. Cyanoacrylate monomers "autopolymerizable" in contact with water are finding many applications in medicine and dentistry.

Allylidene esters of acrylic and methacrylic acids (diesters from acrolein) and their polymerization have been reported by Soviet workers (82). Extreme reactivity in polymerization limits their industrial psssibilities. Alkyl esters of alpha-allyl acrylic compounds are other unusual monomers reported in Russia (83). They homopolymerized on standing. The diallyl esters of methylene malonic esters, as well as the monoallyl methylene malonic compounds, will be expected to polymerize.

DIALLYL ITACONATE POLYMERS

Diallyl itaconate (DAI) has been used in copolymers but has not achieved large production. The monomer does not homopolymerize very fast on storage, but it can be inhibited by 0.01% hydroquinone. DAI monomer $n_D^{20} = 1.4580$ was made by esterification (84). The monomer is trifunctional, the vinyl group being most reactive. The following properties were given in a data sheet of Pfizer and Company:

	$CH_2=CCOOH$ $\overset{|}{C}H_2COOA$	$CH_2=CCOOA$ $\overset{|}{C}H_2COOA$
Boiling point, °C at mm	--	130(9)
Refractive index, n_D^{20}	--	1.466
Melting point, °C	39	<20
Water solubility	slight	insoluble
Miscible solvents	acetone and methanol	acetone and hexane

Moldable thermosetting copolymers of diallyl itaconate with styrene and with MMA have been developed (85). Copolymers from equal weights of DAI and styrene had high heat-distortion temperatures but were rather brittle. A promising monomer mixture for casting contained 20% DAI along with styrene, TAC, allyl methacrylate, and MMA. Diallyl itaconate has been copolymerized with styrene and allyl methacrylate at moderate temperatures to a semi-cured state (86). Then the mass was cured in a desired shape by heating or irradiating in the presence of an organic peroxide. Soluble precopolymers from 4.5 parts styrene with 0.05 part allyl methacrylate prepared with azobis at 60°C (13 hr) were then graft copolymerized with larger proportions of vinyl acetate (87). Copolymers of ethyl methacrylate with minor proportions of DAI were made by heating with BP for 27 hr at 55°C and for 24 hr at 85°C (88).

Diallyl itaconate has been frequently copolymerized with unsaturated polyesters by heating with peroxide initiators (89). Small additions of diallyl itaconate were said to accelerate emulsion polymerization of vinyl acetate (90). Tawney studied the preparation of soluble precopolymers from diallyl itaconate with allyl or methallyl chloride (91). In one example, equal weights of DAI and methallyl chloride with 2% BP were heated at 60°C for 60 hr. The copolymer that was reprecipitated from acetone by adding hexane showed 41% yield and a composition of 18.8% methallyl chloride units and 81.2% DAI units. Soluble precopolymers were also prepared from DAI, styrene, and methallyl chloride by long heating at 60°C with 4% BP (92). Methallyl chloride copolymerized somewhat faster than allyl chloride. Slower in copolymerization were 2,3-dichloropropene and 2-(chloromethyl)allyl chloride. Additions of allyl chloride or alcohol also delayed the onset of gelation in the polymerization of diallyl citra-conate CH_3CCOOA (93). Cured coatings and plastics from
                        $HCCOOA$
such fusible copolymers have not yet attained commercial success. DAI has been added to polyvinyl chloride and to polypropylene to accelerate crosslinking under high-energy radiation (94). Excessive main-chain scission could thus be avoided.

Dimethallyl itaconate (bp 150°C at 4 mm) was prepared by alkaline alcoholysis of ethyl itaconate. After heating for 3.5 hr at 100°C with 1% BP, a soft resin was recovered. Attempts to prepare the monomer by direct esterification with sulfuric acid catalyst gave a black, viscous mass, apparently by way of rearrangement of methallyl alcohol to isobutyraldehyde.

Allylic derivatives of acrylamide and methacrylamide have received surprisingly little study in polymerizations. Some monomer properties and references are given in Table 25.3.

TABLE 25.3

Allylic Acrylamide Derivatives

Compound	bp, °C	Refractive index	References
N-allylacrylamide	108(3)	$n_D^{25} = 1.4850$	Trossarelli, Makro. Chem. 100, 147 (1967) (cyclization)
N,N-diallylacrylamide	110(3)	--	Borden supplies
N-allylmethacrylamide	140(4) or 80(2)	$n_D^{25} = 1.4817$	U.S. 2,311,548 (Du Pont); CA 37, 4502 (1943)
N,N-diallyl-methacrylamide	105(10)	--	Borden supplies
N-(allyloxymethyl)-methacrylamide	112(0.3)	$n_D^{20} = 1.4819$	Mueller, Makro. Chem. 57, 27 (1962)

References

1. M. A. Pollack, I. E. Muskat, and F. Strain, U.S. 2,273,891 (PPG); cf. U.S. 2,912,418 and 2,441,516 (U.S. Rubber).
2. P. O. Tawney, U.S. 2,643,991 (U.S. Rubber).
3. S. G. Cohen et al, J. Polym. Sci. 3, 278 (1948).
4. C. E. Rehberg et al, J. Am. Chem. Soc. 65, 1003 (1943).
5. A. G. Chenicek and M. A. Pollack, U.S. 2,275,466 (PPG).
6. W. Kawai, J. Polym. Sci. A 1, 4, 1191 (1966).
7. M. A. Pollack and A. G. Chenicek, U.S. 2,296,823 (PPG); J. E. Bludworth, U.S. 2,256,544 (Celanese); cf. U.S. 2,250,520.
8. W. H. T. Davison et al, J. Chem. Soc. 2607 (1953).
9. J. L. Speier et al, J. Am. Chem. Soc. 79, 974 (1957); E. P. Plueddemann et al, Mod. Plastics 39, 135 (August 1962); also June 1963.
10. E. R. Blout and B. E. Ostberg, J. Polym. Sci. 1, 230 (1946).
11. S. G. Cohen, B. E. Ostberg, D. B. Sparrow and E. R. Blout, J. Polym. Sci. 3, 264-82 (1948).
12. G. M. Bristow, Trans. Faraday Soc. 54, 1239 (1958).
13. Brit. 649,173 (Du Pont); cf. U.S. 2,912,418 (Peterlite); cf. Brit. 818,417.
14. H. Melville et al, J. Polym. Sci. 34, 199 (1959).
15. C. E. Barnes, U.S. 2,189,735, 2,278,637, 2,308,581 (Du Pont).
16. W. I. Patnode, U.S. 2,181,739 (G.E.); Brit. 532,022 (Norton).

17. I. E. Muskat et al, U.S. 2,320,356, 2,320,533, 2,322,310, and 2,332,461 (PPG); cf. U.S. 2,423,583 (Polaroid).
18. Netherlands 6,510,693 (Am. Cyan.) (1966).
19. Brit. 820,120 (Du Pont).
20. W. G. Deichert et al, U.S. 3,318,975 and 3,387,988 (Am. Cyan.).
21. L. Seidl, Brit. 937,619; CA 59, 15010 (1963).
22. M. N. Gilano and I. W. Martenson, U.S. 3,376,138-9.
23. J. J. Johnson and P. M. K. deGooreynd, U.S. 2,912,418 (Petrolite).
24. E. R. Blout and B. E. Ostberg, J. Polym. Sci. 1, 230 (1946).
25. G. M. Bristow, Trans. Faraday Soc. 54, 1064 and 1239 (1958).
26. H. Schreiber, Ger. 1,014,741 (R & H); CA 54, 1921 (1960).
27. S. G. Cohen et al, J. Polym. Sci. 3, 264, 269 (1948); W. H. Calkins and W. M. Edwards, Ger. 1,127,591 (Du Pont); CA 58, 630 (1964).
28. W. Bauer, Ger. 656,642 (October 1928); Ger. 652,821, 684,533, and 675,808 (R & H); Other references in C. E. Schildknecht, Vinyl and Related Polymers, Wiley, 1952.
29. E. Schnebel, a pharmacist of Darmstadt, first added Michler's ketone in 1938, the first tertiary amine activator.
30. E. M. Wolff, Austral. Dental J. 7, #6, 439 (December 1962).
31. R. L. Bowen et al, J. Am. Dental Soc. 64, 378 (March 1962) and 66, 57 (January 1963).
32. H. Schreiber, Ger. 1,014,741 (R & H); CA 54, 1921 (1960).
33. H. T. Tillson, U.S. 3,219,610.
34. W. S. Simpson, J. Appl. Polym. Sci. 15, 967 (1971).
35. J. C. H. Hwa, U.S. 2,963,453 (R & H).
36. M. Baer, U.S. 3,461,188 (Monsanto); cf. A. M. Clifford, U.S. 2,279,293 (Wingfoot).
37. A. T. Bullock et al, CA 77, 88942 (1972).
38. J. P. Bruce, U.S. 3,354,112; cf. U.S. 3,193,825 (Dow).
39. E. C. Britton, U.S. 2,160,941 (Dow).
40. K. Tessmar, Ger. 931,733 (R & H); CA 50, 12546 (1956).
41. K. Tessmar, Ger. 919,140, 963,476, and 1,031,517 (R & H); CA 53, 2694, 20914 (1959); CA 54, 13744 (1960).
42. Brit. 886,610 (Baker Chem.); CA 57, 1092 (1963).
43. L. Tossarelli et al, J. Polym. Sci. B 5, 129 (1967) and C 8, 4713 (1969); J. P. J. Higgins et al, J. Polym. Sci. 6, 3007 (1968).
44. D. M. Wiles and S. Brownstein, J. Polym. Sci. B 3, 951 (1965).
45. G. G. Odian and B. S. Bernstein, U.S. Atomic Energy Comm., CA 58, 6977 (1964); CA 61, 1963 (1964); cf. Brit. 820,120 (Du Pont).
46. A. Charlesly, Atomic Radiation and Polymers, Pergamon, 1960.
47. E. M. Beavers and J. L. O'Brien, U.S. 3,002,004 (R & H); cf. U.S. 2,562,849 (Shell).
48. H. D. Finch and E. A. Youngman, U.S. 2,998,447 (Shell).
49. G. M. Bristow, Trans. Faraday Soc. 54, 1064 and 1239 (1958).
50. M. M. Renfrew, U.S. 2,335,133 (Du Pont).
51. C. T. Walling and R. H. Snyder, U.S. 2,500,265 (U.S. Rubber).

52. G. F. D'Alelio and T. R. Hoffend, J. Polym. Sci. Al 5, 323 (1967).
53. H. Kamogawa et al, J. Polym. Sci. C #23(2) 655 (1968).
54. C. E. Rehberg and C. H. Fisher, U.S. 2,456,647 (U.S.); J. Org. Chem. 12, 226 (1947); cf. Org. Syn. Coll., Vol. III, 147 (1955).
55. C. E. Rehberg et al, J. Am. Chem. Soc. 65, 766 (1943).
56. A. Neuman et al, U.S. 2,778,848 (R & H); H. Lautenschlager and H. H. Friederich, Ger. 1,046,030 (Badische).
57. Brit. 578,266 (PPG); CA 41, 2274 (1947).
58. R. H. Snyder, U.S. 2,441,515-6 (U.S. Rubber).
59. M. D. Barnett and G. B. Butler, J. Org. Chem. 25, 309 (1960); cf. M. Raetzsch and L. Stephan, CA 76, 46545 (1972).
60. R. C. Schulz, Makromol. Chem. 44-46, 281 (1961); cf. G. B. Butler et al, J. Polym. Sci. A3, 3413 (1965).
61. R. K. Schlatzer, U.S. 3,476,722 (Goodrich).
62. J. R. Caldwell et al, U.S. 2,893,970 (Eastman).
63. Brit. 844,231 (Dow); CA 55, 5031 (1961); G. Odian and B. S. Bernstein, J. Polym. Sci. A2, 2835 (1964).
64. J. R. Hatton et al, Polymer 8, 41 (1967).
65. M. Donati and M. Farina, Makromol. Chem. 60, 233 (1963); cf. C. T. Walling and R. H. Snyder, Can. 484,688 (Dominion Rubber); CA 50, 1370 (1956).
66. S. Bywater et al, Can. J. Chem. 44, 695 (1966).
67. G. F. D'Alelio and T. R. Hoffend, J. Polym. Sci. A1, 323 (1967).
68. Brit. 884,706 (R & H); CA 58, 2517 (1963).
69. R. N. Chadha, U.S. 3,284,422 (Grace).
70. J. C. H. Hwa and O. H. Leoffler, Ger. 1,070,381 (R & H); CA 55, 15847 (1961).
71. C. E. Rehberg et al, J. Am. Chem. Soc. 66, 1723 (1944); J. Org. Chem. 12, 228 (1947).
72. W. Kawai, J. Polym. Sci. A1, 4, 1191 (1966).
73. L. Tossarelli et al, Makromol. Chem. 100, 147 (1967); J. Polym. Sci. C 8, 4713 (1969).
74. R. H. Martin, U.S. 2,996,486 (Monsanto).
75. R. Dowbenko and R. M. Christenson, Belg. 610,933 (PPG); CA 57, 16506 (1962).
76. E. H. Specht et al, U.S. 2,773,063 (R & H).
77. Brit. 947,610 (Bayer); CA 60, 13414 (1964).
78. M. Baer and M. F. Vignale, U.S. 3,079,428 (Monsanto).
79. R. H. Martin, U.S. 2,996,486 (Monsanto).
80. J. D. Nordstrom, J. Polym. Sci. A1, 7, 1349 (1969).
81. G. F. Hawkins and H. F. McCurry, U.S. 3,254,111.
82. I. A. Arbuzova et al, CA 50, 16667 (1956).
83. O. S. Stepanova, CA 58, 1341 (1963).
84. C. E. Rehberg and C. H. Fisher, U.S. 2,712,025 (USA).
85. E. E. Parker and T. L. St. Pierre, Mod. Plastics 38, 133 (June 1961).
86. Brit. 1,009,421 (PPG).

87. J. W. Breitenbach and H. Edelhauser, U.S. 3,247,174 (Kreidl).
88. G. F. D'Alelio, U.S. 2,310,731 (G.E.).
89. G. F. D'Alelio, U.S. 2,441,799, 2,279,885 (G.E.); Ger. 966,705; CA 54, 6204 (1960).
90. S. B. Luce, U.S. 2,611,762 (Swift).
91. P. O. Tawney, U.S. 2,556,989 and 2,569,959 (U.S. Rubber).
92. P. O. Tawney, U.S. 2,569,960 (U.S. Rubber).
93. P. O. Tawney, U.S. 2,649,437 (U.S. Rubber).
94. G. G. Odian et al, U. S. Atomic Energy Comm. report, CA 55, 25345 (1961).

26. DIALLYL MALEATE AND DIALLYL FUMARATE

Diallyl maleate (DAM), diallyl fumarate (DAF), and related compounds are highly reactive trifunctional monomers useful in small proportions in preparing branched and crosslinked copolymers and in accelerating rates of copolymerization and curing. Their toxicity and irritation to the skin seem to vary among different individuals. When stored, the purified monomers may undergo premature polymerization to crosslinked homopolymer gels unless stabilized by potent antioxidant inhibitors and molecular oxygen.

The chief use of diallyl maleate has been in copolymers for protective coatings (e.g., with vinyl esters, acrylic monomers, and/or drying oils). The cured copolymers have improved heat resistance and reduced sensitivity to water and organic solvents. DAM has also found use in pharmaceutical synthesis. Diallyl fumarate has been added along with diallyl phthalates in fiber-reinforced laminates. Soluble precopolymers of DAF with diallyl phthalates can be prepared which are thermosetting on molding in presence of peroxides. Diallyl maleate is discussed first.

Because of the high polymerization reactivity of diallyl maleate, the monomer is best made in the presence of efficient inhibitors and solvents. For example, maleic anhydride may be warmed with allyl alcohol in the presence of camphor sulfonic acid and bronze powder in ethylene dichloride (1). The monomer is soluble in common alcohols, ketones, and aromatic hydrocarbons, but insoluble in water and in petroleum ether. It has moderate toxicity to rats--$LD_{50} = 0.30$ g/kg compared with 3.2 g/kg for diethyl maleate. Properties of DAM and related monomers are given in Table 26.1.

Diallyl maleate reacts by addition with many compounds containing active hydrogen atoms (e.g., alcohols, amines, inorganic acids, nitroparaffins, mercaptans, and malonic ester) to give more saturated derivatives. DAM was reported to have value as an insecticide (2).

Dimethallyl maleate was prepared by passing gaseous methallyl chloride into aqueous sodium maleate at 85°C

TABLE 26.1

Diallyl Maleate, Fumarate
and Related Monomers

Compound	Properties, etc.	References
Diallyl maleate HCCOOA HCCOOA	bp 112°C(4); $n_D^{20}=1.4664$ or 1.469; viscosity = 4.3 cps (20); $d_{20}^{20}=1.0773$; surface tension, 33 dynes/cm (20°C)	FMC data sheet; U.S. 2,251,765 and 2,281,394
Dichloroallyl maleate	bp 160°C(5)	U.S. 2,127,660 (Dow)
Allyl-2-nitro- butyl maleate	bp 163°C(3)	U.S. 2,425,144 (R & H)
Diallyl fumarate HCCOOA AOOCCH	bp 140°C(3); $n_D^{25}=1.4670$; viscosity = 3.0 cps; $d^{25}=1.0516$	U.S. 2,712,025 (USA); Sartomer data sheet
Dichlorallyl fumarate	bp 165°C(5); mp 34°C	U.S. 2,448,531 (Eastman)
Dimethallyl maleate	bp 148°C(7); (co- polymers)	U.S. 2,221,663 and 2,472,661 (Du Pont); U.S. 2,296,823 (PPG)
Monomethallyl maleate	(Copolymers)	Ger. 745,031 (R & H); U.S. 2,160,940 (Dow); U.S. 2,254,384
Butyl methallyl maleate	bp 180°C(30); $n_D^{20}=1.4555$	U.S. 2,296,823 (PPG); 2,411,136 (Monsanto)
Dimethallyl fumarate	bp 144°C(7); (co- polymers)	U.S. 2,296,823 (PPG); U.S. 2,221,663 and 2,346,612 (Du Pont)
Bis(2-allyloxy- ethyl) fumarate	(Copolymers)	U.S. 2,594,825 and 2,636,874 (U.S. Rubber)

in the presence of an emulsifier (3); in another method,
methallyl alcohol was reacted with diethyl maleate in the
presence of sodium ethylate and p-phenylene diamine (4).
   Diallyl maleate is one of the fastest curing of commercial ethylenic monomers. Once polymerization of the
trifunctional monomer has begun, it can occur very rapidly,
producing highly crosslinked polymer masses (Fig. 26.1).
Impurities in samples of DAM may act as regulators. On
storage, commercial DAM can form crumbly, soft crosslinked
polymers that only slowly become hard and brittle. The
commercial monomer may not polymerize actively when heated
in the presence of inhibiting oxygen until temperatures of

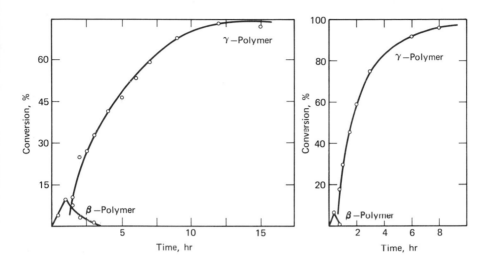

Fig. 26.1. Polymerization of diallyl maleate on heating at 70°C,
with 1% BP catalyst gives insoluble gel, or gamma-polymer already
at low conversions. Syrup and prepolymer are not readily prepared
without added regulators. Alex Schulthess, Promotionsarbeit,
Zurich ETH, 1960.

Fig. 26.2. Polymerization of a sample of diallyl fumarate on
heating at 70°C with 1% BP occurred faster than that of the diallyl
maleate in Fig. 26.1 to form crosslinked polymers. Alex Schulthess,
Promotionsarbeit, Zurich ETH, 1960.

150°C or above are reached. When the compound is heated
at 50 to 70°C with peroxide catalyst such as BP and in the
absence of molecular oxygen, homopolymerization proceeds
rapidly. Gelation of FMC monomer with 1% BP occurred in
3 to 5 min at 100°C. The polymerization can be arrested
at the soluble prepolymer stage (5) but not as readily as
in the case of diallyl phthalates. Aldehydes or other
regulating agents may be added to raise the yield of prepolymer (6). Dimethallyl maleate also can form soluble
prepolymers (7). One commercial sample of DAM became
highly viscous but free of gel during storage for 4 years.
Prepolymer could be precipitated by adding methanol.
    Kropa heated DAM without catalyst to 143°C during 43 min
when exothermic polymerization occurred with rapid rise of
temperature to 163°C (8). The solution was then cooled
and vacuum distilled to form a viscous solution of pre-

polymer that could be copolymerized with unsaturated
polyester (ethylene glycol maleate) by heating with 0.5%
BP. Although most of the examples of this patent show
applications of polyester copolymers in coatings, reinforced thermoset moldings also were disclosed. Thermoset
coatings, sheets, and moldings of DAM homopolymer have
been disclosed in patents (9), but these products are too
brittle for practical use (even with reinforcement) because of their tight crosslinked structures. Soluble
homopolymers of dimethallyl maleate of comparatively low
molecular weight were believed to contain lactone rings
(10). DAM has been used commercially as a crosslinking
and branching minor comonomer with vinyl ester, styrene,
and acrylic monomers. In patents DAM often has been disclosed as one of a list of possible polyfunctional comonomers (11).

As minor crosslinking comonomers, DAM and DAF have found
application in latices made by emulsion copolymerization
and in casting resin syrups for thermoset fiber-reinforced
structures. Minor proportions of DAM have been used for
manufacture of branched and crosslinked vinyl acetate
copolymer latices for use as adhesives and water paints
(12). Low proportions of DAM may raise rates and molecular weights obtainable under given conditions of polymerization without causing polymer insolubility (13).
Thus copolymers of vinyl chloride-vinyl acetate with up
to 0.1% DAM have been patented (14). In latices, crosslinking does not prevent deposition of films and impregnation of fibrous materials. Controlled polymer branching
and crosslinking frequently improve physical properties in
coatings and in adhesives.

Fisher and co-workers found DAM to be an effective comonomer with ethyl acrylate and acrylonitrile for preparing vulcanizable acrylic ester rubbers (15). This group's
pioneer work on emulsion copolymerization with peroxide
and persulfate initiators, along with mercaptans, contributed substantially to the development of acrylic rubbers
and coatings. As little as 1% added DAM permitted vulcanization with sulfur and accelerators. Vulcanizates
prepared from 91% ethyl acrylate, 6% acrylonitrile, and
3% DAM were resistant to solvents and to aqueous solutions.
Copolymers obtained from monomer mixtures containing 10%
or more DAM were hard and inelastic and could not be
milled.

Styrene-DAM crosslinked polymers were studied by Dow
(16), but there has been little application. D'Alelio
prepared a precopolymer by heating at reflux 83 g styrene,
157 g DAM, 2.4 g BP, and 240 g toluene (17). The cold
viscous solution was poured into excess ethanol to precipitate the prepolymer which, after drying, was a mold-

able, free-flowing white powder. Soluble precopolymers
of dimethallyl maleate and styrene were evaluated in
baking enamels (18). In one example, the following mixture was heated at 140°C:

Dimethallyl maleate	179 parts
Dimethallyl adipate	179
Styrene	83
Xylene	294
Benzoyl peroxide	2.2

When the desired viscosity range was reached, 0.04 part
hydroquinone was added and the batch was cooled. Coatings
on steel (baked 15-30 min at 400°F) were resistant to
alkaline solutions and to outdoor exposure.

Thermosetting copolymers of styrene, acrylonitrile, and
DAM were developed as electrical insulating materials (19).
DAM and monoallyl maleate may be formed and copolymerized
in one process from heating maleic anhydride, allyl alcohol, and styrene with BP in acetone (20). Popcorn or
crosslinked growths of styrene-DAM copolymer formed in
bulk at 80°C during 1 to 3 days (21). Graft copolymers
of n-butyl acrylate and MMA upon styrene-DAM copolymer
made in emulsion were then blended with vinyl chloride
polymers for improving impact resistance (22).

Maleic anhydride and maleate esters do not copolymerize
readily with methyl methacrylate or methacrylic acid, although copolymers have been patented (23). However, the
more Lewis-basic monomer 2-allyloxyethyl methyl maleate
copolymerized in bulk with MMA using 0.7% BP at 66°C for
40 hr followed by 2 hr at 130°C (24). A crackfree plastic
sheet of good abrasion resistance was obtained.

Early research by Kropa on the copolymerization of unsaturated polyesters with DAM used relatively large proportions of the latter (25). Recent commercial applications add smaller amounts; for example, 8% DAM along with
other comonomers (26). DAM, like diallyl fumarate, has
been copolymerized with oil-modified alkyds in coatings
(27). Coatings for food cans have been developed from
unsaturated polyesters and DAM (28). Other copolymerizations are listed in Table 26.2.

Recently sulfur-containing polyesters were made by
hydrogen migration or polymerization of dithiols with DAM
(29). Both 1,2- and 1,3-dithiols were useful. The polyester products could be crosslinked by ionic addition of
dithiol to maleate double bonds or by radical-initiated
copolymerization with styrene.

DAM was found to crosslink preformed polymers such as
polyvinyl acetate (30), polyvinyl chloride (31), cellulose acetate sorbate (32), polyethylene and polypropylene

TABLE 26.2

Other Copolymerizations of Diallyl Maleate

Compounds	Patents
Styrene, etc. (ion exchange resins)	D'Alelio, U.S. 2,340,110 (G.E.); cf. U.S. 2,403,213
Styrene and maleic anhydride	Seymour, U.S. 2,647,886
Vinyl chloride	Hopff, U.S. 2,187,817 (I.G.); Martin, U.S. 2,898,244 (Monsanto); Wolf, U.S. 2,608,549 (Goodrich)
Vinylidene chloride	Britton, U.S. 2,160,940 (Dow); Matsuoka, CA 53, 8715 (1959)
Chloroprene, etc.	Mighton, U.S. 2,392,756 (Du Pont)
Isobutylene	Rust, U.S. 2,459,501
Vinyl acetate	Fordyce, U.S. 2,444,817 (Monsanto)
Dialkyl fumarates and maleates	D'Alelio, U.S. 2,431,374 (Prophylactic)
Trifluoroethylene	Landrum, U.S. 2,951,783 (3M)
Unsaturated polyesters	D'Alelio, U.S. 2,441,799 and 2,305,224 (G.E.)
Drying oils	Sorenson, U.S. 2,280,862; cf. U.S. 2,343,483
Resin esters	Rust, U.S. 2,398,668-9 (Montclair Res.)
N-Vinyl pyrrolidone	Forchielli, U.S. 2,831,836 (GAF)

(33), and diallyl isophthalate prepolymer (34). Under influence of heat or high-energy radiation, however, co-monomers other than DAM, where the ethylenic groups are further separated from each other, often have proved to be better for curing polymers with addition of peroxides. DAM dissolved in atactic polypropylene was polymerized by heating with BP (35).

Most of the copolymerizations of DAM listed in Table 26.2 were in radical-initiated systems. Ionic polymerizations have been little used. However, Grignard reagents such as ethyl magnesium bromide can form colored viscous liquid prepolymer syrup (36). Dimethallyl maleate reacted more rapidly to give soluble polymers or crosslinked gels.

OTHER MALEIC MONOMERS

Among related polyfunctional allyl monomers are bis-(allyl lactate) maleate (37), diallyl chloromaleate (38), and di-2-chloroallyl maleate (39). Allyl vinyl maleate has been prepared by heating monoallyl maleate with vinyl acetate in the presence of mercuric acetate and sulfuric acid (40). Mixed esters of methallyl and methyl butenyl maleate and fumarate gave soluble polymers of low

molecular weight; these were believed to contain cyclic
units (41).

Monoallyl alkyl or hydrogen maleates and related monomers have better possibilities for control of branching
and crosslinking but relatively little work has been
reported. Rust copolymerized monoallyl maleates with
drying oils, styrene, vinyl acetate, and acrylic monomers
(42). Butyl methallyl maleate was polymerized by heating
with 1% benzoyl peroxide (43). Cyclohexyl allyl maleate
formed polymers of relatively high refractive index
($n_D^{25} = 1.492$), and for that reason copolymers were suggested as components of wax polishes (44). Monoallyl
esters of butenedioic acids have been esterified by
hydroxy-terminated polyesters (45). Soluble prepolymers
from monoallyl maleates have been reported from Russia
from bulk polymerization with BP at 60°C (46). Estimates
of polymer unsaturation suggested about 60% cyclic chain
segments. Monoallyl glycol maleate derivatives were
copolymerized with styrene in acetone solution using
benzoyl peroxide catalyst at 60°C (47). Allyl glycol
maleate and fumarate were studied (48).

A number of nitrogen-containing monoallyl derivatives
of maleic and fumaric acid have been investigated. For
example, allyl 2-nitrobutyl maleate, a pale yellow oil,
was heated with 2% BP at 100°C to give a transparent
solid polymer (49). This monomer was distilled near 160°C
at about 2 mm. Monoallyl maleamate derivatives such as
$NH_2CONHCOCH=CHCOOA$ have been copolymerized with vinyl-type
monomers (50).

N-allyloxymethyl maleimide was prepared (51). The monomer, distilled at 80°C at low pressure, was polymerized
by heating at 50°C for 16 hr with 1% BP. Copolymers with
unsaturated polyesters and with triallyl cyanurate were
promising in glass cloth laminates. The N-allyl monoamide
of maleic acid (mp 97°C) was polymerized in the solid
state using high-energy radiation (52). Low yields of
polymer were obtained. Polymerization continued after
irradiation was terminated. Synthesis and polymerization
of N-allylmaleimide was studied at the University of

Akron (53). This monomer CH-C⟨O / NA / CH-C⟨O was made by heating

N-allylmaleamic acid at 130°C. From free radical polymerizations in tetrahydrofuran at 55°C for 200 hr the
products were highest in cyclic units in the main chain
when initial concentrations of monomer were low.

DIALLYL FUMARATE

Because of its great tendency to polymerize, diallyl fumarate (DAF) usually has been prepared in solution in the presence of special antioxidant stabilizers. Fumaric acid and allyl alcohol can be reacted in benzene in the presence of p-toluene sulfonic acid or allyl alcohol with fumaryl chloride (54). High temperatures must be avoided. The toxicity, irritating odor, and lachrymating action of DAF are limitations to use. The compound also can produce skin rash. The monomer supplied by Sartomer contains 60 ppm hydroquinone as inhibitor. DAF is insoluble in water but soluble in common organic solvents, including hydro-carbons. Shrinkage in homopolymerization is about 20%. Crosslinked polymer gels are formed rapidly at even low conversions on heating with radical-forming initiators (Fig. 26.2 on page 639).

Dimethallyl fumarate has been prepared by reaction of methallyl chloride with aqueous sodium fumarate solution at 80 to 100°C (55). Rothrock in 1937 prepared dimeth-allyl fumarate by alcoholysis of dimethyl fumarate in the presence of sodium methylate (56). He observed that heating purified DAF with BP formed rubbery insoluble polymer at once. However, heating monomer that contained high-boiling impurities gave first a viscous syrup. After addition of peroxide catalyst, the latter could be cured at 100°C to form hard coatings. This was one of the early two-stage polymerizations of a synthetic monomer, a fea-ture widely explored in subsequent years. Linseed oil, of course, forms viscous liquid prepolymer solutions, but these prepolymers are not readily isolated and quickly cured.

Diallyl fumarate polymerizes very easily to insoluble polymers, as does dimethallyl fumarate (57). Photopoly-merization occurs readily in UV light, especially with 5% BP. Rates of homopolymerization of DAF and other diallyl esters at 60°C with BP catalyst were reported (58). On storage, uninhibited DAF gives a rubbery, crumbly gel that becomes hard on heating. In 1937 Bradley suggested application of polymer syrup in coatings (59). He noted that the polymers were not readily saponified by alkali. By addition of chloroalkenes such as methallyl chloride, polymerization of DAF with BP at 88°C can give soluble polymers (60). DAF catalyzed by BP showed two exotherms, the second near 158°C (61).

DAF serves as a modifying comonomer in the preparation of crosslinked vinyl acetate polymer latex adhesives and as an accelerating comonomer with diallyl phthalate in potting compositions for electrical resistors. DAF is often used when delayed cure or crosslinking is desired

because of the lower reactivity of allyl groups compared
with acrylic or vinyl double bonds. Fumarate 1,2-disub-
stituted double bonds are known to be more reactive in
copolymerizations with styrenes than maleate double bonds.

Styrene was copolymerized with DAF in bulk in the presence of BP at 50 to 110°C to form clear castings free of
popcorn structures (62). Tawney and Kuderna made soluble
terpolymers of styrene, DAF, and allyl chloride by heating with peroxide catalyst (63). The following mixture
was heated at 80°C for 100 min to form a syrup of 100
poises (64): 180 p styrene, 70 p DAF, 810 p MMA, 8.9 p
ethylhexyl mercaptan, and 9.3 p lauroyl peroxide.

A 52% yield of soluble copolymer was indicated. To
700 parts of the syrup were added 300 parts of MMA and 10
parts BP for glass-fiber-reinforced moldings. Styrene-DAF
soluble copolymers were prepared by heating with BP and
carbon tetrachloride as regulator (65). These could be
copolymerized with unsaturated polyesters. Styrene-
methacrylic acid-DAF precopolymers were used in fiber-
reinforced moldings (66). Copolymerizations of 1.5 to 25%
DAF with styrene have been evaluated in Russia in heat-
resistant coatings. Copolymerization of DAF with styrene
in benzene at 80°C with benzoyl peroxide initiator gave
evidence of cyclization from low monomer concentrations
such as 0.30 molar (67).

Soluble copolymers of DAF with diallyl maleate have been
made in BrCCl$_3$ solution with BP catalyst at 60°C (68).
References to other copolymerizations are given in Table
26.3.

DAF does not polymerize readily under ionic initiation
although viscous liquid discolored products may be formed.
DAF has been suggested as an accelerator of the curing of
unsaturated elastomers (69).

TABLE 26.3

Other Copolymerizations with Diallyl Fumarate

Comonomers	References
Styrene, p-chlorostyrene	Zanaboni, U.S. 3,218,299 (Monte)
Styrene, MMA	Ital. 630,731 (Monte); CA 57, 16894 (1962)
Vinyltoluene	Kramer, U.S. 3,281,497
Vinyl acetate	Bradley, U.S. 2,238,030 (Am. Cyan.); Brit. 1,038,228 (Esso); CA 65, 18402 (1966)
Vinyl chloride	Stafford, U.S. 3,351,604 (G.E.)
Vinylidene chloride	Britton, U.S. 2,160,940 (Dow)
Alkyl acrylates	Bolgiano, U.S. 3,361,691 (Armstrong)
Dibutyl fumarate	Symcox, Brit. 1,072,723 (Monsanto)
Unsaturated polyesters	Joo, U.S. 3,207,815 (Pure Oil); Brit. 962,882 (Lab. Italiani); CA 61, 8487 (1964)

References

1. E. L. Kropa, U.S. 2,249,768 (Am. Cyan.); cf. U.S. 2,275,467 (PPG).
2. W. Moore and R. O. Roblin, U.S. 2,325,790 (Am. Cyan.).
3. M. A. Pollack and A. G. Chenicek, U.S. 2,296,823 (PPG).
4. H. S. Rothrock, U.S. 2,221,663 (Du Pont).
5. M. A. Pollack, I. E. Muskat, and F. Strain, U.S. 2,273,891 and 2,370,578 (PPG).
6. B. Phillips and W. M. Quattlebaum, U.S. 2,543,337 (Carbide); cf. P. O. Tawney, U.S. 2,560,495 (U.S. Rubber).
7. H. S. Rothrock, U.S. 2,221,663 (Du Pont).
8. E. L. Kropa, U.S. 2,443,740 (Am. Cyan.), appl. June 1944.
9. I. E. Muskat et al, U.S. 2,379,247; U.S. 2,320,536 (PPG); D. A. Swedlow, U.S. 2,456,093 (Shellmar).
10. G. B. Butler and M. D. Barnett, PB 145,435; CA 55, 24083 (1961).
11. FMC, Sartomer, and Borden data sheets (cf., U.S. 2,430,564 and 2,444,817).
12. cf. P. Gordon, U.S. 2,430,564 (Am. Waterproofing); R. G. Fordyce, U.S. 2,444,817 (Monsanto); B. E. Sorenson, U.S. 2,343,483 (Du Pont); S. B. Luce, U.S. 2,611,762 (Swift).
13. cf. A. Voss and W. Heuer, U.S. 2,200,437 (I.G.).
14. G. Gatta and G. Benetta, Ital. 611,733 (Edison); CA 55, 21670 (1961).
15. W. C. Mast, L. T. Smith, and C. H. Fisher, Ind. Eng. Chem. 36, 1027 (1944); cf. U.S. 2,449,612 (Dow); India Rubber World 119, 596 (1949).
16. J. W. Zemba et al, U.S. 2,311,607, 2,331,263, 2,311,615, and 2,341,175 (Dow); cf. U.S. 2,443,915 and 2,482,087.
17. G. F. D'Alelio, U.S. 2,403,213 (Prophylactic); cf. U.S. 2,443,915 (LOF).
18. R. E. Holman, U.S. 2,472,661 (Du Pont).
19. H. F. Minter, U.S. 2,542,827 (Westinghouse).
20. N. C. Foster, U.S. 2,482,087 (Westinghouse).
21. R. L. Letsinger and S. B. Hamilton, J. Am. Chem. Soc. 81, 3009 (1959).
22. R. M. Myers and D. L. Dunkelberger, Ger. Offen. 2,130,989 (R & H), CA 77, 62753 (1972).
23. J. Johnson et al, U.S. 2,912,418 (Peterlite); cf. U.S. 2,910,456 and Brit. 818,471.
24. H. T. Neher et al, U.S. 2,514,895; cf. U.S. 2,514,786 (R & H).
25. E. L. Kropa and T. F. Bradley, U.S. 2,280,242, appl. April 1942; E. L. Kropa, U.S. 2,443,738-41, appl. June 1948; U.S. 2,473,801 (Am. Cyan.).
26. H. J. Beck and H. Dannenbaum, Ger. 1,166,467 (Beck); CA 60, 16064 (1964).
27. J. C. Petropoulos and L. E. Caldwell, U.S. 2,626,250 (Am. Cyan.); J. B. Rust and W. B. Canfield, U.S. 2,530,315-6 (Ellis-Foster).
28. D. G. Patterson, U.S. 2,305,224 (Am. Cyan.).
29. A. A. Oswald and W. Naegele, Makromol. Chem. 97, 258 (1966).
30. B. S. Garvey, U.S. 2,155,590 (Goodrich).

31. G. Guenther, Ger. 1,069,314.
32. W. B. Hewson, U.S. 2,749,319 (Hercules).
33. Brit. 844,231 (Dow); W. F. Busse, U.S. 2,868,001 (Du Pont).
34. C. A. Heiberger and J. L. Thomas, U.S. 3,113,123 (FMC).
35. J. Barton and J. Pavlinec, Plaste Kautschuk 15, 397 (1968).
36. J. W. King, J. W. Hafey, K. J. Wildonger, and C. E. Schildknecht, unpublished.
37. C. E. Rehberg and C. H. Fisher, U.S. 2,452,209 (USA).
38. A. M. Clifford and J. D. D'Ianni, U.S. 2,464,488 (Wingfoot).
39. R. F. Dunbrook, India Rubber World 117, 745 (1948).
40. T. W. Evans et al, U.S. 2,612,491 (Shell).
41. G. B. Butler et al, J. Am. Chem. Soc. 81, 5946 (1959).
42. J. B. Rust, U.S. 2,503,772; U.S. 2,570,385; cf. U.S. 2,478,015 (Montclair Res. and Ellis-Foster).
43. E. B. Luce, U.S. 2,411,136 (Monsanto).
44. J. H. Hunsucker, U.S. 2,862,900; U.S. 2,852,495 (Holcomb).
45. Brit. 796,006 (U.S. Rubber); CA 52, 21235 (1958).
46. I. A. Arbuzova et al, CA 59, 6523 (1963).
47. H. F. Minter, U.S. 2,524,921 (Westinghouse).
48. Brit. 933,697 (Allied); CA 60, 1700 (1964).
49. H. A. Bruson and G. B. Butler, U.S. 2,425,144 (R & H).
50. P. O. Tawney et al, J. Org. Chem. 25, 56 (1960).
51. C. H. Alexander, U.S. 3,151,182 (U.S. Rubber).
52. F. F. Shcherbina and I. P. Fedorova, Polym. Sci. USSA 10, 56 (1968).
53. T. M. Pyriadi, CA 75, 6403 (1971).
54. T. F. Bradley, U.S. 2,295,513; cf. U.S. 2,249,768 and 2,295,513 (Am. Cyan.); U.S. 2,220,855 (Dow).
55. M. A. Pollack and A. G. Chenicek, U.S. 2,296,823 (PPG); cf. U.S. 2,346,612 (Du Pont).
56. H. S. Rothrock, U.S. 2,221,663 (Du Pont), appl. August 1937; cf. Reference 5, Chapter 27.
57. C. J. Mighton, U.S. 2,392,756 (Du Pont).
58. D. A. Kardashev et al, CA 40, 4559 (1946).
59. T. F. Bradley, U.S. 2,295,513; U.S. 2,311,327 (Shell).
60. J. G. Kuderna and R. H. Snyder, U.S. 2,498,084 (U.S. Rubber).
61. J. Panlik et al, CA 60, 1542 (1964).
62. R. E. Davies and A. R. Esterly, U.S. 2,493,948 (Catalin); cf. H. A. Wright, U.S. 3,259,595 (Koppers).
63. P. O. Tawney and J. G. Kuderna, Can. 493,853; CA 50, 1371 (1956).
64. Ital. 630,731 (Monte); CA 57, 16894 (1962).
65. Brit. 578,736 (Monte); CA 53, 23106 (1959); cf. Brit. 858,916 (Monte), CA 55, 14993 (1961).
66. P. Zanaboni and C. Bargensi, U.S. 3,218,299 (Monte); cf. CA 53, 23106 (1959).
67. K. Noma and M. Niwa (Kyoto), CA 77, 88887 (1972).
68. F. J. Foster, U.S. 2,547,696 (U.S. Rubber).
69. J. F. Smith, U.S. 3,297,626 (Du Pont); D. G. McRitchie, U.S. 3,312,757 and Brit. 1,047,053 (Raybestos).

This chapter discusses some bifunctional compounds with reactivity in radical polymerization generally lower than that of monomers which are activated by resonance. The allyl vinyl compounds bear electron-donating or weakly electron-attracting substituents. Examples are the following:

CH$_2$COOCH$_2$CH=CH$_2$
CH$_2$COOCH=CH$_2$

CH$_2$-CH=CH$_2$
SO$_2$
CH=CH$_2$

CH$_2$-CH=CH$_2$
O
CH=CH$_2$

Allyl vinyl succinate   Allyl vinyl   Allyl vinyl ether
                               sulfone

Viscous liquid or semisolid polymers of relatively low molecular weight have been prepared from a number of allyl vinyl compounds by cationic polymerizations. In commercial radical polymerizations, they may be added in minor proportions as regulators.

Little has been published about allylstyrenes and allyl-oxystyrenes, which polymerize reluctantly by radical initiation. Preparation of p-allylstyrene (bp 60°C at 3 mm) was described (1). A polymer from o-allylstyrene containing six-membered rings in the main chain was reported by radical polymerization, and another with five-membered rings by cationic polymerization (2). Allyl (o-vinylphenyl) ether (bp 66°C at 0.7 mm) and the para-isomer (bp 77°C at 0.7 mm) gave soluble polymers with pendant allyl groups when polymerization in CH$_2$Cl$_2$ at -78°C was initiated by boron fluoride etherate (3). With azobis at 70°C, 20% monomer in tetrahydrofuran formed soluble polymers, but 50% solutions formed insoluble, crosslinked polymers. The monomers were copolymerized successfully with ethyl vinyl ether and with styrene by using BF$_3$ catalysts. The homopolymers underwent Claisen rearrangement in diethylaniline when heated for 3 hr at 215°C under nitrogen. Crosslinked polymers were formed containing phenolic hydroxyl groups (confirmed by acetylation). From p-AOCH$_2$C$_6$H$_4$CH=CH$_2$ (bp 69°C at 0.1 mm and

648

$n_D^{20} = 1.5413$) only low polymers resulted from sodium naphthalene catalyst in tetrahydrofuran at -40°C (4).

An allyl derivative of a 2-substituted styrene was observed to polymerize long ago. When the allyl compound $C_6H_5CH=CH-CH=CHCOOA$ was heated for a week at 210°C, it passed through a viscous syrup to form an amber solid (5). The polymer was purified by precipitation from acetone solution by adding methanol. It could be hydrolyzed to a polymeric acid. Alkenylstyrenes were copolymerized with styrene (6). The allylstyrene monomer, 2-allyl-4-vinylanisole, polymerized readily by radical initiation to a heat-resistant, yellow solid (7).

ALLYL VINYL ESTERS AND ETHERS

Allyl vinyl esters of succinic, adipic, and sebacic acid were prepared by reaction of acetylene with the half-allyl esters in the presence of a mercuric compound and BF_3 or HF (8). Dioxane may be used as solvent, and a little hydroquinone may be added to minimize formation of brown low polymers (9). Ethylidene bis(allyl succinate) is formed as a byproduct. Allyl vinyl diglycolate was polymerized first to prepolymer and then to a crosslinked state (10). Vinyl allyloxyacetate (bp 78°C at 28 mm) was prepared from allyloxyacetic acid and vinyl acetate in the presence of mercuric acetate and sulfuric acid (11). It polymerized on heating with 5% BP at 85°C to give a clear, yellowish solid. Vinyl undecenoate, which is $CH_2=CHOCO(CH_2)_8CH=CH_2$, was copolymerized in minor proportions with vinyl chloride, vinylidene chloride, acrylonitrile, butadiene, styrene, and methyl acrylate (12). The nature of the acyl group of the vinyl ester had little influence on the Q and e values in copolymerization.

The allyl ester of vinylacetic acid was made by dropwise addition of vinylacetyl chloride to a stirred mixture of allyl alcohol and dry ether containing pyridine at 0 to 5°C (13). The following monomer properties were reported:

Allyl vinylacetate: bp 58°C at 27 mm; $n_D^{22} = 1.4313$
Methallyl vinylacetate: bp 74°C at 27 mm; $n_D^{22} = 1.4351$

Bulk polymerization of the monomers by heating with benzoyl peroxide at 100°C or azobis at 75°C gave only viscous liquid low polymers. This confirms that these are in fact diallyl compounds and allyl vinyl in name only.

Allyl vinyl ether can be made by alcoholysis of alkyl vinyl ethers in the presence of a mercuric salt (14). It was made also by reacting potassium hydroxide with 2-bromoethyl allyl ether (15). Heating the monomer with sodium rearranged it into propenyl vinyl ether (boiling at 62°C). Allyl vinyl ether vapor undergoes Claisen-type

rearrangement on heating at 255°C to form allyl acetaldehyde or 5-pentenal. Rhône-Poulenc studied synthesis of allyl vinyl ether and use of the viscous liquid prepolymers
as air-curing coatings. Allyl alcohol was best vinylated
with acetylene and alkaline catalyst at 95°C, a much lower
temperature than needed in typical Reppe-type vinylations
(16). Properties of allyl vinyl ether and some miscellaneous allyl vinyl compounds are given in Table 27.1.
Allyl vinyl ether, boiling at 67°C, is rapidly hydrolyzed
by dilute acids to acetaldehyde and allyl alcohol. It
very slowly undergoes Claisen rearrangement even at 80°C.
Allyl 1-phenylvinyl ether rearranged on heating to form
allylacetophenone (15).

Vinyl polymerization of allyl vinyl ether to give viscous prepolymers was carried out at -25 to -60°C in
toluene using boron fluoride-etherate catalyst. Oxidative
polymerization of the soluble vinyl polymer in the presence of cobalt salts with heating gave glossy, hard
insoluble coatings. However, rates of air cure were
slower than desired in synthetic "drying oils." Stereoregular polymers were reported from polymerization of
allyl vinyl ether in toluene at -78°C with BF₃ etherate
catalyst (17). Adding nitromethane as a polar solvent
reduced the tacticity of the polymer formed. The isotactic polymer in acetic acid reacted with HBr apparently
forming a ternary copolymer of vinyl alcohol, vinyl
acetate, and vinyl bromide units. A similar polymerization procedure at -78°C gave solid polymers from methallyl
vinyl ether which were reported to be isotactic (18). At
0°C only viscous liquid polymers of low tacticity were
obtained.

Allyl vinyl ether was copolymerized with ethyl vinyl
ether in petroleum ether at -5°C, using aluminum sulfate-
sulfuric acid catalyst (19). The rubberlike copolymer
contained 9.3% allyl vinyl ether units. Allyl vinyl ether
was copolymerized with larger proportions of 2-chloroethyl
vinyl ether in heptane solution with aluminum isopropoxide
catalyst added portionwise at 0°C (20). The mixture was
reacted overnight, quenched by ammoniacal ethanol, and
then treated with a solution of phenolic antioxidant in
methanol. The rubbery copolymer was cured by heating with
a tetramethylthiuram disulfide, sulfur, and zinc oxide
formulation. Methallyl vinyl ether, boiling at 89°C, was
prepared by alcoholysis, and its rearrangement on heating
was studied (21).

The allyl and methallyl vinyl ethers of ethylene glycol
were made by reaction of the sodium alkenoxide with vinyl
2-chloroethyl ether (22). They were polymerized in the
presence of boron fluoride etherate at -70°C to give
soluble low polymers containing unsaturation. Heating

TABLE 27.1

Miscellaneous Allyl Vinyl Compounds

Compound	bp, °C	Refractive index	References, etc.
Allyl vinyl ether	67	$n_D^{20}=1.4109$	Watanabe, JACS $\underline{79}$, 2828 (1957)
Methallyl vinyl ether	88	$n_D^{20}=1.4256$	Watanabe
Chloroallyl vinyl ether	100	$n_D^{20}=1.4502$	Watanabe
Allyl(o + p-vinyl) phenyl ether	66(0.7) 77(0.7)	--	Kato, J. Polym. Sci. Al, $\underline{6}$, 2993 (1968) (polymers by BF_3 etherate)
N-Allyl vinyl sulfonamide	--	--	Goethals, Makromol. Chem. $\underline{108}$, 312 (1967) (radical polymerization)
Allyl N-vinyl carbamate	38(0.1) mp 11°C	--	Schulz, Monatsh. $\underline{92}$, 303 (1961); Makromol. Chem. $\underline{44-46}$, 281 (1961) (radical cyclopolymerization)
N-Allyl 2-vinyl-oxyethylcyanamide	100(2)	--	DeBenneville, U.S. 2,727,068 (R & H) (polymer by BF_3)
Allyl crotonate	64(22)	$n_D^{22}=1.4452$	Butler, J. Org. Chem. $\underline{25}$, 309 (1961); D'Alelio, U.S. 2,260,005 (G.E.)
Methallyl crotonate	77(22)	$n_D^{22}=1.4491$	Butler; (low polymer oils)
Allyl p-vinyl benzoate	106(10)	--	D'Alelio, U.S. 3,335,119 (DalMon)

the latter with peroxides slowly formed crosslinked in-fusible polymers. As much as 20% BP was added in cross-linking prepolymer of this allyl ether at 65°C during 24 hr. Viscous liquid low polymers were made by Butler by addition of BF_3 etherate at -60°C to the following monomers (23):

$104°C(2)$; $n_D^{25} = 1.5187$ $112°C(1)$; $n_D^{25} = 1.5335$

Heating the polymers at 100°C for 24 hr with peroxides gave harder insoluble products.

Allyl vinyl ethers of glycerol and other polyols have been prepared, and air-drying properties of prepolymers made by cation methods have been tested without promise (24). The rearrangement of allyl-3-allyloxy-2-butenoate at 135°C gave a ketone derivative along with polymer(25). By refluxing vinyl acetate with allyloxyacetic acid in the presence of mercuric acetate and a little tannic acid as polymerization inhibitor, vinyl (allyloxy) acetate was prepared (26). The compound $AOCH_2(OH)CHCH_2OCH_2CH_2OA$ polymerized with Lewis-acid catalysts at low temperatures (27).

Allyl vinyl acetals $AO\overset{H}{C}(R)OCH=CH_2$ were reported to polymerize on heating, and the viscous liquid products were stated to give vinyl alcohol-allyl alcohol copolymers by hydrolysis (28). Such monomers, made by heating allyl 2-chloroethyl acetals with KOH boiled as follows:

A vinyl formal, 120°C   A vinyl butyral, 55°C(7 mm)
A vinyl acetal, 52°C(17 mm)   A vinyl benzal, 130°C(20 mm)

Allyl isopropenyl ketone (bp 47°C at 11 mm; $n_D^{19} = 1.4712$) was made from 1-vinyl-2-isopropenylacetylene. It did not give homopolymers of high molecular weight (29).

Allyl vinyl carbonate was made by reaction of vinyl chloroformate and allyl alcohol in pyridine at 0°C (30). The vinyl chloroformate was made by pyrolysis at 500°C of the dichloroformate of ethylene glycol.

Allyl vinyl sulfonate (bp 62°C at 0.2 mm) was prepared by reacting 2-chloroethane-1-sulfonyl chloride with allyl alcohol in methylene chloride solution with pyridine at 0°C (31). Polymerization in bulk by heating at 45°C with azobis gave crosslinked polymers, whereas polymerization in benzene solution gave soluble polymers believed to contain six-membered cyclic units. Increased cyclization at lower polymerization concentrations was indicated by IR data. Allyl vinyl sulfone was epoxidized to give a 50% yield of glycidyl vinyl sulfone (32). Examples were given of copolymerization of allyl glycidyl sulfone with vinyl chloride and with other monomers under radical conditions.

References

1. G. F. D'Alelio, J. Polym. Sci. A 1 5, 1245 (1967).
2. K. Yokota and Y. Takada, Chem. High Polym. (Tokyo) 26, 317 (1969).
3. M. Kato et al, J. Polym. Sci. 6, 2993 (1968).
4. G. F. D'Alelio and T. R. Hoffend, J. Polym. Sci. A 1 5, 1249 (1967).
5. F. F. Blicke, J. Am. Chem. Soc. 45, 1562 (1923).
6. G. Greber and G. Egle, Makromol. Chem. 59, 174 (1963).

7. S. Kawai and I. Suzuki, CA <u>47</u>, 3606 (1953).

8. Brit. 595,061 (PPG); CA <u>42</u>, 8819 (1948).

9. C. D. Hurd et al, J. Am. Chem. Soc. <u>78</u>, 104 (1956).

10. Netherlands 70,514 (Shell); CA <u>47</u>, 3040 (1953).

11. R. S. Barker and L. N. Whitehill, U.S. 2,448,246 (Shell).

12. C. S. Marvel and W. G. DePierri, J. Polym. Sci. <u>27</u>, 39 (1958).

13. G. H. Jeffery and A. J. Vogel, J. Chem. Soc. 658 (1948); M. D. Barnett and G. B. Butler, J. Org. Chem. <u>25</u>, 309 (1960).

14. W. H. Watanabe and L. E. Conlon, U.S. 2,760,990 (R & H); and J. Am. Chem. Soc. <u>79</u>, 2828 (1957).

15. C. D. Hurd and M. A. Pollack, J. Am. Chem. Soc. <u>60</u>, 1905 (1938); cf. J. Org. Chem. <u>3</u>, 550 (1939); F. W. Schuler and G. W. Murphy, J. Am. Chem. Soc. <u>72</u>, 3155 (1950).

16. R. Paul et al, Bull. Soc. Chim. France 121 (1950); CA <u>44</u>, 6672 (1950); M. L. A. Fluchaire and G. Collardeau, Fr. 943,821; U.S. 2,603,628; U.S. 2,546,431 (Rhône-Poulenc).

17. Y. Heimei et al, CA <u>72</u>, 1954 (1970).

18. H. Yuki et al, CA <u>74</u>, 126113 (1971).

19. K. Herrle et al, U.S. 2,825,719; cf. U.S. 2,830,032 (Badische).

20. R. Heck, U.S. 3,025,275-6 (Hercules).

21. R. F. Webb et al, J. Chem. Soc. 4092 (1961).

22. G. B. Butler and J. L. Nash, J. Am. Chem. Soc. <u>73</u>, 2538 (1951).

23. G. B. Butler, J. Am. Chem. Soc. <u>77</u>, 482 (1955).

24. F. W. Hoover, U.S. 2,518,321 (Du Pont).

25. J. W. Rolls et al, J. Org. Chem. <u>28</u>, 3521 (1963).

26. R. S. Parker and L. N. Whitchill, U.S. 2,448,246 (Shell).

27. L. N. Bauer and H. T. Neher, U.S. 2,692,256 (R & H).

28. I. A. Arbuzova et al, CA <u>61</u>, 1949 (1964).

29. I. N. Nasarov et al, CZ I, 1244 (1942); cf. CA <u>39</u>, 1620 (1945).

30. F. Strain and F. E. Kueng, U.S. 2,370,589 (PPG); cf. U.S. 2,384,143 and 2,377,111.

31. E. J. Goethals et al, J. Polym. Sci. B <u>4</u>, 691 (1966); Makromol. Chem. <u>108</u>, 312 (1967) and <u>115</u>, 234 (1969).

32. D. L. MacPeek et al, U.S. 3,220,981 (Carbide).

28. ALLYL PHOSPHORUS COMPOUNDS

The allyl phosphorus compounds comprise a great number of types bearing different amounts of oxygen. Little is known about the toxicity of many of them. Care should be exercised in research with these challenging compounds, since explosions have been reported during distillations and polymerizations on several occasions.

Diallyl arylphosphonates first received attention in the United States for flame-resistant plastics, and a little later triallyl phosphate and related monomers were evaluated in fabric flameproofing and ion-exchange resins. Interesting researches on other allyl phosphorus compounds were carried out in England and in Russia, but much remains to be learned about the polymerizations. The high reactivity of some of these compounds as reducing agents must be kept in mind in studies of polymerization. Allyl phosphites and phosphines generally resist polymerization.

Some additional Russian references to those included here have been given by Gefter (1). A number of the phosphorus-containing allyl monomers have not been prepared in sufficient purity for critical comparison of their polymerizabilities and the properties of the polymers. It is probable that future research will lead to important applications in flame-resistant and hydrophilic copolymers.

Many of the names for organic phosphorus compounds used in the older literature are different from those adopted by the American Chemical Society in 1952 (2). The following allyl derivatives illustrate names accepted:

O=P-OA with OA, OA — Triallyl phosphate (ester of phosphoric acid)

P-OA with OA, OA — Triallyl phosphite (ester of nonexistent phosphorous acid)

O=P-OA with OA, H — Diallyl phosphonate (ester of phosphonic acid)

P-OA with OA, H — Diallyl phosphonite (ester of nonexistent phosphonous acid)

O=P-OA with OA, C_6H_5 — Diallyl phenylphosphonate

P-OA with OA, C_6H_5 — Diallyl phenylphosphonite

$$O=P\!\!\begin{array}{c}\nearrow OA \\ -H \\ \searrow H \end{array}$$ Allyl phosphinate (ester of phosphinic acid) $$P\!\!\begin{array}{c}\nearrow OA \\ -H \\ \searrow H \end{array}$$ Allyl phosphinite (ester of nonexistent phosphinous acid)

The hydrogen atoms attached to oxygen are acidic and can be replaced by a metal ion to form a salt or by an organic radical to form an ester. The hydrogen atoms attached to phosphorus are not acidic. When an allyl or alkyl radical of a phosphite, phosphonite, or phosphinite is replaced by hydrolysis with hydrogen, then the hydrogen normally moves immediately to phosphorus creating a P=O group. Thus the partial hydrolysis of a triallyl phosphite gives a diallyl phosphonate. The free phosphorous, phosphonous, and phosphinous acids are unstable and unknown.

Other approved names include allylphosphonic dichloride $(AP(O)Cl_2$ and allylphosphonous dichloride $APCl_2$. Because of variations in nomenclature in the literature, formulas of the allyl phosphorus compounds are emphasized.

DIALLYL PHENYLPHOSPHONATE

Diallyl phenylphosphonate (DAPP) is useful in raising refractive index, flame resistance, and adhesion in thermosetting copolymers; it was studied by Toy of Victor Chemical Works (later merged with Stauffer Chemical Co.). The compound can be made by slow addition of $C_6H_5P(O)Cl_2$ to a mixture of allyl alcohol and pyridine cooled to 5°C during 4 hr (3).

$$\text{C}_6\text{H}_5\text{PCl}_2 + 2HOCH_2CH=CH_2 \longrightarrow \text{C}_6\text{H}_5\text{P}(OCH_2CH=CH_2)_2 + 2HCl$$

After bringing the mixture to room temperature during an hour, water was added and DAPP recovered from the oily layer. The monomer was distilled at 128°C with copper powder as inhibitor (3). The monomer also was prepared by action of $C_6H_5P(O)Cl_2$ upon sodium allylate. Related monomers are listed, with their properties, in Table 28.1.

DAPP is slightly soluble in water (2 g/liter) and is soluble in common organic solvents. It can be stored for years with little change. A commercial sample after seven years' storage was a colorless, smoothly viscous solution miscible with methanol (4). However, with 4.5% BP it gave a gel of crosslinked polymer after 40 days at room temperature. Diallyl hydrogen phosphonate, without the benzene ring, also polymerizes readily, but DAPP was favored for development by Victor because of availability of phenylphosphonic dichloride.

When DAPP containing 2% dissolved BP is heated under nitrogen at 90°C, it passes through a viscous prepolymer

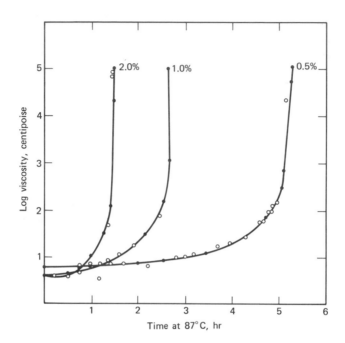

Fig. 28.1. Diallyl phenylphosphonate with three different concentrations of benzoyl peroxide polymerized at 87°C to give highly viscous solutions of prepolymer before crosslinking to a gel. For complete polymerization of such diallyl monomers, 4% or more of peroxide initiators is often required. A. D. F. Toy and L. V. Brown, Preprint ACS Div. Paint, Varnish and Plastics 75 (September 1947).

stage (Fig. 28.1) and then gives hard, clear, crosslinked polymer castings after 20 hr. Polymers were colorless to light yellow. When DAP containing 3% BP was polymerized under air at 100°C, there was a sudden temperature rise to 165°C, and a brittle gumlike polymer was formed that could not be converted to a hard solid by further heating (5). In general, temperatures below 100°C and low catalyst concentrations favored formation of hard, fully cured polymers. Once gelation had been attained, however, higher temperatures could be used to finish cure. A viscous liquid prepolymer of 20 poise could be stored for several months before curing with peroxide at 130°C for 2.5 hr. Besides flame resistance and insolubility, the cured homopolymers were found to have the following properties:

Colorless to amber
$n_D^{28} = 1.573$
$d_4^{28} = 1.273$
Heating 11 days at 116°C gave
  darkening
Water absorption in 24 hr, 0.66%

Tensile strength, 5300 psi
Flexural strength, 6000 psi
Impact (Izod), 0.24 ft-lb/in.
Rockwell hardness, M-95
Heat distortion, 216°F

Cured polymers are insoluble in solvents but swollen in aromatic hydrocarbons.

Diallyl phenylphosphonate, first supplied under the name V-Lite by Victor Chemical, copolymerizes by radical initiation with MMA, vinyl acetate, DAP, DA succinate, DA oxalate, and with unsaturated polyesters (6). When 25% or more phosphonate monomer units are present in copolymers, self-extinguishing, nonflammable crosslinked plastics are possible (7). Mixtures of MMA and DAPP were copolymerized with peroxide catalyst to give refractive indices matching those of glass fibers and therefore permitting nearly transparent glass fiber-reinforced plastics. Such a result was obtained from 68.4% DAPP and 32.6% MMA by volume with 2.5% BP heated for 18 hr at 75°C with glass fibers in a sheet cell. Polymerizations of DAPP in aqueous dispersion were undertaken (8). In copolymerizations with vinyl acetate or MMA it was necessary to increase the peroxide catalyst concentrations when higher proportions of DAPP were used. In copolymerizations with diallyl oxalate, the larger proportions of DAPP gave cured copolymers that were less brittle. More than 10% DAPP with DAP or with unsaturated polyesters gave flame-resisting copolymers. The copolymers tend to become yellow when heated for full cure and they discolor slowly on outdoor exposure. With lauryl methacrylate, DAPP catalyzed by BP showed copolymerization reactivity ratios $r_1 = 19.5$ and $r_2 = 0.07$; diallyl butylphosphonate gave $r_1 = 18.7$ and $r_2 = 0.09$ (9). Figures 28.2 and 28.3 show the copolymer compositions at low conversion and also the products much higher in allyl monomer units formed at higher conversions. The reluctance of DAPP to copolymerize is also shown in reactivity ratios with MMA ($r_1 = 23$, $r_2 = 0.13$) and with styrene ($r_1 = 29$, $r_2 = 0.03$) (10). DAPP was copolymerized with acrylonitrile at 75°C in bulk using azobis as initiator (11). Diallyl alkylphosphonates were copolymerized with sulfur dioxide at low temperatures (12). Dimethallyl phenylphosphonate and related monomers and prepolymers were copolymerized with diethylene glycol maleate and fumarate polyesters (13).

Diallyl benzyl- and butylphosphonates were heated for 20 hr at 115°C with di-t-butyl peroxide, producing clear solid polymers (14). Diallyl phosphonate $(AO)_2 HPO$ gave

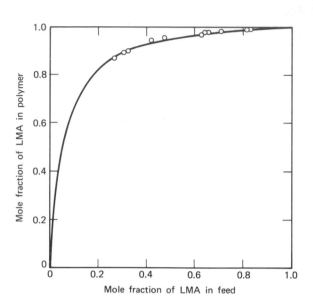

Fig. 28.2. Copolymer composition curve for copolymerization of
lauryl methacrylate (LMA) with diallyl phenyl phosphonate in benzene with benzoyl peroxide at low conversion. Diallyl n-butyl
phosphonate reacted similarly. K. I. Beynon, J. Polym. Sci. A <u>1</u>,
3352 (1963).

hard brittle polymers (15). Diallyl cyclohexylphosphonate
was heated up to 115°C during 7.5 hr with the following
percentages of benzoyl peroxide (16):

0 %	No change
1.0%	Viscous liquid
2.0%	Clear, stiff gel
5.0%	Moderately hard solid

   Diallyl phenyl- and related phosphonates have been
studied in plastic lenses (17), in optical cements (18),
in flameproofing of fabrics (19), and in aqueous dispersions with melamine-formaldehyde resins (20). The
polymerizations of most diallyl aryl- and alkylphosphonates pass through a viscous syrup stage before gelation.
   Toy suggested that some cyclic polymerization may occur
in the formation of soluble prepolymers from diallyl
phenylphosphonate and related monomers. Spooncer polymerized allyl phenyl allylphosphonate in 2-chloroethyl

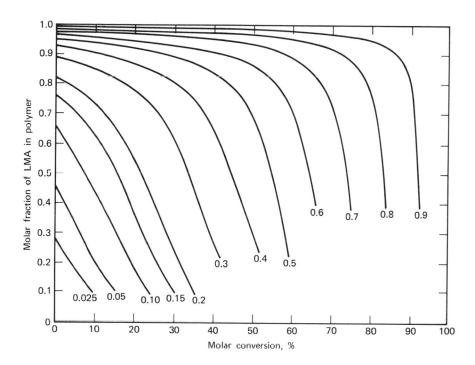

Fig. 28.3. Instantaneous copolymer compositions at different conversions from different starting mixtures of lauryl methacrylate (LMA) and diallyl phenyl phosphonate. The copolymerizations were carried out in benzene with benzoyl peroxide as initiator. K. I. Beynon, J. Polym. Sci. A $\underline{1}$, 3354 (1963).

ether solution at 150°C with t-butyl peroxide to form soluble polymers of molecular weight 26,000 (21). This was explained also by partial cyclization.

OTHER ALLYL PHOSPHONATES

Harman and Stiles prepared bis(diallyl phosphonomethyl) ether $(AO)_2 P(O)CH_2OCH_2P(O)(OA)_2$ (bp 107°C at 1 mm), a tetrafunctional monomer (22). It polymerized to a clear infusible solid on heating for several hours with 1% t-butyl peroxide at 110°C. Impurities and air may greatly affect radical polymerizations of polyfunctional monomers. Some samples of this monomer polymerized explosively. Co-polymerizations with different ethylenic monomers were suggested using radical initiation.

Diallyl phenylphosphonate in chloroform solution was epoxidized by peracetic acid at 30°C (23). A tetra-functional allylphosphonate monomer AOOCCH$_2$CHCH$_2$COOA was
OP(OA)$_2$
made by reacting the allyl half-ester of maleic acid with P(OA)$_3$ (24). The compound was of doubtful purity and it did not reach gelation until it had been heated with benzoyl peroxide for 65 hr at 100°C. Flame-resistant copolymers were made with MMA.

Diallyl allylphosphonate with 1.5% BP has been poly-merized by Kennedy in Britain by heating at 100°C for 3 hr or more (25). The polymer could be partially saponified by boiling aqueous alkali to form an ion-exchange resin. Kennedy also copolymerized diallyl phosphonate (AO)$_2$P(O)H with CH$_3$CH=NC$_2$H$_5$ by heating with BP (26). The product was partially saponified to give an ion-exchange resin. Hooker had supplied (AO)$_2$P(O)A as a water-white liquid with a sharp, unpleasant odor. It is insoluble in water but miscible with common organic sol-vents. The compound homopolymerized slowly on heating to form crosslinked polymers. Copolymers showed good flexibility and resistance to crazing. Diallyl allyl-phosphonate should not be heated above 130°C since violent reactions may occur. The monomer is considered to be mildly toxic. Allylphosphonate esters and salts have been observed to isomerize to propenyl compounds under the influence of transition metal catalysts (27).

Diallyl chloromethylphosphonate has been used as a co-monomer for imparting flame resistance. It has been supplied by Victor Chemical Works as a pale yellow liquid (bp 87°C at 1 mm; d_{25}^{25} = 1.1870). It is soluble in common organic solvents and slightly soluble in water. It was polymerized and copolymerized using radical initiators to give comparatively hard, strong polymers with good optical properties and solvent resistance. The polymers were self-extinguishing, as were copolymers containing 20 to 30% phosphonate units. A typical polymerization suggested by a Victor data sheet used 2% BP and heating for 16 hr at 85°C out of contact with air. Copolymers were formed with vinyl acetate, MMA, DAP, and unsaturated polyesters. Properties of allyl phosphonate monomers, along with other references, are presented in Table 28.1.

Copolymerization of CH$_3$PO(OA)$_2$ was carried out with ethylene using azobis at 400 atm and 80°C (29). Copoly-mers containing 4 to 11% phosphorus showed evidence by infrared of branches containing allyl double bonds. Mix-tures of diallyl methylphosphonate and ethylene were co-polymerized at high pressures using azobis to form polymers of varying elasticity (30).

TABLE 28.1

Properties of Allylic Phosphonates, $O=P\overset{\displaystyle OA}{\underset{\displaystyle R}{\diagdown OA}}$

Phosphonate	bp, °C	Refractive index	References, etc.
Diallyl phenyl [a]	128(1)	$n_D^{28} = 1.5106$	Toy, U.S. 2,425,765 (Victor)
Diallyl hydrogen (Formerly called diallyl phosphite)	58(0.5)	$n_D^{20} = 1.445$ $n_D^{20} = 1.4478$	Kennedy, J. Appl. Chem. 8, 459 (1958) Cade, J. Chem. Soc. 2266 (1959)
Diallyl allyl	72(0.45)	$n_D^{20} = 1.4622$ $n_D^{25} = 1.463$	Cade Hooker data sheet
Dimethallyl allyl	143(3)	$n_D^{28} = 1.5057$	Toy, U.S. 2,425,765
Diallyl tolyl	128(3)	$n_D^{28} = 1.5097$	Toy
Dimethallyl tolyl	139(4)	$n_D^{27} = 1.5065$	Toy
Dimethallyl benzyl [a]	163(2)	$n_D^{25} = 1.5053$	Toy
Diallyl isobutenyl [a]	115(4)	$n_D^{26} = 1.4666$	Toy, U.S. 2,425,766 (Victor)
Dimethallyl isobutenyl [a]	115(2)	$n_D^{25} = 1.4668$	Toy
Diallyl styryl	158(1)	$n_D^{26} = 1.5442$	Toy (polymerized on distillation)
Dimethallyl styryl	164(2)	$n_D^{25} = 1.5391$	Toy (polymerized on distillation)
Diallylchlorophenyl	139(2)	$n_D^{25} = 1.5208$	Toy, JACS 70, 188 (1948)
Diallyl acetyl	71(0.25)	$n_D^{20} = 1.4534$	Cade, J. Chem. Soc. 2266 (1959)
Diallyl chloromethyl	98(1)	$n_D^{25} = 1.4674$	Toy, U.S. 2,714,100 (Victor); Victor data sheet
Diallyl bromomethyl	76(0.1)	$n_D^{20} = 1.4842$	Cade
Diallyl trichloro-methyl	118(0.5)	$n_D^{20} = 1.478$	Kosolapoff, JACS 69, 1002 (1947)
Dimethallyl β-cyanoethyl	127(0.5)	$n_D^{25} = 1.4653$	Toy, U.S. 2,844,558 (Victor)
Allyl vinyl ethyl	--	--	Zykova, CA 76, 46559 (1972)
Tetraallyl methy-lene bis	142(0.1)	$n_D^{20} = 1.4675$	Cade
Tetraallyl thio bis	--	--	Harman, U.S. 2,630,450 (Shell) (hard opaque polymers

[a] These compounds polymerized most readily to hard, glassy polymers (3% BP at 88°C for 18 hr), Toy and Cooper, JACS 76, 2191 (1954)

Diallyl chlormethylphosphonate was prepared by the following reactions (28):

$$PCl_3 + HCHO \longrightarrow ClCH_2POCl_2 \xrightarrow[\text{amine}]{\text{allyl alcohol}} ClCH_2PO(OA)_2 \text{ , etc.}$$

In one patent example, 67 g of dichloride was added dropwise during 1.25 hr to a mixture of 250 ml benzene, 85 g triethylamine, and 48.7 g allyl alcohol cooled at 0°C. Cold water was added to dissolve the ammonium salt that had formed. From the washed and dried benzene layer, the diallyl monomer was recovered by vacuum distillation in the presence of cupric resinate as polymerization inhibitor. Purified $ClCH_2PO(OA)_2$ polymerized overnight at 80°C with 3% BP under nitrogen gave harder polymers than were obtained when 2% BP was used. Slow polymerizations under air form relatively soft, yellow solids or viscous solutions. Copolymers with an unsaturated polyester, Laminac 4201, were self-extinguishing when removed from a flame. The diallyl monomer, with an equal weight of 2,2'-dichlorodiethyl ether and 1% BP at 100°C, gave viscous prepolymer solution from which white, granular solid prepolymer was isolated by precipitation with hexane. The prepolymer could be used as a thermosetting molding material or could be deposited upon cloth from solution and cured to give fire-resistant fabrics.

Diallyl phosphonate $(AO)_2P(O)H$ gave hard glasslike castings quickly by free radical bulk polymerizations (31). The monomer can be made by adding PCl_3 in toluene slowly to a 1:1 mixture of allyl alcohol and toluene maintained at -20 to -30°C (32). Hydrogen chloride was removed by a stream of dry air at 0°C and by adding ammonia. After filtering off ammonium chloride and adding hydroquinone as polymerization inhibitor, the diallyl phosphonate was recovered by vacuum distillation at 80°C (2 mm). In water solution the compound acts as a fungicide (33).

Hooker and Weston have supplied $(AO)_2P(O)H$. It is a colorless liquid of pungent, unpleasant odor boiling at 62°C at 1 mm and having $n_D^{25} = 1.444$. It is slightly soluble in water and is soluble in ethanol, hexane, benzene, and acetone. Tests with rats showed diallyl phosphonate to be toxic (LD_{50} of 178 microliters/kg).

Explosions were encountered in the vacuum distillation of diallyl phosphonate under carbon dioxide, apparently from polymerization (34). The monomer adds bromine readily. Phosphonoformate and acetate were prepared and also found to add bromine readily. Attempts to form formanilides and p-formotoluides gave glassy solids, presumably polymers.

Kennedy and co-workers at the British Atomic Energy Research Laboratories made interesting studies of polymerizations of different allyl phosphorus compounds in their research on crosslinked copolymers as complexing agents for uranyl ions (35). In agreement with Toy and Cooper, they concluded that polymerizability of diallyl phosphonic esters is enhanced by electronegative or electron-attracting groups attached to phosphorus. The monomers $(AO)_2 P(O)CCl_3$ and $(AO)_2 P(O)OCH=CCl_2$ polymerized violently in bulk with peroxide catalysts. They could only be homopolymerized with control in aqueous suspension. After these two highly reactive monomers, the following order of decreasing reactivity in polymerization was observed: $(AO)_2 P(O)H > (AO)_3 P=O \simeq (AO)_2 P(O)OCH_3 \simeq (AO)_2 P(O) \cdot CH_2 COOC_2 H_5 > (AO)_2 P(O)A >> A_3 P=O \simeq (AO)_2 P(O)OH \simeq (AO)_2 P(O) \cdot N(CH_3)_2$. The last four compounds did not give solid homopolymers, whereas the others formed hard, brittle polymers on heating with benzoyl peroxide. The following compounds were found to inhibit allyl polymerizations catalyzed with BP: $(AO)_2 P(O)N(CH_3)_2$, $(AO)_2 P(O)CHN(C_2 H_5)_2 \cdot CH=CH_2$, and $(AO)_3 P$. Although $(AO)_2 P(O)H$ polymerized readily, neither $A_3 PO$ nor $(AO)_2 P(O)OH$ did so. In general, the distribution coefficients of heavy metal salts between the liquid monomers and water followed a sequence reverse to that of the relative polymerizability.

Kennedy and co-workers found that conventional phenolic inhibitors, basic nitrogen-containing allyl compounds, as well as inorganic bases such as sodium carbonate were effective inhibitors for preventing violent polymerizations during distillation. Complete exclusion of oxygen during distillation and prevention of peroxide formation also helped to avoid explosions during distillations when 0.5% anhydrous sodium carbonate was added.

The authors suggested that electron-attracting or acid-strengthening substituents attached to phosphorus enhance polymerization by favoring formation of the resonance-stabilized allyl ions $CH_2 \doteq CH \doteq CH_2{}^+$. The British workers made crosslinked allyl phosphorus polymers of controlled permeability and complexing properties, but details of the polymerization conditions were not published. By empirical choice of time, temperature, and initiator concentration, allyl phosphorus polymers of varying DP and crosslinking characteristics could be prepared. Allyl compounds had advantages over vinyl compounds in control of crosslinking to obtain linear segments of about 20 monomer units. Suspension polymerization of a number of the monomers in water at 100°C gave acidic polymers because of partial hydrolysis. Several of the monomers, such as triallyl phosphate, diallyl phosphonate, and diallyl carbethoxymethylphosphonate, gave only about 50%

conversion to insoluble polymers under the conditions used for polymerization. Diallyl hydrogen phosphate gave only about 25% conversion to crosslinked polymer.

Beads of triallyl phosphate and diallyl allylphosphonate polymers could be saponified by refluxing with 20% aqueous NaOH during 3 hr. The beads were then washed with dilute sulfuric acid to convert them to the acid form. Insoluble crosslinked polymers were made having water uptake of 4 to 5 g/g of dry resin for the sodium form and exchange capacity of 4 to 4.5 mequiv./g of dry resin. Alkaline hydrolysis of polymers from diallyl monomers such as diallyl phosphonates produced water-soluble basic polymers. It was reported that hydrolysis of the allyl phosphate and phosphonate ester polymers with HCl or HBr formed soluble allyl alcohol polymers. Since diallyl phosphonamide and diallyl α-aminophosphonate did not polymerize readily enough for preparing crosslinked polymers for studies of heavy metal ion complexing, similar polymers were made indirectly from diallyl hydrogen phosphonate polymers (36).

Raff and co-workers of Washington State University polymerized a number of allyl phosphorus monomers within western white pine wood using ^{60}Co radiation (37). The wood samples ($0.64 \times 0.64 \times 4.0$ cm) were evacuated for 15 min and then were left in the impregnating monomer solutions for 8 hr at atmospheric pressure. In order to accelerate formation of crosslinked polymers within cells of the wood, 5% of vinyl-type crosslinking agents was added. In one series of experiments the following data were recorded:

Phosphorus monomer (95%)	Crosslinking monomer (5%)	Monomer in wood, %	Polymer in wood, %	Char, %
0	0	0	0	3.9
Dimethallyl methallyl-phosphonate	Ethylene di-methacrylate	76.2	51.0	30.3
Diallyl chloromethyl-phosphonate	Divinylbenzene	101.0	33.8	31.6
Diallyl allyl-phosphonate	Ethylene di-methacrylate	94.9	41.5	33.0
Diallyl phosphonate	Ethylene di-methacrylate	105.5	52.6	57.5
Triallyl phosphate	Ethylene di-methacrylate	88.5	73.7	34.4

The allyl phosphonate monomers were superior to phosphates in that the polymers were more resistant to leaching out of the wood in water. The phosphonate polymers also were more stable to oxidation, hydrolysis, light, and heat. The treated wood showed promising flameproof

properties. The chars were found to contain only 40 to
75% of the original phosphorus. After 14 days in water,
the dried wood still showed unchanged flame resistance.
   Monoallyl phosphonates have been prepared (38), but
they have stimulated little interest because of low reactivity in polymerization. Cyanoallyl compounds such as
$CH_2=C(CN)CH_2PO(OR_2)$ are not allylic in behavior but act
as acrylonitrile derivatives (39). The diethoxy compound
polymerized readily on warming with 0.3% acetyl peroxide.
Dialkyl chloroallylphosphonates gave copolymers but not
homopolymers (40). Diethyl and di-n-butyl allylphospho-
nates did not homopolymerize but were copolymerized with
acrylonitrile, MMA, and vinyl chloride by activated
aqueous persulfate systems (41). Diethyl allylphosphonate
$APO(OEt)_2$ has been reacted with polyesters bearing terminal hydroxyl groups by heating under nitrogen for 2 hr at
160°C in the presence of 0.1% EtONa (42). The resulting
allylphosphonate-polyesters could be polymerized to
soluble products containing 7.8% phosphorus. Allyl isopropenyl alkylphosphonates have been polymerized to solids
of low DP by heating with azobis at 70 to 90°C (43).
Monoallyl aryl esters of alkylphosphonic acids gave sulfone polymers by reaction with sulfur dioxide at -60°C
(44). No polysulfone formed at +20°C.

TRIALLYL PHOSPHATE

   Triallyl phosphate $(AO)_3PO$, or TAP, polymerizes comparatively fast with free radical initiation and copolymers have had considerable interest, especially for
flame-proofing fabrics, ion exchange resins, and plastics.
The monomer was made by Cavalier by reaction of Ag_3PO_4 and
allyl iodide (45). It is a colorless, pleasant-smelling
liquid, insoluble in water but soluble in ether and in
ethanol. It adds 6 atoms of bromine readily. It was
prepared also by addition of $POCl_3$ to allyl alcohol in
toluene-pyridine at -30 to 0°C (46). TAP was made by
slow, portionwise addition of $POCl_3$ and triethylamine to
allyl alcohol in a solvent (47). In a more recent process
$POCl_3$ was added slowly to a mixture of warm alcohol and
alkali alcoholate (48). TAP can be prepared also by oxidation of triallyl phosphite, e.g., by air or H_2O_2 (49).
Cavalier prepared a number of salts of monoallyl and
diallyl phosphoric acids and compared their melting
temperatures (45).
   Purified triallyl phosphate supplied by Aldrich boiled
at 90°C at 0.1 mm and had refractive index $n_D^{20} = 1.4496$.
Commercial TAP from Borden boiled at 80°C and 0.5 mm with
$n_D^{22} = 1.4508$.   TAP is an irritant but is considered only
moderately toxic based on oral LD_{50} of 1700 microliters/kg

of rat body weight. Hooker warns against heating TAP
above 130°C, where violent spontaneous decompositions may
occur. Early gelation by polymerizations at lower temperatures may cause exothermic overheating, leading to the
dangerous range. The author observed a commercial sample
of TAP pass through a viscous syrup to a clear, soft but
brittle gel on storage 6 years. The prepolymer was not
precipitated by excess methanol or acetone. The ability
of TAP to copolymerize with a wide variety of monomers
and to form soluble prepolymers makes it interesting for
nonflammable textile and paper coatings, plastics, and
oil modifiers. Purified TAP polymerizes quite readily on
heating with peroxide catalysts as well as by irradiation.
Thus TAP containing 5% BP gave hard, colorless polymer by
7 hr exposure to a mercury arc lamp (4). However, the
surface in contact with air was soft from oxygen inhibition.

Atomic Energy Research laboratories in England investigated TAP polymers (50). By heating $(AO)_3PO$ with aqueous
sodium hydroxide and acidifying, $(AO)_2P(O)OH$ was prepared;
it decomposed when distillation was attempted. On heating
for 6 hr at 100°C with 1.5% BP, $(AO)_3PO$, $(AO)_2P(O)A$, and
$(AO)_2P(O)CH_2COOC_2H_5$ gave predominantly crosslinked polymers. Hydrolysis during polymerization in aqueous suspension could be avoided by buffering at pH 8. Trimethallyl phosphate gave only viscous oily polymers by
suspension with BP. However, it could be converted to a
solid crosslinked polymer by exposing to ^{60}Co radiation
for 3 days. This polymer was acidic from radiation
damage. Crosslinked polymers of $(AO)_3PO$ and $(AO)_2P(O)A$
were refluxed with 20% aqueous sodium hydroxide for 3 hr
and then reacted with normal sulfuric acid to form cation-
exchange resins of capacity about 4.5 milliequiv/g.

Ion-exchange resins based on TAP polymers were studied
as selective adsorbents of uranium ions (51). In one
example, 10 g TAP with 0.15 g BP in a sealed glass tube
was heated for 3.5 hr at 100°C for partial polymerization
to a solid. The polymer mass was crushed to powder and
refluxed 30 min with ether. This extraction was repeated
four times to remove soluble fractions. The purified
polymer was treated for 3 hr with boiling 20% aqueous KOH
and then was washed with water and dried. The analysis
was that of a sodium diallyl phosphate polymer. The
crosslinked polymer was converted to the acid form by
treating with 1.0 N sulfuric acid. When the ion-exchange
resin was treated with uranyl sulfate at pH 1.0, it showed
a capacity of 3.0 milliequiv. U^{6+}/g. The order of affinity
at pH 1.0 for cations was $Th^{4+}>Fe^{3+}>U^{6+}>La^{2+}>Ba^{2+}>Na^+$.
The adsorbed uranyl ions could be eluted by aqueous

ammonium carbonate. The essentially nonionic crosslinked polymers of TAP, diallyl ethyl phosphate, and diallyl methyl phosphate showed even more affinity for metal ions; this was attributed to Werner-type complexing. Sodium diallyl phosphate polymers showed higher affinity for beryllium ions compared with sulfonated resins, suggesting that complex formation occurred with phosphate groups (52). Beryllium and uranium ions were quantatively adsorbed in the presence of disodium EDTA. On addition of sodium carbonate, U^{6+} ions were eluted and beryllium ions were retained in the resin bed.

A method of flameproofing textiles with bromine derivatives of triallyl phosphate developed as far as a pilot plant stage. Walter and co-workers polymerized TAP by heating it upon fibers with peroxide catalyst (53). For example, a solution of 50 parts TAP, 1.5 parts perbenzoic acid, and 150 parts benzene was used to impregnate cloth such as cotton, which was heated at 90°C for 30 min in nitrogen atmosphere. The 33% allyl unsaturation remaining was then reacted with a solution of bromine in carbon tetrachloride. Washing with dilute alkali and with water and drying at 110°C gave flameproof cloth resistant to laundering and dry cleaning.

TAP prepolymer was made in solution using an inhibitor to arrest polymerization short of 67% (54). The prepolymer could be precipitated by adding hexane. In one process the prepolymer was dissolved in $(ClCH_2CH_2)_2O$, and the residual double bonds were partiallyl brominated before the textile was impregnated. A final heat treatment cured the brominated polymer. Difficulties were encountered in completing the allyl polymerizations in aqueous dispersions.

Allyl amino-phosphorus compounds were also investigated for use in flameproofing textiles (54). Tetrabromodiallyl ammonium phosphate was made by dropwise addition of Br_2 to $A_2P(O)NH_2$ in solution in carbon tetrachloride-isopropanol. Allyl alcohol was reacted with $POCl_3$ in the presence of toluene and pyridine to give $A_2P(O)Cl$ which, by reaction with ammonia, gave $(AO)_2P(O)NH_2$.

The Southern Regional Laboratory of the U.S. Department of Agriculture investigated prepolymers and telomers from TAP for flameproofing cotton and other textiles. Bromomethylated TAP polymers were prepared in aqueous dispersion (55). In one example, 252 parts TAP, 163 parts $CHBr_3$, 12.5 parts $K_2S_2O_8$, 85 parts polyvinyl alcohol, 22.5 parts $NaHCO_3$, and 850 parts water were heated at 90°C for 2.5 hr. Cotton fabric was dipped into the polymer dispersion until the dried cloth carried 25% polymer solids. The treated cloth was heated for 4 to 8 min at

140°C for curing. This flameproofing resisted laundering.
However, tear resistance, color, and flexibility of the
cloth were impaired. Polymerizations of TAP in CCl_4
occurred on heating with peroxides, but in CBr_4 it was
necessary to add anhydrous $AlCl_3$ to activate the peroxide
polymerization (56). The soluble polymer could be precipitated by pouring into ether. The telomerizing agents
CH_2ClCH_2Cl and $CHCl_2CHCl_2$ were less promising in giving
flameproofing polymers. The bromoform-triallyl phosphate
polymers were evaluated in flameproofing of textiles. In
one modification the aqueous dispersions were blended
with melamine-formaldehyde resins (57). Allyl phosphorus
compounds and their evaluation in flameproofing were
reviewed (58). Recently cotton fabrics were flameproofed
by impregnating with TAP and N-methylol acrylamide, followed by application of ionizing radiation (59). Triallyl
phosphate has also been copolymerized with styrene and
with unsaturated polyesters (60).

Diallyl chlorophosphate (bp 65°C at <.01 mm) was found
to be extremely sensitive to heating (61). Even careful
vacuum distillation left viscous polymer residues. The
polymer contained only 0.93% chlorine, suggesting complex
polymerization with loss of halogen. Diallyl ethyl phosphate, because of the high reactivity of the allyl group,
was made at 0°C from the novel reaction of diallyl phosphonate, carbon tetrachloride, ethanol, and triethylamine.
Vacuum distillation at 72°C(2 mm) left a tan-colored
viscous polymer residue. Little has been published about
mono- and diallyl phosphates. Allyl phosphates have been
observed as products of fermentation (62). Minor proportions of diallyl hydrogen phosphate were disclosed to
copolymerize with lauryl methacrylate in high-boiling
hydrocarbon solution (0.5% BP, 18 hr, 70°C)(63). Such
long-alkyl methacrylate copolymers with allyl phosphates,
phosphinates, or phosphonates were suggested as lubricating oil additives.

TAP was found to crosslink preformed polymers such as
acrylate ester polymers on heating with peroxides (64).
Interesting studies of graft copolymerization of TAP and
diallyl phosphate on polyethylene were carried out by
Radiation Applications laboratories for the U.S. Atomic
Energy Commission (65). Choice of solvents greatly
affected grafting rates. This was attributed to a
Trommsdorff effect from insolubilization of the growing
graft copolymer chains and also from swelling of the base
polymer to give the monomer access to grafting sites.

Kropa copolymerized TAP with diethylene glycol fumarate
polyester with BP initiator; heating up to 120°C during
4 hr gave substantially insoluble, infusible copolymers
(66). TAP was copolymerized with vinyl acetate (67).

Cast sheets of methyl chloroacrylate-TAP copolymer have been studied (68). Copolymers of TAP with styrene were made by Dow (69). Copolymers of TAP with long-alkyl methacrylate esters of molecular weights 50,000 to 10^6 were proposed as oil modifiers (70). Copolymers of acrylonitrile with minor proportions of 2-chloroethyl or 2-bromoethyl allyl or methallyl phosphates were studied as flame-resistant products (71).

Triallyl phosphate has been copolymerized with a number of types of unsaturated ester polymers (72). In one example, an hydroxyl-terminated polyester was prepared from maleic anhydride and an excess of ethylene glycol. This was reacted with $POCl_3$ to give a phosphate-maleate polyester, which was copolymerized with TAP in the presence of glass cloth reinforcement using 3% BP at 100°C. Ethyl β-hydroxyethyl maleate was copolymerized with TAP by heating with 3% BP at 125°C. In another example, a low polymeric ester with terminal allyl groups was prepared by heating ethylene glycol, allyl alcohol, and maleic anhydride under an inert atmosphere. This was reacted with $POCl_3$ to form a phosphate-maleate polyester containing allyl groups that could be copolymerized with TAP.

Small proportions of TAP have been copolymerized with acrylic acid for the preparation of water-soluble polymers of specially high viscosity for mucilage applications (73). In one patent example, the following solution was heated under nitrogen for 20 hr at 70°C:

Acrylic acid	98-99.5 parts
TAP	0.5-2.0
Caprylyl peroxide	0.2
Benzene	900

The copolymer precipitated as formed in benzene. By neutralization with dilute ammonium hydroxide to pH 7, the branched acrylic acid copolymers gave 0.2% aqueous solutions having the following very high viscosities: 0.5% TAP units (21 poises); 1.0% TAP units (125 poises). References to other allyl phosphate esters are given in Table 28.2.

ALLYLIC PHOSPHITES

Of the allylic phosphites there has been greatest interest in triallyl phosphite $(AO)_3P$. This compound can be made by dropwise addition of PCl_3 in ether to an ether solution of allyl alcohol and $C_6H_5N(CH_3)_2$ (74). Triallyl phosphite also was made by reacting triphenyl phosphite with allyl alcohol and a little sodium allylate (75). It was suggested as an insecticide. Allyl phosphites were

TABLE 28.2

Allylic Phosphate Esters and Related

Compound	bp, °C (at mm)	Refractive index, etc.	References
$(AO)_3 PO$	80(0.5)	$n_D^{25} = 1.4472$ $n_D^{24} = 1.4520$	Pianfetti, U.S. 3,020,303 (FMC); Cadogan, Brit. 937,609; CA 59, 1469 (1963)
$(AO)_2 P(O) OH$	--	(Copolymers with methacrylates)	Kirby, CA 57, 2508 (1962); U.S. 3,166,505 (Shell)
$(AO)_2 P(O) OCH_3$	60(0.5)	--	Kennedy, J. Appl. Chem. 8, 459 (1958)
$(AO)_2 P(O) OC_2H_5$	72(2)	$n_D^{20} = 1.4350$	Steinberg, U.S. 2,666,778 (USA)
$(AO)_2 P(O) OC_6H_5$	102(0.5)	$n_D^{25} = 1.4957$ (Hard polymer)	Toy, JACS 76, 2192 (1954); cf. Yuldashev, CA 57, 12707 (1962)
$AOP(O)(OC_6H_5)_2$	155(0.3)	$n_D^{21} = 1.5388$	Miller, J. Chem. Soc. C 1837 (1968)
$(AO)P(O)(OCH_2 CHCH_2)_2$	--	--	Mueller, U.S. 2,826,592 (Shell)
$(AO)_2 P(O) ONH_4$	mp 134	--	Meise and Machleidt, Ann. Chim. 693, 76 (1966); CA 65, 7172 (1968)
$(AO)_2 P(O) NHR$	--	--	Walter, U.S. 2,574,515-518 (Martin)
$(MAO)_3 PO$	135(5) 92(0.1)	$n_D^{25} = 1.4454$ $n_D^{19} = 1.452$	Britton, U.S. 2,160,942; 2,176,416 (Dow) Kennedy, CA 53, 3032 (1959)
$(MAO)_2 P(O) OC_2H_5$	--	$n_D^{25} = 1.4390$	Britton
$(CH_2=CClCH_2O)_3PO$	133(1)	$n_D^{20} = 1.4866$	Reinhardt, JACS 74, 1093 (1952)
$A(OCH_2CH_2)_2 OP(O)(OM)_2$	--	(Copolymers with acrylonitrile)	Caldwell, U.S. 3,449,303 (Eastman)
Allyl thiophosphate	--	(Pesticides)	Regel, U.S. 3,513,233 (Chemagro)

670

TABLE 28.3

Some Allylic Phosphites

Compound	bp, °C (at mm)	Refractive index, etc.	References
$(AO)_3 P$	34 (0.06) or 84 (12)	$n_D^{20} = 1.4589;$ or $n_D^{20} = 1.4610$	Cade, J. Chem. Soc. 2269 (1959); Aldrich data
$(AO)_2 POH$	--	Unstable, forming $(AO)_2 P(O)H$	See phosphonates
$P(OCH_2CHClCH_2OA)_3$	--	--	Brit. 803,082 (Carlisle Chem.); CA 53, 8704 (1959)
$AOPOCH_2$ $\ \ \ \overset{|}{O} \overset{|}{C}H_2$	132	$n_D^{20} = 1.5025$	Arbuzov, CA 45, 1512 (1951)
$(MAO)_3 P$	72 (0.3)	$n_D^{25} = 1.458$	Hooker data sheet

patented as stabilizers for rubber and other polymers (76). Properties of allyl and methallyl phosphites are given in Table 28.3.

Triallyl phosphite has been supplied by Hooker (data sheet 1966). The colorless commercial liquid has an unpleasant odor (bp 84°C at 12 mm; $n_D^{20} = 1.459$). About 0.1% dissolves in water, and the monomer is miscible with ethanol, hexane, benzene, and acetone. The compound is toxic, having LD_{50} of 178 microliters per kg rat body weight. Hooker's data sheet stated that homopolymers of $(AO)_3 P$ had not been obtained.

Allyl phosphites can undergo the Michaelis-Arbusov rearrangement on heating to form phosphonates (77): $(AO)_3 P \longrightarrow (AO)_2 \overset{}{\underset{O}{P}} A$. In one example, $(BrCH_2CH_2O)_3 P$ was heated near 200°C using a thin-film reactor to form $(BrCH_2CH_2O)_2 P(O)CH_2CH_2Br$ (78). Substituted allyl phosphites were also studied (79). Triallyl phosphite can be oxidized to triallyl phosphate by heating with ethylene oxide under pressure (80). The double bonds were not attacked.

Cade of British Atomic Energy Research prepared triallyl phosphite and diallyl phosphite, finding that the physical properties for the latter differed from those previously reported (81). Alkyl phosphites are active reducing agents and may be oxidized under air with UV light. Cadogan and co-workers discovered that with 2800 Å of UV light, triallyl phosphite is oxidized to triallyl phosphate; but with the mercury resonance radiation (2537 Å) triallyl phosphite undergoes an exothermic reaction,

giving black polymers (82). Long irradiation of $(AO)_3P$
with the hot mercury light (3600 Å band) does not produce
polymerization, even when benzoyl peroxide is added as
photosensitizer (4). Triallyl phosphite does not homopolymerize under conventional conditions of free radical
polymerization. On heating with benzoyl peroxide, $(AO)_3P$
was oxidized with little polymerization (83). When the
monomer ($n_D^{22} = 1.4572$) in 2-chloroethyl ether was heated
at 150°C with t-butyl peroxide, a viscous solution resulted; addition of hydrocarbon gave a precipitate of
taffylike polymer of uncertain composition.

Copolymers of vinyl chloride with minor proportions of
triallyl phosphite may be made by aqueous persulfate- or
peroxide-initiated suspension systems (84). Outstanding
heat resistance was expected from internal stabilization.

There have been few other publications on copolymerization with triallyl phosphate (85, 86). Sodium methallyl
phosphite (3%) has been copolymerized with acrylonitrile
in aqueous bisulfite-persulfate at 40°C (87). The copolymer from solution in 67% nitric acid was spun to give
dyeable fibers.

OTHER ALLYL PHOSPHORUS COMPOUNDS

Many of these compounds are reducing agents, and they
may be expected to react with peroxide and other oxidizing catalysts. Phosphinate esters have the structure
$RR'P(O)OR''$, and such esters are known where allyl groups are
attached directly to phosphorus or oxygen. Allyl bromide
was reacted with $C_6H_5P(OCH_3)_2$ to form $C_6H_5P(O)(A)OCH_3$
(bp 111°C at 2 mm) (88). Attempts were made to polymerize
this compound, as well as the n-butyl ester, by heating
with t-butyl peroxide for 24 hr at 115°C. Allyl octyl-
phosphinate (134°C at 1 mm; $n_D^{20} = 1.4536$) was prepared (89).
To sodium allylate suspended in allyl alcohol was added
$(n-butyl)_2P(O)Cl$ at 0 to 5°C. After 30 min at 55°C, allyl
dibutyl phosphinate (95°C at 1 mm) was recovered. Viscous
low polymers were obtained from these and related compounds by heating for 5 hr at 175°C with t-butyl peroxide.
A number of esters of allylphosphinic acids, $AP(O)R(OR)$,
have been prepared by Liorber and co-workers at Kazan (90).

The allyl ester of dimethyl phosphinate $(CH_3)_2P(O)OA$
(bp 98°C at 14 mm) and the corresponding methallyl compound (bp 104°C at 13.5 mm) with 1% BP only gave red-brown
viscous solutions on heating for 25 hr at 80°C followed
by 25 hr at 115°C and 65 hr at 140°C (91). Ultraviolet
radiation for 100 hr without added peroxide gave no evidence of polymerization. Diallyl phenylphosphinate (bp
138°C at 2 mm) was prepared, as well as diallyl vinyl
phosphinate (bp 123°C at 1-2 mm) (92). The allyl ester

of ethyl allyl phosphinic acid gave a waxy solid on heating for 8 hr with 2% BP at 80 to 120°C (93). The allyl
ester of ethylvinylphosphinic acid was reported to polymerize rapidly. Diallyl phenylphosphonite $C_6H_5P(OA)_2$ (bp
100°C at 0.3 mm) and diallyl alkylphosphonites have been
considered for polymerization and copolymerization (92).
These compounds are relatively unstable. The compound
$AP(O-n-C_4H_9)_2$ was reported to boil at 59°C at 1.5 mm with
$n_D^{20} = 1.4500$ (95). Popov in Russia studied conjugation in
these and other allylic phosphorus-oxygen compounds from
infrared spectra (96).

Allylmagnesium chloride reacted with $POCl_3$ under cooling
to give triallylphosphine oxide A_3PO (bp 78°C at 0.1 mm
and mp of 17°C (74). Heating the monomer with t-butyl
peroxide at 200°C for 2 hr gave a yellow glassy polymer
(97). A mixture of A_3PO and unsaturated polyester (30 to
70%) was copolymerized in bulk with 1.5% BP (0.5 hr at
100°C) (98). Copolymerizations of lauryl methacrylate
with reluctant diallylphenylphosphine oxide at 80 and
140°C showed different reactivity ratios (99). Diallyl-
and vinylarylphosphine oxide derivatives were disclosed
vaguely, with suggestions for flame-resistant copolymers
(100). On heating with radical initiator, diallylphenylphosphine oxide gave soluble, low polymers believed to
contain cyclic segments (101).

ALLYLIC PHOSPHINES

Allyl phosphines are highly reactive and unstable compounds, and little has been published on their properties
and reactions. The tertiary phosphine $C_6H_5(C_2H_5)PA$ (bp
115°C at 15 mm) was made by cathodic reduction of solutions of quaternary phosphonium salts (102). Phenyl
monosodium phosphide in ether has been reported to react
with allyl chloride to give $C_6H_5P(A)H$ and other allylic
compounds (103). Surprisingly polymerization was observed
when the product was allowed to stand at 0°C. Heating
$CH_3P(A)H$ with B_2H_6 and tributylamine for 21 hr at 200°C
was said to give a linear liquid phosphinoborine polymer
suggested for use as a dielectric or hydraulic fluid
(104). Among other monoallyl phosphines prepared were
$AP(C_6H_5)_2$ (105) and allyl bis(2-cyanoethyl)phosphine
(bp 179°C at 1.6 mm; $n_D^{25} = 1.4980$) (106).

Triallylphosphine (bp 69°C at 13 mm) was prepared by
Jones and co-workers of Cardiff by reaction of allyl
magnesium bromide with phosphorus trichloride (107).
Diallyl was a byproduct. The same workers prepared tri-
methallylphosphine (bp 112°C at 15 mm) and phenyl dimeth-
allylphosphine (bp 148°C at 3 mm; $n_D^{15} = 1.5485$). The
allyl and methallyl phosphines oxidized rapidly in air

but were stable to heating an hour at 250°C in the absence
of air. The allyl phosphines reacted with $HgCl_2$ to give
$PR_3 \cdot HgCl_2$ addition compounds, with CS_2 to yield red-colored
products, and with CH_3I to give methyl phosphonium iodides.
Methods were proposed in Russia for the preparation of
allyl tertiary phosphines, for example, reaction of triphenyl phosphine with allyl acrylate or allyl methacrylate
(108).

The highly reactive lower alkyl phosphines take fire
spontaneously in air, but some aryl phosphines can be
handled with precautions. Butler and co-workers reacted
phenyl dichlorophosphine or diphenyl chlorophosphine with
allyl or methallyl magnesium halides to give tertiary-
allylarylphosphines (109). The latter were reacted with
alkyl halides in acetone solution at room temperature
under dry nitrogen to form the following allyl and methallyl phosphonium halide salts:

Salt	mp, °C
$[A_2(C_6H_5)PCH_3]^+ \bar{Br}$	99
$[A_2P(C_6H_5)_2]^+ \bar{Br}$	176
$[A(MA)P(C_6H_5)_2]^+ \bar{Br}$	164
$[(MA)_2P(C_6H_5)_2]^+ \bar{Br}$	180
$[A(CH_2=CH)P(C_6H_5)_2] \bar{Br}$	171
$[A_2P(C_6H_5)_2]^+ \bar{Cl}$	160

The diallyl aryl phosphonium halides were heated in water
or dimethyl formamide with azobis under nitrogen at 60°C
for 3 to 34 days; the resulting polymers had low intrinsic
viscosities (below 0.1 in ethanol). The polymers had
melting points somewhat higher than those of the monomers,
and they showed little unsaturation. Butler suggested
that cyclopolymerization occurred in the case of diallyl
diphenyl phosphonium bromide. The hydroscopic polymers
seemed to retain a molecule of water per phosphonium
group. The low polymer of diallyl diphenyl phosphonium
bromide was heated with alcoholic KOH to form a brown
gummy solid, believed to be a polymer of diallyl phenyl
phosphine oxide. When dry it softened at 260°C. MCB
Corporation supplies $AP(C_6H_5)_3Br$ melting at 223°C.

Little work has been reported on sulfur analogs of allyl
phosphorus compounds such as thiophosphonates (110) and
thiophosphinates (111). Formation of P=S bonds from
P-S-A bonds can occur by Michaelis-Arbusov rearrangement
on heating. Polyfunctional allyl phosphorine derivatives
have been copolymerized with acrylic and styrene monomers
(112).

References

1. E. L. Gefter, Organophosphorous Monomers and Polymers, Pergamon, 1962.
2. Chem. Eng. News 30, 4515 (1952). For the interpretation here I am indebted to A. D. F. Toy.
3. A. D. F. Toy, U.S. 2,425,765 (Victor); J. Am. Chem. Soc. 70, 186 (1948); cf. U.S. 2,186,360 and 2,945,052 (U.S. Rubber).
4. C. E. Schildknecht, unpublished.
5. A. D. F. Toy and L. V. Brown, Ind. Eng. Chem. 40, 2276 (1948); U.S. 2,497,637-8 (Victor); cf. E. C. Shokal and L. N. Whitehill, Brit. 645,222 (Shell); CA 45, 4267 (1951).
6. D. E. Warren, CA 41, 6767 (1947); A. D. F. Toy, U.S. 2,485,677, 2,538,810, 2,586,834, 2,586,884-5 (Victor).
7. A. D. F. Toy and Brown, Ind. Eng. Chem. 40, 2276 (1948).
8. J. R. Costello and T. P. Traise, U.S. 2,867,597 (Victor).
9. K. I. Beynon, J. Polym. Sci. A 1, 3343 (1963).
10. S. Hashimoto and I. Furukawa, Chem. High Polym. (Japan) 21, 647-651 (1964).
11. T. Skwarski (Poland), CA 73, 131357 (1970); cf. CA 78, 5336 (1973).
12. I. N. Fazullin et al, CA 74, 112479 (1971).
13. A. D. F. Toy and L. V. Brown, U.S. 2,586,885 (Victor); cf. U.S. 2,577,281 (Lockheed).
14. D. Harman and A. R. Stiles, U.S. 2,601,520; cf. U.S. 2,659,714 and 2,681,920 (Shell).
15. A. D. F. Toy and R. S. Cooper, J. Am. Chem. Soc. 76, 2191 (1954); G. E. Walter et al, U.S. 2,574,516.
16. A. J. Castro and W. E. Elwell, J. Am. Chem. Soc. 72, 2275 (1950).
17. A. O. Hungerford, U.S. 3,038,210 (Parmelee Plastics).
18. J. J. Lugert, U.S. 2,675,586 (Eastman); E. Carnall and J. J. Lugert, U.S. 2,768,108 (Eastman).
19. A. D. F. Toy et al, U.S. 2,735,789 (Victor); J. R. Costello and T. P. Traise, U.S. 2,841,507 (Victor); A. D. F. Toy et al, U.S. 2,867,547 (Victor).
20. B. W. Lew, U.S. 2,726,177 (Atlas).
21. W. W. Spooncer, Belg. 608,759 (Shell); CA 60, 15666 (1964).
22. D. Harman and A. R. Stiles, U.S. 2,632,756 (Shell).
23. C. W. Smith et al, U.S. 2,856,369 (Shell).
24. N. E. Boyer and R. R. Hindersinn, U.S. 3,376,274 (Hooker).
25. J. Kennedy, Brit. 825,767 (U.K. Atomic Energy); CA 54, 11565 (1960).
26. J. A. Cade, J. Chem. Soc. 2266 (1959); CA 54, 263 (1960).
27. P. I. Pollak and H. L. Slates, U.S. 3,597,510 (Merck).
28. A. D. F. Toy and K. H. Rattenbury, U.S. 2,714,100 and 2,918,449 (Victor).
29. M. K. Atakozova et al, CA 69, 107,115 (1968).
30. V. A. Kargin et al (Moscow), CA 72, 56029 (1970).
31. A. D. F. Toy and R. S. Cooper, J. Am. Chem. Soc. 76, 2193 (1954).
32. G. E. Walter et al, U.S. 2,574,516 (Martin).
33. W. E. Craig and W. F. Hester, U.S. 2,495,958 (R & H).

34. E. I. Shugurova and G. Kamai, J. Gen. Chem. 21, 658 (1951); CA 45, 8970 (1951); CA 51, 3440 (1957).
35. J. Kennedy, E. S. Lane and B. K. Robinson, J. Appl. Chem. 8, 459 (1958); CA 53, 3033 (1959); cf. J. Kennedy and Davies, Chem. Ind. (London) 308 (1956).
36. J. Kennedy and G. E. Ficken, J. Appl. Chem. 8, 465 (1958).
37. R. A. V. Raff, I. W. Herrick, and M. F. Adams, Forest Prod. J. 16, 43 (February 1966).
38. A. E. Arbuzov and V. M. Zoroastrova, CA 45, 1512 (1951); G. M. Kosolapoff, J. Am. Chem. Soc. 73, 4040 (1951); Boyer, U.S. 2,866,807 (Va-Carolina); E. A. Perren et al, Brit. 707,961 (Natl Res. Dev.); J. F. Allen and O. H. Johnson, J. Am. Chem. Soc. 77, 2871 (1955); S. Melamid, U.S. 2,843,586 (R & H); L. Burger and R. M. Wagner, CA 53, 15943 (1959); T. Skwaski and Wodka, CA 76, 46561 (1972).
39. J. B. Dickey and H. W. Coover, U.S. 2,721,876; cf. U.S. 2,780,616 (Eastman).
40. H. W. Coover, U.S. 2,827,475; cf. 2,632,768 and 2,636,027 (Eastman).
41. H. W. Coover and J. B. Dickey, U.S. 2,636,027; T. Skwaski and T. Wodka, CA 68, 115003 (1968).
42. V. I. Kodolov and S. S. Spasskii, CA 68, 87638 (1968).
43. A. I. Razumov et al, CA 69, 107,106 (1968); I. A. Krivosheyeva et al, J. Polym. Sci. USSR 11, 287 (1969).
44. I. N. Faizullin et al, CA 72, 121,987 (1970).
45. J. Cavalier, Ann. Chim. Phys. [7] 18, 444-507 (1899); CZ I, 102 (1900); cf. Compt. Rend. 124, 91 (1896).
46. L. N. Whitehill and R. S. Barker, U.S. 2,394,829 (Shell); explosions from overheating.
47. A. D. F. Toy and J. R. Costello, U.S. 2,754,315 (Victor).
48. J. O. Pianfetti and P. L. Janey, U.S. 3,020,303 (FMC).
49. I. Hechenbleikner, U.S. 2,851,476 (Shea Chemical); J. I. G. Cadogan, Brit. 937,609 (Albright and Wilson); cf. A. J. Burns and J. I. G. Cadogan, CA 59, 1469 (1963); Brit. 937,560 (Hooker).
50. J. Kennedy, U.S. 2,882,248 (USA).
51. J. Kennedy, Brit. 777,248; CA 51, 15839 and 16044 (1957).
52. J. Kennedy and V. J. Wheeler, Anal. Chim. Acta 20, 412 (1959); CA 53, 21403 (1959).
53. G. E. Walter, I. Hornstein, and C. A. Sheld, U.S. 2,660,542-3 (Martin); cf. Chem. Eng. News 31, 326 (1953).
54. G. E. Walter et al, U.S. 2,574,515-7 (Martin).
55. J. G. Frick and J. W. Weaver, U.S. 2,686,768-9; J. Polym. Sci. 20, 307 (1956); cf. U.S. 2,892,803, 2,810,701 (all USA); Textile Res. J. 25, 24 (1955); 26, 136 (1956).
56. J. W. Weaver, U.S. 2,778,747 (USA).
57. I. Hechenbleikner, U.S. 2,832,745 (Shea Chemical).
58. R. C. Laible, Chem. Rev. 58, 823 (1958).
59. T. D. Miles and A. C. Delasanta, U.S. 3,592,683 (USA).
60. U.S. 2,186,360 (Dow); U.S. 2,443,735 (Am. Cyan.); Farbe und Lack 62, 181 (1956).

61. G. M. Steinberg, J. Org. Chem. 15, 637 (1950); U.S. 2,666,778 and CA 49, 6300.
62. S. Baba (Tokyo), CA 45, 9114 (1951).
63. P. Kirby, Ger. 1,124,620 and U.S. 3,208,942 (Shell), CA 57, 2508 (1962).
64. Brit. 930,761 (Bayer); Ger. appl. 1960; CA 59, 10302 (1963).
65. G. G. Odian et al, CA 50, 4929 (1956) and 57, 13979 (1962).
66. E. L. Kropa, U.S. 2,443,740 (Am. Cyan.), examples 24 and 25 (appl. 1944); cf. U.S. 2,409,633 (Am. Cyan.); U.S. 2,583,356 (U.S. Rubber).
67. J. W. Haworth, Brit. 675,783 (Brit. Oxygen); CA 46, 11778 (1952); Brit. 706,577 (Brit. Oxygen).
68. H. D. Anspon, U.S. 3,234,192 (GAF).
69. E. C. Britton et al, U.S. 2,186,360 (Dow).
70. P. Kirby, U.S. 3,208,943 (Shell).
71. F. E. King et al, U.S. 2,999,085 (Celanese).
72. W. F. Bruksch and L. H. Howland, U.S. 2,583,356 (U.S. Rubber).
73. F. A. Wagner, U.S. 3,426,004 (Goodrich).
74. J. Kennedy et al, J. Appl. Chem. (London) 8, 459 (1958); Brit. 778,077 (U.K. Atomic Energy); I. H. McCombie et al, J. Chem. Soc. 380 (1945).
75. I. Hechenbleikner et al, U.S. 3,056,823 (Hooker); C. F. Baranauckas et al, Belg. 611,708 (Hooker); cf. U.S. 2,852,551 (Shea).
76. J. R. Mangham and T. M. Melton, U.S. 3,027,348 (Va.-Carolina).
77. A. Arbusov, J. Russ. Phys.-Chem. Soc. 38, 687 (1906); CZ II 1640 (1906); A. H. Ford-Moore, J. Chem. Soc. 1465 (1947).
78. R. A. Davis and E. L. Larsen, U.S. 3,483,279 (Dow).
79. A. N. Pudovik and I. M. Aladzheva, CA 59, 13798 (1963).
80. C. B. Scott, U.S. 2,909,555 (Collier).
81. J. A. Cade, J. Chem. Soc. 2269 (1959); cf. J. Am. Chem. Soc. 76, 2191 (1954); cf. Gerrard, J. Chem. Soc. 1464 (1940).
82. J. I. C. Cadogan, M. Cameron-Wood, and W. R. Foster, J. Chem. Soc. 2549 (1963).
83. R. C. Laible et al, J. Appl. Polym. Sci. 1, 376 (1959).
84. B. D. Halpern, U.S. 3,069,400 (Borden); S. Koyanagi, Japan 71-20,492 (Shin-Etsu).
85. H. Schmidt, Ger. 1,098,709 (Hoechst); CA 56, 6166 (1962).
86. R. R. Hindersinn and N. E. Boyer, Ind. Eng. Chem. Prod. Dev. 3, 141 (1964).
87. N. Fujisaki et al, Japan 15,088 (1963); CA 59, 15402 (1963).
88. A. R. Stiles and D. Harman, U.S. 2,711,403 (Shell); cf. H. Reinhardt et al, Ber. 90, 1656 (1957).
89. D. Harman and A. R. Stiles, U.S. 2,659,714 (Shell).
90. B. G. Liorber et al, CA 64, 15914 (1966); CA 65, 736 (1966); cf. CA 61, 8335 (1964).
91. H. Reinhardt et al, Chem. Ber. 90, 1656 (1957); cf. Mat'kova CA 55, 7898 (1961).
92. E. L. Gefter, CA 59, 1677 (1963); CA 56, 10184 (1961).
93. G. Kamai and V. S. Tsivunin, CA 54, 7539 (1960).

94. R. C. Morris et al, U.S. 2,577,796 (Shell).
95. M. I. Kabachnik et al, CA 58, 9127 (1963).
96. E. M. Popov et al, CA 58, 5165 (1963).
97. H. C. Fielding, Brit. 864,086 (ICI); CA 55, 19343 (1961).
98. R. S. Ludington and W. W. Young, U.S. 3,009,897 (Westinghouse).
99. K. I. Beynon, J. Polym. Sci. A 1, 3357 (1963).
100. R. S. Cooper, U.S. 3,035,096 (Stauffer).
101. K. D. Berlin and G. B. Butler, J. Am. Chem. Soc. 82, 2712 (1960);
     cf. J. Org. Chem. 25, 2006 (1960); W. W. Spooncer, U.S. 3,160,593
     (Shell).
102. L. Horner and A. Mentrup, Ger. 1,114,190 (Hoechst); CA 57,
     2256 (1962).
103. E. Steininger and M. Sander, Angew. Chem. 75, 88 (1963); CA 59,
     1677 (1963).
104. A. B. Burg and R. I. Wagner, U.S. 3,071,553; cf. U.S. 2,926,194
     and 3,071,553 (Am. Potash).
105. M. C. Browning et al, J. Chem. Soc. 693 (1962); CA 57, 9448
     (1962).
106. M. Grayson and P. T. Keough, U.S. 3,005,013 (Am. Cyan.).
107. W. J. Jones et al, J. Chem. Soc. 1446 (1947).
108. G. M. Vinokurova and S. G. Fattakhov, J. Gen. Chem. (USSR) 36,
     70-73 (1966).
109. G. B. Butler et al, Preprint Organic Coatings and Plastics 29,
     244 (1969); cf. Rothstein, J. Chem. Soc. 4002 (1953); Keough,
     J. Org. Chem. 29, 631 (1964).
110. I. N. Faizullin and L. E. Shvltsova, CA 76, 72827 (1972);
     cf. D. Harman, U.S. 2,630,450 (Shell).
111. A. E. Arbusov and K. V. Nikonorov, CA 43, 3801 (1949).
112. H. R. Allcock and W. M. Thomas, U.S. 3,329,663 (Am. Cyan.).

29. ALLYL SILICON COMPOUNDS

Although few industrial applications have come from research on allyl compounds containing silicon, some interesting chemistry has been disclosed. The allyl chlorosilanes are discussed first, since the allyl-containing silicone polymers derived therefrom were evaluated early by General Electric before the commercial development of vinyl-modified curable silicones. Vinyl silicone copolymers generally are more stable to alkali and to heating in air than allyl silicones. Allyl silanes were studied both in America and Russia. Allyl alkoxysilanes were developed in the United States as finishes for glass fibers in reinforced thermosetting plastics. There has been little work on allyl silicates and chlorosilicates since that of Peppard and co-workers. A number of allyl silicon compounds are available from Pierce Chemical Company, Rockford, Illinois, and Rotterdam, Holland.

Practical applications of allyl silicon compounds have been limited by their high reactivity in side reactions competing with monomer synthesis and polymerization, as well as by low rates of polymerization and difficulties in preparing really high-molecular-weight thermoplastic polymers, either ethylenic type polymers $--CH_2CH--$ or siloxane polymers $--Si(A)_2O--$. $CH_2Si\lessgtr$ Polyfunctional allyl silicone or siloxane compounds can form crosslinked polymers by free radical initiation. However, residual allyl groups in these polymers often are unstable when heated in air and this leads to discoloration. The resonance energy of the allyl radical is believed to lower the dissociation energy of the allyl-silicon bond and to promote reactivity.

Ziegler-type catalysts have given soluble stereoregular polymers of moderate molecular weight from allyl alkyl silanes. Other complex ionic catalyst systems such as Grignard reagents may be expected to polymerize allyl silanes. Interesting copolymers form with maleic anhydride in a number of cases. The highly reactive allyl silicon compounds make a challenging area for further academic research. Research on allyl silicon compounds has not been sufficiently complete for drawing many generalizations.

679

ALLYL CHLOROSILANES AND THEIR POLYMERS

The literature on allyl chlorosilanes and their polymers was reviewed until 1955 by George, Prober, and Elliott (1). Allyl chlorosilanes comprise about 60% of the product of the vigorous reaction of allyl chloride with silicon-copper at 230 to 300°C. The principal volatile product is $ASiCl_3$ (2). This compound also can be made by dehydrochlorination, but the propenyl compound then predominates (3). Temperatures below 165°C with $AlCl_3$ catalyst were used to minimize silicon-carbon cleavage and polymerization. Hurd found large differences in reactivity of the chlorides with copper-silicon in the order $ACl>CH_3Cl>CH_2=CHCl$. He reacted $ASiCl_3$ with CH_3MgBr in ether to give $A(CH_3)SiCl_2$.

Allyl trichlorosilane was made by reacting allyl chloride vapors with finely divided silicon at about 300°C (4). By halogenation of propyl compounds, followed by dehydrohalogenation, allyl chlorosilane derivatives were prepared (5). Hurd observed that A_2SiCl_2 formed nonvolatile polymers when distillation was attempted at 130°C. The compound slowly polymerized and darkened on storage. Tyran obtained a brown, brittle polymer in low yield by long heating at 70°C with BP (6).

Allyl and vinyl chlorosilanes have sharp olefinic odors; they fume in moist air and slowly liberate HCl. In the hydrolysis, alcoholysis, and other reactions of allyl halosilanes, an acid acceptor such as pyridine may be added in order to minimize silicon-carbon bond cleavage. Such cleavage contributes to the difficulties in preparing linear allyl-containing polysiloxanes of high molecular weight.

Hurd disclosed the hydrolysis of allyl and diallyl dihalosilanes to give oily low siloxane polymers (or cross-linked polymer gels):

$$
\begin{array}{ccc}
A & A & A \\
ClSiCl + H_2O \longrightarrow & HOSiOH \longrightarrow & --OSi-- + H_2O \\
CH_3 & CH_3 & CH_3
\end{array}
$$

When these unsaturated silicones are heated with peroxides or in air, slow curing occurs by crosslinking through allyl groups. However, introduction of allyl or vinyl groups into silicones gave inferior oxidation resistance and thermal stability compared with unmodified methyl and phenyl silicones (7). Diallyl siloxane low polymer oil obtained by aqueous hydrolysis of A_2SiCl_2 polymerized to clear solid films in air at room temperature in a few days. The large amount of residual unsaturation in such polymers results in poor oxidation resistance with discoloration. Heating the films at 100°C for a week caused disintegration.

Research of General Electric concentrated on synthesis of curable copolysiloxanes containing largely methyl groups with smaller proportions of phenyl, allyl, and/or vinyl groups. Silicone rubbers containing about 1% or more vinyl groups for curing were promising because of low compression set, hardness, and relatively short curing time. Silicone rubbers were also studied using minor proportions of CH_3ASiCl_2, A_2SiCl_2, or $ASiCl_3$ in cohydrolysis with $(CH_3)_2SiCl_2$ in ether-ice-water mixtures. Properties of these allyl silicone rubbers were favorable except that their oxidation on heating was rapid when more than 20% of the polymer side groups were allyl. An important advantage of vinyl groups over allyl groups is resistance to oxidation. When an experimental siloxane polymer containing 80% diphenyl and 20% diallyl silicone units was baked at 200°C for 48 hr, the product was gel-like at 200°C and a fairly tough solid at room temperature. Properties of glass-fiber-reinforced cured silicones are given in the Materials Selector 1973 published by Reinhold.

Hurd and Roedel reported copolymerization of allyl methyl siloxane liquid polymers by long heating with butyl methacrylate, MMA, or vinyl pyridine in the presence of 1% t-butyl perbenzoate. The methacrylate polymers exhibited the best clarity and hardness without brittleness. Copolymerization with styrene gave white, opaque products. Copolymers from allyl methyl siloxane low polymers and o-diallyl phthalate were harder and tougher than unmodified, cured polysiloxanes. Polysiloxanes of low molecular weight bearing allyl groups such as $ASi(CH_3)_2-[OSi(CH_3)_2]_nR$ (where $n = 1$ to 4) were isolated as fractions by vacuum distillation (8). Burkhard prepared $ASi(CH_3)_2OSi(CH_3)_3$, $n_D^{20} = 1.4061$, and $[ASi(CH_3)_2]_2O$, $n_D^{20} = 1.4280$ (9).

The condensation of methyl hydroxysilanes to form silicone-type polysiloxanes occurs so rapidly that the silanols ordinarily cannot be isolated. However, higher alkyl- and allyl-substituted silanols are more stable. The compound $A_2Si(OH)_2$ was reported to melt at 81°C (10). From hydrolysis of $AC_6H_4Si(CH_3)Cl_2$ the diol $AC_6H_4Si(CH_3)(OH)_2$ was prepared (11). Allylphenylsilanediol and allylbenzyl-silanediol have been made by hydrolysis of the diacetoxy-silane derivatives (12). Such aromatic silanediols do not polymerize readily by condensation.

Scott and Frisch prepared allyl aryl chlorosilanes from allyl chlorosilanes and appropriate Grignard reagents (13). Being careful to avoid contact with air, they were able to isolate A_3SiCl, along with the mono and diallyl compounds, from reaction of allyl magnesium bromide in ether with silicon tetrachloride. They studied IR absorption bands of allyl chlorosilanes. In progressing from mono to tetraallyl compounds, the IR bands near 8.5 and 11.0 microns shifted position toward longer wavelengths, whereas the bands near 13 microns, attributed to stretching, shifted to shorter wavelengths.

Methallyl dichlorosilane and dimethyl dichlorosilane in 1:9 molar ratio were cohydrolyzed to give clear oily siloxane copolymers of low molecular weight (14). A number of other laboratories besides those of General Electric contributed to the study of allyl halosilanes. Kropa made diallyl dichlorosilane by reacting allyl magnesium bromide and silicon tetrachloride (15). The Grignard solution was added slowly to SiCl$_4$ in ether with good agitation. The A$_2$SiCl$_2$ recovered was poured slowly into water to prepare silicone-type viscous liquid polymers. These could be set to a gel by heating with benzoyl peroxide for 2 hr at 100°C, but they required 24 hr of heating at 100°C followed by heating at 135°C to attain a solid polymer of constant hardness. Copolymerizations of allyl chlorosilanes, allyl silicols, and allyl-containing silicones were suggested with unsaturated polyesters and with a number of vinyl, acrylic, and styrene monomers.

Swiss and Arntzen of Westinghouse pointed out the limitation of dimethyl siloxane copolymer elastomers (uncrosslinked) in their swelling and crumbling in hydrocarbon solvents such as toluene (16). They showed that allyl-bearing silicone copolymers could be cured by heating with peroxides to give better solvent resistance. Such silicone polymers could be prepared by hydrolysis of mixtures of CH$_3$Si(OC$_2$H$_5$)$_3$, (CH$_3$)$_2$Si(OC$_2$H$_5$)$_2$, ASi(OC$_2$H$_5$)$_3$, xylene, and sulfuric acid catalyst; for example, by adding the mixture dropwise to equal parts of cold water and ethyl borate. A 60% solution of the siloxane copolymer in toluene was spread on a metal surface. The coating could be cured at 120°C overnight to a tack-free, hydrocarbon-resistant condition. However, curing was accompanied by discoloration.

As reported by Bailey and Pines, in the hydrolysis of ASiCl$_3$, cleavage of allyl-silicon bonds competes with polymerization, and aqueous alkali may cause complete cleavage (17). Vinyl silane derivatives are more stable to alkali. Allyl trichlorosilane in ethanol with quinoline gives ASi(OC$_2$H$_5$)$_3$. The latter undergoes disproportionation in the presence of NaOC$_2$H$_5$ to form Si(OC$_2$H$_5$)$_4$ and A$_2$Si(OC$_2$H$_5$)$_2$.

Addition polymerization of allyl halosilanes has not been promising. For example, A$_2$SiCl$_2$ with 25% benzoyl peroxide was heated under nitrogen for 64 hr at 70°C (18). Evaporation gave only a small residue of brown, brittle solid. Allyl trichlorosilane is even less reactive toward additions to the double bond, including polymerization,

TABLE 29.1

Allyl and Methallyl Halosilanes

Compound	bp, °C (at mm)	Refractive index	References
ASiCl$_3$ [a]	118	$n_D^{20} = 1.4449$	George, Chem. Rev. **56**, 1065 (1956)
ASiHCl$_2$	98	--	Hurd, JACS **67**, 1813 (1945)
ASiCH$_3$Cl$_2$ [a]	120	$n_D^{20} = 1.4419$	Hurd
ASi(CH$_3$)$_2$Cl [a]	112	$n_D^{20} = 1.4295$	Hurd
ASiF$_3$	46(20)	--	Shaw, Brit. 637,739
ACH$_2$SiCl$_3$	143	$n_D^{20} = 1.4548$	Bailey, Ind. Eng. Chem. **46**, 2363 (1954)
ACH$_2$=CHSiCl$_2$	65(57)	$n_D^{25} = 1.4602$	Scott, JACS **73**, 2599 (1951)
AC$_6$H$_5$CH$_2$SiCl$_2$	85(1)	$n_D^{25} = 1.597$	Scott
AC$_6$H$_5$SiCl$_2$ [a]	101(8)	$n_D^{25} - 1.5351$	Scott
A$_2$SiCl$_2$ [a]	166 or 84(50)	--	Hurd; also U.S. 2,386,793 (Du Pont)
A$_2$SiF$_2$	46(20)	--	Shaw
A$_3$SiCl	98(30)	$n_D^{25} = 1.4779$	Scott
A$_3$SiF	67(20)	--	Shaw
ASiBr$_3$	185	$n_D^{20} = 1.5350$	Chernyshev, CA **57**, 8601 (1962)
MASiCl$_3$	141	$n_D^{20} = 1.4504$	Shaw; also Plueddemann, U.S. 2,642,447 (L-O-F)
(MA)$_2$SiCl$_2$	--	--	Hyde, U.S. 2,480,822
(MA)$_3$SiCl	123(20)	--	Peppard, JACS **68**, 70 (1946); U.S. 2,394,642 (PPG)

[a] Supplied by Pierce Chemical Company

than is allyl trimethyl silane. In other reactions, ASiCl$_3$ behaves somewhat similarly to CH$_2$=CHSiCl$_3$. Allyl trichlorosilane has been reacted with 3 moles of glycidyl methacrylate to give ASi[OCH$_2$CHClCH$_2$OC(O)C(CH$_3$)=CH$_2$]$_3$, a liquid stable at -10°C which polymerized on warming to form a crosslinked product (19).

Properties and references to allyl halosilanes are given in Table 29.1. Infrared spectra of allyl chlorosilanes and tetraallyl silane are shown in Fig. 29.1.

ALLYL ALKOXYSILANES

The mono-, di-, and triallyl ethoxy compounds are best known. The allyl alkoxysilanes are colorless, liquid organosilicon esters with faint camphorlike odors. They

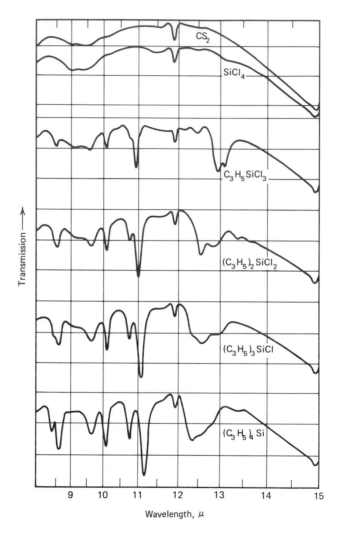

Fig. 29.1. Infrared spectra of allyl chlorosilanes and tetraallyl silane (10% in CS₂). Note the increasing absorptions and shifts of wavelength with introduction of more allyl groups. R. E. Scott and K. C. Frisch, J. Am. Chem. Soc. 73, 2599 (1951).

are soluble in common organic solvents and only slightly
in water, in which they hydrolyze rapidly. Allyl triethoxysilane was obtained by alcoholysis of allyl trichlorosilane in the presence of pyridine as HCl acceptor
(20). In the absence of added base, yields are low
because the hydrogen chloride formed may add to the
double bond of the $ASi(OC_2H_5)_3$ and also may lead to
cleavage of the carbon-silicon bond:

$$ASi(OC_2H_5)_3 + HCl \longrightarrow CH_3CHClCH_2Si(OC_2H_5)_3 \longrightarrow$$

$$CH_3CH=CH_2 + ClSi(OC_2H_5)_3$$

Mixtures of mono-, di-, and triallyl ethoxysilanes were
obtained by reaction of allyl chloride or allyl bromide
with ethyl orthosilicate in the presence of magnesium (21):

$$ACl + Si(OC_2H_5)_4 + Mg \longrightarrow ASi(OC_2H_5)_3 + MgOC_2H_5Cl$$

An improved procedure employed 1008 g of finely divided
magnesium turnings in a flask equipped for Grignard reactions; to this was added a "starter" consisting of 15 ml
of ethyl bromide in 25 ml of dry ether (22). Next was
added 2080 g of $Si(OC_2H_5)_4$ in 4500 ml of ether. With
cooling by an ice bath, 3822 g of methallyl chloride was
added dropwise during 14 hr. The mixture was then refluxed for 2 hr and cooled, and the solid were filtered
off. After ether had been distilled off, vacuum distillation yielded the following: 234 g of $MASi(OC_2H_5)_3$,
604 g of $MA_2Si(OC_2H_5)_2$, and 660 g of $MA_3SiOC_2H_5$. A
similar experiment using allyl chloride was described in
which the largest yield was the monoallyl triethoxysilane.
Allyl alkoxysilanes were prepared by promoted Grignard
coupling reactions without solvent (23). The allyl
ethoxysilanes were estimated by addition of bromine
(bromate-bromide) during 30 min in the dark.

Peppard and co-workers prepared $(C_2H_5O)_2SiA_2$ by adding
cold, dry allyl alcohol to a cold mixture of $(C_2H_5O)_2SiCl_2$
and pyridine (24). An ice bath was used to control the
strongly exothermic reaction. The preparation of allyl
methyl esters by this procedure was unsatisfactory,
probably because of greater susceptibility to hydrolysis.
A large amount of gelatinous solid was formed. Hexaallyl
and hexamethallyl disilicates were prepared from the
corresponding trialkyl chlorosilicates. Hydrolysis of
the diallyl and dialkenyl dichlorosilicates in the presence of pyridine gave replacement of halogen without
replacement of alkoxy groups. The clear, viscous liquid
products showed nearly the correct analysis for low
polymers $--SiO(OR)_2--$. Hydrolysis eventually proceeds
to completion with formation of silica gel.

Some physical properties of allyl alkoxysilanes are

given in Table 29.2. The reactions of allyl and methallyl alcohol with commercially available tetraethyl
orthosilicate were found by Peppard and co-workers to be
sensitive to HCl, $SiCl_4$, and other impurities. Alcoholysis of di- and trialkoxysilanes were accelerated by adding
$SiCl_4$ in catalytic amounts. Heating $(C_2H_5O)_4Si$ with
methallyl alcohol in dioxane with portionwise addition
of a little $SiCl_4$ gave a mixture of mono-, di-, and trimethallyl ethyl silicates.

Allyl ethoxysilanes can undergo disproportionation in
the presence of sodium alkoxides; for example (25):

$$2ASi(OC_2H_5)_3 \xrightarrow{NaOCH_3} A_2Si(OC_2H_5)_2 + Si(OC_2H_5)_4$$

Mercaptans add in the expected reverse way to allyl alkoxysilanes under free radical conditions (26). Bromine
in carbon tetrachloride adds to the double bond of
$ASi(OC_2H_5)_3$ without further reaction. However, in the
presence of acetic acid nearly 4 equivalents of bromine
react, the allyl-silicon bond is split, and addition to
the allyl double bond occurs (27).

Purified allyl ethoxysilanes may polymerize slowly at
room temperature becoming increasingly viscous and finally
forming clear, glassy solids (28). The monomers polymerize more rapidly on heating with free radical initiators such as t-butyl peroxide (29).

Allyl and methallyl alkoxysilanes may hydrolyze to
silanols or "silicols" HOSi(A)(R)OH, which can condense
to form water and unsaturated siloxane low polymers
capable of curing by heating with peroxide catalysts (29,
30). Thus a solvent-resistant, heat-stable siloxane
polymer suggested for electrical insulation was obtained
by dropwise addition at 20 to 25°C of 100 ml of 5% H_2SO_4
to 36 ml of $MASi(OC_2H_5)_3$, 60 ml of $CH_3Si(OC_2H_5)_3$, and
97 ml of $(CH_3)_2Si(OC_2H_5)_2$ in 450 ml of toluene (31). The
organic layer containing prepolymer was separated and
treated by stirring with successive 100-ml portions of 50,
60, 70, and 75% aqueous H_2SO_4 for half-hour periods, in
order to promote further condensation. After the organic
layer had been treated with ice and a liter of water, the
polymer was recovered by distilling off the toluene. A
segment of the copolymer can be represented:

$$\begin{array}{cccccc} & --O & CH_3 & CH_3 & O-- \\ & --Si-O-Si-O-Si-O-Si-O-- \\ CH_2=C(CH_3)CH_2 & CH_3 & CH_3 & CH_3 \end{array}$$

The use of allyl phenyl siloxanes to prepare curable siloxane copolymers also was suggested (32).

TABLE 29.2

Allyl and Methallyl Alkoxysilanes

Compound	bp, °C (at mm)	Refractive index	References
$A_3SiOC_2H_5$	129(84)	$n_D^{25} = 1.4569$	Swiss, U.S. 2,595,728 (Westinghouse); Scott, JACS 72, 2600 (1951)
$A_2Si(OC_2H_5)_2$	190(741)	$n_D^{25} = 1.4316$	Swiss; also Brit. 624,362 (Westinghouse)
$A_2Si(O\text{-}iso\text{-}C_4H_9)_2$	--	--	U.S. 2,637,718 (Ellis)
$ASi(OC_2H_5)_3$	176(740) or 100(50)	$n_D^{25} = 1.4063$ or $n_D^{20} = 1.4073$	Swiss; also U.S. 3,008,975 (Carbide); Adrianov, CA 33, 1266 (1939); Pierce Chemical Co.
$ASiCH_3(OC_2H_5)_2$	154(736)	$n_D^{25} = 1.4097$	Swiss, U.S. 2,595,729; Brit. 624,363 (Westinghouse)
$ASi(CH_3)_2(OC_2H_5)$	123(743)	$n_D^{25} = 1.4100$	Swiss
$ASi(OC_2H_5)_2C_6H_5$	120(10)	$n_D^{25} = 1.4822$	Brit. 624,364 (Westinghouse); Grafstein, JACS 77, 6650 (1955)
$(MA)_3SiOC_2H_5$	112(12)	$n_D^{20} = 1.4675$	Bunnell, U.S. 2,632,755 (LOF); Brit. 663,770; CA 46, 11228 (1952)
$MASi(OC_2H_5)_3$	102(40)	$n_D^{25} = 1.4122$	Bunnell
$(MA)_2Si(OC_2H_5)_2$	122(40)	$n_D^{25} = 1.4387$	Brit. 624,361 (Westinghouse)
$MA(A)Si(OC_2H_5)_2$	78(13)	$n_D^{20} = 1.4370$	Bunnell
$(MA)Si(OC_2H_5)_3$	101(40)	$n_D^{25} = 1.4122$	Brit. 624,361
$MASi(C_6H_5)(OC_2H_5)_2$	144(20)	$n_D^{20} = 1.4840$	Bunnell
$MASi(C_6H_5)_2(OC_2H_5)$	185(17)	$n_D^{20} = 1.5259$	Bunnell
$MASi(C_6H_{11})(OC_2H_5)_2$	133(20)	$n_D^{20} = 1.4560$	Bunnell
$MASi(CH_2C_6H_5)(OC_2H_5)_2$	106(4)	$n_D^{20} = 1.4973$	Bunnell
$MASi(C_4H_9)_2OC_2H_5$	98(16)	$n_D^{20} = 1.4270$	Brit. 663,770

The compounds $A_2Si(OC_2H_5)_2$ and $A_3SiOC_2H_5$ polymerized on heating with BP or di-t-butyl peroxide, going through viscous solutions to clear, glassy solids (33). Bromine in acetic acid added rapidly to the monomers with splitting of silicon-allyl bonds. Hydrolysis of allyl ethoxysilanes by dilute sulfuric acid occurred readily.

Although $MASi(OC_2H_5)_3$ did not homopolymerize readily when heated with benzoyl peroxide near 100°C (34), it copolymerized with maleic anhydride violently when heated in the presence of BP. The product was hard and glasslike. When the copolymerization was carried out in dioxane solution, a viscous syrup of copolymer solution formed in 10 min. Films deposited from dioxane solution were baked at 125°C to give clear, colorless, but brittle films. Allyl methallyl diethoxy and related silanes also were copolymerized with maleic anhydride. Allyl ethoxysilanes have been copolymerized using free radical catalysts with fluoroolefins (35) and with unsaturated polyesters (36). Methallyl and dimethallyl ethoxysilanes such as $(MA)_2Si(OC_2H_5)_2$ have been homopolymerized and copolymerized with DAP or CR-39 by heating with 2% di-t-butyl perbenzoate (37). Diallyl disiloxanediols, obtained by hydrolysis of corresponding diacetoxysilanes, formed copolysiloxanes curable through allyl groups (38).

Diallyl diethoxysilane and methallyl dichlorosilane were examples of silicon compounds used in treating glass fibers (39). They may form strengthening bonds between glass (through the SiO group) and impregnating unsaturated polyester resins (through the C=C group). Diallyl dialkoxysilanes were reacted with carboxyl-terminated polyesters (40).

Relatively few compounds of the types $A(R)Si(OR)_2$ and $A_2Si(R)(OR)$ have been studied. The patent literature tells how $A(CH_3)SiO(C_2H_5)_2$ was made by adding a dilute solution of allyl magnesium chloride in ether to an ether solution of $CH_3Si(OC_2H_5)_3$ in ether followed by distillation (41). The monomethyl compound, as well as $ASi(CH_3)_2OC_2H_5$, could be copolymerized with diallyl phthalate by heating with 3% BP at 70°C to give hard, transparent films.

Allyl phenyl diethoxysilane was prepared by the following procedure (42). To 2 moles of $ASi(OC_2H_5)_3$ in 171 ml of dry ether containing 3 g of magnesium turnings was added slowly during 9 hr a dilute solution of 1.33 moles of C_6H_5MgBr in dry ether. Distillation gave $A(C_6H_5)$-$Si(OC_2H_5)_2$, boiling at 120°C at 10 mm. The monomer was suggested for condensation copolymerization with siloxanes to give thermosetting silicone polymers. Allyl acyloxy and allyl alkyl acyloxy silanes have been prepared (43). Allyl triacyloxy silanes were added to resin latices for

sizing glass fibers used for polymer reinforcement (44).
Allyl compounds developed as finishes or coupling agents
for glass fibers in reinforced plastics were discussed by
P. W. Erickson and co-workers of Naval Ordnance Laboratory
(45). A reaction product of vinyl trichlorosilane and
chloroallyl alcohol was best with unsaturated polyester
resins; it was good with epoxy resins, but inferior with
phenolics (46). Products from heating allyl trichlorosilane with phenol or meta-cresol are excellent finishes
with epoxy resins and phenolic resins. The product NOL-24,
derived from m-cresol of structure $CH_2=CHCH_2Si(Cl)_2O$⟨◯⟩$-OH$
was regarded as most useful. There was some difficulty
in the hydrolysis of the finish on the glass cloth, with
regeneration of odorous cresol. Allyl trichlorosilane is
a fairly good finish for glass fibers with epoxy resins.
Other allyl-silicon compounds evaluated as glass-fiber
finishes, particularly with unsaturated polyester-styrene,
included diallyl diethoxysilanes (47) and allyl siloxanol
derivatives (48). Except for the last, hydrolysis apparently precedes action as coupling agent between glass
fibers and plastic (49).
Allyl-containing siloxane polymers can be prepared from
the cohydrolysis of allyl and alkyl alkoxylsilanes (although the use of allyl chlorosilanes is preferred) (50).
Methyl borate and sulfuric acid may be used as catalysts
in the following reactions involving intermediate hydrolysis to silanols that condense to polymers:

$$A_2Si(OC_2H_5)_2 + R_2Si(OC_2H_5)_2 \xrightarrow{H_2O} --Si-O-Si-O-SiO--$$
$$\phantom{A_2Si(OC_2H_5)_2 + R_2Si(OC_2H_5)_2 \xrightarrow{H_2O} --}A_2R_2R_2$$

The liquid low polymers thus formed can be cured by heating with peroxides (51); also they can be copolymerized
with ethylenic monomers.
Relatively little work has been reported on polymerization of allyloxy silanes. Nagel and West prepared
$CH_2=CHSi(OA)_3$ (bp 153°C at 6.5 mm; $n_D^{25} = 1.4631$) (52).
Reaction of allyl alcohol with $C_6H_5SiCl_3$ in pyridine gave
$C_6H_5Si(OA)_3$, which partially polymerized to a gummy solid
when distilled (53). A mixture of $C_6H_5SiH_2OA$ and
$C_6H_5SiH(OA)_2$ formed a rubbery polymeric residue (54).

ALLYL SILAZANES

Even bifunctional allyl silazanes were resistant to
polymerization on heating with 1% di-t-butyl peroxide at
139°C (55). Analogous vinyl compounds were more reactive
in free radical polymerization under these conditions. A

TABLE 29.3

Allyl Silazanes

Compound	bp, °C	Refractive index	Preparation
$(CH_3)_2 ASiNHCH_3$	64(98)	$n_D^{26} = 1.4250$	aminolysis of $(CH_3)_2 ASiCl$
$CH_3 ASi(NHCH_3)_2$	87(80)	$n_D^{25} = 1.4409$	aminolysis of $CH_3 ASiCl_2$
$[(CH_3)_2 ASi]_2 NH$	115(45)	$n_D^{27} = 1.4481$	ammonolysis of $(CH_3)_2 ASiCl$
$[(CH_3)_2 ASi]_2 NCH_3$	130(45)	$n_D^{26} = 1.4580$	$(CH_3)_2 ASiNHCH_3$ heated to 170°C with $(NH_4)_2 SO_4$
$[(CH_3 Si]_2 NA$	177	$n_D^{27} = 1.4372$	cf. Andrianov, CA 59, 7551
$[(CH_3) ASiNCH_3]_3$	104(0.2)	$n_D^{23} = 1.4888$	$CH_3 ASi(NHCH_3)_2$ heated to 145°C with $(NH_4)_2 SO_4$

number of allyl silazanes were prepared by Elliott and
Breed (55), (Table 29.3).

Elliott and Breed found the C=C stretching frequency in
the IR spectrum at 1575-1590 cm^{-1} in vinyl silazane,
compared with 1630-1645 cm^{-1} in allyl silazanes. The
higher frequency from the allyl compounds is consistent
with the work of Smith (56).

On long heating with ammonium sulfate, allylamino trimethyl silane formed N-allyl hexamethyl disilazane (57).
Additions of silanes to olefinic compounds were studied
also.

ALLYL SILICATES AND HALOSILICATES

Peppard, Brown, and Johnson made critical studies of
allyl silicates and halosilicates (58). Slow portionwise
addition of 9 moles allyl alcohol to 3.6 moles silicon
tetrachloride with cooling gave a mixture of 6.1 g of
$AOSiCl_3$, 130 g of $(AO)_2 SiCl_2$, 449 g of $(AO)_3 SiCl$, and
112 g of $(AO)_4 Si$. By varying the reacting ratio, the
proportions of products were controlled. If $SiCl_4$ is
added in reverse order, a large proportion of $(AO)_4 Si$ is
formed and much $SiCl_4$ remains unreacted. Another method
of preparing $(AO)_4 Si$ involves alcoholysis of methyl orthosilicate without catalyst or of ethyl orthosilicate with
allyl alcohol in the presence of $SiCl_4$ as catalyst (58).
Tetraallyl silicate was made by reaction of allyl alcohol
with $SiHCl_3$ (59).

Tetraallyl orthosilicate, $(AO)_4 Si$, is a clear, colorless liquid ester with almost no odor. It is soluble in
common organic solvents and has low solubility in water.
In water it slowly hydrolyzes to siloxane polymers. It
boils at 116°C(13 mm), has $n_D^{20} = 1.4336$, and specific
gravity at 17°C = 0.9842 (60). The compound slowly adds

4 moles of Br_2 in carbon tetrachloride without cleavage
of the silicon-carbon bonds (60). For complete reaction
24 hr in the dark was required.

Homopolymers of $(AO)_4Si$ are hard, transparent, heat-
and scratch-resistant solids which are insoluble in all
solvents. In spite of their high functionality, the
tetraallyl and tetramethallyl orthosilicates do not polymerize very fast, and the brittle crosslinked polymers
have not found much use. Rothrock obtained a soft but
brittle gel by long heating of tetramethallyl silicate
with ascaridole (61). Large additions of $SnCl_4$ in $CHCl_3$
produced violent reaction of $(MAO)_4Si$ forming a brittle,
brown resin. By heating $(AO)_4Si$ at 250°C for 20 hr at
3000 atm, a scratch-resistant, clear, brittle polymer was
obtained (62). Heating methallyl orthosilicate at 250°C
for 40 hr in an evacuated sealed tube gave a similar but
less brittle polymer, insoluble in solvents and not
softened below 300°C. Copolymerizations of these tetrafunctional monomers with MMA and other monomers were
carried out by heating with peroxide catalysts (61, 62)
or by use of high temperatures and pressures (61).

Polymerizations of mixed allyl silicate esters by heating with acetone peroxide gave liquid to solid products
(63). Addition of up to 1% methallyl silicate to unsaturated polyesters improved bonding to glass fibers (64).
Hexaallyl disilicate acted as a crosslinking agent for
polysiloxanes (65).

Triallyloxy silane $(AO)_3SiH$ $(n_D^{20} = 1.4284)$ was prepared
from dry allyl alcohol and $HSiCl_3$ in benzene in the
presence of HCl (66). It polymerized at the boiling
point, 188°C, but properties of the polymer were not reported. Surprisingly, $(AO)_3SiCH=CH_2$ (bp 90°C at 7.5 mm;
$n_D^{25} = 1.4380)$ was not observed to polymerize (67).

Peppard, Brown, and Johnson heated diallyl dichlorosilicates with $AlCl_3$ to obtain white solid polymers that
subsequently became brown. Allyl ethyl dichlorosilicate
was obtained in low yield by adding a cold solution of
allyl alcohol in benzene to a cold solution of $C_2H_5OSiCl_3$
and isoquinoline in benzene. Allyl cyclohexyl dichlorosilicate was obtained in better yield (without adding
base to prevent disproportionation). Only 13 mole % allyl
dicyclohexyl chlorosilicate was formed at the same time.
Alkyl chlorosilicates hydrolyze in water slowly to give
clear viscous solutions, but alkyl fluorosilicates such
as $(C_2H_5O)_3SiF$ and A_3SiF react instantly with water to
give polymer gels. Hydrogen fluoride catalyzes the
elimination of alkoxy groups. Physical properties of
allyl chlorosilicates and related compounds are given in
Table 29.4. Allyl triphenyl orthosilicate was prepared
by heating $(C_6H_5O)_4Si$ with allyl alcohol for 10 hr at 140°C
(68).

TABLE 29.4

Some Allyl and Methallyl
Chlorosilicates and Silicate Esters (58)

Compound	bp, °C (at mm)	Compound	bp, °C (at mm)	n_D^{20}
$AOSiCl_3$	122, 36 (32)	$(AO)_3SiOCH_3$	116 (32)	1.4252
$(AO)_2SiCl_2$	81 (32)	$(AO)_2Si(OCH_3)_2$	95 (34)	1.4110
$(AO)_3SiCl$	115 (32)	$AOSi(OCH_3)_3$	70 (34)	1.3919
$(MAO)SiCl_3$	141, 53 (32)	$(AO)_3SiOC_2H_5$	121 (34)	1.4230
$(MAO)_2SiCl_2$	89 (20)	$(AO)_3SiOCH(CH_3)_2$	75 (2)	1.4204
$(MAO)_3SiCl$	123 (20)	$(AO)_3SiOC_6H_{11}$	105 (3)	--
$AO(C_2H_5O)SiCl_2$	66 (32)	$(MAO)_3SiOCH_3$	128 (18)	1.4320
$AO(C_6H_{11}O)SiCl_2$	127 (32)	$[(AO)_3Si]_2O$	161 (10)	1.4394
$(AO)_3SiF$	99 (32)	$[(MAO)_3Si]_2O$	178 (4)	1.4414

Compounds of the type $CH_2=CH(CH_2)_nOSi(CH_3)_3$, where n = 3 and 9, were polymerized with low yields by catalysts made from AlR_3, $TiCl_3$, and $AlCl_3$ (69). Reaction times of 8 to 64 hr at 50°C were used. Hydrolysis of the polymers gave polymers bearing primary hydroxyl groups that were only soluble in alcoholic solutions of mineral acids.

ALLYL SILANES AND RELATED COMPOUNDS

Tetraallyl silane (A_4Si) was made by Kropa by reaction of silicon tetrachloride with an excess of allyl chloride and magnesium in ether and benzene (70):

$$SiCl_4 + 4CH_2=CHCH_2Cl + 4Mg \longrightarrow Si(CH_2CH=CH_2)_4 + 4MgCl_2$$

The liquid monomer (bp 103°C at 15 mm; $n_D^{20} = 1.4864$) polymerized after heating above 100°C for a day, forming a clear solid mass. It was copolymerized with unsaturated polyesters, vinyl esters, styrene, and other ethylenic monomers (70, 71). Pyle prepared A_4Si by dropwise addition of $SiCl_4$ in ether to AMgBr in ether (72). On heating, A_4Si and its mixtures with styrene polymerized more slowly than did styrene. Alkali metal carbonates and oxidizing agents accelerated the slow thermal homopolymerization; acids acted as inhibitors (73). Rochow was able to distill SiA_4 without polymerization and confirmed acceleration of polymerization by potassium carbonate (74). Tetramethallyl silane was distilled at 270°C, apparently without polymerization (75). Methallyl silanes were made by reacting organosilicon compounds with diisobutene at 400 to 600°C (76).

Allyl alkyl and allyl phenyl silanes can be made by reacting dilute allyl magnesium bromide with alkyl or aryl halosilanes; for example (77):

$$AMgBr + (C_6H_5)_3SiF \longrightarrow (C_6H_5)_3SiA + MgBrF$$

A detailed procedure for preparation of allyl Grignard reagent has been supplied by Peninsular Chemical Research, Inc. Yakovlev reacted $(C_2H_5)_2SiCl_2$ in increments with AMgBr to obtain $A_2Si(C_2H_5)_2$, which was said to polymerize under the action of BP and other catalysts (78).

Sommer and co-workers made allyl trimethyl silane using allyl magnesium bromide (79). It was also made by reacting allyl trichlorosilane with methyl magnesium halide. Acid catalysts gave oily polymers. Even at 0°C, $ASi(CH_3)_3$ adds HI in the normal way. Chlorine and hydrogen bromide also add to the double bond, but some cleavage of the silicon-allyl bond of $ASi(CH_3)_3$ occurs with bromine and hydrochloric and sulfuric acids. Iodine at 0°C vigorously cleaves allyl silanes forming allyl iodide (80). Recently $ASi(CH_3)_3$ was made by reaction of $(CH_3)_3SiCl$ with allyl chloride in the presence of zinc and an activator (81).

With $AlCl_3$ catalyst an oily low polymer was formed from $ASi(CH_3)_3$ (bp 300°C at 15 mm; $n_D^{20} = 1.4760$). Peroxides attack the allylic hydrogen atom of $ASi(CH_3)_3$ on long heating at 90°C (82). Allyl silanes have been used as coupling agents between glass fibers and resins (83).

Swiss and Arntzen obtained no polymer by long heating of $ASi(CH_3)_3$ with 4% benzoyl peroxide (84). Under pressure of 5500 atm with di-t-butyl peroxide at 130°C for 6 hr, Petrov and co-workers were able to obtain solid, crosslinked polymers from CH_3SiHA_2, $(C_6H_5)_2SiA_2$, $(CH_3)_2SiA_2$, CH_3SiA_3, SiA_4, and tetramethallyl silane (85). Four different monoallyl alkyl silanes gave only oily trimers to pentamers. Topchiev and co-workers concluded that allyl silanes containing silicon-hydrogen bonds are stable in the absence of catalysts, even on heating at 100 to 150°C (86). Radiation-produced polymers of allyl diethyl silane and allyl ethyl phenyl silane were viscous liquids of molecular weights between 1000 and 5000. Polymers of $A_2SiHC_2H_5$ and A_3SiH were infusible insoluble solids. Polymers made by use of beta and gamma rays showed no IR bands for double bonds. Polymers prepared with benzoyl peroxide had weak residual double-bond absorptions and a predominantly vinyl-polymer structure. Polymers made by use of platinum/carbon catalyst showed no silicon-hydrogen bonds. Structures of copolymers with styrene and acrylonitrile were studied. In the case of A_3SiH, the rates of copolymerization with styrene fell off sharply with increasing A_3SiH in the monomer feed. From 1:1 molar ratio, crystalline copolymers were obtained which retained silicon-hydrogen bonds. Using ^{60}Co radiation, copolymers can be prepared at room temperature (87).

Crystalline polymers were prepared from allyl trimethyl silane in Italy, the United States, and Japan. Natta, Mazzanti, Longi, and Bernardini polymerized the monomer at 70 to 80°C using a catalyst from aluminum triethyl and titanium tetrachloride in molar ratio of 2:1 (88). The crude polymer was fractionated by solvent extraction. The most crystalline fraction melted at 350 to 360°C and gave an x-ray fiber pattern indicating an identity period of 6.5 Å, with helical structure and ternary symmetry of an isotactic polymer. However, these polymers were more soluble in solvents than most highly isotactic polymers, and considerable stereoirregularity was suspected. The same paper reported crystalline polymer from allyl silane (mp 127°C) prepared at 60 to 70°C using violet titanium trichloride. The freshly prepared polymer was completely soluble in boiling n-heptane, but because of the high reactivity of S-H bonds crosslinking occurred by oxidation in air. Water and alcohol also promoted insolubilization.

Longi and co-workers also disclosed terpolymerization of up to 5% of an allylsilane with 20 to 35% propylene and ethylene using Ziegler-Natta catalyst (89). Copolymers of propylene with minor proportions of an allylsilane were evaluated in crosslinked fibers (90). In one example, a solution of $ASiH_3$ and propylene in n-heptane was reacted 5 hr at 70°C in presence of catalyst from $TiCl_3$ and $(C_2H_5)_2AlCl$. The copolymerization was quenched by adding a mixture of acetone and methanol. There was recovered a copolymer of crystal mp 170°C, 62% crystallinity, and intrinsic viscosity 5.6 in tetrahydronaphthalene. From the copolymer product only 4% was extractable by boiling n-heptane. Heating the copolymer fibers at 120°C for 5 hr in a mixture of n-butanol and ammonium hydroxide caused crosslinking through reactive S-H links with little change in normal crystallinity.

Campbell obtained crystalline, high-melting homopolymers from allyl trimethyl silane using a suspension of catalyst prepared from lithium aluminum tetradecyl and titanium tetrachloride (91). Murahashi and co-workers prepared crystalline polymer fractions from allyl, vinyl, as well as 3-butenyl trimethyl silane; they employed Ziegler-Natta type catalyst at 70°C (92). Amorphous polymer fractions were removed by extractions with toluene followed by ether. In Russia, catalysts from reaction of aluminum triethyl with titanium tetrachloride were employed with allyl phenyl dimethyl silane to obtain polymers, and fractions of these melted near 300°C (93). Diallyl dimethyl silane gave liquid and rubbery polymer fractions, whereas diallyl methyl phenyl silane yielded a polymer fraction melting near 400°C. Details of polymerization procedures were not disclosed.

Diallyl dimethyl silane was polymerized in heptane at
30°C using Ziegler-type catalyst. The soluble polymers
formed were believed to contain cyclic recurring units
--CH₂ [cyclohexane ring with] Si(CH₃)₂   (94). In most reactions some crosslinked
polymer was formed. Polymers prepared using di-t-butyl
peroxide melted at 80-100°C, as did the Ziegler polymers,
but the former showed more evidence of residual double
bonds by IR analysis (95). From A₂Si(C₆H₅)₂ by Ziegler
catalyst low polymers melting at 125-155°C were prepared.
The polymers reported from both of these monomers have
low molecular weights (intrinsic viscosity in benzene
0.13 and lower) and probably have some branched chains.
Irradiation of A₂Si(CH₃)₂ with UV light for 27 hr gave an
80% yield of an oily polymer (96). Attempts were made to
relate the UV spectra of diallyl, triallyl, and tetraallyl
silanes to the tendency toward cyclopolymerization (97).

Butler and Stackman prepared some diallyl and dimethallyl silanes containing cyclic radicals but they were
only able to obtain polymers of very low DP with intrinsic
viscosities of 0.12 and below (98). Allyl magnesium
halide solution was added very slowly to dichlorosilanes
to prepare the monomers with properties given in Table
29.5. The solubility of the low polymers in solvent was
attributed to cyclopolymerization. The diallyl monomers
were treated with a dark brown catalyst suspension in
n-heptane at 85°C, obtained by reaction of Al(C₂H₅)₃ with
TiCl₄. The dimethallyl compounds did not respond to the
Ziegler-Natta type catalyst, but gave low polymers on
heating with 2 to 5% di-t-butyl peroxide at 135°C. The
solid polymers melted in the range of 90-110°C.

Curry and Harrison made allyl dimethyl silane under
high dilution conditions by way of a Grignard reaction,
starting from dimethyl chlorosilane, allyl bromide, and
magnesium (99). The monomer was heated under reflux with
0.06% platinum-on-carbon catalyst as developed by Wagner
(100). A colorless, moderately viscous polymer resulted.
Evidence obtained in NMR studies revealed that the
polymer had the structure --Si(CH₃)₂CH₂CH₂CH₂Si-- by
1,3-polymerization. When allyl dimethyl silane was
treated with the same catalyst in toluene solution, a
crystalline dimer melting at 110°C was obtained which was
believed to have the structure:

$$(CH_3)_2Si \underset{CH_2CH_2CH_2}{\overset{CH_2CH_2CH_2}{\diagup \diagdown}} Si(CH_3)_2$$

Double bonds in allyl monoalkyl and dialkyl silanes were

TABLE 29.5

Allyl Silanes   (98)

Compound	bp, °C (at mm)	Refractive index	Polymer intrinsic viscosity
$A_2CH_3(C_6H_5)Si$	124(15)	$n_D^{20} = 1.5200$	0.07–0.12
$A_2(CH_2)_4Si$	101(35)	$n_D^{24} = 1.4857$	0.05–0.11
$A_2(CH_2)_5Si$	104(24)	$n_D^{24} = 1.4888$	0.04
$A(CH_3)_2(CH_2=CH)Si$	111(760)	$n_D^{22} = 1.4315$	--
$(MA)_2(CH_3)_2Si$	79(25)	$n_D^{20} = 1.4525$	0.04
$(MA)_2(C_6H_5)_2Si$	148(0.5)	$n_D^{22} = 1.5650$	0.03
$(MA)_2CH_3(C_6H_5)Si$	138(12)	$n_D^{22} = 1.5180$	0.03
$(MA)_2(CH_2)_5Si$	95(2)	$n_D^{24} = 1.4907$	0.04

determined by first destroying Si–H bonds by reaction with KOH in methanol, after which bromine in methanol was added quantitatively (101).

Mono-, di-, tri-, or tetrafunctional allyl silanes co-polymerized reluctantly under free radical conditions with styrene, acrylic, and vinyl monomers (102), but useful products have not been developed. Allyl silanes were copolymerized with ethylene (103) and with 1-pentene (104). Allyl silanes have been polymerized and copolymerized with acrylic and other comonomers in Russia (105).

Diallyl dialkyl silanes react with dimercaptans by hydrogen migration polymerization to give oily or soft solid low polymers (106):

$$HS(CH_2)_6SH + CH_2=CHCH_2Si(CH_3)_2CH_2CH=CH_2 \longrightarrow$$

$$--S(CH_2)_6SCH_2CH_2CH_2Si(CH_3)_2CH_2CH_2CH_2--$$

Buffered persulfate emulsion systems may be used. This type of reaction seems to occur by successive free radical chain transfer. Other reactions of polyfunctional allyl silanes will be found in references of Table 29.6.

TABLE 29.6

Other Allyl and Methallyl Silanes

Silane	bp, °C (at mm)	Refractive index, etc.	References
Triallyl vinyl	70(6)	$n_D^{25} = 1.4790$	Nagel, J. Org. Chem. 17, 1379 (1952)
Triallyl ethyl	106(34)	$n_D^{20} = 1.4723$	Yakovlev, CA 47, 6340 (1953)
Triallyl methyl	89	--	Pyle, U.S. 2,448,391 (G.E.)
Diallyl dimethyl	60(22) or 68(50)	$n_D^{20} = 1.4410$	Marvel, J. Polym. Sci. 9, 53 (1952); Pierce Chem. Co.
Diallyl diethyl	92(34)	$n_D^{20} = 1.4594$	Yakovlev, CA 44, 1016 (1950)
Diallyl ethyl	147(740)	$n_D^{20} = 1.4510$	Petrov, CA 49, 7510 (1955)
Diallyl diphenyl	140(2)	$n_D^{20} = 1.5750$	Petrov
Allyl trimethyl	86	$n_D^{20} = 1.4069$	Burkhard, JACS 72, 1078; Brit. 624,363 (Westinghouse)
$A(CH_2)_n SiR_3$	--	--	George et al, Chem. Rev. 56, 1079 (1956)
Allyl triphenyl	178(0.3)	--	Gilman, JACS 79, 4560 (1957); Johnson, J. Org. Chem. 26, 4092 (1961)
Dimethallyl (tri-chlorosilyl-propyl), etc.	--	--	Topchiev, CA 59, 1670 (1963)
4-Allylphenyl trimethyl	125(32)	--	Frisch, U.S. 2,759,959 (G.E.)
1,3-Diallyl di-silylmethylene	85(11)	$n_D^{20} = 1.4624$	Greber, Makromol. Chem. 52, 184 (1962)
3-Butenyl	--	polymerized	Oppegard, U.S. 3,026,213 (Du Pont)
$MASi(C_2H_5)_3$	189	$n_D^{20} = 1.4505$	Petrov, CA 46, 11102 (1952)
$(MA)_2 Si(CH_3)H$	63(17)	$n_D^{20} = 1.4450$	Petrov, CA 48, 1247 and 5078 (1954)
$(MA)_3 SiCH_3$	232	$n_D^{20} = 1.4772$	Petrov, CA 48
Allyl silicon nitriles	--	--	Sommer, U.S. 2,906,767 (Dow)

References

1. P. D. George, M. Prober, and J. R. Elliott, Chem. Rev. 56, 1065 (1956).
2. D. T. Hurd, J. Am. Chem. Soc. 67, 1813 (1945); U.S. 2,420,912; Brit. 620,692 (G.E.).
3. D. L. Bailey and A. N. Pines, Ind. Eng. Chem. 46, 2363 (1954); U.S. 2,736,736 (Carbide).
4. Brit. 597,367 (ICI), CA 42, 4601 (1948); cf. Brit. 618,608 (ICI).
5. C. Pape, Ann. 222, 373 (1884).
6. L. W. Tyran, U.S. 2,532,583 (Du Pont).
7. D. T. Hurd and G. F. Roedel, Ind. Eng. Chem. 40, 2078 (1948).
8. E. Huseman, Ger. 1,102,152 (Metzinger et al), CA 57, 9881 (1962).
9. C. A. Burkhard, J. Am. Chem. Soc. 72, 1078 (1950).
10. R. Okawara et al. (Osaka), CA 48, 5081 (1954).
11. K. C. Frisch, U.S. 2,759,959 (G.E.).
12. K. C. Frisch et al, J. Am. Chem. Soc. 74, 4584 (1952).
13. R. E. Scott and K. C. Frisch, J. Am. Chem. Soc. 73, 2599 (1951).
14. C. A. Burkhard, U.S. 2,604,486 (G.E.).
15. E. L. Kropa, U.S. 2,465,731 (Am. Cyan.).
16. J. Swiss and C. E. Arntzen, U.S. 2,595,728 (Westinghouse), appl. December 1943 and March 1945.
17. D. L. Bailey and A. N. Pines, Ind. Eng. Chem. 46, 2363 (1954).
18. L. S. Tyran, U.S. 2,532,583 (Du Pont).
19. R. Brown, Polym. Lett. 1, 208 (1963).
20. C. A. Burkhard, J. Am. Chem. Soc. 72, 1078 (1950); D. L. Bailey and A. N. Pines, Ind. Eng. Chem. 46, 2363 (1954).
21. K. Adrianov and M. Kamenskaya, J. Gen. Chem. USSR 8, 969 (1938); CA 33, 1266 (1939); cf. J. Swiss et al, U.S. 2,595,728-9 (Westinghouse).
22. R. H. Bunnell, U.S. 2,632,755; Brit. 663,770 (L-O-F); CA 46, 11228 (1952).
23. J. Nagy, Dissertation, Budapest, CA 54, 1377 (1960); 58, 4593 (1963).
24. D. F. Peppard, W. G. Brown, and W. C. Johnson, J. Am. Chem. Soc. 68, 70 (1946).
25. D. L. Bailey, U.S. 2,723,985 (Carbide).
26. C. A. Burkhard, J. Am. Chem. Soc. 72, 1078 (1950).
27. E. Larrson, CA 50, 16662 (1956).
28. Data sheets of Linden Labs, State College (University Park), Pa.
29. J. Swiss and C. E. Arntzen, U.S. 2,595,728-9 (Westinghouse); D. L. Bailey, U.S. 2,723,985 (Carbide); cf. L. W. Tyran, U.S. 2,532,583.
30. D. T. Hurd and G. F. Roedel, Ind. Eng. Chem. 40, 2078 (1948).
31. Brit. 624,361 (Westinghouse); CA 44, 2287 (1950).
32. J. Swiss and C. E. Arntzen, U.S. 2,595,730 (Westinghouse).
33. E. Larrson (Lund University), CA 50, 16663 (1956).
34. R. H. Bunnell, Brit. 663,770 (L-O-F); U.S. 2,632,755.
35. L. W. Frost, U.S. 2,596,967; CA 47, 4365 (1953).
36. F. J. Sowa, U.S. 2,605,243; CA 46, 11770 (1952).
37. Brit. 624,361 (Westinghouse); CA 44, 2287 (1950).

38. K. C. Frisch et al, J. Am. Chem. Soc. 74, 4584 (1952); U.S. 2,678,938 (G.E.).
39. L. P. Biefeld, U.S. 2,683,097 and 2,723,210 (Owens-Corning); T. R. Santelli et al, U.S. 3,046,243.
40. L. P. Biefeld, U.S. 2,763,573 (1956).
41. J. Swiss and C. E. Arntzen, U.S. 2,595,729, Brit. 624,363 (Westinghouse).
42. J. Swiss and C. E. Arntzen, Brit. 624,364 (Westinghouse).
43. C. A. MacKenzie and M. Schoffman, U.S. 2,623,832 (Ellis); A. Likovets et al, CA 57, 2242 (1963).
44. G. E. Eilerman, U.S. 2,994,619 (PPG).
45. P. W. Erickson, I. Silva, and H. A. Perry, Preprint booklet, ACS Div. of Paint, Plastics, and Printing Ink Chem. 14, No. 1, 19-31, March 1954; U.S. 2,720,470 (USA).
46. cf. J. Bjorksten and L. L. Yaeger, Mod. Plastics 29 (11), 124 (1952).
47. H. Steinbock, U.S. 2,563,288 (Owens-Corning).
48. U.S. 2,472,799, 2,507,200, 2,567,110 and 2,582,215.
49. E. Plueddeman et al, Mod. Plastics 39, 135 (August 1962).
50. J. Swiss and C. E. Arntzen, U.S. 2,595,727-30 (Westinghouse).
51. D. T. Hurd and G. F. Roedel, Ind. Eng. Chem. 40, 2078 (1948); D. L. Bailey, U.S. 2,897,222 (Carbide); M. M. Olson, U.S. 2,945,003 (PPG).
52. R. Nagel and H. W. Post, J. Org. Chem. 17, 1382 (1952).
53. R. Filler, J. Org. Chem. 19, 544 (1954).
54. V. A. Prokhorova and V. O. Reikhsfeld, CA 60, 541 (1964).
55. R. L. Elliott and L. W. Breed, J. Chem. Eng. Data 11/4, 604 (October 1966).
56. A. L. Smith, Spectrochim. Acta 10, 87 (1960).
57. J. L. Speier et al, J. Am. Chem. Soc. 78, 2278 (1956).
58. D. F. Peppard, W. G. Brown, and W. C. Johnson, J. Am. Chem. Soc. 68, 70-78 (1946); cf. B. Helferich and J. Hausen, Ber. 57, 795 (1924); U.S. 2,566,365 (3M).
59. I. Joffe and W. E. Post, J. Org. Chem. 14, 421 (1949).
60. Data Sheet, Linden Laboratories, State College (University Park), Pa.; also available from Pierce Chemical Company.
61. H. S. Rothrock, U.S. 2,276,094 (Du Pont); cf. E. E. Lewis, CA 43, 3235 (1949).
62. P. J. Garner, U.S. 2,396,692 (ICI).
63. F. Strain et al, U.S. 2,394,642 (PPG).
64. R. Steinman, U.S. 2,513,268 (Owens-Corning); cf. U.S. 2,529,214 (Am. Cyan.).
65. H. G. Brod and O. Schweitzer, Ger. 1,035,358 (Degussa); CA 54, 17958 (1960).
66. I. Joffe and H. W. Post, J. Org. Chem. 14, 421 (1949).
67. R. Nagel and H. W. Post, J. Org. Chem. 17, 1382 (1952).
68. F. Weigel, Ger. 954,245; CA 53, 11306 (1959).
69. U. Giannini et al, Polym. Lett. 5, 527 (1967).
70. E. L. Kropa, U.S. 2,338,161 and 2,443,740-1 (Am. Cyan.).
71. C. A. McKenzie and J. B. Rust, U.S. 2,438,612 (Montclair Res.).

72. J. J. Pyle, U.S. 2,448,391 (G.E.).
73. Brit. 616,420 (G.E.).
74. E. G. Rochow, U.S. 2,538,657 (G.E.).
75. A. D. Petrov, CA 48, 5078 (1954).
76. R. H. Krahnke et al, U.S. 3,631,085 (Dow Corning).
77. A. D. Petrov and L. L. Shchukovskaya, CA 50, 3275 (1956);
     cf. R. Nagel and H. W. Post, J. Org. Chem. 17, 1379 (1952).
78. B. I. Yakovlev, CA 44, 1016 (1950).
79. L. H. Sommer et al, J. Am. Chem. Soc. 70, 433 and 2872 (1948).
80. D. Grafstein, J. Am. Chem. Soc. 77, 6650 (1955).
81. M. Lefort, Ger. appl. (Rhône-Poulenc); CA 74, 142686 (1971).
82. G. Sosnovsky and H. J. O'Neill, Compt. Rend. 254, 704 (1962);
     CA 56, 14316 (1962).
83. R. E. Vanderbilt, U.S. 3,301,739 (Esso); cf. U.S. 3,337,391 (Esso).
84. J. Swiss and C. E. Arntzen, U.S. 2,595,729 (Westinghouse).
85. A. D. Petrov et al, CA 49, 15728 (1955).
86. A. V. Topchiev, N. S. Nametkin, and L. S. Polak, J. Polym. Sci.
     58, 1349 (1962).
87. A. V. Topchiev et al, CA 58, 4659 (1963).
88. G. Natta, G. Mazzanti, P. Longi, and F. Bernardini, J. Polym.
     Sci. 31, 181 (1958).
89. P. Longi et al, U.S. 3,644,306 (Monte).
90. Ital. 792,644 (Monte), CA 70, 20658 (1969).
91. T. W. Campbell, U.S. 2,958,681 (Du Pont).
92. S. Murahashi, S. Nozakura, and M. Sumi, Bull. Chem. Soc.
     Japan 32, 670-674 (1959).
93. N. S. Nametkin, A. V. Topchiev, and S. G. Durgar'yan, J. Polym.
     Sci. 52, 51 (1961).
94. C. S. Marvel and R. G. Woolford, J. Org. Chem. 25, 1641 (1960).
95. G. B. Butler and R. W. Stackman, J. Org. Chem. 25, 1643 (1960).
96. K. Gutweiler and H. Niebergall, U.S. 3,070,582 (Shell).
97. G. B. Butler and B. Iachia, J. Macromol. Sci.-Chem. A 3, 803 (1969).
98. G. B. Butler and R. W. Stackman, J. Macromol. Sci.-Chem. A 3,
     821 (1969).
99. J. W. Curry and G. W. Harrison, J. Org. Chem. 23, 1219 (1958).
100. G. H. Wagner, U.S. 2,637,738 (Carbide).
101. A. P. Kreshkov et al, CA 63, 2391 (1965).
102. J. J. Pyle, U.S. 2,448,391 (G.E.); Brit. 641,268; A. D. Petrov
     et al, CA 51, 4979 (1957); Y. Iwakura et al, J. Polym. Sci. A 1
     6, 1633 (1968).
103. Ital. 606,018 (Monte), CA 56, 1609 (1962).
104. G. D. Cooper and A. R. Gilbert, U.S. 3,125,554 (G.E.).
105. A. V. Topchiev et al, J. Polym. Sci. 58, 1349 (1962); N. S.
     Nametkin et al, CA 73, 15364 (1970).
106. C. S. Marvel and H. N. Cripps, J. Polym. Sci. 9, 53 (1952);
     C. A. Burkhard, U.S. 2,604,487 (G.E.); J. Am. Chem. Soc. 72,
     1078 (1950); cf. Brit. 688,408 (G.E.).

30.   ALLYL BORON COMPOUNDS

Allyl and methallyl borates and boranes have had little commercial application. Polymers from allyl carboranes have been studied by the Olin Company and the U.S. Army as solid propellants and heat-resistant materials. Allyl borazoles, investigated by American Cyanamid, are examples of heat-resistant polymers based on cyclic units. Only a little exploratory work has been reported on allyl derivatives of arsenic and other metalloids. Many of these compounds are highly reactive and unstable. Triallyl borate and other allylic esters of boric acid are discussed first.

Triallyl borate (A_3BO_3) was prepared by heating 1 part of B_2O_3 with 4 parts of allyl alcohol at 130°C (1); NH_2SO_3H may be used as catalyst at reflux with benzene (2). This monomer (178°C; $n_D = 1.4332$) reacted immediately with 6 atoms of bromine per molecule. Triallyl borate was also made by refluxing boric acid (H_3BO_3) with excess allyl alcohol in benzene (3). Heating the monomer at 130°C with air bubbling through produced gelation by polymerization. Triallyl borate was prepared by alcoholysis of $(CH_3O)_3B$ and by reaction of sodium borohydride with allyl alcohol in the presence of acetic acid at room temperature (4). Triallyl borate is useful as an allylating agent.

Rothrock prepared trimethallyl borate by refluxing methallyl alcohol with boric acid (5). The crystalline monomer melted near 30°C and would not homopolymerize readily with benzoyl peroxide, sulfur dioxide, or tin tetrachloride. Trimethallyl borate also was made by heating methallyl alcohol and B_2O_3 in toluene with anhydrous cupric sulfate (6). The monomer was polymerized in bulk at 66°C and at 100°C using portionwise addition of benzoyl peroxide (7). Properties and other references to borate ester monomers are given in Table 30.1.

Triallyl and trimethallyl borate were prepared by reacting the alcohols with boron trichloride in methylene chloride or n-pentane at -78°C (8). These esters hydrolyze in water rapidly. Hydrogen halides cleaved the

compounds to give allyl halide, dihalopropane, and boric
acid. Bromine added to A_3BO_3 readily. The ester showed
little tendency to copolymerize with MMA or acrylonitrile
on heating with BP at 60°C (9). Russian workers reported
that triallyl borate polymerized best in phosphoric acid
solution at 60°C, forming colorless, glasslike masses of
molecular weight about 4600 (10). In the presence of
metallic sodium, triallyl borate became a viscous mass
within 5 days. Copolymers could not be formed with
styrene using as catalysts sodium or dicumyl peroxide,
nor with MMA or acrylonitrile using benzoyl peroxide at
60°C.

Mikhailov and Tutorskaya made compounds of type $AB(OR)_2$
and A_2BOR such as $AB(OCH_2CH_2CH_2CH_3)_2$ (bp 100°C at 15 mm;
$n_D^{20} = 1.4230$). These compounds were readily hydrolyzed by
water. The Russian workers obtained different properties
for A_2BOCH_3 from those reported by Rothstein and Saville
(12). The latter found allyl boranes or borines of the
type ABR_2 to be less stable than corresponding mercury
compounds and they could not be isolated readily (12).
Allyl decaborane was made from reaction of $B_{11}H_{13}MgBr$ with
excess allyl bromide (13). Allyl quaternary ammonium
hexaborates were prepared from allyl ammonium iodide and
boric acid (14).

Triallyl borane A_3B (bp 60°C at 3 mm; $n_D^{22} = 1.4500$) can
be made by portionwise addition of boron fluoride etherate
to allyl magnesium bromide solution (15). Russian workers
observed some polymerization during distillation at 60°C.
Triallyl boron undergoes allylic rearrangement which can
be followed by NMR (16). Addition of tetrahydrofuran
retarded rearrangement. A_3B ignites spontaneously in air
and is otherwise extremely reactive (17). However, only
4 equivalents of bromine add readily. Compounds of the
types $ABOR_2$ and A_2BOR were prepared by Topchiev (18). The
residue from preparation of one of the diallyl compounds
was a colored polymer that oxidized rapidly in air. Reaction of A_3B with propionic acid gave $A_2BOOCCH_2CH_3$, along
with a dark polymer. Triallyl boron was prepared from
reaction of allyl bromide, BF_3 etherate, and magnesium in
ether (19). On contact with a mixture of nitric acid and
oxygen triallyl borane formed a solid yellow polymer that
was not soluble in common solvents. Heating A_3B with
carbon tetrachloride, tetrahydrofuran or butanol gave
partially soluble polymers melting above 250°C. Adding
5 mole % A_3B to MMA accelerated polymerization, but addition to styrene retarded polymerization. From A_3B Topchiev and co-workers made diallylboroacetate by reaction
with allyl alcohol. Properties and references to allyl
borates and other allyl boron compounds are given in

TABLE 30.1

Allyl Boron Compounds

Compound	bp, °C	Refractive index, etc.	References
Triallyl borate	76(15)	$n_D^{21}=1.4276$	Gerrard, J. Chem. Soc.
	66(5)	$n_D^{20}=1.4276$	3287 (1956); Eastman catalog
Trimethallyl borate	225 or 118(19) (mp, 30)	--	Gerrard; Scattergood, JACS 67, 2150 (1945); Copolymers with MMA, U.S. 2,276,094 (Du Pont)
$(AC_6H_4O)_3B$	188(0.05)	$n_D=1.5600$	Gerrard, CA 51, 13797 (1957)
$A_2B(OC_2H_5)$	49(7)	--	Topchiev, CA 58, 3452 (1963)
A_2BCH_3	82(15)	--	Rothstein, J. Chem. Soc. 2990 (1952)
Allyl carborane	--	copolymers	Clark, U.S. 3,121,117 (Olin)
Allyl triphenyl boron derivatives	--	--	Seyferth, J. Org. Chem. 26, 4797 (1961)
Allyl decaborane	--	--	Palchak, U.S. 3,299,144 (Olin)
Diallyl vinyl-benzene boronate	--	copolymers	Hoffmann, U.S. 2,931,788 (Am.Cyan.); Crosnos, U.S. 2,934,556
Diallyl nonyl boronate	77(1.2)	--	U.S. 3,152,166 (Gulf Res.)
Diallyl phenyl boronate	69(0.7)	$n_D^{25}=1.5011$	U.S. 3,152,166
Bisdiallyl boron oxide	--	--	Rothstein, J. Chem. Soc. 2988 (1952)
Diallyl diboronate esters	--	--	Bamford, Brit. 913,385 (ICI)
Triallyl ethers of dioxaborinane	--	--	Lund, U.S. 3,325,528 (Allied)

Table 30.1. It is possible that allyl-boron complexes may be intermediates in hydroboration of alkenes (20).
Olin chemists have prepared copolymers of 1-allyl-o-carborane $B_{10}H_{10}CHCA$ as well as 1-vinyl-o-carborane by long heating under nitrogen with t-butyl peroxide at 170°C (21). One of these products had a molecular weight of 11,500. A vinyl carborane gave polymer of molecular weight 71,000 by use of phenyl lithium (22). Polymers from unsaturated carboranes have interest in solid pro-

pellant compositions and heat-resistant materials for
space vehicles. The following unsaturated carborane
monomers with novel closed cage structures have been
available from Olin (about \$4/g):

Monomer	mp, °C	Solubility
1-Vinyl-o-carborane,[a] $C_4H_{14}B_{10}$	79	In common organic solvents
1-Allyl-o-carborane, $C_5H_{16}B_{10}$	65	In common organic solvents
1-Isopropenyl-o-carborane, $C_5H_{16}B_{10}$	46	In common organic solvents
1-Isopropenyl-o-carborane 2-carboxylic acid, $C_6H_{16}B_{10}O_2$	175	In acetone, ethanol, chloroform

[a]For o-carborane Chemical Abstracts prefers 1,2-dicarbadodeca-
borane(12).

The vinyl compound could be homopolymerized best by phenyl
lithium; aluminum trichloride, on the other hand, was
found to give low polymers from the allyl and the isopropenyl carboranes (23). The vinyl and isopropenyl carboranes copolymerized reluctantly with some vinyl monomers
by radical initiation (24). Preparation of a vinyl-
modified carboranylenesiloxane polymer was described (25).
Ventron Corp. has supplied small amounts of allylcarborane
A($B_{10}C_2H_{11}$) and isopropenylcarborane carboxylic acid
$CH_2=C(CH_3)(C_2B_{10}H_{11})COOH$.

Triallyl-N-triphenyl borazine or borazole C_6H_5N
$$\begin{array}{c} A\ C_6H_5 \\ B-N \\ \diagup \qquad \diagdown \\ \qquad\qquad BA, \\ \diagdown \qquad \diagup \\ B-N \\ A\ C_6H_5 \end{array}$$

melting at 99°C, can be prepared by reacting boron trichloride with aniline to obtain β-trichloro N-triphenyl
borazole (26). Allyl magnesium bromide in ether may be
added dropwise to a suspension of the trichlorotriphenyl
borazole to give the triallyl triphenyl borazole (TTB).
This compound was polymerized by heating at 150°C for 6 hr
with pinane hydroperoxide. The material produced holds
promise as a neutron shield (27). Copolymers were made
by heating with radical catalysts and fumarate esters,
ethyl acrylate, styrene, or unsaturated polyesters. TTB
does not polymerize readily under conditions that give
high rates of polymerization of styrene (28). Forcing
conditions with 8% or more peroxide at 130°C gave white
polymer powders that flowed at 180°C. However, after
cooling and reheating, the polymers do not soften up to
300°C and are largely crosslinked. Monomers analogous
to TTB, but with groups other than phenyl on the nitrogen

atoms, were disclosed (29). Divinyl and diallyl borazines were prepared in England, but stable polymers of promising heat resistance were not reported (30).   Some other references on allyl boron compounds were reviewed by Mikhailov (31).

Some allylic compounds of other metalloids are included here with the boron compounds. Triallyl arsenic and allyl alkyl arsines can be prepared from reacting arsenic halides with allyl magnesium bromide (32). Allyl dichloroarsine and allyl arsonic acid (bp 42°C at 4.5 mm) were studied (33). Trimethallyl arsine (bp 114°C at 15 mm), as well as aryl diallyl arsines, were found to oxidize in air (34).

Triallyl arsenite (bp 106°C at 11 mm; $n_D^{20} = 1.4794$) is a mobile, odorless liquid that darkens in light and is immediately hydrolyzed in water (35). Arsenic trichloride was reacted with allyl alcohol and with substituted allyl alcohols in pyridine to give easily hydrolyzed esters that showed no tendency to polymerize in the presence of BP or KHF_2 (36). Tetraallyl arsonium hydroxide (needles melting at 72°C) (37) was made, as well as quaternary allyl arsonium salts (38).   Allyl esters of trivalent arsenic and antimony did not homopolymerize on heating with benzoyl peroxide, but the arsenic compounds formed copolymers with methyl methacrylate and with styrene (39). Allyl dibutyl antimony was obtained by reacting allyl bromide with lithium dibutyl antimony (40). Triallyl antimonate was added as a flame retarder to polyesters (41).

References

1. C. Councler, J. Prakt. Chem. [2] 18, 376 (1871); cf. H. G. Kuivila et al, J. Am. Chem. Soc. 73, 123 (1951).
2. E. P. Irany, U.S. 2,523,433.
3. S. A. Ballard, Brit. 595,502 (Shell); CA 42, 3998 (1948).
4. H. C. Brown et al, J. Am. Chem. Soc. 78, 3613 (1956).
5. H. S. Rothrock, U.S. 2,276,094 (Du Pont).
6. L. H. Thomas, J. Chem. Soc. 820 (1946).
7. B. N. Rutovskii and N. S. Leznov, CA 44, 1007 (1950).
8. W. Gerrard et al, J. Chem. Soc. 3285 (1956).
9. P. Losev et al, CA 51, 4045 (1957).
10. B. I. Mikhant'ev and S. A. Kretinin, CA 60, 10798 (1964); P. Losev et al, CA 51, 4045 (1957).
11. B. M. Mikhailov and F. B. Tutorskaya, CA 53, 6990 (1959).
12. E. Rothstein and R. W. Saville, J. Chem. Soc. 2987 (1952).
13. J. Maurel et al, Bull. Soc. Chim. France 1953 (1959); CA 55, 16396 (1961).
14. S. D. Ross et al, U.S. 3,118,939 (Sprague Electric).

15. B. M. Mikhailov and F. B. Tutorskaya, CA 53, 6990 (1959); E. A. Weilmuenster, U.S. 3,109,029, cf. U.S. 3,022,350 (Olin).
16. V. S. Bogdanov et al (Moscow), CA 68, 100,409 (1968).
17. A. V. Topchiev et al, CA 52, 12752 (1958).
18. A. V. Topchiev et al, CA 54, 7533 (1960); 55, 7272 (1961); cf. Ref. 12.
19. A. V. Topchiev et al, CA 58, 3452 (1963).
20. Cf. H. C. Brown and N. Ravindran, J. Org. Chem. 38, 182 (1973).
21. J. W. Ager, S. L. Clark, and T. L. Heying, U.S. 3,146,260; CA 61, 11840 (1964); cf. L. G. Wiedemann, Report N67-37852 Army Weapons Command, June 1967; cf. CA 78, 16797 (1973).
22. H. L. Goldstein and T. L. Heying, U.S. 3,109,031 (Olin).
23. Data Sheet on Isohedral Carboranes, Olin Chemicals Div.
24. S. F. Reed, J. Polym. Sci. 9, 825 (1971) and 10, 1557 (1972).
25. S. Papetti et al, Preprints ACS Organic Coatings and Plastics Chem. 25, 298 (September 1965).
26. S. J. Groszos and S. F. Stafiej, J. Am. Chem. Soc. 80, 1357 (1958); U.S. 2,892,869 and 2,917,543 (Am. Cyan.).
27. S. F. Stafiej et al, U.S. 2,954,402 and 2,954,361 (Am. Cyan.).
28. J. Pellon et al, J. Polym. Sci. 55, 153 (1961).
29. S. J. Groszos and S. F. Stafiej, U.S. 2,954,401 and 2,892,869 (Am. Cyan.).
30. I. B. Atkinson et al, Polymer Preprints 13, #2, 770 (1972).
31. B. M. Mikhailov (Zelinski Institute, Moscow), CA 78, 4300 (1973).
32. L. Maier et al, J. Am. Chem. Soc. 79, 5884 (1957).
33. C. K. Banks et al, J. Am. Chem. Soc. 69, 927 (1947).
34. W. J. Jones et al, J. Chem. Soc. 1446 (1947).
35. E. J. Salmi and E. Laaksonen, CA 41, 5394 (1947).
36. G. Kamai and R. K. Karipov, CA 57, 4529 (1962).
37. E. Mannheim, Ann. 341, 223 (1905).
38. L. Horner and H. Fuchs, Tetrahedron Lett. 1573 (1963); CA 59, 15309 (1964).
39. G. Kaman and N. A. Cadaeva, Chem. & Chem. Tech. (Moscow) 4, 601 (1959).
40. S. Herbstman, J. Org. Chem. 29, 986 (1964).
41. L. Williams et al, Brit. 837,696 (Spence Ltd.).

31. ALLYL COMPOUNDS WITH METALS

The so-called complexes formed through interaction of
π-bonds (pi-bonds) of allyl compounds with unfilled orbitals of transition metals have been studied during the
last decade especially in England and Germany. The
π-allyl compounds with nickel and palladium halides and
with chromium and zinc have given some promise as catalysts for polymerization of olefins, dienes, and certain
acrylic monomers. However, the polymers obtained generally are low in molecular weight and have not been very
useful to date. The sigma-bonded allyl compounds with
the alkali, aluminum, tin, and other metals seem less
fascinating than π-complexes to physical chemists; but
they also should be considered as chemical intermediates
and catalysts in polymerizations. The remarkable Ziegler-
Natta or coordination polymerization systems discussed in
Chapter 3 and elsewhere in this book may involve intermediate π-bonding of monomers to metals. Isomerizations
of olefins catalyzed by transition metals are believed to
occur through π-allyl complex intermediates (1). The
allyl compounds with transition metals are discussed
first.

In contrast to such sigma-allyl compounds as sodium
allyl, where each allyl group supplies one electron to
make a covalent bond, in π-allyl complexes each allyl
group may be considered to contribute three electrons
toward forming a stable electron shell. The conjugated

allyl ligand $>\overset{\cdot}{C}-\overset{\cdot}{C}-\overset{\cdot}{C}<$ donates electrons to unoccupied

electron orbitals of the transition metals to form π-
bonds. Apparently there are some cases intermediate
between sigma and pi bonding, as well as examples in
which allyl groups appear to act as bridges between metal
atoms. Banthorpe has reviewed the π-complexes as reaction intermediates (2).

It has long been known that heavy metal ions (e.g., Ag$^+$
in water) can form complexes with olefins. In some cases
only complexes of low stability are formed and these can
promote solubility of olefins, especially at lower

temperatures. In other cases stable allyl complexes
melting up to 100°C are obtained. The π-complexes often
have yellow to red or brown colors. In one example, an
ethanolic solution of rhodium trichloride and 1,5-cyclo-
hexadiene was refluxed to form a stable yellow crystalline
solid $C_8H_{12}RhCl$ (3). Stability of such compounds falls
off in the order Cl>Br>I>SH.

Allyl carbonyl complexes of manganese and of cobalt can
be made by reacting the sodium metal carbonyl with allyl
chloride to give a sigma-complex which, in light or when
heated, may form a π-complex (4):

$$NaMn(CO)_5 + C_3H_5Cl \longrightarrow (CO)_5Mn\cdots C_3H_5 + NaCl$$

$$UV \downarrow or\ 100°C$$

$$mp\ 53°C,\ yellow\ (CO)_4Mn\cdots C_3H_5 + CO$$

This allyl-manganese compound seems to have a sandwich-
type structure. The odor of the tetracarbonyl complex
suggests that of camphor. The analogous yellow methallyl
complex was made by McClellan and co-workers, whose paper
reviews early researches in the field (5). Some π-allyl
iron tricarbonyl halide complexes and derivatives were
prepared (6). The π-complexes of compounds with noble
metal halides may have an allyl sandwich structure (7).
Thus allyl alcohol reacted with palladium dichloride to
give a surprisingly stable compound which melts at 175°C
with decomposition (8). A palladium halide complex with
allylbenzene was prepared (9). A π-allyl palladium
chloride complex catalyzed carbonylation of allyl chloride
(10). Alkylation of olefins was believed to involve intermediate π-allyl palladium complexes (11).

Of interest to synthetic organic chemists are the π-
allyl complexes of metals applied by Wilke and others as
catalysts (12). Nickel dibromide reacted with allyl
magnesium chloride to give bis(π-allyl)Ni. When palladium
chloride in ether was mixed with 0.35N AMgCl and stored
at -78°C for 6 hr, there was formed a 50% yield of yellow
diallyl palladium. Wilke also prepared dimethallyl
nickel, triallyl complexes of Fe, Co and Cr, diallyl
platinum, diallyl molybdenum dimer, and tetraallyl tungsten. A number of the complexes, especially diallyl
nickel, were used as catalysts by Wilke for cyclopolymerization of butadiene to form cyclododecatriene, and
these gave some promise in polymerizations of 1-olefins.
The π-allyl compounds of transition metals were reacted
with halogen or halogen halides to give more reactive
catalysts for polymerization and oligomerization (13).
Adducts with Lewis acids and bases also were used for
cyclooligomerization of olefins (14). Butadiene oligomers

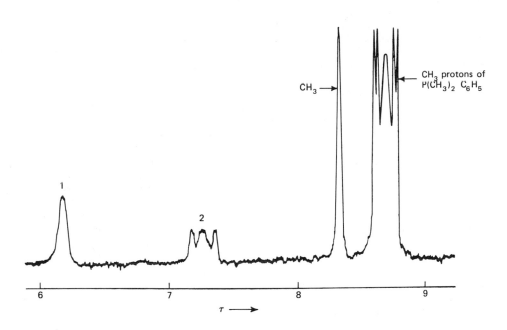

Fig. 31.1. NMR resonance pattern of the palladium complex
[Pd(MA)(P(CH₃)₂C₆H₅)₂]B(C₆H₅)₄ in CDCl₃ solution. Syn- and anti-
protons of the methallyl group produced resonances at 1 and 2,
whereas the methyl protons gave the sharp singlet. To the methyl
protons of the phosphine group was attributed the complex group of
resonances. J. Powell and B. L. Shaw, J. Chem. Soc. A 775 (1968).

were prepared by use of diallyl platinum, but APtCl was
ineffective (15).

Cationic allylic complexes of palladium have been prepared recently in which, judging from NMR data, the
methallyl ligands are symmetrically π-bonded to palladium
(16). Complexes of the type Rh₂Cl₂A₄ can be prepared by
oxidative hydrolysis of Rh₂Cl₂(CO)₄ in aqueous methanol
with allylic halides (17). Allyl palladium complexes
were made from Pd^II salt and an allyl compound treated
with an olefin in alcohol solution (18). Olefin complexes of platinum and palladium have been reviewed (19).
Methallyl palladium complexes are among those studied by
NMR (Fig. 31.1). Slightly basic solvents such as N,N-
dimethyl acetamide were reported to favor formation of
π-allyl palladium complexes (20). The dimer (APdCl)₂ is
available from Ventron Corporation.

The π-allyl nickel halide catalysts in diene polymerization were studied by Porri and Natta (21). In benzene
solution π-ANiI and π-ANiBr gave mainly trans-1,4-poly-
butadiene; but allyl nickel chloride and allyl nickel
acetate gave largely cis-1,4-polybutadiene (22). In other
solvents, all three gave the latter polymer. In Ziegler
and other transition metal catalysts employed for stereopolymerization of 1-olefins, π complexes may be reactive
intermediates. Porri and co-workers used π-NiBr for
selective polymerization of cyclobutene and norbornene
without ring rupture (23). The molecular weights of the
polymers of butadiene and of cyclic olefins prepared by
π-allyl nickel catalysts were rather low (24).

Teyssie and co-workers, using π-allyl complexes of
nickel or cobalt salts, were able to prepare predominantly
cis- or trans-1,4-polybutadienes, depending on the anion
present (25). Complexes of chromium, molybdenum, and
niobium promoted 1,2-polymerization of butadiene. Polymers
of higher molecular weight were possible by use of mixtures of benzene and nitrobenzene as the polymerization
solvent. Butadiene was polymerized by a living polymer
mechanism by use of π-ANiI in solvents (26). Electron
acceptors such as p-chloranil were added to π-allyl nickel
halide catalysts (27). Styrene was dimerized at 50°C by
$(\pi\text{-ANiOC(O)CF}_3)_2$ (28). Allyl anions may replace π-cyclo-
pentadienyl and other ligands, for example:

$$\pi\text{-}(C_5H_5)_2Ni + AMgX \longrightarrow \pi\text{-}(C_5H_5)Ni(\pi\text{-}A) + C_5H_5MgX$$

Preparation of colored catalyst solutions of diallyl
nickel at 20 to -78°C in ether or other solvents was discussed by Wilke (29). Precautions must be taken to exclude air and moisture. Bridged π-allyl compounds of
molybdenum and tungsten were prepared and suggested as
catalysts for cyclopolymerization of butadiene and other
diolefins (30). The synthesis of allylbenzene by reaction
of allyl bromide with benzene and nickel pentacarbonyl
was believed to occur via a π-allyl nickel bromide complex
(31).

ICI laboratories prepared the complexes PdA_2, ZrA_4,
MoA_4, CrA_3, and $Cr(MA)_3$ (32). Only the last two catalyzed
polymerization of acrylonitrile, isoprene, and MMA; none
initiated polymerization of styrene or vinyl chloride.
Triallyl chromium was prepared by reaction of chromium
trichloride with allyl magnesium chloride in ether for an
hour under argon at -30°C (33). Pentane was added and
the complex filtered off. The π-allyl chromium polymerized MMA slowly in heptane solution at 40°C. A catalyst
suspension made from CrA_3 and $TiCl_4$ was dispersed in
hexane and used to polymerize propylene at 40°C (34). A

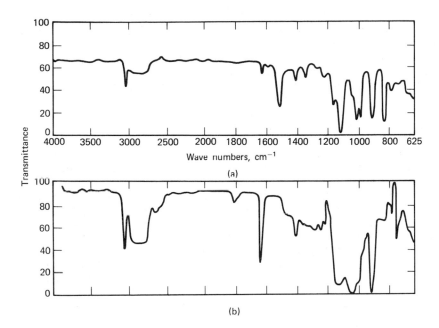

Fig. 31.2.   Infrared spectra: (a) Zr(π-allyl)₄; (b) its reaction product with methyl methacrylate. Polymerization did not occur as it did when Cr(π-allyl)₃ and MMA were mixed with less evidence of complex formation (judging from IR data).   D. G. H. Ballard, W. H. Janes, and T. Medinger, J. Chem. Soc. B, 1170 (1968).

partially stereoregular polypropylene of reduced viscosity 3.3 was obtained. Allyl π-complexes of tungsten, molybdenum, and zirconium were also claimed in the patent as effective. Tetraallyl zirconium was used as catalyst for polymerization of ethylene to linear polymer (50°C, 60 atm, 3 hr) (35). Infrared spectra of Zr(π-allyl)₄ and its reaction product with methyl methacrylate are shown in Fig. 31.2.

Few uses have been developed yet for allyl complexes of transition metals other than as catalysts. Allyl manganese complexes have been suggested as gasoline antiknock agents (36). Pi-allyl complexes of palladium were proposed as oil and gasoline additives and as antioxidants (37). Ruthenium tetracarbonyl trimer was heated with allyl halides to give allyl ruthenium tricarbonyl halides, suggested as catalysts in carbonylation reactions (38).

An allylidene complex formed with triphenyl phosphine and molybdenum carbonyl (39).

ALLYL ALKALI METAL AND MAGNESIUM COMPOUNDS

Allyl sodium can be made from sodium and diallyl ether (40). Allyl lithium can be prepared by cleavage of allyl phenyl ether with lithium in tetrahydrofuran at -15°C. Allyl and methallyl lithium were made from reaction of lithium phenyl with the alkenyl triphenyl tin compounds (41). An order of decreasing reactivity was observed based on molecular weights of the polymers that were formed when the organolithium compounds were used as polymerization initiators: BuLi, C_6H_5Li, $C_6H_5CH_2Li$, allyl Li, vinyl Li (42). Some vinyl and allyl compounds of alkali metals and of magnesium can be prepared directly from the metals and allyl halides (43). Allyl lithium was prepared from allyl magnesium bromide and lithium (44). Morton's Alfin catalyst for styrene and diene polymerization is believed to contain allyl sodium (45). Allyl sodium has been used with titanium tetrachloride for polymerization of ethylene at 200 atm (46).

Alkali and alkaline earth allyloxides or allylates $CH_2=CHCH_2OM$ can be prepared by metals replacing hydrogen of alcohols. Sodium allylate has been used, for example, in the initiation of anionic polymerization of acrylonitrile (47).

The usual conditions for preparing Grignard reagents from magnesium metal and alkyl halides in ether do not give good yields of allyl magnesium halide; instead allylic coupling forms principally diallyl (48). Gilman and McGlumphy of Iowa State College obtained good yields of allyl magnesium bromide by employing high dilutions with ether, gradual mixing of reactants, and rapid stirring (49). In one experiment 0.5 mole allyl bromide in 285 cc of diethyl ether was added slowly to 3 equivalents of 30-mesh granular magnesium with rapid stirring during 2 hr. They obtained a yield of 90.3% allyl magnesium chloride. Slower mixing during 4 hr gave 93.6% yield of AMgBr. Gilman and McGlumphy reacted allyl magnesium bromide with benzyl chloride to give 1-phenyl-3-butene and with benzophenone to form allyldiphenylcarbinol. AMgBr and AMgCl solutions in ether or tetrahydrofuran at 1 or 2 molar concentrations are available commercially from Pfaltz and Bauer, Ventron, Peninsular Chemresearch, and others.

Recently the preparation of allyl magnesium chloride was described by slow addition of allyl chloride to a well-stirred suspension of magnesium powder in diethyl ether (50). In another example of synthesis of a Grignard

solution, 153 g of magnesium turnings were agitated in
360 ml of ether containing a few crystals of iodine and
under a reflux condenser cooled at 0°C (51). To this was
added dropwise a mixture of 351 g of allyl bromide in 2.6
liters of ether during 3 hr. The mixture was further
refluxed for two additional hours and then was reacted
with acrolein to make allyl vinyl carbinol.

Reaction of AMgCl in ether with tetrafluoroethylene
during 17 hr at room temperature gave higher yields of
ACF=CFA than did AMgBr (52). The diallyl difluoroethylene
polymerized to a tar even at -78°C. Vinyl, methyl, and
ethyl magnesium bromides were much less reactive than
AMgCl with $CF_2=CF_2$. Rearrangement of allylic Grignard
compounds of the type $ACH(CH_3)MgCl$ were studied (53). Reaction of allyl magnesium chloride with 2-methylquinoline
at 25°C gave 1,2-addition, but at higher temperatures
largely 1,4-addition (54).

Rothrock prepared triallyl aluminate by heating metallic
aluminum with toluene-allyl alcohol 1:1 along with mercuric chloride (55). After filtering off sludge, the
liquid mixture was distilled at room temperature under
reduced pressure to give a viscous oil that was still
soluble in toluene. Heating this monomer $(AO)_3Al$ at 210°C
for 8 hr gave a stiff polymer gel that was insoluble in
common solvents. A mixture of 5% triallyl aluminate with
MMA was heated at 65°C for 3 days; it formed a nearly
colorless, hard but brittle, insoluble polymer. On exposure to boiling water, the surface of the polymer became
opaque, apparently from hydrolysis. The polymer then
could be polished to give a bright, white surface.

Rothrock made trimethallyl aluminate by alcoholysis of
triethyl aluminate with methallyl alcohol. It also produced a gel under similar conditions of heating at 210°C
and formed insoluble copolymers with MMA. Allyl aluminum
diethyl and diallyl aluminum ethyl can be prepared by
reacting allyl alkali compounds with aluminum halides (56).
Allyl alcohol was reacted with aluminum isobutyl at -10°C
to form diisobutyl allyloxy aluminum (bp 115°C at 1 mm)
(57).

OTHER ALLYL METAL COMPOUNDS

Allyl and methallyl alkyl germaniums were made from the
alkenyl germanium trichloride and RMgCl, but they resisted
polymerization (58). The compounds $(CH_3)_2GeA_2$ and CH_3GeA_3
polymerized at high pressure and 120°C with t-butyl peroxide to form clear, glassy polymers (59). Low polymers,
possibly of cyclic structure, were reported from dimethyl
diallyl germanium using Ziegler catalyst (60). Tetraallyl
germanium (bp 107°C at 15 mm; $n_D^{20} = 1.5030$) gave solid,
insoluble polymers on heating with BP (61). The monomer

was cleaved by carboxylic acids and was attacked by halogens, halogen halides, and mercaptans. Allyl tributyl germanium (bp 117°C at 2 mm; $n_D^{20} = 1.4664$) did not polymerize. When R_3GeH was added to allyl alcohol, there was formed $R_3Ge(CH_2)_3OH$ in the presence of benzoyl peroxide or platinum (62). Triphenyl germane added to allyl triphenyl germane when heated with BP (63).

Jones and co-workers made monoallyl and diallyl alkyl stannanes using Grignard reagents (64). These compounds are oxidized rapidly by air. Numerous allyl compounds of tin have been studied, but most of these have not yielded homopolymers of high molecular weight. Korshak and co-workers could not homopolymerize allyl trialkyl tin compounds; however, they copolymerized them with styrene at 120°C under 6000 atm using t-butyl peroxide catalyst (65). Allyl compounds of silicon polymerized more readily than did those of germanium or tin. Koton found the allyl derivatives of tin to be more stable than those of lead (66). The former acted as inhibitors for radical polymerization of styrene or MMA. Allyl triphenyl lead decomposed at room temperature in contact with Ziegler catalyst.

Tetraallyl tin was made by reaction of $SnCl_4$ and AMgBr in ether (67). It was decomposed by air or water and reacted with HCl to give propylene. With nitric acid tetraallyl tin took fire. When heated with azobis to 160°C, tetraallyl tin formed an infusible solid containing 55.1% tin (68). When heated at 170°C, it gave ASnOOH and other products. Allyl tin and lead compounds retarded radical polymerization of acrylic and styrene monomers (69).

Tetraallyl tin and related compounds have been studied as modifiers of Ziegler-type catalysts for polymerization of l-olefins (70). Tetraallyl tin with chlorides of Ti, V, or Al polymerized propylene in n-heptane at 80°C and 9 atm (71). SnA_4 was added as modifier in polymerization of dienes (72). Bis(allyloxy) dibutyl tin and other allyl tin compounds have been tested as stabilizers for vinyl chloride polymers (73).

Reaction of lead chloride with $p-AC_6H_4MgBr$ was reported to give $(AC_6H_4)_4P$ together with polymer (74). Methallyl triethyl lead was prepared (75). Allyl triphenyl lead and diallyl diphenyl lead are less stable than analogous tin compounds (76).

Rothrock made tetramethallyl stannate by alcoholysis of tetraethyl stannate with methallyl alcohol under reflux (77). Ethanol and excess methallyl alcohol were removed under reduced pressure, leaving a yellow amorphous solid insoluble in toluene and in dioxane. A mixture of 19.5 parts of MMA, 0.5 part of the solid monomer, and 0.2 part

TABLE 31.1

Allyl Compounds Containing Tin

Compound	bp, °C	Other properties	References
Allyl trimethyl tin	129	$n_D^{25} = 1.4741$	Seyferth, JACS 79, 515 (1957)
Allyl triethyl tin	77(10)	--	Jones, J. Chem. Soc. 1446 (1947)
Diallyl diethyl tin	100(17)	--	Jones
Diallyl dibutyl tin	146(17)	--	Jones
Allyl tributyl tin	155(17)	--	Jones
Allyl triphenyl tin	--	mp, 74°C	Gilman, J. Org. Chem. 20, 763 (1955)
Diallyl diphenyl tin (unstable)	174(5.5)	$n_D^{20} = 1.6025$	Gilman
Allyl phenyl tin compounds	--	--	Henry, JACS 82, 558 (1960)
Diallyl dibromo tin	79(2)	mp, 103°C (polymerized)	Shishido, J. Org. Chem. 26, 2301 (1961)
Triallyl butyl tin	--	--	Rosenberg, JACS 81, 972 (1959)
Tetraallyl tin	70(1.5)	$n_D^{32} = 1.533$	Vijayaraghavan, J. Indian Chem. Soc. 22, 138 (1945)
A₂SnO		unstable	Baum, J. Polym. Sci. Bl 9 , 517 (1963)
ASnOOH	--	--	Meyer, Z. Anorg. Chem. 68, 106 (1910)
Allyl diphenyl tin acetate	--	decomp. at 260°C	Rosenberg, JACS 81, 972 (1959)

of benzoyl peroxide was allowed to stand for 24 hr and then was heated for 3 days at 65°C. A hard, tough translucent polymer formed which was insoluble in toluene, dioxane, and butyl acetate. The product softened at 118°C. A copolymer with styrene was made by heating with 1% BP at 50°C for 48 hr. Moldings were clear and strong.

Properties of some tin-containing allyl compounds are shown in Table 31.1.

The complex vinyl ferrocene [⬠]⟩··Fe··⟨[⬠] $CH=CH_2$ polymerized with azobis at 80°C to give benzene-soluble polymers melting near 280°C (78). Resonance prevents the ring hydrogen from acting allylic. Bis(allylcyclopentadienyl) iron (bp 123°C at 1 mm) was prepared (79). The corresponding manganese complex was made and was suggested as

an additive to fuel and lubricating oils (80). Allyl
ferrocene compounds have not given high polymers. Addition of isopropenyl ferrocene greatly retarded free radical polymerizations of methyl methacrylate and styrene
(81).

Rothrock prepared tetramethallyl titanate (bp 174°C at
6 mm) by reaction of tetraethyl titanate with methallyl
alcohol at reflux temperature (82). Although it did not
homopolymerize, it reacted with either vinyl acetate
monomer or vinyl acetate polymer to give crosslinked
products. Tetraallyl titanate (TAT) is a yellow viscous
liquid (bp 185°C at 12 mm; $d_4^{25} = 1.1138$) which readily
hydrolyzes in moist air and is soluble in common organic
solvents (83). On heating, TAT becomes gelatinous and
finally forms a hard, glassy, insoluble polymer. On
distillation at low pressure at 240°C, allyl alcohol and
a little crotonaldehyde were formed. A Japanese patent
gives the boiling point of TAT as 185°C at 12 mm. TAT
has been applied to glass fibers used in polymer laminates (84). Soluble homopolymers of high molecular weight
have not been reported. TAT is limited as an industrial
crosslinking and adhesive agent because of hydrolysis
during handling and storage.

Allyl orthotitanates were copolymerized with styrene,
MMA, and other monomers (85). Tetramethallyl titanate
(bp 174°C at 6 mm) did not homopolymerize on heating or
on addition of $SnCl_4$ or BF_3 (86). It gave insoluble
polymers on heating with vinyl acetate. Korshak and co-
workers made an allyloxy titanocene melting at 58°C which
polymerized on heating at 100°C with BP for 10 hr to a
dark orange product which melted at 160°C with decomposition (87). Orange-colored copolymers of moderate molecular weight formed with styrene and with MMA. Allyl
Grignard reagents react with dicyclopentadienyl titanium
monochloride to give paramagnetic purple crystals of
$(C_5H_5)_2TiA$ melting at 118°C (88). An analogous allyl
compound of trivalent vanadium formed a brown solution
but did not crystallize.

Allyl zinc bromide behaves somewhat like allyl Grignard
reagents in hydrolysis, carbonation, and reaction with
carbonyl compounds, but it is less reactive (89). Cadmium
diallyl dithiocarbamate was suggested as a rubber vulcanization accelerator (90). Polymers were obtained by heating triallyl chloromolybdate with radical catalysts (91).

Diallyl mercury was obtained by action of concentrated
KCN on AHgI or by reduction of AHgI by sodium stannite
(92). It is a nondistillable oil of density 2.1 g/ml,
soluble in ether and benzene, and slightly soluble in
ethanol. It decomposes slowly at room temperature and
faster on heating to give mercury and diallyl. Diallyl

mercury reacts with Br_2 or Cl_2 to give AHgX. Allyl phenyl mercury was prepared from phenyl mercuric chloride and allyl magnesium bromide (93). Allyl mercury halides slowly decompose at room temperature. The vinyl compounds are more stable. Allyl mercuric halides are cleaved by aqueous perchloric acid giving propylene (94). Allyl bromide stirred with mercury in light yielded 30% AHgBr in 19 hr (95). Polymers formed as a byproduct, and with allyl chloride Hg_2Cl_2 and polymers formed.

Products of thermal decomposition of allyl mercuric acetate were studied (96). Allyl mercury mercaptides were suggested as comonomers and as fungicides (97). Substitution reactions of allylic mercury compounds were studied (98).

References

1. J. F. Nixon and B. Wilkins, CA 78, 4348 (1973).
2. D. V. Banthorpe, Chem. Rev. 70, 295 (1970); cf. J. W. Faller et al, J. Am. Chem. Soc. 93, 2642 (1971).
3. J. Chatt, Nature 177, 852 (1956); cf. J. Inorg. Nucl. Chem. 8, 515 (1958); J. Chem. Soc. 2939 (1953); E. O. Fischer and W. Hafner, Z. Naturforsch. 10b, 140 (1955).
4. H. D. Kaesz et al, Z. Naturforsch. 15b, 682 and 763 (1960); CA 55, 20924 (1961).
5. W. R. McClellan et al, J. Am. Chem. Soc. 83, 1601 (1961).
6. R. F. Heck and C. R. Ross, J. Am. Chem. Soc. 86, 2580 (1964); U.S. 3,338,936 (Hercules).
7. J. Smidt and W. Hafner, Angew. Chem. 71, 284 (1959); cf. R. Huettel and J. Kratzer, Angew. Chem. 71, 456 (1959); E. O. Fischer and G. Buerger, Z. Naturforsch. 16b, 702 (1961).
8. J. C. W. Chien and H. C. Dehm, Chem. Ind. (London) 745 (1961); CA 55, 23056 (1961).
9. A. D. Ketley, U.S. 3,493,591 (Grace).
10. W. T. Dent et al, J. Chem. Soc. 1588 (1964).
11. B. M. Trost and T. G. Fullerton, J. Am. Chem. Soc. 95, 292 (1973).
12. G. Wilke and B. Bogdanovic, Belg. 631,172; Ger. 1,194,417; U.S. 3,379,706 (Studienges. Kohle); Angew. Chem. 73, 756 (1961) and 75, 10 (1963); Wilke, U.S. 3,422,128; U.S. 3,511,863; U.S. 3,522,283 (Studienges. Kohle).
13. G. Wilke, U.S. 3,424,777, U.S. 3,536,740 (Studienges. Kohle); cf. P. L. Maxfield, U.S. 3,475,471 (Phillips).
14. G. Wilke, U.S. 3,468,921 (Studienges. Kohle).
15. A. I. Lazutkina et al (Novosibirsk), CA 77, 5855 (1972); cf. CA 78, 4352 (1973).
16. J. Powell and B. L. Shaw, J. Chem. Soc. A 774 (1968).
17. J. Powell and B. L. Shaw, J. Chem. Soc. A 583 (1968).
18. D. Medema, U.S. 3,398,168 (Shell); cf. A. D. Ketley, U.S. 3,479,379 (Grace); D. G. Brady, CA 78, 16748 (1973).

19. F. R. Hartley, Chem. Rev. 69, 799 (1969); Angew. Chem. 84, 657 (1972).
20. H. C. Bach, U.S. 3,584,020 (Monsanto).
21. L. Porri, G. Natta et al, J. Polym. Sci. C 16, 2525 (1967); J. F. Harrod and L. R. Wallace, Macromolecules 2, 449 (1969); cf. E. F. Magoon et al, U.S. 3,483,269 (Shell).
22. F. Dawans and P. T. Teyssie, Polym. Lett. 7, 111 (1969); cf. T. Matsumoto and J. Furukawa, J. Macromol. Sci.-Chem. A6, 281 (1972); F. Dawans et al, CA 78, 16532 (1973).
23. L. Porri et al, Chim. Ind. (Milan) 46, (4), 428 (1964); CA 61, 1944 (1964); cf. V. A. Kormer et al, J. Polym. Sci. A 1, 10, 251 (1972).
24. J. C. Marechal et al, J. Polym. Sci. A 1, 8, 1993 (1970).
25. P. T. Teyssie et al, Macromolecular Preprints, p. 118, IUPAC Boston, July 1971; J. Polym. Sci. A 1, 8, 979 and 1993 (1970).
26. G. Henrici-Olivé et al, J. Organometal Chem. 39 (1), 201 (1972).
27. A. V. Alferov et al, U.S. 3,468,866 (Moscow).
28. F. Dawans, Tetrahedron Lett. 22, 1943 (1971).
29. G. Wilke et al, Angew. Chem. 78, 156 (1966); English ed. 5 (2), 151 (1966).
30. H. D. Murdoch, U.S. 3,483,238 (Am. Cyan.).
31. K. Sato et al, J. Org. Chem. 37, 464 (1972).
32. D. G. H. Ballard et al, J. Chem. Soc. B 1168 (1968); CA 76, 25646 (1972).
33. S. O'Brien et al, Brit. 1,091,296; U.S. 3,436,383 (ICI).
34. W. Herwig et al, U.S. 3,501,415 (Hoechst); cf. H. Naarmann et al, U.S. 3,454,538 (Badische).
35. W. H. Janes and S. O'Brien, Brit. 1,112,889 (ICI); CA 69, 10918 (1968); cf. E. A. Demin et al, CA 77, 5848 (1972).
36. W. R. McClellan, U.S. 2,990,418 (Du Pont).
37. R. G. Schultz, U.S. 3,446,825 (Monsanto).
38. P. Pino et al, U.S. 3,546,264 (Lonza).
39. A. Greco, CA 78, 4333 (1973).
40. F. Galbraith and S. Smiles, J. Chem. Soc. 1234 (1935); cf. A. A. Morton et al, J. Am. Chem. Soc. 69, 950 (1947); R. Letsinger et al, J. Am. Chem. Soc. 74, 400 (1952).
41. D. Seyferth and M. A. Weiner, J. Org. Chem. 26, 4797 (1961); R. Waack and M. A. Doran, J. Phys. Chem. 68, 1148 (1964).
42. R. Waack and M. A. Doran, Polymer 2, 365 (1961); CA 58, 11473 (1963).
43. D. Braun, Angew. Chem. 73, 197 (1961).
44. T. E. Londergan, U.S. 2,734,091 (Du Pont), CA 50, 15588 (1956); cf. Seyferth and Weiner, J. Org. Chem. 24, 1395 (1959).
45. A. A. Morton et al, J. Am. Chem. Soc. 74, 5434 (1952).
46. Brit. 796,912 (Badische); CA 52, 19243 (1958).
47. A. Zilkha et al, J. Chem. Soc. 928 (1959).
48. R. Lespieau, Ann. Chim. et Phys. (8), 27, 149 (1912); F. Cortese, J. Am. Chem. Soc. 51, 2266 (1929).
49. Henry Gilman and J. H. McGlumphy, Bull. Soc. Chim. 43, 1322-8 (1928); CA 23, 1870 (1929).

50. L. H. Shepherd, U.S. 3,597,488 (Ethyl); cf. U.S. 2,959,598 (Metal and Thermit).
51. J. C. H. Hwa and H. Sims, Org. Syn. 41, 49 (1961); cf. L. W. Butts, J. Org. Chem. 5, 171 (1940).
52. P. Tarrant and J. Heyes, J. Org. Chem. 30, 1485 (1965).
53. J. D. Roberts et al, J. Am. Chem. Soc. 82, 2646 (1960).
54. J. J. Eisch et al, CA 78, 4102 (1973).
55. H. S. Rothrock, U.S. 2,258,718, appl. 1939; Brit. 545,114 (Du Pont).
56. P. Pansini and G. Bartolini, Fr. 1,336,568 (Monte); CA 60, 3008 (1964).
57. L. I. Zakharkin and L. A. Savina, CA 57, 12520 (1962).
58. A. D. Petrov et al, CA 51, 4938 (1957).
59. V. V. Korshak et al, CA 51, 13815 (1957).
60. G. S. Kolesnikov et al, CA 54, 17940 (1960).
61. P. Mazerolles and M. Lesbre, Compt. Rend. 248, 2018 (1959); cf. J. Satge, CA 57, 5941 (1962).
62. M. Lesbre and J. Satge, Compt. Rend. 247, 471 (1958).
63. H. Gilman and C. W. Gerow, J. Am. Chem. Soc. 79, 342 (1957).
64. W. J. Jones et al, J. Chem. Soc. 1446 (1947).
65. V. V. Korshak et al, CA 53, 15959 (1959).
66. M. M. Koton et al, CA 54, 22436 (1960).
67. K. V. Vijayaraghavan, J. Indian Chem. Soc. 22, 137 (1945).
68. M. M. Koton and T. M. Kiseleva, CA 52, 7136 (1958).
69. M. M. Koton et al, J. Polym. Sci. 52, 237 (1961).
70. T. T. Li, U.S. 3,163,629 (M & T).
71. Ital. 792,997 (Monte); CA 70, 20472 (1969).
72. W. J. Trepka and R. J. Sonnenfeld, U.S. 3,536,691 (Phillips).
73. G. P. Mack et al, U.S. 2,700,675; U.S. 2,626,953 and 2,684,973 (Advance Solvents).
74. E. A. Puchinyan and Z. M. Manulkin, CA 61, 677 (1964).
75. F. Glockling and D. Kingston, J. Chem. Soc. 3001 (1959).
76. P. Austin, J. Am. Chem. Soc. 53, 3514 (1931); H. Gilman and J. Eisch, J. Org. Chem. 20, 763 (1955).
77. H. S. Rothrock, U.S. 2,258,718 (Du Pont).
78. F. S. Arimoto and A. C. Havens, J. Am. Chem. Soc. 79, 6295 (1955).
79. M. A. Lynch and J. C. Brantley, Brit. 785,760 (Carbide).
80. E. J. DeWitt et al, U.S. 2,839,552 and 2,964,547 (Ethyl).
81. M. Howard and S. F. Reed, J. Polym. Sci. A1, 2085 (1971).
82. H. S. Rothrock, U.S. 2,258,718 (Du Pont).
83. N. M. Cullinane et al, J. Appl. Chem. (London) 1, 400 (1951); cf. CA 45, 7951 (1951) and 46, 7994 (1952); Nature 164, 710 (1949).
84. R. E. Clayton et al, U.S. 3,337,391 (Esso).
85. S. A. Kretinin et al, CA 60, 13326 (1964); S. Hashimoto and I. Furukawa, CA 61, 5853 (1964).
86. Brit. 545,114 (Du Pont).
87. V. V. Korshak et al, CA 59, 14117 (1963).

88. H. A. Martin and F. Jellinek, Angew. Chem. 76, 274 (1964).
89. M. Gaudemar, Compt. Rend. 246, 1229 (1958).
90. W. S. Murray, U.S. 2,831,823 (Du Pont).
91. M. Bloom, U.S. 3,052,658 (Hughes Aircraft).
92. K. V. Vijayaraghavan, CA 36, 4477 (1942) and 38, 2006 (1944).
93. M. S. Kharasch and S. Swartz, J. Org. Chem. 3, 405 (1938).
94. R. D. Bach and P. A. Scherr, J. Am. Chem. Soc. 94, 220 (1972).
95. O. A. Reutov et al, CA 46, 10120 and 48, 12692.
96. D. J. Foster and E. Tobler, J. Org. Chem. 27, 834 (1962).
97. A. L. Flenner, U.S. 2,967,191 (Du Pont).
98. P. D. Sleezer et al, J. Am. Chem. Soc. 85, 1890 (1963).

APPENDIX

DATES OF PATENTS

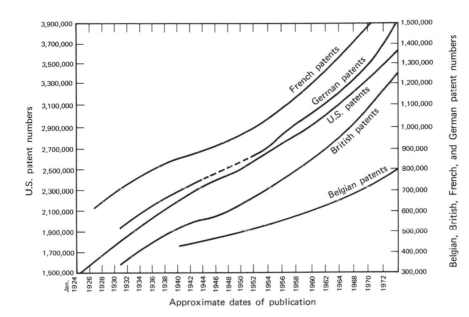

Approximate dates of publication

ABBREVIATIONS

A	Allyl
A.C.S.	American Chemical Society
Azobis	Azobisisobutyronitrile
Belg.	Belgian Patent
t-BHP	t-Butyl hydroperoxide
BP	Benzoyl peroxide
bp	Boiling point
Brit.	British patent
CA	Chemical Abstracts
Can.	Canadian patent
Ger.	German patent
Japan	Japanese patent
MMA	Methyl methacrylate
MA	Methallyl
mp	Melting point
n_D^{20}	Refractive index, 20°C
U.S.	United States patent

COMPANIES FREQUENTLY CITED

Algemene Kunstzijde Unie (A.K.U.), Arnhem, Holland
Allied Chemical and Dye Corporation, New York, N. Y.
American Cyanamid Company, Wayne, N. J.
Badische Anilin und Soda Fabrik (BASF), Frankfurt, Germany
Bayer Farbenfabriken, Leverkusen, Germany
Borden Company, New York, N. Y.
Calgon Corporation (Division of Merck), Pittsburgh, Pa.
Canadian Industries, Ltd., Montreal, Quebec
Celanese Corporation of America, New York, N. Y.
Chemische Werke Huels, Huels, Germany
CIBA-Geigy, Basel, Switzerland
Dow Chemical Company, Midland, Mich.
Dunlop Tire and Rubber Corporation, London, England
E. I. du Pont de Nemours and Company, Wilmington, Del.
Eastman Kodak Company, Rochester, N. Y.
Farbwerke Hoechst, Frankfurt-Hoechst, Germany
FMC Corporation, San Jose, Cal.
GAF Corporation, New York, N. Y.
General Electric Company, Schenectady, N. Y.
B. F. Goodrich Company, Akron, Ohio
Goodyear Tire and Rubber Company, Akron, Ohio
Gulf Oil Corporation, Pittsburgh, Pa.
Hercules, Inc., Wilmington, Del.
Hoffmann-LaRoche, Basel, Switzerland
I.G. Farbenindustrie, formerly Frankfurt, Germany
Imperial Chemical Industries, Ltd., London, England
Johnson and Johnson, New Brunswick, N. J.
Kuraray Company (formerly Kurashiki), Osaka, Japan
Libbey-Owens-Ford Company, Toledo, Ohio
M and T Chemicals, Greenwich, Conn.
Mitsubishi Industries, Osaka, Japan
Mobil Oil Corporation, New York, N. Y.
Monsanto Chemical Company, St. Louis, Mo.
Montecatini Edison S.p.A., Milan, Italy
Phillips Petroleum Company, Bartlesville, Okla.
PPG Industries (Pittsburgh Plate Glass), Pittsburgh, Pa.
Polymer Corporation, Sarnia, Ontario
Rohm and Haas Company, Philadelphia, Pa.
Sartomer Industries, Essington, Pa.
Shell Development Company, Houston, Texas
Standard Oil, Indiana, Chicago, Ill.
Standard Oil, New Jersey (Esso, later Exxon), New York, N. Y.
Stauffer Chemical Company, New York, N. Y.
Sumitomo Chemical Company, Osaka, Japan
Union Carbide Corporation, New York, N. Y.
Uniroyal, Inc., New York, N. Y.

INDEX